Boats

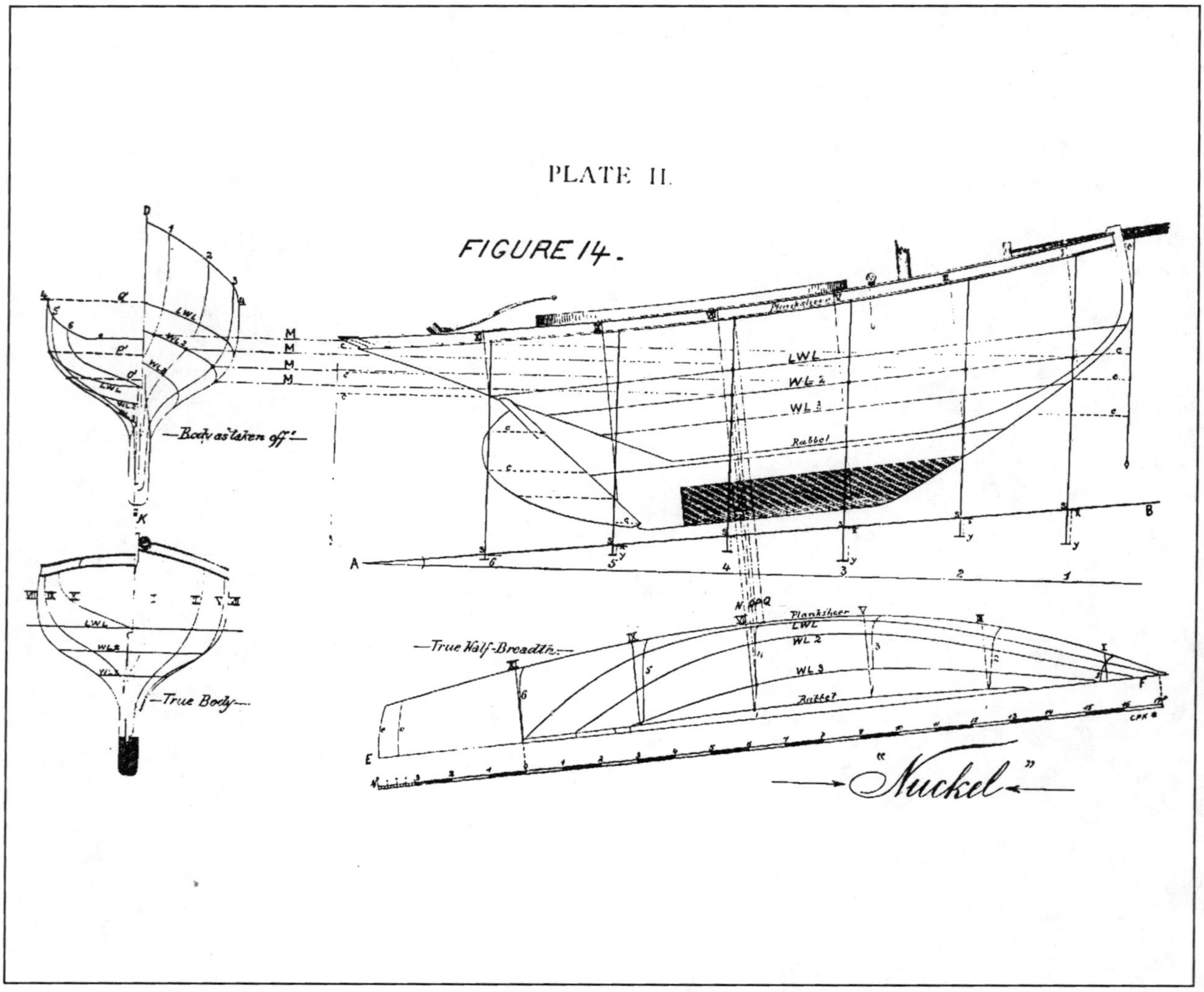

Lines taking in the 1880s. From *Small Yachts* by C. P. Kunhardt, published in 1891.

Boats

A Manual for Their Documentation

Museum Small Craft Association

Willits Ansel Maynard Bray
Kevin J. Crisman D. W. Dillion
Benjamin A. G. Fuller
Paula J. Johnson Paul Lipke
William Peterson Peter Schmid
David A. Taylor Peter T. Vermilya
Garth Wilson

Paul Lipke, *Project Manager/Editor*
Peter Spectre, *Editor*
Benjamin A. G. Fuller, *Editor*

American Association for State and Local History

Nashville, Tennessee

The Museum Small Craft Association gratefully acknowledges the significant financial assistance of its members, the National Trust for Historic Preservation, the National Maritime Initiative of the National Park Service, the Institute of Museum Services, the Council of American Maritime Museums, the American Association for State and Local History, and Mystic Seaport Museum.

To Howard Chapelle, John Gardner, J. Richard Steffy, and those who follow in their footsteps

Produced by Zenda, Inc.
Design: Griffin Norman/IKON
Cover photo by Peter T. Vermilya

Library of Congress
Cataloging-in-Publication Data

Boats : a manual for their documentation / Museum Small Craft Association, Willits, Ansel . . . [et. al.] : Paul Lipke, project manager/editor, Peter Spectre, editor, Benjamin A.G. Fuller, editor.
p. cm.
Includes bibliographical references (p.) and index.
ISBN 0-942063-17-1 : $37.95
1. Boats and boating--Conservation and restoration.
I. Ansel, Willits Dyer. II. Lipke, Paul. III. Spectre, Peter H. IV. Fuller, Benjamin A.G., 1945- . V. Museum Small Craft Association.
VM 149.B68 1993 94-11098
623.8'202'075--dc20 CIP

Printed on acid-free paper

Contents

Preface /vii

Acknowledgments /xi

1 **An Introduction to Documenting Boats**
Benjamin A. G. Fuller
Defines documentation, examines the reasons, and considers the levels / 1

2 **Stabilization: Managing the Boat**
Peter T. Vermilya
Methods for preserving boats / 9

3 **The Boat as Record**
Willits D. Ansel
How a boat changes through its lifetime / 19

4 **Accuracy and Verification**
D. W. Dillion
Determining the quality of the job and how long it may take / 27

5 **Tools for Measuring, Drafting, and Lofting**
D. W. Dillion
The tools needed for the job / 35

6 **Taking Off Construction**
D. W. Dillion
Studying and recording a boat's construction, with illustrations of boat parts and where they are found / 51

7 **Taking Off—The Measuring Process**
D. W. Dillion
Step-by-step guidance to measuring a boat's shape / 115

8 **Organizing and Plotting Field Notes**
D. W. Dillion
Keeping clear, useable notes / 155

9 **Recording Sails, Sailing Rigs, and Rigging**
D. W. Dillion
Recording the sailing rigs of watercraft, including illustrations of types and terms / 179

10 **Selective Documentation: A Guide to Quick Measuring**
Maynard Bray & Paul Lipke
Recording quickly the information needed to describe a particular boat / 209

11 **Drawing and Lofting from Field Notes**
D. W. Dillion
The techniques needed to produce finished drawings / 223

12 **An Archaeological Approach**
Kevin J. Crisman
Recording boats from their remains and learning from their parts / 303

13 **Sketching and Photographing Boats**
Peter Schmid & Paul Lipke
The photographs and sketches that should be part of the recording process / 321

14 **Historical Photographs**
William N. Peterson
Finding historical photographs of boats and learning from them / 331

15 **Beyond the Boat: Documenting the Cultural Context**
Paula J. Johnson & David A. Taylor
Recording the cultural, social, and economic environment of the boat; ethnographic fieldwork methods / 337

16 **Computer Applications for Documentation and Analysis**
Garth S. Wilson
Promises and problems in using a computer to help record, draw, and analyze a boat's shape / 357

Contributors / 371

Appendix A: *List of Helpful Organizations / 373*

Appendix B: *Union List of North American Watercraft: Descriptive Summary and Reporting Forms / 379*

Appendix C: *Informant Consent Form and Artifact Loan Form / 383*

Bibliography: *A guide to published boat documentation projects / 387*
Benjamin A. G. Fuller

Index / 403

Editors' note: Unless otherwise noted, all illustrations within a chapter are the work of the chapter author or authors.

Acknowledgments

The first order of appreciation must go to the authors of *Boats* for their selfless contributions of time, energy, and expertise. In chapter order they are: Benjamin Fuller, Peter Vermilya, Willits Ansel, D. W. Dillion, Maynard Bray, Paul Lipke, Kevin Crisman, Peter Schmid, William Peterson, Paula J. Johnson, David A. Taylor, and Garth Wilson. Peter Spectre's work as co-editor was invaluable. The second order of thanks must go to the Museum Small Craft Association's Board of Officers for endless patience and perseverance: Anne Witty, Hallie Bond, Dick Wagner, Peter Vermilya, Garth Wilson, Bill Doll, John Summers, and Michael Alford.

No order of importance is implied in the listing that follows. Any omissions (all unintended) are the sole fault of the author.

At the offices of our copublishers, the American Association for State and Local History, we were helped immensely by Joanne Jaworski, Mary Bray Wheeler, Peter La Paglia, Larry Tise, and John-Paul Richiuso. Their good efforts were augmented by those of Patricia Hogan at Zenda, Inc. James P. Delgado and Kevin Foster of the National Maritime Initiative of the National Park Service provided moral, financial, and technical support, as did Marcia Myers, Michael Naab, Lynn Hickerson, Lisa Wormser, Constance Beaumont, and Bridget Hartman at the National Trust for Historic Preservation. The Institute of Museum Services' Dan Lucash and Rebecca Danvers could not have been more helpful. The same could be said of the staff of Mystic Seaport Museum, especially J. Revell Carr, Mike Rachlin, Margaret Pellegrino, Jerry Morris, Paul O'Pecko, Helen Packer, Katia Maloney, Mary Anne Stets, Peggy Tate-Smith, and Dana Hewson. Spread around the country but united in their helpfulness were John Carter, Nathan Lipfert, Bob Webb, Philip Shelton, Sam Manning, Richard Lunt, Leslie Fuller, J. Richard Steffy, Peter Neill, Richard Anderson, Sam Johnson, Anne Bray, Cynthia Curtis, Joseph T. Butler, Isacco Valli, Joel White, Lance Lee, James Mays, and the many people who tested the manuscript in the field and improved it with their constructive comments.

Four people know better than anyone, what this project has cost in personal sacrifice. They are Suzanne Alford, Leslie Fuller, Marcelle Lipke, and Margaret Vermilya. If this work proves to be of any value, they should receive the heartfelt appreciation of the entire maritime preservation community.

Last but not least, this book could not have been completed without the financial support of the National Trust for Historic Preservation, the National Maritime Initiative of the National Park Service, the Professional Services Program of the Institute of Museum Services, the Council of American Maritime Museums, the American Association for State and Local History, the Mystic Seaport Museum, the North Carolina Maritime Museum, the Center for Traditional Louisiana Boatbuilding at Nicholls State University, the Manitowoc Maritime Museum, the San Francisco National Maritime Historic Park, the Center for Wooden Boats, Benjamin A. G. Fuller, D.W. Dillion, Paul Lipke, and other private contributors.

—Paul Lipke

Preface

Any book advocating the documentation of history can hardly afford to deprive its readers of an understanding of the context in which it was produced. This is especially true for this work since the reader cannot fail to notice a very broad (some would say peculiar!) range in the level of detail between different chapters. This breadth results from a complex series of events that occurred between the initial idea and publication.

In 1980, the idea for a manual on small craft documentation was put on paper by an informal group of museum professionals meeting as the "Small Craft Curators Conference" (SCCC). Even then, the concept had been around for more than five years. Assistance was then sought from a number of sources including the Maritime Preservation Task Force of the National Trust for Historic Preservation. In 1983, that task force's Standards Committee produced draft Guidelines for Documentation, with essays by Maynard Bray, Richard Lunt, and Sam Manning.

Shortly thereafter, Congress requested the National Park Service to "conduct a survey of historic maritime resources. . ." (Congressional Record, Oct. 10, 1984, p. 11922). A new program, the National Maritime Initiative was created in the National Park Service to carry out this request to inventory, evaluate, and recommend standards for maritime resource preservation.

This request helped shift the Task Force's emphasis towards large vessels because it coincided with a strong push to nominate large vessels to the National Register, and to document them to the standards of the National Park Service's Historic American Building Survey/Historic American Engineering Record (HABSHAER). This shift eventually resulted in the 1988 publication of *Guidelines for Recording Historic Ships* (Anderson, 1988). While *Guidelines* concentrates primarily on Library of Congress standards for drawings, documentation, and presentation, rather than on the methods by which these are produced, it does cover some of the field work involved.

In light of these events, an ad hoc committee's 1985 review of the 1983 manuscript set out additional areas that needed to be covered and some that needed expansion. The manual needed to address the issues surrounding the documentation of small watercraft, a subject *Guidelines* could not address. It needed to encompass recording to a higher level of detail than *Guidelines* generally advocated. It needed to be completed quickly; too much time had already gone by, boats were being lost without being recorded at alarming rates. It was determined that a large number of institutionally supported contributors, each having responsibility for a relatively small section of the book, would produce text quickly at the lowest cost. To keep *Boats* manageable in size and avoid unnecessary repetition, wherever the work would refer to the reader to *Guidelines.*

> Photocopies are available from
> Technical Information Center
> (DSC-PGT)
> National Park Service
> Denver Service Center
> P.O. Box 25287
> Denver, Colorado 80225-0287
> (303) 969-2130
> order number: 999/0-378; include check
> or money order for $18.60

Getting this revised plan for the "DocManual" underway required the formalization of the SCCC into a professional organization: the Museum Small Craft Association (MSCA). Incorporated in 1987, the MSCA sought and received funding from the Critical Issues Fund of the National Trust for Historic Preservation, the Institute for Museum Services, Mystic Seaport Museum, the National Park Service, the Council of American Maritime Museums, and several private sources. National Maritime Initiative help for both the manual and to produce a national watercraft inventory for all historic watercraft came to the Museum Small Craft Association through a cooperative agreement.

While the work's content was fairly fixed early on by the need to fill in the gaps in the available literature, the level of detail to include was (and will certainly continue to be) a source of controversy. For example, one participant would want each measuring method to be described in minute detail, another would say, "Just give them the basics." Most of the authors made their contributions on a volunteer basis while simultaneously meeting substantial professional obligations to their employers; under such circumstances their vociferous, expert opinions in such matters could hardly be casually over-ruled.

Because of this history, *Boats* devotes most of its pages to the intimate, step by step "how to's" of capturing the "telling details" that form the physical evidence for reconstructing a boat's working life, and presenting this life in text, lines, rigging, and construction plans. We have assumed the reader has boats to care for and little expertise in such matters, but is highly motivated to invest the necessary time and energy. Methods are included even if they have significant drawbacks in some applications, since they may provide the perfect answer to others. The work is specially bound to ease use in the field. The contributors have also attempted to provide enough background material to enable the reader to assess the accuracy and suitability of new and/or different methods. The guiding philosophy in the work's composition is one of inclusion rather than exclusion.

In this vein, a basic understanding of topics such as field sketching, boat stabilization, cultural context, and archaeological reconstruction of deteriorated boats are considered essential in meeting the needs of the work's broad constituency. These "basic" chapters are summaries; they point the reader towards helpful institutions and/or more complete works on the subject while clarifying issues and tricks of the trade specific to documenting small boats. For example, the chapter on field sketching assumes you know how to produce perspective sketches (or can learn through standard reference works on the subject) but could use specific guidance on how to sketch boats, a topic not covered in any detail in any work of which we are aware.

While maritime museum professionals began the drive to assemble this book, amateur and professionals in the areas of ethnography, sport diving, and state and local history have proven to be important, if less obvious, audiences with an interest in boat documentation. Clearly, more boats survive than there are experienced personnel available to make documentation decisions about them. A frequent telephone call to maritime museum staff, from both professional and amateurs, is: "I have a boat, what do I do?" The manual seeks to provide a framework for answers. There are no geographical limits to such inquiries, either. While the work was in production, serious inquiries were received from as far afield as Sweden and Polynesia.

The consensus within the maritime preservation community has long been that, at least in the United States, we have fallen far behind the historic building preservation movement in the preservation of our heritage. While the preservation and/or adaptive reuse of buildings has become widely valued and applied, the maritime sector has had difficulty overcoming the greater logistical, financial, and cultural obstacles to the preservation of ships and boats. Matters are made worse in two ways. First, watercraft are far less durable than buildings, with the result that in many cases documentation is the only way of assuring their permanence. Second, to document the structure you must overcome the greater technical difficulties of recording complex curved shapes and their supporting framework.

Under these challenges, on top of the usual lack of funds, staff, storage space, and/or time, organizations and individuals are forced to turn down donations of boats and/or opportunities to document which they know it would otherwise be better to accept. Alternatively, they have been known to accept a boat (and the endless curatorial obligation it entails) because one small detail is deemed important. This work attempts to reduce the domination of such absolute choices—to save in its entirety or not, to document in full or not—and replace them with a wider range of options to be evaluated with expertise and sober judgment.

Decisions to save part of a boat or some limited aspect of the information it embodies can, of course, create political and curatorial difficulties of their own. In such an instances, a full narrative explanation of the decisions should be made part of the permanent record. At present, however, some preservationists fail to recognize that for every boat they "save," many with significant value are turned down and then disappear without any effort being made to save critical features and/or information. It is hoped this narrow view will be broadened by the wide range of methods and values expressed here.

There have been only three issues on which total consistency and unanimity was achieved among the project participants. First, it is harder to write about the measuring and drafting process than it is to do it. In general, the actual measuring is relatively fast and straightforward; the drafting takes more skill and much more time. Second, the reader's input is highly desired. It is hoped a time will come when a second, revised edition can be produced. Finally, we urge the reader to think of boat documentation as a wide range of activities that will help make sense of boats and the dynamic cultural processes that underlie their creation and use.

—Paul Lipke
—Ben Fuller
August 30, 1992

Chapter 1

An Introduction to Documenting Boats

Defines documentation, examines the reasons, and considers the levels

Benjamin A. G. Fuller[1]

Small boats were once essential to transportation in America. Early in the nineteenth century, the design of small boats began to take on regional flavors with special combinations of construction methods and styles. As models and methods were adapted to particular economic and geographical conditions, small boats became true folk or vernacular artifacts.

Boats also migrated from area to area, outside of places where they were originally developed. They moved as part of the working complement of a ship, such as whaleboats and dories, or as somewhat generic ship's boats of various sizes.

Builder's knowledge moved with builders. Moving into new areas, they brought designs and methods learned in their apprenticeships. In some cases their knowledge was put to use; in others builders adapted their skills to the methods in use in their new homes. In the Chesapeake, where the need for boats outstripped the production capacity of available boat builders, settlers adopted native craft—dugout canoes—using new tools and joining techniques to enlarge them.

By the early part of this century, small craft began to reflect a growing national recreational and commercial market. Critical to this emerging national market were improvements in rail transportation and the development of national distribution systems (mail order) by pioneers such as Sears and Montgomery Ward. Yet regional builders still brought local aesthetics and techniques to bear. These regional forms could and were distributed nationally. In response to a wider market, many builders refined their methods into a form of mass production. Today's boats are almost exclusively mass market, mass production boats, not the products of local building traditions.

Small craft are, then, highly complex artifacts, with variations in form and technique directly linked to their temporal and geographical context. Their study leads to questions of technical skill, competence, aesthetics, and invention or diffusion of knowledge.

The study of boats can move beyond the artifact to questions of economics, function, markets, and social organization as the role of the boat in its community (users, owners, and builders) is examined. Indeed, the community may well be defined by the boat rather than by geographic boundaries. Equally worthy of study are the community of owners of a boat type that is nationally or regionally distributed, and the community of work boats owners based in a single harbor.

Thus documentation can take on many forms. It can be the close examination and recording of a single boat. It can be defining types and then doing a regional survey to determine their distribution, ownership, and use. Documentation may also focus on how a boat serves to define a community. It can be a record of what owners, operators, and creators think about them. It can concentrate on context, the world in which the boats live.

What Do We Mean by Small?

In his seminal survey, *American Small Sailing Craft* (1951), Howard I. Chapelle defined small boats as forty feet and under and stated that traditional craft in that range were good candidates for recreational boats. However, there are long, light boats, longer than forty feet, that qualify as small by virtue of their displacement or weight, a more accurate guide to size. Five tons and under is considered by the U. S. Coast Guard to be a small boat when issuing numbers. Some even feel that any boat that takes more than six or eight people to carry is not small!

Based on preservation and collection policies of museums, which can handle small craft easier than large, the Historic American Engineering Record considers thirty feet and over as eligible for listing as they are less likely to be preserved in a museum collection. The International Congress for Maritime Museums in its Historic Ship Register used fifty feet or twenty tons as a standard. Whatever the criterion, this manual focuses on techniques suitable for boats generally smaller than these standards, boats that usually can be built and operated by a crew of not more than two or three.

Why a Specialized Manual?

Watercraft documentation manuals are not new. A century ago, one of this country's first yachting writers, C. P. Kunhardt, saw the need and wrote about methods for recording a boat's shape in *Small Yachts*. More recently others have written articles or chapters of books on watercraft measurements. What is new about this book is that it provides a single source where techniques and methods are presented. Using it will allow readers to understand the issues involved and the ways to solve problems.

This manual emphasizes recording boat construction and shape and putting this information on paper. It raises questions of a boat's context and history, and reviews the boat documentation literature. The manual points out aspects of recording and fieldwork methods that are used in other disciplines and which can be applied to maritime artifacts. It provides, as well, an introduction to the study of boat-related artifacts and original verbal and written information sources in watercraft documentation.

Most of the book is designed as a field manual for the person doing the work. But it will also help individuals start, plan, and supervise a project, as it sets out the questions that should be considered before beginning a project, such as equipment needed, time and costs involved, personnel coordination, and the use of museums and other cultural institutions.

Technical information for recording watercraft tends to be the domain of a few specialists, generally historically oriented naval architects, preservation oriented boat builders, or technically oriented museum curators and folklife specialists. Even for these, information about techniques that others have used is poorly distributed. Just as there are many construction methods used to build acceptable watercraft, there are different recording techniques, suitable for different situations. Familiarity with one does not mean familiarity with all. For the experienced, the manual can give insight into a new or different technique.

More importantly, more boats survive than there are experienced personnel available to make documentation decisions about them. A frequent telephone query to maritime museum staff, from both professional and amateurs, is: "I have a boat, what do I do?" The manual seeks to provide a framework for answers.

Manuals exist for types of fieldwork ranging from oral history to architectural photography. But there is nothing readily available on the idiosyncratic problems posed by boats. Recording their shape is a technical problem of curves. Detailing construction requires initiation into the arcane world of rabbets, stems, and clamps. There can be purely physical problems of access to a boat half buried on a tidal mud flat, with severe and certain limits on the amount of time available to work. Recording the use of boats requires knowledge of maritime matters and a special vocabulary. The researcher must be familiar with the literature, some of which contains important information on maritime cultures, techniques, and methods. Boat documentation also requires knowledge of geographical and cultural environments.

In short, this manual begins to build a cadre of individuals knowledgeable about boat documentation.

The Challenge of the Boat

Watercraft are mobile objects that as a class are second only in size to architectural and civil engineering structures. Some examples represent the largest man-made moving structures. Manuals and techniques for the study of buildings are common, for as part of our built environment, they are all around us. When considering boat documentation as an exercise in measuring, drawing, and recording, the procedure resembles superficially that for architectural recording. An easy assumption is that skills needed in the two fields are parallel and transferable.

Recording boats, however, is made complex by the boats themselves. Watercraft are curved shapes, which need special skills to create. The problem is one of translating the flat and square surfaces of materials and paper to three-dimensional curves. Measuring requires similar skills to render these back into flat surfaces and to understand and delineate their construction. These skills are less common than architectural skills, just as boats are less common than houses. These methods are basically the same for all watercraft, but are often required to be applied in greater detail by the goals of a small craft documentation project. However, commonly needed by both are drafting skills, whose basics can be gained in a basic drafting course. Marine drafting, however, requires the recorder to go far beyond the straight shapes of the architectural and mechanical delineator to the world of special tools: splines, curves, battens, and ducks.

With watercraft, studying a large vessel is

equivalent or analogous to studying an industrial structure. Large ships, like large buildings, were built with gangs, many people, often specialists in particular jobs, always professionals. Their preservation and recording generally requires similar professional skills, with significant budgets and a team lead by a specialist. Their scale also makes recording details difficult. And in fact, for most large vessels built since the turn of the century, plans exist, especially for those built of steel and iron.

Small craft were originally built by one or two men, or at best, a small crew. Until the late nineteenth century, their designers were their builders. Not until recreational boating came on the scene did the professional naval architect turn his attention to boats. For the naval architect, small craft design needed a market to become profitable. Not only did the recreational market become large enough and competitive enough to require boat building production lines, it also made designing highly competitive, especially when customers turned to racing.

Thus plans of small craft built prior to World War I are rare, and those of regionally specific work boats are virtually nonexistent. Even when plans may have once existed, boat yard fires often consumed them. Half models (carved scale models representing only one half of the hull) were used as design tools by builders, but these, too, often found their way into wood stove fires.

In documenting and studying small craft, fine attention to detail is often required. Construction features can differ among boat types, even when the boats are similar below the waterline. Similarly, in architectural recording, attention to detail is generally greater in documenting a modest house or architecturally inspired mansion than it is for a factory structure, a chemical plant, or a ship. When documenting small craft, a level of detail greater than practical for the study of a large vessel is the rule, not the exception. Reproduction of historic small craft is often economically feasible, and indeed, often necessary to understand the building process. A plan drawn with sufficient detail to permit reproduction is an achievable standard.

Recording small craft requires modest personnel and financial resources, resources similar to those used in their creation. Solo workers or teams of a few, if properly trained and with some experience, can do the job. Professionally trained naval architects or marine engineers are not necessarily needed, and indeed much of the best work in the field has been done by those for whom small craft recording skills were developed without formal training.

Recorders of buildings usually have the additional advantage of time over boat recorders. Buildings are generally meant to last more than one or two generations. This means that there are generally more of a certain type available for study, and recorders can afford to be selective. Because they are built to last, saving their details does not necessarily require documentation; it usually only requires good maintenance.

Watercraft are far less durable than buildings. As John Gardner, one of the pioneers in small craft preservation, says: "Wood was meant to rot." Similarly, iron and steel were meant to rust. Ships and boats were designed with a working life span of twenty-five years or so, a generation. They operate in an environment whose hazards range from storms that overwhelm vessels to shoals that rip out their bottoms. They are rusted and rotted by rain water from above and sea or lake water from below. When they sink, they are conveniently erased. Until the early years of this century, watercraft were primarily built for moving passengers and goods, for fishing, or for war. More recently, going to sea for pleasure has become a primary purpose for building watercraft. When watercraft become obsolete by competing technologies or new designs, motivation for their owners to continue to maintain them disappears. Documentation, in many cases, is the only way of ensuring their permanence.

Confronting the Boat

The first and most important decision to be made in a documentation project is whether to do one at all. To answer the question requires a more detailed examination of the various levels or types of documentation projects possible.

The most important question is, why? What is the record for? What is the project's objective? Do you want to nominate the boat to the National Register? Add it to a national inventory or union list of historic small craft? Replicate or build a new one? Is the boat important for family or local history reasons? Do you think it might have regional or national importance? Do you need to understand it to be able to understand better the culture of its makers and users? Or do you have an economic interest? Perhaps you are a boat owner. Perhaps you are a builder interested in developing a new hull design based on a historical shape or want the same shape in a different material.

For National Register nomination, a boat needs to be "significant" and possess "integrity." "Significance" means association with historical events that are major parts of our history. Specific information about the boat's relationship to these themes needs to be provided. A boat may also be significant due to its association with a historically important person, or it may be significant due to its architectural, artistic, or engineering distinctiveness.

"Integrity" is recognized through location, design, setting, materials, workmanship, feeling, and association. Besides having documentable historical context, strong candidates for the National Register are craft that are operational or at least floating. Watercraft less than fifty years old, those owned by religious groups and used for religious purposes, and replicas and collections need special justification for Register nomination. In the nomination process, the boat needs a written, detailed description and account of its significance, coupled with photographs and possibly drawings. (For more information, see National Register Bulletin No. 20.)

Small craft need to be especially outstanding examples in order to be considered for the register. Boats that have documentable careers or that possess the quintessential qualities of a type might be eligible. Small craft that are part of a museum collection, now out of context, might otherwise be good candidates. But the collection as a whole, especially if in a historical context, might be eligible. Thus it might be difficult to nominate a twenty-foot dory to the National Register, even if it could be shown to have the essence of doriness, if it were part of a museum display. If the dory were part of the inventory of a historic dory shop, it would have a better chance, while a unique and famous thirty-foot racing boat still afloat and sailed might be an excellent candidate.

A boat documentation project can survey the types found within a region. While this manual emphasizes the tools needed to document in detail an individual boat, regional surveys can be equally valuable. For these, general descriptive information for the craft is ade-

Documentation Levels and Purposes

Documentation Type	*Purpose*
I. Basic description and dimensions	Regional surveys and inventories
II. Photographs, delineations, and basic drawings	National Register nominations and catalogs
III. Accurate drawings of shape and construction	Building/replication drawings, basis for analytical study, study of changes in a boat through time
IV. Analytic study of shape and replication	Study of boat performance and use in its physical environment
V. Detailed contextual study	Study of the boat's role in its social and cultural environment
VI. Comparative physical and cultural study	Tracing migrations of types and people; analysis of match of boat and environment; comparison of folk types and historical study of the relationships between folk and popular type

quate, including dimensions, sketches, or photographs of key features, colors, and historical information on the ownership of the craft. The descriptive elements now being developed for the Museum Small Craft Association's Union List project[2] and in use at Mystic Seaport Museum for its Vessel Authority file are excellent guides to the recording needs of such a survey.

A documentation project can also record the shape and structure of a boat to an extent that exact reproduction is possible. However, accurate drawings alone are not enough, as Barry Thomas (supervisor of the Exhibit Boat Shop at Mystic Seaport Museum) found out in reproducing a catboat built by the Crosby family. As experienced builders, Thomas and his crew could have arrived at a boat that faithfully reproduced their drawings, with methods they were accustomed to using. But those were not the Crosby techniques, methods that evolved over a half century arriving at an unsurpassable economy and efficiency, and which could only be determined by working with the last of the builders.

A "deep" or complete documentation project might attempt to explore the behavior of the boat in all wind and sea conditions in which it lived, doing the various jobs for which it was designed, something that has been of tremendous interest to scholars interested in experimental archaeology.

Finally, the lives of builders and users could be recorded, how they used the boat, what they thought of it, what impact it had on their lives—in short, the cultural context of the boat. Boats are built for a purpose and exist within the strict environmentally defined parameters of seaworthiness, speed, economy, and capacity. Solving the problem of getting afloat is constrained by the environment. An unseaworthy boat sinks; the physics of the medium puts limits on speeds achievable given power and size of a vessel. What is considered satisfactory, however, differs from culture to culture. Investigating unique solutions to the technical problem of doing a particular job on the sea involves the boat's human element: the values and standards used in creating and using it.

Going beyond a particular boat and its context, one could then ask questions of where similar boats and cultures are found, or why neighboring ones are different. The documentation circle, from in-depth to survey, would be complete.

Imminent demise, or in architectural terms, the wrecking ball, is often where a boat documentation project starts. Some one says, "Get it off my land," or a drought reveals a sunken boat in what was a lake bottom. What is the basic information needed before one can decide what to do? If the boat could be headed for a museum, enough information needs to be taken to help make that decision, but once the boat is stabilized, it may be more important to document the way it was built and used. Those sources, the people that built, used, and knew it may still be available, though aging, while its shape and construction can wait. If the boat is in a museum and restoration for display on land or water is being considered, the boat's existing state must be well enough recorded so that information is not lost in the restoration process. Prospects of a restoration to give a boat a new use in life can lend urgency to documentation that might not be felt in a museum project.

Assuming you have decided that documentation is in order and you have some idea of what you might want to do, the next question is "Has it been done?" Have documentation projects recorded the same or similar craft? Have they investigated the questions that interest you? This question must be answered in order to justify the considerable time and expense of a documentation project.

Expert assistance can be found through your local maritime museum. If you are your local maritime museum, find one with documentation experience. The Maritime Preservation Office of the National Trust for Historic Preservation can put you in touch with some one who can help, if you know of no nearby museum. University history, folklore, or anthropology departments and state archaeology or historic preservation offices can also plug you into the information network.

Professional advice will help ensure that you don't do something that already has been done. You may start by wanting to record the shape of the boat, but may find that it is a type whose shape is well understood. Instead, the idiosyncrasies of construction and joinery that render it different from a boat built on the next island, or the boat from the next lake, may be unrecorded and may lead to searching for an explanation of the differences.

Experts can help you review and streamline the recording process. They can help you gain support by letting others know about projects in process. They can help you make decisions about what should be looked at and documented. Experienced people can review data taken, drafts of information, and drawings.

Besides your local maritime museum, the other major source for expert assistance is the library. There you can find case studies, surveys, and monographs—the results of other documentation projects. Seeing what others have done with the products of small craft documentation is essential towards helping you make a decision. As with many projects, the idea is the first step, a trip to the library is the second.

You should next determine the level of planning, preparation, research, and recording for your project. An Adirondack guide boat built by Dwight Grant is a boat whose shape, construction, and context is well recorded. Documenting your boat might entail researching only its history and use. Or your project might record how the boat differs from a "standard." If your guide boat had a builder's plate from a hitherto unknown builder, greater efforts to record shape and construction would be justified.

Once you know why you want to document the boat or boats, set the project's scope, and determine that it has not been done, you can decide whether you can do it. Ultimately, it comes to measuring your resources and skills against the need. If the need is greater than the estimate of your means, get help from the preservation community.

Notes

1. This introduction is an attempt to summarize many of the ideas discussed in the meetings of the Museum Small Craft Association and benefited greatly from a close and careful reading by David A. Taylor, followed by Paula J. Johnson, whose help is gratefully acknowledged.

2. This project is compiling a list of watercraft in the collections of American and Canadian maritime museums and plans to expand this compilation to other museums and private owners and associations of historical watercraft. This information is being kept on computer, and plans call for its distribution on paper, on disc, and on networks, beginning in 1992. Contact the Museum Small Craft Association for more information.

Chapter 2

Stabilization: Managing the Boat

Methods for preserving boats

Peter T. Vermilya

You have a responsibility for the welfare of the boat you are documenting, both before and after the process. Documentation is only one step in preserving the heritage embodied in the boat. It's a mistake to think that once you have documented the boat, you have captured its essence and can discard the boat itself. No amount of recording facts and figures can substitute for the boat itself. When it is gone, so too is the feeling of immediacy, the tactile sense, the sense of relative proportion, the experience only given by the real object itself and not by some two-dimensional substitute. In addition, by discarding the boat, you deprive future generations from the chance of using improved techniques and superior knowledge to do a better job of documentation.

This chapter deals with the first steps to take upon gaining stewardship of the boat and then goes on to outline ways to ensure the boat's lasting welfare.

On Finding the Boat

If you are lucky enough to find the boat "in situ," uncollected, undisturbed by previous plunders or well-intentioned investigators, count yourself among the very fortunate. Your first step will be to make a record of both the boat itself, as found, and its immediate locale as a future reference tool. This effort with camera and notebook should be done with care.

Take photographs of the boat as you find it, whether unspoiled, "in situ," or not. Start with the bow and work your way around the exterior, taking shots at each quarter, as well as the stern. Take shots of the interior in the same methodical manner. Try to take full-length raking shots of the interior in order to inventory loose parts and to establish their relative positions.

Label all loose parts with consecutively numbered tags and make an inventory keyed to these numbers. If the parts are at risk, remove them to safe storage, but take photographs first.

Record the relationship of the boat to the location in which it was found. How far is it to the nearest usable water? To inland transportation networks? Are there allied tools used with the boat lying nearby? Throughout these initial steps your goal is to set the boat within what remains of its cultural and geographical context before it is moved or otherwise changed by the preservation or conservation process.

You will have to use your judgment as to what effort is justified if the boat has no relationship with its immediate surroundings other than one of storage.

Assess the Boat's Condition

Your second major step will be to determine the boat's physical condition. You are doing this for two reasons: to establish a benchmark for future comparison and to help you decide what immediate steps are needed to stabilize the boat. Put simply, to stabilize a boat is to prevent any further deterioration. Keep in mind while you are assessing the boat's condition that you may have to decide whether the boat can be moved safely.

Look first at the overall shape of the boat. Is the boat unfair to the eye—that is, are there marked discontinuities in the lines, are there humps and bumps or flatness where there should be none? If so, the boat may have lost its original shape because the underlying structure is no longer sound. The boat may be loose and flexible because it has dried out. You will have to find out if the boat has lost its structural integrity in order to take proper remedial steps, before either measuring it or moving it.

If you do not feel competent to make such an assessment, get help. Appeal to your qualified local wooden boat builder, to the Museum Small Craft Association, to a member of the Council of American Maritime Museums, or to a qualified marine surveyor. For names of qualified surveyors, contact NAMS (see Appendix A). Because much of your planning will depend on the conclusions you make at this time, effort and money spent at this stage is worthwhile.

Short-Term Steps

Your short-range plan for the boat should cover selecting a work site, or a storage site if work cannot start immediately, selecting a method for getting the boat there, and seeing that the boat is properly supported and protected from the elements.

Should the boat be moved before the documentation effort proceeds further? Do you have a choice? Is the bulldozer imminent? Will you have to move the boat eventually? Is the site adequate for working and is it secure?

There are obvious advantages to doing your measurement work on site. Threats of potential damage to the boat and personnel are reduced if the boat is not moved. Especially if the boat is large or in poor condition, this is not the time to try to learn the art of boat moving. The boat may lose shape during the moving process in ways that are difficult to detect and to reverse, and parts may shake loose and be lost. In addition, if time is short, you can get right to the job of measuring.

On the other hand, moving the boat allows you to choose your work environment. An ideal working space is:

- sufficiently heated and lighted so that personnel can work comfortably
- sufficiently large for the set up and for walking around the boat with some ease
- secure from vandals and environment conditions
- accessible to the project staff.

Moving the boat also buys you time and control and the ability to double-check your work at little cost. A venerable naval architect was hired by a prominent marine museum to measure and draw its small craft collection. Because he lived an hour away from the museum, he would measure one or two boats during a given day and then repair home for a matter of weeks to draw the plans. If he missed a detail or two, it was understandable because of the speed with which he had to work. To delay the drawing process by returning to the boat to get the necessary missing detail or dimension would be too costly. He drew the boats in large part from good data but in small part by guess work. He has since been the target of revisionists, and his entire work is thought to be unreliable. All because his drawing board was an hour away from the boat.

If you can choose a work site that allows you to compare your boat to other well-documented examples of its type, you may be able to concentrate your documentation effort on those aspects of your boat that are different.

Choosing the Best Moving Method for You

If you decide the boat should be moved, your choice of method and whether you elect to move the boat yourself will be guided by factors such as the boat's size, weight, and shape, the distance to be traveled, the surface to be moved over, the available tools, and, most important, your own skill level.

For short distances, hand carrying requires the least skill and is fast and safe. Be sure to distribute personnel around the edge in a way that ensures that all carry an equal amount of weight. Pay attention to individual height and place carriers accordingly. Be sure that they do not grab vulnerable parts of the boat. Have the carriers lift to commands.

A six hundred-pound boat might be moved easily by hand by a dozen well-directed but unskilled people, while the same boat may pose a real problem to a team of two skilled people even if they are using sophisticated tools. Examples of such situations would include moving a boat across uneven terrain or carrying a wide boat through a door while the boat is tilted on its side.

Generally speaking, unskilled personnel should not be asked to carry more than fifty pounds apiece—even less weight if the boat is especially unwieldy or overhead lifts are involved. Have only as many people on hand as you feel you can meaningfully employ. Too many people, standing by idly, soon degenerate into a social affair, and concentration suffers.

Up to a point, wisdom and experience compensate for lack of youthful strength. If the move requires too much physical exertion, you have probably chosen the wrong method. Think every move through, and talk it out with everyone present. Move slowly. One person should be clearly in charge.

For many years, boat yards moved very heavy boats using simple hand-operated machines, which were only a little more advanced than those used by the Egyptians to raise their pyramids. The traditional boat mover used rollers, planks, wedges, block and falls, pry bars and mauls, plus blocking which was never in sufficient supply but which he knew intimately. (Boat movers had the shapes and sizes of their blocking memorized and knew which combinations could be used most effectively). Later, wheeled dollies, come-alongs, and hydraulic jacks eased his labor. Although the skills involved were not complex, and the tools were simple, there was often considerable potential danger to boat and personnel. The traditional boat mechanic gained skill through hands-on work under the direction of those with more experience. Such opportunities are rare today. If you do not have experience using these tools, leave the work to those who do, or choose other ways. Most situations allow the use of modern machinery, such as fork lifts, travel lifts, and vehicles.

If you load the boat into the back of a truck, be sure to block the boat well, both fore and aft and side to side. Take the time to pad the boat well and be sure to tie it down securely. Do not rely on gravity to keep the boat from shifting. If your truck bed is eight feet long, do not attempt to load a boat that is more than twelve feet. Most boats are more stable when put into the bed stern first, the padded transom bearing firmly against the forward end of the box.

All lines should be well padded where they bear on the boat. Pad them with old rags, foam, or rugs taped firmly in place. Tubular foam, sold at most canoe franchises, works well, as does pipe insulation, sold at plumbing supply houses. Wide nylon webbing also works well, especially when combined with a ratchet tightener. Once you are underway, stop frequently to check for chafe and line stretch.

If you plan to load a boat on a roof rack, pad it well to prevent chafing. How long a boat you can safely carry on roof racks depends partly on the distance between the racks as well as on the weight and condition of the boat. If the length of the boat extends too far beyond the racks, it is easy to damage the boat by bending it when tying down the ends.

Pay attention to the suspension system of the vehicle you plan to use. Truck suspensions are usually kinder to boats than trailer suspensions. If the suspension is too jarring, the boat's caulking can be bounced out of its seams.

Most trailers come with adjustable roller poppets, hydraulic poppets, or are flat beds. A "poppet" is the physical element that provides side support, whether the boat is on a trailer, cradle, or on blocking alone. Most rental trailers have adjustable roller poppets. Their usefulness is limited and depends on how accurately they can be adapted to the shape of the boat in question. The flat-bed type is more easily adapted, as the blocking and poppetting can be of any number and shape required. Most museums opt for a flat bed if they frequently move a variety of boat shapes.

Large flat-bed and hydraulic poppet trailers require special equipment for towing and considerable knowledge about positioning and securing the load for safe operation. They should be left to personnel who have had extensive experience in their operation.

If your ability to move the boat safely is marginal, consider getting professional help. Again, your local maritime museum, the Council of American Maritime Museums, or your contacts in local boat yards can provide names of individuals competent in boat moving.

Research the question of moving with care. Consider, for a moment, the difference in perspective and attitude that may exist between the personnel working at the local boat yard and you, a member of the maritime preservation world. The local boat yard operator is in business to survive economically. This fact shapes his decision making. He is in the habit of doing a job in a manner that costs him the least. He may not be willing to take, or even be aware of, the extra measure of care necessary when dealing with an old, fragile boat. Such boats present little economic value to him and may get an equal amount of his respect. He also will usually have a different time perspective—he thinks in terms of getting from year to year whereas you should be thinking in terms of decades, at least. In contracting with a commercial enterprise, look for the exception to this rule.

Short-Term Care

For short-run storage, you can use blocking and minimal poppetting for supports. Often, centerline blocking and a pair of three-legged horses (fig. 2.1) will suffice for a short time. A better system, however, includes two or more pairs of poppets, perhaps adjustable "Brownell" type poppet stands. Poppet pairs should always oppose each other, side to side. The length of time a boat can safely be stored in this manner depends on the boat's structural integrity and how secure the area is.

For a boat with a real structural keel, almost all of the weight should be on the center blocking. Poppets are used only to keep the boat upright.

Because casual support increases the risk of damage to the boat, a more formal cradle system should be assembled and used as soon as possible. Passing traffic can loosen supports. Frost and precipitation can cause the ground to heave or sink and thus creates treacherous footings for a system that once gave evenly applied support.

Whether it is stored inside or out, the boat should be covered, even for short periods of time. Inside, the boat is subject to dirt and vandalism. Outside, the boat can be harmed by rain and drying winds. Covers should be supported by frames strong enough to resist high winds and be ventilated to avoid moisture build-up and to minimize rot. Frameworks with rafters that overlap the sides of the boat are good for ventilation. Screens in openings will prevent animals nesting in the boat.
Do not use plastic for covers because it can cause sweating and the greenhouse effect. Self-sealing cotton duck, treated for mildew, is a good choice for covers because it breaths. Inspect the cover regularly to make sure it is in good condition.

Long-Term Care

Most wooden boats depend on some combination of their wood-to-wood fits, their fastenings, and proper support to keep their shape when stored out of the water. When a boat dries out, its structural members shrink, its wood-to-wood fits are no longer tight, and the boat becomes increasingly loose and flexible. The uncontrolled temperature and humidity levels in most storage environments cause seasonal swelling and drying. As a result, a certain amount of compressive shrinkage often occurs—that is, the wood, having shrunk during dry periods, does not swell to its former size during subsequent humid periods.

A dry boat with loose fits depends on its fastenings and support system to retain its shape. When loose boats are moved, they tend to wear at their fastenings and become even more loose. Metal fastenings have a finite lifespan. After about forty to sixty years for bronze or twenty to thirty for galvanized iron, the fastenings begin to lose their viability. Proper support becomes more and more important as a boat ages, especially if it has been damaged.

To visualize the ways a simple canoe shape will change if stored with improper blocking, fold a piece of notebook paper down the middle along its long axis. Hold the corners that are away from the fold in your fingers, so that the long edges hang out, making a canoe-shaped form. Bend the two ends in farther, and note how the sides bulge out. When a keel sags, the same thing happens—the ends tend to fall in and the sides move farther apart. This often happens when the keel is supported toward the ends but lacks support in the middle. Now pull the paper canoe ends out, and note how the sides pinch in. This illustrates keel hog, often caused by too much blocking near the middle of the keel and none at the ends. Always support the keel in at least three places.

The more complex a boat is—with multiple thwarts, longitudinal members, and bulkheads—the more difficult it is to predict how it will go out of shape, but the principles are the same. Think of wood as a plastic material, one that will flow downhill around its supports, given gravity and enough time.

Arresting Further Change

Stabilization consists of arresting further change to a boat, primarily by providing physical support, cover from the elements, and routine maintenance and by controlling the environment. The best solution for long-term physical support is a cradle.

Cradles. A cradle usually consists of a horizontal platform, which serves as a base to which are attached vertical supporting arms called "poppets." A cradle should:

- support the shape of the boat
- protect the boat from damage (overturning, etc.)
- make it possible to transport the boat
- provide access to the boat for maintenance

Cradles used in museums should also:

- not obstruct the visitor's view
- present a neat appearance

You need a cradle if:

- you cannot easily move the boat by hand and must use tools. The tools should con-

tact the cradle, not the boat. The rule to remember is "Move the cradle, not the boat."
• you must move the boat often. The cradle should protect the boat against frequent flexing.
• the boat's shape makes the boat unstable. Boats should be stored upright, with certain exceptions.

Boats are designed to maintain their shape when immersed in the water. The water supports the boat against the downward pull of gravity with an equal resistance at all points below the waterline. The water does not discriminate between those points that can best resist the upward push of the water and those that cannot.

When you build on-land support systems, you do discriminate. You seek out the major strength members of a boat and concentrate support in these areas. At the same time, you avoid heavy point loads, which might occur when the support is too localized. If you can spread support over two or more strength members, such as the keel, deadwood, and frames, as opposed to the planking and decking, you should do so.

Cradles usually have two longitudinal runners running full length along the bottom. The runners should be separated enough to make the entire cradle-boat combination stable, but close enough that the runners can fit between wheel wells and span dollie frameworks. Try to level the boat's waterline plane with the plane of the cradle.

Cross pieces are firmly attached to the runners. There should be at least three in number, and they should be distributed to suit the load, with special attention paid to heavy keels or engines. The keel rests directly on the cross pieces or on blocking stacks. If the keel is marginal in strength, you may need to provide support along its full length.

X-bracing should be placed on the underside of the cross pieces, inside of the runners. Place wedges under the centerline of the cross pieces if the cradle is going to be left in one place for any time. The cross pieces will sag over time, causing more of the boat's weight to rest on the poppets. If neglected, the poppets will indent the sides of the boat. Do not hang your boat on the cradle's poppetts.

Ideally, cradles should be absolutely stiff to avoid racking or twisting the boat's hull when the cradle is lifted. Unless you use very expensive and large dimension timber or metal stock, absolute stiffness is difficult to achieve. Compromises are usually found. Because some twisting and racking may occur, always apply lifting forces along the centerline of the cradle's cross pieces, not out toward the sides, nor under the runners.

Poppets should be in opposing pairs, usually three in number, to provide better support and to allow one poppet to be removed when working on the hull. Poppets are usually paired with cross pieces and are often angled in toward the hull, crossways, for additional safety.

Various types of poppets are available. The two main categories are stick or full, depending on how they contact the hull. A stick poppet is usually made of dimension stock and contacts the hull at one point. The support is spread over a wider area by means of a poppet pad, usually a square piece of wood equal in area to two or three times the surface area of the top of the poppet. Often the pad is padded with soft canvas, old fire hose, or a similar material. Most boat yards toenail through the poppet into the poppet pad in such a way that the pad can be removed during maintenance to the hull. When the nails are pulled, the poppet pad should never be so tight that it cannot be twisted. Poppets are usually X-braced both fore and aft and crossways to the cross pieces or the runners.
This bracing is usually doubled up when a major move is made. In exhibit situations, X-bracing is kept to a minimum to avoid obstructing vision.

Full poppets are scribed to fit transverse sections of the hull, and several pairs are used to support the hull. The fit must be good, and the edge in contact with the hull should be padded. This type of poppet is often made of plywood and is used by mass-market boat manufacturers for shipping purposes. Full poppets make maintenance difficult if they have not been made easily removable, and they definitely obstruct sight lines.

When used for exhibit purposes in a museum, cradles should be unobtrusive, not a woodworker's personal statement. Muted grey or medium brown paint works well.

Boats with structural keels should be stored upright. Canoes, coracles, kayaks, and rowing shells are often built without a strong keel. Store these types of boats upside down, with their weight resting on the sheer clamps. If you use web slings, make sure they are eight to ten inches wide, and use stretcher bars to keep the hull from being pinched.

If your boat has long ends, if the overall length greatly exceeds the waterline length (as with the deep keel square-meter sailing classes, for example), be sure to support the overhanging ends with centerline poppets.

Maintenance

Your best maintenance tool is vigilance. Set up and maintain a regular schedule of inspection, and keep a log. Look for changes in the boat's shape, support system, and cover, and keep the boat clean.

At first glance, it would seem natural to keep valuable boats as clean as possible. Dirt obscures structural details and cosmetic features. Left in place, it becomes increasingly difficult to remove safely. Dirt absorbs moisture, causing it to stick to structural surfaces and to discolor paint. By keeping structural members moist, dirt will help rot to flourish.

A dirty boat advertises the fact that those people responsible for it do not appear to value it highly enough to take adequate care of it. Worst of all, a dirty boat invites the disrespect of the public and of those people who work with the boat. Unfortunately, few people see much glamour in cleaning boats. Done right, it is tedious, painstaking work. To make the job more manageable, break down a large collection into smaller units, and inspect and clean the groups one at a time. Do not allow food in the storage area, and where visitor visibility is not a factor, use dust covers.

On the other hand, cleaning should not be an automatic activity. Boats can be overcleaned. Cleaning puts the original fabric of the boat at risk. Take care to train your cleaning staff properly. Careless work can knock parts loose, chip paint, and dim the luster of varnish.

Choose your tools with care. Use a vacuum cleaner when you can. Brushes tend to scratch paint when pushed across the surface. Vacuum dust brushes should be as soft as is practical. A damp chamois cloth works well on brightwork. Do not use the dusting sprays. They leave residues that can trap more dirt.

Inspect your boats regularly, and clean them only on an as-needed basis. The best way to decrease the amount of cleaning needed is to control the environment.

The Physical Environment

Perhaps because many people and practices involved in the small craft preservation field come directly from the boat yard, the idea that small craft should be given the same care and stored in the same conditions as any other museum artifact has met with slow acceptance beyond the narrow world of curators. Even where accepted, because small craft are relatively large collection items that consume much costly space and staff time, the idea has met with little success in practice.

A good storage environment for historical small craft starts with a tight envelope—the roof must not leak. If the building lacks temperature and humidity control, it must be ventilated. Treat the space like a warehouse, by providing a complete air turnover every five minutes. Make sure there are no isolated, unventilated sections. Try to keep the relative humidity between 40 and 60 percent and the temperature below 75 degrees Fahrenheit. If the relative humidity exceeds 80 percent for extended periods of time, mildew can become established. If the level falls below 20 percent, the boats will dry out and become brittle. Your ideal goal is a constant state of around 55 percent relative humidity and 68 degrees Fahrenheit. Monitor conditions and chart them over time.

If you have enough space, allow ample access around each boat to prevent visitors from scratching painted surfaces. Do not allow direct sunlight to strike the boats. Sealed concrete floors, with a vapor barrier under them, provide a good surface on which to move boats and the opportunity to control the relative humidity. Gravel and dirt do not provide these conditions, although they may be kinder for boats that go into the water on an annual basis. Security and adequate fire protection are a requirement.

It is good practice to store the parts of a boat with the boat itself. The exception to this rule is for sails and spars. Sails should be cleaned, rolled or folded, and stored in a rodent-free, well-ventilated, dry location. Spars should be stored on special racks designed to give support every four to six feet, depending on their dimensions.

Museums and Small Craft

A museum is often the only practical repository for your historical small craft if keeping the boat in an "as was" condition is the most important goal. Most owners need to use a boat to justify the expense of ownership. Regular use usually requires adapting the boat to a modern lifestyle and standards of safety. Boats that do not suit their owner's needs fall into disuse, are neglected, and gather speed on the down hill slide of deterioration.

If you are considering donating a boat to a museum, keep the following in mind. Most maritime museums have formal collections policies. These policies spell out the limits within which the museums are prepared to accept boats for their collections. The limits refer to factors such as geographical area, period of use, type, function, material, and size. Potential donors should not expect the limits to be compromised.

Most museums have limited space. They divide their watercraft collection between those boats on exhibit and those in storage. Exhibit space is always at a premium, and donors should not expect their boats to always be on exhibit. The storage area for the reserve collection is limited, but it should be accessible to qualified people by appointment. A good measure of a museum is its storage area, an area not generally open to public censure.

Most museums do not spend much of their restoration funds on boats that are not put in the water. Boats carry a record of their own history by virtue of their wear marks, paint layers, repairs, and altered constructions. To pick a point in the life of the boat and to restore the boat to the condition it was in at that time usually results in erasing the historical record. The record is no longer there for succeeding scholars, with more knowledge or better research methods. Restoration is sometimes acceptable if the boat in question is not a one-of-a-kind or the last of a class, or, if not restoring it would mislead the viewing public. A lot can be done with proper interpretation, however.

A museum should be quite clear with you about whether it intends to put your boat in the water for use or not. Most museums make a distinction between the boats meant for use in the water and the boats considered as part of their permanent collection. Boats used in the water are fungible assets, meant to be used up. A boat in the water is a boat at risk, both from accidental damage and routine wear. The more a museum repairs a boat or gives a boat annual maintenance for in-water use, the less value the boat retains as a study item. For this reason, a museum will often build a replica of a boat in its collection, providing not only a boat for comparative use in the water but also a chance to rediscover and maintain boat-building skills.

Most museums accept donations of boats only on an unconditional basis, which allows them to dispose of boats in their collection at will. How they dispose of them is important. Some museums make a policy of first offering the items to similar institutions, then offering the boats for sale to as wide a market as possible, either through auction or advertisement for direct sale.

If you donate a boat to a valid nonprofit museum, you may be entitled to take a tax deduction for it on your income tax return. Do not expect a museum to make an appraisal of your donation for you. The museum can help you find an appraiser, but the appraiser must be a disinterested third party who works directly for you, not the museum. Appraisals in the historical small craft field are not easy to make or defend, because the boats are no longer in production and often there is no market. Tax law changes from year to year, so it is a good idea to get expert, up-to-date information.

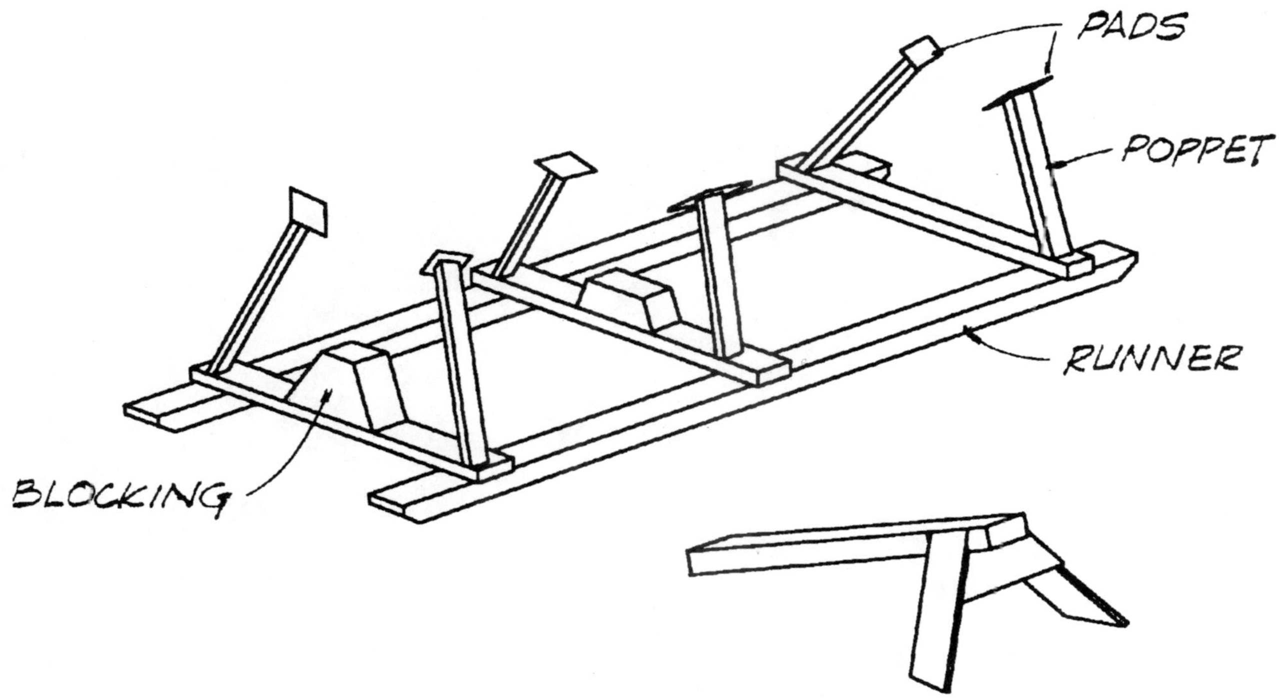

BOAT CRADLE AND THREE LEGGED HORSE

Chapter 3

The Boat as Record

How a boat changes through its lifetime

Willits D. Ansel

The boat itself is the primary source. Before a boat is documented through the tools and methods detailed in later chapters, it should be examined—working from a general, overall view of shape and lines to an increasingly detailed study.

Reading the artifact may be difficult. The process requires more background knowledge than the relatively straightforward measuring and drafting of the lines of a boat. Changes in hull shape, combined with evidence of repairs and modifications, allow the researcher to reconstruct the boat's history. Why the boat was designed and built as it was; how it fared, was repaired, and what it now represents are the basic questions.

Study the materials. Was this or that wood available or commonly used in the role in which it is found? Are the fastenings of the material, size, condition, and placement one would expect if they were original?

Consider the style of workmanship and the logic of the builder. If you see something in an otherwise logical structure that you would not associate with efficient, strong simple building, you can speculate the boat builder probably did not build it that way originally. It is an alteration or repair. Put yourself in the positions of the builder, the repairers, and the users of the boat and ask, "How would I have done it?"

If at all possible, approach the boat armed with information on its purported type, age, and history. It was designed and built to perform a job under certain conditions. What were those conditions? Compromises are inevitably involved in boat design. Though custom and taste may influence it, form and function are complementary in boats. Study the boat from a distance, from different heights, and from different angles. Think about the intentions of those who made it.

Looking at Shape

First, consider the boat's size and hull shape. Is the hull narrow or beamy? Roughly, how large is the vessel? What does its size say about the intended use? What bearing did economics, availability of materials, building conditions, and other factors have on the original size? It was not at all unusual for vessels to be lengthened or otherwise changed in dimension.

Look at the bow from various heights. What does the bow's shape suggest about the boat's performance and use? Is it full or fine forward? How did convention, construction requirements or limitations, or other factors affect this feature? Does the boat have a relatively high volume forward to provide buoyancy because its owner hauled salmon over the bow or because the bow had to be kept up when a harpooned whale sounded? Or does the high-volume bow seem to follow the once popular theory that such bows were faster? Was speed a factor? How was the boat propelled? Has it a deep or cut-away forefoot? The presence of a cut-away forefoot, a more modern variation, could help you date the example you are studying. Is the bow flared above the waterline, a feature that indicates its builder was striving for a dry boat?

Next, look at the sections. Is there a lot of deadrise or are the floors flat, or something in between? Is the turn of the bilge slack or hard? Are the topsides flared or wall-sided, or do they tumble home? Try to relate what you see to the requirements of performance and use.

Move to the side and look at the boat in profile. How much freeboard is there? Why? Was it low because the boat worked in protected waters, and the men fished or hauled nets over the side? Was it high to take rougher water or increase cargo capacity? Think of the boat's story after building, about changes in the shape of the hull, and why those changes were made. Some shapes are inherently stronger than others. Think about construction, materials, stress under way, and modification for different purposes such as those that might be brought on by a switch in fisheries.

How is the sheer: flat, hogged, sweet, extreme? Is the sheer you see now original to the boat? Is the sheer no longer fair? How has it changed? It is pulled up at the chain plates? Consider the factors that cause hogging: weights at the ends, fineness or fullness at the ends, loading, rigging, and strains. How do these factors apply to the boat? Go back to the ends and squat; look for evidence of hogging in the keel.

Move aft and look at the boat from the quarter. Speculate about the boat's performance and the conditions under which it operated.

Would the boat have pulled a quarter wave? Would the boat squat? Was it designed to be launched on and off the beaches stern first? Was it used in heavy following seas? Look at the underwater profile of the hull. Is the keel straight or rockered? If straight, does it have drag? Is the drag great? Does this help date the boat? Sometimes relatively subtle changes reflect the evolution of a type.

As you think of the intentions of the designer and builder as they are reflected in the boat's shape, observe how the shape has been held or lost. What has happened to the boat is recorded in its shape. One dragger's stern was racked and twisted by years of dragging. Another flexed with loading to the point that a stay from mast to stem went slack or taut depending on the load of fish.

The questions to ask about shape are infinite. The reasons for the original shape of the hull constitute the primary issue. How that shape was obtained structurally and how it survived follow.

Looking at Construction

Availability and types of material combined with the local construction processes influence hull form. The shape of the hull is developed structurally through the layout of the deadwood and the principal strength members including the keel, stem, frames, and floors. The sections are formed by the shape of the floors and frames. The plank lines complete the definition of the shape of the hull. Ask yourself how each component was made, what processes were involved, and how the construction of the components affected the whole. Examine each material used for what is says about the period's timber resources, woodworking technology, economy, and quality of workmanship.

For example, are the frames sawn to shape, steam bent, or hewed out of the natural sweep of the tree, or a combination of these? Are the frames in the ends set up square to the keel or do they cant (slant) to follow more easily the narrowing of the hull? Different traditions can be determined by details like these. In sawn and natural sweep frames, look to see if the wood quality was so high that the grain followed the section shape very closely.

Consider the pros and cons of the possible planking methods for a given shape and the practices common to the region where the boat was built. Some boats are easier to plank than others, and some are more easily planked using a particular method such as lapstrake or cross-planked bottom. How does the planking affect shape?

In the process of studying hull shape, you may have seen other things that tell a story. For example, availability of planking stock is often a factor in boat building. Are there butts in the planking of the boat? Did the builder have enough length and sweep in his planking stock to avoid butts completely? From the fastening pattern at the butts, you can tell if the butts are supported by butt blocks between the frames or are simply butted at—and fastened to—the frames. Butts that are not well spaced and those that do not land on either substantial double-sawn frames or butt blocks might indicate the builder's attempt to keep labor costs to an absolute minimum.

After pondering questions of original construction, consider how the boat was used and maintained over its lifetime. Reconstruct the story from study of the boat's structure, hardware, and rig. For example, you might find that rust is bleeding from plank fastenings all over the place or that the boat is iron fastened (which says something about building costs, intended use, and intended life span). But then you might notice there are many fastenings at each intersection of the planking and framing. While examining the fastening pattern and sizes, you might find that the boat has been refastened, not just once, but twice. In fact, there might no longer be any space to insert new fastenings. Short of total rebuilding, this boat reached the end of its useful life when the newest fastenings lost their holding power. From fastening samples and observations below decks, you might determine that the original fastenings were iron-clench nails, the second set, galvanized boat nails, and the third set, common house nails. Each successive repair used a lower grade fastener; a picture of the boat's decline begins to form.

While looking closely at the butts, you might notice the sections are unfair at the turn of the bilge because some planks are pulling away. Either the fastenings have wasted away, or the frames are broken, or both. Make a note to look inside to determine the cause.

Changes to deck layout and structures can be detected. Sawed-off deck beams, discontinuity in deck planking, and old paint lines may indicate modification to houses, hatches, bitts, etc. You have to learn to look for these things; think like a detective and try to see patterns.

Mystic Seaport Museum's *Emma C. Berry* was built as a sloop, rerigged as a schooner, then rerigged again as a sloop. Changes in rig may result from shifts in style, manpower, technology, or other factors. Some rig changes can be detected fairly easily from observation of stumps of masts, mast steps, mast partners, and chain plates. To uncover some changes, you may need to analyze the location, lead angles, and capacity of deck hardware to determine the arrangement and size of the spars, sails, and lines that you find wholly missing.

Go back over the boat more carefully now, moving in for more specific details of when and how the many changes occurred. Go back as many times as needed until you feel you have wrung the boat dry of information. Make sure your notes clearly state what you have seen. Take pictures or make sketches to document the details.

An Example

Let's take as a case study a specific boat, one said to be a Muscongus Bay sloop (fig. 3.1). You have knowledge of this type and related types but no specific knowledge of this particular boat other than the owner's statement that it was originally a work boat and later was bought by his grandfather and used as a pleasure boat by the family in Maine. It has no builder's plate, but tradition says it was built in Bremen, Maine.

You have looked it over from the outside, more or less as described above, and have concluded it fits your conceptions of the general size and shape of the Muscongus Bay sloop.

Figure 3.1. The Muscongus Bay sloop *Ranger*, in the collection of the Maine Maritime Museum. Although the boat is old, the shape of its hull is well preserved. The sheer is fair; there is little or no hog in the keel or sheer. The keel has considerable drag. The run is very hollow, and the quarters tumble home. The profile of the bow is simple. Although many repairs have been made to the planking and the fastenings are badly deteriorated, the plank lines are still fair. The house appears disporportionally long for a work boat requiring a large cockpit. *Maine Maritime Museum, Bath.*

From its lines and general condition, it looks old. Now you begin to look more closely.

You find a lead patch on a butt and a strip of lead fastened with copper tacks along the garboard seam. The butts were leaking, as was the garboard seam, and the owners tried to caulk it first, one supposes. Later it was leaking again, and the owners didn't have the money or time to replace the planking, so they put on the patches.

There are tacks in a row indicating missing lead patches on each side of a seam in the deadwood. This was an early effort to stop leaking along the seam across the rabbet. With further leaking, the patch was removed, and a second stopwater was added outside of the rabbet in the seam in the deadwood. The seam between the new stopwater and the rabbet was then caulked.

While under the hull looking at a false keel, you wonder if it covers a centerboard slot that is now plugged. You know Muscongus Bay sloops had centerboards, so you make a note to look inside for evidence of one. The false keel looks like it was added on; it doesn't fair in with the stem and stern in profile. It seems less worn, and the workmanship is different from the keel itself. The underside of the keel is rounded and worn. The boat has been aground. You also find worm damage behind the rudder stock.

Meanwhile you find dutchmen and cement patches in the deadwood. There are pieces in the forefoot that don't look like original construction (fig. 3.2). Following the logical sequence of construction, the stem

Figure 3.2. Ranger's stem—besides wear, checks, exploded iron fastenings, and rot—shows refastened and split hood ends, patching, and marks of former bobstay fittings. The stem is oak, cut to take advantage of the sweep of the grain; the stock from which it was cut was not particularly clear. The entrance is rather fine. The forward sections appear to have retained their shapes despite weak frames and fastenings. *Maine Maritime Museum, Bath.*

assembly would have been made more simply, with fewer pieces. A builder is highly unlikely to have built a boat that way. You hypothesize there was damage or rot in the forefoot and, working from the outside without removing the garboards, the end of the keel and the heel of the stem were cut back. A block of wood, fastened with drifts, was placed in the gap. The fit is not the same as the workmanship elsewhere. Besides that, the repair piece is yellow birch, whereas the rest of the stem, keel, and deadwood appear to be oak.

After scraping off some paint, you find that the planking shows evidence of refastening and replacement. The replacement pieces are of a different wood from the original. Actually, you suspected planking repairs because you've seen some stealers (short, very narrow planks) in places where any builder with reasonable plank stock would be unlikely to put them. Furthermore, some of the planks are very short; their butts are very close together. Looking at the wear marks and workmanship, you try to determine which repairs were done at the same time, or by the same workman. You want to reconstruct as much of the repair history as possible.

Various plugged holes suggest through-hull fittings for an engine, and then you realize you overlooked the obvious—the stern post was drilled for a shaft, now filled with a large plug. The rudder has had a lot of patching, and there is a pattern of small holes near the waterline on both faces. You suspect this is evidence of copper ice sheathing, so you look for more holes elsewhere on the hull.

Questions about the rig arise as you look at what is left of it. It has two sets of chain plates. Earlier Muscongus Bay sloops had one set or unstayed masts. You look for evidence of moved chain plates, and find plugged fastening holes in the sheer strake on one side. Scraping, you find rust stains and compression marks where the earlier chain plate lay. The strake on the other side looks new and has no similar holes. In the stem are holes and grooves where an earlier bobstay fitting was let in.

The deck has lost its crown, its wide pine decking has seams that are wide open. The toe rail does not look original because the shaping is too "yachty." The original is unlikely to have survived anyway.

There is no bridge deck as was common in Muscongus Bay sloops. The cabin afterbulkhead is vertical tongue-and-groove stock rather awkwardly fastened to a nailing strip on an transverse beam. It does not look like a logical way to have built it if it was original. The cabin sides have been lengthened. You suspect the bridge deck, where lobster traps were emptied and baited in the boat's working days, was removed and the cabin lengthened in order to increase cabin space. Dropping below, you confirm this hypothesis by observing cut ends of deck beams where the bridge deck used to be and butt blocks where the cabin sides were lengthened (fig. 3.3).

Pulling up the cabin sole hatches reveals

Figure 3.3. This is the underside of *Ranger*'s deck forward of the bulkhead between the cockpit and the cabin. The cut-off deck beam with the knee, paint lines on the ceiling, and butt block on the cabin side are evidence that the cabin was lengthened at the aft end. In the process, the bridge deck was eliminated, and the bulkhead was moved aft. Comparing paint layers confirmed the relative age of the bulkhead. *Maine Maritime Museum, Bath.*

the cap on the keel covering the centerboard slot. Old floor timbers that once butted against the centerboard case are still in place, plus new ones that go across the keel (fig. 3.4). There are sister frames where the old frames are eroded on the face and split apart by rusting iron fastenings. The old frames are of small scantlings and bent on the flat—construction features common to Muscongus Bay sloops and related types. Perhaps these repairs and alterations were made when the boat was converted to a yacht; you will have to try to date them.

Paint lines and layers help date structural members and define earlier layouts. Deck beams are useful sources of information, often having been painted with each painting-out and being the landing place for bulkheads and stanchions. Scraping the side of a beam and noting numbers of layers and colors can provide a useful key in establishing the age of other parts. Wear marks, old fastenings, and lines of removed fastenings, cleats, and nailers may also provide evidence of change.

Figure 3.4. The story of aging, repair, and modification is clear in this view of *Ranger* from below. A slot for a centerboard was plugged and capped. Floors were placed across the keel and drifted through into a false keel on the bottom of the original keel. The original frames were light, bent on the flat, and notched to the keel. The plank fastenings were iron-clench nails. Sister frames were a later repair, done after the removal of the centerboard trunk. *Maine Maritime Museum, Bath.*

Seats were put on each side of the cockpit during conversion, but there is evidence of a transom across the after end of the cockpit. The location of the transom is fixed by paint lines on the rudder box and cleats on the ceiling (fig. 3.5). The height of the cockpit deck was not changed, as shown by wear and paint lines. Standing in the cockpit, you notice it is a good working height if you are handling lobster traps. Wear marks on the ceiling under the cockpit seats were left by traps and barrels during the boat's working days. Under the deck is a solid shelf rather than a clamp, another construction detail typical of the Muscongus Bay type.

Under the hatch in the cockpit deck are the engine beds. They are roughly fitted on the floors and extend past the stern end of the capped centerboard slot. This arrangement indicates that the engine was installed after the centerboard was removed.

Now you can even generate some hypotheses. The boat was built as a working Muscongus Bay sloop, in a workmanlike manner using standard construction, workmanship, and materials for the time and place of its origin. The rig was simple and typical of the period—a single set of shrouds, probably a single large jib with one bobstay on a relatively short bowsprit. The boat had a centerboard, which was later plugged, and a false keel was added. Later, an engine was installed; the rig may or may not have been removed at this time.

The boat suffered the usual amount of wear and tear as a working fishing boat. There were some groundings, and the boat worked in ice. Seams leaked and were first recaulked and then patched over. At some point, repairs were made at a location where yellow birch was a boat-building material, presumably farther east. The boat was bought by a summer resident and converted to a yacht. The new owner took out the engine and enlarged the rig. It now had a jib and staysail on a longer bowsprit and two sets of shrouds. The new owner also had alterations made in the deck layout and cabin. The direct evidence for these changes confirms and expands upon what the owner has told you.

Finally, the style of workmanship and the logic of the boat builder tell a story. If you see something in the structure that you would not associate with efficient, strong, simple building, you can say that the boat builder probably did not build it that way originally. It is an alteration or repair. You must put yourself in the position of the builder, the individual making repairs or changes, and the users of the boat and ask, "How would I have done it?"

It helps if the person studying the boat has done some building, repairs, surveying, or

Figure 3.5. The aft end of *Ranger*'s cockpit shows little modification other than the removal of an athwartships transom. Paint lines on the rudder box and ceiling indicate the transom's existence and location. There is evidence of repair to frames and planking. The beds of an engine, now gone, are in place. Some ceiling planking was replaced in the course of the repairs. In general, the layout and structure of this end of the cockpit appear original. *Maine Maritime Museum, Bath.*

has sailed or used a similar boat, but, of course, you can read a boat without this experience. Your goal is to be observant, questioning, logical, and imaginative, and, in the process, the boat yields its story.

Figure 3.6. Detail drawings of *Ranger* by D. W. Dillion at a scale of one inch to one foot. The original drawings have been reduced for publication. *Maine Maritime Museum, Bath.*

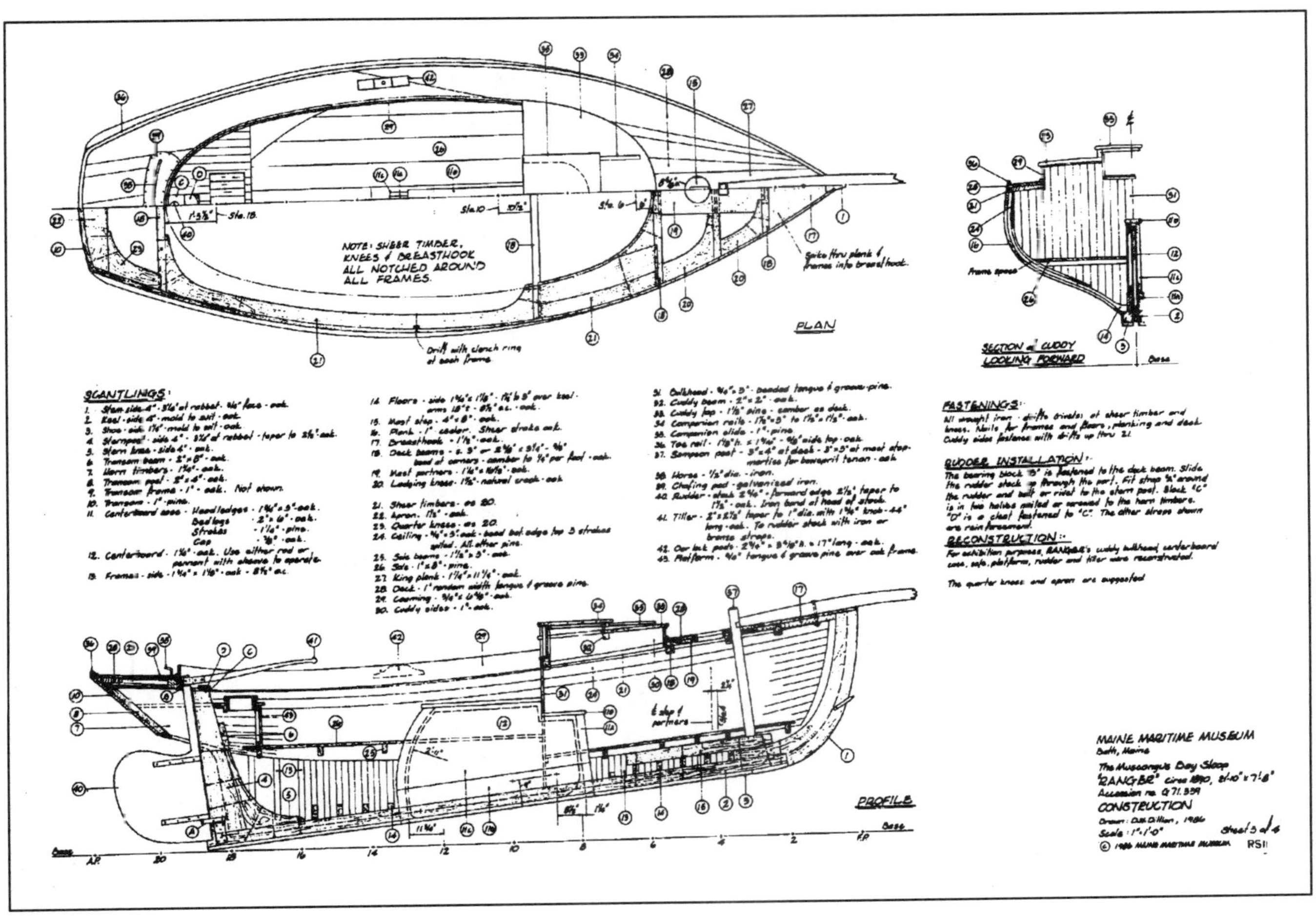

Chapter 4

Accuracy and Verification

Determining the quality of the job and how long it may take

D. W. Dillion

There are many forms of accuracy, many things that contribute to it, and many reasons why it is important. Proving that work is accurate is an ongoing process.

On one occasion I measured a fifteen-foot boat in four hours. As a result of rushing the job, I omitted the critical measurement shown in fig. 4.1. Good work takes time, so when there is time, take it.

In the spring of 1988, I measured a boat that had a shape unfamiliar to me. I plotted my measurements at 1-1/2" = 1'-0" scale as I took them from the boat. When I laid the lines down on the drawing board, the shape was not what I thought it should be. I went back to the boat to check. That shape, the very slight reverse curve shown in fig. 4.2, was there just as the notes and measurements said; not at just one section, but for at least fourteen feet of the boat's twenty-eight-foot length and also port and starboard. Because I did not "know" the boat, I was trying very hard to impose my perception of it on the drawing. You need to "know" the boat to understand its shape and construction.

My friend, the late Bob Baker, was a highly respected builder, designer, restorer, and historian in Westport, Massachusetts. When he could, he would set up a boat where he practically had to trip over it in order to get in and out of his shop. This was his way of getting the boat into his mind's eye before he started to work on it. He got to "know" the boat. What's more, he was a great detective when it came to figuring out what had happened to a boat in its lifetime. He had an intuition based on many years of observation and work.

For many of us, accuracy is the substitute for intuition when trying to trace evidence of repair work or rebuilding and modification. For example, new frames put into the boat as repairs may not be exactly the same as the originals, even when the original fastening types and holes are used. If there is a mixture of original and replacement frames in the boat, accurate measurement and inspection will help determine what is original or a replacement and sometimes will aid you in figuring out what a joint is, even though you can't see all of it.

What degree of accuracy is acceptable? How is accuracy defined? Years ago I owned a pocket chronograph. It was accurate to within five seconds of U. S. Naval Observatory time over a forty-eight hour period. That is 0.003 percent error. Apply that 0.003 percent to a twenty-foot boat. Twenty feet is 240 inches; 0.003 percent of 240 inches is 0.0072 inches, or slightly thicker than this page. That degree of accuracy is unnecessary when taking the length of a twenty-foot boat. Measuring 20'-0" as 19'-11 7/8" or 20'-1/8" is not a gross error, but measuring a thwart or knee as 7/8" when it is 13/16" is.

To the man who built the boat, accuracy was simple to define. If he cut two pieces of wood to fit together and they did, his work was accurate. Accurate fit was once defined to me this way: "If your business card fits into the joint, pretty good; if the cutting edge of a sharp knife won't start into the joint, damned good!" On the other hand, the degree of accuracy of the measurements taken from a boat will be defined by numbers and the relationship of points plotted from those numbers. The ranges of accuracy shown in table 1 are appropriate when measuring length, beam, sectional, and profile shapes and locations of hardware, cockpits, masts, etc. The ranges have not been worked out by applying percentages to the measurements; instead, they reflect the experience of measuring and what can be done by competent workers.

The ranges of accuracy given in table 1 are not suitable for measuring the scantlings, hardware, or machinery in a boat. Measuring a thwart or knee as 7/8" when it is 13/16" is really an error. What then is acceptable when measuring scantlings, hardware, and machinery?

Anything made of wood is subject to variations. Humidity affects the dimensions of wood from day to day and some woods more than others. The greatest dimensional changes due to climate will be across the grain (fig. 4.3). The greatest dimensional variation due to workmanship is caused by the way rough-cut lumber is made into usable boards. The workman hand planing a rough-sawn board to make a thwart probably did not worry a great deal about whether it was slightly thicker or thinner than the 7/8" that he planned. It may not only be slightly thicker or thinner overall, but it also may not be uniform in thickness from end to end. A board that has been planed to thickness in a planer will be much more uniform.

How do you measure these things accurately? Measure wood in several places on the same part and average small differences. Record the large differences that do or may affect shape. A range of ± 1/32" should be expected and should allow for the slight variations caused by finishing and refinishing.

Manufactured metal items—fastenings, hardware, and machinery—are usually very uniform in their dimensions. Most small boat hardware is simple in form and does not require measuring to 0.001" (1/1000"). Such items should be measured accurately enough to be reproduced or picked out of a catalogue. If there is any manufacturer's identification on the hardware, record it and the basic dimensions in a sketch, and take a good close-up photograph of it.

The Measuring Team

The selection of the individuals who will measure the boat is crucial. Three is a good number, with four the absolute maximum. Having more than four workers causes confusion; there may not even be room enough for them to work. Decide who will do what job. Recording a boat should not be treated as a learning experience, even though it frequently turns out that way. If you have a mix of experience levels on the team, assign jobs according to ability and experience. Talk the project over, and agree among yourselves who does what job. Make it a team effort.

First, evaluate the person who is going to do the job. If that person is a known quantity, a person with credentials, you may be all set. But, if that person has no credentials, training, or experience, the value judgment on that person is yours. Assessing ability is difficult, especially during an interview. Sometimes the enthusiasm shown by the candidate leaves the strongest impression. Enthusiasm, however, does not ensure competence. More important are intelligence and determination. While enthusiasm is a great starting impulse, determination will sustain the enthusiasm, and intelligence will make the observations, ask the questions, and solve the problems.

If you have three people on the team, assign a notetaker, a measurer, and a plotter. The notetaker's job is to make up the forms on which the measurements are recorded and to do the recording. The notetaker should be in control of the procedure by comparing the status of the notes to what is actually being done. The notetaker should always see and hear what the other team members are actually doing not what they say they are doing and not what the notetaker thinks they are doing. When the time comes to sketch and measure the construction detail, you might want to reassign jobs. Perhaps more than one of you is able to do the sketching and measuring.

The measurer's job is to take the measurements that the notetaker wants. The plotter should make up the grid forms necessary to plot the measured points—sections, profile shapes, etc.—and plot them as the measurements are called out by the measurer to the notetaker.

If you have four people, assign the fourth as a checker and helper, to keep everyone else on the ball. This person might help the measurer get lined up either on a point or with the equipment or move lights around so the point being measured is visible.

At some point you may decide to switch jobs. For instance, the notetaker may not be able to sketch the notes or write legibly under field conditions, or the measurer may get tired of crawling around under the boat.

If this is your first time measuring a boat as a team or the first for each of you individually, practice the measuring procedures and notetaking before actually recording anything. There is nothing wrong with taking a maiden voyage. Just don't sink anyone else with the results. Your practice should include setting the boat up so it is ready to measure, then moving it and setting it up again. Do the same thing for the bow profile, the transom, and a station section. Of course, you should follow the instructions in this manual while you practice. Keep practicing until you feel that each of you has your assigned task right. Do not be afraid to check each other to see that you are each doing your job.

The Method

Consult the appropriate chapters of this manual, and decide which method to use. Write into your notes which method you have chosen, and indicate the chapter and pages. Choose the method that suits the boat, the conditions where you will work, the time available, and the capabilities of the team.

Even though you have selected a particular method, do not get locked into using it. If your practice session shows that your choice or a part of it doesn't work for the boat you are measuring, change methods or the part that doesn't work. Maybe part of another method will work. Use it, but note that you have changed or mixed methods.

If none of the team members has measured a boat before, there is only one expert present—the manual. If there are any disagreements, refer to the manual.

Procedure

Determine a rough work schedule based on the method evaluations. Don't assume that if it will take one person eight hours to do the job, four will be able to do the job in two hours. Include time for bringing the boat to the work area, rounding up help to move the boat, and gathering bracing materials to support the boat. I moved one boat from under a barn in order to measure it. Working alone, that took me four hours. Don't forget the time needed to get your measuring equipment and materials. And don't forget coffee breaks, lunch, and kibitzing.

Make sure that everyone is clear about the measuring process described in the method that you have chosen and that they know how to use the tools. Then—and only then—go to work.

Time and cost for a job is another piece of important information towards making a decision about a documentation level. It is not easy to estimate it. The boat's complexity, its environment, and the skills of the documentation team are all critical variables. *Susie*, the 13' Woods Hole Spritsail Boat used in some of my examples, took two days to record and about two weeks to produce three drawings. A simple 15-foot pulling boat could be recorded by one very experienced person in about four hours, with another eight to twelve hours needed to finish the first stage of drawing: rough lines in pencil. For a small semidecked open boat, a complete job with inked drawings might take ten to fifteen working days. Larger boats take longer. *Imp*, a 1929 28-foot Gold Cup racing boat, looks deceptively simple, but in fact just recording its construction and the subtleties of its shape took one man a very full forty-hour week. For a 25-foot, decked Muscongus sloop, a solid month was needed to get to four finished drawings.

These are all estimates for experienced persons working from their own notes. An experienced team can reduce the measuring and recording time. Add time if you are inexperienced: persons with graduate architecture and drafting experience, but with no maritime experience, can take up to twelve weeks to produce three to four inked drawings. Inexperienced delineators may need twice as much time as that of experienced in just doing the drawings.

If you want to record each stage that a greatly modified boat has gone through, allow for it both in the taking off and the drawing. It can take double the normal time just to analyze the evidence. Once recording is complete and the stages sorted out, two to four weeks may be needed to go from pencil drawing to finished inks.

A decision to record accurately the shape and construction of a boat is not to be made casually. Adequate funds and hours need to be budgeted. Motives need to be clear and justifiable.

Do everything you can to familiarize yourself with the boat. Shine a light along the hull in all directions to highlight the shape. Bend a batten on it to show up unfairness. Lie on your back under it, or turn the boat over if you can. Run your hands over the hull first with your eyes open, then with them closed. All of these actions will help you "see" the boat's shape.

Organize Your Notes

Set up the measuring equipment accurately, use the right tools accurately, and keep good records. All team members should participate in setting up the measuring equipment at each location. Be sure that when you take measurements they are in the plane the measuring equipment is set up to define on the boat.

The measurer should call all instructions back to the notetaker, who should then repeat all measurements out loud. If you have any doubt, repeat it all again. The notetaker should watch to see that the measurer is doing what the notetaker thinks he is doing. This works both ways, the measurer should also watch the notetaker. Don't assume that the other person knows what he or she is doing or that he or she is doing it right.

No member of the team should goof off. If there are three of you and only two can work, the third should watch what the other two are doing and act as a sounding board during discussions about methods or the details of the boat.

Record measurements as they appear on rulers or tapes. If your ruler reads 5'-3 7/8", write that. Do not use the form common to a table of offsets and write 5-3-7 instead. A notation like that is taken only from a faired lofting or drawing. Also, someone who isn't familiar with a table of offsets might take 5-3-7 to mean 5'-3 7/16".

Some people measure with a tape by starting 1" or 1'-0" from the end because the end clip is loose. Then they subtract that 1" or 1'-0" from the measurement to compensate. They feel their measurements will be inaccurate because of the loose clip. Most of those clips move about 1/32" to allow for the thickness of the clip when taking inside and outside measurements. Forget that 1/32". It is better to have that 1/32" error instead of the resulting 1" or 1'-0" error if you forget to make the necessary subtraction.

The most common errors in measuring and recording are very simple ones. For example, you might write the wrong inch or foot, or 16" instead of 1'-6". Keep this in mind when measuring, taking notes, and plotting.

I learned some of my boat measuring practices from land surveyors back when everything was done by hand. There were no electronic measuring or recording devices available in 1959. All of our field data was written in a notebook and then taken into the office to be "reduced" (converted to readable form) for plotting onto maps. We did this because even writing the notes can be difficult under field conditions. Don't attempt conversions in the field.

Take overall measurements two or three times to check them. If they differ, average them and use the result. For example, if you measure the length of the same boat three times as 19'-11 7/8", 20'-0", and 20'-0 1/8", the best answer is 20'-0". Or, 19'-11 13/16", 20'-0", and 20'-0 1/4", call it 20'-0" because 19'-11 13/16" is 3/16" short of 20'-0" and 20'-0 1/4" is 1/4" long. The average would be 20'-0 1/32" and 20'-0" is acceptable. If however, you measure the boat as 19'-11 7/8", 21'-0", and 20'-0 1/8", someone is reading the tape wrong, measuring to the wrong place, or reading the tape beginning at 1'-0".

For examples of ways to take measurements, refer to fig. 4.4. The overall measurement must be taken. Taking running measurements will automatically give the overall measurement. The chance of accumulative errors is nil. The errors that do occur are most often in units of one inch or one foot, both of which can be sorted out by trial and error on the drawing board. Additive measurements are most prone to error because of the degree of accuracy possible between points.

Do not expect the total of additive measurements to equal the overall measurements, because additive measurements will be accurate to ± 1/16". This will be close but not exact and therefore will not, when totaled, equal the overall measurement. However, the result should be close, say ± 1/8".

Plot the field notes as the measurements are taken or at least before you move the measuring equipment. If you find an error while plotting, it will be easy to check if the measuring equipment is still in place. If you have the slightest doubt about anything that you have done, whether setting up the measuring equipment or taking a measurement, check it. You should spot-check your work occasionally anyway.

Take photographs that show the shape of the hull. Use portable reflector lights rather than a camera-mounted flash, which may wipe out what you want to show. Adjust the lights to produce the shadows or highlights you want. Photographs taken inside a cabin may require a combination of flash and reflectors. Take photographs so that they show both profile shapes and locations. Try to take photographs from overhead. You can turn a small boat up on its side for the same result. Take a panoramic—overlapping—series of photographs.

Sooner or later everyone gets bored, tired, or fed up with the work. When that happens, take a break. Get a cup of coffee, or walk around the block. When you go back to the boat, make sure that neither it nor the measuring equipment has been moved. Then go back to work and look, look, look, check—then look again.

The procedure I follow includes twelve steps:

1. Walk around the boat and get to know it.
2. Measure the scantlings and begin to sketch the details.
3. Photograph the boat.
4. Set the boat up plumb and level.
5. Organize the notes, including the material from 2, above.
6. Set up the measuring equipment, measure for shape, and plot the measurements.
7. Keep a running list of things that need to be done, including those missed as a part of 2, above.
8. Do the things in 7, above.
9. Go to the drawing board, and lay down the lines and construction plans.
10. Keep a running list of questions about the boat. (My habit over the years has been to "build" whatever I am drawing in my head as I work. This leads me to questions about how the boat is built.)
11. Take prints of drawings and lists of questions back to the boat to get the answers.
12. If the boat is complicated, I may go through steps 9, 10, and 11 more than once.

Verification

Much of the material on procedure contains the essence of verification. By checking what you do and plotting measurements as they are taken, you are verifying that you are working accurately. The best form of verification is to check while you work. An old carpenter's admonition, suitable here, is "Measure twice, cut once".

Final Thoughts

Your work should be so accurate that plotting it at large scale (1-1/2" = 1'-0") or lofting it full size goes smoothly. If plotting or lofting does not go smoothly, check your plotting procedures, especially your measuring and scaling, before you go back to the boat.

Does plotting the notes give the size, shape, and location of things as you remember them? If not, use the photographs to refresh your memory. They will help sort out gross errors, sometimes to within an inch. If this is unsatisfactory, go back to the boat.

Do you have the right number of planks, frames, thwarts, cleats, etc., and do the joints look right?

When you measure a bow profile and then plot the points, you may not get a good visual representation of the actual shape. Because of this, you must combine accurate measuring, good fairing, and a good visual appreciation of the shape. Sometimes only a pattern will do the job. A boat is a complex shape to measure. There are combinations of measurements that even when taken accurately will not give coincident points when plotted. This is a legitimate discrepancy and can be compensated for in the fairing process. (See chapter 11.) Recognize that such discrepancies and out-and-out errors will show up in the best of records, but do the best job you can at each and every step.

Finally, accuracy and verification are part and parcel of the whole process. The result is going to be a historical record that may be used to write a book, build a boat, or support a hypothesis. The work that you do should leave enough clear evidence so that the next person to use your record can follow along.

Table 1

Less than 12" (1'–0")	±1/16" *(except as explained in the text)*
Between 1'–0" and 6'–0"	±1/16" *very close* ±1/8" *good* ±1/4" *no good*
Between 6'–0" and 15'–0"	±1/8" *very close* ±1/4" *good* ±1/2" *no good*
More than 15'–0"	±1/4" *very close* ±1/2" *good* ±1" *no good*

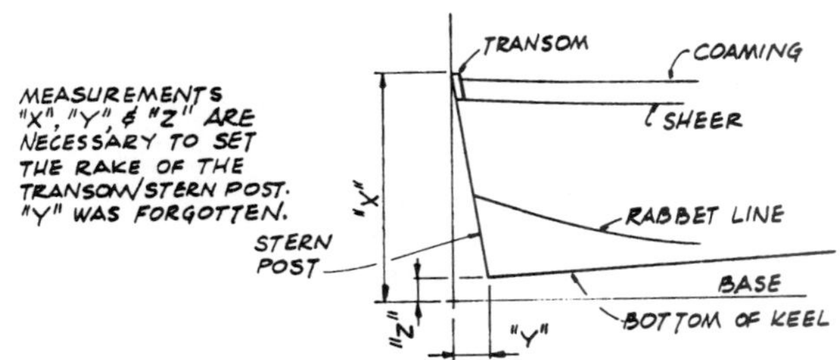

FIG. 4-1

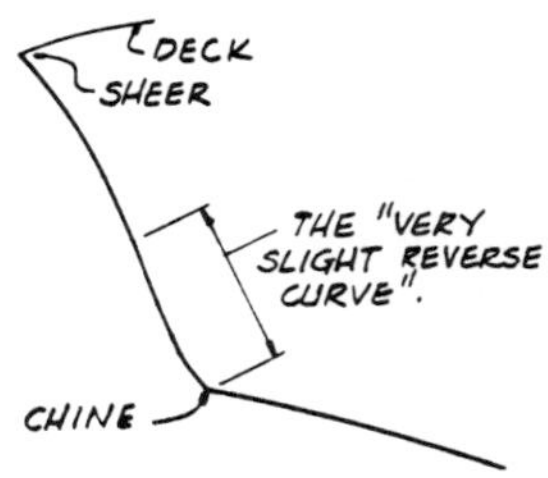

FIG. 4-2

FIG. 4-3

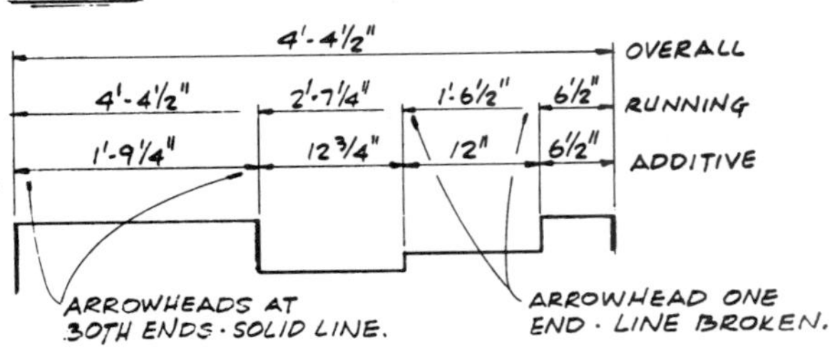

FIG. 4-4

Figure 4.1. Measurements necessary to determine the rake of the transom.

Figure 4.2. Before taking off measurements, it is important to familiarize yourself with the boat thoroughly, so that slight or gentle shape changes such as the "very slight reverse curve" shown here are not missed.

Figure 4.3. Variations in humidity affects the shape of wood, mostly across the grain.

Figure 4.4. Either running or additive measurements can be taken off. Running measurements are easier to take and more accurate than are additive measurements, which are prone to errors.

How do I know if I have done a good job? If after I have plotted all of my measurements and have found no holes, I have a lines drawing that gives, without major fairing or juggling of shapes, a good representation of what I saw when I took those measurements, I feel that I have what I went after. Weaving between plotted points where the slightest nudging of a spline will put it on either point—averaging—I think of this result as good.

To judge the accuracy of a drawing, you must be well enough versed to read lines and construction plans. Your best tool for checking is a pair of dividers; a tick strip will also do the job. Check the fore-and-aft locations of the buttock-waterline intersections in the profile and half-breadth plan; the buttock heights in the profile and body plan; the same for the sheer height in the body plan and profile; and the half breadths in the body plan and half-breadth plan. I don't suggest checking every point—only enough to be sure. If you find discrepancies, ask the draftsman to check the original work or check the original draft yourself. If there are offsets, make sure that they were taken from the original work.

In a table of offsets there are one hundred to two hundred bits of information. The law of averages says there will be mistakes. If you check a final drawing, instead of the original work, you are checking a tracing. Tracings are not always exactly the same as the original and are not as accurate. Offsets and dimensions scaled from tracings and prints almost never agree exactly with the original.

Construction drawings are easier to check by visual comparison if you have either the boat or photographs of it. In addition, check the photographs by laying them out to make a mosaic. They should give you a good perspective view of the boat. They should leave you with the feeling that you have had either a hands-on or a walking tour of the boat. Let that impression govern.

Remember that accuracy is a state of mind combined with hard work. Verification is maintaining a high level of both.

Chapter 5

Tools for Measuring, Drafting, and Lofting

The tools needed for the job

D. W. Dillion

The tools used to measure a boat could fill a small truck if you let things get out of hand. I have gone on expedition with my 1980 Saab so loaded there was barely room for me. On the other hand, I once went to Scotland to record a boat, and my tools fit in a three-quart food-storage box. Aside from wedges and props, the only "tool" that I had to find at the work site was a straight board to use as a level. Keep in mind it is possible to have so many tools that you pay more attention to them than to the boat you are trying to measure.

Where you measure the boat will determine what you need to support the measuring operation rather than the tools you need for the actual measuring. Ideally, you might work in a brightly lighted workshop with a smooth wooden floor. Working in a dimly lighted dirt-floored shed is a very different story. The tools used for measuring will be the same in either location, but in the shed you will need lights and possibly more propping and bracing material. In addition, in both locations you may need equipment to move the boat.

Most measuring or lofting tools are relatively inexpensive, but drafting tools can be quite expensive depending on the desired quality of the finished drawing. Drawing with pencil on vellum (paper) requires the least expensive tools. If you decide to draw in ink on mylar (plastic), you will spend a substantial amount of money. The prices for tools given in the following list were current as of Spring 1992.

If you have tools that approximate those listed and you are familiar with them, don't go out and buy a tool just because it is mentioned here. Just be sure that your tools are accurate—your level must read level, your framing square must be square, and so forth.

In the following list, the tools used for similar operations are grouped together in the approximate order of use. Note that the simplest tools are best because they are easiest to use. Photocopy this tool list, and use it as a checklist while you inspect the boat and the place where you will be measuring it. Imagine the processes of getting to know the boat, photographing it, measuring scantlings and details, and measuring for shape (described in chapters 6 and 7). Then check off the tools you think you will need.

Measuring and recording boats takes a few tools plus a lot of imagination and ingenuity. So be imaginative and ingenious.

Warning: Some of the instructions and techniques for using the tools mentioned in this chapter require marking, drilling, or otherwise treating the boat in an unkind way. These instructions and techniques are generally accepted practices and are legitimate. However, before doing any of these things, get permission from the person responsible for the boat. Also, test what you are going to do on a piece of scrap wood or in a concealed place in the boat.

Tools for Field Use

Flashlight: A six-volt hand lantern will throw a strong beam. Use it to see difficult to reach places or to give adequate light for focusing a camera. Cost: $5 to $10.

Reflector lights: Aluminum reflectors with clamping handles and rated for a minimum of a 100-watt bulb. Use for highlighting shape and for available light photography. Cost: about $10 without bulb.

Extension cords: Use cords you have on hand as long as they are in good condition. You may need more than one plug-in point on the extension. If you buy an extension cord, get one at least 100 feet long. Cost: about $10.

Plug and socket adapters: You may need a screw-in type adapter if the only electric power available is at a light socket from which a bulb can be removed. Also, carry a three-pronged adapter.

Mirror: A small glass or stainless steel mirror about 3" x 5" can be hand-held or fastened to a stick. Use the mirror to look into hard-to-see places. As a last resort, use the rearview mirror from your car.

Photographic equipment: Use a 35mm single-lens reflex camera if you can. The single-lens reflex allows you to frame your picture before you take it. To avoid excessive distortion, use the standard focal-length lens for your camera. For most models it is 50mm or 55mm. Consult your camera's handbook. If you use a variable focal-length lens, set it to the recommended standard.

Wide lens openings (*f*5.6 or wider), slow shutter speeds (1/30th of a second or slower), and camera movement combine to give poor results. A tripod eliminates camera movement but can be awkward to use. Bracing the camera against a wall or the boat is a good alternative.

Black-and-white film and prints are recommended for record photography. Most color films cannot give the fine resolution obtainable with black-and-white. In addition, the processing of color film is not as "adjustable" as black-and-white. Even in a well-lighted shop, it can be difficult to get good photographs using 400 ASA-rated film. Kodak Tri-X or T-Max both give good results. They both can be "pushed" to ASA 1600 or even 3200. Consult your local processor for availability and limitations, but remember that employees of a "fast service" or pharmacy are unlikely to know what you are talking about. Both the Tri-X and T-Max when enlarged to 8" x 10" size show very high resolution. If you have shot the film at an ASA higher than the standard, don't forget to tell the processor that the film has been "pushed."

Order 8" x 10" contact proof sheets for use at the drawing board, and use magnifying glass to read them. Working with contact proof sheets is easier than shuffling through 30 or 40 large pictures as you work. But remember, do not use photography as a substitute for good field work and note taking. Special processing and one contact proof sheet will cost about $10.

If space and lighting conditions combine to thwart your efforts to take close-up pictures of details inside a boat, try using a telephoto lens. A 135mm lens has a minimum focusing distance of about five feet. Place lights as necessary, then shoot over or from behind them.

When surveying a complex boat project or a number of boats, photographs will often prove handy for on-site reference. Since processing for regular film takes time, consider using an instant developing-printing camera, such as a Polaroid, along with the standard camera.

Saw horses: Boats that can be picked up and moved by three or four people can be put on saw horses while they are recorded. Make sure the horses are strong enough to hold the boat and are not wobbly. A height of sixteen or so inches is convenient.

Bracing (propping or shoring): Most boards of 1" x 3" up to 1" x 6" will work as braces or props. They must be sound, reasonably straight, and long enough to reach from the floor to above the sheer. If you have to buy material from a lumber yard, buy strapping or furring strips. Around a boat shop or museum, look for sound scrap material and use that.

Clamps: Bracing must be temporarily fastened to the boat with clamps. The size you use depends on the size of the boat. For small open boats, any clamp with enough reach will work. Whether you use C-clamps or bar clamps, position them so that the handles and ends are over the boat to avoid bumping into them and ruining your leveling work. Small or medium-size spring clamps, which look like overgrown spring clothes pins, can be handy when clamping bracing to hold measuring equipment in place. Don't forget to put padding under all clamps to protect the finish on the boat; or modify the clamps by adding pads to them.

Weights: If you are working in a space with a concrete floor, you will need weights to hold the bottom of bracing in place, because you cannot nail the bracing to the floor. Bricks, lead ballast blocks, or even buckets of paint or floor cleaning liquid work well. A gallon can of paint weighs about ten pounds, and five gallons about fifty. Ingenuity is like having another hand; I have used a five gallon can of floor wax to hold the end of a tape measure while working alone measuring the length of a boat.

Wedges: You will need a number of wooden wedges of various sizes to level both the boat and the measuring equipment (fig. 5.1). Cedar shingles are handy for shimming and padding.

Wood blocks: Have an assortment of wood blocks to put under either the boat or the cradle. Blocking procedures are discussed in chapter 2.

Hollow-core concrete blocks: Measuring staffs or braces can be clamped to hollow-core concrete blocks (fig. 5.2). They can also be used as blocking under a boat cradle.

Pry bars, crow bars, jacks, and levers: All sorts of pry bars, crow bars, and jacks are available for moving a boat. The selection is yours, but be sure to get advice from knowledgeable people for the specific situation before you begin. When you move boats that weigh much over 350 pounds, you have crossed into dangerous territory. A cradle that looks as though it is strong and properly supporting a boat may in fact be neither. If you jack up a cradle to level the boat, it is possible to change the shape of the boat or, worse, force one of the supporting members (poppets) through the planking. Sometimes the cradle will break up, causing the boat to fall and be damaged or destroyed. This has happened more than once even to professionals using the best of equipment, so take special care when using jacks, pry bars, crow bars, and levers.

Pencils: My personal choice for note taking and sketching is a draftsman's fine-line pencil. Fine-line pencils are similar to "automatic" pencils. The lead sizes vary from 0.3mm to 0.9mm. (I prefer 0.5mm and 0.9mm.) The leads of fine-line pencils don't need sharpening, and you can carry a supply of lead in the pencils themselves. Use a medium lead of "F" (softer) or "H" (harder) degree. (These are standard drafting supply catalog designations.) If you have to mark a white-painted boat, use an extremely soft lead of "B" degree. You will be able to erase the mark if you don't grind the lead into the paint. Test it to be sure. Cost: pencils, $3 to $7; leads, $1.50 per dozen.

Don't use ball-point or felt-tipped pens or markers for field notes or sketches. If your work gets wet, the ink will run and blot.

Use colored pencils to highlight recorded data that is to be checked. My habit over the years has been to use red. Whatever color you use, test it on a photocopier to ensure that it will copy.

Erasers: Erasers don't come on all pencils, and those that come in the fine-line pencils are worthless except in an emergency. Buy a small block eraser. White vinyl erasers work well and can be used to take temporary pencil marks off the boat. Cost: $1 or less.

Paper: Plain, white, bond paper works well for note taking and sketches. For making scale sketches, paper printed with a square grid is useful. It can be bought in a number of styles and materials. For field use, buy white bond paper printed with a blue grid of either 8 x 8 grid (1/8" squares) or 4 x 4 grid (1/4" squares) in pad or notebook form. These grids work with scales of 1 1/2" and 3" = 1'-0", which are suitable for field notes (fig. 5.3). Don't use vellum gridded paper if you will be working out in the weather. It shrivels and may disintegrate when wet. Whatever paper you use, leave a clear margin of about 1" width at the left edge for binding and 1/2" on the other three edges so that data is not lost when photocopying.

Clipboard: Writing on a single loose sheet or even a pad of paper can be aggravating, so use a backing for your notepad. Buy a clipboard, or use a piece of plywood or Masonite with rubber bands or spring clips to hold your paper in place.

Folding extension rulers: Most folding extension rulers are 6 feet long. The extension is a sliding brass strip in one end, adding 6 inches to the length of the ruler. The minimum graduations are spaced 1/16 of an inch apart for the full length, with inches marked from 1 inch to 71 inches and stud spacing indicated at 16-inch intervals. A folding extension ruler has several advantages. First, you can use the sliding brass insert as a depth gauge to 6". Second, you can take inside measurements up to 6'-6" (fig. 5.4). This is especially useful when measuring the boat for lines and shape. Third, you can use it for a scale; not for plotting your notes, but when making sketches (fig. 5.8). Finally, you can use the ruler as a caliper (figs. 5.5, 5.6, and 5.7). Genuine calipers are discussed later in this chapter. Cost: about $25.

Carpenters' tape measures: Available in lengths up to 25 feet, with minimum graduations of 1/16 inch, carpenters' tape measures are marked in inches for the total length, marked and highlighted every foot with the inches marked in between, and also marked for 16-inch stud spacing. Cost: $20.

Long tapes: These include carpenter's tapes and surveyor's tapes. Use a carpenters' tape which is graduated in feet, inches, and eighths-of-an-inch, not a surveyor's tape, which is graduated in feet, 1/10 of a foot, and 1/100 of a foot. Long tapes can be bought in steel, cloth, and fiberglass. Avoid cloth tapes because they stretch appreciably. Cost: from $10 for a steel tape to $35 for a 100-foot fiberglass tape.

Seamstress's tape: Handy for measuring along a curved surface, as when measuring plank widths or the circumference of a mast, a seamstress's tape can be held in place by a push pin or tack through one end while you read and record the measurements. Check the seamstress's tape against a ruler or metal tape to see if it stretches when pulled tight. Cost: about $1.

Calipers: For measuring diameters of spars, inside diameters of and thicknesses, calipers are needed when you can't use a ruler or tape measure. You can make your own from heavy wire or plywood, or you can use a folding ruler as shown in figs. 5.5, 5.6, and 5.7. Cost: $10 to $15, depending on quality.

Scale: A plastic architect's scale, 6 inches long in four-bevel style with scales of 1/8", 1/4", 1/2", 1", and 3/8", 3/4", 1-1/2", 3" = 1'-0" is handy for field use. It fits comfortably in a shirt pocket and can be used as a straight edge for sketching. ("Scale" is the generic term used in catalogues and by architects and engineers. Boat builders call it a "scale rule.") Buy the cheapest you can find for field use, because it can get lost, broken, or so dirty that it shouldn't be used on a finished drawing. Use a scale when plotting field notes. For sketching structure and joiner work, a ruler can be used instead of a scale as shown in fig. 5.8. Cost: $5 to $7.

Knife: Use a handyman's knife for poking around if you don't want to ruin your pocket knife. If you are going to fly to the work site, do not carry a pocket knife or a fixed blade knife (bench knife) as they can be considered weapons. Instead, carry a utility knife with the blades stored in the handle. Make sure that it and the rest of your tools are in your checked luggage, not in your carry-on bag.

Screwdrivers: You will probably need a medium-size screwdriver at least. A screwdriver with a long slender blade, 1/8" wide x 4" long, is handy for digging into seams, holes, and dark corners.

Feeler gauges: For probing joints, seams, and other places where a screwdriver or knife blade won't fit, try a hacksaw blade with one end broken off more or less square. Wrap the rounded end with tape to make a handle. A small machinist's ruler with a sliding T-clip and 1/4" wide, 6" long, 1/64" thick also works. You also might find that you can slip one or the other of these tools into a joint or through a plank seam to measure plank thickness. Don't overlook using business cards, credit cards, or automotive feeler gauges as probes.

Eggbeater drill with small drill bits: Sometimes drilling a hole is the only way to determine the thickness of planking. Measuring plank thickness at the transom can be misleading because of wear and the angle that the planks and transom make. Measuring the exposed top edge of the sheer strake can also be misleading. Even in small boats, plank thickness vary and probably did even when the boat was built because of the planing required to fair the hull.

Check with whomever is responsible for the boat before drilling holes or marking on it with anything. Demonstrate what you intend to do before you do it. Before drilling holes, examine the boat carefully. You may find a seam where a feeler gauge can be pushed through or even an existing hole where a fitting has been removed.

If a boat is on display in a museum and will never be used or put in the water, 1/32" or 1/16" holes drilled in concealed places may be acceptable and not require plugging. If the boat will be used, drill a hole that will take a small plug, say, 1/8" in diameter. The plug should be made from soft, dry wood, slightly tapered, driven into the hole. Cost: eggbeater drill, $20 to $40; bits, $1.

Plumb bob: Plumb bobs come in many sizes and shapes. Buy one that weighs at least twelve ounces and has a long, slender point. Good bobs have replaceable points that can be purchased separately. The best plumb bobs have replaceable points, one of which is contained in the top of the bob itself and can be used alone as a miniature plumb bob in tight spaces. Use cotton string for plumb bobs, because it does not stretch with weight as nylon will.

If wind causes the plumb bob to swing, use your body as a wind break, hang the plumb bob inside a can of water, or build a wind screen of two short boards nailed at a right angle to one another. A plumb bob will always hang true and give a "plumb" reference for 360 degrees around it (fig. 5.9). Cost: $7 to $20.

Chalk: Use white chalk for marking the perpendiculars, center line, and stations on the floor. You can also use chalk to mark frame numbers and stations on the boat, as it will wipe off most finishes easily. Check to be sure, though. For drawing lines, sharpen the chalk end to a chisel point, with a knife using a skiving cut or with a piece of sandpaper. Keep the chalk sharp while drawing grid and reference lines. Sharpen several at once and save time (fig. 5.10).

Masking tape: Masking tape can be used to tag numbers on frames and to indicate stations on the boat as you work. (Chalk doesn't show up on white paint.) Do not leave masking tape on the boat for very long, as it can adhere too tightly with the surface and pull paint or varnish off when it is finally removed. You can also use masking tape on the baseline to mark station locations. If you do that, and the baseline string is moved or broken and then repaired, be very sure to check that the tape pieces are still at the true stations. If the string has been disturbed, it is a safe bet that the station marks on the tape are not where you think they are. Check! It will take less time than doing the whole job over again.

Carpenter's chalk line: A length of cotton string impregnated with chalk dust, a chalk line is used to make long, straight lines. The string is stretched tightly along a floor, roof, or wall and then snapped, leaving a line of chalk dust behind. Generally, a chalk line comes in a plastic or metal case into which chalk dust is poured. You may want to use a chalk line to "snap" lines on the floor for measuring grid. (See chapter 7). Cost: $12.

Measuring staffs: Good boards, 1" x 4" or wider, or strips of 3/4" plywood at least 4" wide can be used as measuring staffs. No matter what you use, one edge must be straight, as you will be measuring from that edge. The length depends on the boat you are measuring.

String: Use stranded or braided nylon string (not monofilament) for baselines and reference lines. Nylon string is easy to pull taut and straight, and it is not affected by humidity. Cotton string will sag as the humidity rises. Another characteristic of nylon is that it is safe. If it breaks while you are tightening it, the flying ends are not likely to hurt you. If nylon has a drawback, it is that it is difficult to hold taut while tying. Use #17 or #18. Cost: $2 to $3 for a 100-foot roll.

Do not use wire for baselines and reference lines. A great deal of tension is required to pull wire straight. If wire snaps while under tension, the loose ends will flail and cut—even through light clothing. In addition, you can peel the skin off your shins if you walk into taut wire.

Mason's line level: This device is a small metal or plastic case with hooks at each end that allow it to be hung from a string. A level vial is enclosed in the case. When the bubble in the vial is centered between the lines on the vial, the string (line) that the case is hung from is level (fig. 5.11). Cost: about $5.

To check the level's accuracy, (fig. 5.12) proceed as follows:

1. Tie a length of nylon string between two points. Be sure the string is very taut, and hang the line level at the mid-point of the string. Level the string by raising or lowering one end. When the string is level, mark its height at both ends. Switch the line level end for end. If the bubble is still between the lines on the vial, the string is level and you are done. If the string is not level, proceed as recommended below.
2. If the test in step 1 shows that the level is not accurate, leave the line level in its switched position, and level the string by raising or lowering one end only. Don't move the other end. Mark the height of the end that you have moved.
3. Next, divide the space between the marks made in step 1 and step 2 in half. Make a new mark and align the string with it. The bubble will probably fall between the lines on the vial. Should it not, the hooks on the case will have to be adjusted. Do not remove the level from the string. The bubble will be nearer one end of the vial than the other. Bend the hook at the opposite end of the case down very slightly and gently. When the bubble is between the lines, switch the level end for end to check it. If it checks, you are done. Otherwise start again as in step 1 but only make the adjustments on the hook.

Carpenter's level: Available in various lengths and made of wood or aluminum, most carpenter's levels eighteen inches and longer come with both leveling and plumbing vials (fig. 5.13). Some levels use curved vials that should be read only with the bubble at the top of the curve.

Whichever type of level you use, it must be checked for accuracy as shown in fig. 5.14. Put a wedge under the low end to bring the bubble between the lines. Mark the location of the wedge and the level. Then switch the level end for end. If the bubble settles between the lines in the vial, the level is accurate. If not, adjust the vials according to the manufacturer's instructions. If the vials are not adjustable, shim the low end of the level with tape. Be sure to rest the level on the shims when using it. The same principle applies to checking the plumbing vials. With the carpenter's level checked, you can compare the mason's line level to it as in fig. 5.14. Cost: $20 to $30.

Leveling board: A line level can be combined with a straight board to make a leveling board—a substitute for a carpenter's level. This is a handy trick when you have to travel light. There is no guarantee that the line level will read the same when hung by the hooks and when laid down on top of a surface. To check it, lay the line level on top of a true reading carpenter's level. If the bubbles agree, fine. If not, wrap plastic tape around the low end of the line level until the bubble agrees with the carpenter's level (fig. 5.15).

Carpenter's framing square: A steel carpenter's framing square is useful for squaring station lines off the baseline. The factory-trimmed corner of a piece of plywood or hard board, or a sheet of paper folded into quarters are good substitutes. If you use a framing square, check it and, if necessary, adjust it as shown in fig. 5.16. Lay the tongue along a straight line. Draw a fine pencil line along the blade. Flip the square as shown. The blade should line up on the pencil line. Correct any misalignment by either closing or spreading the angle between the tongue and blade. Use a center punch and hammer to dent the square where shown; do this cautiously. Several dents may be necessary to do the job. Cost: $25.

Wooden yardsticks: Often available free from local lumber yards or hardware stores, wooden yardsticks are great tools. They don't rust and can be glued, nailed, or taped to a measuring frame, a leveling board, or a level. To top it off, breaking one won't break your heart—it was free! I have a number of them, all of which check very closely against my very expensive English-made steel ruler.

Sliding bevel gauge: Used to pick up angles in odd places, such as the angle of the garboard plank to the keel or of a seat back to the seat, a sliding bevel gauge is a handy tool. If you do not have a gauge, you can measure an angle by folding a piece of paper to fit the angle or fitting a folding rule to the angle and tracing the angle onto your notepad (fig. 5.17). You can make your own bevel gauge by breaking 3 inches off each end of a hacksaw blade and riveting or bolting through the end holes to hold the pieces together. Cost: $15.

Combination square: Useful for squaring off from the measuring staff or frame to a point on the boat, a combination square is accurate only over a maximum distance of 8 inches (fig. 5.18).

Strap hinges: If you build any kind of a measuring frame, strap hinges are helpful for holding the frame or bracing in place. A half dozen of these combined with spring clamps or a few small nails will make setting up and bracing measuring frames easier (fig. 5.19).

Hammer: A 16-ounce hammer will do for nailing bracing and measuring equipment in place. Cost: $5 to $10.

Saw: A regular cross-cut hand saw will suffice for cutting bracing. Cost: $15 to $25.

Imagination and ingenuity: Develop these qualities, use them and make them work for you.

While all the tools mentioned here will be handy as you work in the field, it is often not practical to carry them all to the boat's location. On my trip to Scotland, mentioned earlier, I could carry only the bare minimum of tools and hope that I could find whatever else I might need when I got there. This is what I carried in the three-quart food storage box (10" long x 6" wide x 3-1/2" deep):

- Camera flash unit and extra batteries
- 6-foot folding rule
- 50-foot tape measure
- Seamstress's tape measure
- Plumb bob
- String
- Line level
- Chalk
- Pencil
- Scale
- Eraser
- Push pins
- Knife

All of the above went in my duffle bag as checked luggage. My camera and film went with me as carry-on luggage.

Tools for Drafting or Lofting

Drafting table: A professional drafting table, while useful, is not absolutely necessary. There are many reasonable substitutes. One of the best is a flush, hollow-core door on saw horses. A good height for either will be just about at your waist band. You might like to have it slope slightly; say about 5 inches higher at the back than the front.

Vinyl covering materials are available from drafting supply houses for the table top and make a very nice surface on which to draw. A sheet of white paper under your drawing will accomplish the same thing for one or two jobs. The working surface itself should be at least 30 inches wide by 60 inches long and must be flat in all directions. If you buy a hollow-core door to use as a drafting table, you can save money by buying one that is damaged on one side. Use the good side for a drafting surface. A hollow-core door will cost $30. The vinyl covering material for a 30 inch by 60 inch surface will cost about $40.

Lofting surface: Any flat, smooth surface—the shop floor or plywood sheets nailed to the shop floor or set up on saw horses—will work as a lofting surface. The surface should be painted white or very light gray. Your lofting will be done directly on the painted surface.

Drafting material (media): Use drafting vellum or mylar for your drawings. Drafting vellum is a very high-quality paper usually of 100 percent rag content. It has an extremely long life when properly cared for, but it is dimensionally unstable. Humidity variations will cause vellum to shrink or stretch as much as a quarter inch in 36 inches overnight.

Mylar is a polyester film that is dimensionally stable and extremely durable, even when wet. Anything drawn on it will stay put unless rubbed off. Mylar is generally available in two thicknesses, 0.003" (3 mil.) and 0.004" (4 mil.), both with a matte surface on one or two sides. The matte side is the drawing surface. The 4 mil. double matte (matte on two sides) is recommended for archival work. Cost: Mylar, 24" x 36" sheet $4; vellum, 30 inch x 20 yard roll, $25.

Regular leads and pens and inks are used to draw on vellum, while drawing on Mylar requires special leads, pens, and inks.

Straight edge: You need a straight edge at least as long as the drawing you will make. Do not use a short straight edge to make a long line from several short ones; there is a very good chance that the resulting line will not be straight. The straight edge can be plastic, stainless steel, or a draftsman's parallel edge fitted to your drawing board.

Whichever type you use, check to see that the straight edge is really straight, as shown in fig. 5.20. Lay the straight edge on a sheet of drawing material, and draw a very fine line along it. (A fine line is as close to a knife cut width as you can get.) Mark the corners, then either flip the straight edge top to bottom or switch it end for end, and align the corners. If there is a gap between the line and the edge of the straight edge or if the edge covers part of the drawn line, the straight edge is not straight. Exchange it for another, or straighten it very carefully with a file and sharpening stone. Be careful not to change the sectional shape of the straight edge. Cost: $55 for a 48-inch long parallel straight edge; $50 for a 48-inch stainless steel straight edge.

Lead holder: This mechanical device that looks like a pencil holds leads, bought separately and inserted into it. You sharpen the lead only, not the "pencil." Cost: $3 to $5.

Leads: Available in varying degrees of hardness, leads are either a clay-graphite mixture (commonly called "lead") or a polymeric composition (referred to as "leads for film"). "Lead" is used on all papers but usually not on Mylar unless you want to achieve fine lines. The penalty paid for drawing on Mylar with leads is very faint linework. The benefit is that you achieve fine lines, which give very accurate drawings, especially if you use the hardest (9H).

"Lead for film," although very durable, does not readily produce very fine lines. But it is great for lettering on Mylar if you don't want to letter in ink.

The degrees of hardness for clay-graphite leads are standard and range from 6B (super soft) to 9H (super hard). I use 9H for laying down lines on Mylar. Lead of that hardness holds its sharpness fairly well if you have a light hand. The designations vary from manufacturer to manufacturer for leads for film, but in most cases are cross referenced to the hardness of standard leads. Cost: about $6 per dozen.

Lead pointer: A device for sharpening the lead in a lead holder, a pointer is unnecessary. Use fine sandpaper instead. For good, fine lines, the lead must be sharp. The tapered length of the point should be about a quarter inch. When it looks lethal, it is sharp. Cost: $15.

Ink and pens: See the comments above on drafting materials (media). Unless you have previous experience drafting in ink, limit your use of ink to drawing the straight lines for the grid. To draw the grid lines, you need only one pen. A Pentel brand Ceranomatic pen in either 0.13mm or 0.18mm size does the job. These small sizes don't seem to stand up to long-term use on Mylar, but at about $12 each, they are inexpensive enough to be throwaways.

For serious drafting in ink, you will need three to six pens ranging in sizes from 0.13mm to 0.5mm with tungsten carbide or jewel tips. Proper maintenance of the pens requires an ultrasonic cleaner and cleaning fluid. The pens cost $25 to $30 each; cleaning fluid, $5 for a half pint (or use Windex); and the ultra sonic cleaner, up to $100. Ink costs about $2 per ounce. If the cost of necessary pens, etc., does not discourage you, by all means take the time to learn to work with ink. The results can be very satisfying and worthwhile.

Scale (scale rule): A good 12-inch long architect's scale of wood with plastic faces and very fine graduation lines is a necessary tool. A 6-inch scale will be too short for laying out long measurements on the drawing. If at all possible, compare a number of scales, and buy the one with not only the finest graduation lines but also the smallest increments. Cost: $12 to $15.

The styles shown in fig. 5.21 come graduated in a number of scales. The triangular scale, "A," is probably the most frequently used. "B" is easy to pick up; "C" is difficult to pick up; "D" is a good compromise because it has a wide selection of scales and does not take up the space that "A" does. The dashed lines at "A" represent a wooden triangular scale.

Scales on the edges of:

"A," the triangular scale, are 3/32" and 3/16", 1/8" and 1/4", 3/8" and 3/4", 1/2" and 1", and 1 1/2" and 3', all equal to 1'-0" (or with 2" and 4" equal to 11'-0" instead of 3/32" and 3/16");

"B," the opposite bevel scale, and "C", the two bevel scale, 1/8" and 1/4", 1/2" and 1", or 3/8" and 3/4", and 1 1/2" and 3" all equal to 1'-0";

"D" the four-bevel scale, 1/8" and 1/4", 1/2" and 1", 3/8" and 3/4", and 1 1/2" and 3", all equal to 1'-0".

Ruler: The same folding ruler used to measure the boat can be used at the drawing board. It can be used to rough out the view areas on the drawing. More importantly, it can be used to check what appear to be measuring errors in the field notes by comparing the ruler and the notes. The most common error made with the ruler is to take a measurement with the ruler partially folded. If you want to buy a good aluminum or steel yardstick, do so. An aluminum yardstick costs about $10; a steel one, about $25.

Triangles: You will need at least an 8-inch 45-degree triangle and a 10-inch 30-60-degree triangle. Both should have plain square edges as shown in fig. 5.22. Smaller triangles that work well for drawing details are 4-inch 45-degree and 6-inch 30-60-degree. Cost: $5.

Protractors: Used to plot or lay out angles, protractors must be laid down so that the two 0/180-degree marks fall on a straight line. The scribed center mark is used for lateral positioning only and is aligned over the point from which the angle is to be drawn. The end of the center mark on the cross bar may not fall on a straight line between the 0/180-degree marks (fig. 5.23). Cost: $5.

Ships' curves: Generally used when laying down a boat's lines, ships' curves will be used when tracing the "rough" lines plan in ink—especially if you have faired the lines using splines (fig. 5.24). Cost: varies from $1.50 to $10, depending on size and shape.

Compasses and dividers: These come in a number of styles and sizes as shown in fig. 5.25. Bow pencils and bow dividers are approximately 3 1/2 inches overall in length, and compasses and dividers are 6 inches.

Compass "A" is available from several manufacturers in a "quick adjust" model. The diameter of circles drawn can be increased by using the extension arm "B". The old-style compass "C" is quick to adjust, has a range of about 6-inch radius and, when used with extension bar "D", has a range of about 9 inches. The bow divider "E" is useful for small work and is quick to adjust, as it can be squeezed and the adjusting wheel spun. The same is true for the bow pencil "F". The standard sized dividers "G", have a useful reach of about 5 inches. "H" shows the proper sharpening of the leads in compasses and bow pencils.

Of the tools shown in fig. 5.25, the compasses "A" and "C" are the most useful. Cost: "A" and "C", $15 to $20; dividers, $10; bow dividers, $15; bow pencil, $15.

Beam compass: Consisting of two separate clamping heads that are moved along a beam, a beam compass is used to draw circles and arcs of greater radius than possible with compasses and extensions. The most likely uses are in laying out sail plans and the large-radius arcs of a curved, raked transom, (fig. 5.26). Cost: $20 to $30.

Trammel heads (yard stick compass): Similar to a beam compass, trammel heads are fitted to a wooden yardstick or similarly sized strip of wood, (fig. 5.27). Cost: $10 to $15.

Battens and splines: These are used for laying down and fairing lines on both the drawing board and loft. Wooden battens are usually made as needed from selected stock in a number of sizes, from 1/8" x 1/4" to 1" x 1 1/2". Vinyl splines, 5/32" x 5/16" with one grooved edge, can be bought in lengths up to 6-feet. For use with a modern technical drafting pen, modify the shape as shown in fig. 5.28.

For tight curves, use piano wire, which comes in sizes from 0.015" to 0.125", or use plastic-coated braided picture wire. Wire must be supported from two directions to bend it to the shape desired. For this reason, it can only be used with a pencil for rough drawings. Cost: $5 to $10.

Ducks: Lead weights used to hold battens and splines in place while laying down lines, commercially made ducks are shaped rather like whales, weigh about three pounds each, and cost about $15. I cast my own from melted wheel weights and save a lot of money. If you know someone who has a lead-melting pot, consider making your own (fig. 5.29).

Templates: Thin plastic sheets with circles, squares, ellipses, triangles, and other shapes cut out of the plastic, templates are handy for drawing shapes you use frequently (fig. 5.30). You can cut your own shapes into the template (fig. 5.31). Cost: $.50 to $5.

Lettering guide: The Ames Lettering Guide shown in fig. 5.32 is indispensable for making guide lines to aid in lettering. It can also be used to draw shading lines. Cost: $1.50.

Magnifying glass: A small magnifying glass or jeweler's loupe is handy for reading photographs. Cost: $10.

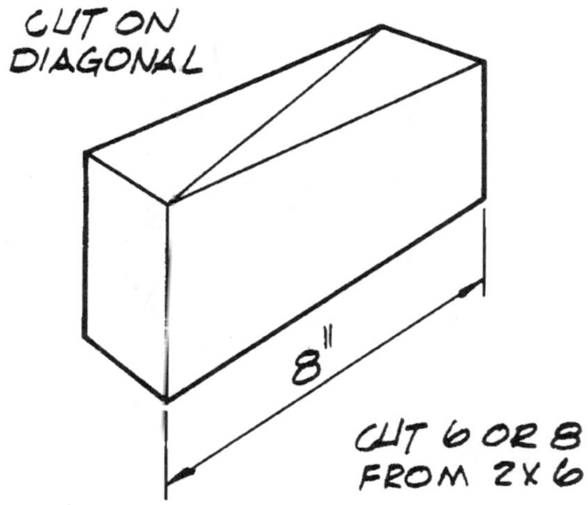

FIG. 5-1 WEDGES

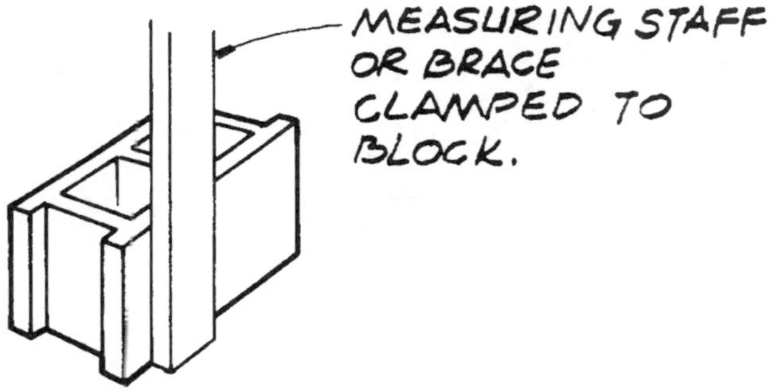

FIG. 5-2 CONCRETE BLOCK

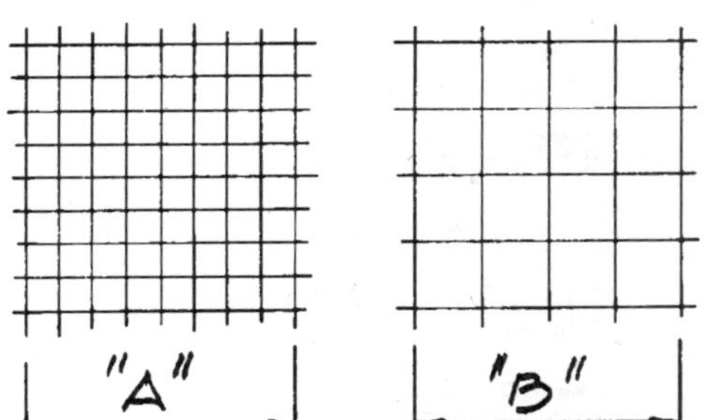

"A" = 8 1/8" SQUARES EACH WAY
OR 8 x 8 GRID.
"B" = 4 1/4" SQUARES EACH WAY.
OR 4 x 4 GRID.

FIG. 5-3 GRID PAPER

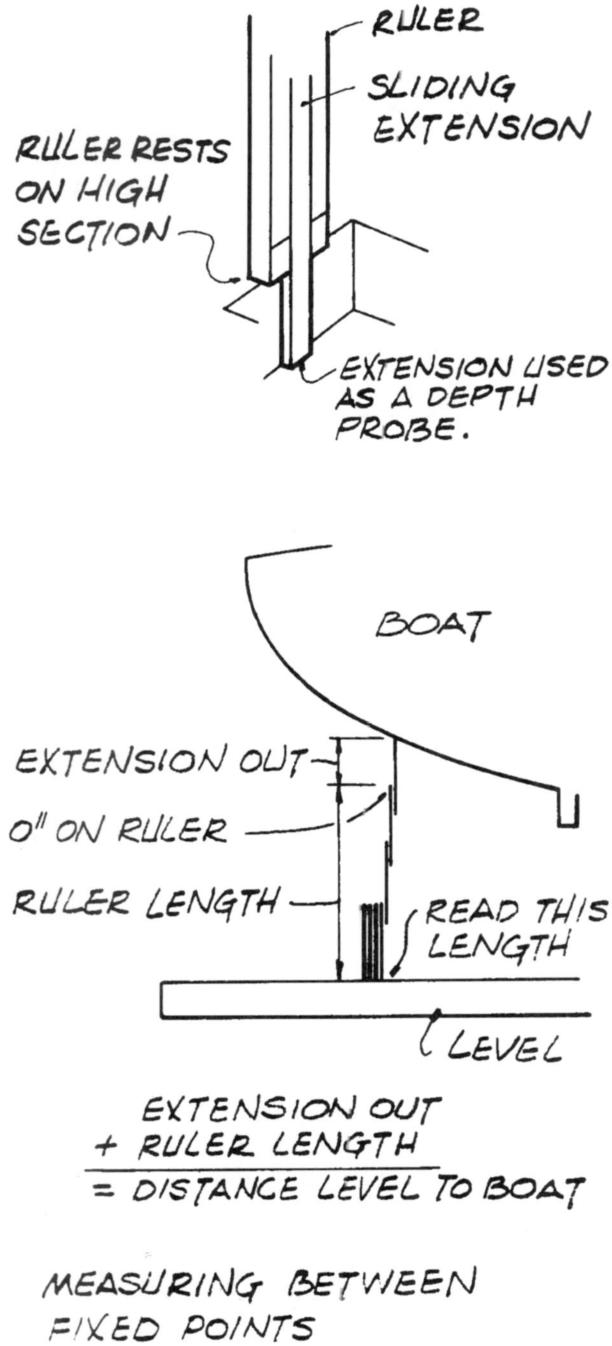

FIG. 5-4. THE FOLDING RULE
USING THE
SLIDING EXTENSION

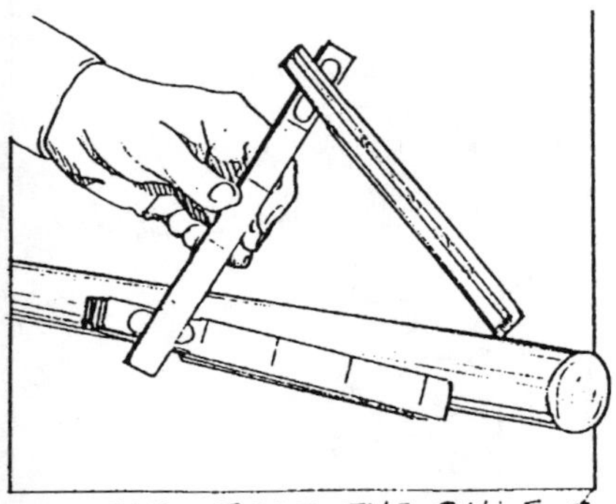

FIG. 5-5 FOLD THE RULE & TAKE THE SIZE.

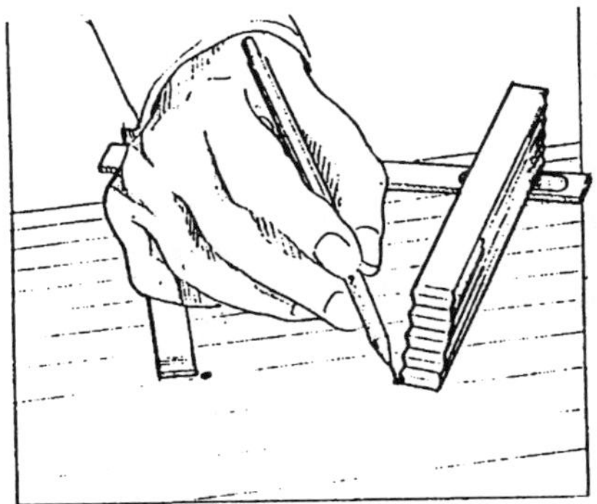

FIG. 5-6 MARK THE SIZE ON PAPER.

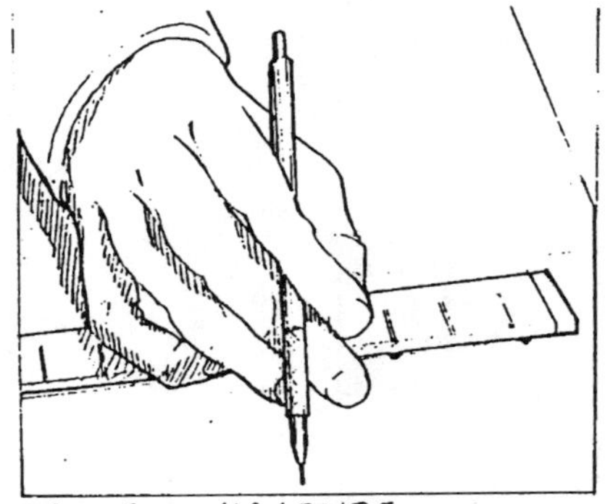

FIG. 5-7 MEASURE

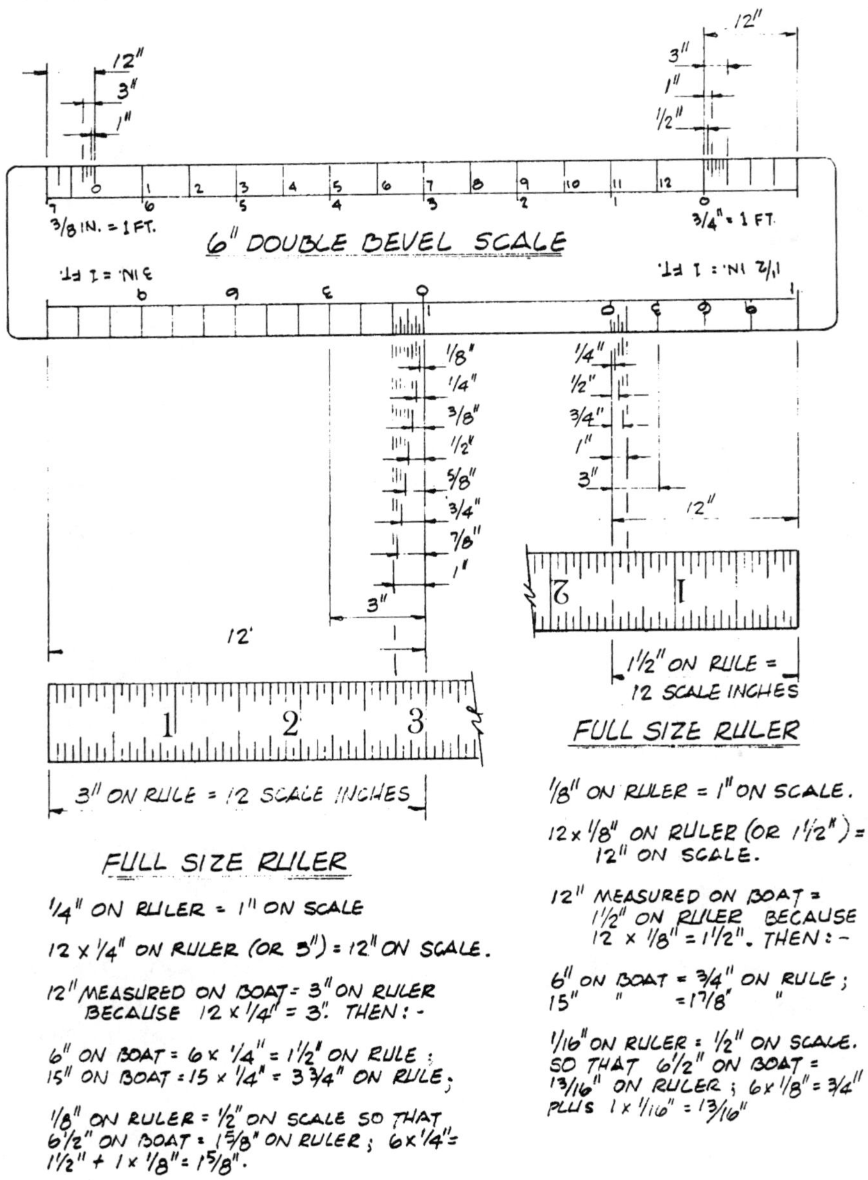

FIG. 5-8 THE SCALE AND RULER AS A SCALE

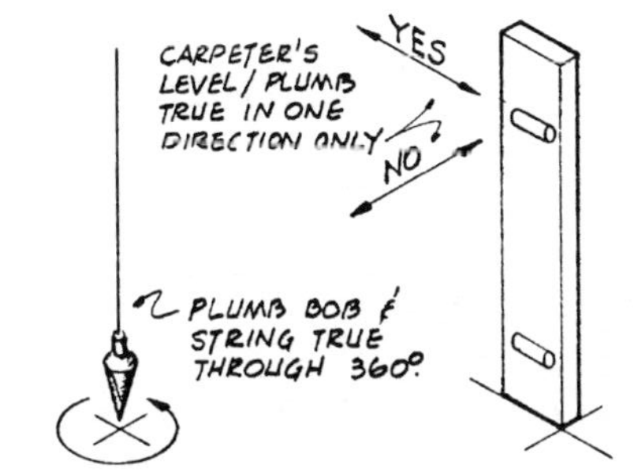

FIG. 5-9 "PLUMB"

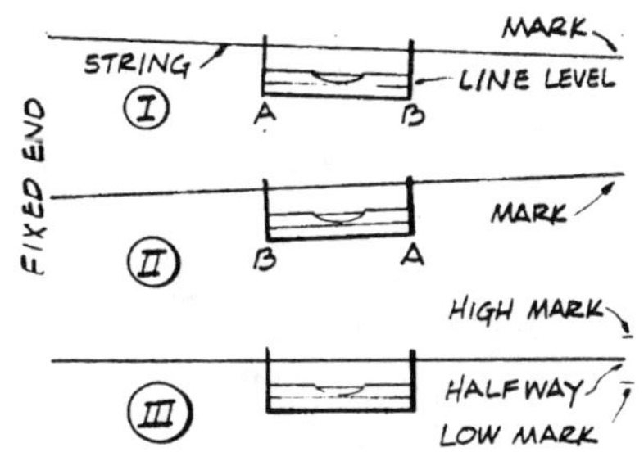

FIG. 5-12 CHECKING LINE LEVEL

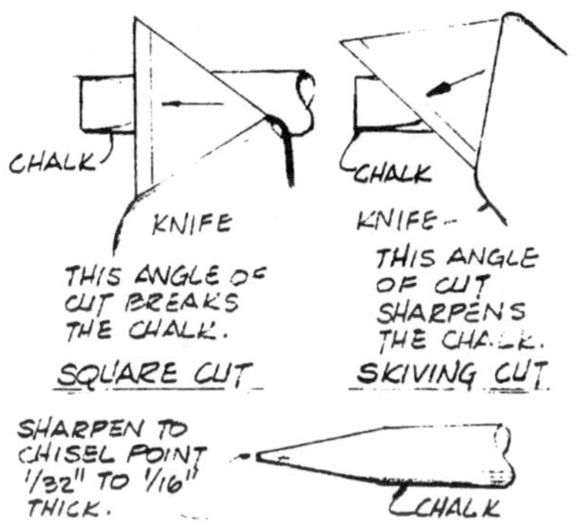

FIG. 5-10 SHARPENING CHALK

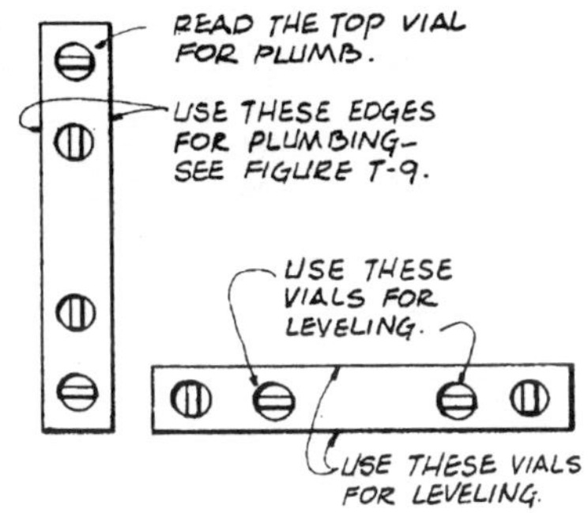

FIG. 5-13 THE CARPENTERS' LEVEL

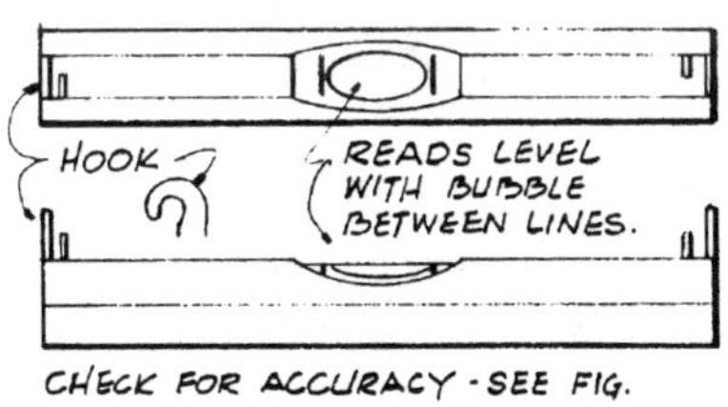

FIG. 5-11 MASON'S LINE LEVEL

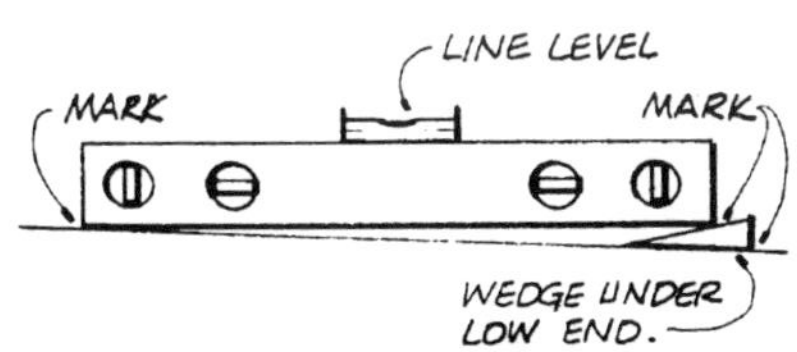

FIG. 5-14 CHECKING THE CARPENTERS' LEVEL

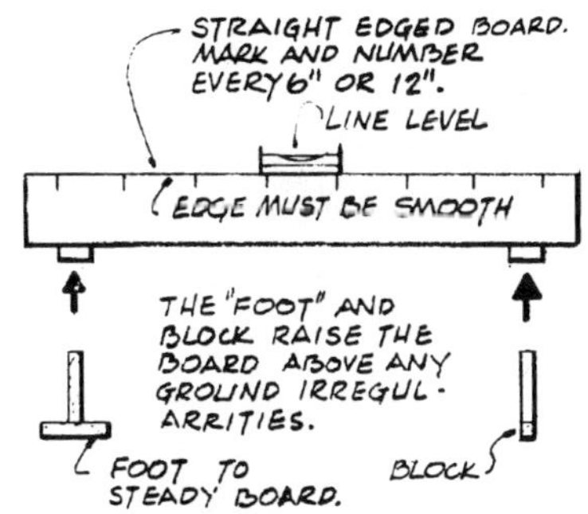

5-15 LEVELING BOARD

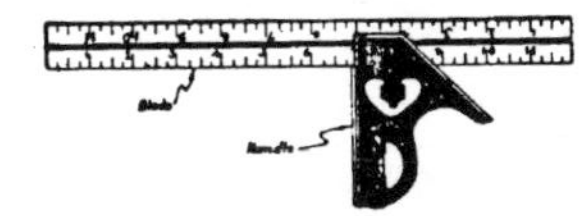
FIG. T-18 COMBINATION SQUARE

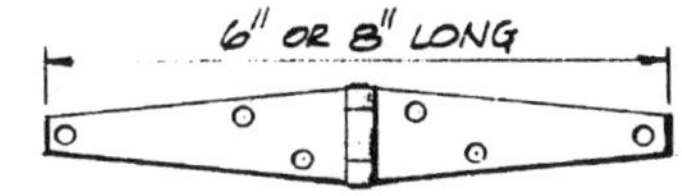

FIG. 5-19 STRAP HINGE

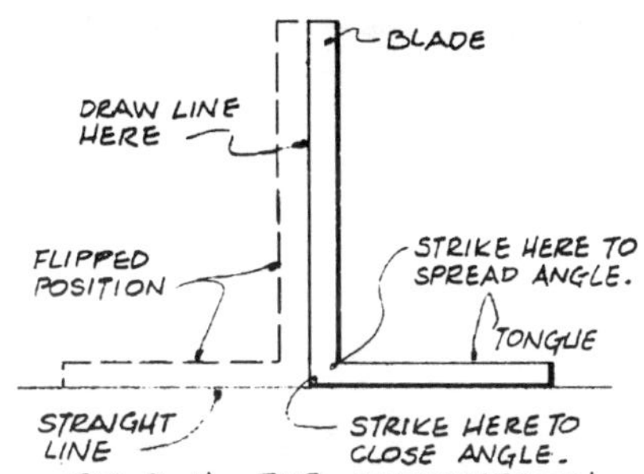

FIG. 5-16 THE CARPENTERS' FRAMING SQUARE

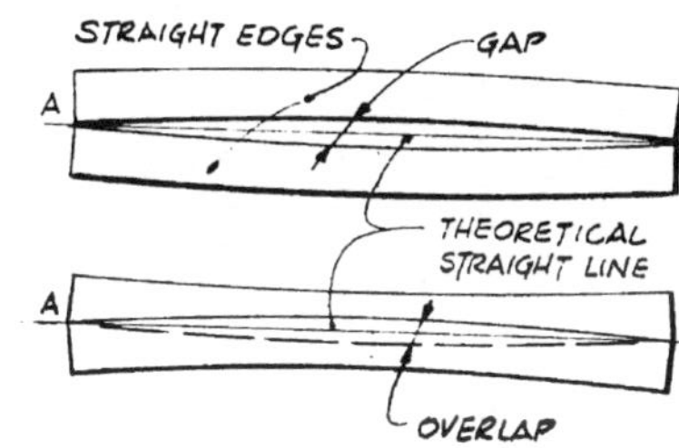

FIG. 5-20 CHECKING A STRAIGHT EDGE

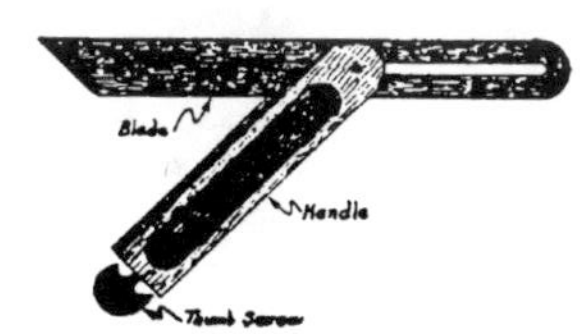
FIG. 5-17 SLIDING BEVEL GAGE

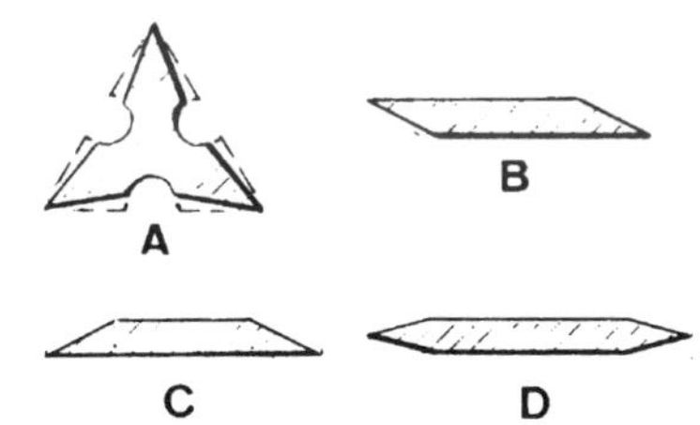

FIG. 5-21 SCALES

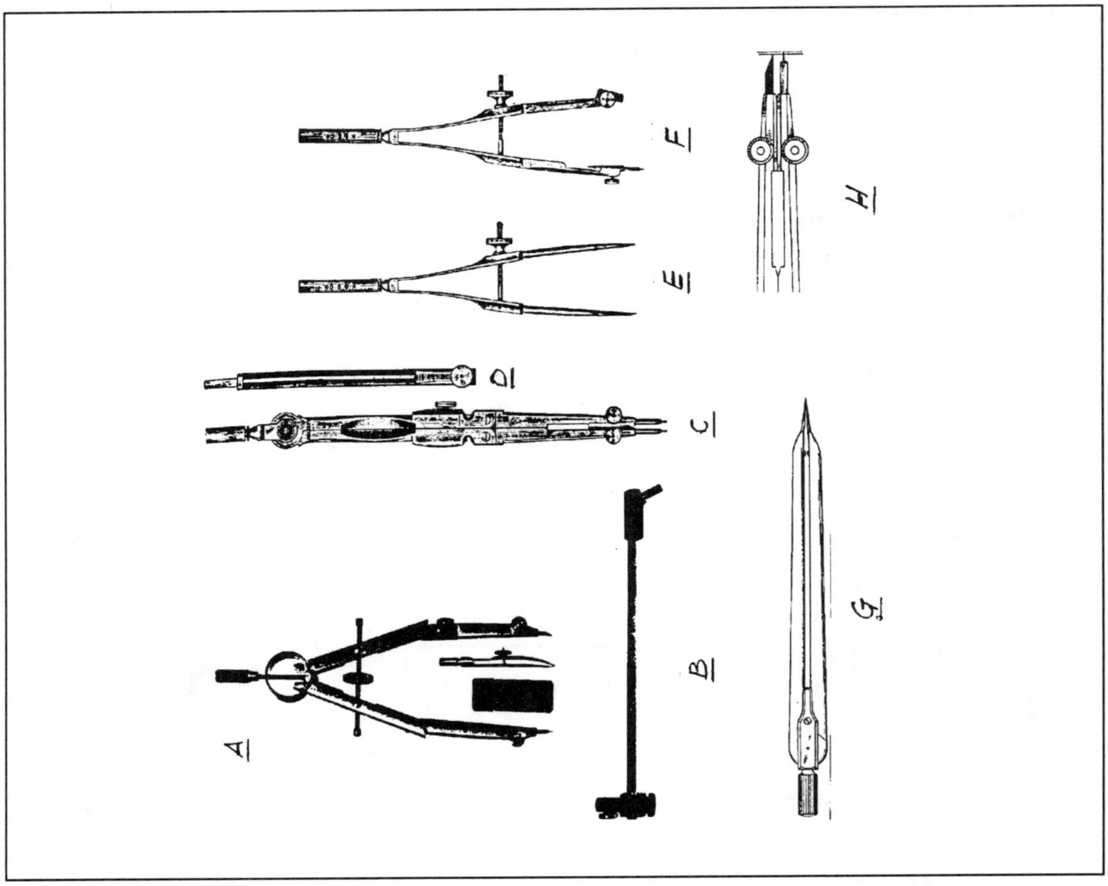

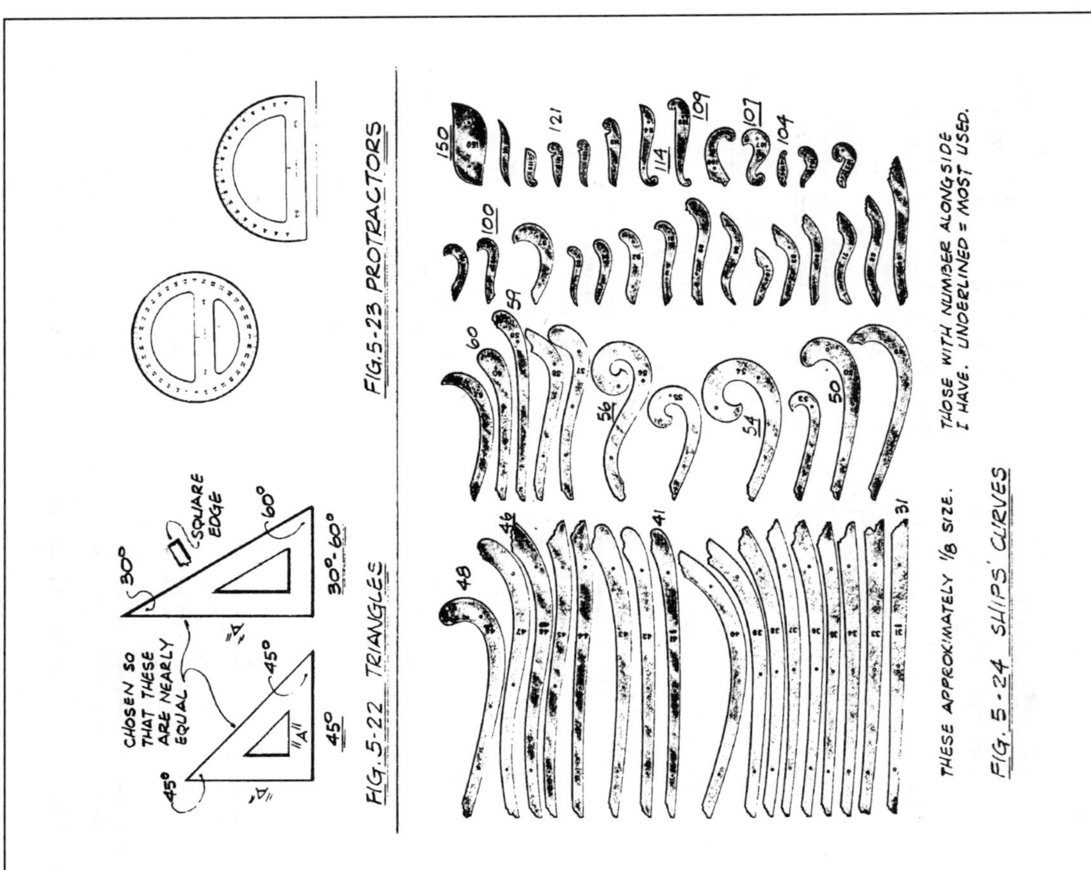

FIG. 5-22 TRIANGLES

FIG. 5-23 PROTRACTORS

FIG. 5-24 SHIPS' CURVES

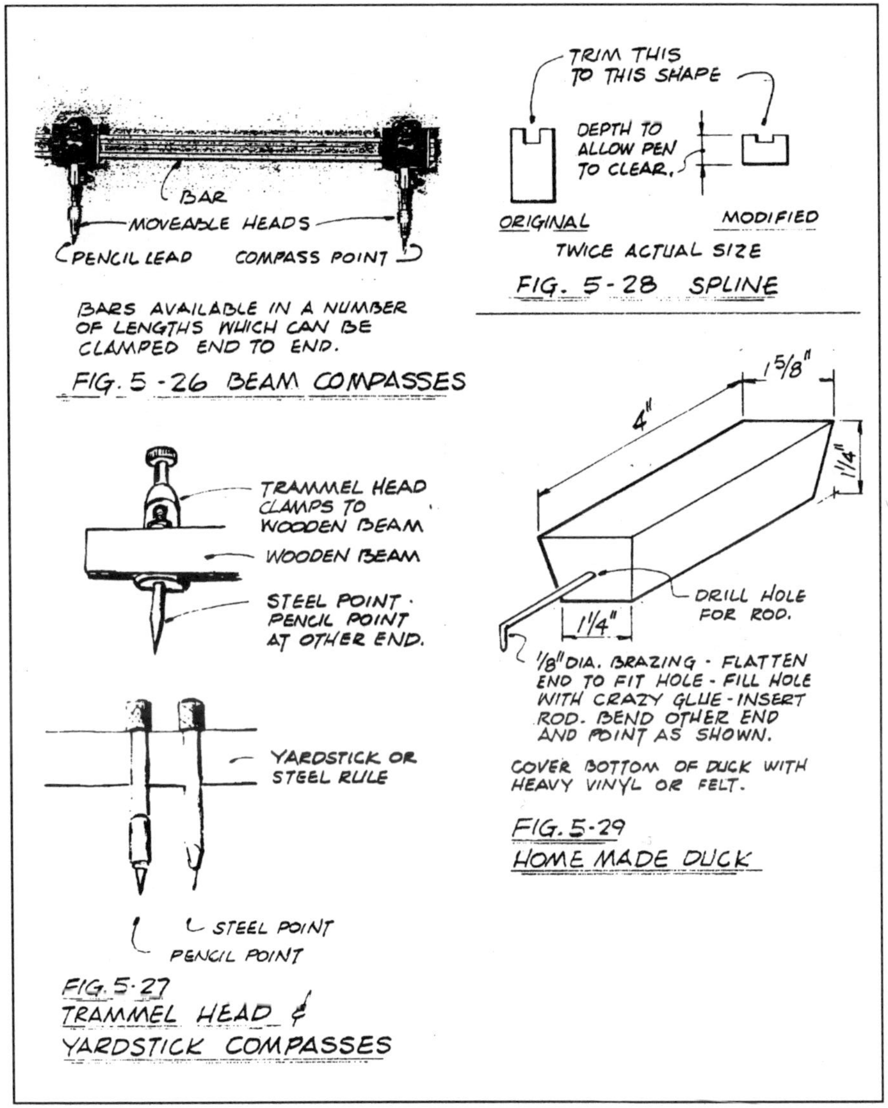

FIG. 5-26 BEAM COMPASSES

FIG. 5-27 TRAMMEL HEAD & YARDSTICK COMPASSES

FIG. 5-28 SPLINE

FIG. 5-29 HOME MADE DUCK

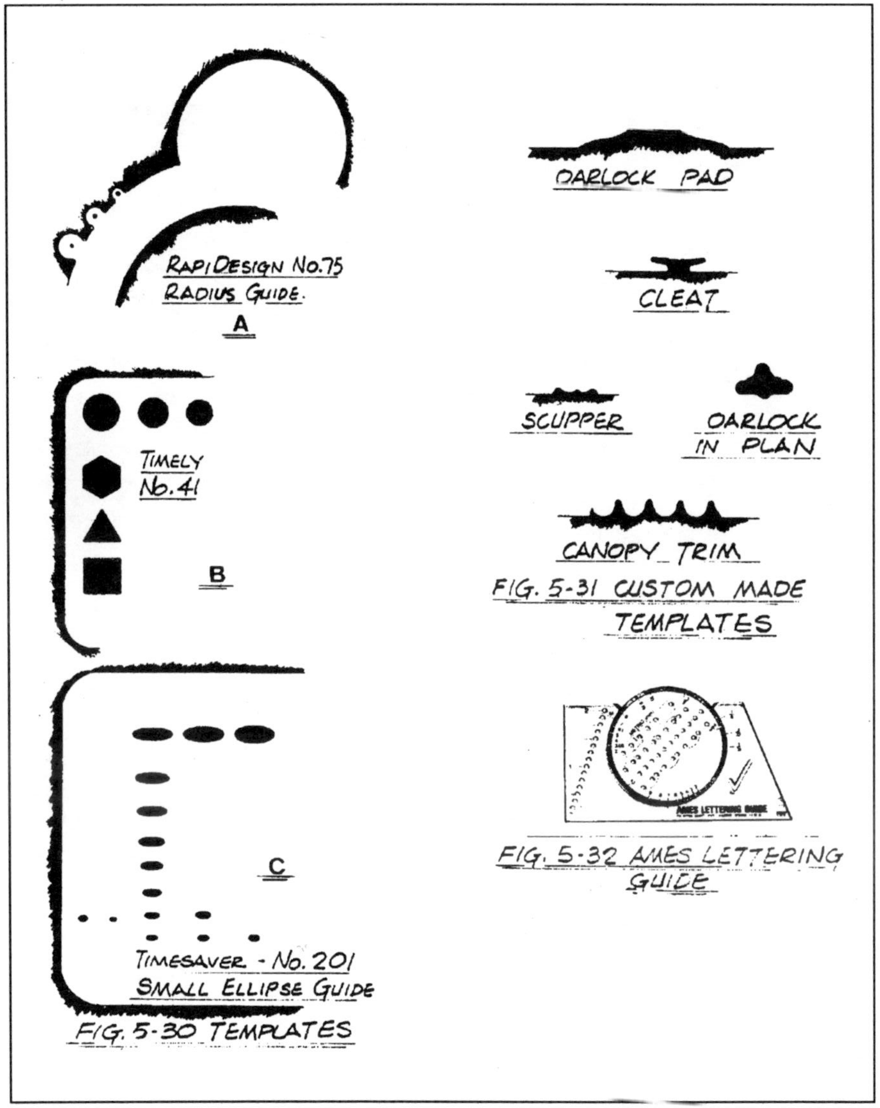

FIG. 5-30 TEMPLATES

FIG. 5-31 CUSTOM MADE TEMPLATES

FIG. 5-32 AMES LETTERING GUIDE

Chapter 6

Taking Off Construction

Studying and recording a boat's construction, with illustrations of boat parts and where they are found

D. W. Dillion

This chapter serves as a guide to recording the way a boat was built and suggests what *might* be found when taking off construction—not what *will* be found. The purpose of this chapter is not to tell how a boat should be built. It is important to consider a boat's construction before taking off its lines because by doing so, you will begin to understand if and how the boat has changed during its lifetime. What you are actually considering is the boat's structural soundness—its condition—which in turn affects its shape. Understanding the effect of condition on shape helps you make adjustments when laying down a lines plan.

The notes you develop while taking off construction should be an archaeological record showing the original work and all evidence of wear, damage, repair, and modification. When time and budget limitations prevent you from making finely detailed sketches, substitute verbal descriptions. Because many boats are taken off in order to make drawings for the recreational boatbuilder to use, the archaeological record is usually overlooked or ignored. The complete information does not have to be included in a set of plans but should be in the notes, which are the record of the boat and which should be kept for future reference.

Taking off a boat's construction and joinerwork means to measure and record the size, shape, and material of its parts and how they are joined and fastened; measuring their location in relationship to the boat's shape; and ultimately developing drawings that detail the boat's construction and joinerwork. Taking off demands attention to several elements:

- Construction (structure or structural): those pieces necessary to give a boat its shape; hold its shape; support other components, such as engines and masts, necessary to the operation of the boat; and keep out water; any member which cannot be removed without weakening the boat.
- Joinerwork (joinery): those things that make a boat comfortable or attractive.
- Condition: the soundness or structural integrity of the boat—deterioration and weakness versus strength.
- Shape or hull form (model): the external form of the boat—visualized as the impression it makes in the water.
- Type: the family to which a boat belongs—such as canoe, peapod, or skiff. Model is the shape of the boat as it relates to type. Considerable confusion arises due to boats being referred to as the same type even though differing in hull form, construction, usage, rig, and locally used name. It follows, then, that an adequate description of a boat demands more than a simple "that's a sailboat." Figures 6.2 through 6.9 illustrate the point. Further discussion and illustration will be found in books listed at the end of this chapter, particularly *Mystic Seaport Museum Watercraft* (Bray, 1986).

In order to explain the process of taking off construction, illustrations of two boats appear in this manual. *Susie* (figs. 6.1, 6.11, and 6.12) was built by Edward E. Swift at Woods Hole, Massachusetts, in 1896. *Susie* is called a "Woods Hole Spritsail Boat," because that is where the type, which carries a single spritsail, originated. When she was measured, *Susie* was owned by Mrs. Robert H. Baker of Westport, Massachusetts. Robert Baker bought *Susie* from the Swift family in the early 1960s and sailed her for a number of years. *Susie* is now in the collection at Mystic Seaport Museum.

Amy J. (figs. 6.60, 6.61, 6.62, and figs. 11.1, 11.4, 11.7, and 11.11 through 11.27), a flat-bottomed skiff, was designed to illustrate processes described in this manual.

The boat itself is the primary source for the information you are recording. The best way to know its construction is to take it completely apart. Since this would destroy the boat, you must instead closely examine the boat and identify its pieces, their size and shape, and how the joints are fitted and fastened. You are not trying to determine how the builder did the job of building or even if he did a good job. You may never know how the boat was built—the actual mechanical operations of setting up, cutting, fitting, and joining the pieces that are the boat. However, the more you learn of the actual building process, the more you will decipher about the boat. This requires experience or research, or both, some of which can be gained from the books listed in the bibliography. Be cautious about accepting what you read as gospel. What is shown may be an adaption, inaccurate, or not detailed enough. It is possible, however, that a detail found in a reference book will help solve a puzzle. Other, perhaps better, sources are plans made for museums of boats in their collections. Also, look at, study, ask, and talk about boats.

At this point, an admonition is appropriate. No matter how many books you read, or boats you study, or how much illustration is provided, you will never see or know all of the variations in construction. It has been said, "No boat builder would do that!" This may be a true statement—at least until you see the next boat! Be alert and study, study, and study some more.

Steps in Taking Off

When you have completed reading this chapter, it should be obvious to you that there is a simple rule governing the taking off process: "If you see it, describe it and locate it." While that sounds simple, you will be aided in taking off by first reading or reviewing chapter 8, "Organizing and Plotting Field Notes," and chapter 7, "Taking Off." Also study the figures in chapters 6, 7, and 8. Notice the organization of the field notes and the various techniques of notation. Refer also to figures illustrating *Susie* and *Dog Star.* Notice the form used to show the scantlings and fastenings.

Scantlings

Following the form of the scantlings list in fig. 8.9, begin a scantlings takeoff. Start on the outside of the boat and list all the pieces you see, including hardware and fittings. Do not record sizes or materials yet. Concentrate on the pieces.

Avoid upsetting the boat or any measuring equipment to be used for the lines takeoff. Complete as much of the scantlings takeoff as possible before setting the boat up. In this way, you will concentrate on the boat and not on avoiding the measuring equipment.

The following definitions will help you take off the scantlings:

- ***Scantlings***: a listing of the size of the wooden parts of the boat, including the name of the part and the species of wood used. Technically, scantlings do not include hardware and fittings.
- ***Measurements***: what you take off the boat. They already exist. Dimensions are the numbers on a drawing that tell the size(s) of a thing to be made.
- ***Siding***: the thickness dimension of wood; that which is the first cut when logs are sawn into planks; and the first dimension that you look for when you go to the lumber stack.
- ***Molding***: the pattern shape lifted from the loft floor and taken to the lumber stack to find a piece of wood of an appropriate shape and grain pattern.

When you think that the list is complete, repeat the process, this time fleshing out the list with sizes, shapes, materials, and fastening locations. Make patterns of irregular shapes, label the patterns, and note their existence in the list. A completed scantlings takeoff should include:

- Name of part—accepted and/or regional (both would be preferred so that regionalisms are preserved).
- Measurements (sizes).
- Material.
- Shape, noted in measured sketches or patterns.
- Fastening size and location. Look for and record fastening patterns and their relationship to joint lines. A "fastening pattern" is the arrangement or dimensioned layout of the fastenings.
- Shape of joints where they are visible.
- Sketches showing a half cross-section of the boat, the size and shape of openings, framing around the openings, deck framing, centerboard case, and cockpit layout, etc. Leave plenty of room on your sketches to record the measurements that locate these things.

The Bits and Pieces

Taken a piece at a time, a boat is not a complex structure. However, when assembled, the pieces can seem complex and confusing. So, let's begin the study of the bits and pieces of boats with a one-piece boat, the Louisiana log pirogue (fig. 6.10). The pirogue is misleadingly simple—just a hewed-out log. When the builder has the outside of the pirogue hewed to the desired shape, he must have a way of maintaining the desired hull thickness—say a half or three-quarters of an inch. To do this, he inserts gauge pins (dowels) of the appropriate lengths into the hull at predetermined locations and then hews the hull out to the ends of the gauge pins.

A record of any boat includes defining both the outboard and the inboard. This means taking the outboard shape of the pirogue—an easy task—and then establishing the relationship between the outboard and the inboard to find the hull thicknesses. Only then can a proper drawing of the pirogue be made. It is possible to drill a great many holes to determine the thicknesses, but that would severely damage the pirogue and is unnecessary. A much more practical, and acceptable, way to measure the outboard and inboard simultaneously is shown in fig. 11.46.

Knowing that gauge pins were probably used, look for them along the centerline, either inside the hull or outboard. If all is bare wood, the pins should stand right out due to a difference in grain from the rest of the hull. If the pirogue is painted, and in good condition, look for hairline cracks in the paint outlining the pin. Work forward or aft from the first pin found to find a second pin, then to port or starboard to establish a pattern. Record the locations of the pins when taking off the lines, and take body sections at the gauge pins.

The plank pirogue (fig. 6.10b) presents a very different problem. It has two stems, two side planks, a bottom with outside chine logs, and a thwart (seat). Since planks generally have a constant thickness, drilling one or two holes in the bottom at the centerline will give that thickness. The side plank thickness can be measured at the sheer. (There are actually several more pieces in a plank pirogue, but this description suffices for this example.)

Since recording a boat is very much a visual exercise, what *you see* and what *you record*, illustrations of what you might expect to find in a boat are the best guidance.

The illustrations and captions included with this text describe what has been found in vernacular ("traditional" or "folk art" boats) and the more traditional of naval architect-designed boats. Some of the material shown is very regional and rarely encountered elsewhere. Let the material encourage your suspicion and curiosity and help you decipher unexpected details. The captions not only identify the content, but also contain definitions, sources for details in the illustrations, and in some cases, the geographical origins. The illustrations include:

- The geography of the boat, showing where the parts are and general terminology (figs. 6.1 through 6.16).
- Descriptions of the backbones from very small to the largest boats covered by this manual (figs. 6.17 and 6.18).
- Details found in the backbone—stems, keels, centerboards, and centerboard cases, transoms, sterns, and floors (figs. 6.19 through 6.28).
- Hull construction—flat-bottomed, V-bottomed, and round hull; frames, floors, knees, and clamps (figs. 6.29 through 6.33).
- Planking—smooth and lapstrake and bottom planking (figs. 6.34 through 6.37).
- Deck and deckhouse construction (figs. 6.38 through 6.45).
- Canoes—wood-canvas and wood planked (fig. 6.46).
- Fasteners (fastenings) found in boats (fig. 6.47).
- Joints (figs. 6.48 and 6.49).
- Rudders/steering (figs. 6.50 through 6.53).
- Mechanical and electrical systems (figs. 6.54 through 6.59).
- Locating and recording the boat's parts—taking off (figs. 6.60 through 6.72).
- Analyzing the boat (figs. 6.73 through 6.76).

Joinerwork

Taking off joinerwork requires no more skill than taking off either the lines or the construction. Because every boat owner has an "ideal" arrangement for the interior of a boat and modifies what the builder or naval architect suggests, there are myriad variations in the arrangements of interior joinerwork.

Joinerwork is divided into two groups: deck joinerwork and interior joinerwork. Deck joinerwork includes toe rails and bulwarks; coaming; watertight cockpit; benches, seats, and thwarts; and deckhouse, companionway, and hatches. Interior joinerwork includes virtually everything inside the boat from the frame and floors in, except bilge stringers and ceiling—cabin sole, bunks, head enclosure, lockers, cupboard, tables, benches, ladders, and bulkheads.

Identifying and naming joints may prove the most difficult part of recording joinerwork. Several common joints are included in the illustrations.

Mechanical and Electrical Systems

In a boat recording project, the mechanical and electrical systems are often ignored. To do a proper job, however, you must look for and record the following:

- Engine installation: engine, battery, starting switch, fuel tank, fuel pump (if separate from engine), fuel piping, lubricating oil tank, throttle controls, cooling system, exhaust system, smoke/fire detector, and gauges (ammeter, oil pressure, temperature, and fuel level)
- Drive line: propeller shaft and the hardware along in from the engine coupling to the outboard end of the shaft
- Sanitary facilities: head, holding tank, wash basin, shower, freshwater tank, and piping for them
- Food preparation: galley sink, stove, refrigerator, and lockers for utensils and food
- Ventilation and heating: stove, air conditioning, and fixed ventilators
- Automatic fire extinguishing equipment
- Steering gear
- Winches: electrically or manually powered for the anchor and halyards
- Electrical systems: engine starting and battery charging and possibly a separate battery with its own wind driven generator to operate an automatic bilge pump, cabin lighting, and navigation lights

Before recording mechanical and electrical systems, decide how detailed the finished drawings will be. Measuring an engine to illustrate it faithfully is excessively time consuming. Instead, record manufacturer's names, model numbers, and/or specifications from data plates fastened to equipment, or cast into hardware. Listing that data on the drawings may be adequate. Photograph all equipment, hardware, and data plates if possible.

If you decide to show equipment and hardware on the drawings, try to obtain copies of "catalog cuts" (Manufacturer's catalogs) from a marine hardware dealer, boatbuilder, museum library, or even the manufacturer. The copies can be enlarged or reduced on a photocopier to suit the scale of the drawing and then traced onto the drawing.

Record the locations of mechanical and electrical equipment and hardware and the routing of pipes and wires while you are recording the locations of bulkheads, bunks, lockers, etc. For example, if a wire for a bow light runs along the clamp, write that in your notes. If an exhaust fitting passes through the hull, record where it is on the outside of the hull with a distance below the sheer and from the nearest station section.

Measured sketches of systems are not necessary. Instead, make schematics and supplement them with detailed sketches and/or photographs. (A schematic is a simple single line sketch, not drawn to scale, showing the locations of beginnings, changes of direction, and end points, and of equipment and hardware. See fig. 6.59.)

Locating and Recording: Taking Off

The way that you go about locating and taking off the shape and size of individual parts depends on the size of the boat. A twelve-foot, flat-bottomed skiff and a thirty-foot sailboat require different approaches.

A skiff can be set up, and its lines and construction taken off, without your feet leaving the ground, because it is possible that no part of the boat will be above chest height. But, a skiff is fairly light and may be disturbed while you work around it. It must be in and remain in the position that it was in when the lines were taken off so that the detail take-off can be related to the lines take-off.

Stretch a tape measure the length of the boat between staffs at forward perpendicular (F.P.) and aft perpendicular (A.P.) or from the stemhead to the top of the transom to locate the details inside the boat (fig. 6.60). It makes no difference whether the tape is set up with the 0-inch mark at F.P. or the A.P., or even if the tape is level, but be sure to record in your notes how the tape measure was set up (level or not) and where the 0-inch mark was. Any measuring error caused by the tape sagging will be negligible, so don't worry about it.

To measure locations, hold a plumb bob alongside the tape with the point touching the edge of a thwart, for example, and the string touching the tape, and record the distance from 0" off the tape. To measure the location of details that are not on the centerline (oarlocks, for example), lay a straight board across the sheer, align it with the oarlocks, and measure the location of the board. Or tie a piece of string between the oarlocks and measure the distance from the 0-inch mark to the string. A carpenter's level can be substituted for the plumb bob by using the plumbing vial in the level.

A thirty-foot boat, on the other hand, can have the outboard and inboard taken off simultaneously if the boat is well supported. To do this, the relationship between the outboard and inboard must be established using a few simple methods.

Referring to chapter 7, rig a baseline, erect staffs at the perpendiculars, and rig a top reference line to use as the reference for all measuring—distances and heights—down to and inside the boat. Rig an auxiliary reference line, and if necessary, auxiliary vertical staffs, inside the boat to aid measuring what is otherwise inaccessible (fig. 6.67). The auxiliary reference can be used for height and athwartship measurements, but be careful not to deflect (bend) the string. When an auxiliary reference line cannot be set up directly below the top reference line, erect plumbed staffs and record their relationship to the top reference line (figs. 6.67, 6.68, and 6.69).

Once a system of reference is set up, the actual measuring is very straightforward, as long as you don't forget to record the relationship between the reference lines. If you have already take off the lines, you should have no difficulty with the rest of the job.

Much of what you record will be obvious, easily described and measured. But what about things hidden such as joints? Take off enough data to give good clues, which can be combined with experience, records of similar boats, or knowledge of local boatbuilding practices, to enable you to reconstruct the missing data on a drawing. It is *reasonably* safe to assume that *most* joints are made with straight cuts. It follows that if you accurately locate what appears to be the ends or corners of a joint, you can reconstruct that joint by drawing a straight line between them. However, if after drawing the joint, it does not look right, chances are that it is not.

Hints and Comments

- Do not depend on photographs, even those that include rulers, to provide accurate measurements. If the ruler and the object, or any part of it, are not in the same plane (distance from the camera), it is impossible to take accurate measurements from the photograph.
- Make patterns of thwart knees, breasthooks, seat beams, seat backs, and odd items of trim and hardware, such as hinges and latches. It is quicker to make patterns than to make a sketch with complex measurements. Remember to record thicknesses. Patterns can be photographically reduced and then traced onto the drawings.
- Do not overlook the way the planks lay on the hull. Observe the plank seams, especially in the last several feet of their length. This is where fairing the lines is critical to the planking process. The reasons for this are dealt with at length in chapter 11 and figures 11.36 and 11.37.
- Thinking in descriptive terms, ask, "How do I describe this?" "How did he do this?" "Why did he do this?" There is a logic to the way boats are constructed. Once you begin to see that logic, the rest follows.
- Rely on simple one-, two-, or three-view sketches with the measurements clearly indicated. Perspective sketches are not necessary and can be very awkward to show measurements clearly on.
- The flow or direction of the grain in wood can indicate bent or natural crook. Generally, grain that very closely follows the outline of parallel sided pieces of wood indicates a bent piece. Grain that swirls and crosses the edges indicates that the piece was cut from a "natural crook." Rubbing chalk into the wood's surface will highlight the grain.
- Don't waste time locating both edges of a

thwart, for example. With one edge located and with the width recorded, you have all of the fore and aft information available. Measure the vertical location of the thwart by measuring from either the sheer or a top reference line (figs. 6.60 and 6.61).

- Some points will be difficult to locate from reference lines. Triangulate their location from easily accessible parts.
- Wood identification is for the knowledgeable. While local folks may give the indigenous name, which is important to record, it may be necessary to send a sample to a laboratory for proper identification.
- The parts of even the simplest boat can be deceptive. Things that look straight are frequently curved; those that appear to have a constant thickness may taper; and those that appear to be original or a replacement may be exactly the opposite! Be careful!
- When locking for joints, especially in the backbone, keep in mind that if suitably shaped timber was available for curved members, it was used. This meant fewer joints, less work, and less potential for leaks.

Analyzing the Boat

Correctly analyzing a boat's condition and its effect on the boat's shape is critical to producing a reasonably accurate representation of the boat as it was when built. This means that you have to determine if the boat is fair, and if it is not, why not.

Boats are subject to changes that affect their condition, and their shape. Age, abuse, neglect, poor construction methods and materials, and the interaction of boat and water contribute to change, and all must be considered when analyzing how a boat's shape has changed. Boats are not like houses, they have no straight visual reference lines—corners, doors, and window openings, or roof lines—so must be studied within themselves.

Major changes in shape, visible on the outside of the boat, can usually be traced to structural failure which will be readily apparent inboard. Other changes, not so apparent, will take some cogitation fueled by assuming that:

- The boat was built true and fair.
- The bottom of the keel should be straight or rockered (curved upward).
- The rabbet line should be a slightly upward curving line, even with a straight bottomed keel.
- Neither the keel nor the rabbet line should hog (curve downward). There are exceptions, so be sure of the boat type.
- The sheer should be fair, a continually sweeping, curved line, even though not exactly identical port and starboard. A slight difference port and starboard does not necessarily indicate a change.
- Plumb bobs hung from the center of the stemhead and the center of the transom will align with each other and the centerline of the keel when viewed under the boat. If the plumb bob strings can not be made to align, either the whole boat is twisted from end to end, the keel is bent, or the stem or sternpost is misaligned in relation to the keel.
- The boat's girth (distance from sheer to rabbet and equaling slightly more than the sum of the width of the plank widths) at any body section cannot change unless there is a reason—excessive caulking is one.
- Sections at the ends of a twisted, hogged, or sagged boat will show little or no change in shape.
- In a round-hulled boat, there are no angles or awkward changes in shape—either along or across the planks. That is, the edges and ends (butts) of planks lie flush with each other.
- In a flat-bottomed or V-bottomed hull with straight sections, the planks should lie straight with edges flush across the planks, and smoothly, and flush at the butts, along the length of the plank.
- Planking fits snugly to frames, floors, and deck beams.
- Lapstrake planks do not fit snugly to frames and floors unless they are jogged to fit.
- Plank seams will be approximately the same width throughout the boat.
- All joints should fit tightly.
- Parts, such as knees, breasthooks, and floors, cut from grown or natural crooks,

should retain their original shape, unless broken or cut to extremely slender proportions.

- "Grown" or "natural crooks" are pieces of wood selected for use because the grain approximates the shape of the part to be cut from them.
- The outline shapes of flat transoms do not change. While the outline shapes of curved transoms do not change, they may flatten out—check the transom's frame for loose joints and fasteners, which will indicate that it has flattened.
- Any part of a boat can change its shape if sufficient damaging force is applied.
- Assumptions made about hull planking apply also to deck planking except at king planks, mast beds, covering boards, or other apparently intentionally thickened areas.
- Hull planking will be laid out so that a long plank port will be mirrored by a long plank starboard. Look for butts and scarfs in the garboard, broad, and sheer planks to be located opposite each other. This consistency should also be evident in the deck planking. Laying out the planks this way lets the builder make one plank pattern for use on both sides of the boat. An exception to this style of planking is found in production-built canoes—such as Rushton, in which the planks are of a fairly constant width and cut to a pattern. The scarfs were located to suit the planking stock and were fastened with the forward end of the scarf inboard and fastened to a frame (fig. 6.36).
- The same wood was used for all hull planks. The sheer strake is frequently different—oak or mahogany with the remainder cedar—rarely the garboard might be oak.
- Deck planking is all of the same wood, with covering boards, king planks, etc., of a different wood.
- A mixture of frames that differ in sectional size and shape usually indicates repairs or rebuilding, except where there is a definite pattern of bent and sawn frames, which is common to some boat types. A single broken frame does not necessarily indicate any more than that a single frame is broken. Look very closely for other signs of damage or structural failure in the area of the broken frame. The single frame may have broken because of short grain across the frame. On the other hand, several frames in a row that are broken along approximately the same line means damage and that the shape of the hull must be examined closely for fairness in all directions.
- Wood used for frames will be the same throughout the boat.
- Woods are often a matter of regional preference—red oak for a stem in Maine, white oak south of Boston, maple in the Canadian Maritimes, or spruce in the Adirondacks. Consequently, expect the same wood to be used for the same job throughout the boat, allowing for exceptions as noted above.
- Fasteners of the same type and material were used for the same job throughout the boat. For example, rivets for plank fastening, screws may have been used for the garboard to keel and hood ends or clench nails for the planks, and the sheer strake riveted. If there are variations, they should have rhyme and reason.
- Fastenings of different metals were used in different eras and locales, and in some locales the type of fastening used was governed by the size of the boat.

As a rule of thumb, fastenings—plain wrought (wrot) iron—have in salt water life expectancy of up to 40 years; galvanized iron, about 40 years; galvanized steel, 20 years; silicon bronze, 25 years; and manganese bronze, 50 years. Iron contains no carbon; steel is carburized iron. The impurities in iron, usually silica, give it its long life expectancy. Galvanized steel came into general use between 1900 and 1920. Bronze and brass are copper-based alloys; hence their longevity. All metals have a longer—even indeterminate—life in fresh water, above the water line in salt or fresh water, or where there is no active electrical current.

- A delaminated (looking like layers of heavy paper), rusting fastening is iron. A swollen, rusting fastener is steel.
- Copper-based fastenings erode and lose material at a rate of approximately 0.001

inch per year where active electrical current leakage is present.

- All fastenings below the waterline should be in approximately the same state of deterioration or erosion. The same is true for all those above the waterline. Do not compare those above the waterline with those below or those in the interior with those in the exterior.
- A line of fastenings in the hull or deck indicates a frame or deck beam, possibly hidden; along the deck or carlin or such.
- An otherwise inexplicable mixture of fasteners, in type or material, suggests repair or modification.
- An unusual number of fastenings showing in the plank outboard may indicate a scarf or a butt joint with a butt block inboard that is hidden by ceiling. Look for the same condition on the opposite side of the hull to indicate original work; a difference may indicate a repair. In production-built boats, especially canoes, the scarfs may be located willy-nilly.
- Fastenings should not be visible through joints.
- Fastening type, style, shape, and material help distinguish repairs from original work, so examine them closely. Galvanized common wire nails alongside copper rivets or clench nails, bronze and iron drifts, bolts or nails shout repair. Round headed nails used with square or rectangular headed nails, also indicate alterations.
- Fastenings can be used as indicators of age; sequence of construction, modification, and repair; and geographical origins.
- All of the preceding assumptions are based on collective experience, and any assumption that does not stand up—as in the case of a single broken frame—may mean absolutely nothing to the boat as a whole or, it may mean isolated damage. The boat for example may have been run onto a rock.
- A structure's present shape can be very different from its original shape, and there may be little or no evidence that it has changed.
- Changes occurring slowly over a long time and those resulting from poor storage methods are difficult to deal with and require extra thought.
- If a change in shape is evident in one area of a boat, look for a change opposite that area.
- Small details offer strong clues. Deck beams and carlins are frequently beaded on the bottom corners; risers, clamps, ceiling, and floorboards on one, and sometimes two corners. If the beading is not continuous, or if it changes size or shape, you are looking at repairs.
- There should be a definite rhythm in the boat's construction—location of frames, floors, deck beams, and fastenings—and that planking on the hull, bottom and deck will be symmetrical port to starboard. Any break in that rhythm is cause to suspect repairs or rebuilding.

Where to Start?

Keeping the preceding assumptions in mind, look for a hull that is not fair and figure out why it is not fair. (Fair is a technical definition and results from a graphic solution using straight and curved lines and their intersections to define an irregularly shaped curved surface. Any line drawn on a faired hull will be fair. (Refer to chapter 11.) Sweet is very subjective, as what is attractive in a person, a painting, or scenic view.

When there is enough evidence of change—it is a change, either in shape or arrangement.

Indications of change are illustrated in figs. 6.74, 6.75, and 6.76.

Evidence and Conclusions

After you have spent time with the boat, you know a good deal about it and have ideas about what has changed and the effect on the boat. Those ideas should be put together to describe the boat's condition and shape. So let's sketch out a hypothetical case showing what you have found in a fifteen-foot transom sterned rowing boat:

1. Stem at sheer damaged, flattened, grain looks spongy and squashed.
2. Joint between stem and knee opened up at face.
3. Stem to keel joint opened up.
4. Keel straight in all directions.
5. Port bilge noticeably slacker than star-

board, no broken frames in area of slack bilge, all seams unfair, sheer noticeably higher and not fair.

6. Port riser unfair, sweeps up at slack bilge, thwart knee fits snugly to thwart and plank.
7. Starboard knee fits snug to plank at sheer but one-half inch gap between knee and plank, knee fastened at sheer and to thwart, thwarts fastened to riser. Two other thwarts in boat have knees scribed (fitted snug) to plank and fastened to riser.
8. Joints in deadwood wide open with light visible through them, same for stern post to deadwood joint.

With the preceding evidence in mind, you might reach the following conclusions:

A. Items 1 and 2 indicate that stem has been driven aft—examine the sheer to see if it is bulged or unfair abaft the stem. Look from the sheer down and the keel up, along the side of the hull, and from fore and aft. Also, try to determine if the planks have been forced aft. A gap between the hood ends and the rabbet line is a clue.

Solution:

- Reconstruct the bow profile. Make patterns of the separate parts of the bow and try to fit them together to agree with your observations. This will move the stem face forward and lower the sheer. The sheer should fair in both the profile and the half breadths.
- A stem-keel joint that is open (item 3) usually indicates that the boat has dried out.
- Even though the keel is straight in all directions (item 4), keep an eye on it while adjusting and fairing the body plan and buttocks. It may have had rocker.
- The slackness, unfairness, and gaps noted in items 5, 6, and 7 may indicate that the boat rested on or against an object and slowly settled out of shape.
- The open joints in the deadwood (item 8) are normal for a boat that has dried out.

The job of detecting the details of construction and any changes in the boat's shape and structure is yours. This manual can only point to and suggest what you might find. So, look, observe, and record; then check what you have done.

The overriding rule in taking off is: *It is impossible to take off too much data.*

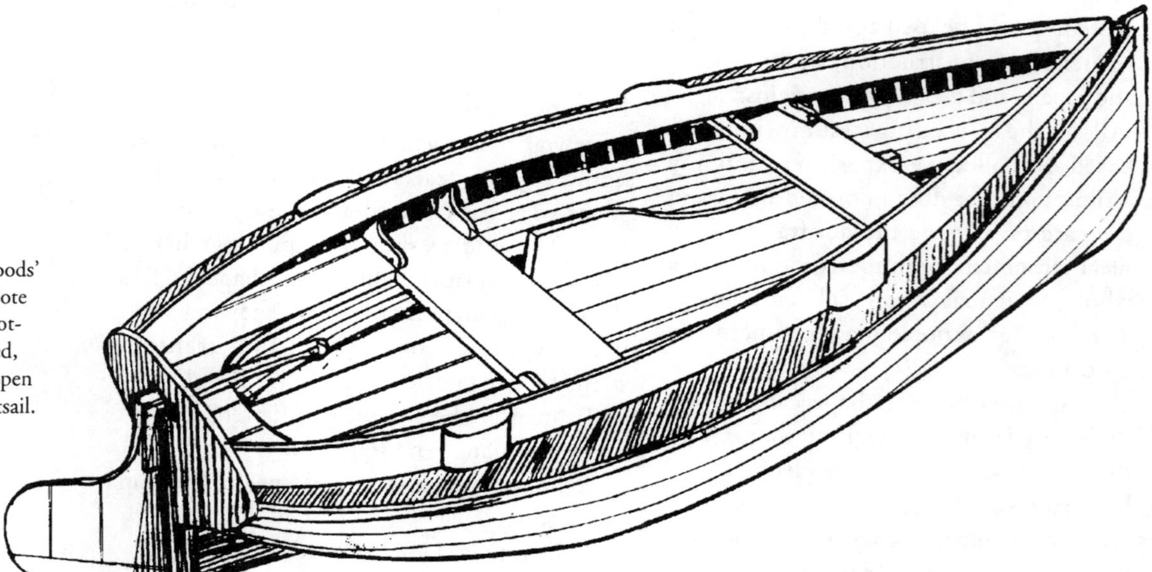

Figure 6.1. Susie, a Woods' Hole Spritsail Boat. Note that *Susie* is a round bottomed, transom sterned, batten seam planked open boat with a single spritsail. *Peter Schmid.*

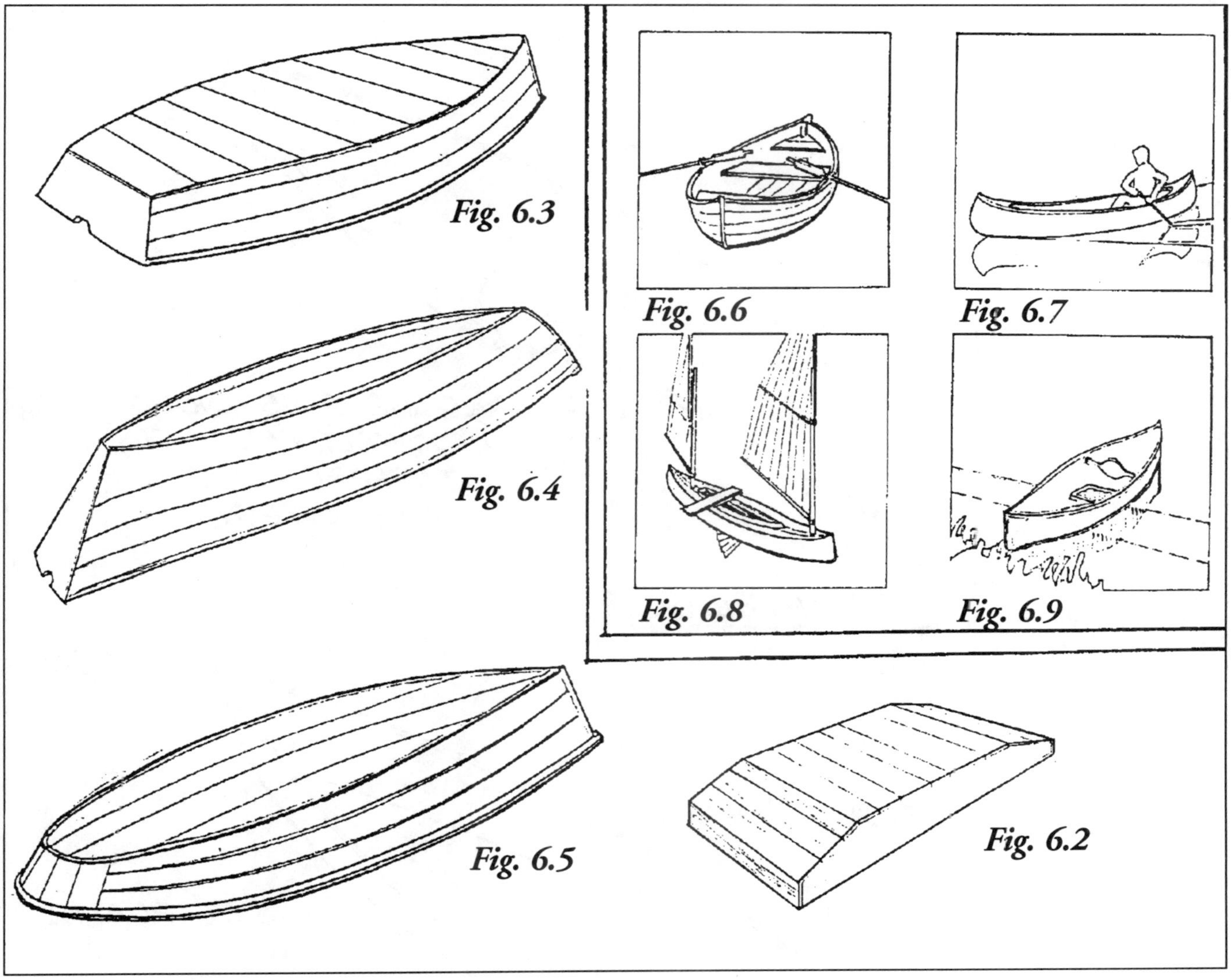

Figure 6.2. A scow with plumb flat sides and a cross-planked bottom that sweeps up bow and stern to rectangular transoms. *Peter Schmid.*
Figure 6.3. A flat-bottomed skiff with flared single or multi-planked sides and cross-planked bottom. Boats of this type can be either double ended or transom sterned. *Peter Schmid.*
Figure 6.4. A dory with flared single or multi-planked sides and flat-fore-and-aft planked bottom. *Peter Schmid.*
Figure 6.5 A New Haven Sharpie with flared multi-plank sides and flat-fore-and-aft planked bottom. Boats of this type are usually round sterned. *Peter Schmid.*
Figure 6.6. A double-ended rowing boat with stems bow and stern, but not necessarily symmetrical, and with carvel (smooth) planked or lap-strake planked. *Peter Schmid.*
Figure 6.7. A traditionally shaped canoe with canvas covered wood or lapstrake planked with regular lap or feather-edged lap (smooth skinned). *Peter Schmid.*
Figure 6.8. A decked sailing canoe. *Peter Schmid.*
Figure 6 9. An Adirondack guide boat—almost always double ended and feather-edged lapstrake (smooth skinned) planked. *Peter Schmid.*

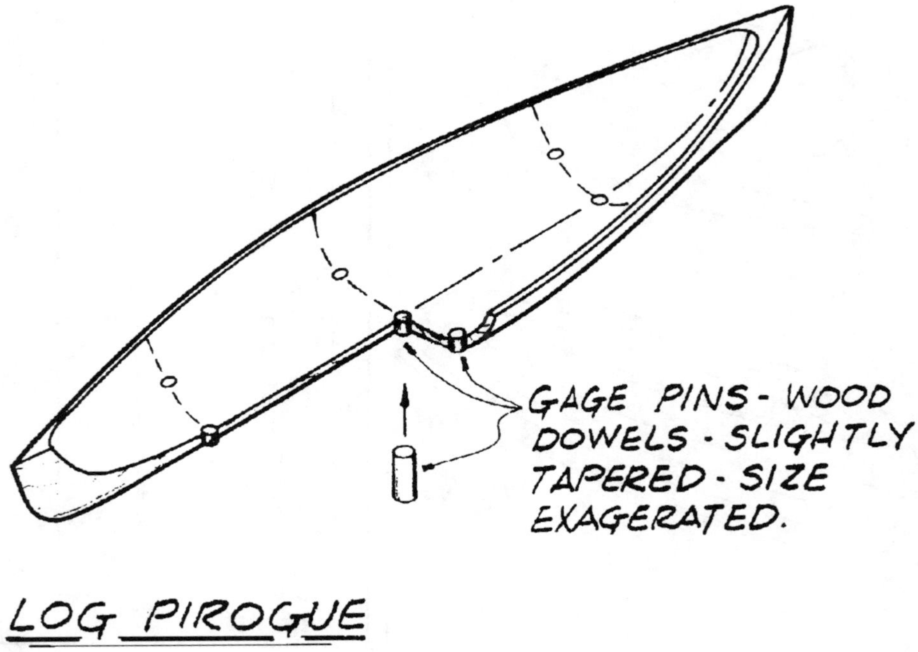

Figure 6.10a. A log pirogue.

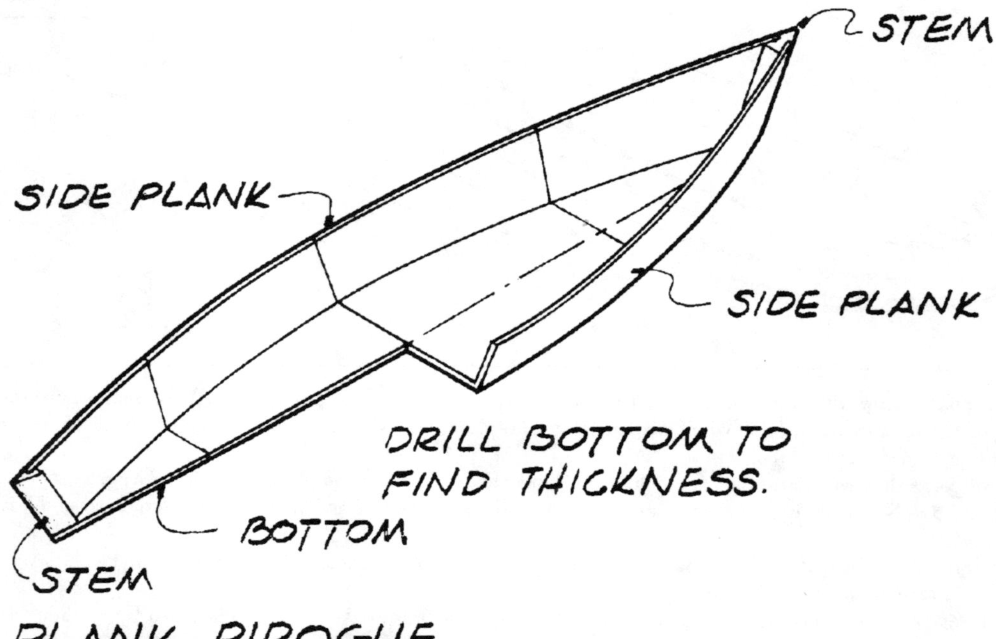

Figure 6.10b. A plank pirogue.

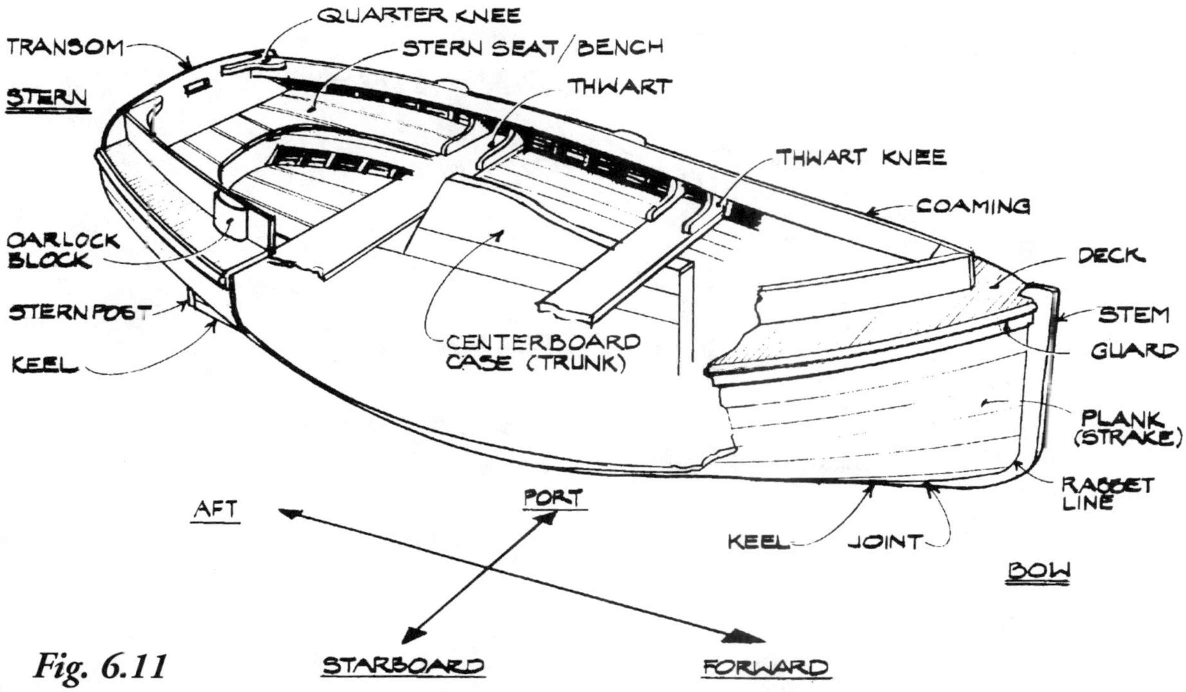

Fig. 6.11

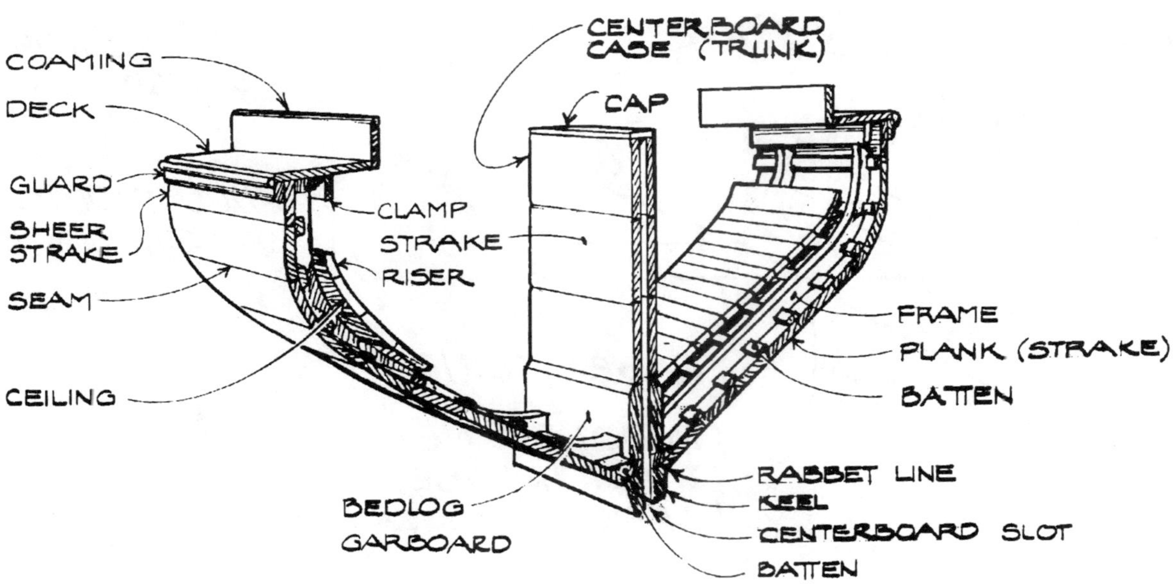

Fig. 6.12

Figure 6.11. (top) The "Geography" of the boat, *Susie. Peter Schmid*

Figure 6.12. (bottom)Susie, a cutaway section amidships. *Susie's* two-piece guard is characteristic of Woods Hole Spritsail Boats. *Peter Schmid*

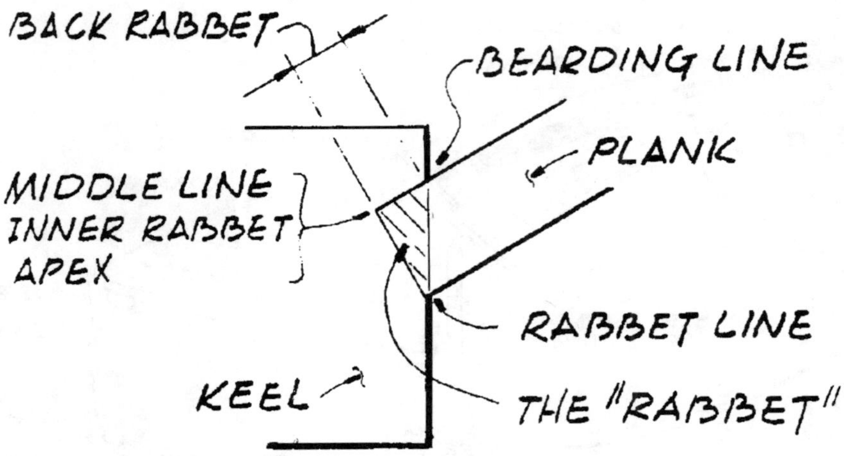

Fig. 6.13

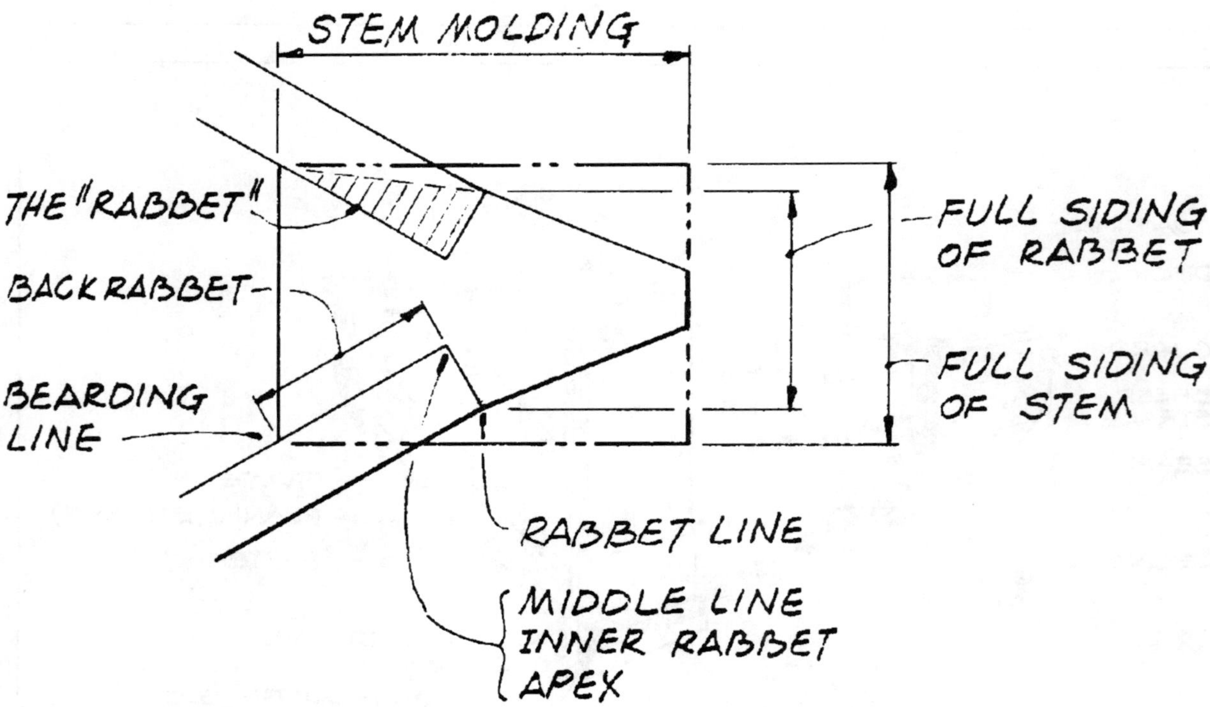

Fig. 6.14

Figure 6.13. (top) The parts of the rabbet in the keel.

Figure 6.14. (bottom) The parts of the rabbet in the stem.

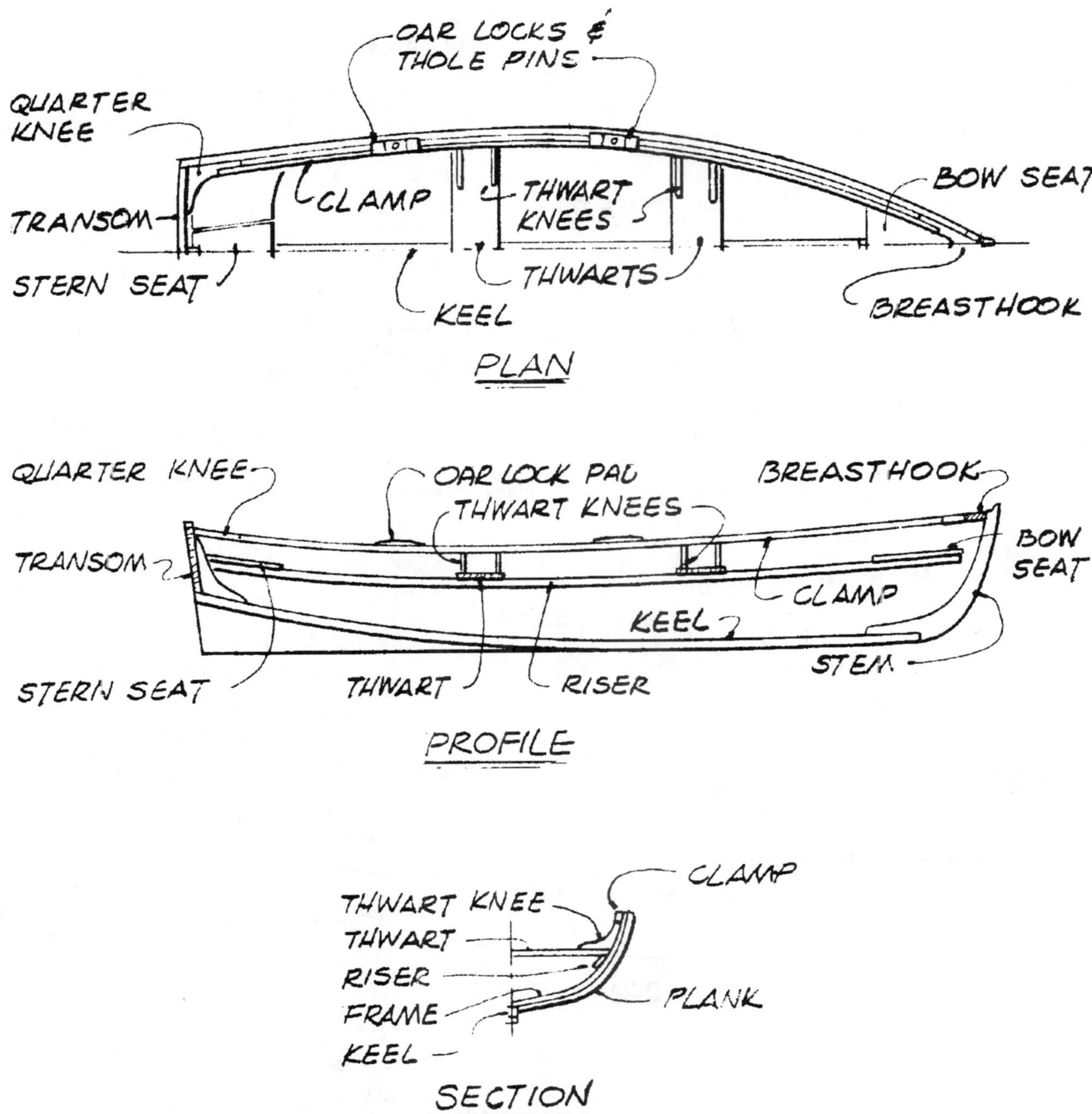

Figure 6.15. A key plan of an open rowing boat.

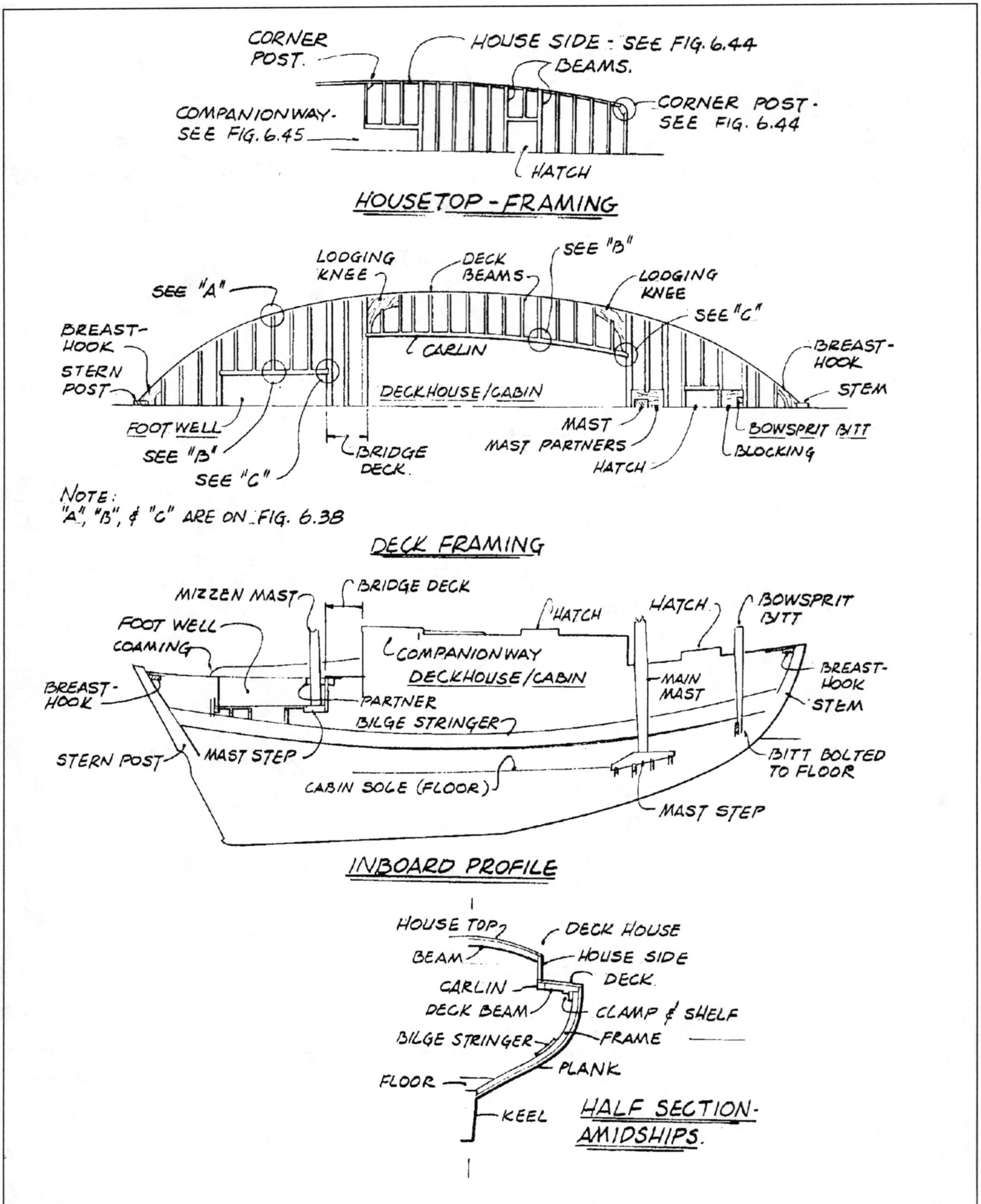

Figure 6.16. A key plan of the thirty-foot cruiser *Dog Star*.

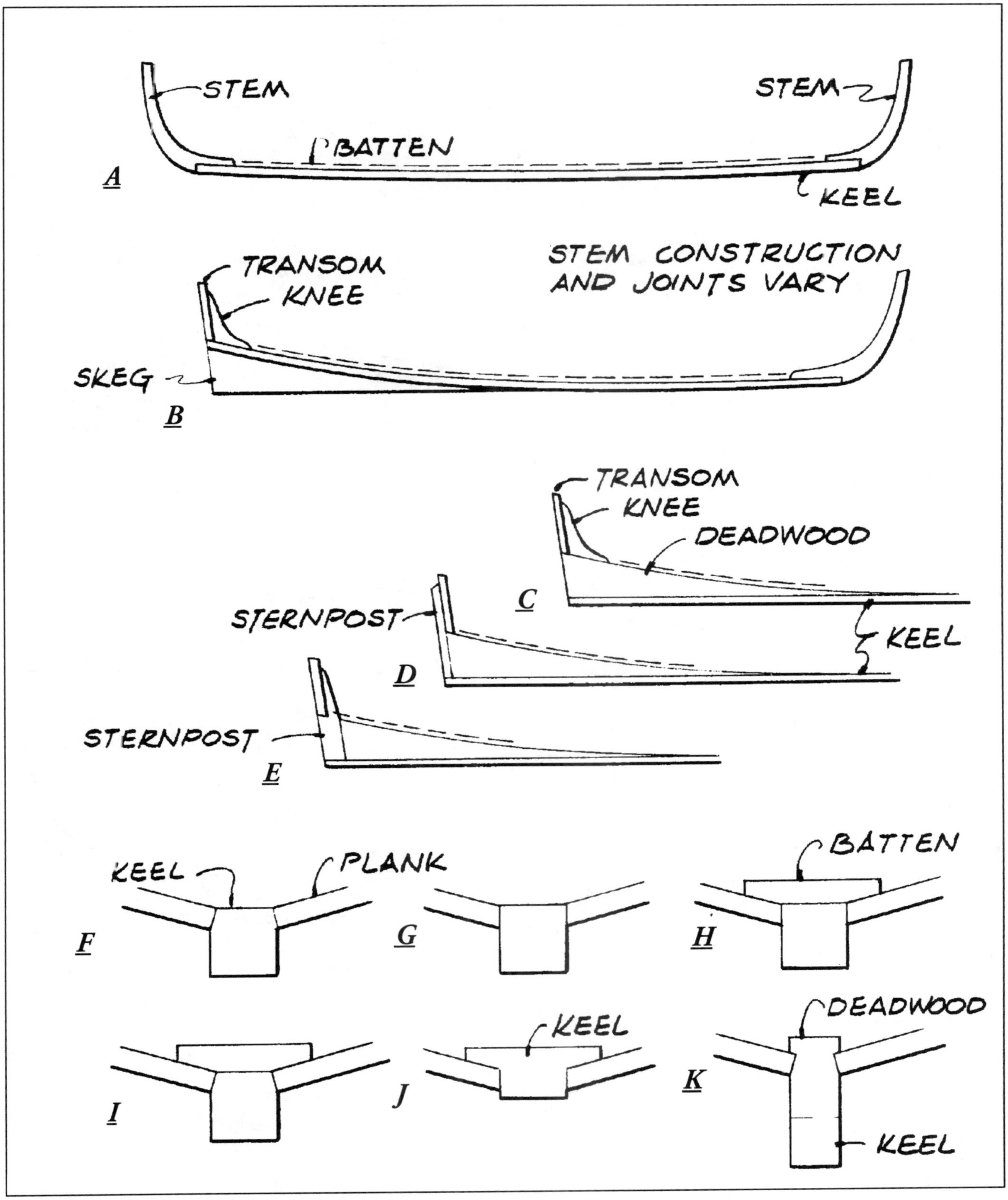

Figure 6.17. Variations in the backbones of very small boats. (A) Double-ended boat with a rockered keel (bent upwards from amidships). (B) Transom-sterned with sprung (bent) keel. (C) Transom-sterned with straight keel, deadwood, and knee. (D) Transom-sterned with straight keel, deadwood, and outboard sternpost. Frequently found on small sailing boats. (E) Transom-sterned with deadwood and stern post/knee in combination. (F) Keel beveled for square edged plank. (G) Keel left square; plank edges beveled. (H and I) As f and g but with a batten giving a bearing for the plank. (J) Plank keel with a plumb cut rabbet. Found in canoes, Delaware Duckers, etc. (K) In some boats the deadwood stands proud to support the plank. Of those backbones shown, F, G, I. J, and K are the most common.

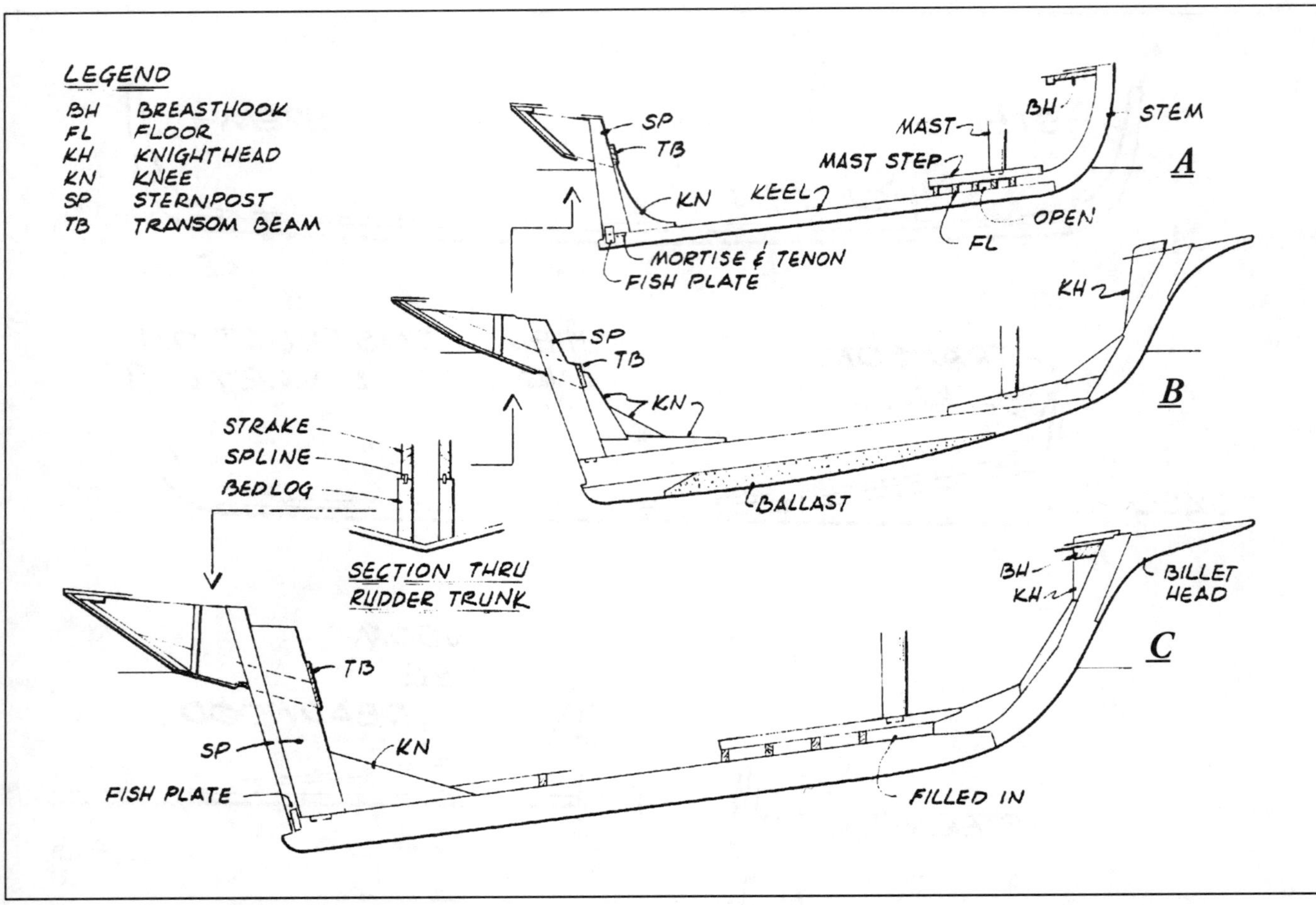

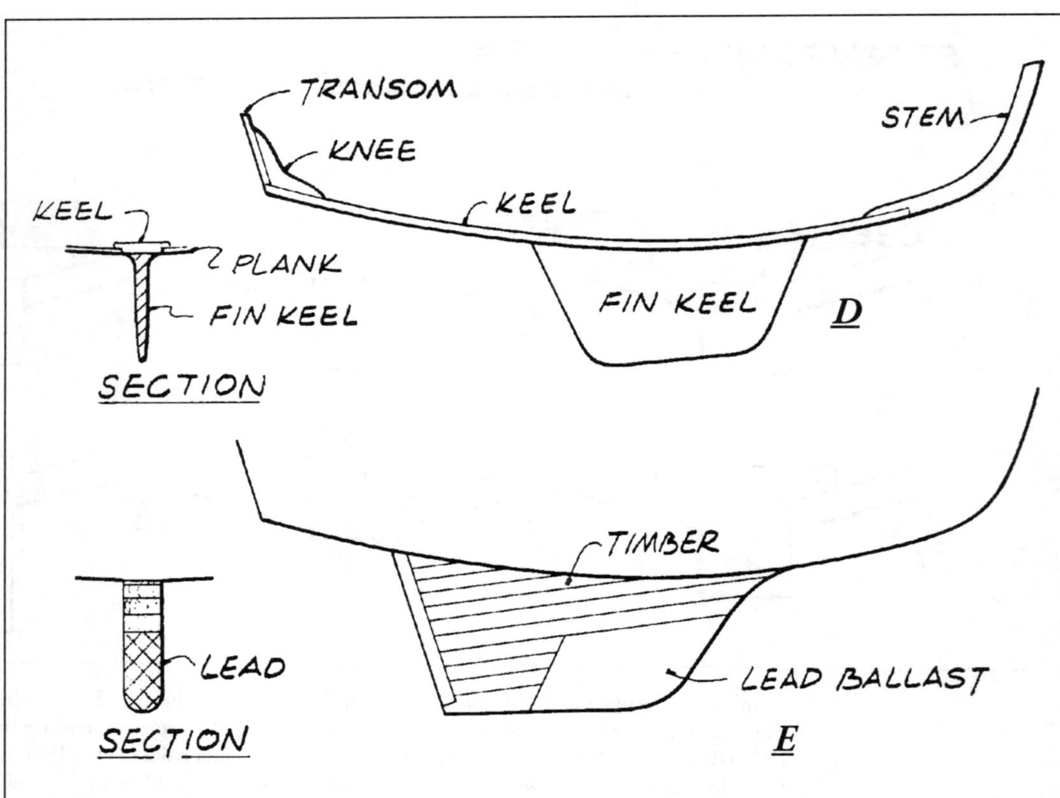

Figure 6.18. Backbones for larger boats. (A) *Ranger,* a twenty-two-foot Muscongus Bay Sloop. (B) *Little Hattie,* a twenty-eight-foot Friendship Sloop. (C) *Estella,* a thirty-five-foot Friendship Sloop. (D) A sailboat with sprung keel and bolted on cast-iron fin keel. (E) A sailboat with timber ballasted fin keel. Note the similarities in construction of the sterns of A, B, and C—the rudder trunk construction and transom beam. The major difference is the number of pieces used to do the same job. Bedlogs are also called horn timbers. Knight heads are heavy planks placed vertically alongside the stem. In all areas except along the keel in a, where the middle line is at the top of the keel, the entire rabbet is cut into the backbone.

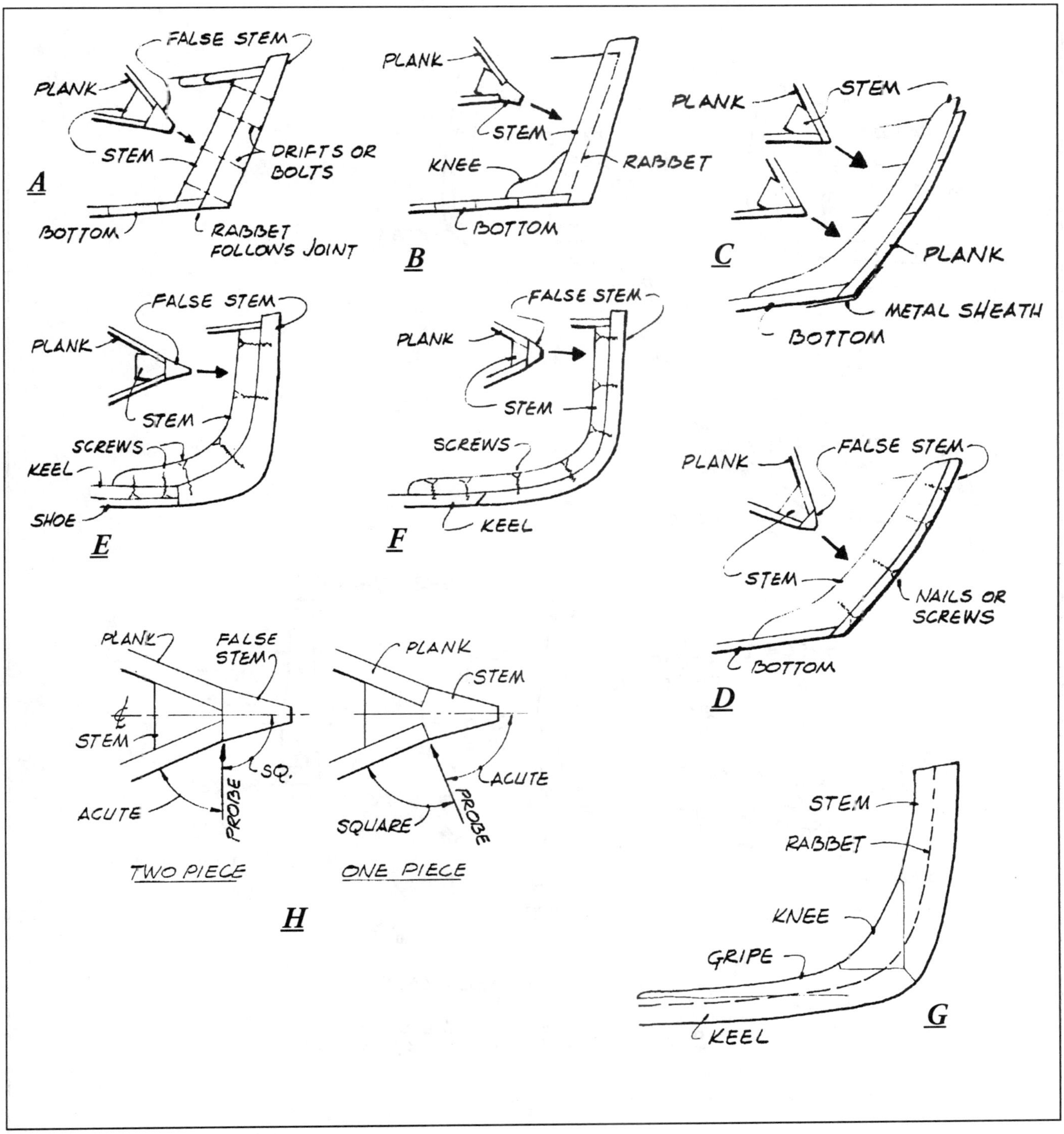

Figure 6.19. Stems: (A) A two-piece stem—the pieces joined either before or after planking. Look for fasteners visible inboard to help determine if there are two pieces. If the two are nailed together, no fastenings will be visible inboard. Look for bungs or putty outboard to tell if there are fasteners in the stem. Otherwise, probe the rabbet with a slim knife or other probe (see fig. 6.19 H). (B) A single-piece straight stem with the hood ends cut square (at 90 degrees to the surface of the plank). There are no fastenings to be found, and the probe inserted at the rabbet (fig. 6.19 H) is square to the plank. (C) Three-plank round hulled dory with a stem but no false stem. Instead of the latter, the planks extend forward of the stem and are mitered to each other. (D) Normal dory construction: a stem sawn from a natural crook and the false stem fastened on after the hood ends are trimmed square across the boat. (E) A Rangeley Lakes Boat—the stem and false stem (called a "cutwater" by the builders) are sawn from natural crooks. The cutwater is fastened in place after the boat is planked. (F) Technique used in lap strake canoes. Two pieces of stock, held together (not fastened to each other), are simultaneously bent over a form of the innermost shape of the stem. The hull is planked with the stem in place. The false stem is then added. (G) Stem found in a transom-sterned Mackinaw boat at Isle Royale National Park, Michigan. (H) Probing the rabbet: If the probe stands square to the plank, it is almost certain that the stem is one piece; if there is a noticeably acute angle to the plank, it is a two-piece stem with hood ends cut square across the boat. ("Hood ends" are the ends of the planks where fitted into the stem.) Occasionally, the false stem-to-stem joint is visible at the stemhead or it may be possible to see daylight through the joint along the rabbet line. Don't be fooled by splits and checks in the wood.

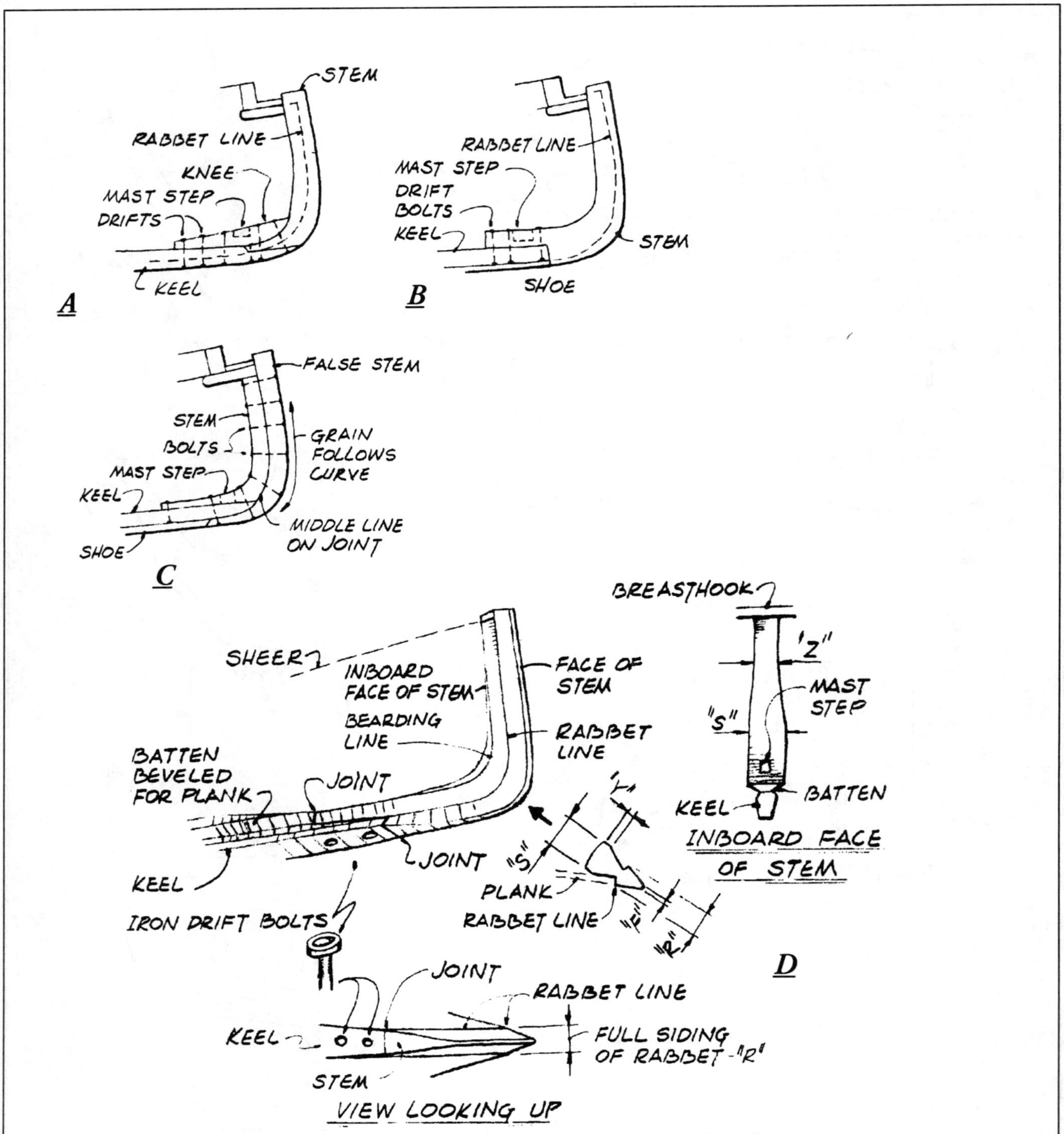

Figure 6.20. Woods Hole Spritsail Boats: (A) *Explorer,* built by Crosby at Osterville, Massachusetts, ca. 1890. (B, C, and D) all by Swift at Woods Hole, Mass. B is *Spy,* 1893; C is unnamed and unfinished, 1913; D, is *Susie,* 1896. A, B, have the mast step mortised into the backbone. *Susie's* mast step (D) is in the cockpit sole (floorboard). A, B, and D have one-piece stems from natural crooks. The unnamed boat C, was the mystery among Swift's boats: The joints between the stem, keel, and false stem are hidden by the planking. Fortunately, Swift's patterns still exist. When they were fitted together, the construction became clear. In this case, even though the fastenings inboard and the grain in the bent stem indicated two pieces, without the patterns to indicate the joints, any other assumption would have been speculative.

Note a stem-to-keel joint in D. An error can be made if the full siding of the rabbet is taken to be the full siding of the stem. The full siding of the stem occurs only where the stem stands clear of the plank abaft (aft of) the bearding line as indicated by "Y" in the section. Don't forget to look inside the boat. The full siding of stem rabbets (the thickness port to starboard) frequently tapers, so take several measurements. Record where they were taken; either height above base or at which plank seam. Attribute variations of a sixteenth or even an eighth of an inch to finishing and refinishing.

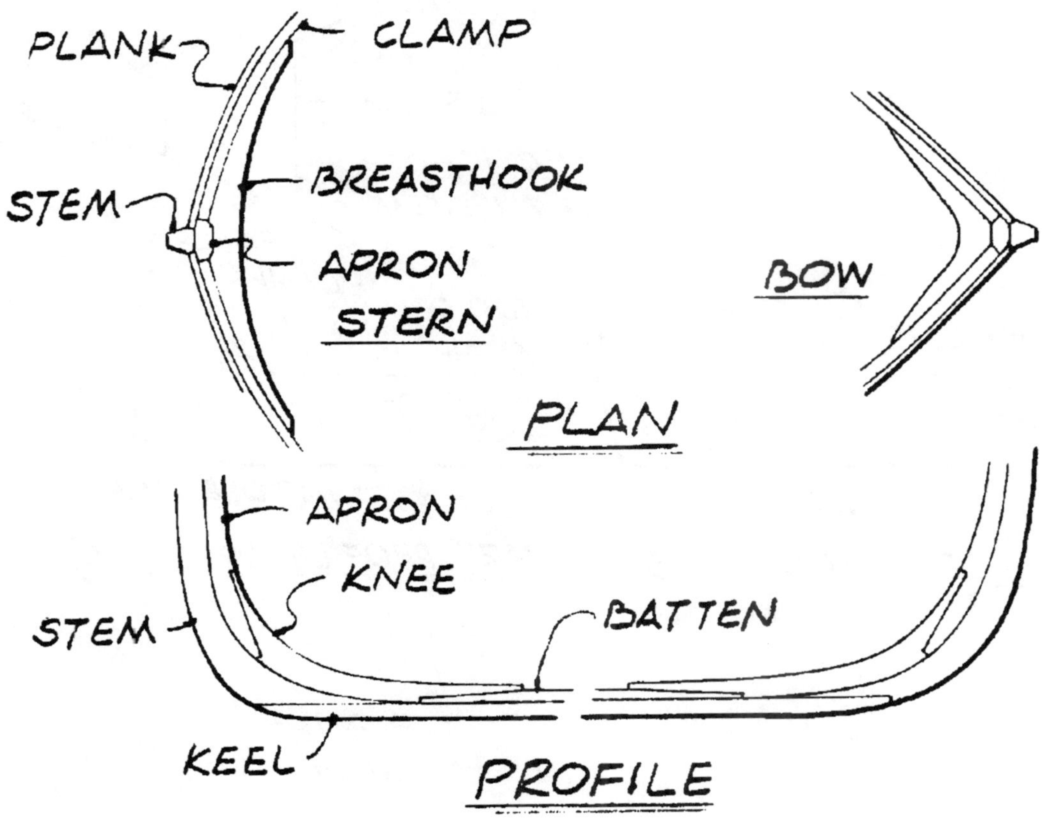

Fig. 6.21

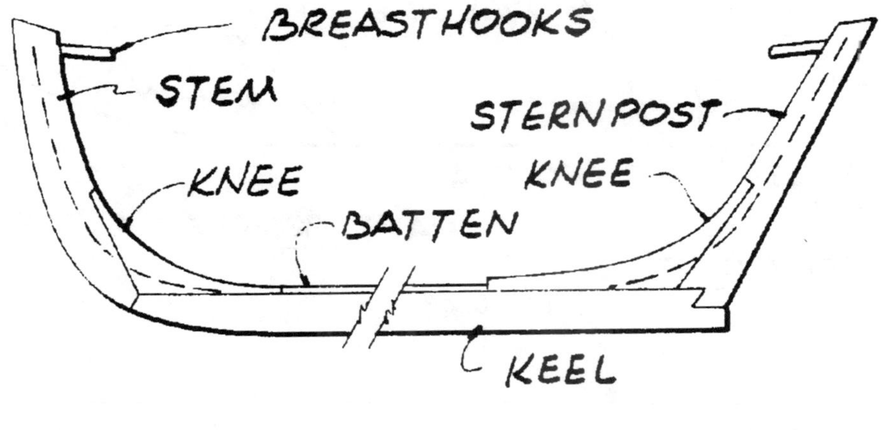

Fig. 6.22

Figure 6.21. A Scottish double-ender with very different ends. The stem-to-keel joints differ bow and stern. The middle line follows the joint line between the stem and the apron and knee and the keel and the keel batten.

Figure 6.22. A Maine-built double-ender with a straight sternpost. The rabbet is cut into the solid timber of the stem knee and sternpost knee.

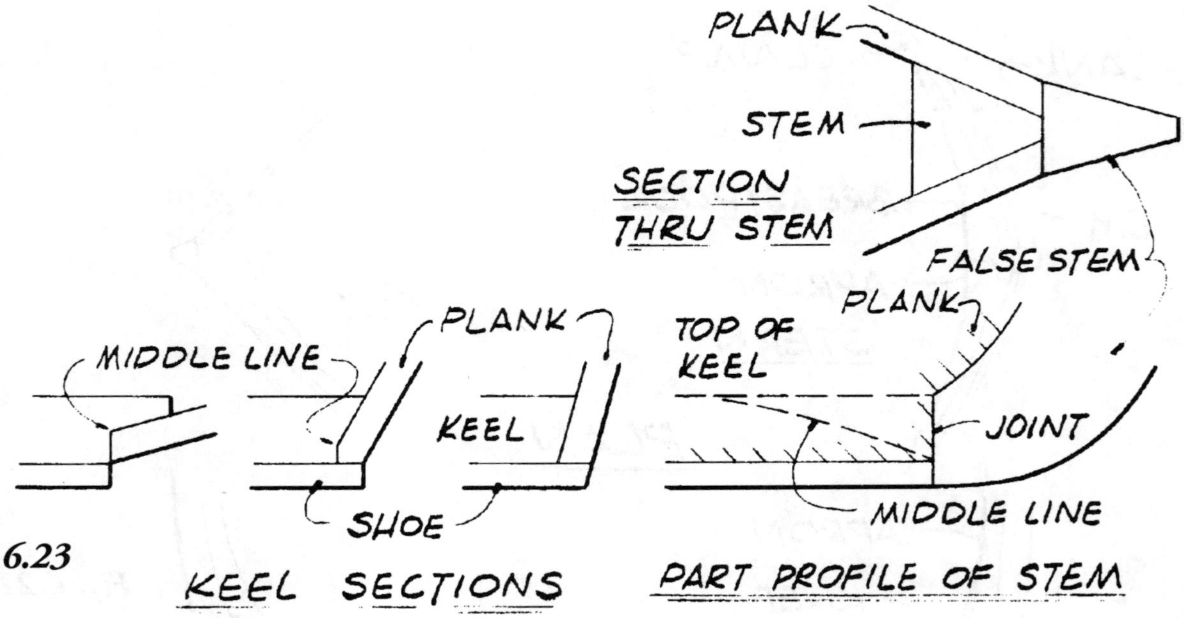

Fig. 6.23

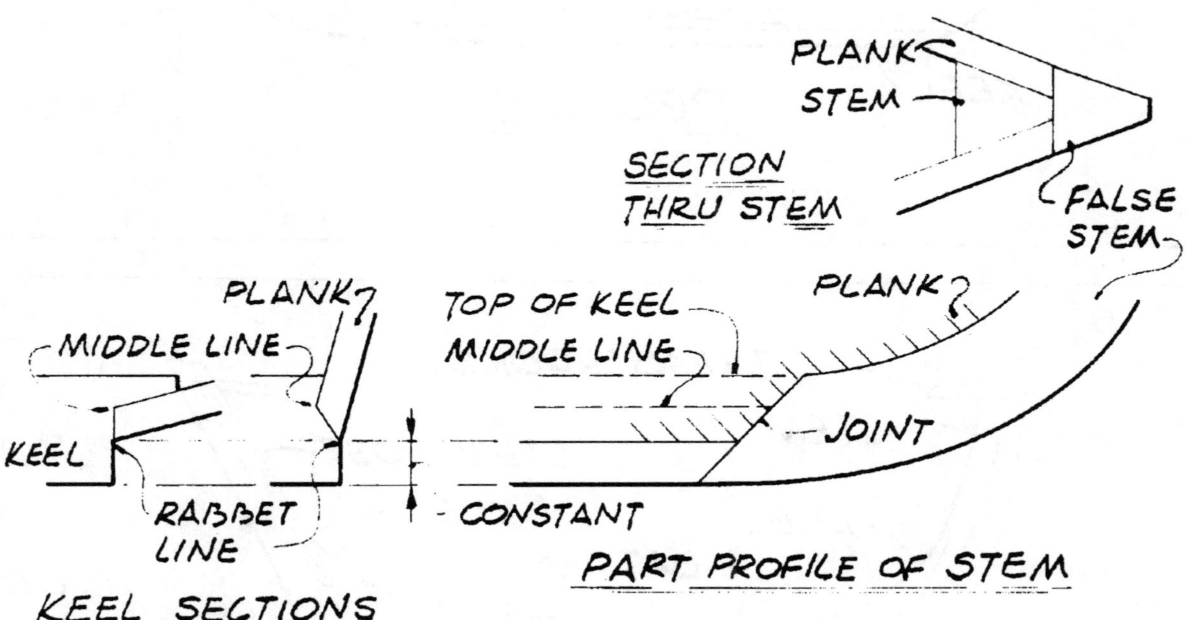

Fig. 6.24

Figure 6.23. Rabbets at the plank keel. Knowledge of regional construction, in this case Rangeley, Maine, is necessary if the boat is in good condition and the rabbet is not exposed. Also, as the garboard rotates toward the vertical, the middle line curves downward.

Figure 6.24. Rabbets at the plank keel. The shape of this rabbet permits the use of a probe to find the angle between the hidden edge of the garboard and the keel.

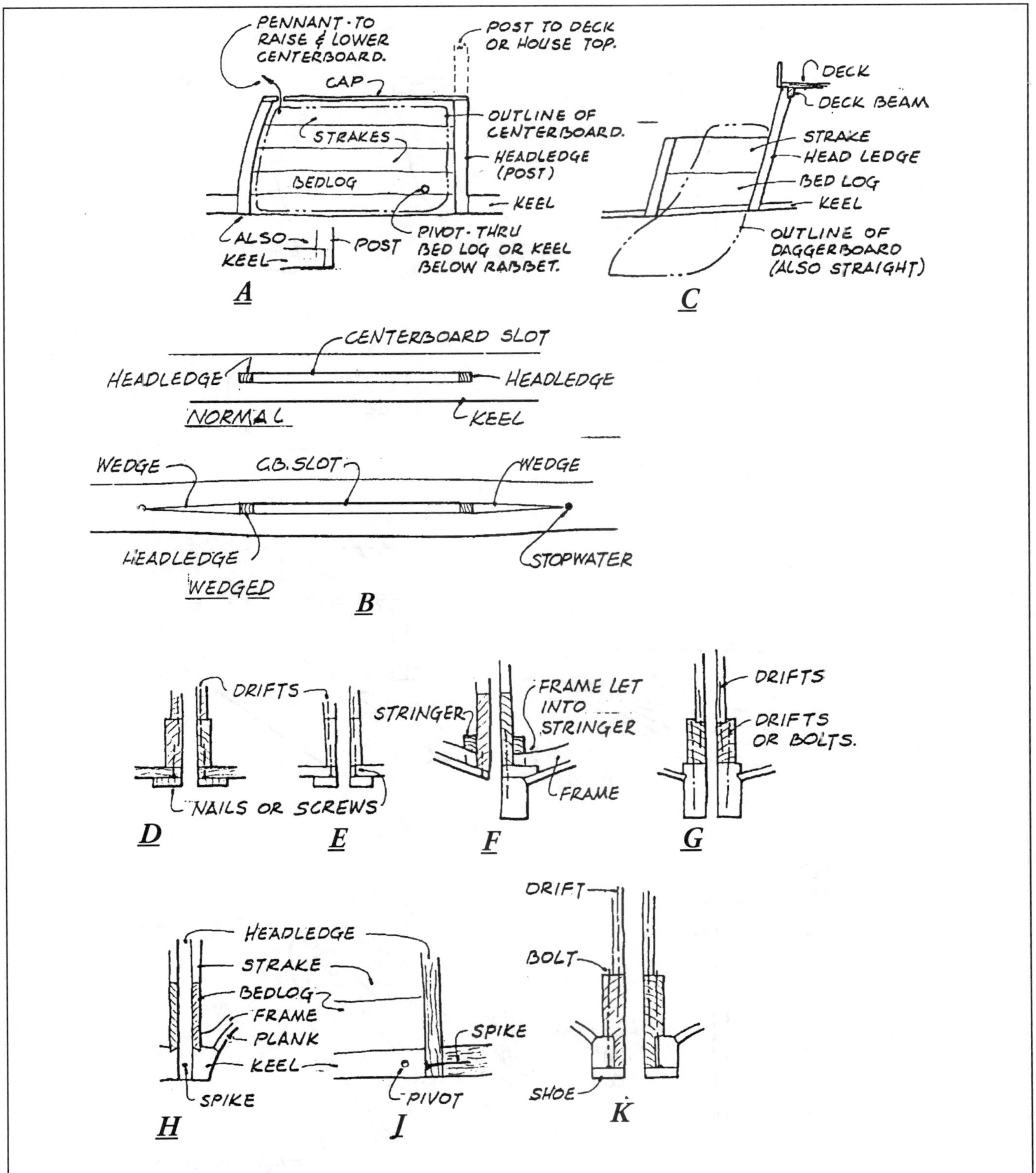

Figure 6.25. Centerboards, daggerboards, and leeboards. (A) Centerboard case: section along the centerline of the boat Different post/keel joints are shown. The bedlogs are likely to be hardwood (oak, yellow pine, Douglas fir); the strakes of softwood (white pine or cedar). The fastenings of the bedlogs to the keel (or bottom) may be nails, screws, drifts, or bolts; to the headledges with nails, screws, or possibly bolts. The strakes are fastened to the bedlogs and each other with drifts; and to the headledges with nails or screws. Examine the centerboard slot closely. It may be tapered. (B) Two ways of cutting the centerboard slot. Normal—a rectangular slot with a constant width and square ends. Sometimes holes were drilled in the keel where the ends of the case were to be. Then a saw cut was made between the holes. Wedges were driven into the saw cut to spread the keel to the desired slot width. The centerboard case was then built in the normal fashion. (C)Daggerboard case. (D and E) Centerboard and daggerboard cases in flat-bottomed boats. (F) Centerboard case with the slot alongside the keel. (G) Centerboard case with the slot through the keel (normal). (H) Bedlog half dovetailed into the keel. From a catboat by Gil Smith, Patchogue, Long Island, New York. (J) Section of H taken along the centerline showing headledge spiked to the keel. (K) Bedlog half lapped through the keel. From a boat built by Chris Brown, Staten Island, New York, for Commodore Ralph Munroe.

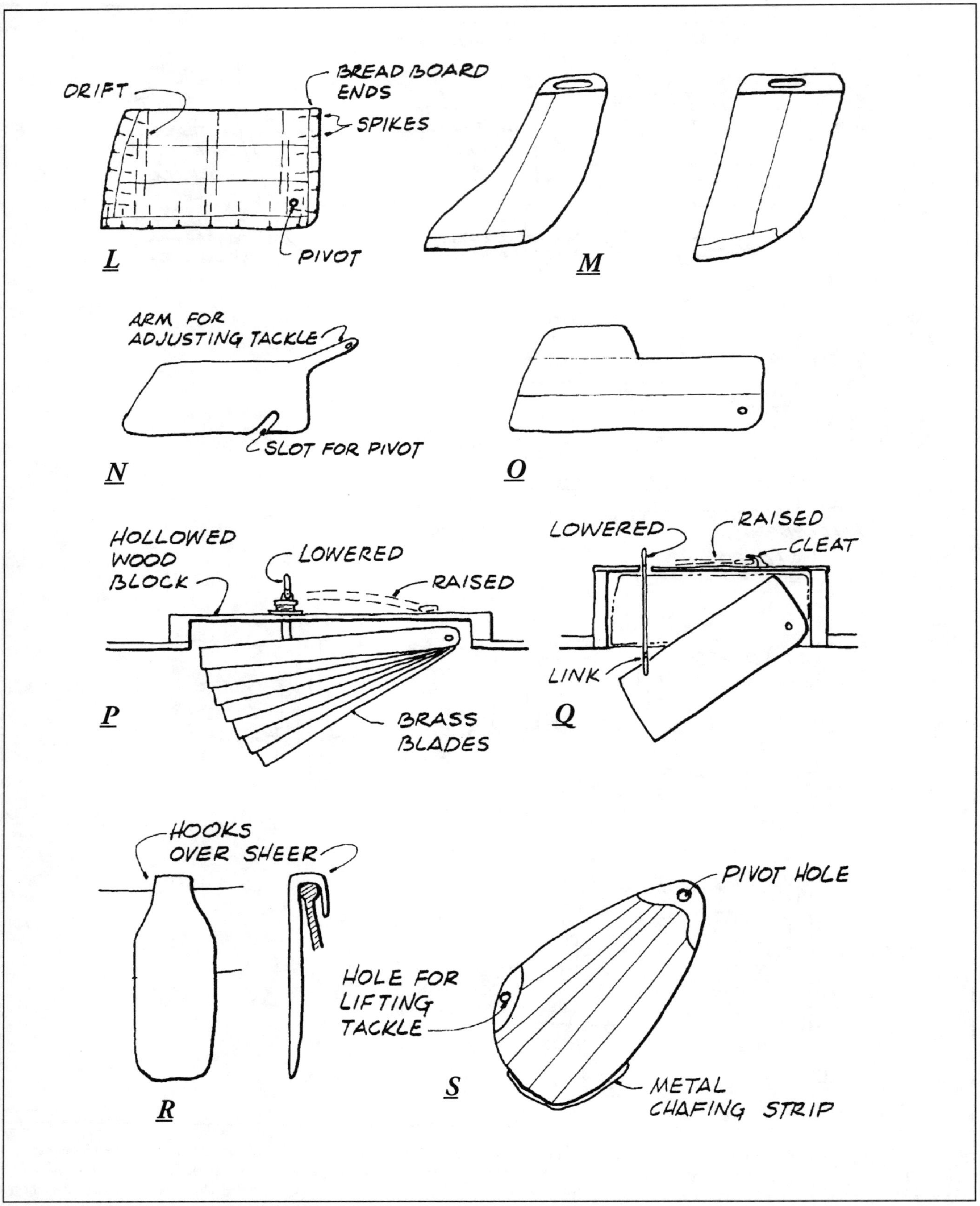

Figure 6.25 cont. (L) Typical centerboard construction. (M) Two daggerboards built up from boards; also may be of fiberglass-covered plywood. In an old boat fiberglass is a replacement. (N) Removable centerboard for a sailing dinghy—of metal or fiberglass. (O) Simple centerboard made up of boards. (P) Metal fan centerboard—the Radix type. (Q) Rod operated centerboard. (R) Simple leeboard for canoe or skiff. (S) Leeboard for a yacht or workboat. Look for hoisting gear, a method of holding the board close to the hull, and a sturdy pivoting arrangement.

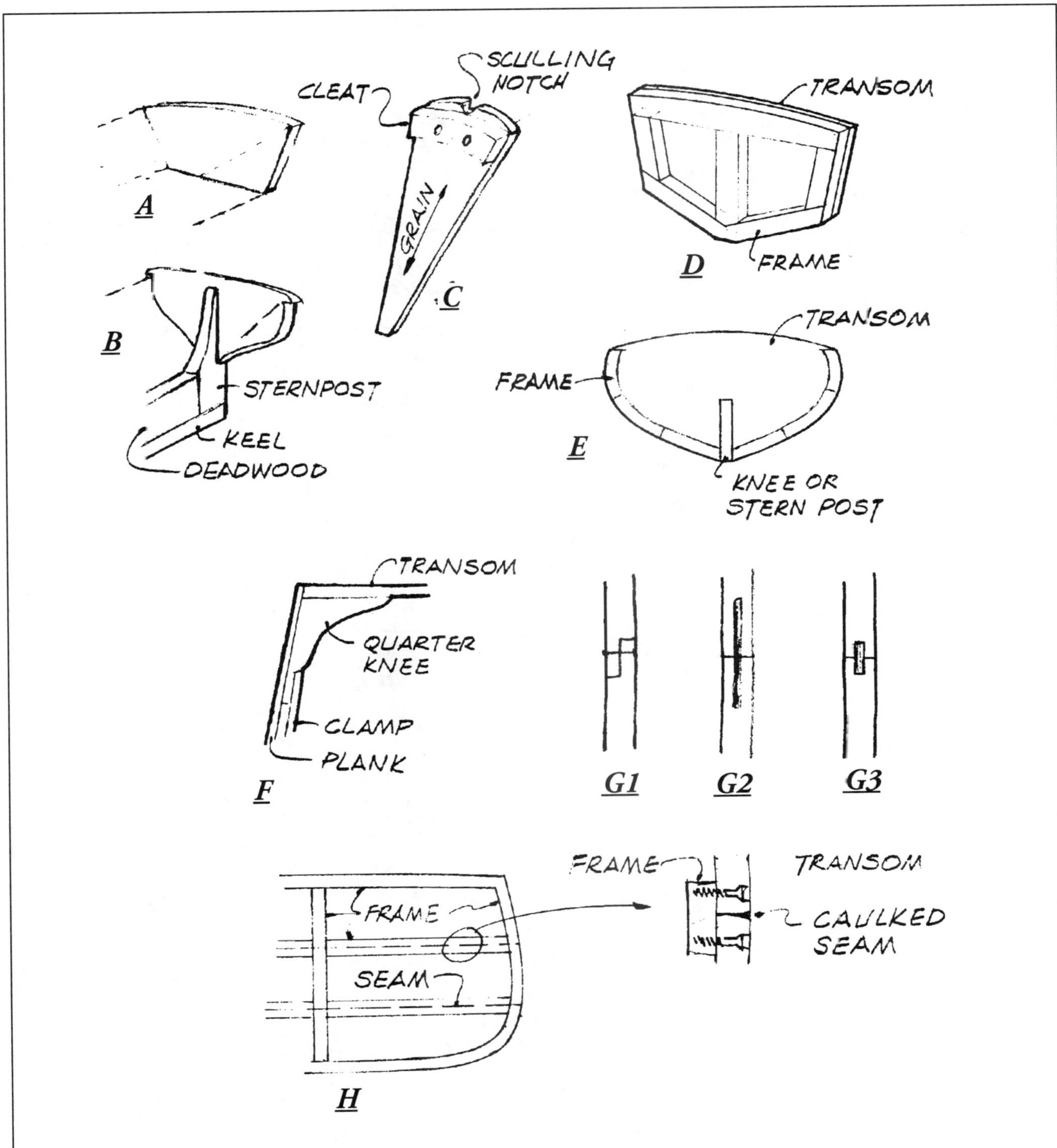

Figure 6.26. Transoms: (A) Simple plank transom as found in skiffs, prams, and some sharpies. (B) A form in small rowing and sailing boats. (C) The traditional dory transom. The cleat provides cross grain reinforcement. The holes are for lifting beckets (rope handles). (D) Generally found in light-weight sailing dinghies where the transom may be 1/4" plywood. If found on a transom that appears thick enough to hold fastenings, the frame may be a repair. Examine the plank ends for old holes or fastenings. Possibly the transom is a replacement, and the frame provides places for the fastenings. (E) Transom with a perimeter frame to allow fastening into face grain instead of end grain. As in the comments at D, above holes or fastenings, may indicate that the perimeter frame is a repair. (F) The quarter knee reinforces the plank-to-transom connection. In decked boats, the knee may be out of sight, snugged up under the deck. (G) Transoms with more than one plank may be half-lapped and nailed (G1), joined with wooden dowels or metal drifts (G2), or with a spline or feather (G3). The seams in (G2) and (G3) would be caulked. (H) A large transom. Planks are fastened to interior frame members and the seams caulked.

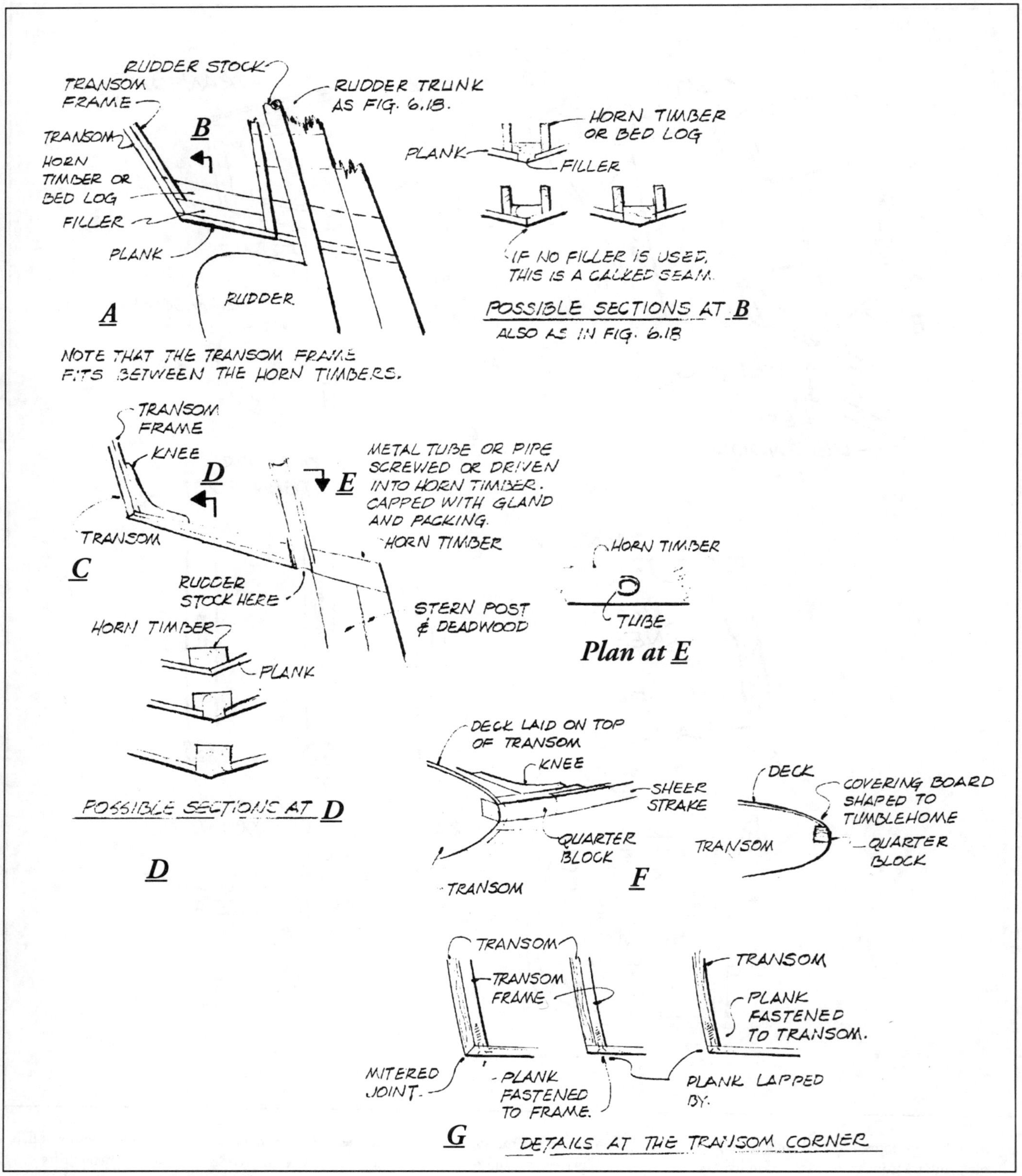

Figure 6.27. Transom Sterns: (A) A section taken along the centerline of a boat at the transom and sternpost showing a built-up rudder trunk. (B) Possible sections through A. Taken at the arrow B. (C) A section taken along the centerline of a boat built with a full horn timber and a metal tube instead of a built-up rudder trunk. The tube may be capped with a gland and packing below deck or extended to the deck and supported by the deck framing. (D) Possible sections through C. Taken at the arrow d. (E) A plan view showing the tube and the horn timber. Taken at the arrow E. (F) Tumblehome transoms have special treatments; the sheer strake is scarfed into a large block of wood—the quarter block and the tumblehome forward of the transom is worked into it. Sometimes, the top two strakes have thicker plank scarfed onto the after end to accommodate the tumblehome. The covering board may or may not be faired into the tumblehome. (G) Several plank-to-transom joints.

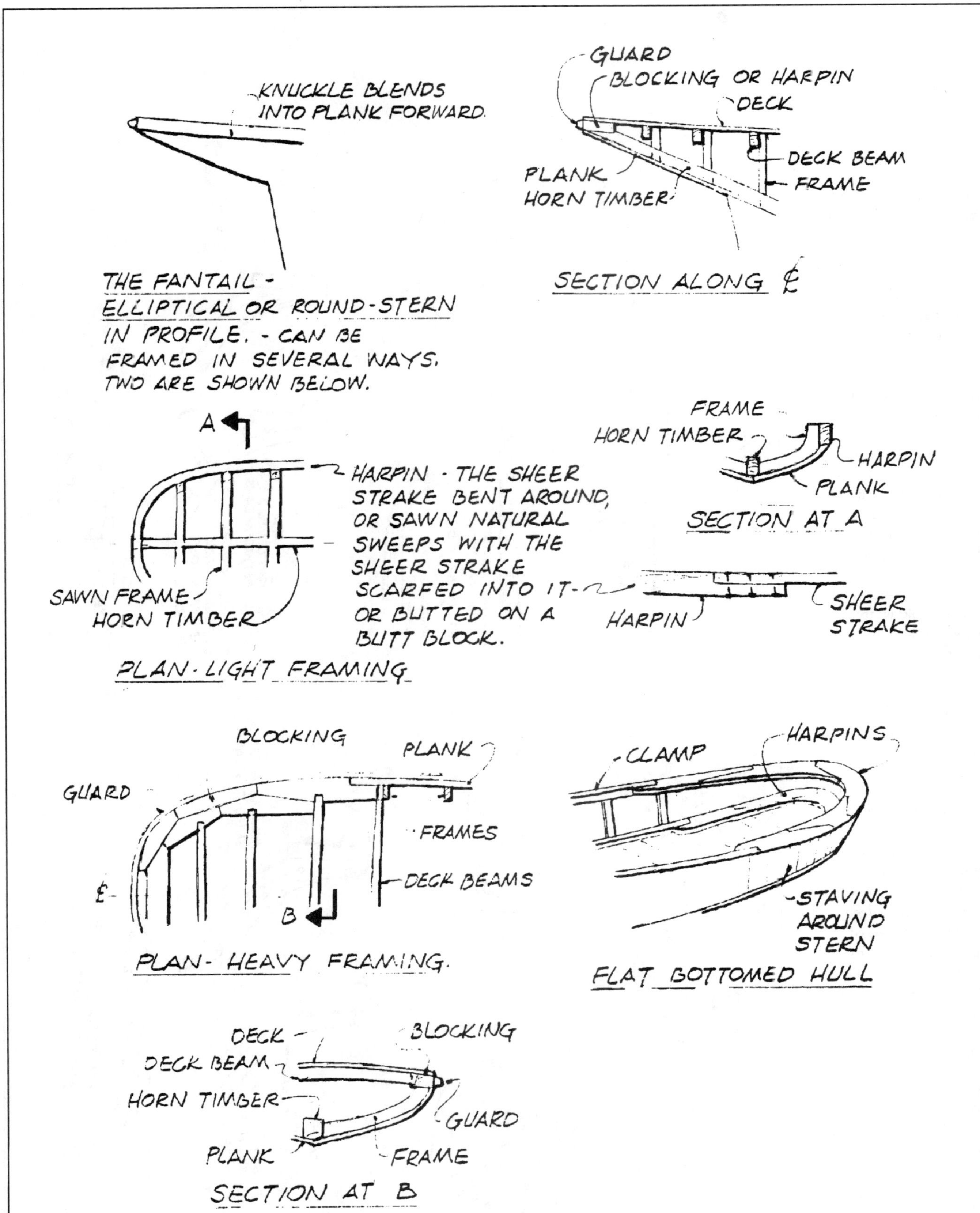

Figure 6.28. Round and elliptical sterns. The harpins in the flat-bottomed boat may be either scarfed up from flat stock as shown, from square or rectangular section pieces of natural crook, or bent.

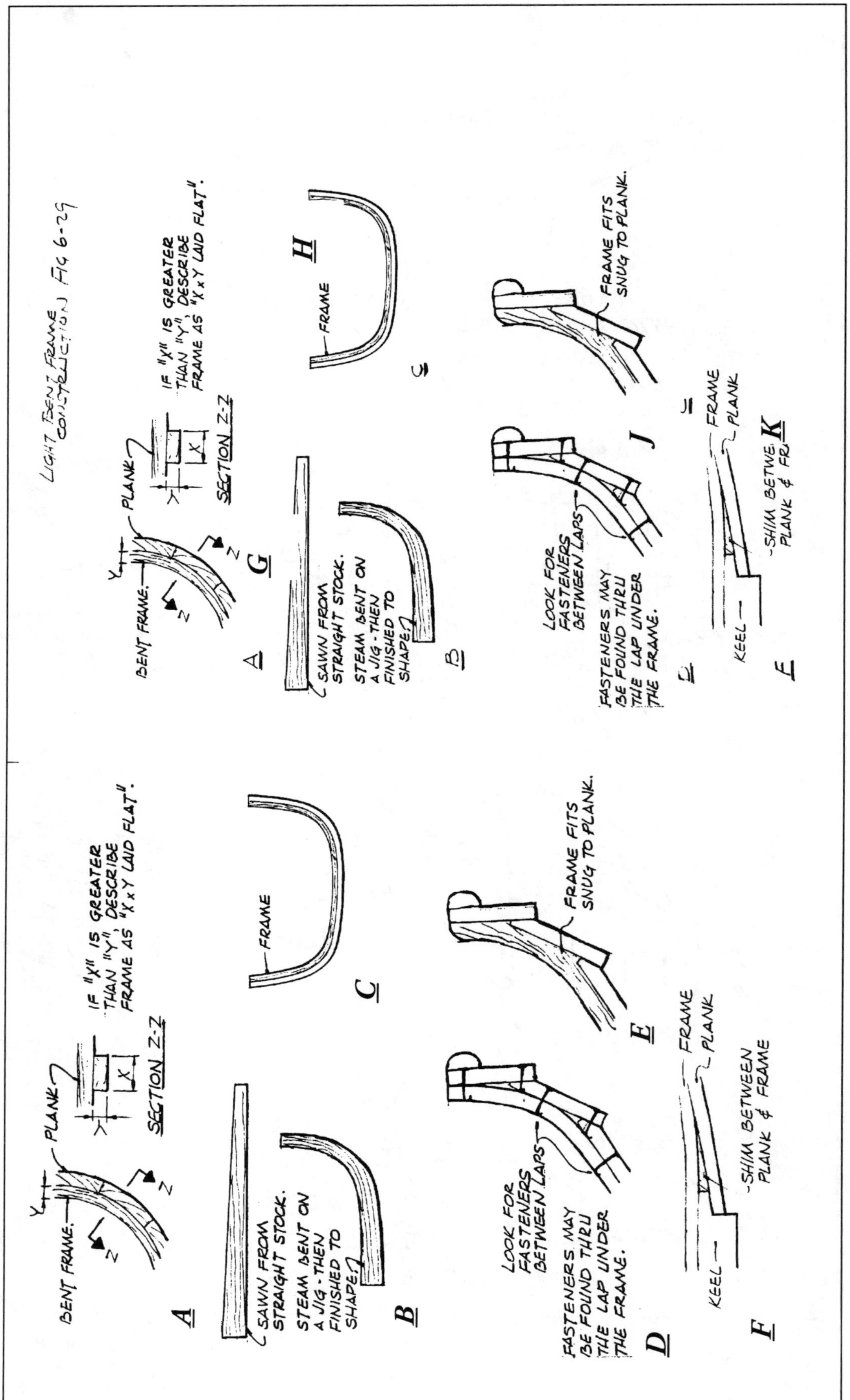

Figure 6.29. Light bent frame construction: (A) Bent frame with carvel (smooth) planking. (B) Tapered bent frame occasionally found in Whitehall or other stock boat (production) construction. (C) Frame bent in one piece from sheer to sheer. Common in boats four feet and under in breadth. (D) Bent frame with lap strake planking. The frames do not always lay against the planks in a fair sweep. Frequently, the frames contact the planks between the laps and are fastened there. Examine, view, and photograph lapstrake planking, from the bow and stern, to notice and record the frames' shape. Also closely examine the frames inboard and the planks outboard to locate fastenings through the frames other than at the lap. (E) Jogged frame with lapstrake planking. The frame may be from natural crook or bent stock. (F) Look for wedges fitted between the frames and planking or a long strip of feather-edged planking stock fitted alongside the keel to support the frames and plank. (G) Frames are frequently "boxed" into the keel or deadwood. (H) Detail of boxed frames found in a catboat built at the Crosby Yard, Centerville, Massachusetts. (J) Treatment of frames and floors at the keel. (K) A bent frame that has been slit to make it more flexible.

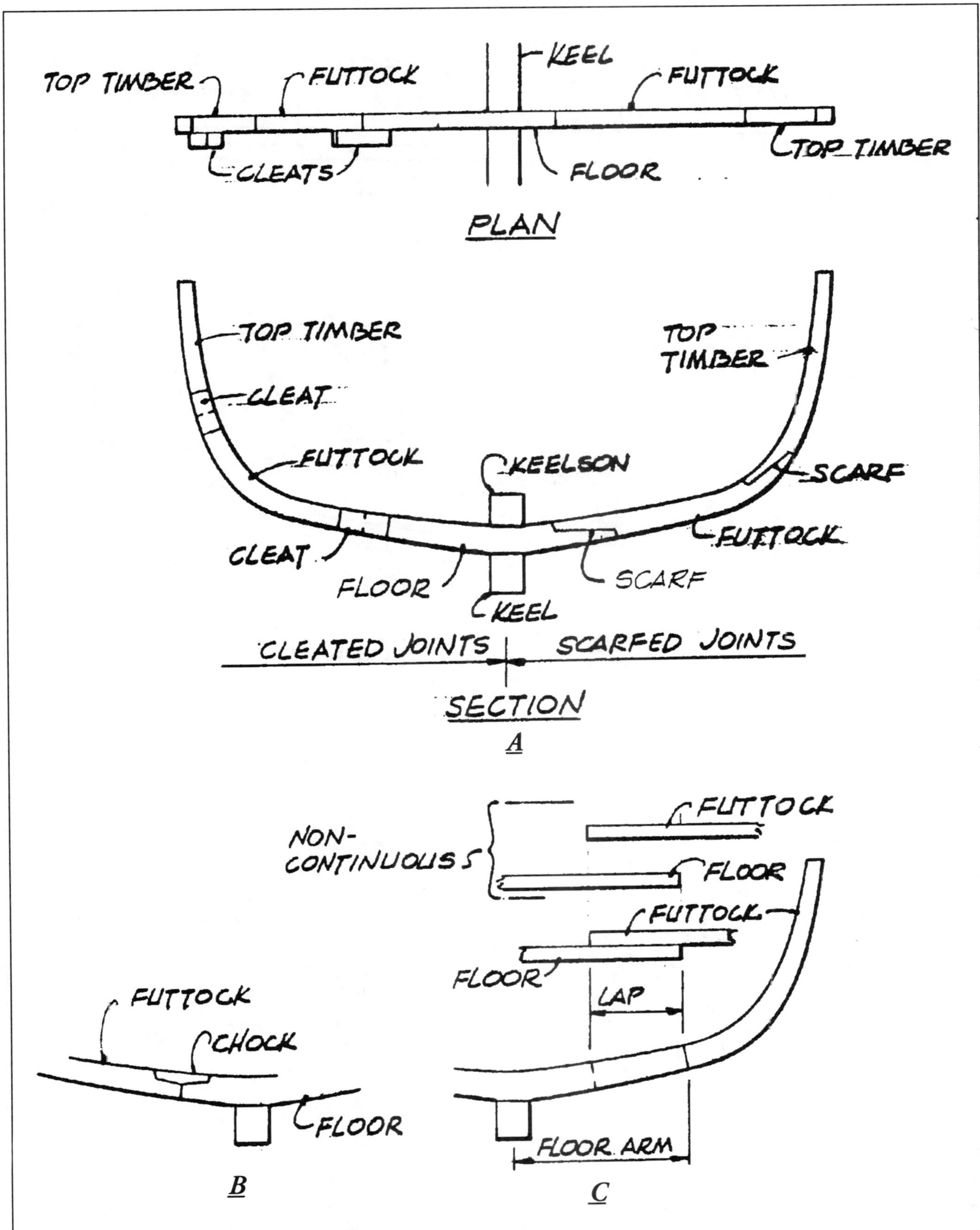

Figure 6.30. (above and on next page) Sawn frames: (A) Single sawn frame with cleated or scarfed joints. Only one type of joint should be found in the frames of any boat. If the joints are a mix, suspect that repairs have been made. (B) Single sawn frame chocked joints. (C) Single sawn frame with lapped joints. Whether the floor and futtock are fastened together or are spaced apart (noncontinuous), they are considered lapped because they support and connect the plank.

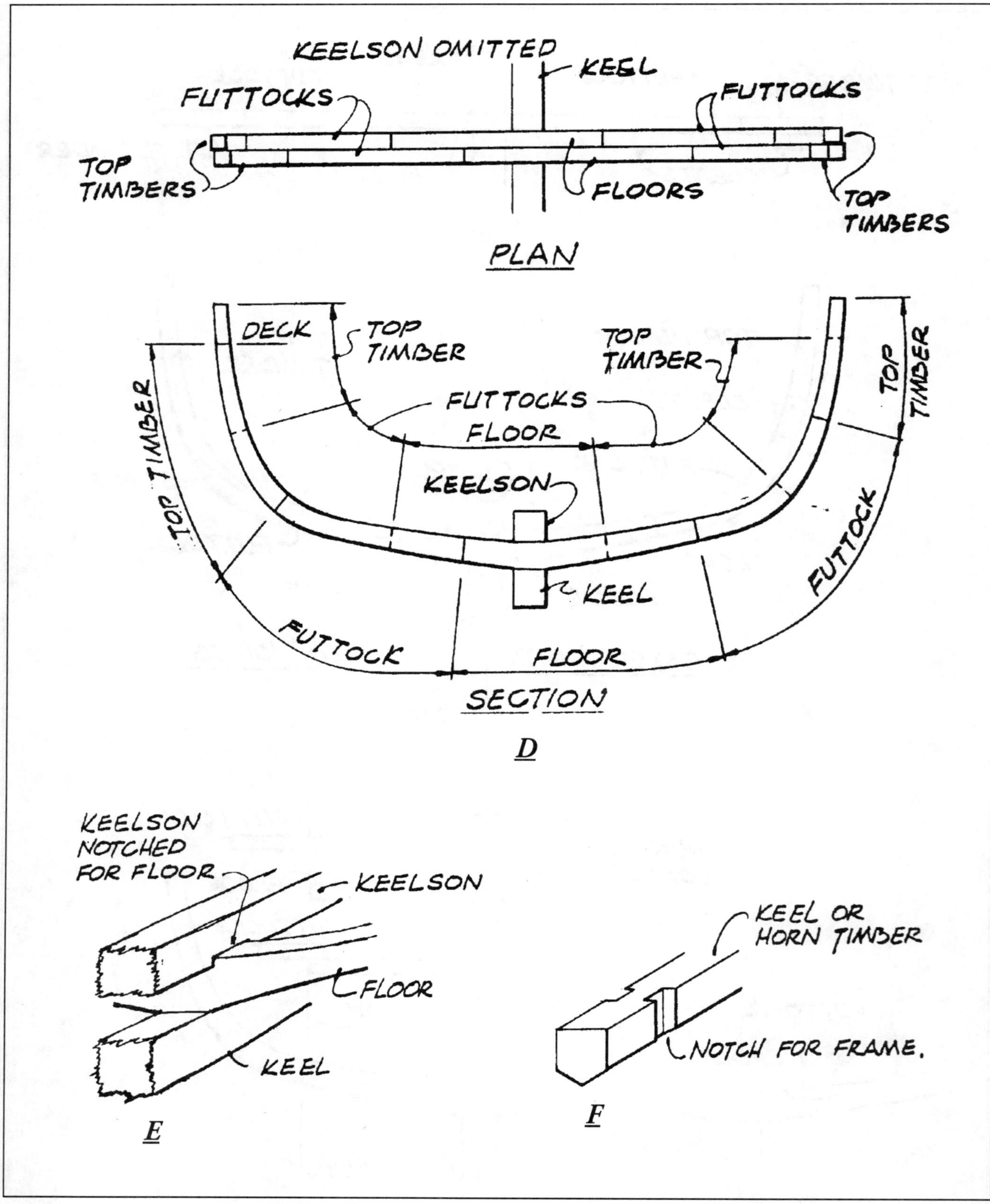

Figure 6.30 cont. (D) Double sawn frame. The portion above the deck is the bulwark stanchion or timberhead, and the timberhead port is forward and the starboard is aft or vice versa. Note: Each frame consists of a floor and a top timber and one or more futtocks depending on the size of the boat. The floors each have a short and long arm arranged so that they alternate with long arm to port, then to starboard, back to port, and so on. (E) The keelson notched for the floor. Sometimes the floor is also notched. Also look for the keel and floor to be a notched joint. (F) Vertical notch in backbone for frame.

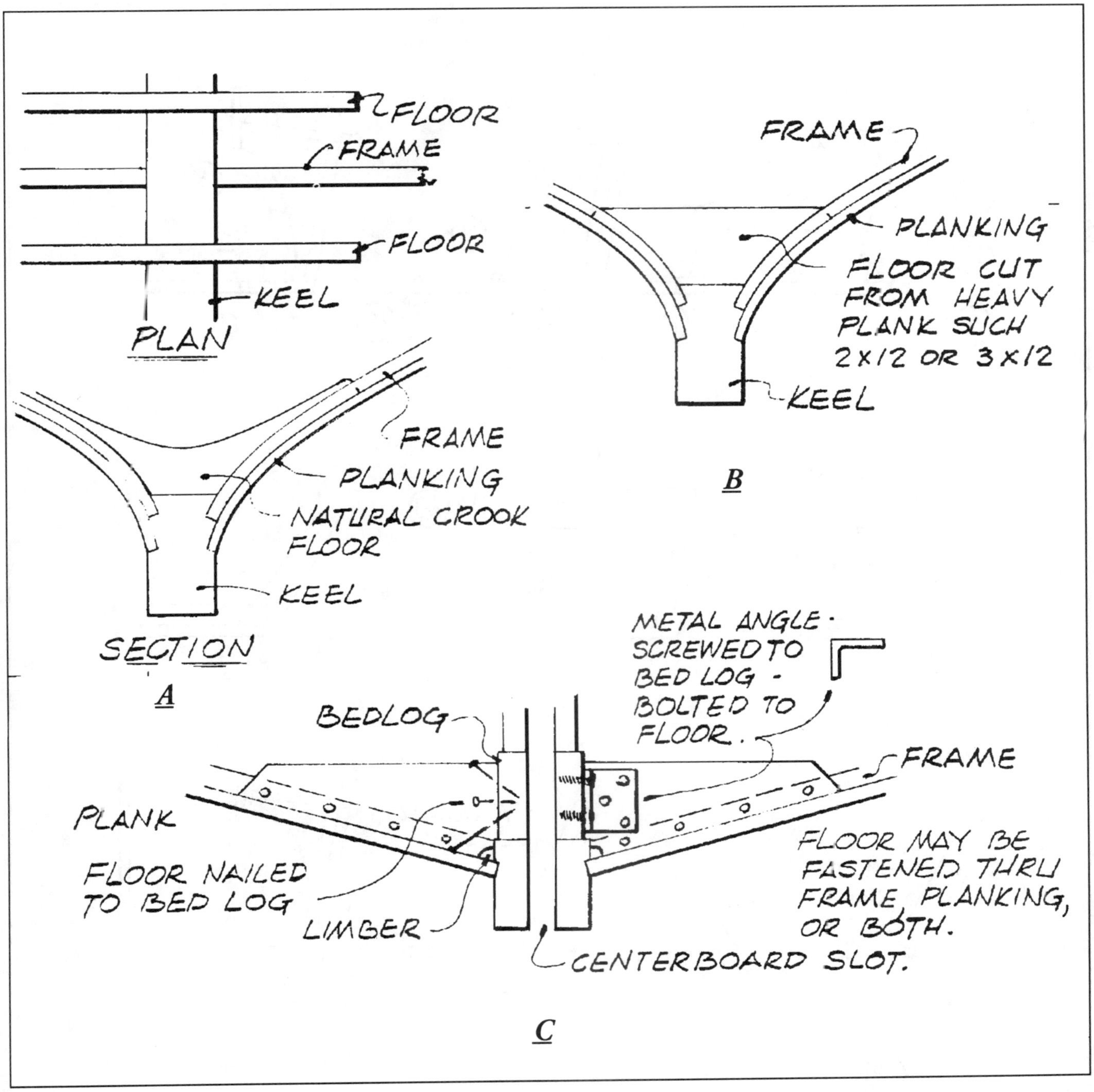

Figure 6.31. Floors: (A) Arrangement of natural crook floors and bent frames common to the Friendship Sloops of Maine. (B) Bent frame with a "plank" floor. (C) Two methods of joining floors to a centerboard case.

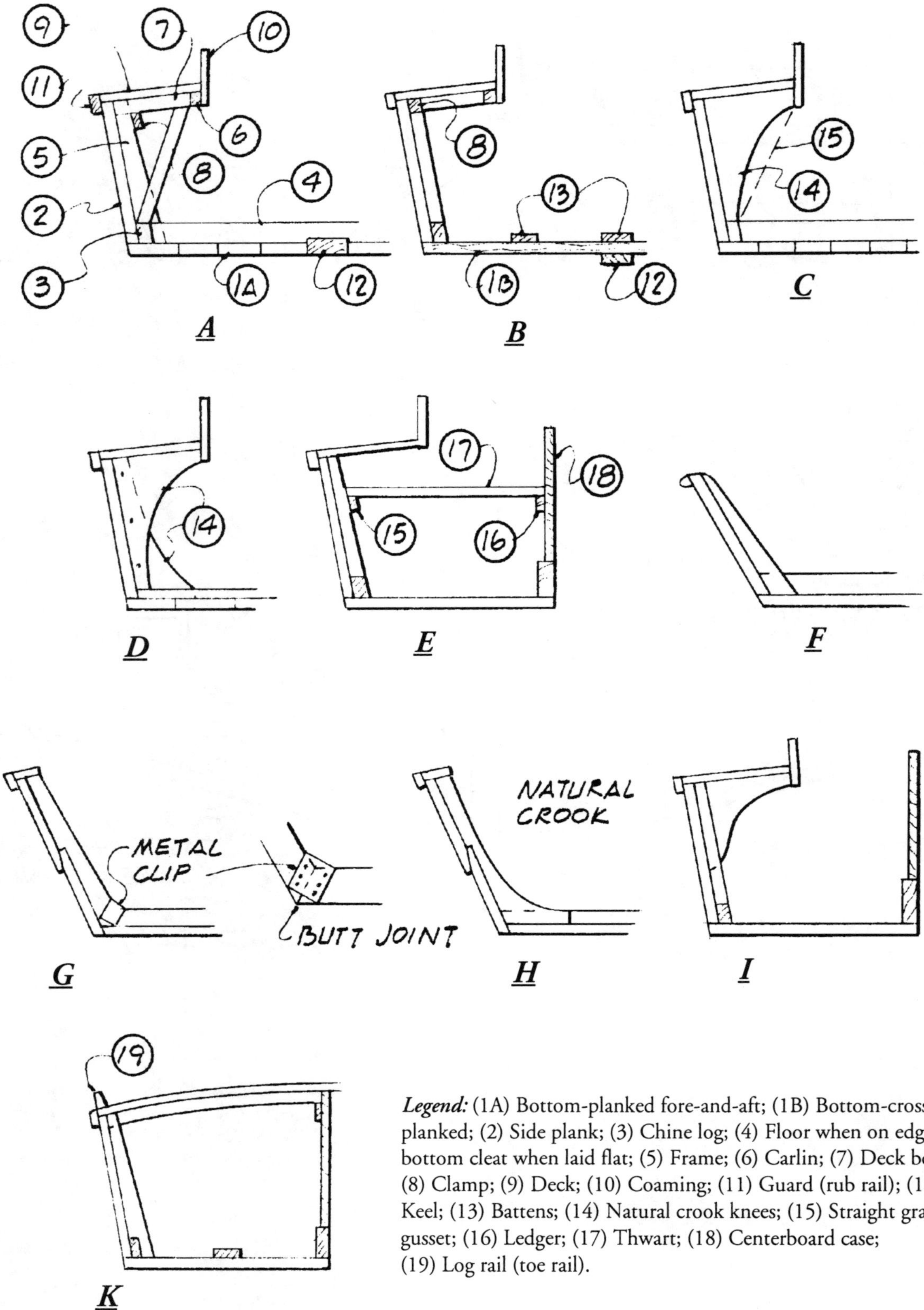

Legend: (1A) Bottom-planked fore-and-aft; (1B) Bottom-cross planked; (2) Side plank; (3) Chine log; (4) Floor when on edge or bottom cleat when laid flat; (5) Frame; (6) Carlin; (7) Deck beam; (8) Clamp; (9) Deck; (10) Coaming; (11) Guard (rub rail); (12) Keel; (13) Battens; (14) Natural crook knees; (15) Straight grained gusset; (16) Ledger; (17) Thwart; (18) Centerboard case; (19) Log rail (toe rail).

Figure 6.32. Chine construction on flat-bottomed boats: A through K are typical flat-bottomed hull sections. What you find when examining a boat may be a composite of the various details shown. Many of the details found in flat-bottomed boats will be also found in V-bottomed hulls. D was found in a fifteen-foot gunning skiff, G and H are typical of Grand Banks dory construction.

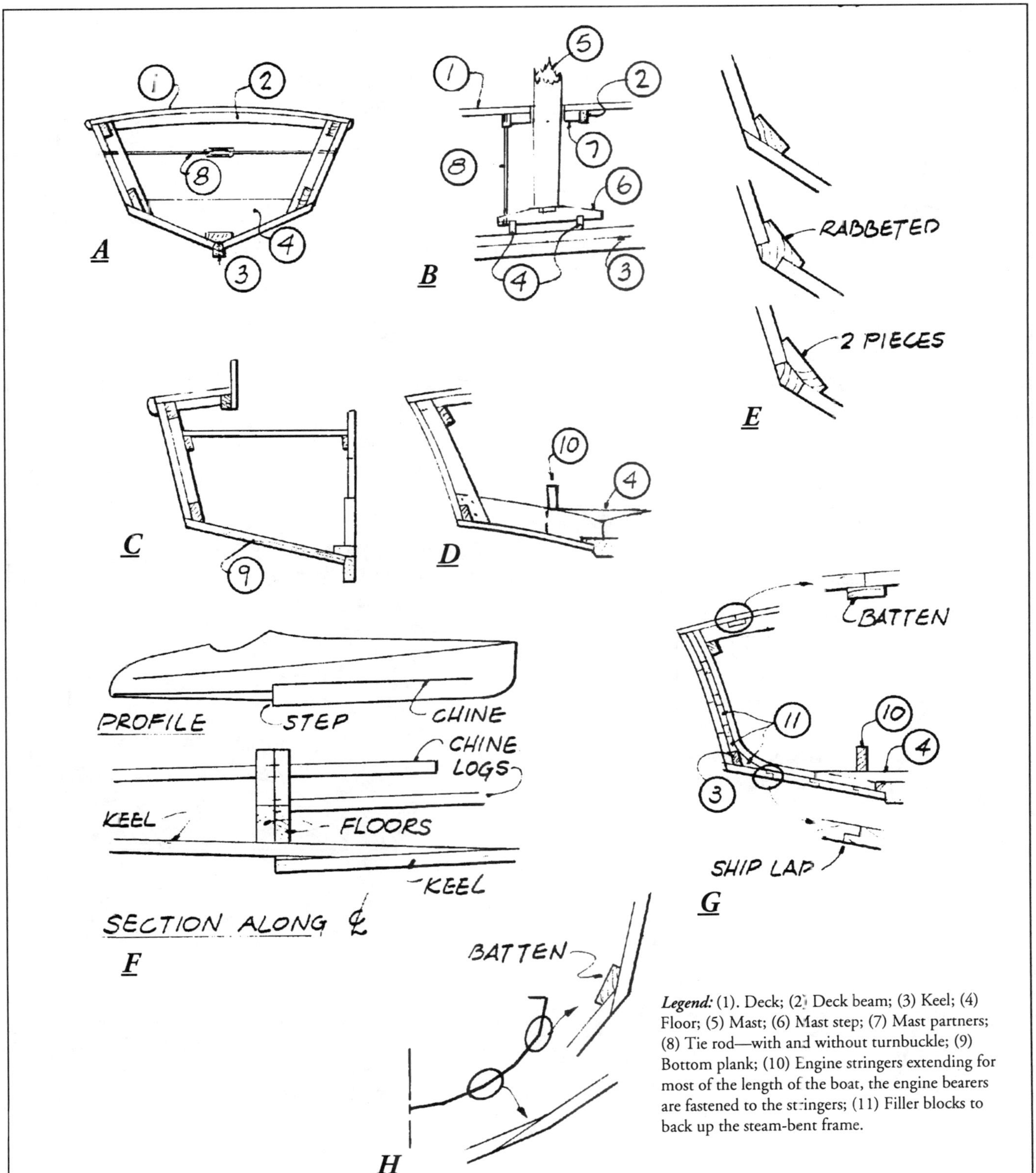

Legend: (1). Deck; (2) Deck beam; (3) Keel; (4) Floor; (5) Mast; (6) Mast step; (7) Mast partners; (8) Tie rod—with and without turnbuckle; (9) Bottom plank; (10) Engine stringers extending for most of the length of the boat, the engine bearers are fastened to the stringers; (11) Filler blocks to back up the steam-bent frame.

Figure 6.33. Chine construction in V-bottom boats: (A and B) Tie rods used to strengthen the bow of a sail boat. (C) The same construction as figure 6.32C, but used in a V-bottomed boat. (D) Typical powerboat section. The frames and floors are usually sawn from straight stock. (E) Different chine logs. (F) In power boats with a step, the chine logs from abaft the step are carried forward several frame spaces for structural continuity. (G) Light-weight powerboat construction combining bent frames, sawn floors, batten-seam deck and side planking, and ship-lapped bottom planking. (H) Multi-chine hulls have battens backing the seam or use a glued-lap planed off smooth. Round-hulled dories are not considered multi-chine because the laps project.

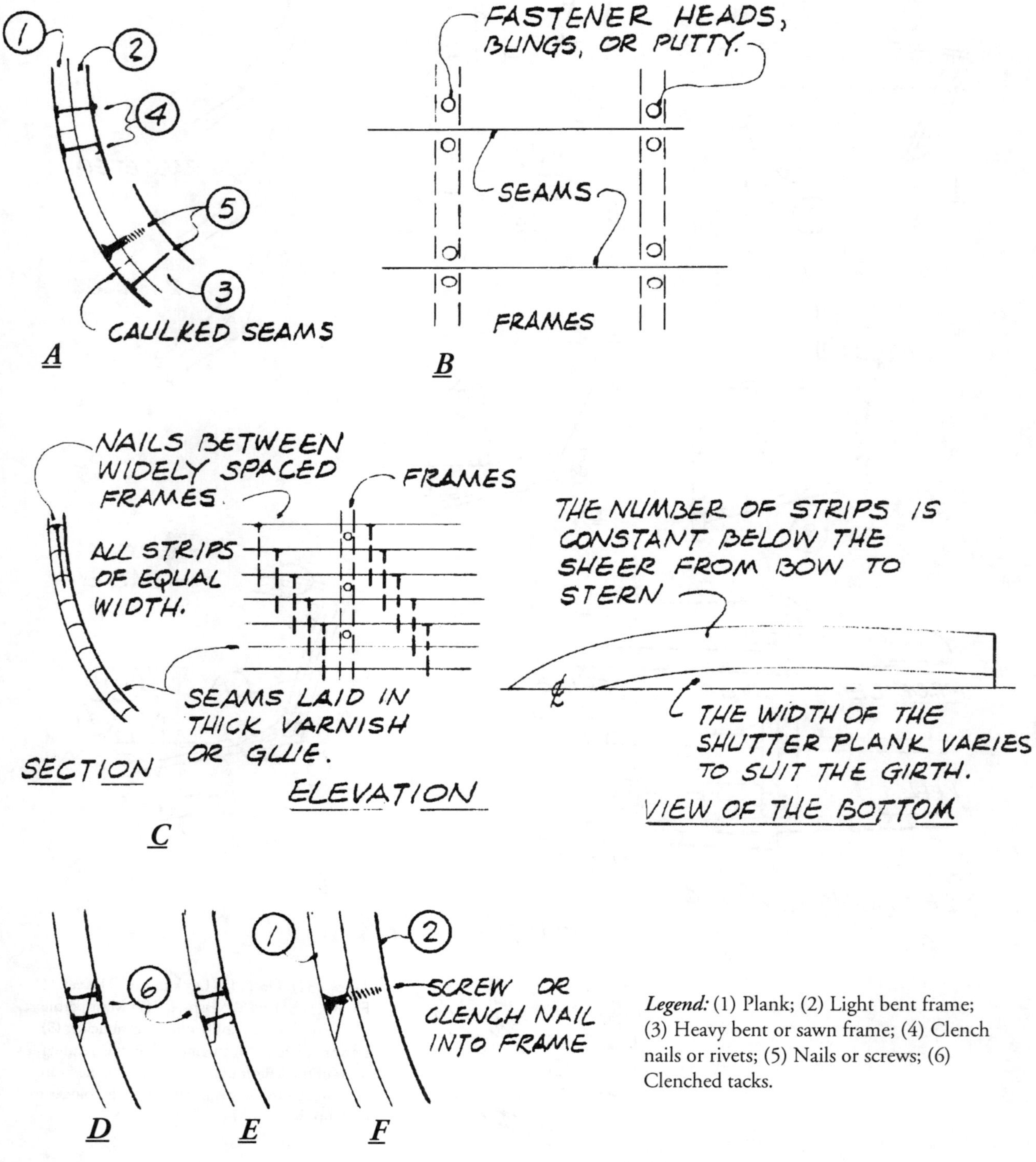

Legend: (1) Plank; (2) Light bent frame; (3) Heavy bent or sawn frame; (4) Clench nails or rivets; (5) Nails or screws; (6) Clenched tacks.

Figure 6.34, above and next page. Planking methods that give a smooth hull: (A) Carvel—may be fastened with rivets, clench nails, screws, or in heavy frames, nails. (B) Outboard view of carvel planking showing fastener locations. (C) Strip planking—used in Western Europe and North America. The example shown was found at Isle Royale National Park in Lake Superior, where examples of both round hulls and flat-bottomed skiffs were found strip planked. (D) Feather lap. This can sometimes be detected by the wavy seam lines. (E) Modified feather lap with a slight shoulder, which allows the seam line to remain crisp. (F) Fastening the feather lap. The two feather laps, D and E, are a form of lapstrake planking.

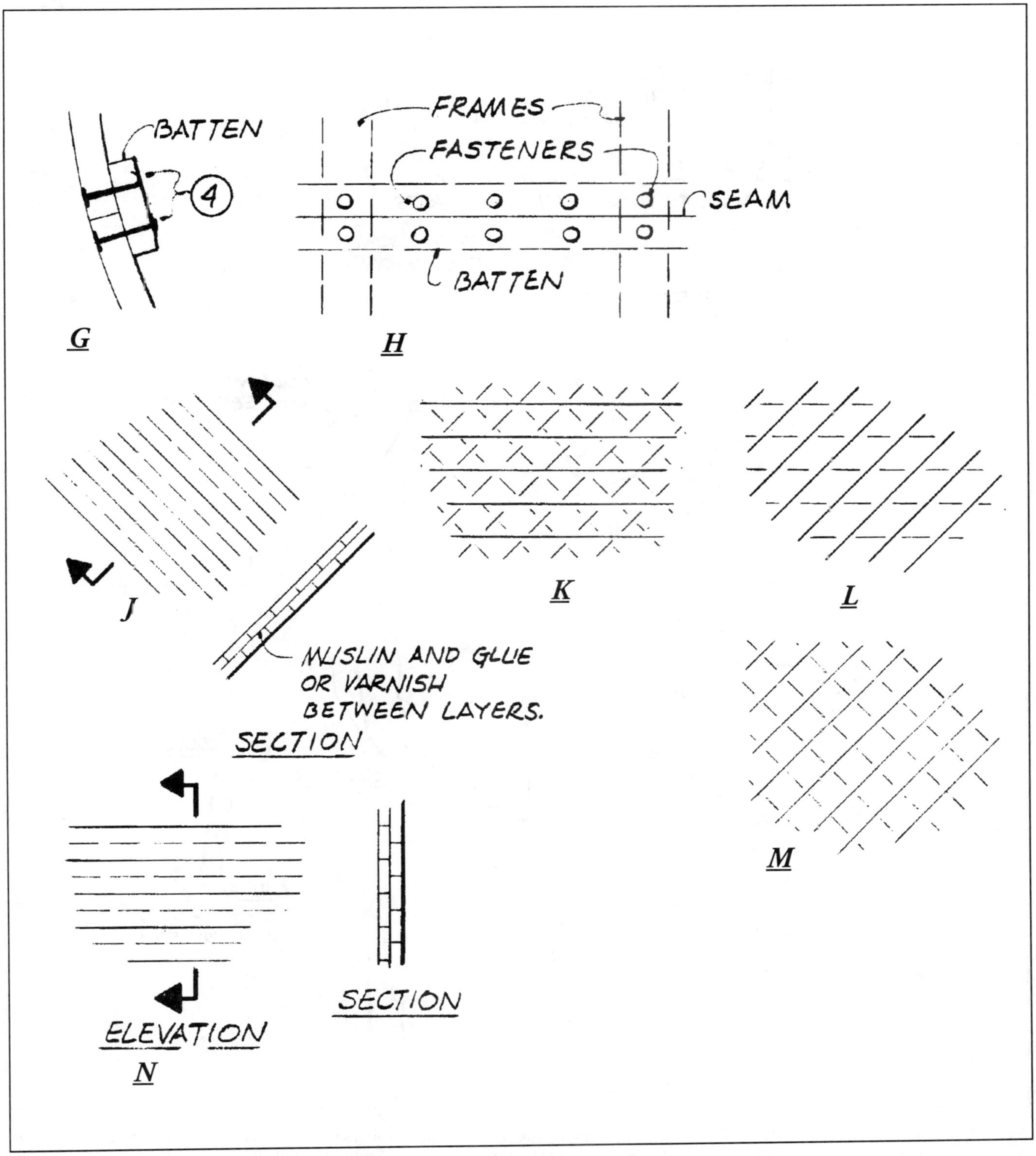

Figure 6.34 cont. (G) Batten-seam. (H) Outboard view of batten-seam planking. (J) Ashcroft planking—two layers laid at approximately forty-five degrees to the keel. (K) Double planked—one at forty-five degrees to keel, the other fore-and-aft. (L) Triple planked—two layers at forty-five degrees to the keel, one fore-and-aft. (M) Double planked—two layers at forty-five degrees to the keel. (N) Double planked—both layers laid fore-and-aft.

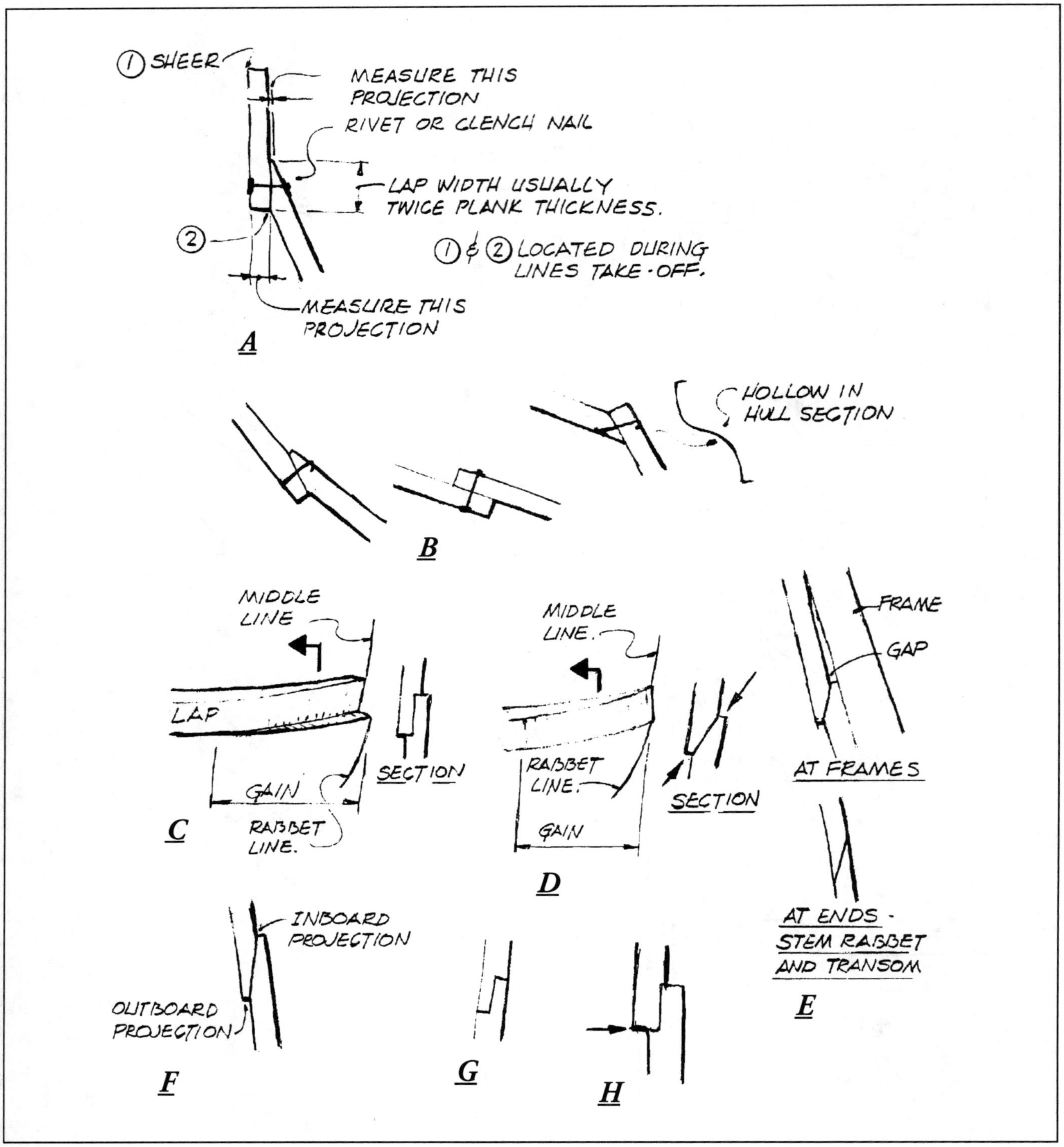

Figure 6.35. Lapstrake planking: (A) The lap bevel taken off the lower plank of the two. (B) Variations that may be found—possibly all in the same boat. (C) The ship-lap or rabbeted gain. "Gain" is the length of plank taken to bring the outboard surfaces of the planks flush with each other at the stem and transom. (D) The beveled gain. The angle of the lap bevel is increased towards the hood end until the plank faces lay flush with each other and the stem; or simply flush at the transom to see how the laps were cut there. They may be the same at the bow. (E) The "Dory Lap." The bevels are cut so that with garboard lying full on the frame, the next plank touches the frame only at its upper inboard corner. Knowing the length of the gains at the hood ends is as critical to reproducing the shape of a lapstrake boat as the bevels on the plank. The length of the gain can be difficult to determine if it is cut as at D. A probe slid into the lap as indicated by the arrows at D and H will help find the shape of the gain—beveled or ship-lap—and in the case of the ship-lap, where the rabbet for the lap begins. With the beveled gain, begin to look for the beginning of the gain where the total of the inboard and outboard projections equals or is slightly less than the plank thickness; at the hood ends, not amidships. Turning the boat upside down sometimes makes the length of the gains more apparent. Observe the way the frames lay to the planking. The builder of the boat probably fitted the plank laps so that the planks lay fair on the molds and showed fair lines along the plank edges. When the frames were bent in, they fit to the planks rather than the mold shape. If the frames are not fair shapes, the lines must be drawn to the throat outboard (not recommended) or to the top inboard corner of plank (recommended). Refer to fig. 7.21 for the points to measure when taking off the lines of a lapstrake planked boat.

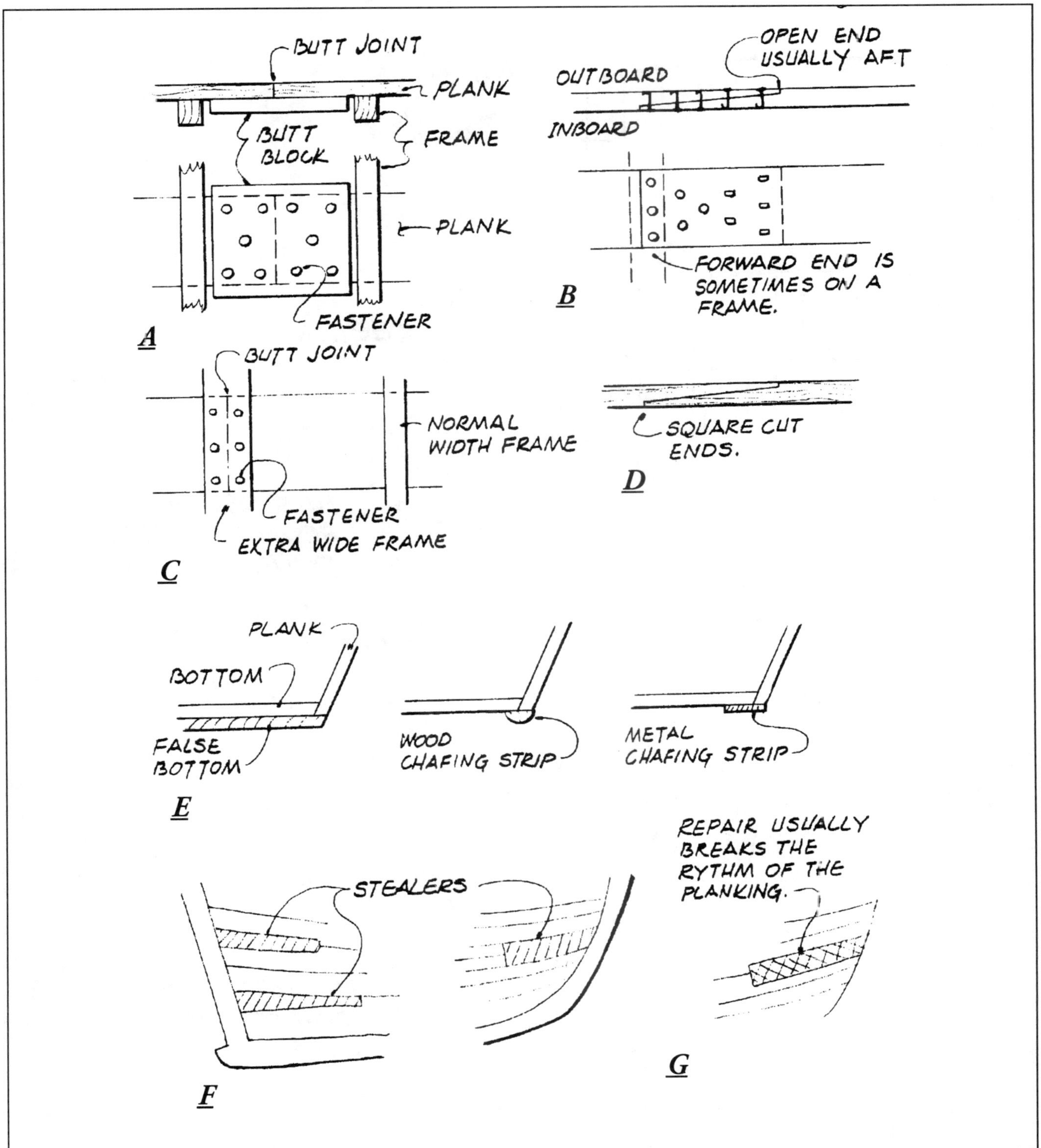

Figure 6.36. Miscellaneous planking details: (A)Planks butted on a butt block. (B) Scarf joint used in lapstrake and feather-lap planking. (c) Planks butted on a wide frame. When an extraordinarily wide frame appears among a sequence of noticeably narrower sided frames, look for the planks to be butted on the wide frames. (D) Nibbed scarf—used in planks, coamings, and other places where a clean-ended scarf is desired. (E) Several ways of protecting the bottom corner of the garboard plank. (F) Stealers—used to prevent the width of a plank from varying radically from narrow to excessively wide. (G) Plank repair. The first and most obvious difference between a repair and a stealer is location. There will be similar work port and starboard if it is a stealer, but only one side if it is a repair. Also, the style of work, fastenings, or wood used for the suspected plank may be noticeably different.

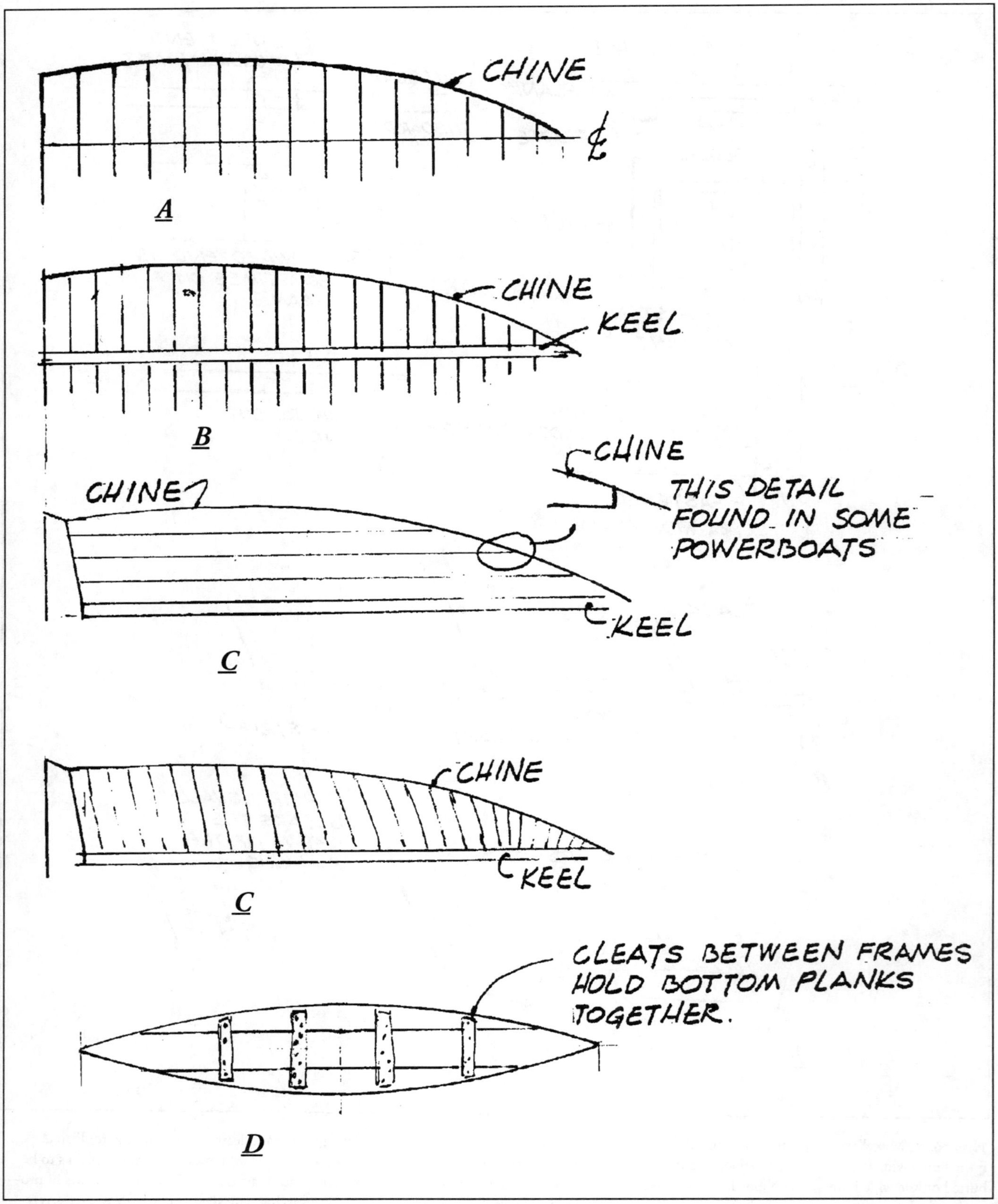

Figure 6.37. Bottom planking: (A) Flat bottom—cross-planked chine to chine. (B) V-bottom—cross-planked chine to keel. (C) V-bottom—cross-planked to keel in a herringbone pattern. (D) Dory bottom—planked fore-and-aft, viewed from inboard.

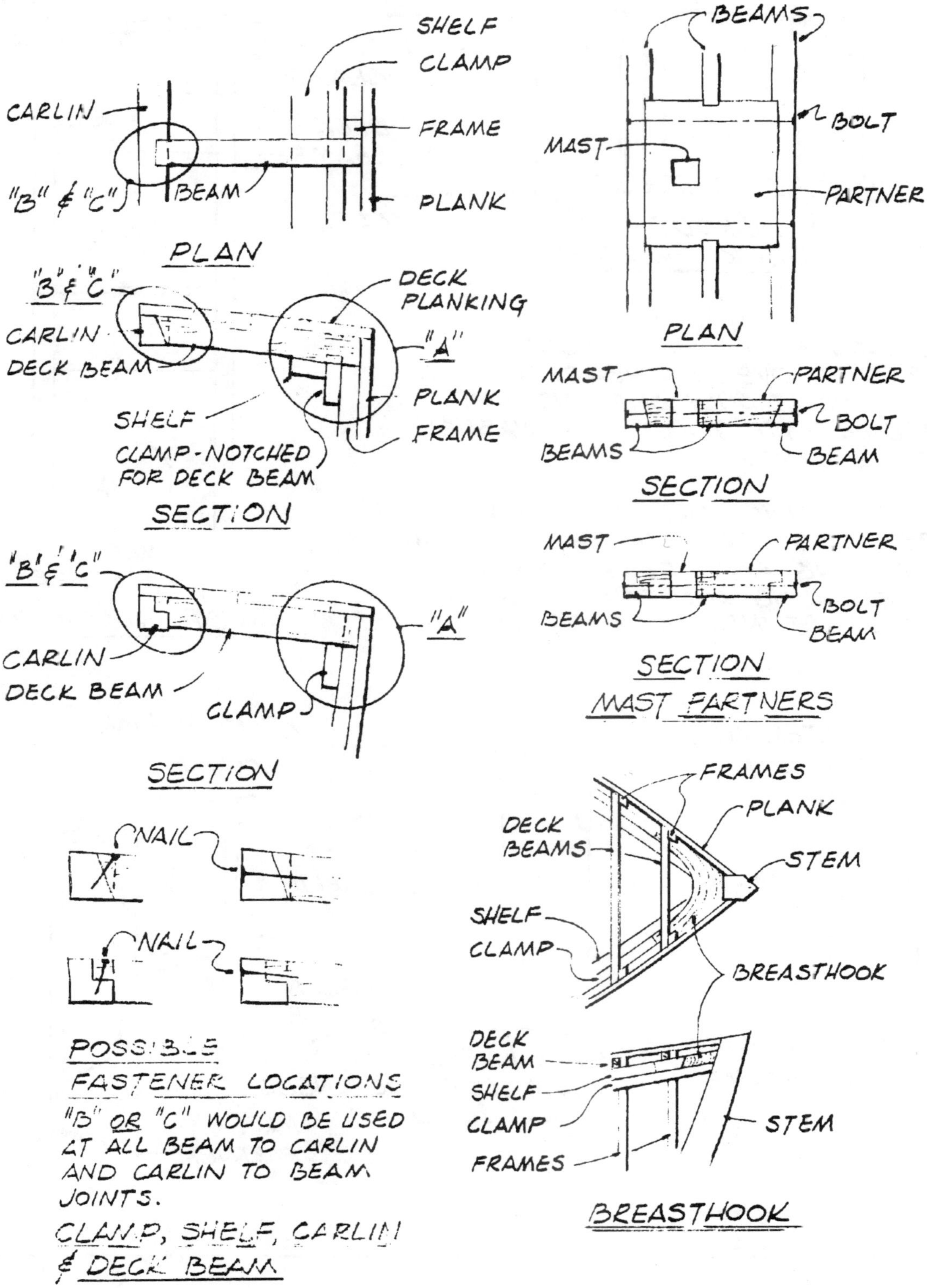

Figure 6.38. Deck framing based on Rhodes's *Dog Star.*

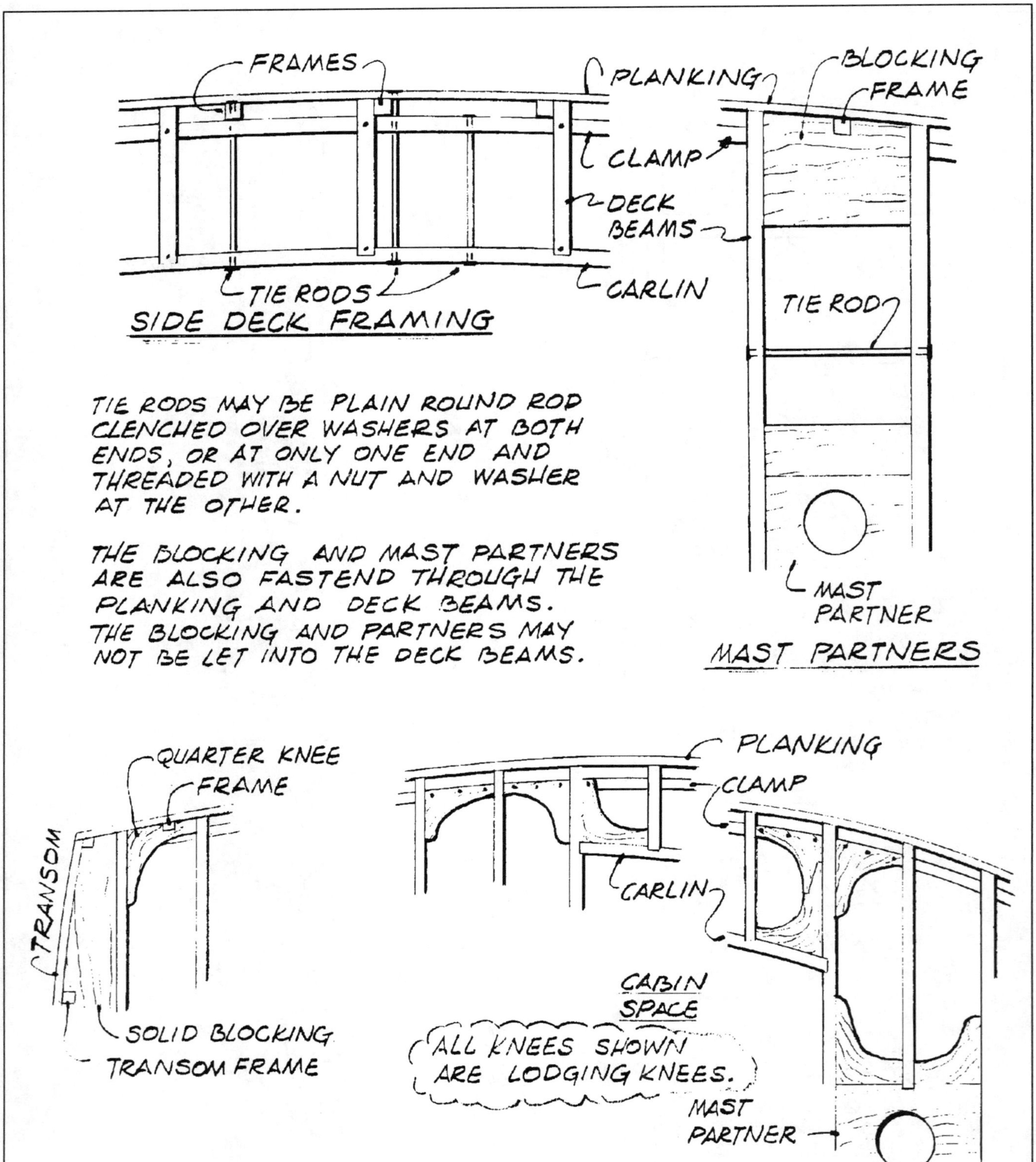

Figure 6.39. Methods of deck framing using tie rods and knees.

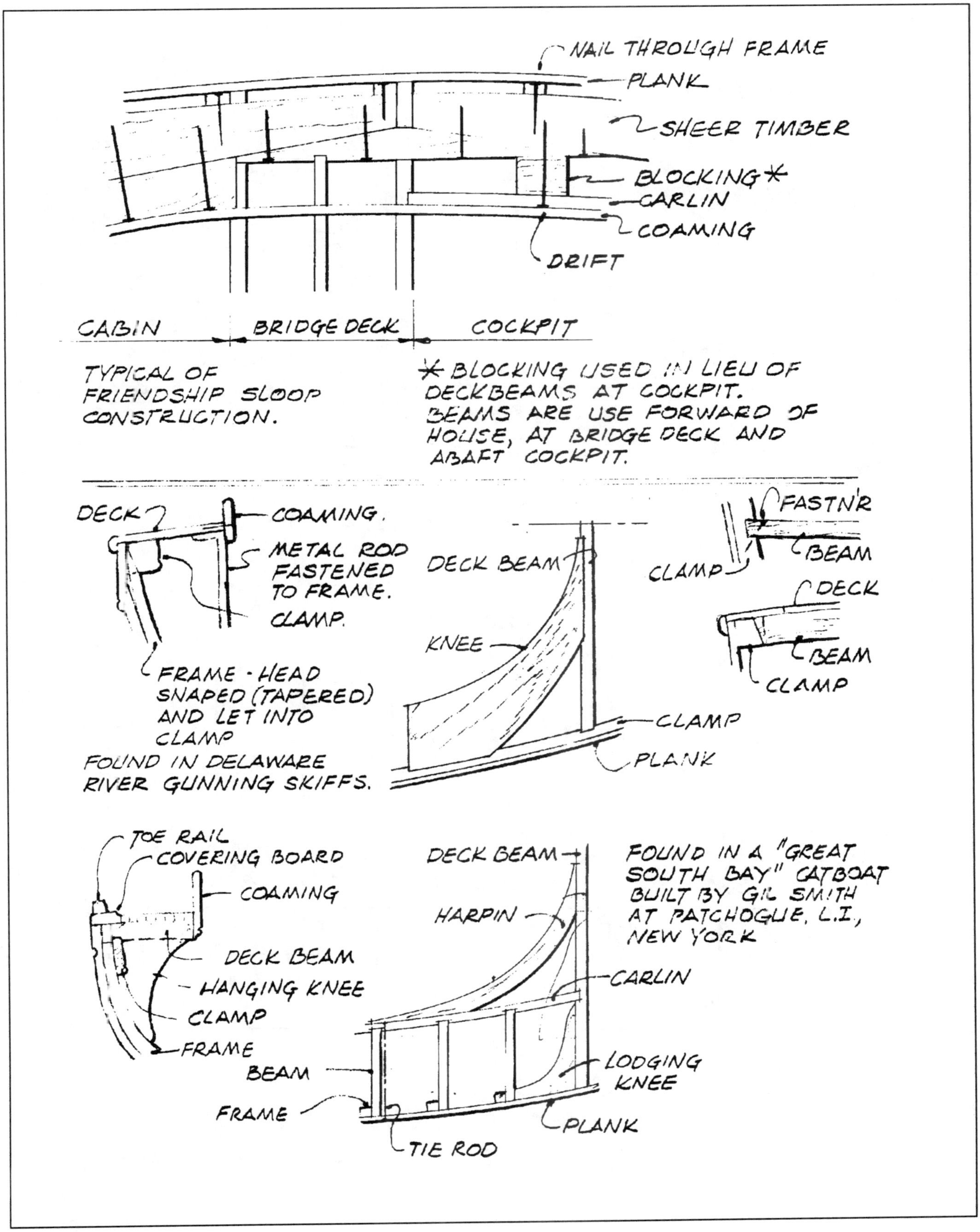

Figure 6.40. Miscellaneous deck framing details.

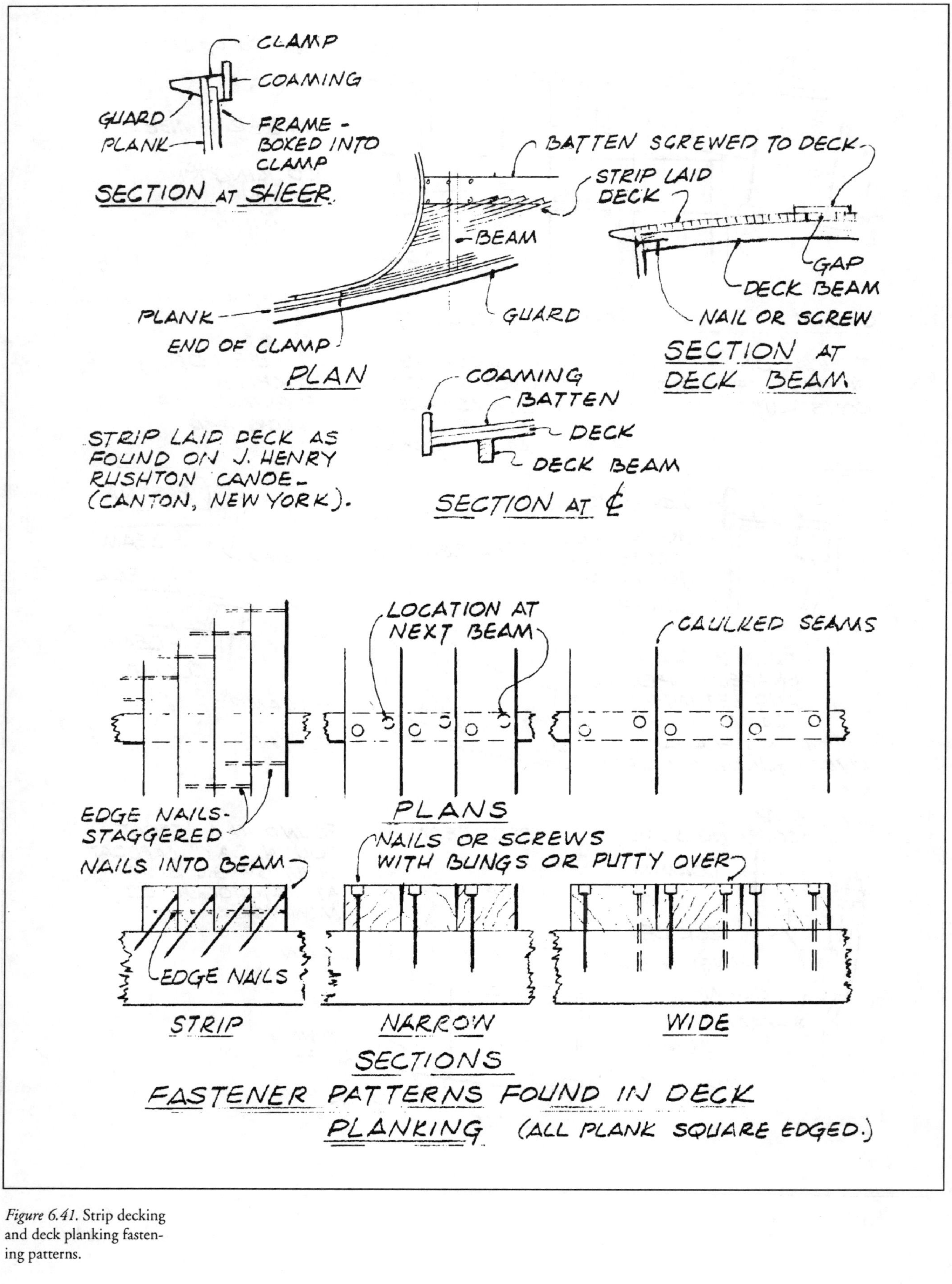

Figure 6.41. Strip decking and deck planking fastening patterns.

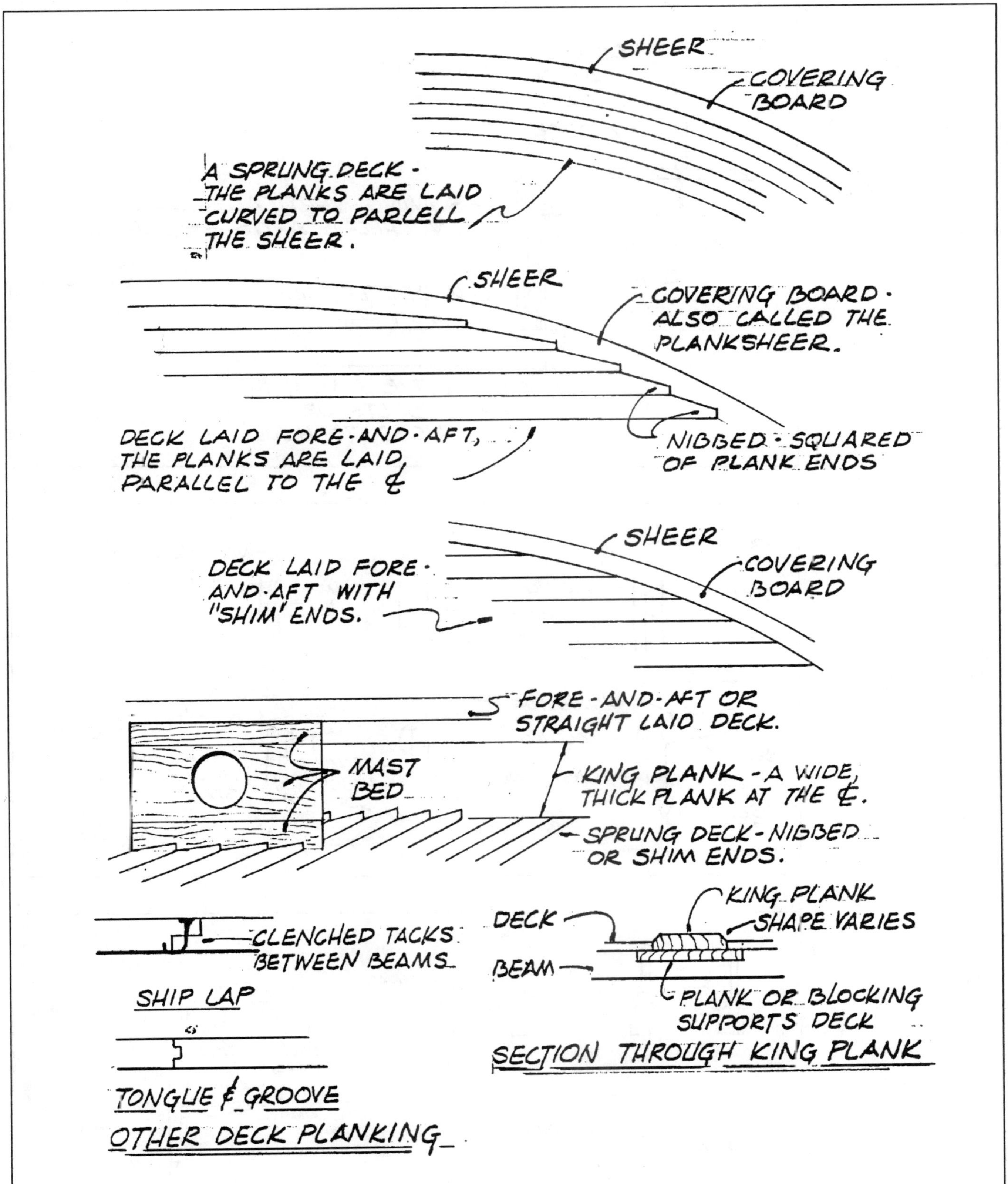

Figure 6.42. Deck planking. Also found but not illustrated are canvas over ship-lapped, tongue-and-groove planking, or plywood.

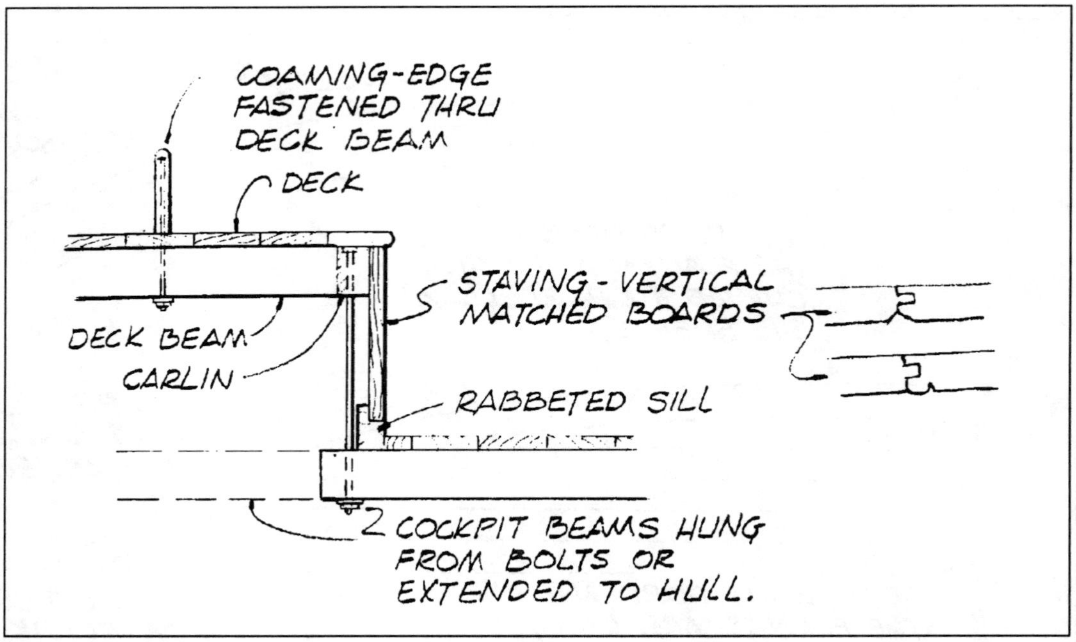

Figure 6.43. Footwell or cockpit construction details.

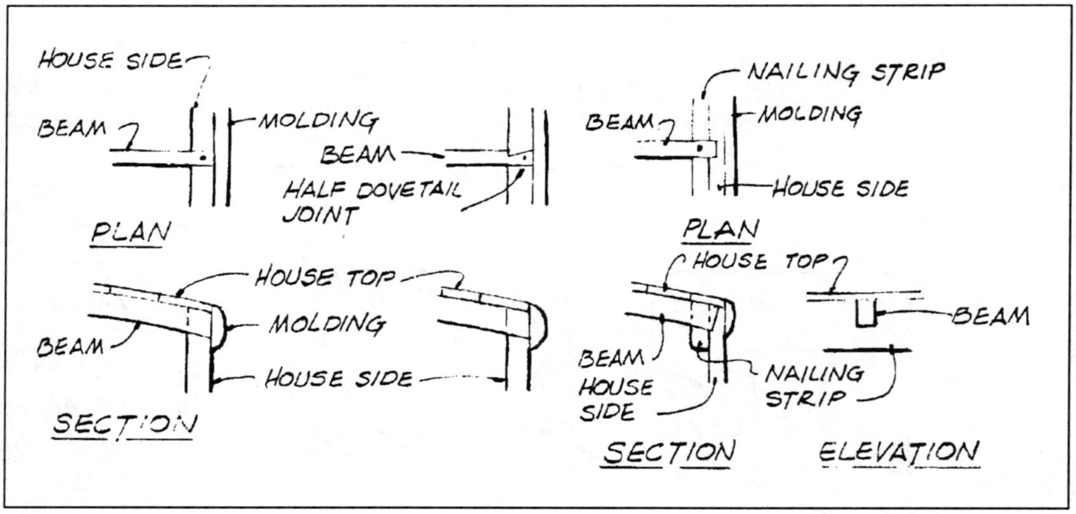

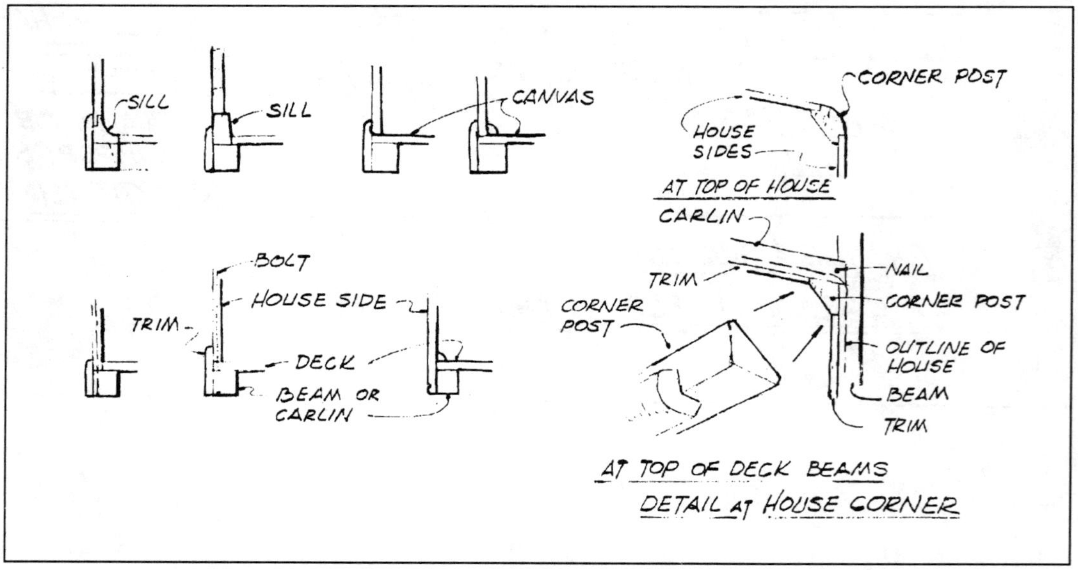

Figure 6.44. Deckhouse construction details. (right and below)

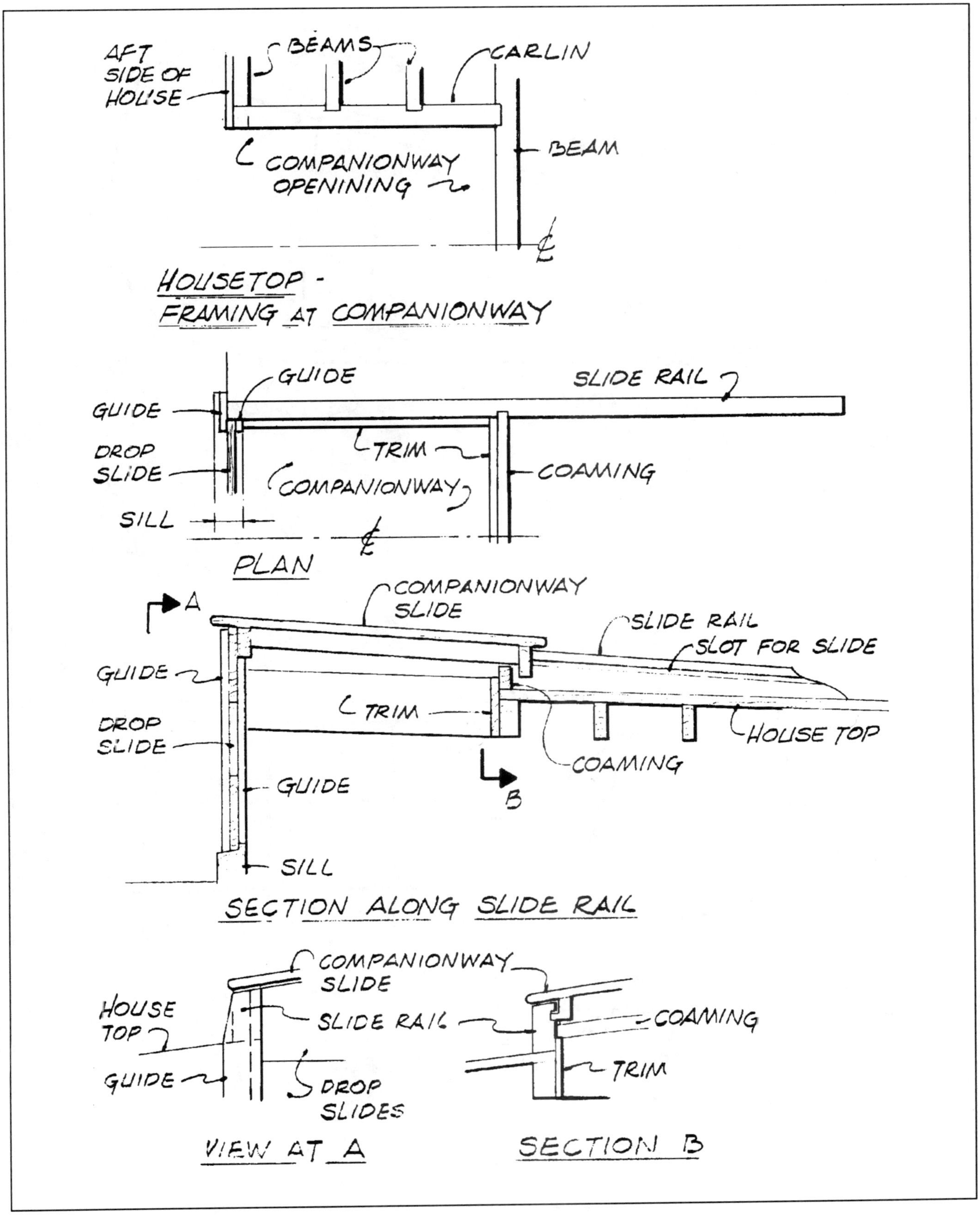

Figure 6.45. Companionway construction details.

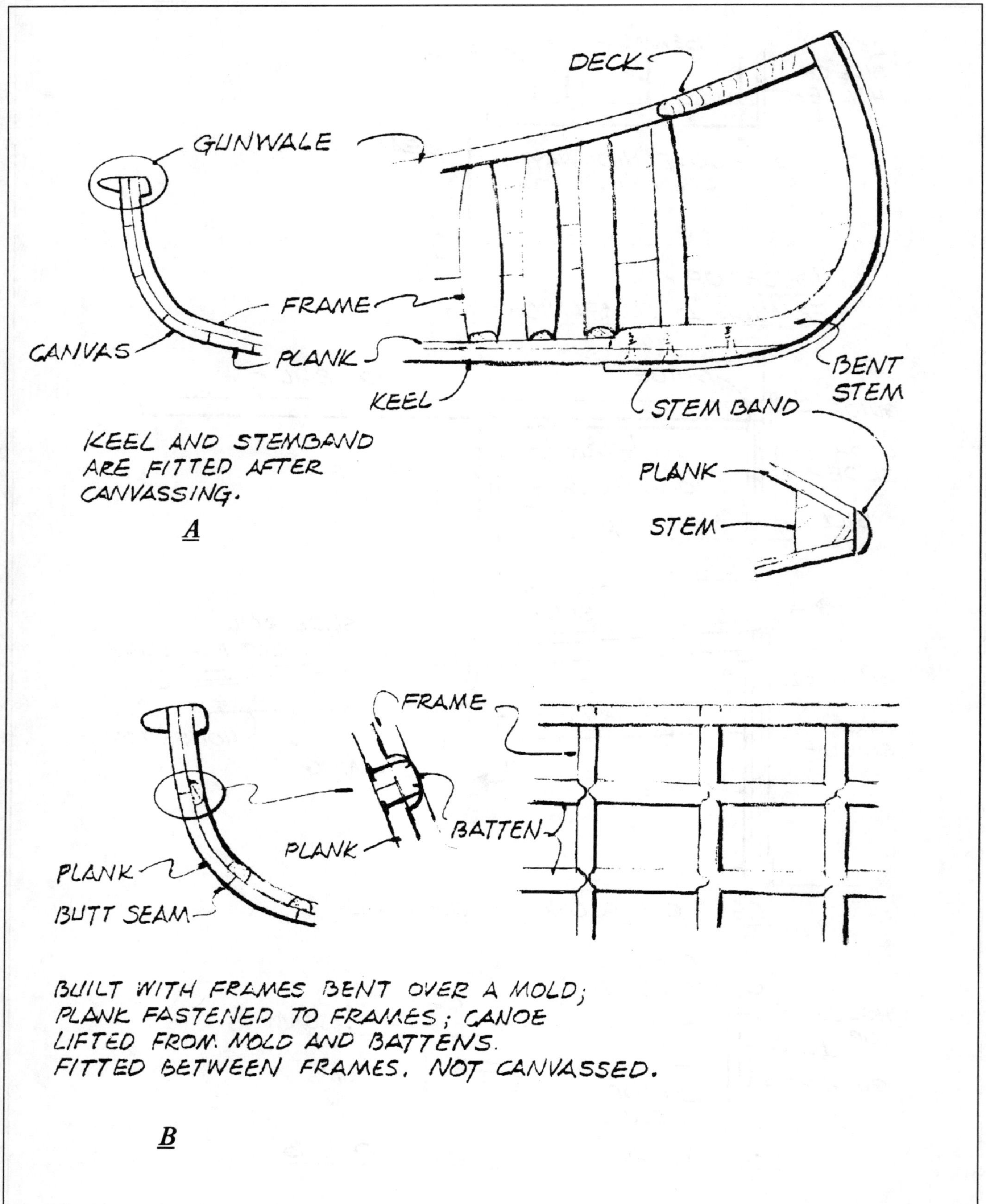

Figure 6.46. Canoes: (A) The most commonly encountered American canoe—canvas over wood construction. (B) Smooth planked construction method used in the Peterborough Canoe (Canadian). Canoes sometimes are of all wood construction; lapstrake, feather lap, and strip planked as shown in figs. 6.34 and 6.35. For a study of native North American boats, refer to *The Bark Canoes and Skin Boats of North America*, by Adney and Chapelle.

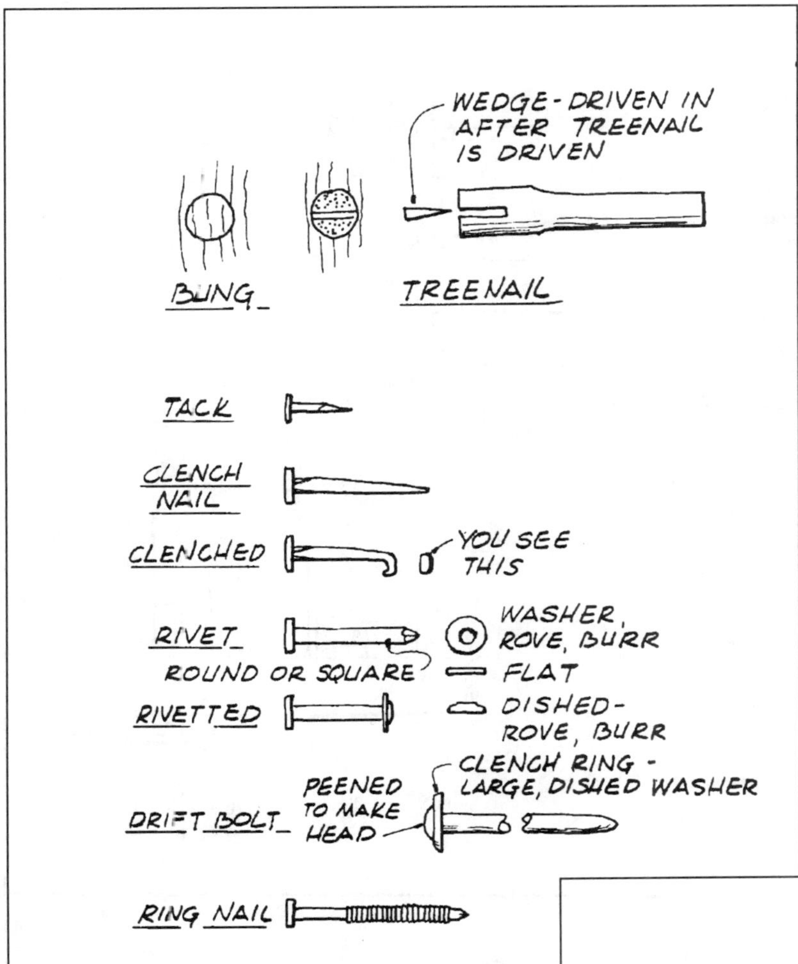

Figure 6.47. Fasteners: Treenails (pronounced trunnels) are not generally found in "boats" as defined by this manual with Scandinavian boats being an exception. The light circular marks that appear on the planking are most likely to be bungs (wooden plugs) covering fastenings. You can distinguish between a treenail and a bung by examining the grain pattern. A treenail shows end grain; a bung is cut from face grain wood and is placed to approximate the grain of the plank that it is set into. If possible, remove several bungs to reveal the fastenings underneath, but get permission first. Only the heads of most fastenings will be visible. Judge sizes (diameters, not lengths) by comparing the heads of samples to the heads of those found. The heads of screws and bolts are related to the size. For example, a bolt head measuring 7/16" across flats, is on a 1/4"-diameter bolt or machine screw. A pocket full of samples and screw and bolt size tables will help. The ratio of size of the head to the diameter of screws and bolts has been fairly standard for about one hundred years. This does not work with nails and rivets because the head sizes are somewhat inconsistent.

Bolts are easier to deal with. The exposed threaded ends can be measured—the diameter and thread—by running a known-size nut onto the thread. When you find a bolt and can't find the other end (head), suspect a hanger bolt. (text continued below left)

Usually both ends of rivets and clench nails are visible—at least of a sufficient number of them to establish a pattern for their location. The heads of rivets and clench nails are roughly similar in appearance with the older styles being roughly rectangular. Modern rivets frequently have a round head. The clenched end of a nail can be mistaken for a small lump of metal embedded in the wood. The end of a rivet is unmistakable—a washer with a small lump in it. The head of a drift bolt is similar but much larger. Copper tacks are driven both straight and clenched. The clench is sometimes tiny. One Adirondack guide boat builder allowed 1/16th of an inch for the clench.

The following is a list of fastenings according to the metal used. Copper: tacks, nails, clench nails, and rivets. Bronze: nails, drifts, drift bolts, and spikes (very large nails). Iron (wrought iron or Swedish iron which has no carbon and a high silica content): nails, clench nails, rivets, drifts, drift bolts, threaded bolts, and screws. Steel and stainless steel: screws, nails, and bolts.

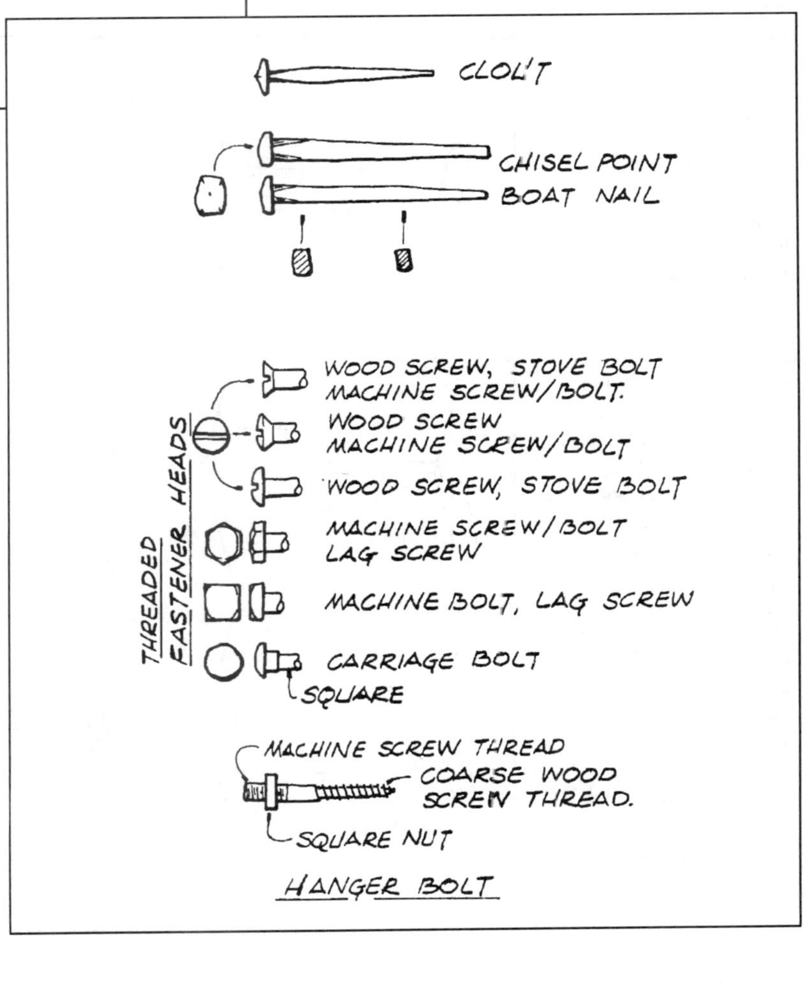

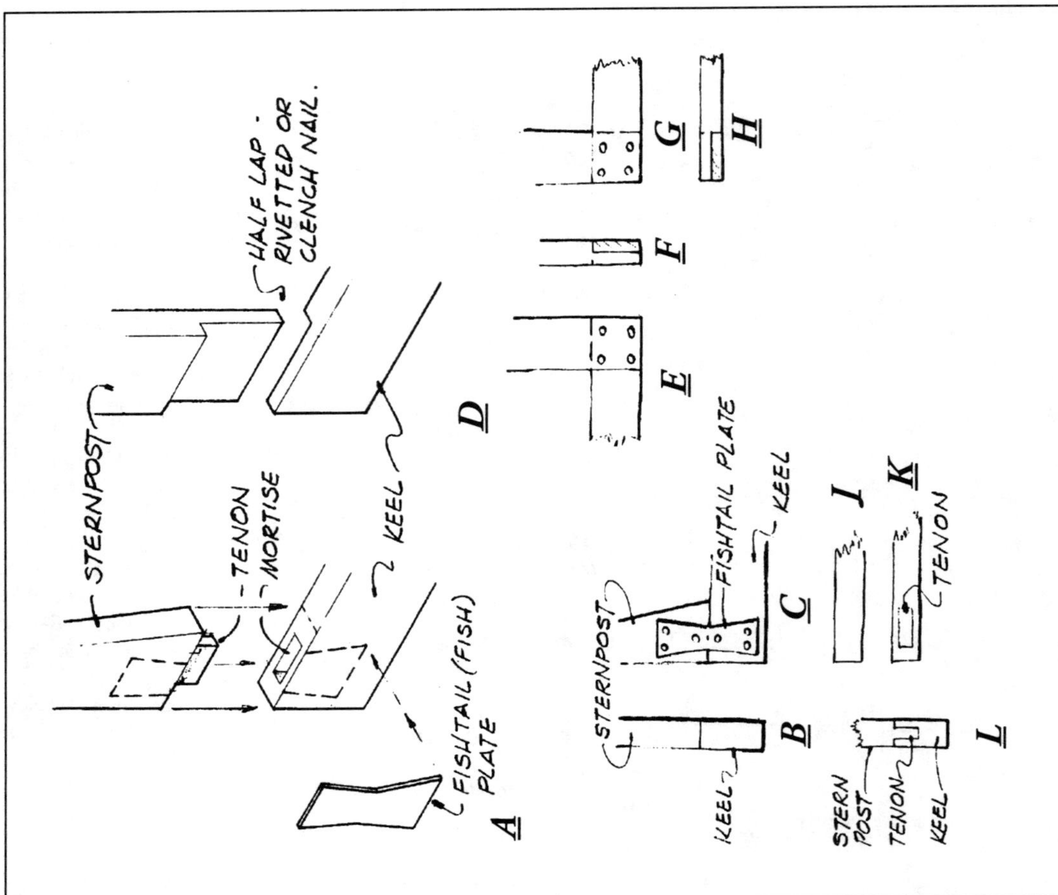

Figure 6.48. Joints: (A) The blind mortise-and-tenon joint; the fishtail plate, fitted on both sides of joint then riveted through the keel and post. Sometimes the plate is let into (recessed) into the wood. (B) Stern view of A. (C) Side view of a looking from starboard to port. (D) The half-lap joint—frequently found in under-bench locker framing and sometimes found at the bow and stern of European-built boats. (E) View of D looking from port to starboard. (F) Stern view of D. (G) View of d looking from starboard to port. (H) View of d looking up from underneath the boat. (J) View looking up from underneath at B and C. (K) View looking up where the tenon passes completely through the keel—a through tenon. (L) View from aft with the tenon stopped in the keel—sufficient to lock the two together and keep the post from twisting out of alignment.

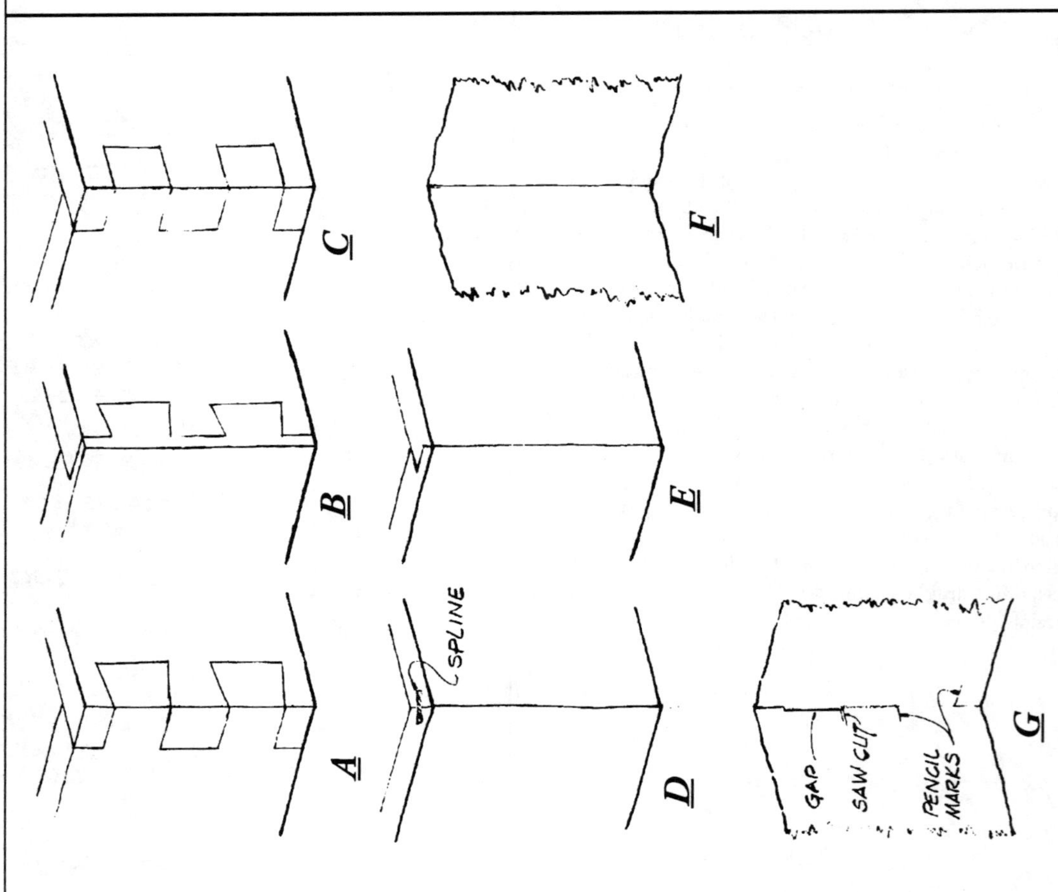

Figure 6.49. Joints as seen in finished work: (A) Dovetail joint (correctly the through-dovetail joint), a general use joint. (B) Blind dovetail joint, most frequently used in drawer fronts. (C) Box or finger joint, another general use joint. (D) Simple miter joint, made both with and without the spline (feather). (E) Half-lapped miter joint. (A through E are all shown looking from the outside of the joint.) (F) If the top and bottom edges of the joined planks cannot be seen, all joints look like this when viewed from the inside. In all cases, look for layout or tool marks, or even a gap, to help identify the joint (G).

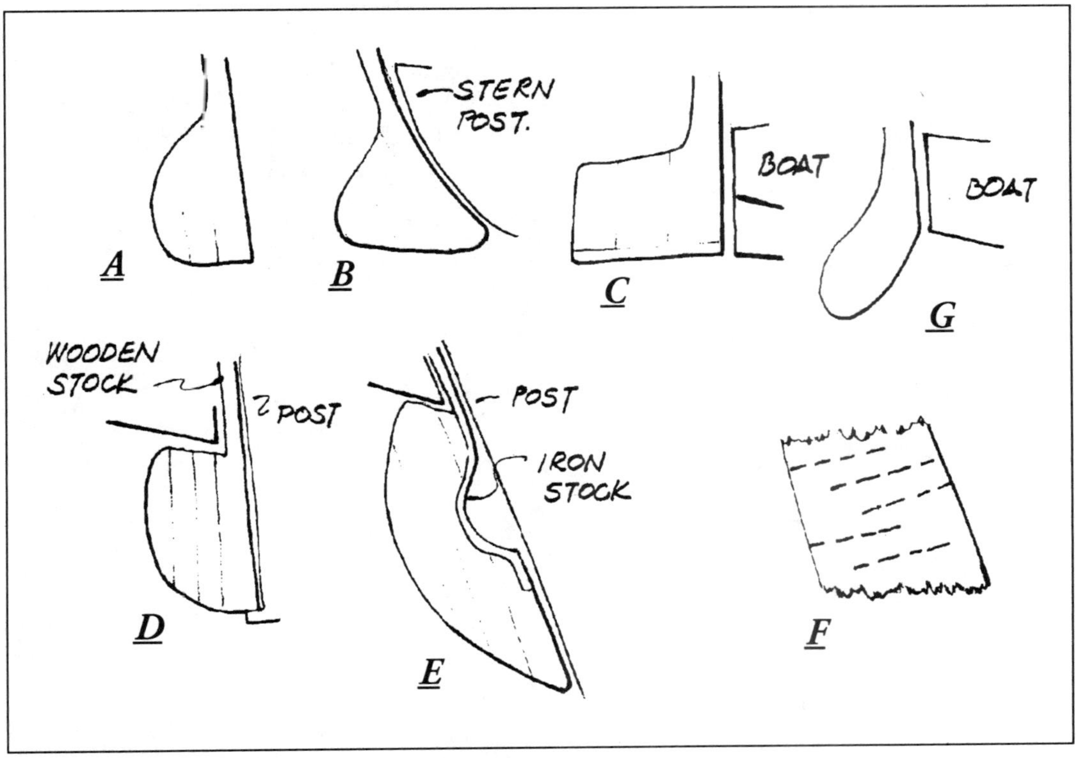

Figure 6.50. Rudders. Wooden rudders (A through E) are made up of a number of planks which are edge fastened with wooden dowels, long spikes, or metal drifts. Dinghy rudders (G) are made of plywood, metal, or fiberglass.

Figure 6.51, right and following page. Attaching rudders to the boat. (A) The most common small boat hardware. To hang the rudder while the boat is floating, some boats have a pintle as in a for the lower fitting, with a loose pintle, (B), which is put into place last. Other methods used to solve the same problem are (C) using a continuous rod for a pintle or (D) fitting a long, upright pintle, which can be engaged first, and a shorter pintle to the rudder which engages the second. (C) is common among small, open boats and sailing canoes. (D) was found on a Bermuda-built pilot's gig of the late nineteenth century. (F) was used on New Haven sharpies; the pin allowed the rudder to be at two depths. (G), the simplest of all, had the pintle and gudgeon bent up from round iron rod. It was found on a New Jersey Barnegat Bay Sneakbox.

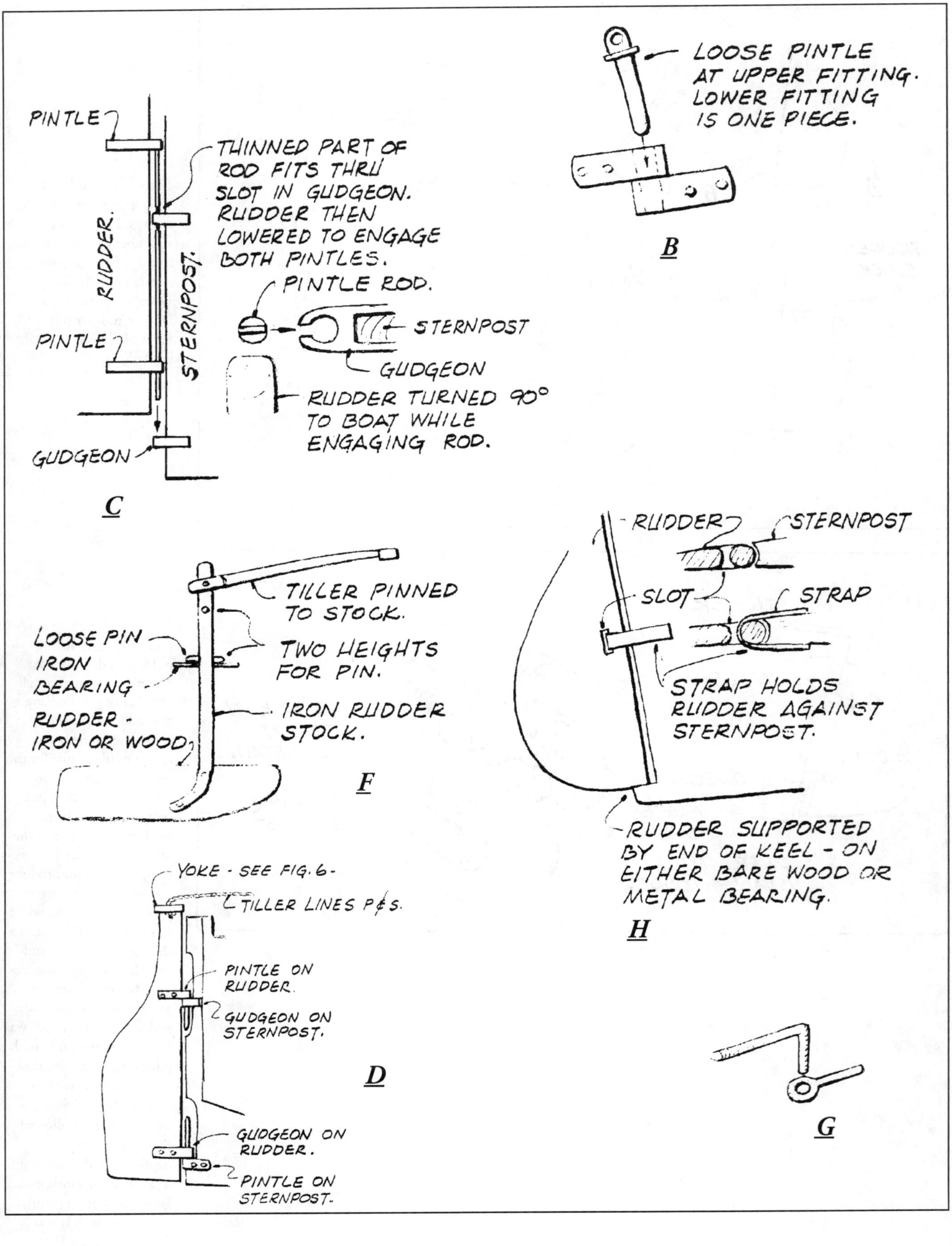
LOOSE PINTLE AT UPPER FITTING. LOWER FITTING IS ONE PIECE.
B
PINTLE
THINNED PART OF ROD FITS THRU SLOT IN GUDGEON. RUDDER THEN LOWERED TO ENGAGE BOTH PINTLES.
RUDDER.
STERNPOST.
PINTLE ROD.
STERNPOST
PINTLE
GUDGEON
RUDDER TURNED 90° TO BOAT WHILE ENGAGING ROD.
GUDGEON
C
TILLER PINNED TO STOCK.
LOOSE PIN
IRON BEARING
TWO HEIGHTS FOR PIN.
RUDDER - IRON OR WOOD
IRON RUDDER STOCK.
F
RUDDER
STERNPOST
SLOT
STRAP
STRAP HOLDS RUDDER AGAINST STERNPOST.
RUDDER SUPPORTED BY END OF KEEL - ON EITHER BARE WOOD OR METAL BEARING.
H
YOKE - SEE FIG. 6 -
TILLER LINES P&S.
PINTLE ON RUDDER.
GUDGEON ON STERNPOST.
D
GUDGEON ON RUDDER.
PINTLE ON STERNPOST.
G

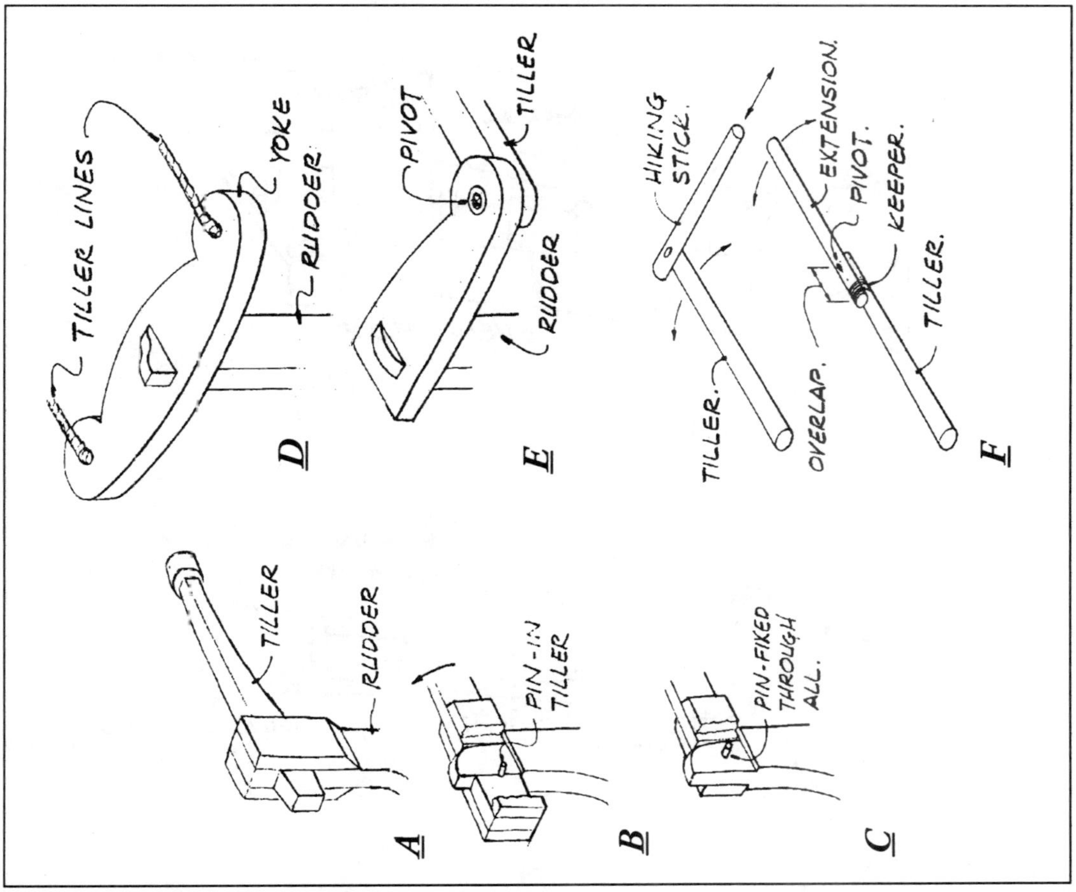

Figure 6.52. Basic steering. An oar held overboard is the simplest of steering methods. Next is a tiller fitted to the rudder in various ways. (A) The tiller, not fastened in place, stuck through a fattened up rudder. (B) The pin locks the tiller in place on the rudder, while at the same time allowing the tiller to be either swung up or removed. (C) The pin both fastens the tiller to the rudder and allows it to be swung upward, so that the helmsman can steer while standing. (D) Rudder fitted with a yoke.The tiller lines lead forward into the boat. Pulling on one line or the other steers the boat. (E) Pulling or pushing on the tiller steers the boat (F). The hiking stick. "Hiking" is to sit as far out on the edge of the boat as possible to balance the boat and keep it upright. The hiking stick permits the helmsman to control the tiller without leaning inboard. When the hiking stick is locked in alignment along tiller with a keeper, the boat can be steered from a position farther forward in the boat.

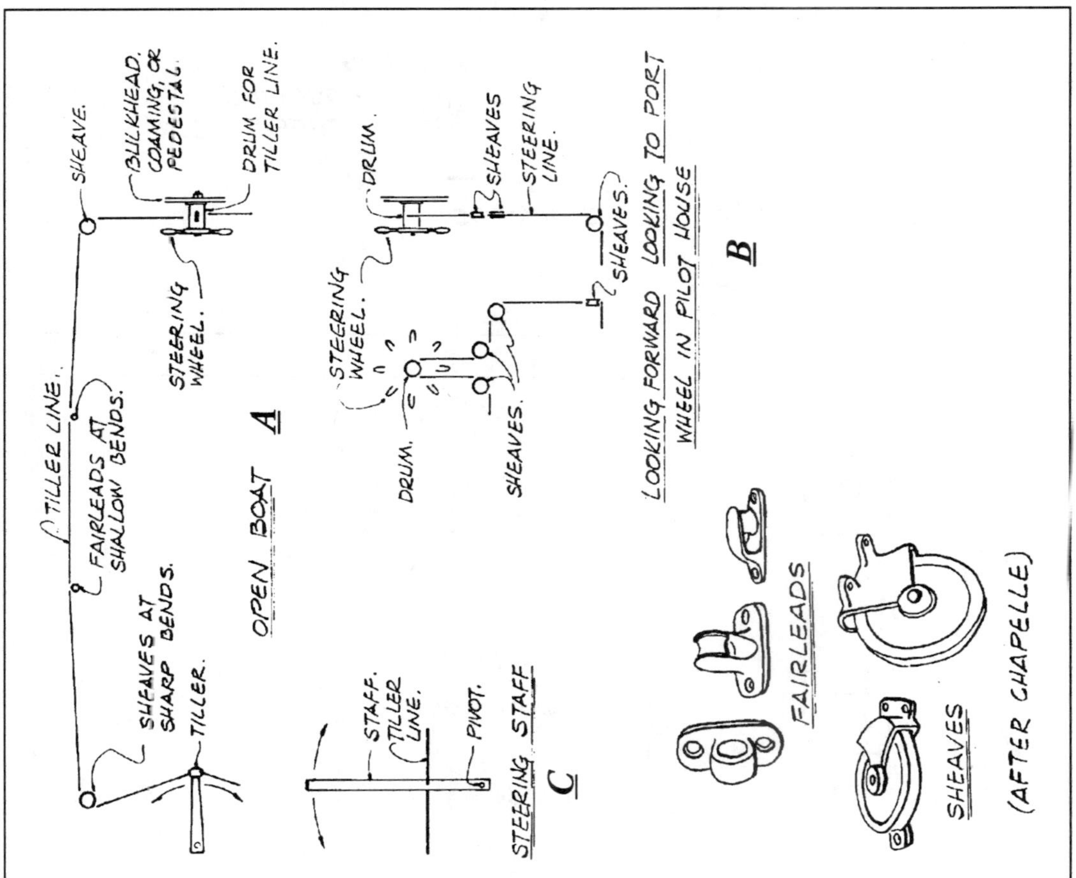

Figure 6.53. Wheel steering. (A) A commonly found arrangement. The steering wheel with a drum is fastened to coaming or pedestal. The tiller line is clamped to the drum and several turns taken, each to port and starboard, around the drum. The lines are then led to the tiller. The same system, with more complicated lead, is used in boats with a pilot house (B). A simplification (C) where a pivoted staff is substituted for the wheel and drum. Note that wire rope (cable) or chain may be found in place of natural fiber rope.

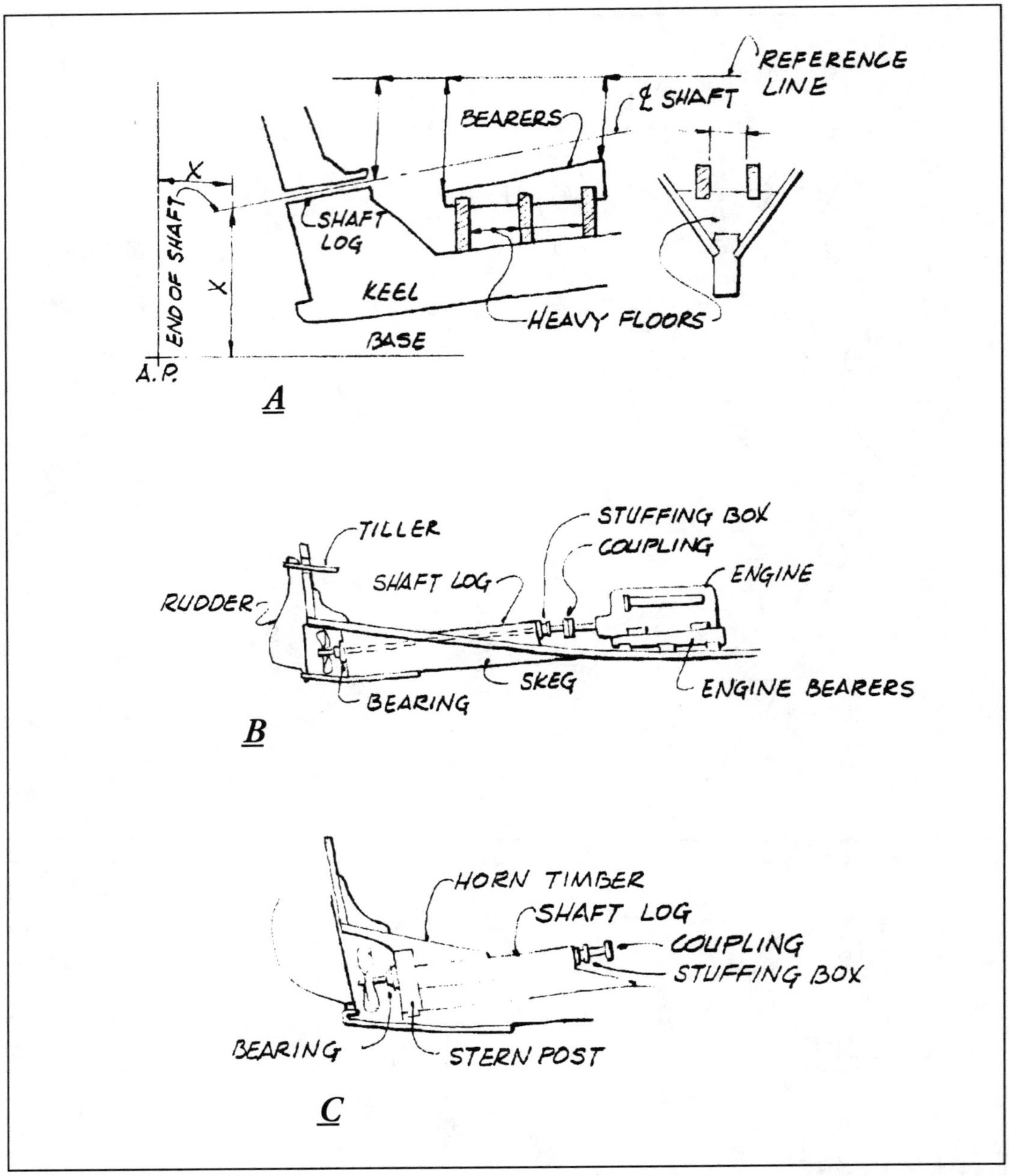

Figure 6.54. Shaft logs: (A) In a heavily built sailboat, the engine bearers are usually set onto heavy floors and the shaft log simply a hole bored through the backbone. Take the measurements marked "X" to locate the propeller shaft and engine bearers relative to the rest of the boat. (B) In a boat with a sprung keel, the shaft log may be a wooden block or a metal fabrication fastened to the keel inboard. (C) A true shaft log built into the boat's backbone.

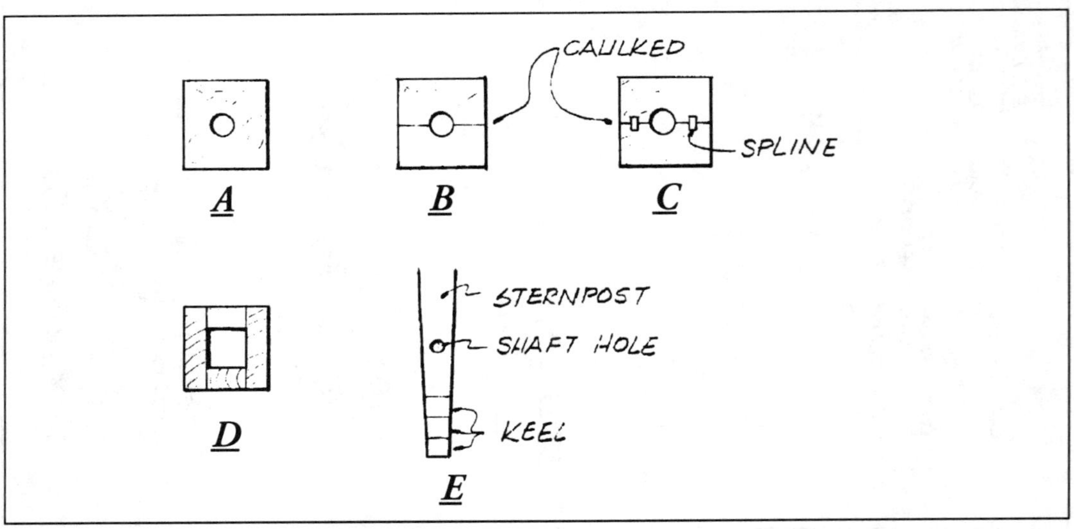

Figure 6.55. Shaft log sections: (A) Most common—generally built into the backbone. (B and C) Used in larger boats. (D) Representative built-up type. (E) View of a sternpost looking forward. The dotted lines indicated an off-center propeller shaft—sometimes original, sometimes a modification.

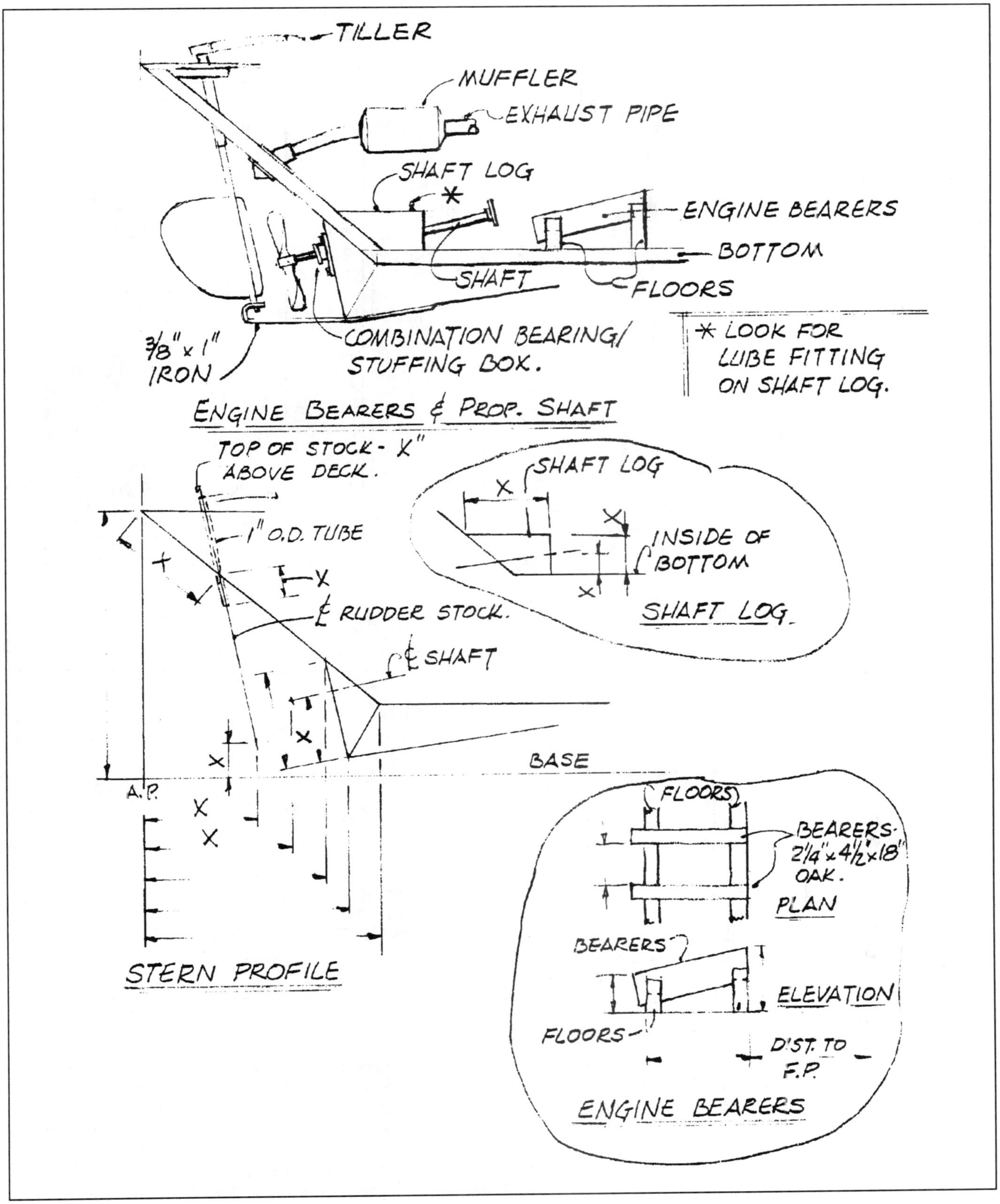

Figure 6.56. The engine compartment arrangement in a Toppan Power Dory built at Medford, Massachusetts, in 1910. Shown are the measurements and sketches taken off to record the stern profile, propeller shaft, and engine bearer locations.

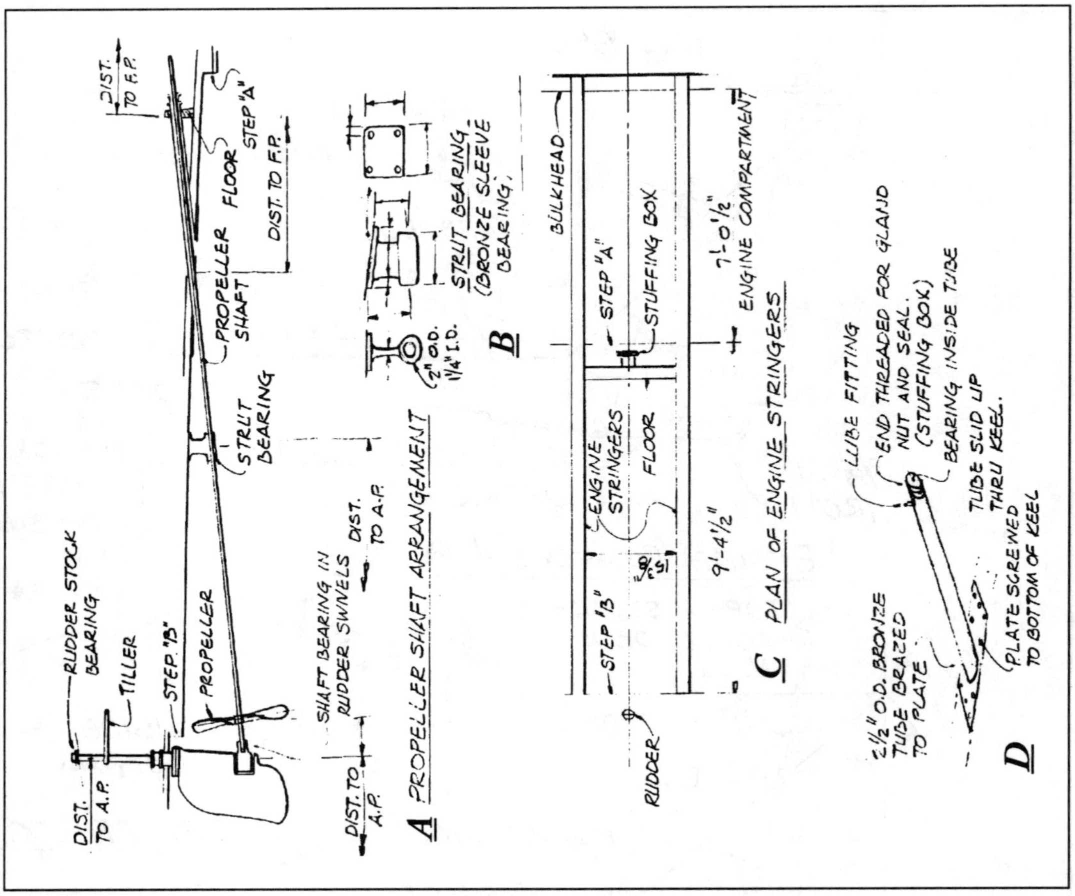

Figure 6.57. Propeller shaft arrangement as found in a 1929 Gold Cup Racer: (A) Semi-schematic sketch showing the fittings along the shaft. (B) Measurements to take from the strut bearing. (C) The engine stringers are used to distribute the engine's weight in a lightly built boat. In this boat, the distance between them accommodated the original engine, and the stringers also served as the bearers. When a different engine was fitted, new bearers were bolted to the stringers. (D) The propeller shaft tube. Sometimes the tube is installed with the plate fastened to the inside of the keel.

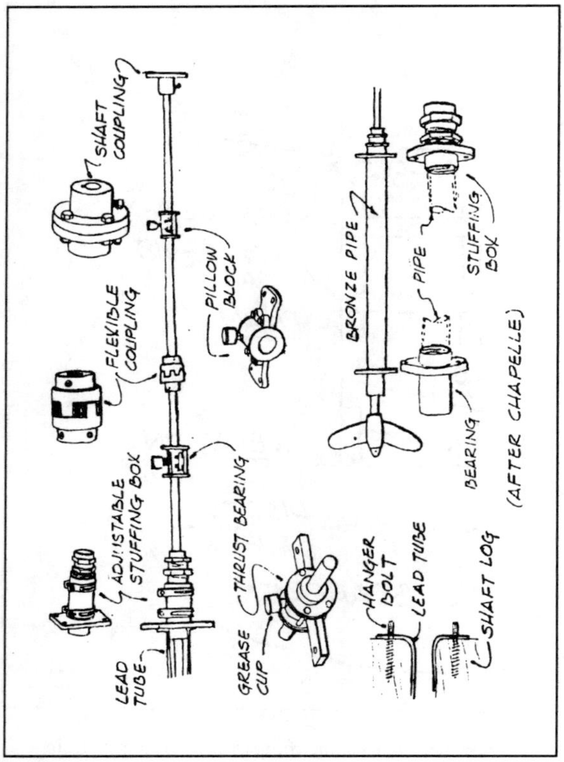

Figure 6.58. Representative propeller shaft fittings.

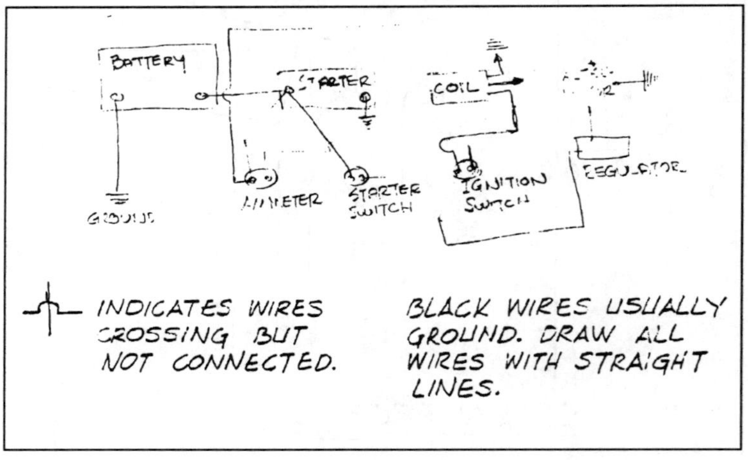

Figure 6.59. Electrical systems: The electrical system in a boat can be simple. The easiest and best way to record an electrical system is to make a schematic drawing of the system as shown here. Show the true locations of the equipment and measured layouts of instrument panels and any equipment groupings in the take-off notes. Record all manufacturer's names and catalog numbers. Record all wire codings (colored pattern as "red with black stripes") and, if possible, the wire size (sometimes marked on the wire). Wire runs may prove impossible to trace. If this happens, the color coding of wiring may prove helpful in determining the connections, if not the actual routing. Older boats that have been rewired may have several generations of wiring in place but not all connected.

Electrical hardware and equipment that may be found: Battery; master switch; starter switch or button; generator or alternator; voltage regulator; fuse box; bilge pump; cabin lighting; navigation/running lights—bow, stern, port, starboard, and masthead; radio; binnacle light; instrument panel: fuel gauge, ammeter, oil pressure; winches; water pumps; ventilation: cabin, engine compartment, and bilges.

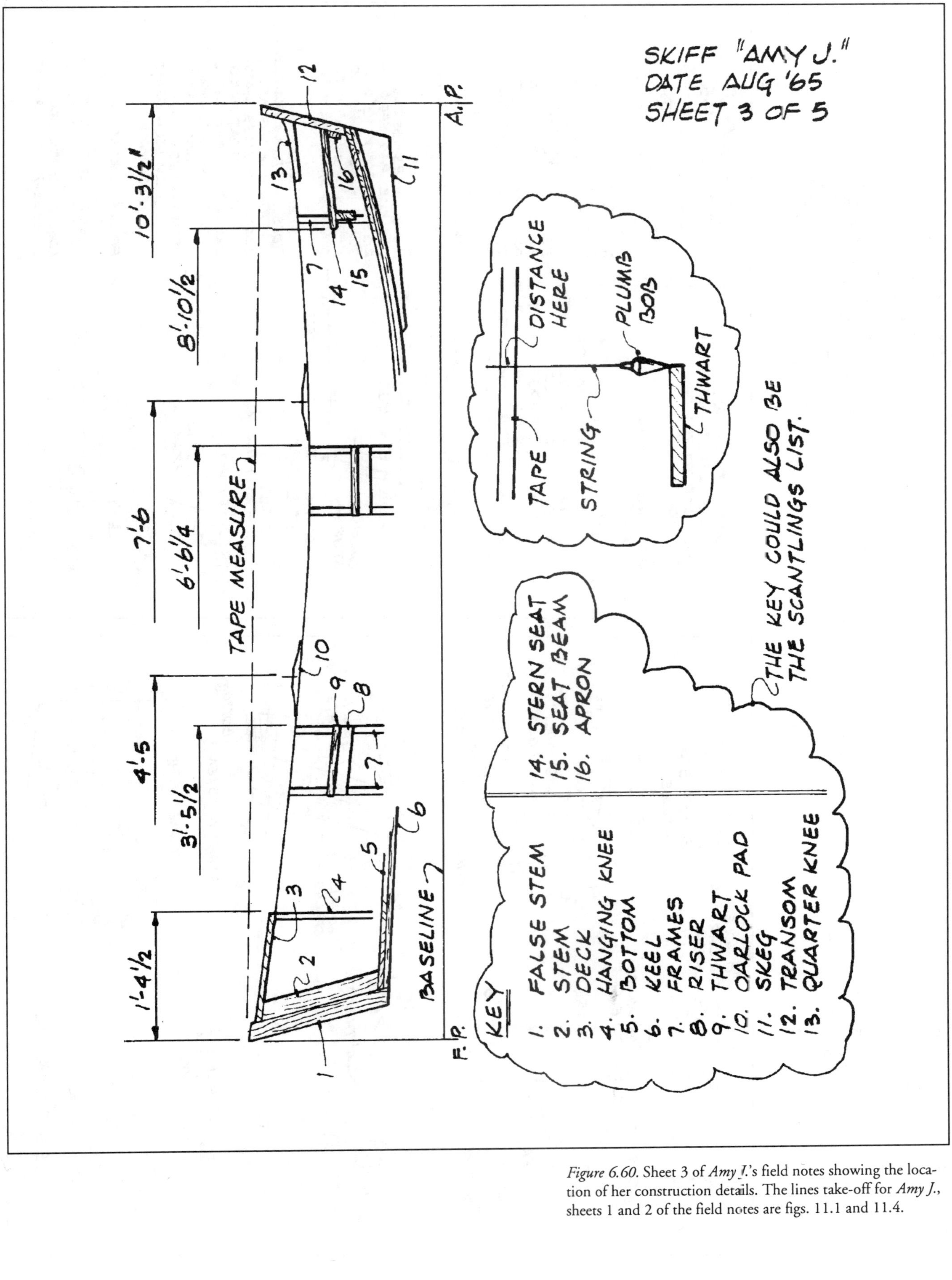

Figure 6.60. Sheet 3 of *Amy J.*'s field notes showing the location of her construction details. The lines take-off for *Amy J.*, sheets 1 and 2 of the field notes are figs. 11.1 and 11.4.

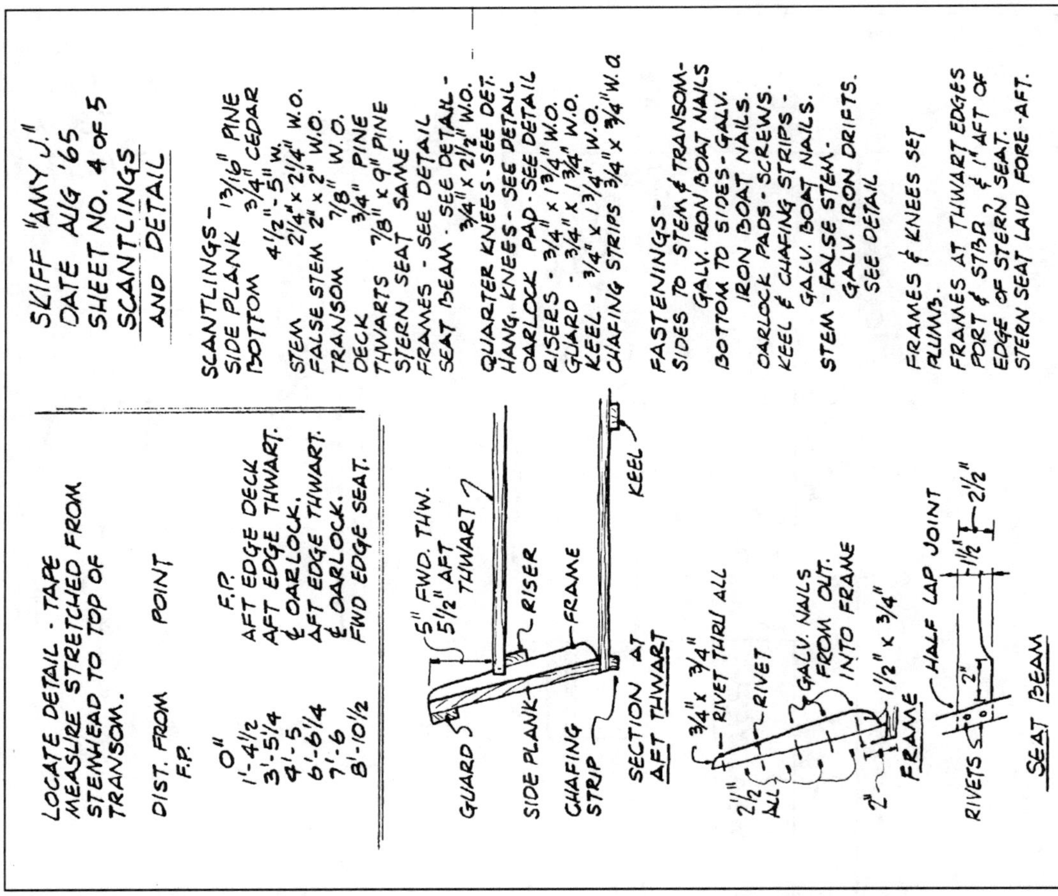

Figure 6.61. Sheet 4 of *Amy J.*'s field notes. Pay particular attention to the different forms for recording the location of construction details. This form takes less space, and when used for a larger vessel can be much neater than making a profile sketch as in figure 6.60.

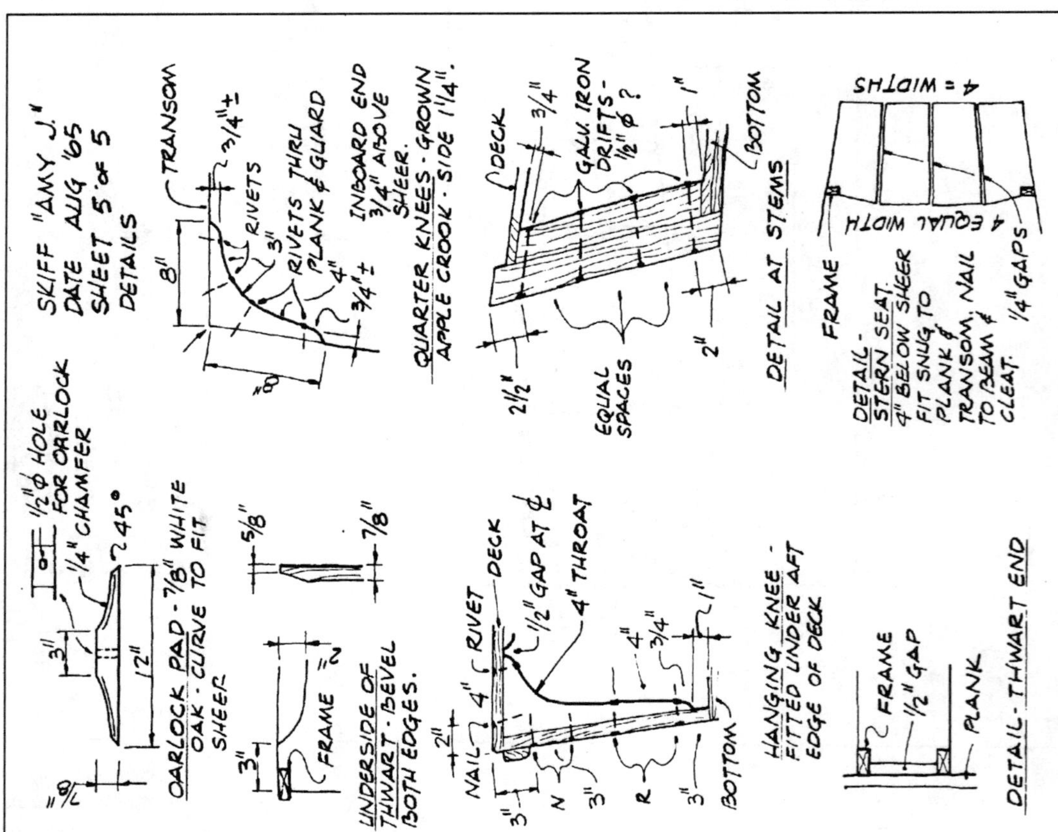

Figure 6.62. Sheet 5 of *Amy J.*'s field notes. The form of notation for measurements and descriptions is quite standard even though several different styles are used.

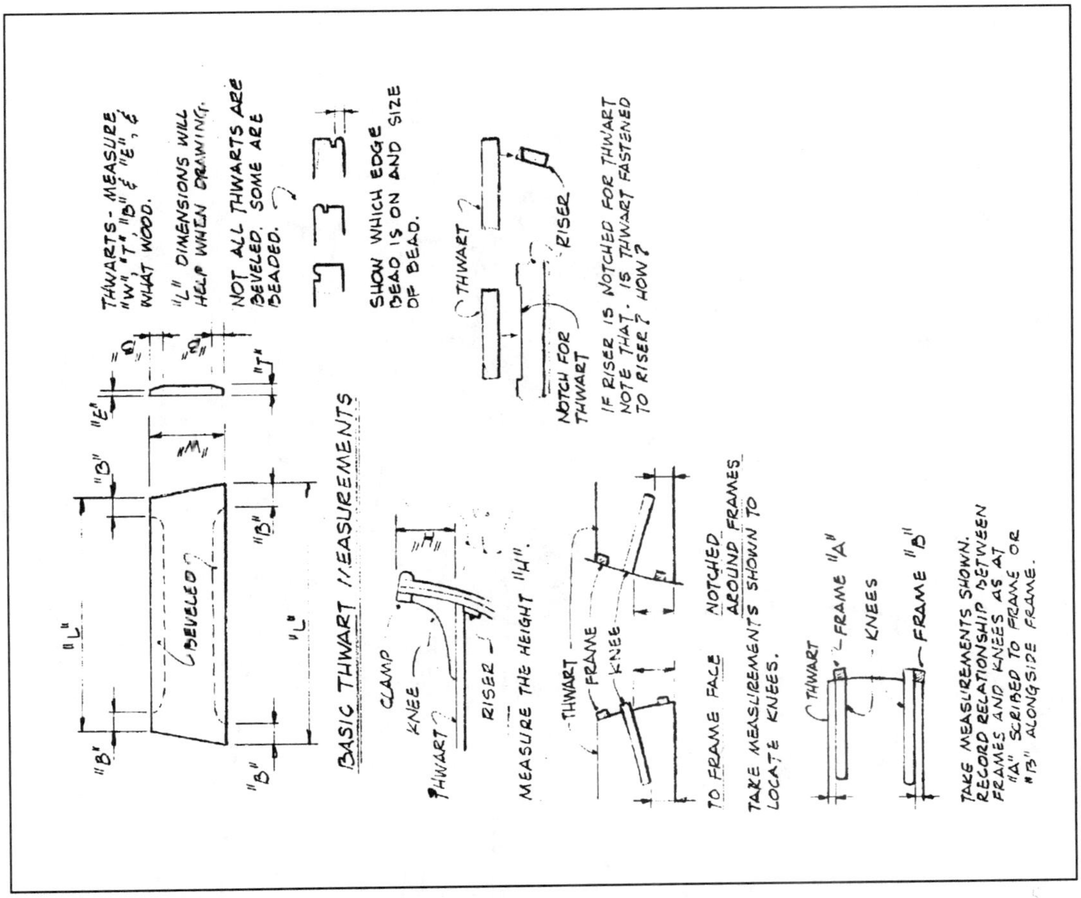

Figure 6.63. An example of thwart details and measurements to record. Aside from the variations in the thwart edges, make note of the locations of the thwarts and knees in relation to the frames. These are critical to having a replica appear correct. When measuring the thwarts, measure the total length of each edge "L". One half of "L" can be drawn in the construction plan to save time.

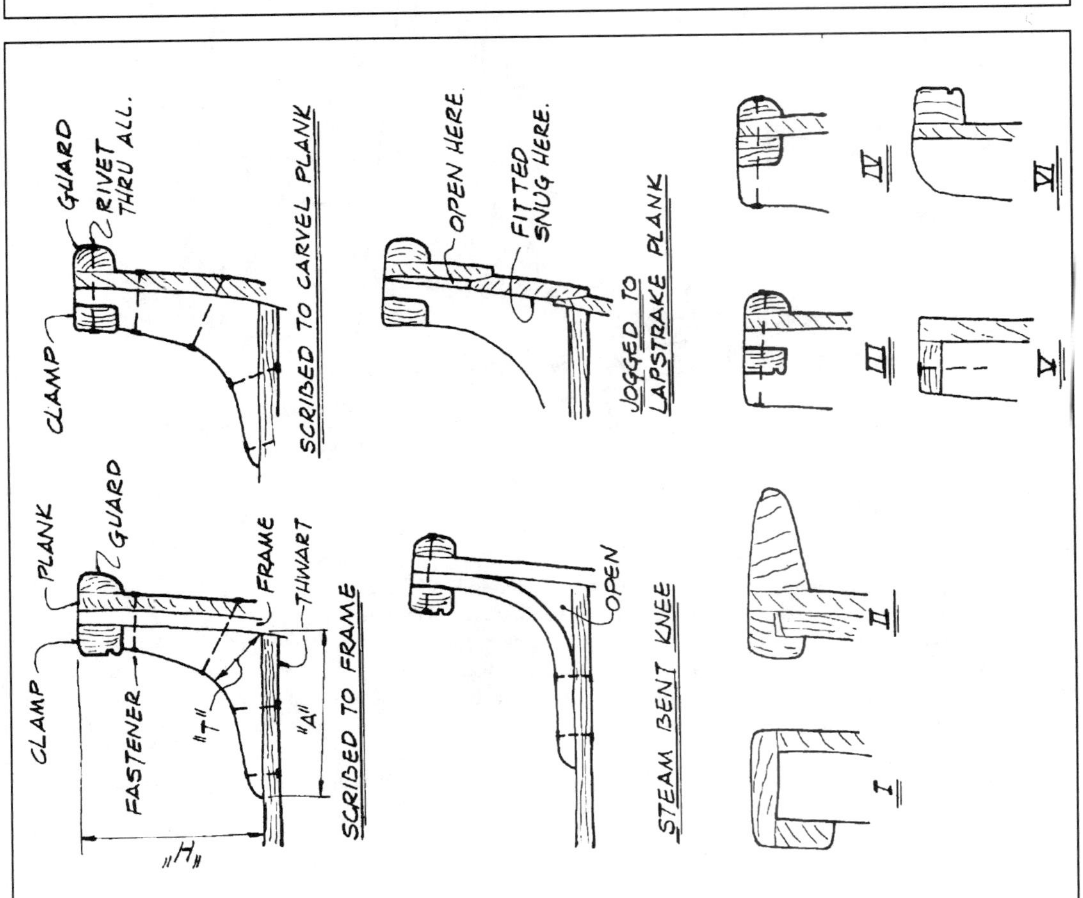

Figure 6.64. Variations in thwart knees, clamps, and guards in open boats. The measurements shown, "A", "H", and "T", are the length of the arm along the thwart, the height from the sheer to the top of the thwart, and the throat (taken at 45 degrees). The four large views are generally representative arrangements. The partial views, while found in boats from a specific geographical locale, are not limited to a single region. (I) Abaco dinghy, Abaco Island, The Bahamas. (II) Canoes by J. H. Rushton, Canton, N.Y. (III) a frequently found variation. (IV) yacht tender by Fife, Scotland. (V) fancy dory from the North Shore of Massachusetts. (VI) Adirondack guideboat, New York.

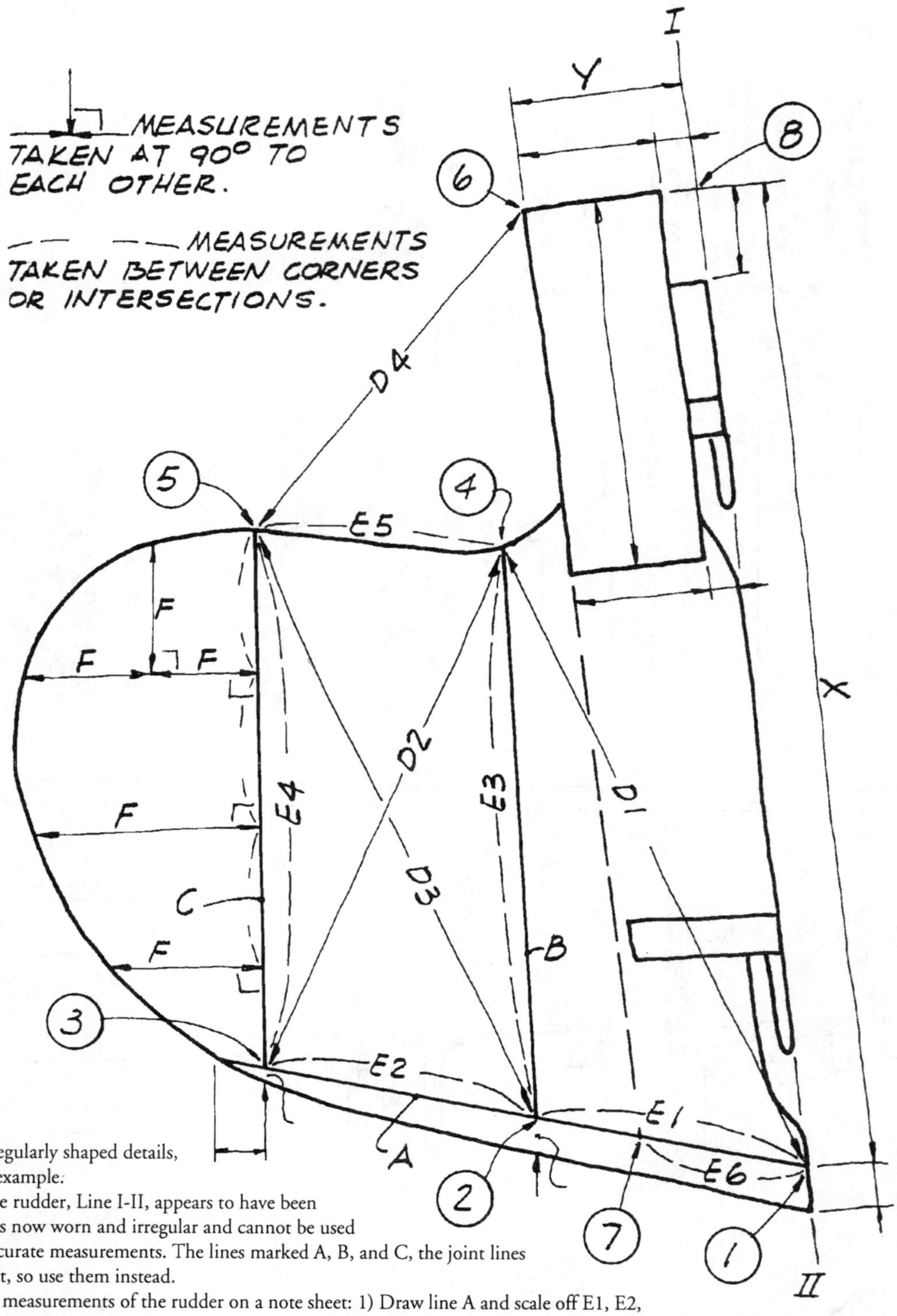

Figure 6.65. Measuring irregularly shaped details, using Susie's rudder as an example.

The forward edge of the rudder, Line I-II, appears to have been built as a straight line but is now worn and irregular and cannot be used as a reference for taking accurate measurements. The lines marked A, B, and C, the joint lines between planks, are straight, so use them instead.

To take and record the measurements of the rudder on a note sheet: 1) Draw line A and scale off E1, E2, and E6 along it. 2) Using points 1 and 3 as centers, draw arcs with radii equal to D1 and D2 to intersect at 4, then draw line B from 2 to 4. Use E3 to check the length of B. 3) Using 3 as a center, draw radius E4; from 4, draw radius E5. The intersection is 5. Draw line C and use D3 to check the accuracy of your work. 4) With 7 as the center, draw the radius equal to the distance between 6 and 7. From 5, draw the radius D4 to locate 6. 5) Draw an arch equal to X with 1 as the center and the arc equal to Y using 6 as center. The intersection is 8. Point 8 was located by laying a framing square with the blade aligned with the forward edge of the rudder with the arm across the rudder head.

These steps complete the straight line necessary to define the basic outline of the rudder. The measurements F are located as necessary to define the curved shape. Other measurements can be taken where needed using the same method.

As an exercise, follow the above steps, using a compass to pick up the measurements from the figure. If you were to make a pattern of the rudder, the lines A, B, and C must be traced onto it. Should there be no straight lines to use, draw them on the object being measured, or the pattern, and go from there.

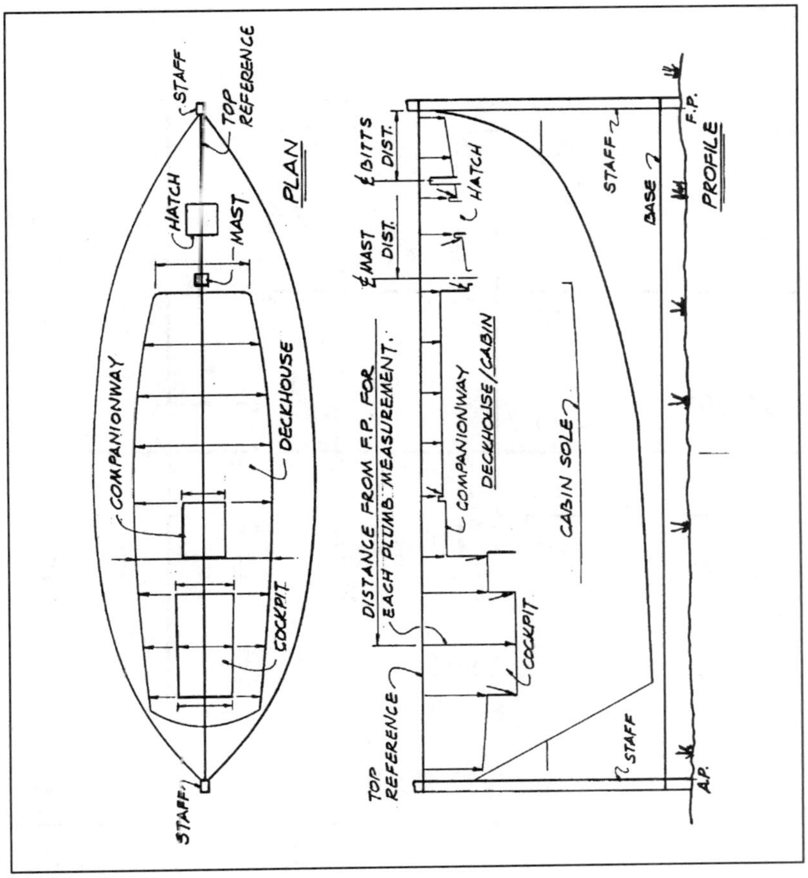

Figure 6.66. Using the top reference line. Rig a top reference line; stretch a tape measure alongside it from F.P. to A.P. - see chapter 7. Take plumb measurements down from the top reference line to the centerline of the deck, the top of the deck house, and the cockpit sole—indicated by arrows. Record the distance from F.P. to each height measurement. Also take athwartship measurements square across the boat (see the arrows in the plan view).

Where bungs for deck beam fasteners are visible, use them as a guide to locate deck beams. Take heights at every second or third beam. As the beam spacing may be irregular, all deck beams must be located.

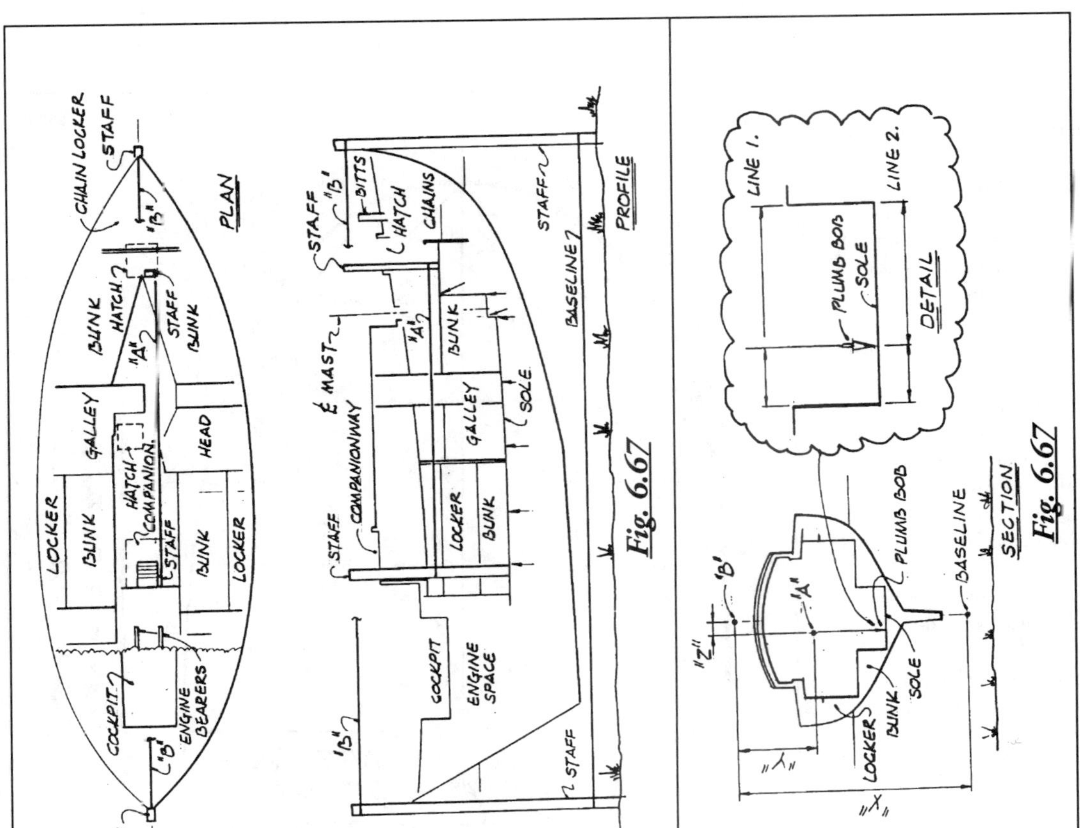

Figure 6.67. Measuring inside cabin space. Rig a reference line (A) inside the boat, attached it if possible, to the boat and directly below parallel to the top reference line (B). If this cannot be done, erect staffs to support A to port or starboard of B, making sure that A is parallel to B horizontally and vertically. Record the relationship between A and B as in figure 6.68. Staffs must be plumb fore-and-aft and athwartships. Measure the profile of the cabin sole by taking plumb measurements below A. Stretch a tape measure alongside A registered with the tape alongside B so that for example, 5'-0" on A is directly below 5'-0" on B.

Figure 6.68. Establish the relationship between the two reference lines by taking measurements "X", "Y", and "Z" as shown. Running or incremental measurements can be taken athwartships as in the detail.

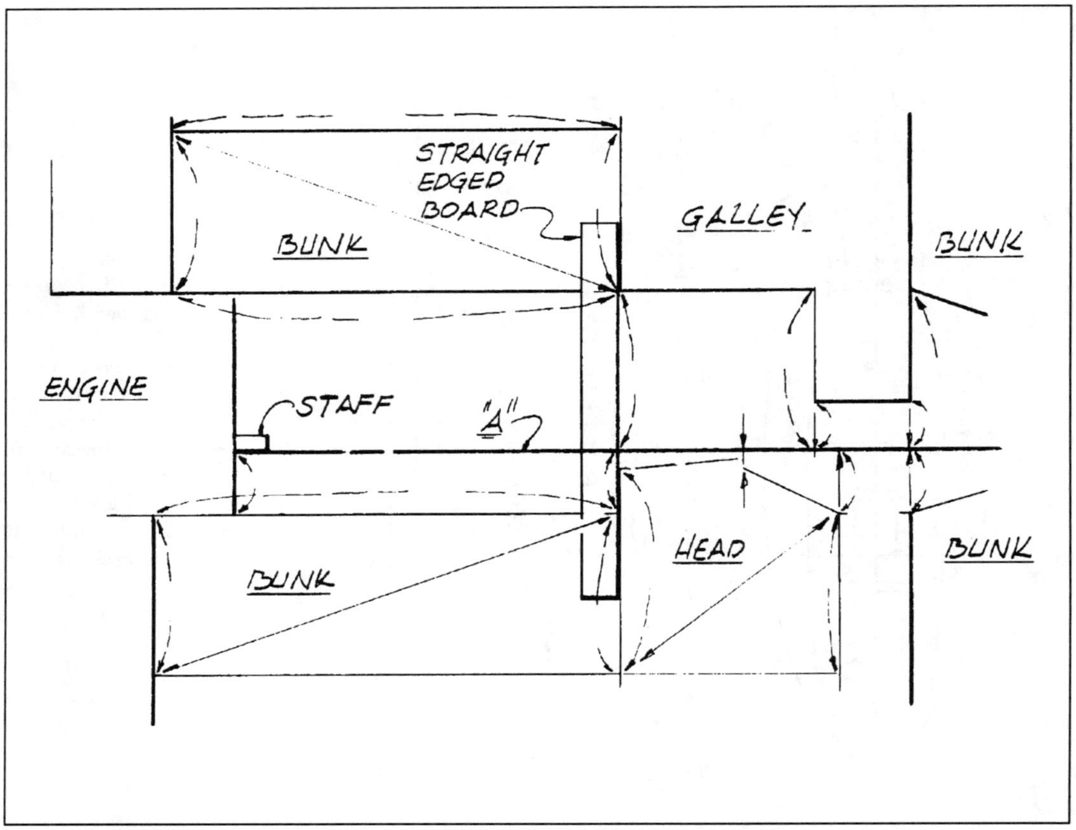

Figure 6.69. Locate interior detail by relating to A or a staff; a bulkhead by using a straight board to extend the line of the bulkhead to "A" and recording the distance from the F.P. A measurement from "A" to the corner of the bulkhead locates the corner. The arrows—straight and curved lines—show dimensions to take to get the gross outlines for the interior arrangement. The exact features in the head and galley should be sketched on separate note sheets. Take diagonal measurements in any apparently square or rectangular space to prove that it is or is not made up of parallel and perpendicular lines.

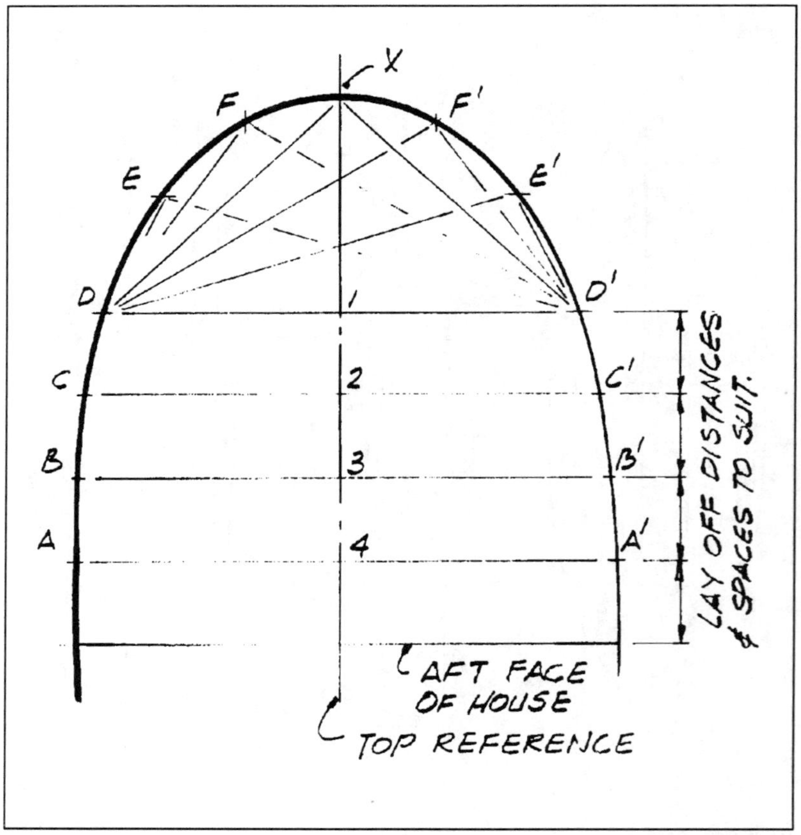

Figure 6.70. Location and measurement by triangulation. Begin by plumbing down from the top reference line and marking point "X". Locate an athwartships line, as at A-A1, at a body section station (refer to chapter 7). Then mark off the additional points of B to D and B1 to D1, equally spaced, or at stations, along the coaming with the last points where the curve of the coaming becomes significant. Mark the points E, F, E1, and F1 dividing the coaming into six approximately equal parts from D to D1. Measure the distances between all of the points as shown by the lines.

Plotting measurements taken by triangulation is straightforward. With D and D1 as center points, swing arcs equal to D-E, D1-E, D-F, D1-F, and so on right around the curve. A line faired through the intersections of the arc is the shape of the coaming. The best shape will be gotten by plotting the coaming shape from D1 around to D and on to point A.

Do not use less than five points to measure the shape of a curve. The results may give an inaccurate, nonrepresentative, curve.

Check the symmetry of the coaming by hanging a plumb bob from points 1, 2, 3, and 4 and measuring from the plumb bob to the coaming.

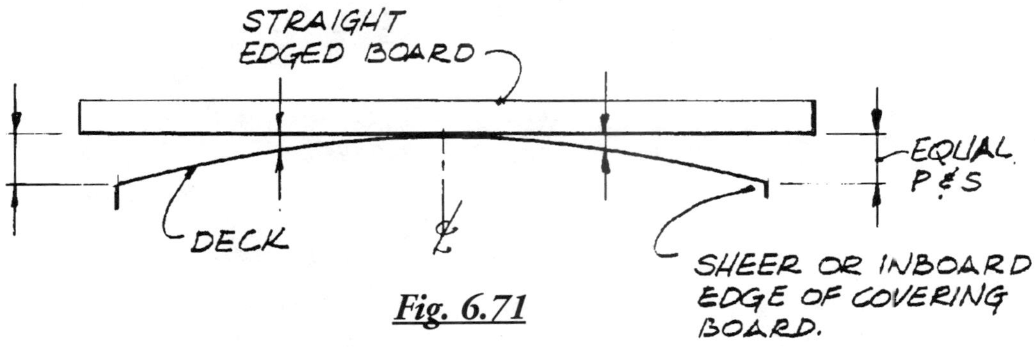

Fig. 6.71

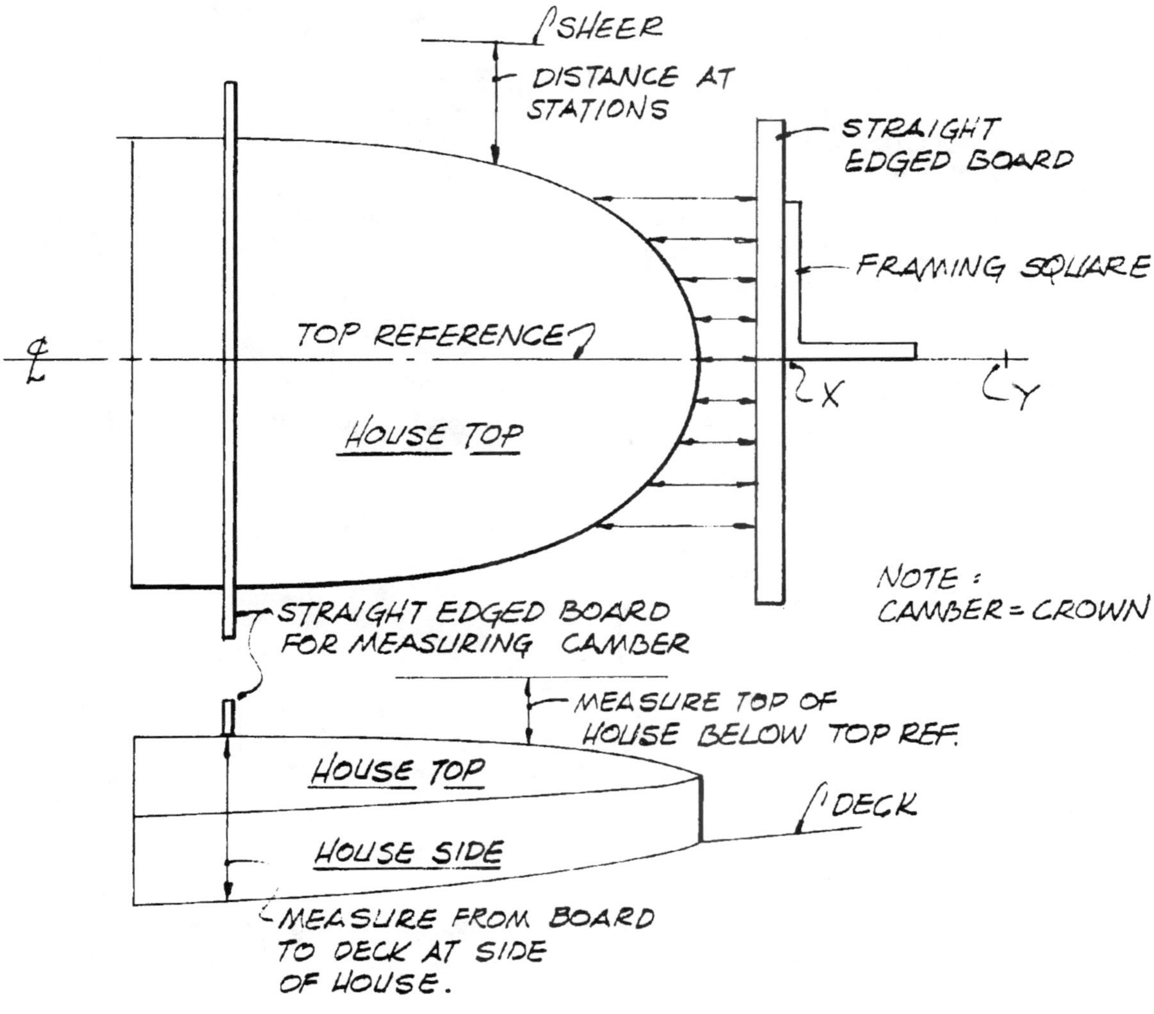

Fig. 6.72

Figure 6.71. Measuring deck camber. Set up a straight-edged board as shown while checking that it is at a body section or deck beam and square to the base line or top reference line. Also, be certain that it touches the deck at the centerline and is equal heights above the deck at the sheer. Take a minimum of the four measurements indicated by the arrows. Measure the camber at several locations, since all the deck beams may not have the same camber. Also inspect the deck framing. Broken beams may mean that the camber has changed.

Figure 6.72. Measuring a deckhouse. To measure a curved forward face, plumb down from the top reference line; mark two points, "X" and "Y" on the deck; mark a straight line between them and lay a straight-edged board across the deck square to the line. Take measurements as shown. Measure the house top profile by taking measurements from the top reference line; measure the camber as described in figure 6.71; measure the deck profile along the house sides down from the straight-edged board.

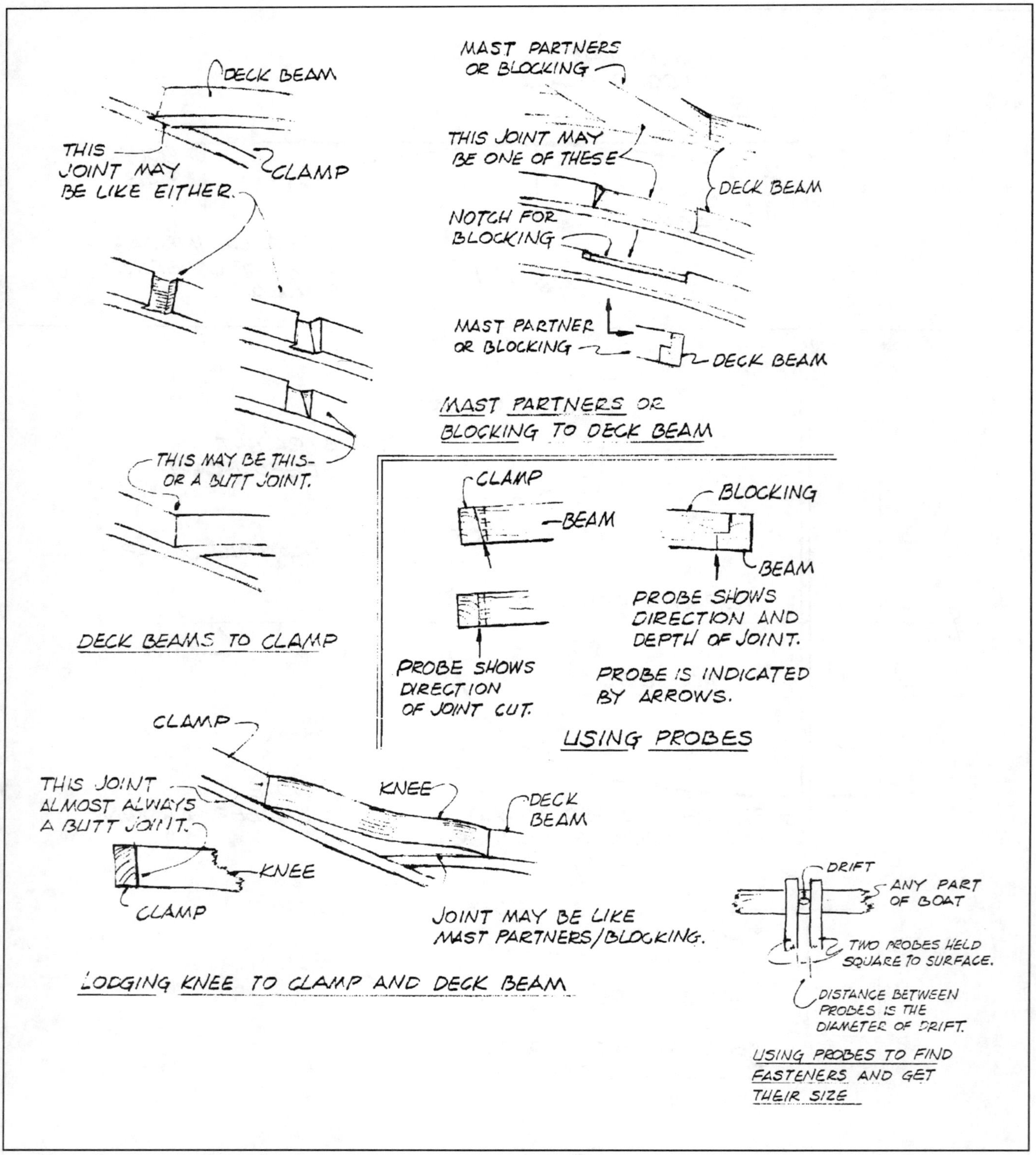

Figure 6.73. When a line where two pieces of wood meet is all that can be seen, use a probe to find the depth and direction of the surfaces of the joint. The probe can be a hacksaw blade, feeler gauge, or, if carefully used, a thin knife blade. Examine cracks, splits, and checks in wood closely. They can be mistaken for joint lines. Try to follow the "joint line" from one side of the wood to the other. Look carefully at grain. It should not be the same either side of joint line. It will be if the line is a fracture.

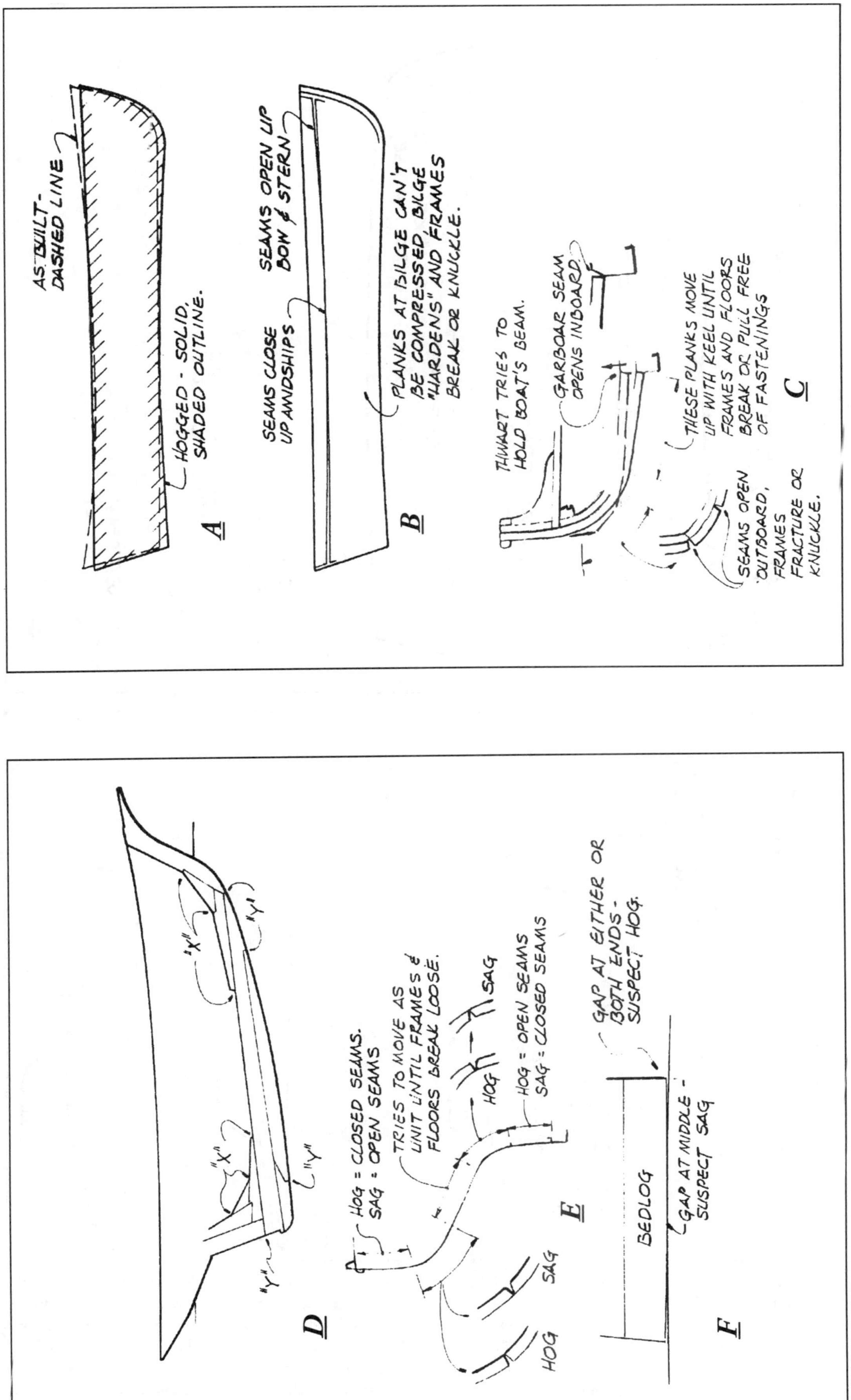

Figure 6.74. Hog and sag. Boats change shape in many ways and for many reasons; not all of them apparent. (A) A boat that was built with a straight or rockered keel is "hogged" when the ends have settled downward. The sheer does not always change to the same degree as the keel. Several things may happen: boat gets longer at sheer and centerline of deck, tries to get narrower, sheer strake tumbles in and breaks the thwart knees, flat spot in the sheer strake when viewed from above (in plan), butts in hull planking near the sheer open up, butts in the deck planking open up, joints in clamp/shelf open up, and the sheer may or may not rise amidships equal to the amount of hog. (B) Planks above the bilge squeeze together and the seams close up. The seams open up bow and stern, and the hood ends of the planks try to pull back from the rabbet line and forward from the transom. (C) Partial sections through a hogged hull. (D) A deep hull with a heavy backbone. In any boat, look for the joints to be opened at "X" when hogged and "Y" when sagged. (E) Section through a deep hull. (F) Examine the seam between the centerboard case bedlog and the top of the keel. If the seam is open at the ends, the boat has probably hogged, if it is open at the mid-length, the boat has probably sagged.

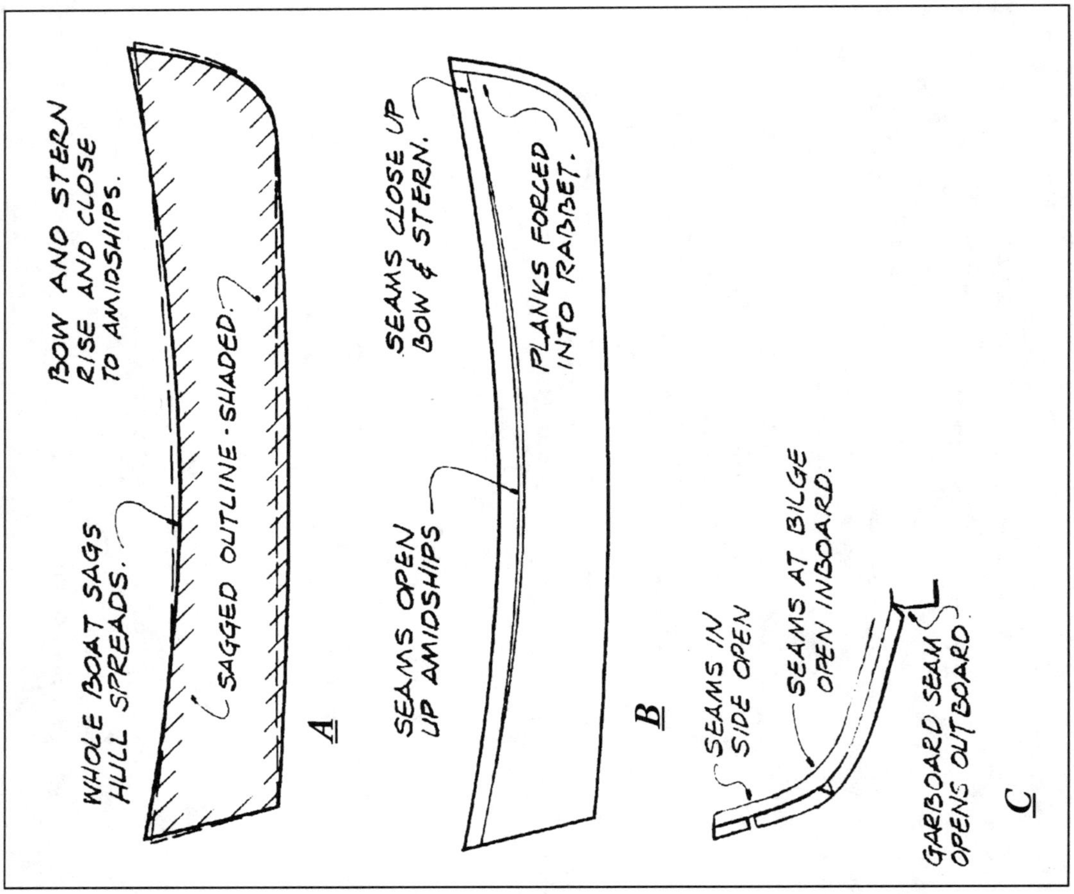

Figure 6.75. Sag. (A) A number of things happen when a boat sags: it may get shorter, or it may get wider—look for signs of tension (stretching) in athwartships members, thwarts fallen inside of risers, thwart knees pulled away from plank/frame or even broken, joints in athwartships deck framing opened up, seams between deck planks of widely varying width near the sheer, gaps between bulkhead and planks, ceiling, and clamp due to planks being forced to a wider beam. (B) Plank seams above the bilge open up amidship and close up bow and stern. Hood ends forced into rabbet and planks try to bulge outward to increase beam. (C) Partial section of a sagged hull.

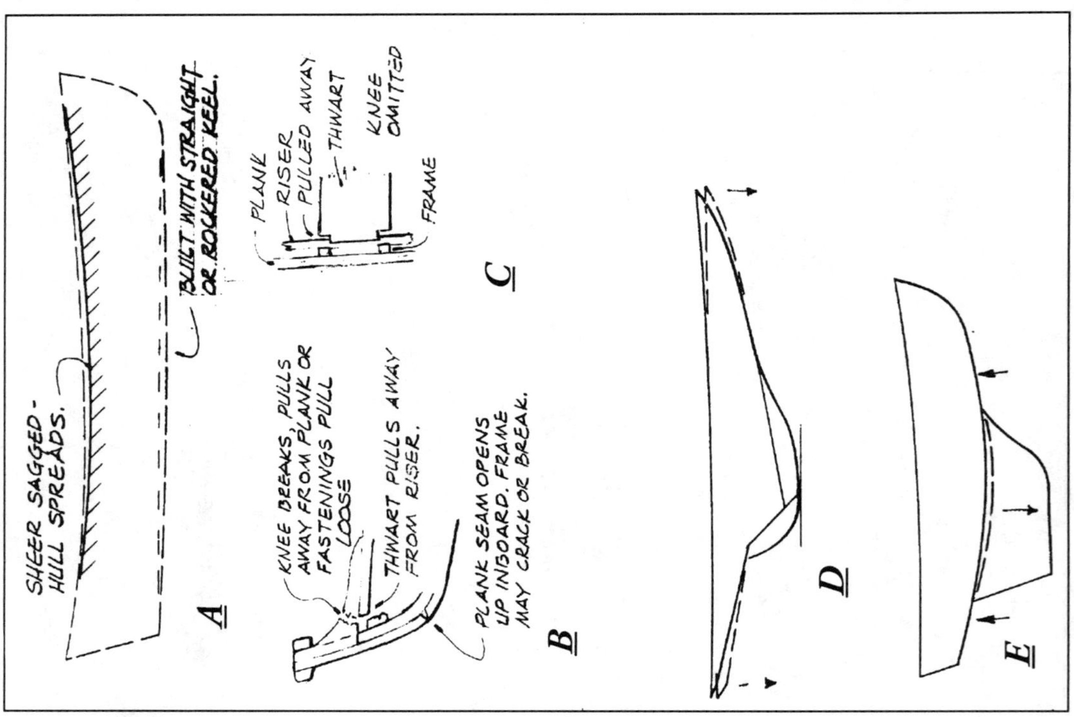

Figure 6.76. Other hull changes. (A) A boat that has been left open to the weather can fill with snow, ice, or debris and spread so that the sheer changes but there is no visible effect on the backbone. (B) Partial section showing the effects of conditions described in A. (C) Plan view of a thwart notched to fit around frames and has pulled away from the frames and riser—a result of the condition described in A. The deck beam-clamp joints in a decked boat do the same thing when the boat spreads open. This is a structural failure. (D) Boats with long overhangs will droop at the ends, if the ends are not supported and develop a hogged sheer. With the ends well supported, but the midsection not, the boat will sag amidships. (E) A fin keel boat supported forward and aft of the fin, will, if built with a light sprung keel, sag at the fin (arrows). In all cases, look at the plank seams. If they are not fair, the direction in which they are not fair is the direction in which the boat has changed its shape.

Chapter 7

Taking Off—
The Measuring Process

Step-by-step guidance
to measuring
a boat's shape

D. W. Dillion

The goal of measuring—taking off—is to record data that can be used to represent accurately the boat's shape, either pictorially or as a replica. This data is taken from the boat in an "as found" or "unreconstructed" form; therefore all irregularities and inconsistencies must be recorded. A complete and accurate take-off will automatically accomplish that.

Steps in taking off the lines are:

1. Select the method
2. Choose which side of the boat to measure
3. Plumb and level the boat
4. Locate the perpendiculars at the ends of the boat
5. Set up the baseline and/or reference line
6. Locate the stations at which the body sections will be taken
7. Take off the body sections
8. Take off the bow profile and detail
9. Take off the stern profile and detail
10. Don't get so wrapped up in the boat and the preceding nine steps that you forget to record what you are measuring and how you are doing it.

Basic geometry, the relationship of right angles and planes, is the foundation for taking off. In taking off a boat's lines, a system of leveled and plumbed planes is used. Because the planes are at right angles to each other, they are easy to set up for taking off and to reconstruct on the drawing board or loft floor. In each of the methods for taking off described in section 7 of this chapter, the measuring equipment is set up to define one plane in reference to the other two planes.

What are these planes? How do they relate to the boat and the taking off process? Imagine standing in a boat that is floating in the water (fig. 7.1). Face forward with one foot on either side of the keel. The centerline plane ABCD will cut you in half from front to back and top to bottom. The base plane EFGH will pass under you and the boat from front to back and left to right as in fig. 7.2. The body section plane IJKL will pass through you from right to left and top to bottom.

The centerline plane ABCD appears in the lines drawing of the boat, and the shapes and dimensions related to it are symmetrical about it unless otherwise noted on the drawings. The base plane EFGH is at a right angle to ABCD. Generally the base plane is located just below the boat or at the design (or painted) waterline. The body section plane IJKL is at a right angle to both the centerline plane and the base plane. The body section plane is moved forward and aft along the boat as needed (fig. 7.3).

Taking off involves taking information from the boat using a system of reference lines that fall in the planes shown in figures 7.1, 7.2, and 7.3. The most easily defined and used line is AB, at the intersection of the centerline plane and the base plane as shown in figure 7.4. This line is the baseline. The centerline is any line in the centerline plane that is also parallel to the base plane.

However, this is theory. In practice, obstructions may prevent setting up a line under the boat that is in the centerline plane. When this situation arises, set up a level line parallel to the centerline plane and use that. Avoid confusion by calling any line that is not in the centerline plane a reference line (fig. 7.4).

Using a single line, either a baseline or a reference line, under the boat is the usual way of setting up to measure. There will be circumstances where two lines will be convenient. For example, a baseline can be set up and used for referencing the vertical location of the measuring equipment, and a reference line used for all lateral (port to starboard) locations (figs. 7.4 and 7.5).

No matter which line you use, you must accurately record the location of the measuring equipment. (See sections 4 and 5 of this chapter for information on setting up baselines and reference lines.)

If visualizing these planes and how they relate to each other is difficult, photocopy fig. 7.6, cut out the three pieces, and assemble as noted on the drawings.

The body section planes are used to take off the cross sectional shape of the boat's hull as described in section 7 of this chapter and then plotted as in chapter 11.

1. Select the Taking Off Method

The team as a group should select the method from section 7 of this chapter. Read the evaluations and study the figures that illustrate each method. Read the tool lists.

Compare the method chosen for the boat, where the boat will be measured, and the team's ability. Then decide which method to use and make photocopies of the appropriate pages of the manual. Finally, practice taking off using the selected method.

2. Choose Which side to Take Off

It is customary to take off one half of a boat; either port or starboard. Therefore, you must decide which side to take off. You must also decide if there is any reason, such as a twisted hull, to take off full body sections.

Try to recognize where the boat's shape has changed from its original, as-built shape, during its lifetime. Stand far enough away so the whole boat can be seen. Is the sheer the same port and starboard? Probably it is not, though the sheer port and starboard should have very similar shapes. Compare the sectional shape of the hull, port and starboard. Look for the same things on both sides—flat spots, curves that change the wrong way, or knuckles. Look at curved shapes from several angles, and move slightly from side to side and up and down as you do. Irregularities are more visible in a foreshortened curve, and as you move your head, your peripheral vision will pick them up.

As you look, talk about what you see. Can you find reasons for shapes that don't look right? Flat spots and knuckles may indicate broken frames or a broken clamp or plank. Perhaps the boat leaned against a wall for a long time and slowly changed its shape without anything breaking. All of these things can contribute to the boat changing its shape. Perhaps the boat was not built fair, straight, or symmetrical. Some of these things will become obvious when you plumb and level the boat.

Start the notetaking now by measuring the scantlings and the general arrangement of the boat. Also start photographing the boat. These steps will help you see things and ask questions about the boat that you might not otherwise.

Work slowly and examine the outside of the boat closely. Shine lights over it; run your hands over it; lie down under it and look at the bottom. Sight along the keel to see if it is straight in both directions, neither hogged nor twisted. Look for broken frames, sprung planks, a hogged keel, seams that have been excessively caulked—especially the garboard seam. Measuring for twist is illustrated in section 7.1; correcting for it, in chapter 11. Hog will be measured while taking off the body sections, and correcting for it is discussed in chapter 11.

Discuss the method you have chosen again. Do all the team members still agree that it is appropriate for the boat, you, and the circumstances? If there is a reason to change or vary the method, now is a good time to consider it.

3. Plumb and Level the Boat

Plumbing the boat is positioning it so the centerline plane splits the boat as nearly as possible into equal halves from bow to stern. Leveling the boat is positioning it forward to aft so that either the bottom of the keel or the painted waterline is parallel to the baseline as described in section 3.2 and illustrated in figs. 7.7 and 7.8.

Plumbing and leveling the boat can be time consuming, even tedious. However, the time taken will keep you from having to deal with out-of-plumb or out-of-level conditions on the drawing board or loft floor. If, after studying the circumstances, you decide that the boat cannot be leveled or plumbed, or both, consider using a declivity board as described in section 8 and illustrated in fig. 7.46. Before you try to use a declivity board, read all of chapter 7 so that you have a good idea of what you are about.

Warning: Make sure the boat is in a stable position. A round- or V-bottomed boat on saw horses must be braced or propped so that it cannot capsize.

3.1. Plumbing the Boat. There are three basic hull profiles: double ended, transom sterned, and counter sterned (transom sterned with an overhang). Each must be dealt with differently.

To plumb a double-ended boat, hang plumb bobs at the stemheads, how and stern, so that the strings split the stem faces or the stembands in half (fig. 7.9). Tie the plumb bob string off to a convenient fitting, jam it in a crack, or tape it in place. Then heel the boat to port or starboard to align the plumb bob strings with the centerlines. Keep a constant check to see how much the plumb bob strings move in relation to the centerlines.

When the strings line up on the center lines for the full height of the stems, the boat is plumbed. Don't expect to be successful plumbing the boat on the first try. More than likely, the strings won't line up because the boat is twisted. It is also possible that the stems were not accurately located when originally fastened to the keel; they could be cocked to port or starboard. Careful examination and measuring should reveal which is the case.

To compensate for twist, take off either full body section at each station or the half section plus the sheer height and half breadth on the opposite side as described in section 7.1, fig. 7.30. What you do depends on whether or not you decide to compensate for the twist, and that decision should be based on whether or not the project warrants the necessary work.

Sometimes it is possible to take the twist out of a very flexible boat. With a person at each end of the boat, place your hands on the sheer, one on the port side and one starboard side; push down with one hand and hold the opposite side from moving with the other hand. Do not use your full body weight to start. Use the least force possible. If you can remove the twist this way, substitute weights for your hands. Be very sure that the only effect is to take out the twist and that no other part of the boat is flexed.

To plumb a transom-sterned boat, begin by marking the centerline on the transom as shown in fig. 7.10 to give a visual reference when sighting the plumb string. Because transoms are rarely symmetrical, some averaging will be necessary when you do that.

Measure across the top of the transom from top inside of plank to top inside of plank and mark the halfway point. Do the same at the bottom of the transom and the sternpost. Don't mark the skeg if there is one, because it may be slightly out of line. A stern post can be used because it is an integral part of the backbone and is less likely to have been forced out of alignment. On the other hand, a skeg may be slightly misaligned, because it was fastened to the keel after the backbone was assembled, or because of heavy use or from outright damage.

Treat the transom of a counter-sterned boat as described above. To aid the plumbing process, mark the centerline on the sternpost. If the rudder can't be removed, mark the centerline of the overhang as close to the rudder as possible and on the sternpost or keel below the rudder (fig. 7.10).

If the boat has a painted waterline, try plumbing the boat in reference to it. Set up a leveled reference line under the boat at midlength and square to the centerline. Heel the boat as necessary to set the waterline at equal heights above the reference line (fig. 7.7). You should still use plumb bobs at the bow and the stern as references for aligning the backbone. There is no guarantee that the painted waterline can be brought 90 degrees to the centerline plane. If it cannot, plumb the boat by using the plumb bobs and centerlines bow and stern.

Do not take the plumb bobs off the boat.

You may wonder why the boat is plumbed in this manner, rather than by using a thwart, for example, to level the boat athwartships. In truth, using the thwart is a practical way of doing the job.

In several instances where it was possible to check the accuracy of the boat's construction against a system of plumbs and levels, it was found that a thwart, for example, was not only a different distance below the sheer port and starboard, it was also not level across the boat. If, instead of plumbing the boat using the centerline plane, the boat was leveled athwartships using the thwart as a reference, the boat may have been inaccurately measured; that is, a twisted boat measured without the twist taken into consideration. By eliminating as many irregularities as possible when setting up the boat, the work on the drawing board and loft floor will be easier. If the boat is set up properly, some of the data taken off will not need correction to make it right.

3.2. Leveling the Boat. Set up a temporary baseline under the boat by putting down two concrete blocks, one forward and one aft of the boat, with a length of nylon string stretched between their tops. Be sure the string does not touch anything in between. Locate the string under the boat by eye so that it is approximately under the center line. The string need not be exactly on the centerline, but it should be situated where it will be possible to measure up plumb from it to either the bottom of the keel or to the painted waterline at the bow and stern. Put a line level on the string exactly halfway between the blocks. Level the string by putting shims or a wedge between the string and the block at the low end.

To level the boat in relationship to the temporary baseline, raise the low end of the boat. Keep making adjustments until the height of the bottom of the keel (at the stern to keel joint forward, and the sternpost to keel joint aft) or the painted waterline above the temporary baseline is equal at the bow and stern (fig. 7.8). If the boat is on saw horses, use wedges or wedges and blocking to adjust the height. Wedge flat-bottomed boats port and starboard so that the wedges support the chines. If the boat is in a cradle, wedge and block the cradle under the poppets where they support the boat. Make smaller changes than seem to be needed. If the bow is 1/2 inch higher than the stern, don't raise the stern 1/2 inch. Raise it 3/8 inches instead. This is because the boat is not supported at the ends, but at points somewhat in from the ends, and that is where it pivots (fig. 7.11).

Plumbing and leveling is a balancing act between keeping the plumb strings centered and the keel or waterline level. Each time you move the boat to level it or plumb it, the other component will change. Don't be discouraged. Each change will be smaller than the preceding one. As long as you keep thinking about what you are doing, you will achieve the desired results. Do not remove the temporary baseline when you have the boat plumbed and leveled. You may still need it.

4. Locate the Perpendiculars

The forward perpendiculars (F.P.) at the bow and the aft perpendicular (A.P.) at the stern are both perpendicular to the base plane. They are used as references for the horizontal dimensions that define the profile shape of the bow and stern of the boat.

When taking off, the perpendiculars will be located where convenient, but to make measuring the ends of the boat possible, the perpendiculars must be located in the centerline plane. At the bow, the perpendicular may be at the stem face, against the stem band, a mooring eye, or a billet head. At the stern it may be at the top of the transom or against the rudder hardware. The perpendiculars used for taking off do not necessarily relate to those used by the designer or builder.

4.1. On a Floor. Lower the plumb bobs so that the points are within 1/8 inch to 1/4 inch of the floor. Start at the bow. Get right down close to the floor, and sight aft to the plumb bob hanging from the stern. Make a thin chalk mark an inch or so long under the bow plumb bob point and aimed at the aft plumb bob. Getting close to the floor allows your eye automatically to align three points: the chalk mark and the two plumb bob points. Then move to the stern and while looking forward, repeat the process, making a chalk mark under the point of the stern plumb bob. These two chalk marks are on the centerline of the boat at the ends. Move around to one side and mark across the centerline exactly under the plumb bob point. Go forward and do the same thing at the bow. Each cross mark is at the intersection of a perpendicular and the centerline. The plumb bob strings are therefore the perpendiculars.

4.2. On a Floor or on Ground. Instead of plumb bobs, use a carpenter's level with a plumbing vial and two straight-edged boards (staffs) long enough to reach from the floor to well above the boat. Put the staffs in place at the bow and the stern with the straight edges against the boat and lined up on the center of the stem and the centerline mark on the transom. Clamp the staffs to the boat at the top and to concrete blocks at the bottom so that they are approximately plumb. Use the carpenter's level to plumb very carefully the staffs in both directions. When they are plumb, the edges of the staffs against the boat are the perpendiculars.

5. Set Up the Baseline

There are several ways to set up a baseline or reference line, to set up the perpendiculars, and then, to measure between the perpendiculars.

5.1. On a Floor. If the perpendiculars are marked on the floor, it is a simple matter to set up the baseline. Put staffs in place at the bow and the stern, with the straight edges against the boat and lined up with the center of the stem and the centerline mark at the top of the transom and with the marks for the perpendiculars on the floor. Clamp the staffs to the boat at the top. Either toe-nail the staffs to the floor or clamp them to concrete blocks. When the staffs are properly set up, trace on the floor around them and the concrete blocks with chalk, so that you can tell later on if the staffs have been moved.

Tie a length of nylon string around one staff. Stretch it tight and tie it around the other staff. Make sure that the string is taut enough to "twang" when plucked. Don't tie the string more than an inch or two above the concrete blocks. Put a line level on the string exactly halfway between the perpendiculars. Slide one end of the string up or down along the staff to level the string (fig. 7.12). Mark the location of the baseline on the staff with a pencil line or a push pin alongside the string so that you can tell later if the baseline has been moved up or down.

Remove the line level from the baseline or push it to one end, so that it is against a staff or concrete block. If the level is at the end of the baseline, it won't pull the baseline down and cause an unaccountable vertical measuring error. Trick: Use a yardstick or a folding rule to push the line level along the baseline to avoid crawling under the boat.

A word of caution: The string must be properly lined up when it is tied around the staffs so it is in the centerline plane (fig. 7.13).

If a level or leveling board is used, set the baseline close to the floor or ground. The top of the level or leveling board can be at the baseline. If you use a measuring frame, be sure that the cross arm will fit either below the baseline or between it and the boat.

5.2. On Ground. Raise the plumb bobs so the points are about 8 inches above the ground. Place a concrete block about a foot forward of the F.P. and another about a foot abaft the A.P. Tie a length of nylon string between the blocks, from top to top, so the string is above the ground, not touching anything in between, approximately under the plumb bob points, and taut.

Lower the plumb bobs until the points are almost at the same height as the string. Move the blocks to one side or the other until the string is exactly under both points. When you are satisfied that the string is under the points, get down and take a good, close look from both ends of the boat. You are halfway to having a baseline. What you have is a centerline.

Put a line level on the string exactly halfway between the two concrete blocks. Level the string by wedging or shimming the low end of the string up from the top of the block as necessary. Raise and lower the plumb bobs to keep the points close to the string but not touching it. When the string is level and directly under the plumb bob points, you have set a baseline. Don't forget to push the line level to one end of the string or to remove it from the string.

Learn to check visually that the two plumb strings and the baseline are all in the centerline plane. Stand forward of the boat. Move your body around until the plumb bob string lines up on the stem the way it did when the boat was plumbed. When the plumb bob string is lined up on the stem, the baseline and the other plumb bob string will disappear behind it. Get down close to the ground so the bottom of the keel is in your line of sight. If the boat is not twisted, the plumb lines will split the keel right down the centerline. A centerline on the floor will also disappear behind the plumb bob string.

This visual checking will come in handy after you have left the boat unattended for a while. It is possible that the boat or one of the references was moved while you were away from it. When you come back to the boat, visually check to see that the perpendiculars, baseline, centerline, and reference lines all are where they should be and that they align as described above. If they are not, put the boat and any line that was moved back where it was. Sometimes you will have to start over again from the very beginning.

Having the boat set up over a floor and using staffs at the perpendiculars is a definite advantage in this situation. With the F.P. and A.P. marked on the floor and the baseline marked on the staffs, it is a simple matter to put things back where they were. Realign the staffs exactly with the F.P. and A.P. marks and in their original places. Plumb the staffs; replace any wedges, blocking, and bracing; check the baseline for location and level; then go back to work.

5.3. With Staffs in Place. Tie a length of nylon string around one staff and stretch it tight. Then tie the string around the other staff, and level the string. That does it.

5.4. Setting Up Reference Lines. When there is not a clear path under the boat to set up a baseline, set up a reference line off to one side as shown in figs. 7.4 and 7.5.

If you are working on a reasonably level floor, locate the perpendiculars as described in section 4. Then, using F.P. and A.P. as centers, swing equal arcs of a convenient radius as shown in fig. 7.14. Then, with a chalk line, snap a line on the floor so that is just touches both arcs. This line, the reference line, will be parallel to the centerline of the boat. Using a carpenter's framing square, and referring to fig. 7.15, square from the reference line to the perpendiculars and mark their location on the reference line. The distance between the marks is the length between perpendiculars but not necessarily the length of the boat.

Unless the floor is straight and level in all directions, this chalked reference line can only be used for locating the measuring equipment athwartships. If, however, the floor is straight and level in all directions, use the chalked line for athwartships and vertical locating of the measuring equipment—a reference line horizontally and a baseline vertically.

When a chalk line cannot be used for a baseline or a reference line, set up a string reference line. Hang plumb bobs at the centerline, bow and stern so that the points almost touch the ground. Set a concrete block down at each end of the boat—a foot or so out beyond the bow or stern—and stretch a length of nylon string (the reference line) between the tops of the blocks. Using a ruler, and with the plumb bob strings as center points, move the blocks about until the string is equidistant from the plumb bob strings. The trick is to hold the ruler so that it barely touches the plumb bob string, swinging it in short arcs over the reference line while moving the blocks (and string) to one side or the other, until the distances from the plumb strings to the reference line are the same at F.P. and A.P. (fig. 7.16). When the reference line is equidistant from both plumb bob strings, level it. Be careful when measuring from the plumb bob strings not to push them out of plumb. Using a carpenter's framing square as a guide, square from F.P. and A.P. and mark their locations on the reference line (fig. 7.15).

The reference line is used in place of a baseline to locate the measuring equipment athwartships, forward to aft, and vertically, with the remaining control for locating the measuring equipment being used to set it up plumb.

To help you keep oriented while you work, put a centerline or reference line on the floor under the boat and mark the station locations, or any reminder notations along it. The centerline or reference line can be either a chalk line or a tightly stretched length of nylon string. To mark a centerline on the floor in chalk, stretch a chalk line tightly along the floor so it passes through the marks for F.P. and A.P., then snap the line. Be sure that the line of chalk dust is straight. A reference line is marked the same way but offset from the centerline and parallel to it as described in section 5.4.

Alternatively, stretch a piece of nylon string along the floor between the marks for F.P. and A.P. Hold the string in place with weights, such as concrete blocks. Make sure that the string lies straight; then put chalk marks along both sides of it roughly where you think the stations will be. Each time you set up to measure a body section, check to see that the string is between the chalk marks and is still straight.

A centerline or reference line on the floor in addition to a baseline is optional. Either one can be used as a reference for all horizontal measurements if, and only if, the line is straight horizontally. It does not matter that it is not straight vertically. You will not use it for vertical reference—only horizontal.

A reference line above the boat, in addition to a baseline, as shown in fig. 7.12, is useful when the boat is decked or has a great deal of interior detail to locate. Measure up on the staffs and mark off equal distances above the baseline at both F.P. and A.P. Tie a nylon string between the staffs at the marks. Since the baseline is level, the top reference line will be level because it is parallel to the baseline. Vertical (or plumb) measurements can be taken down into the boat for height locations. A tape measure stretched alongside the reference line is used to measure fore and aft locations as described in chapter 6.

5.5. Baseline Above the Boat After Plumbing and Leveling. When taking off the lines from the inside of the planking, rig the baseline above the boat (fig. 7.12). Set up, plumb, and level the boat. Place straight-edged staffs at each end of the boat, plumb them to establish the perpendiculars, and anchor them in place. Stretch a string between the staffs above the boat, and level the string using a line level. Then mark the string's location on the staffs.

A baseline set up above the boat can be used when one can't be rigged below the boat because of obstructions. Be very careful when setting this line up, because it will be necessary to locate the measuring equipment by measuring out and down around the boat, and errors can sneak into the measurements taken.

5.6. Baseline Above the Boat Used for Plumbing and Leveling. For measuring the inside of open boats, or where the thwarts can be removed, or for measuring boats that are upside down, setting a baseline above the boat is a good way to go about the job. For this procedure to work, the boat's backbone must be accessible. If access to the backbone is difficult, look for an opening in the deck, such as a hatch or mast partners. If you can't get to the backbone inside the boat, this procedure won't work.

Start by setting the boat up plumb and level by eye. Place staffs at the ends of the boat. Clamp the staffs to concrete blocks, and plumb them fore and aft and athwartships. Brace the upper ends of the staffs to prevent movement. The staffs do not have to touch the boat. Tie and level a string between the staffs about 15 inches above the boat for the baseline.

To set up the boat, plumb it by using four points on the backbone for reference:

1. the center of the stem face at the sheer
2. the center of the backbone inside the boat as near the forward end of the keel as possible
3. the same as near the aft end as possible
4. at the centerline at the top of the transom (or center of the stem face in a double-ender).

To plumb the boat, hold the plumb bob string so that the string touches the baseline. Lower the plumb bob until it touches the top of the keel. Move the boat to port or starboard until the plumb bob point lines up with the center of the keel. Do this at both ends of the boat. When the centerline of the keel is lined up under the baseline, repeat the procedure at the stem head and the top of the transom or sternpost. Start by lining up the top of the keel first, because there will be fewer moves involved by the time you are done.

To level the boat, set the top of the keel, the bottom of the keel, or the painted waterline the same distance below the baseline; one or another, not all three. Use the keel for leveling only if it has no drag and you are certain that it is the same thickness at both ends.

Using the top of the keel as a control is the most straightforward way to level the boat. Raise one end of the boat so that the same two points on the keel used for plumbing are equidistant below the baseline. Don't forget that the boat must also be plumb. Check the points at the stemhead and the top of the transom to be sure they are still under the baseline. When everything is lined up, the boat is plumb and level.

To use the painted waterline or the bottom of the keel, plumb the boat as described above. At the bow, place the top of one end of a carpenter's level on the painted waterline and measure the distance up to the baseline. Do the same at the stern. Raise the low end of the boat until the distances are equal. Check to be sure that the boat is still plumb.

With a boat that is fairly easy to move, set the staffs for the perpendiculars first, and move the boat in between them.

5.7. Measure Between the Perpendiculars. To determine the distance between the perpendiculars, measure between the cross marks on the floor or the inboard edges of the staffs. Don't be concerned if the floor is not level.

Consider this for example, if you measure accurately the level distance between the plumb bob strings, and find it to be 30'-0", the distance along a floor that is 4 inches out of level will be 30'-0.02" or an error of less than 1/32 inch, which is certainly acceptable.

To measure along a string baseline, fold pieces of masking tape over the baseline directly below the plumb bob points and mark where the plumb bob points are. Measure between the marks to get the length between the perpendiculars. Marking F.P., A.P., and the stations on the string baseline and reference line accurately is a bit tricky. Try this: fold pieces of masking tape about 3 inches long over the string where the framing square, or plumb bob strings, cross the tape. Then mark the masking tape. Mark the masking tape after putting it on the baseline and avoid errors due to tape measure and string moving about. Squeezing the tape measure and string together also helps.

To measure from F.P. to A.P., put the 0-inch mark of a tape measure at F.P. and, using a very small clamp, clamp the tape measure to the string, go aft and read the distance to A.P. Be aware that if the perpendiculars are not at the actual ends of the boat, for example, if they are at the stem face and top of the transom, the length between the perpendiculars is not be the length of the boat.

6. Locate the Stations

Taking off the body sections constitutes the bulk of the taking off the lines and gives the greatest quantity of information about the boat's shape. To help you select the body section locations, think about what you found out about the boat's shape while you were deciding which side to measure.

Hint: Study the lines plans of boats that have similar shapes and proportions to the one being measured to get a sense of the relationship between shape, proportion, and station spacing.

A good way to indicate the body section locations is to walk alongside the boat and mark the hull with chalk or tape where you want to measure the sections. Studying the plank lines will help you see the changes in shape. Look for quick shape changes. Space body sections close together at these places no matter the length of the boat. If you question whether to take a body section at any location, take it there. It is better to take too many sections than too few. Do not be concerned about exact or equal spacing. The most important thing is to concentrate on the shape of the boat.

You may find that one of the body section locations falls at an obstruction, such as a saw horse. Don't be foiled. Simply take a section either just forward or aft, or both, of the obstruction.

On most boats you can space the stations fairly widely apart near the middle of the waterline length—say 24 inches to 30 inches apart—because in the middle 50 percent of the boat, the shape does not change very quickly. The 25 percent of the hull at the ends is where the shapes are difficult to develop and require closer station spacing.

6.1. Symmetrical (Double-ended) Boats. A boat that appears to be symmetrical at the ends—having the same shape forward and aft—should have section measurements taken either to prove or to disprove the apparent symmetry. Many boats called "double-ended" are obviously not symmetrical. They should be treated as transom-sterned craft as described in section 6.2.

Locate the body sections for a symmetrical boat with the first section at the boat's mid-length. Use that section as a control for locating the others. Divide the length of the boat in half. Mark that distance off on the baseline or the reference line. The remaining stations should be equally spaced forward and aft. The midships station is "0." Those forward of "0" are "A," "B," "C," etc.; those aft, "1," "2," "3," etc.

6.2. Transom-Sterned Boats. It is customary that the stations for transom-sterned boats begin at the forward perpendicular with 0'-0". You should still mark the hull where you think the sections should be taken. Average the distances to get an even spacing. Then mark off the station locations along the baseline or reference line beginning at the forward perpendicular. Let the odd dimension fall at the stern. Begin the station numbering with "0" at F.P. and assign numbers in sequence as "1," "2," "3," etc.

6.3. Other Cases. Body sections should be measured at specific locations on some boats. For example, power boats with stepped hulls take sections at the steps, and use them as controls for locating the other sections.

If there is a possibility that the boat might be hydrostatically analyzed later, divide the measured waterline length into ten equal sections. Take additional sections as seems necessary.

6.4. Marking the Body Sections. To mark the station locations on a string baseline or reference line, hold the 0-inch end of the tape measure at the F.P. Fold pieces of tape over the string and mark them as described in section 5.7.

Mark the stations on a chalked line on the floor in a similar fashion. Set the end of a tape measure at the F.P., and mark the station locations off along the chalked line.

6.5. The Station Reference. To facilitate locating the body section measuring equipment, establish a port-to-starboard reference under the boat. If you are working on a floor and have chalked the base or reference line on the floor, use a carpenter's framing square as a guide and mark one or all of the station lines on the floor under and across the boat.

In other circumstances, lay down a straight edged board at a station near midships. Plumb down from the selected station mark on the baseline or reference line to the board to align it with the station mark. Square the board to the baseline or reference line using a framing square. Hold the arm against the baseline or reference line—very gently—and let the blade touch the board. Do this port and starboard. Check with the plumb bob to see that the edge of the board is still correctly aligned. Mark the station edge of the board clearly to avoid using the wrong edge when locating the other stations. Weight it or nail it down so that it can not be moved accidentally. Leave the board in place until all of the body sections are measured and plotted (fig. 7.17).

To locate the body section measuring equipment, measure forward or aft from marked edge of the board the distance to the section to be measured and set up the measuring equipment (fig. 7.17).

6.6. Locating the Stations with the Baseline above the Boat. Mark the desired station locations on the hull as described in section 6.1. Hold the 0-inch end of a tape measure at the face of the stem. Mark off a distance slightly less than half the length of the boat aft along the sheer, port and starboard—arcs A and A'.

Then, working forward from the stern and with 0-inch mark of a tape measure at the center of the transom or the sternpost, repeat the process—arcs B and B' (fig. 7.18).

Divide the distance between the marks, C and D port, and C' and D' starboard, locating points E and F. A line drawn between E and F is square to the centerline of the boat at its mid-length (fig. 7.18).

To locate the station reference below the boat (described in section 6.5), plumb down from points E and F and draw the line or lay down a board.

Marks along the sheer showing all of the station locations are not necessary for setting up the measuring equipment, but they are handy to keep yourself located as you work.

Make up a note sheet as in fig. 7.19. Show all of the information indicated.

7. Measuring the Body Sections

Taking off the body sections provides the greatest quantity of information about the boat's shape. There are only two ways that information can be taken off: by taking measurements or by making patterns. Controls must be established so that either form of data can be properly referenced to the baseline and centerline. If you are going to draw the boat, take measurements. If you are going to the loft floor directly from the boat, either measurements or patterns will do the job.

The general assumption is that since waterlines and buttocks appear in lines plans, they must be used when taking off the body sections. That is not necessarily the case. There are two reasons not to use waterlines and buttocks when taking off. First, it can be difficult for the inexperienced to visualize where the waterlines and buttocks should cut through the boat so the resulting take-off accurately reflects the shape of the boat. Secondly, it may be possible to locate the measuring equipment in a constant relationship to either the boat or the baseline, so that vertical corrections must be recorded and then used when plotting the measurements much as described in section 7.1 and illustrated in figure 7.20. Keeping track of the vertical correction to locate the measuring equipment is confusing in itself without adding a correction for the waterlines.

The greatest concern, then, is to take off data that will accurately reproduce the section when plotted. The solution is simple. Take the measurements at the plank seams. For a lapstrake boat, take measurements at every plank; do the same for a small (up to 20 feet) carvel-planked boat. Measure at every seam, or every second or third seam for a larger boat depending on the shape being measured. Then, in all cases, measure the plank widths. Set the 0-inch end of a measuring tape at the rabbet or the sheer, and read off the locations of all the plank seams. Determining plank widths is then a matter of subtracting the smaller measurement from the larger (fig. 7.21). This process will give data not only for section shape but also for the plank widths. For a lapstrake boat, knowing the plank widths is absolutely vital to achieve an accurate reproduction of the boat because of the visual effect of the plank lines. For a carvel boat, knowing how the boat was planked (lined off) will help in the building of a reproduction. In either case, taking off the plank lines will achieve a historically accurate job and will make the builder's job easier. When a plank seam is not accessible for measuring, try to measure the plank width at the location or very near it.

With the boat plumbed and leveled, the baseline or reference line in place, and the stations located, you are ready to start taking off the body sections.

7.1. Method 1—Level and Plumb Bob. This method may be the easiest to understand as it uses only plumb and level measurements. The one complication is the need to take a correction measurement at each body section. The correction measurement must be recorded in the notes and used when plotting the body sections. Without the correction measurement, it will be impossible to plot the body sections in proper relationship to each other or to the baseline. Do not assume that the correction will be the same for all body sections. Several situations are illustrated in fig. 7.20.

Tools needed for taking off include a note sheet for each body section, plumb bob, carpenter's level with yardstick or tape measure attached along the edge, leveling board, folding ruler, blocking and wedges, weights, chalk, scale, triangles, and pencil and eraser.

Figure 7.22 illustrates the form for the sheets and the measurements to be taken off. Show all measurements, the plot of the measurements, and any details on the outside of the hull that occur at or near the station. When plotting the measurements as they are taken off, use two sheets if necessary to permit the plotting and recording of measurements to be done simultaneously.

Figure 7.23 shows several points to look for when measuring a flat-bottomed boat. Figure 7.24 shows where to take measurements on a lapstrake-planked boat and where to take off to establish the lap and the lay of the plank on the hull. Figure 7.25 shows how to compensate for a molded sheer strake when taking the half breadth of the sheer. Figure 7.26 shows planking and keel faired and not faired. These conditions must be observed and noted in order to develop the hull section accurately. Figure 7.27 shows the points to measure on a straight-sectioned V-bottom hull.

At the midship station, square off from the centerline or baseline to locate the station reference line.

Position the level or leveling board so that one edge aligns with the station reference, as shown in fig. 7.28. Block and wedge the level or leveling board so that it clears obstructions. Use the wedge to bring it level.

Begin measuring at either the sheer or the keel. Measure the half breadth of the sheer. Hold the plumb bob string at the sheer (fig. 7.29). Move it forward or aft so that the point aligns with the station line. The figure under the point is the half breadth of the sheer if the 0-inch mark on the rule is aligned with the baseline and if the baseline is also the centerline. Measure the height from the top of the level to where the plumb bob string touches the sheer.

Take the remaining measurements at the seams in similar fashion. Place the plumb bob string on a plank seam. Move it forward or aft along the seam until the point aligns with the station line. Read off the half breadth dimension. Mark the hull at the seam or hold the plumb string in place. Measure up from the half breadth measurement to the mark on the plumb string to get the seam height. Record the measurements in the appropriate places on the form.

As a backup, take as a direct measurement the height from the baseline to the bottom of the keel. Also measure the full siding of the keel at the bottom and at the rabbet line.

If the boat is noticeably twisted, take all of the measurements shown in fig. 7.30.

7.2. Method 2—Leveling Board and Arcs—Triangulation. In addition to the tools used for method 1, you will need a leveling board and, for plotting the notes, a compasses. Organize the note pages as shown in fig. 7.31. Show all measurements, the plot of the measurements, and any detail on the outside of the hull at or near the station. Include a sketch of the leveling board and show how it is marked.

Proceed as in method 1. Establish the station reference at the midship station; then set up the leveling board so it is aligned with the station line and leveled. Plumb up from the station line and make plumb-up marks at the bottom of the keel, the rabbet line, several plank seams, and the sheer to establish proper fore and aft alignment for taking off the body section. The notetaker should be positioned so that he or she can see all of these marks while the measuring is being done. By sighting these marks, the station line, and the end of the ruler as the measurer works, the notetaker can tell the measurer to move the end of the ruler forward or aft as necessary.

Select two marks on the leveling board, and measure from them in turn to the rabbet, seams, and sheer. Take care that the marks selected will give good arc crossings, as shown in fig. 7.31. To save a lot of moving around, take as many measurements from each mark as possible. For example, take the distance from the 12-inch mark to the rabbet and seams 1, 2, and 3; then from the 24-inch mark to seams 1, 2, 3, and 4. Reading the measurement is easier if the 0-inch end of the ruler is held at the plank seam. Then the measurement is read where the ruler crosses the top of the leveling board. Also measure the distance from the baseline to the bottom of the keel, the full breadth of the bottom of the keel, and the full siding of the keel at the rabbet line.

Figure 7.32 shows the position of the leveling board used to take off a boat that is upside down.

7.3. Method 3—Vertical Staff—Level Measurements. The tools needed are those used for method 1, plus a staff long enough to reach from the floor to well above the sheer, clamps, strap hinges, boards for bracing, string, masking tape, and a line level. Follow fig. 7.33.

Establish the station reference and the plumb-up marks on the hull. Place the staff at the station line as shown in fig. 7.33 and use a carpenter's level to plumb the staff fore and aft and athwartships. When the staff is plumbed, place a concrete block alongside it and clamp them together. Then brace the upper end of the staff.

Use either a carpenter's level or a length of string and a line level across from the station on the baseline to the staff. Mark the baseline height on the staff "0". Be very careful to use the same surface of the level for marking the staff that is against the boat. Measure the horizontal distance from the baseline station mark to the 0-inch mark on the staff.

Level across from the bottom of the keel, mark the staff, and record the height above the 0-inch mark. Also measure the distance from the baseline up to the bottom of the keel. Then measure the distance from the staff to the port and starboard sides of the keel. If the keel is tapered in section—wider at the rabbet line than at the bottom—use calipers to measure the full siding of the rabbet.

Level across from the rabbet line to the staff. Measure the height of the level above the 0-inch mark and the distance from the rabbet line to the staff. Do this at each plank seam and the sheer to complete taking off the section.

Hint: Avoid making a clutter of pencil marks on the staff. Put a strip of masking tape along the working edge of the staff, and mark the heights on the tape. When the station take-off is complete and the measurements are plotted, remove the tape. Use a clean piece of tape for each station.

7.4. Method 4—Vertical Staff—Triangulation. The tools for method 4 are the same as for method 3 except that the staff must be marked in 6-inch or 12-inch increments or have a ruler or tape measure fastened to it.

Figure 7.34 shows the vertical staff in place, the method of measuring, and the form for recording the measurements. Make up a note sheet similar to that shown for each body section.

Locate the station reference, put the plumb-up marks on the hull, and erect the staff in place.

Level across to the staff from the baseline, and mark the staff "0". Then mark the staff above and below the 0-inch mark in 6-inch or 12-inch increments. Show the staff with the marks on it in the notes. Since the baseline height will not always be the same distance from the bottom of the staff, put a strip of masking tape along the inboard edge of the station face of the staff to mark on. Replace the tape at each station.

Measure from the baseline to the 0-inch mark on the staff, and record that distance. Then measure from the port and starboard edges of the keel to the staff and the height of the bottom of the keel above the baseline. Use calipers to measure the full siding of the keel at the rabbet line.

Take the measurements as described for method 2 and as shown in fig. 7.34. As an example, you might use a mark 12 inches below the baseline and another 18 inches above it to measure to the rabbet line and seams 1, 2, and 3.

The preceding paragraphs describe the basic methods of digital measuring. There are variations and modifications made to adapt the method to a particular hull section or field condition. A major consideration is to keep the reference edges of the measuring equipment as close to the surface of the hull as possible. The following variations require the same reference lines and planes as used in methods 1, 2, 3, and 4.

7.5. Method 5—Vertical Staff with Horizontal Arm. The same tools are needed as for method 3, plus one or two straight-edged boards longer than the full beam of the boat for the horizontal arms and blocking and wedges to support the arms. The form for recording plumb and level measurements is shown in fig. 7.35.

Establish the station reference, put the plumb-up marks on the hull, and set the vertical staff in place as described in method 3. Clamp one end of the horizontal arm to the staff. Block or wedge the other end to level the arm. Be sure that the staff and the arm are aligned with the station line and in the body plane.

Plumb and level measurements from the horizontal arms and staff must be taken square to the edges of the staff and arms. The quickest way to do this is to use a carpenter's level. To measure athwartships at the deck level, top of the house, and hatches, place another horizontal arm across the top of the boat. Clamp it to the staff, level it, and position it fore and aft so that it is directly above the lower horizontal arm. Be very sure that the straight edges of the arms are the ones leveled, that the distance between those edges is the same at the port and starboard ends, and that both edges are lined up over the station line (fig. 7.36).

Measure the height from the top of the lower horizontal arm to the baseline and to the bottom of the keel, from the baseline to the staff, and the horizontal distance from the baseline to the rabbet line. Plumb down from the seam to the arm and measure from the plumb to the baseline and then the height of the rabbet line above the top of the arm. Follow this sequence right up the curve of the hull to the sheer, marking the locations of seams on the arm or the staff, then measuring from the arm or staff, and recording the distance to the seam.

All measurements from the upper horizontal arm are taken as distances from the staff and below the straight edge of the arm. Be sure to record which edge that is.

7.6. Method 6—Vertical Staff with Horizontal Arm and Diagonal. Tools needed for method 6 are the same as for method 5, plus another straight board 4 to 5 feet long for the diagonal and a small piece of 1/4-inch plywood or hardboard to help in clamping the diagonal in place. The diagonal member is used to shorten the distance from the equipment to the point being measured on the hull. The form of the note page for recording the measurements in this method is shown in fig. 7.37. Prepare a separate page for each body section.

Locate and mark the station reference line and set the staff in place as described in method 3. Then clamp one end of the horizontal arm to the staff. Block or wedge the other end to level the arm. Be sure that the staff and the arm are aligned with the station line as shown in fig. 7.36. Plumb up from the station line, and mark the bottom of the keel, the rabbet line, and the seams to locate the body section plane on the hull.

Because of the arrangement of the vertical staff, the horizontal arm, and the diagonal, all measurements must be taken square to the staff, arm, and diagonal. Do not measure the same point twice, from both the horizontal arm and the diagonal. Not only is that extra work, but minor inaccuracies of measurement can give two slightly different locations for the same point, leading to confusion when plotting the notes.

Use a combination square to take measurements from the diagonal or as a guide to keep a ruler square to the edge of the diagonal. The end of the diagonal at the top of the horizontal arm should be marked "0". To take the measurement, place the 0-inch end of the ruler on the plank seam, and hold it there. Swing the ruler until it is square to the diagonal, and record the distance from the seam to the diagonal. Then record the distance measured along the diagonal from "0" on the horizontal arm. If appropriate, use an upper horizontal arm as described in method 5.

7.7. Method 7—Sloped Staff. Tools needed for method 7 are those described for methods 3 and 4 except that extra-long bracing may be needed to support the staff. Mark the staff with index marks spaced in 6-inch or 12-inch increments for taking triangulating measurements, or fix a tape measure to it for taking square measurements. The form for square measurements is shown in fig. 7.38; for triangulation, in fig. 7.39.

Locate the station reference and the plumb-up marks on the hull.

There are two ways to position the staff:

1. Set the staff in place with the bottom end on the station line. Align the upper end with the station line using a plumb bob suspended from it. Brace the ends of the staff to keep it firmly located.

2. If you choose to use a vertical brace, line it up on the station line, clamp it and the sloped staff together, line the staff up on the station line, and plumb them fore and aft. Brace and clamp everything in place so nothing will move around while you work.

7.8. Method 8—Outside of Plank Using a Joggle Stick and a Scrive Board. Two joggle sticks are shown in fig. 7.40. They are accurate and very simple to use, because, when traced, they establish direction and distance. Make at least two identical sticks from wood, metal, or plastic. Also make a tracing of the stick as a safeguard against loss or damage.

In addition to the joggle stick, you will need paper strips 8 to 12 inches wide for pattern material and a flat, straight backing (scrive) board 8 to 12 inches wide to tape the paper to. The shape of the scrive board will depend on the shape of the hull section being taken off.

Figure 7.41 illustrates several things: In (A.), taking measurements at the sheer, indicated by the arrow, or any other point above bilge requires an excessively long level or leveling board. The example at (B.), the vertical staff, can prove difficult to use accurately because of the scant angles when measuring the bottom as indicated by the arrow. Using a vertical staff and horizontal arm as in (C.) will make the measuring job easier no matter what method you use because measurements can be taken vertically and horizontally (arrows). The sloped staff, as in (D.), can be set up close to the hull and is convenient for measuring hulls with hollow. The vertical staff, shown again at (E.), can be useful when taking off hull sections with a tall thin shape, but can prove difficult when it is too far away from the point measured (arrow).

In fig. 7.41, (C.) and (D.) are the best choices for use with a joggle stick because the scrive boards can be set close to the boat. Figure 7.43 shows the reference marks that should be put on the patterns. Figures 7.42 and 7.44 show the most desirable solutions for use with the joggle stick.

With the baseline in place, the station reference marked, and the plumb-up marks on the hull, set the scrive board, with the pattern paper taped to it, in place and line it up with the station line and plumb it forward and aft. The scrive board does not have to be leveled or plumbed athwartships. To do so is a matter of convenience only.

Record the boat's name, the station number, and the distance to F. P. on the pattern paper. Draw a level line at the baseline height and a plumb line on the pattern paper. Draw a level line at the baseline height and a plumb line on the pattern paper. Record the distance to the plumb line from the baseline. If a level line cannot be drawn at the baseline height, record the vertical distance from one to the other. The level and plumb lines will be used to locate the pattern paper on the loft floor when plotting the section (fig. 7.42).

Other ways to reference the pattern to the baseline are shown in fig. 7.43. When locating level and plumb lines from the baseline, use whole inch distances, not fractions.

Do not assume that because the joggle stick is a very accurate tool that any part of the take-off can be done sloppily. The patterns made must be accurately referenced to the baseline and reference lines just as in the other methods of taking off.

To take off the body section, place the joggle stick on the paper with its point against the boat. Trace carefully around at least two of the "teeth" and as much of the straight edge as possible with a very sharp pencil.

7.9. Method 9—Inside of the Hull with a Joggle Stick. Make up a scrive board as shown in fig. 7.44. Tailor the size and shape to suit the boat. You will need sheets of paper about the size of the scrive board for patterns.

Use this method when the interior of the boat is unobstructed, and you intend that the lines be to the inside of plank, as in a lapstrake boat, for example. In most cases it will be necessary to take off sections at the ends of the boat using another method, probably method 8.

Set up the baseline above the boat as in section 5.5 and locate the midship station as in section 6.5. Put the scrive board in the boat so that the working face is between frames and lines up with or is parallel to the station marks, port and starboard. In this case, the station marks are references for setting up the scrive board square to the centerline. Plumb the scrive board using a carpenter's level as shown in fig. 7.44, and then block it in place while you work.

Record the boat's name, the station number, and the distance to F.P. on the pattern paper. Draw a plumb line from the baseline down the paper. Draw a level line across the paper an even distance below the baseline, and note that it is a level line and how far it is below the baseline.

Take off the section and the location of the riser, bilge stringer, all points on the keel, the sheer, and the baseline from two directions.

Take off the largest section first. Take as many sections as possible with the scrive board at its original size before cutting it down in size to fit smaller sections.

7.10. Method 10—Measure Inside and Outside Simultaneously. Some boats, such as dugouts and log canoes, have to be measured to both the inside and outside of the hull, because there is no way to determine the hull thickness without drilling many holes. Simply combine methods 8 and 9 as shown in fig. 7.45.

8. The Declivity Board

The declivity board is a wedge of wood fixed to a carpenter's level. When a boat can't be leveled or plumbed for measuring, a carpenter's level with a declivity board attached will aid in setting up the measuring equipment. The principles and their application are illustrated in fig. 7.46.

It is a good idea to use two distinctively different carpenter's levels modified with declivity boards when working on a boat that is both out of level and out of plumb. This will prevent using the wrong one in the wrong place. Cut the declivity wedges as shown, and fasten them to their respective levels. They can then be used to erect the measuring equipment.

Note that the declivity for the centerline must be measured using a straight-edged board and that the staffs at F.P. and A.P. must be set up in the centerline plane. The board used for measuring the centerline declivity will have to be positioned by eye.

9. Measuring the Inside of a Boat

A frequently asked question is: "How do I measure my boat in the water? I don't want to take it out." The answer is sufficiently complex that a simple "Don't" is not sufficient.

Consider these questions:

1. Is it necessary to take the lines from that particular boat? Have the lines already been taken from a similar boat?

2. Is recording the lines of the boat worth the cost of moving it? If the boat is unique, the answer is definitely yes!

3. Do you need the lines, or are the construction details and interior and deck arrangements what you really want?

4. Can the job wait until the boat is hauled for another purpose?

Then consider the following:

1. To take off the lines while the boat is in the water, you either have to do everything from above the water, and feel your way about the underwater portion, or engage a diver to assist in the work.

2. There is no practical reason why a boat 40 feet long or with five tons displacement can't be hauled.

3. If the boat is worth measuring, it is worth doing the job right. When you have the time only to get the lines, establish reference marks relating the measuring equipment to the inside of the boat for use when taking off interior detail after the boat is put back in the water.

4. To develop a proper set of lines, you must take off the underwater profile. The true profile can't be determined from inside the boat.

5. Measuring a boat in the water can give satisfactory results in other ways. Enough data can be taken off the exterior down to the water and off the interior to give outline shapes into which construction detail, arrangements, and deck layouts can be drawn.

10. Evaluations

The following evaluations will help you to determine the particular advantages and disadvantages of each of the taking-off methods and to decide which is more appropriate for your situation.

Method 1. This method is possibly the easiest to understand because all measurements are taken parallel to defined planes, either plumb or level. The resulting data are readily plotted. This method also requires the minimum number of tools, which is a definite advantage if you must travel to the site by air.

The disadvantages lie in the method itself. Measurements taken high up, where the angle between the plumb string and the hull is scant, can be extremely inaccurate unless taken with very great care. Avoid inaccuracies by combining methods 1 and 2 and taking triangulating measurements on the sides of the hull and plumbed measurements under the boat. Be sure to record in your notes that the methods have been combined.

Also, the manipulation of the plumb bob can be both slow and tiring. Sixteen ounces is not a very great weight but, after it has been held for a while at arms' length with the string stretched over a finger tip, it will feel like a ton. The plumb bob can also be slow to use when it can't be shielded from the wind to stop it from swinging.

Recommendation: Use method 1 for practice measuring and for teaching the principles of lines taking. Limit its use to measuring the simpler hull forms or in situations where a minimum of equipment can be carried.

Methods 2 and 4. Carefully executed triangulation is very reliable and accurate. Once the reference marks are made on the hull, the measuring is very quick because much of the slow work with a plumb bob is eliminated. Using a compass to plot the notes does make that part of the work a bit slower for these two methods.

Method 3. This uses very simple and straightforward equipment and procedure. Once the baseline and station reference are located, the work proceeds quickly. The plotting in the field and at the drawing board or loft is also quickly accomplished.

Unfortunately, this method is workable only as long as the hull and the level meet at an angle of not less than 20 degrees. Because of this, measuring sections that have a very shallow rise of floor with this method can give very inaccurate results.

Methods 5 and 6. Between them these methods offer one major improvement over methods 1, 2, 3, and 4, the horizontal arm. The arm allows taking measurements at a more normal angle to the hull whether the measurements are plumb and level, or triangulated. But the diagonal used in method 6 can actually lead to confusion. Besides being an additional piece to keep track of, it adds the potential for recording inaccurate measurements if its location is not properly noted.

Method 7. The sloping staff may prove useful when measuring deep hulled-sections, but, since everything that must be done to set up for the other methods must also be done for method 7, give the others first consideration.

Methods 8, 9, and 10. These methods are very suitable when going from the boat directly to the loft and can also be used when going to the drawing board. The accuracy is pin point.

11. Taking off the Profile from Bow to Sternpost

11.1. The bow profile. The best way to take off the bow profile is to use either a vertical staff, sloping staff, or both in combination. With the vertical staff in place, taking off a nearly plumb bow profile is very straightforward, as shown in figs. 7.47 and 7.48.

Mark the baseline height on the staff. Use a level or combination square to mark the sheer and plank seam heights on the staff; then measure the distances to the stem face and the rabbet line and all of the heights above the baseline.

To help fair the stem face and rabbet line into the keel, take additional measurements up to the bottom of the keel and the rabbet line from the baseline in the area between the top of the garboard and where the stem face fairs into the keel. Or take additional level measurements from F.P. to the stem face and rabbet line. Plot the measurements as they are taken to show the gaps in the take-off.

When the bow profile has been plotted, make a tracing or a photocopy of the plotted outline on which you can record the location of all joints, fastenings, and hardware. Trace the shape of the stem above the sheer (the stemhead). Do not forget to measure the full siding of the stem at the rabbet line at several places—at every other plank seam, at the sheer, and at the stem to keel joint. Also measure the width of the stem face and the stem band.

To take off the bow profile using a joggle stick, set up a scrive board with a length of paper taped to it at the F.P. Mark the baseline and F.P. on the paper; then take off the bow profile, rabbet line, and the locations of all joints. If the staff at F.P. is wide enough, 6 inches or more, use that (figs. 7.41 and 7.43).

Take the profile off a raking stem using a sloped staff as shown in fig. 7.49. The form for recording the measurements can be either that shown in fig. 7.49 or in fig. 7.38 for method 7. To take off the bow profile from a boat that is upside down, see fig. 7.50.

11.2. Measuring along the Keel. Deep-keel hulls require some care in measuring. Frequently there are "knuckles" or "quick" spots that occur between body sections. These require additional measurements to record the profile shape accurately. Take the measurements as distances from F.P. and heights above the baseline or the reference line. Fin-keel boats present similar problems for recording adequate data. Try to measure as much of the shape in straight lines, as shown in fig. 7.51.

11.3. Centerboards and Daggerboards. For a centerboard, record the distance from the forward end of the slot to the nearest station and to F.P., the distance from the aft end of the slot to the nearest station, the length and width of the slot and its relationship to the centerline of the keel (it may be offset from the centerline), the maximum draft of the board, its shape below the keel, and the location of the pivot if it is through the keel below the rabbet line.

For a daggerboard, take off all the data

listed for a centerboard, then push the board down to maximum draft, and draw a plumb line on the board at or referenced to a station and if possible the baseline or the plumb distance to it (fig. 7.52). On a large piece of paper, draw a straight line. Lay the board on the paper so the two lines are one over the other. Trace the daggerboard onto the paper. Measure and record the joints, cross section shape, and any other details. The plumb line will be useful in locating the board properly when the time comes to draw the boat.

12. Measuring Stern Profiles and Transoms

12.1. The Double Ender. To take off the stern profile of a double-ended boat, follow the instructions for measuring a bow profile. The only difference is that if the double ender has an engine, the end of the propeller shaft must be located. Measure from A.P. forward and up from the baseline or other previously located points.

12.2. The Straight Sternpost. Record the rake of a straight sternpost as indicated by dimensions C and D in fig. 7.19 and dimension R in fig. 7.51.

12.3. The Overhanging (Counter) Stern. As shown in fig. 7.53, an overhanging stern is measured as a series of straight lines and their intersection points.

12.4. Flat Transoms. Do not waste time trying to measure a flat, raked transom as though it is a body section. Either take off a pattern or measure the transom's surface. Since some transoms have very subtle shapes that lend greatly to the character of the boat, making a pattern is a good idea. The first step is to record the transom's rake as shown in fig. 7.19, dimensions C and D, and the sheer height. Note that some boats have a tumble home in the last few feet of the sheer and at the transom that makes the sheer indeterminate. In this case take measurements to the underside of the deck.

Making a pattern is easier and quicker than taking off measurements. Tape a large piece of paper on the transom, lined up with or marked to show the centerline and sheer height. The best way to show the sheer height is to draw a straight line from sheer to sheer. Hold the paper firmly, or tape it in place, then rub a fingertip around the outline of the transom (fig. 7.54). Very carefully mark each plank seam and, if possible, the inside of the planking. Measure the sheer height by taking a plumb measurement above the baseline to the sheer—sheer line.

Another way to make a pattern is to tape a piece of paper slightly smaller than the transom to it. Mark the centerline and sheer heights. Then make the pattern using a joggle stick as shown in fig. 7.44.

To take measurements off a flat, raked transom, mark the centerline on the transom. Then draw a line from sheer to sheer. Measure from the centerline and square to it, to each plank seam and down from and square to the line drawn from sheer to sheer (figs. 7.55 and 7.56). Plot the transom in the notes exactly as it was measured. Measure the sheer height as described above.

12.5. Curved, Raked Transoms. The procedures for taking off a curved transom are similar to those for taking off a flat transom. Begin by making sure that the measurements shown in fig. 7.53 have all been recorded. In addition to recording the rake of the transom, also record the crown—the degree of curvature—of the face. The curve is almost always an arc of a circle, and, unless proof to the contrary exists, should be assumed to be an arc, with its radius square to the face of the transom as though it were built as a part of the surface of a large cylinder.

Tape a large piece of paper to the transom, and mark the centerline of the transom on it. Bend a wide, thin, straight-edged board across the transom so it lies flat and the straight edge lines up with the sheer, port and starboard. Draw a line along the board from sheer to sheer (fig. 7.57). This is the basic reference for plotting the transom on the lines plan or loft.

Measure the crown of the transom by tipping the straight edge up so that it is square to the rake of the transom at the centerline. Pivot the straight-edge until the distance to the sheer is the same, port and starboard. This distance is the crown. Record it. Then rub the outline of the transom onto the paper to complete the pattern.

To take measurements from the transom, mark the centerline and the sheer to sheer line on the transom. Mark off a series of lines parallel to the centerline on the transom—arbitrary spaces in even inches that will be easy to scale when plotting the transom on a drawing. Measure and record all of the distances above and below the sheer to sheer line to the outside of plank and the deck (top or underside). This is very similar to the procedure for measuring a flat transom except that all measurements are taken along lines parallel to the centerline.

If the transom has a shape that looks as though it will be difficult to reproduce, use additional lines parallel to the sheer to sheer line for taking measurements.

12.6. The Sheer Height and Half Breadth at the Transom. On a flat, plumb transom, draw a line from sheer to sheer. Measure from the line to the baseline to get the sheer height. Take the same measurement for a flat, raked transom, but be sure that the height is taken plumb—not along the raked face of the transom. The half breadth of the sheer should have been taken off when the transom was measured.

If you are measuring a curved, raked transom, there are four pieces of data to take off that will tie things together for lofting or drawing:

1. The rake of the transom
2. The sheer height
3. The half breadth of the sheer
4. The distance from A.P. to the intersection of the sheer with the transom (this is a check of the crown).

The height, half breadth, and distance from A.P. can be taken by hanging a plumb bob from the sheer at the transom. Square back from the plumb bob string to the baseline and measure aft to A.P. for distance from A.P. to the sheer at the transom. Measure from the baseline to the plumb bob string for the half breadth. Level over from the baseline to the plumb bob, measure up, and get the sheer height.

You have completed taking off the lines. Do not take down the measuring equipment. Leave it in place to aid in taking off construction.

13. Summary

The various methods of taking off have been used many times by many workers. Used as described and with reasonable care, they should provide good results. Keep in mind that (1) all references from the measuring equipment to baselines and reference lines must be accurately recorded, (2) distances from the measuring equipment to the boat should be kept as short as possible, and (3) the methods should be chosen to suit the hull's sectional shape.

Figure 7.1. The base plane and centerline plane.

Figure 7.2. The base plane and centerline plane, and the body section plane.

Figure 7.3. The planes used for taking off.

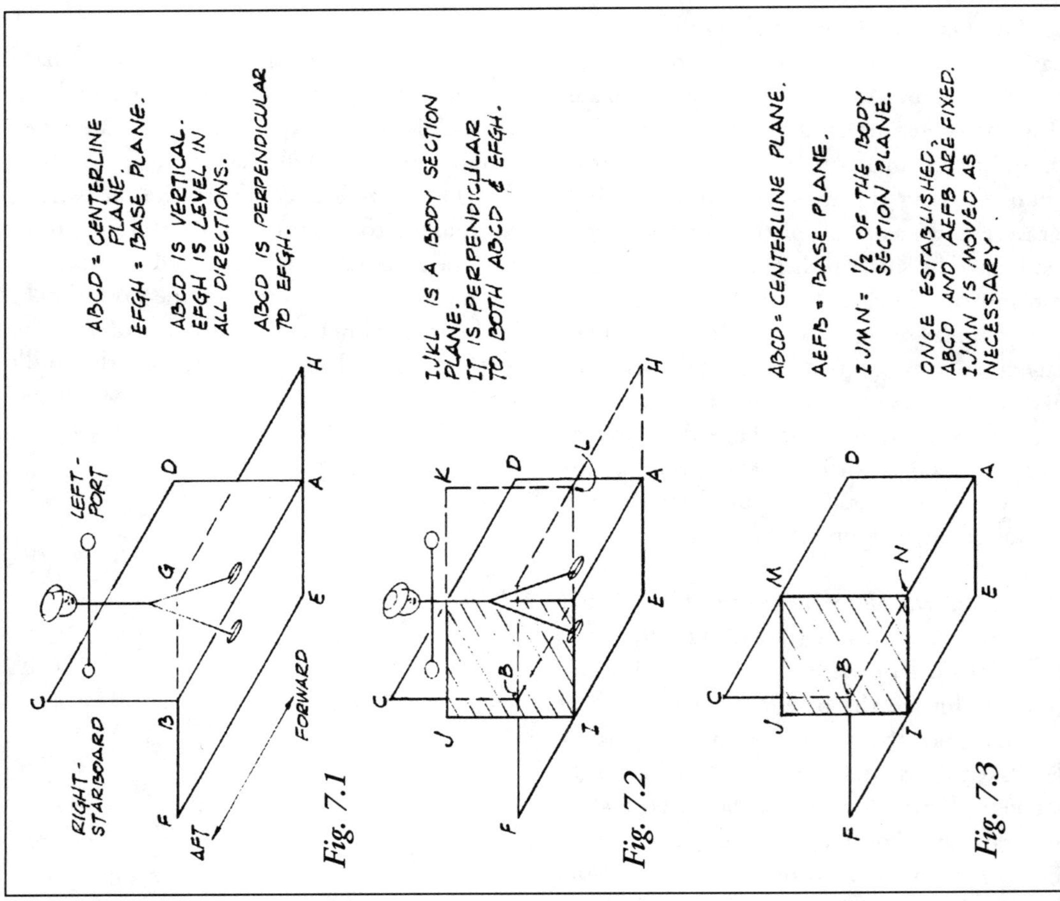

Figure 7.4. A reference line in relation to the centerline plane.

Figure 7.5. A baseline and reference line in relation to the hull.

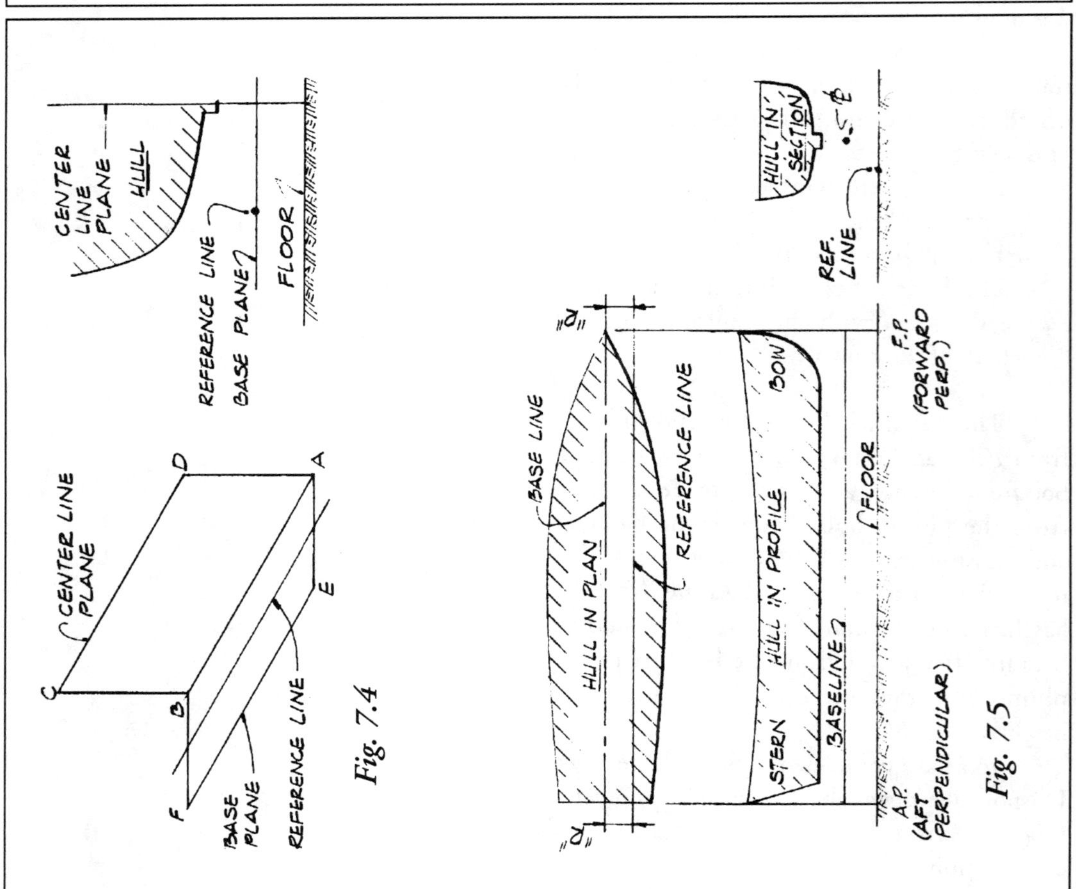

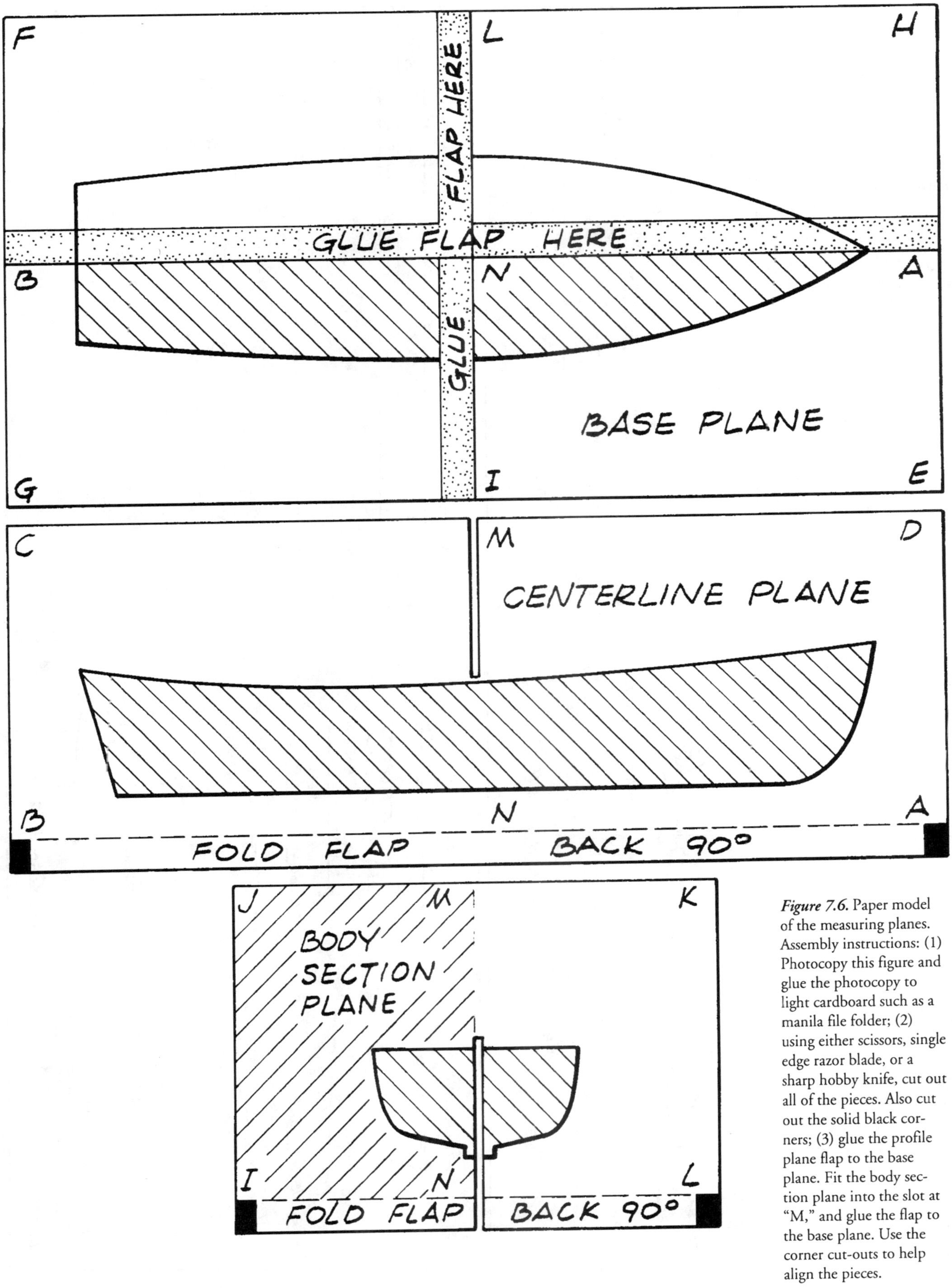

Figure 7.6. Paper model of the measuring planes. Assembly instructions: (1) Photocopy this figure and glue the photocopy to light cardboard such as a manila file folder; (2) using either scissors, single edge razor blade, or a sharp hobby knife, cut out all of the pieces. Also cut out the solid black corners; (3) glue the profile plane flap to the base plane. Fit the body section plane into the slot at "M," and glue the flap to the base plane. Use the corner cut-outs to help align the pieces.

Figure 7.7. Leveling the boat athwartships using the painted waterline.

Figure 7.8. Leveling the boat fore and aft using the bottom of the keel or the painted waterline.

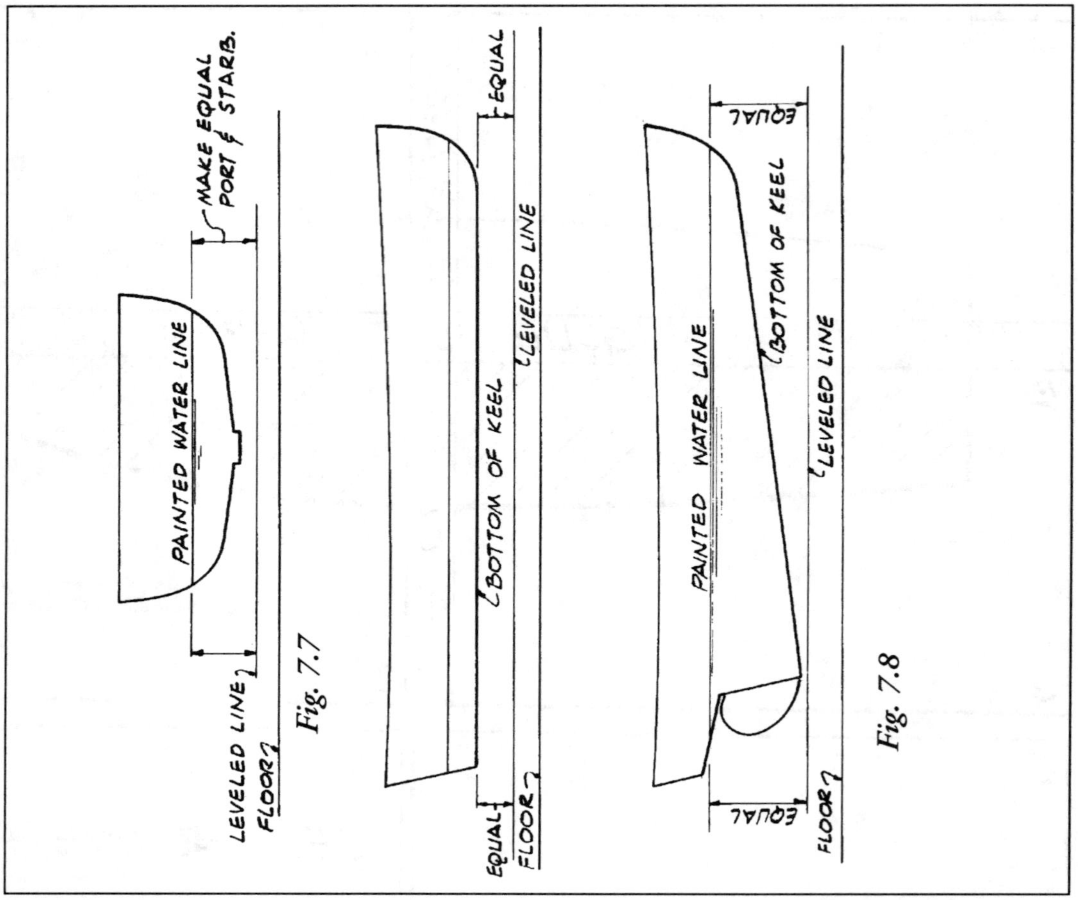

Figure 7.9. Locating the plumb bob at the stem for plumbing.

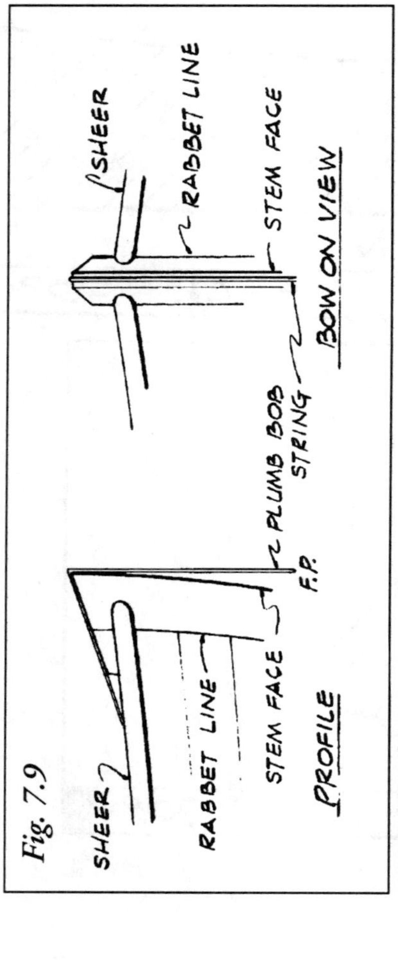

Figure 7.10. Places to mark the centerline on various sterns to aid in plumbing the boat.

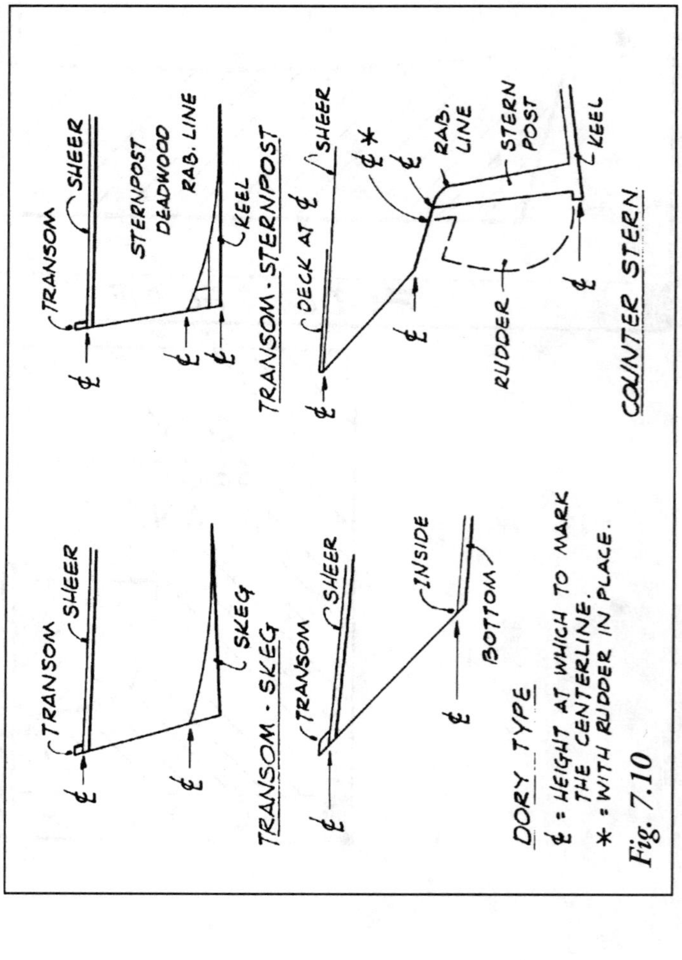

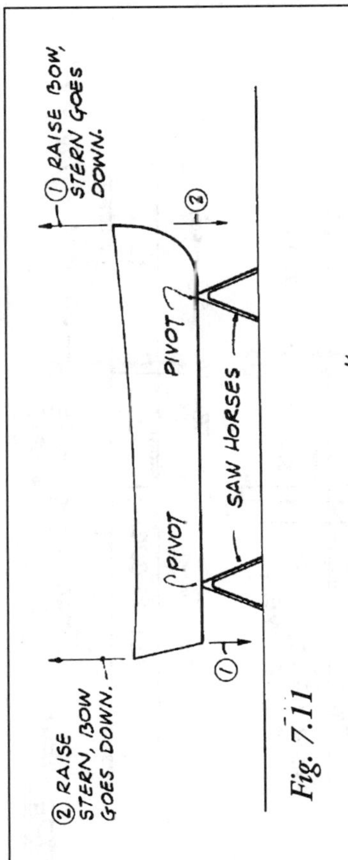

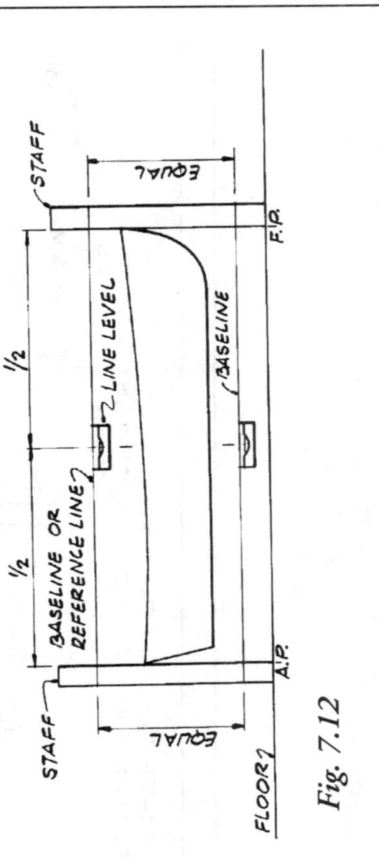

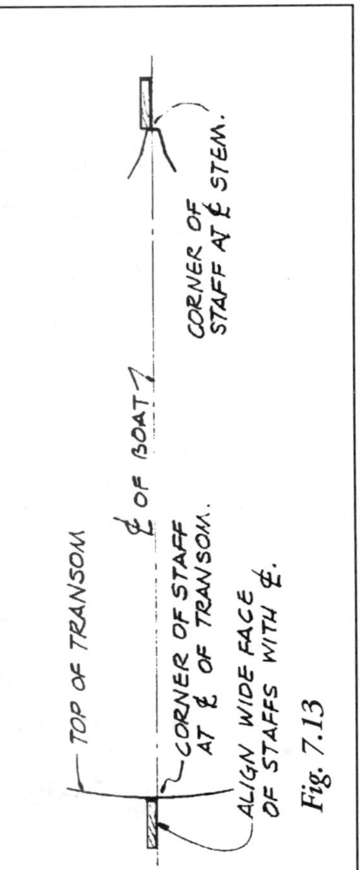

Figure 7.11. Leveling the boat fore and aft. Notice that not only do the ends of the boat move as indicated by (1) and (2), but the movements should be less than required to bring the boat level.
Figure 7.12. Leveling the baseline below or above the boat and rigging a reference line above the boat.

Figure 7.13. The proper alignment of the staffs at the perpendiculars with the centerline.

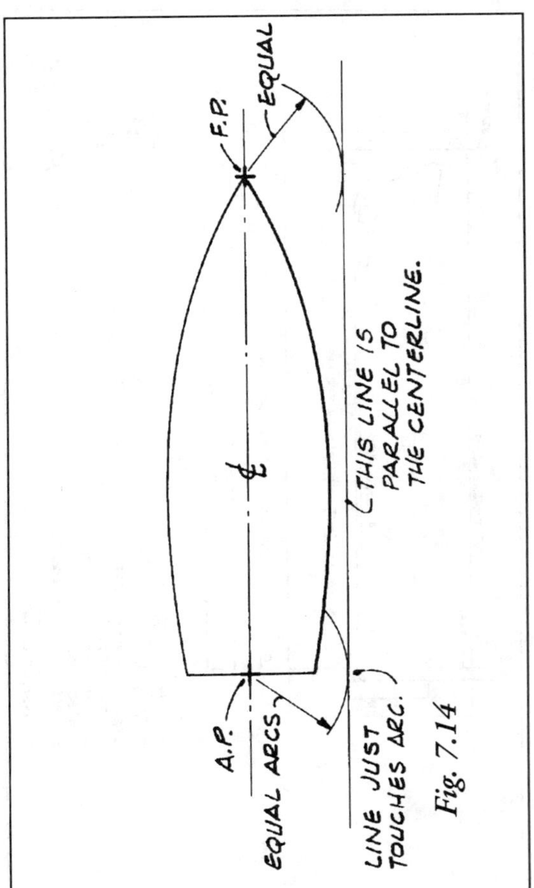

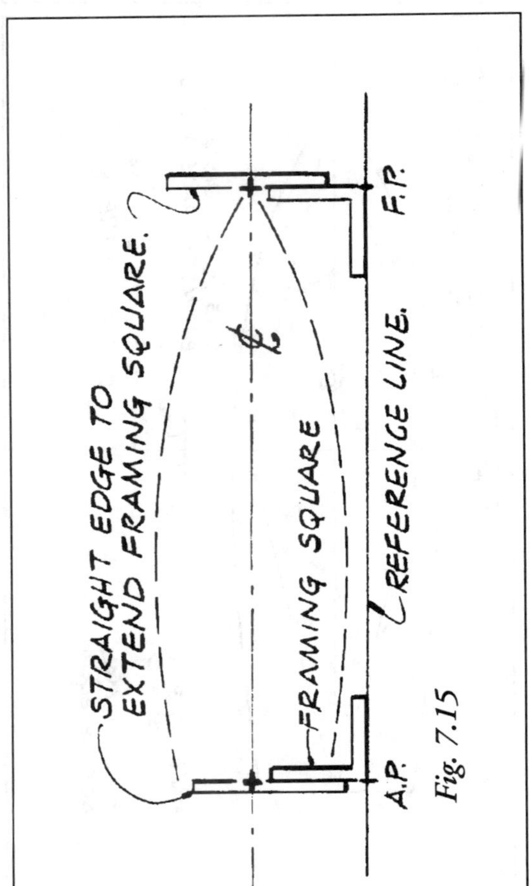

Figure 7.14. Locating a chalked reference line on a floor.

Figure 7.15. Squaring from the perpendiculars to the reference line.

Figure 7.16. Using a ruler to set up a string reference line parallel to the center-line of the boat.

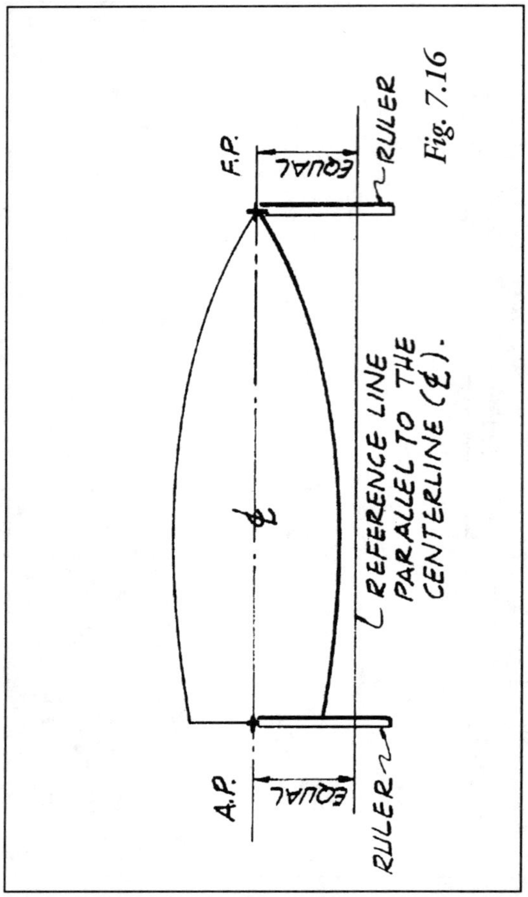

Figure 7.17. Squaring the station reference across the boat.

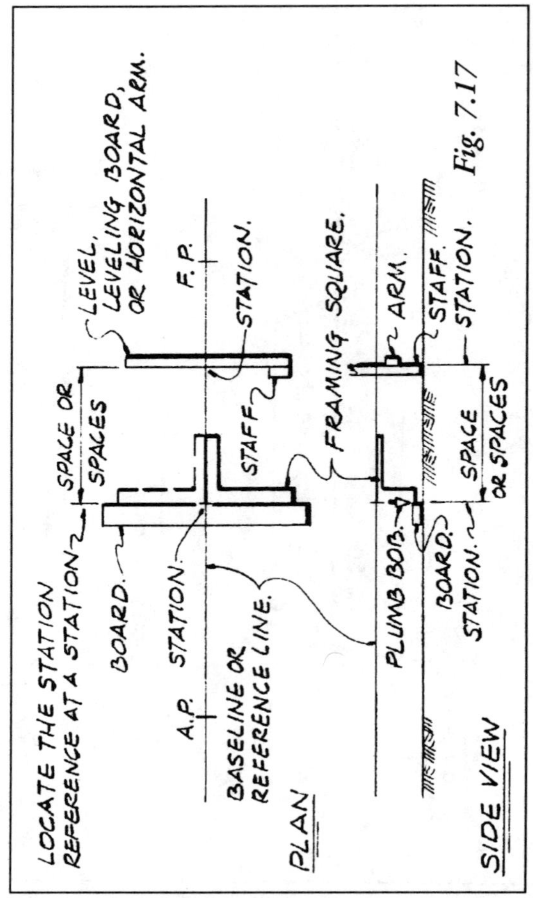

Figure 7.18. Locating a station reference when the baseline is above the boat. This procedure can also be used when you cannot get under the boat to set up a reference.

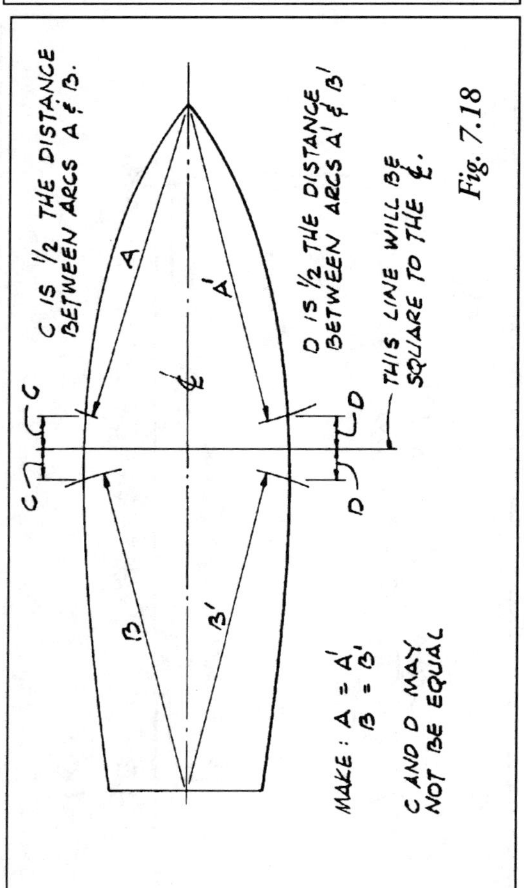

Figure 7.19. Sample of note page showing profile of boat as set up for measuring with necessary data to record.

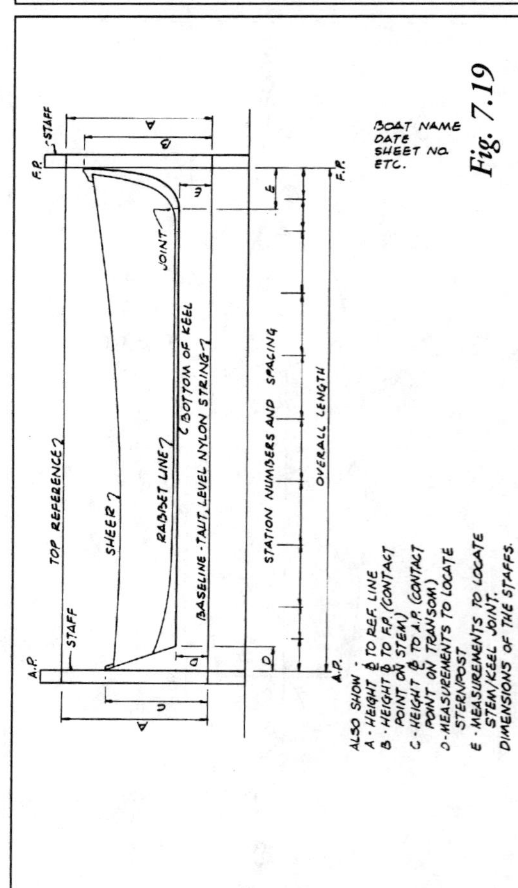

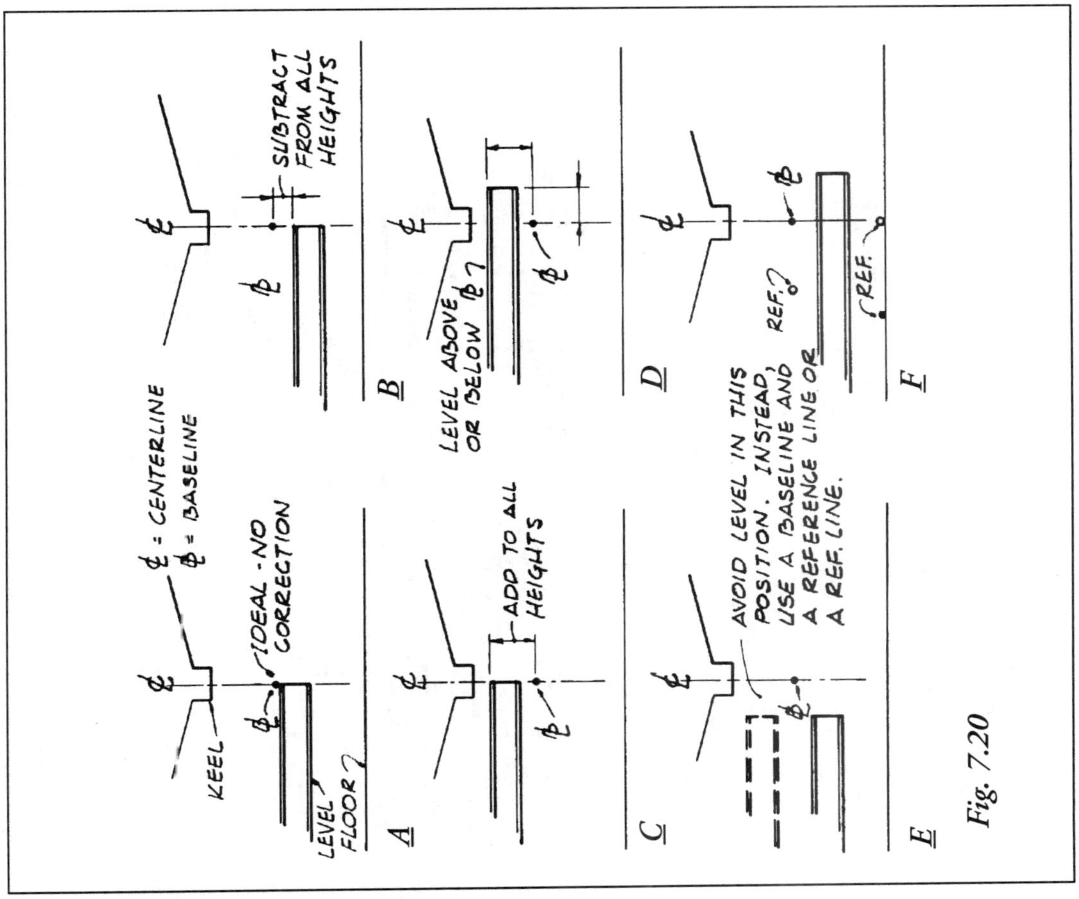

Fig. 7.20

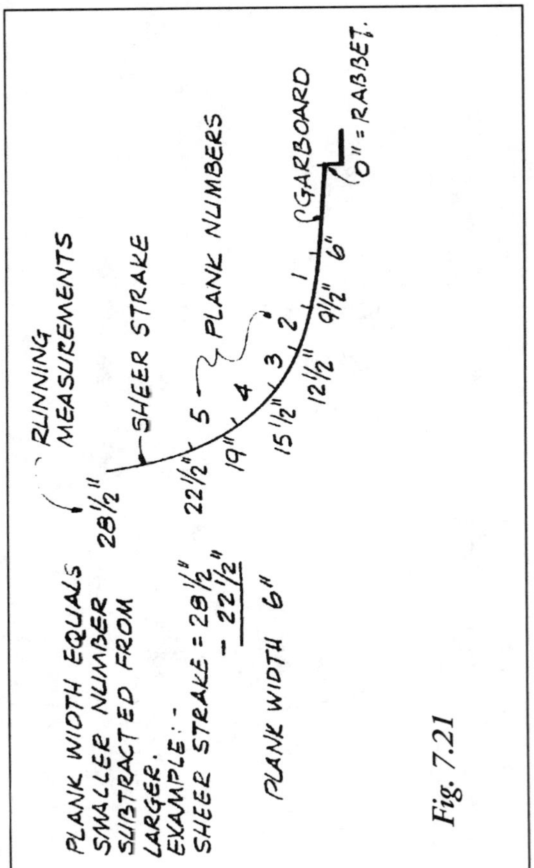

Fig. 7.21

Figure 7.20. Several possible locations of measuring equipment in relationship to baseline or reference line. (A) The ideal situation with the end of the level or leveling board at the baseline—no correction required; (B) if the end of the level or leveling board aligns horizontally with the baseline but is below it, the distance between them is subtracted from all heights; (C) similar to B., but with the level or leveling board above the baseline. The difference between the two must be added to all heights. (D) The end of the measuring equipment sometimes passes horizontally beyond the baseline. Record not only the differences above or below the baseline, but also measurement "X"; (E) if the inboard end of the measuring equipment can not be aligned with the baseline, use a reference line; (F) the reference line can be a string between two concrete blocks or a chalk line on the floor. *Note:* All of the measuring equipment should have a yardstick or tape measure fixed to it.

Figure 7.21. Measuring the plank widths.

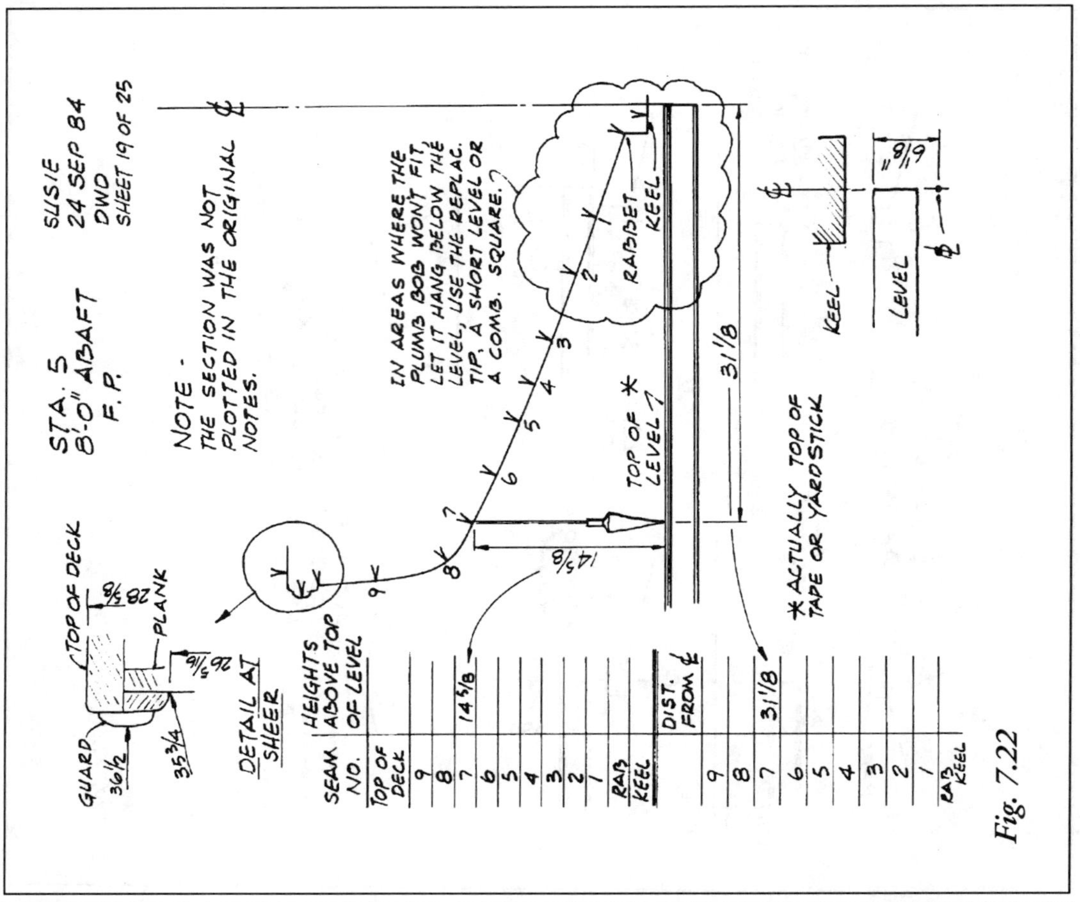

Figure 7.22. Sample note page for method 1, level and plumb bob, showing measurements to take, the table for recording them, and the plot to check the measurements.

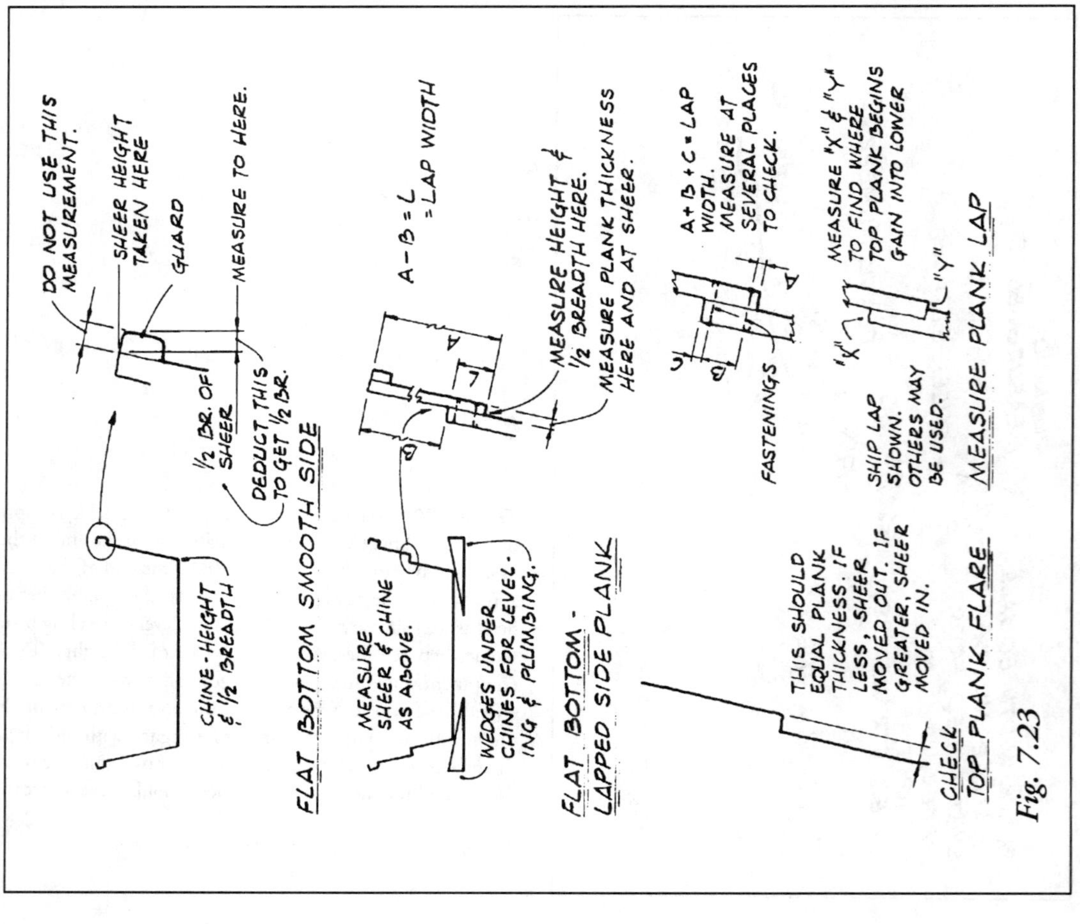

Figure 7.23. The flat-bottomed boat; leveling, and points to measure.

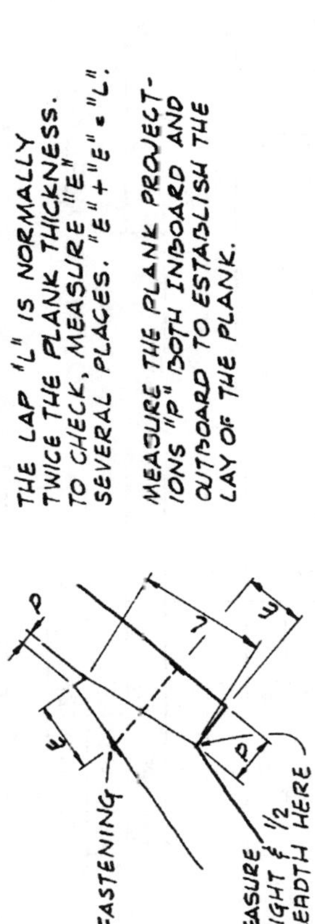

Fig. 7.24

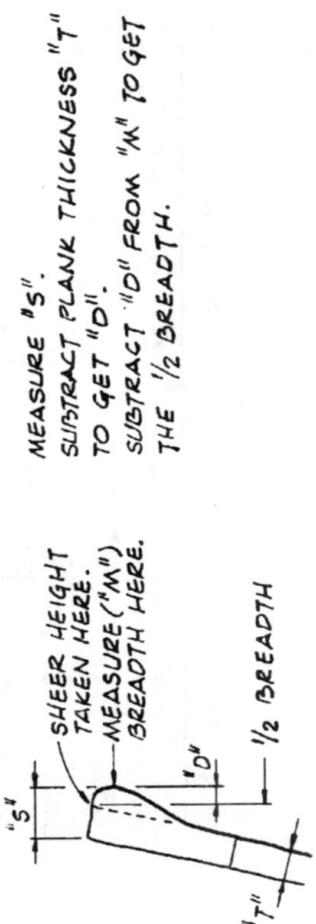

Fig. 7.25

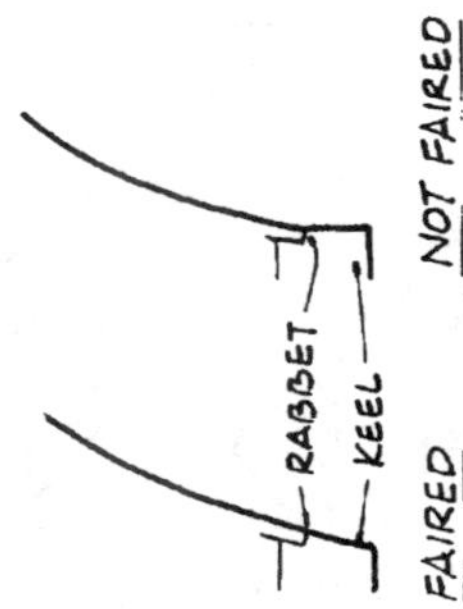

Fig. 7.26

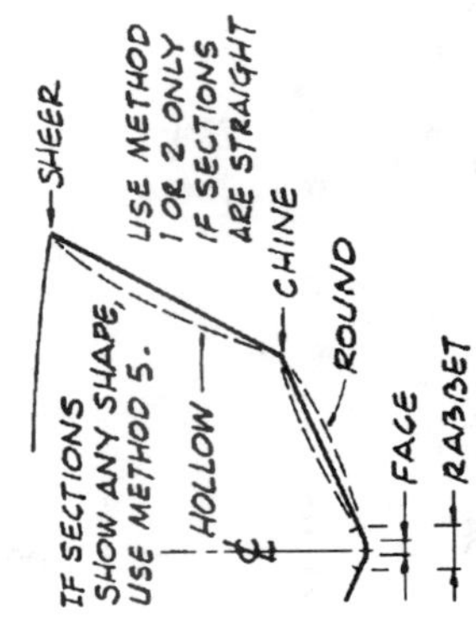

Fig. 7.27

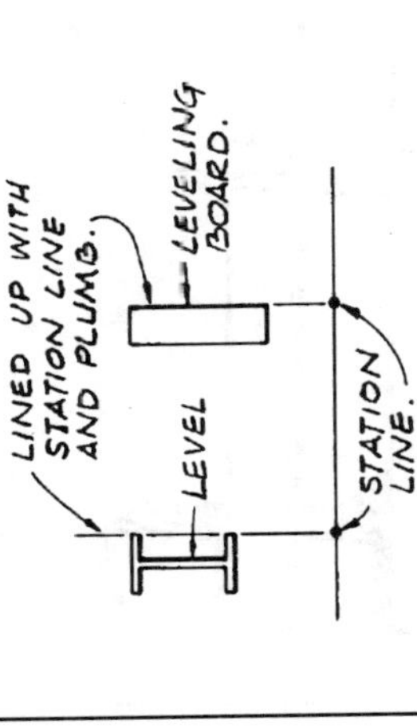

Fig. 7.28

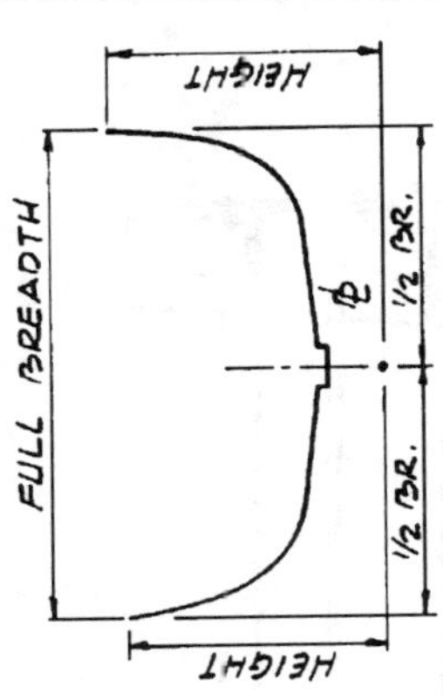

Fig. 7.30

Fig. 7.29

Figure 7.24. Where to take measurements on a lapstrake-planked boat.

Figure 7.25. Compensating for a molded sheer strake.

Figure 7.26. Planking faired and not faired in the keel.

Figure 7.27. Points to measure on a straight-sectioned V-bottom hull—sheer, chine, and rabbet.

Figure 7.28. Align the measuring equipment with the station line.

Figure 7.29. Using the plumb bob and level. *Peter Schmid*

Figure 7.30. Measurements to take if the boat is twisted—the 1/2 br. (half breadth) and the sheer height port and starboard and the full breadth at the sheer.

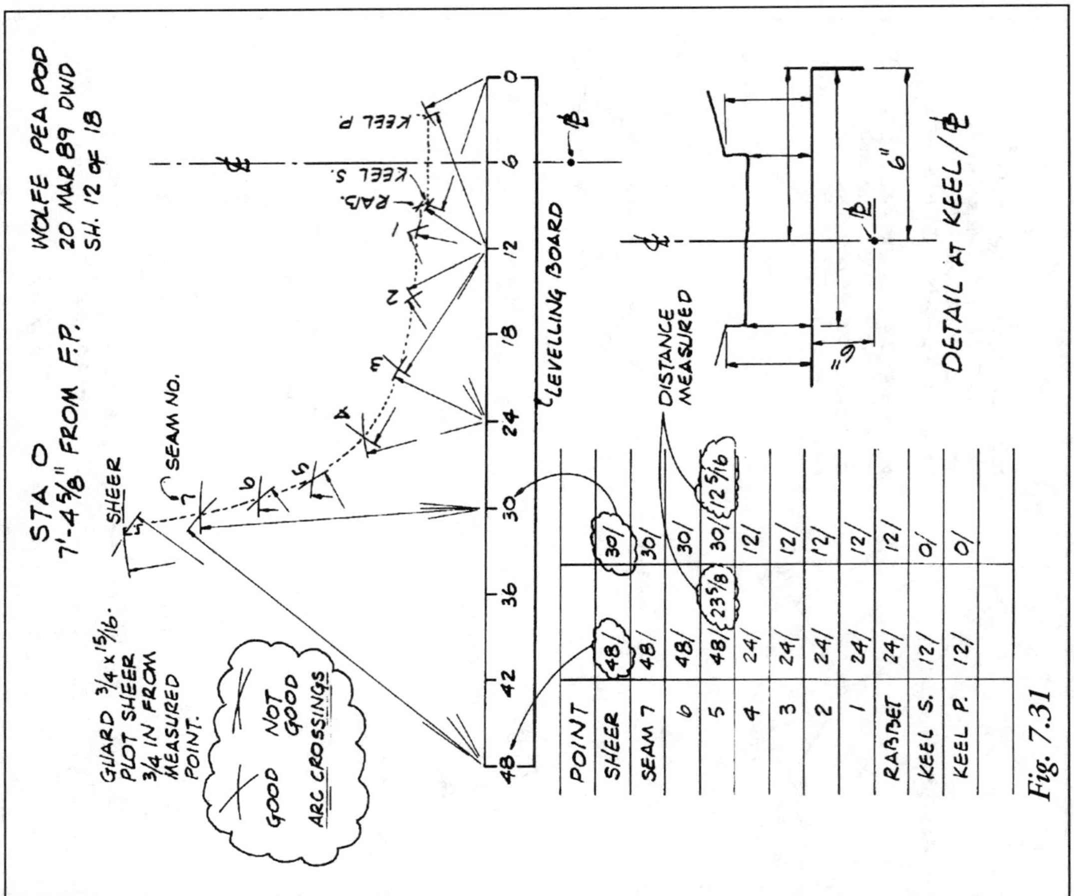

Figure 7.31. Sample note page for method 2, leveling board with triangulated measurements.

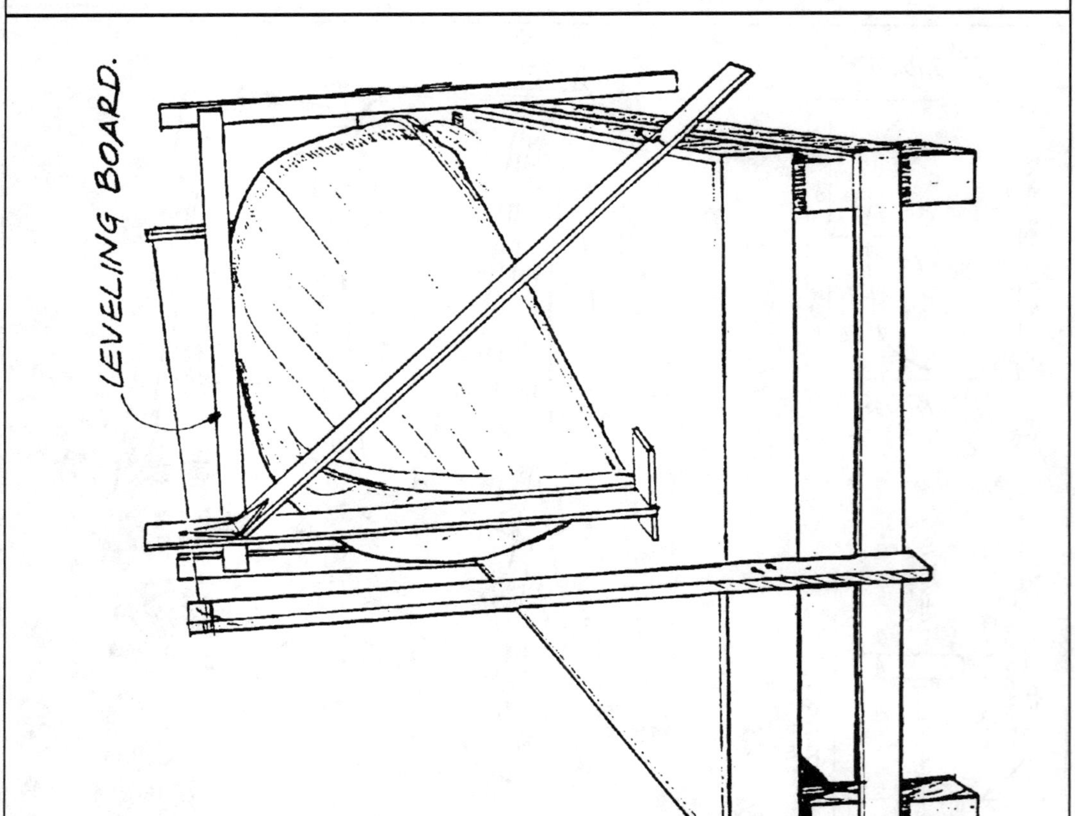

Figure 7.32. Leveling board used to measure a boat that is upside down. Also see figs. 8.16 and 8.17. *Peter Schmid*

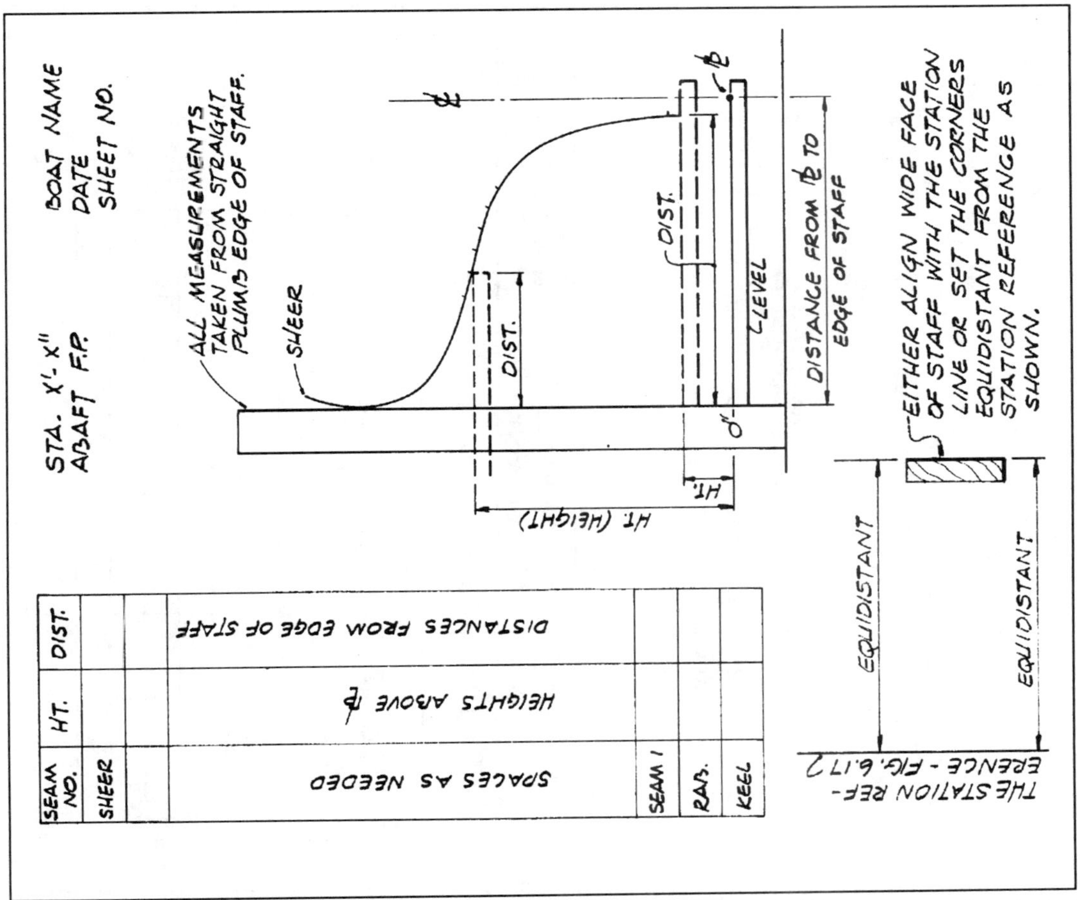

Figure 7.33. Sample note page for method 3, vertical staff with level measurements.

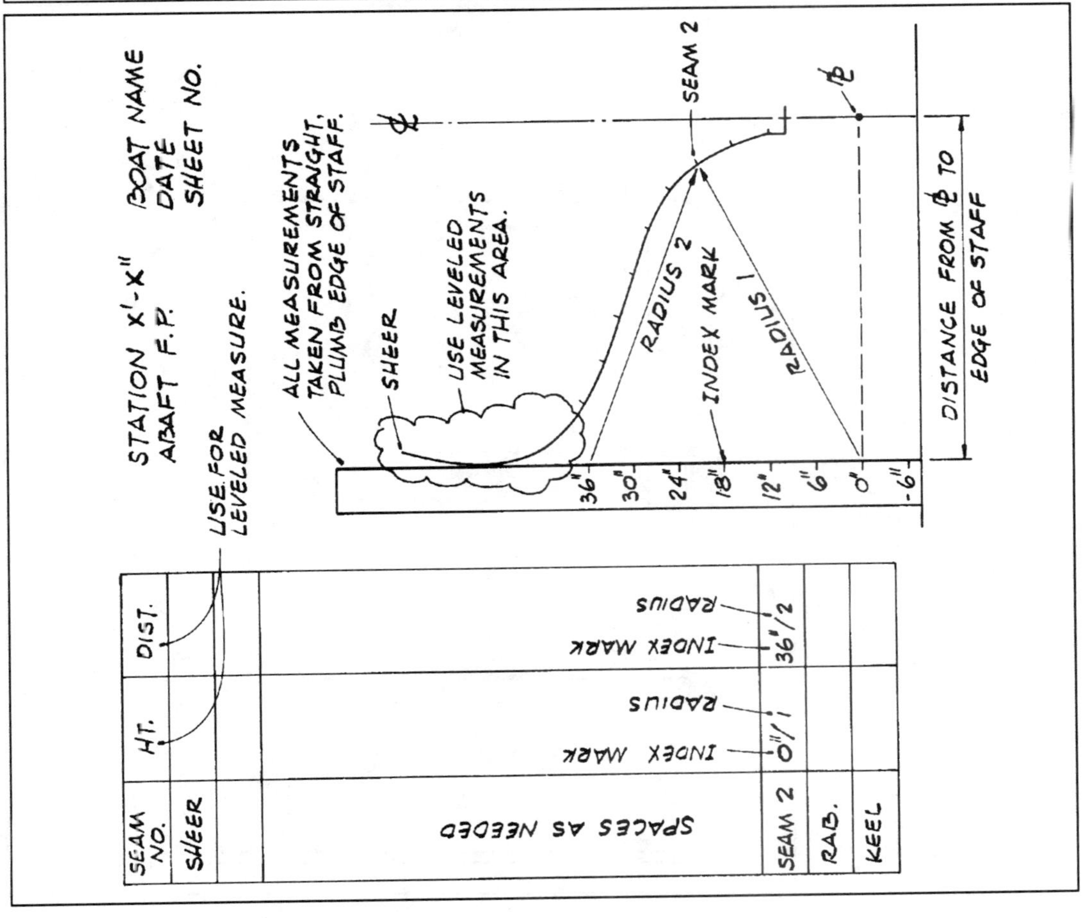

Figure 7.34. Sample note page for method 4, vertical staff with triangulated measurements.

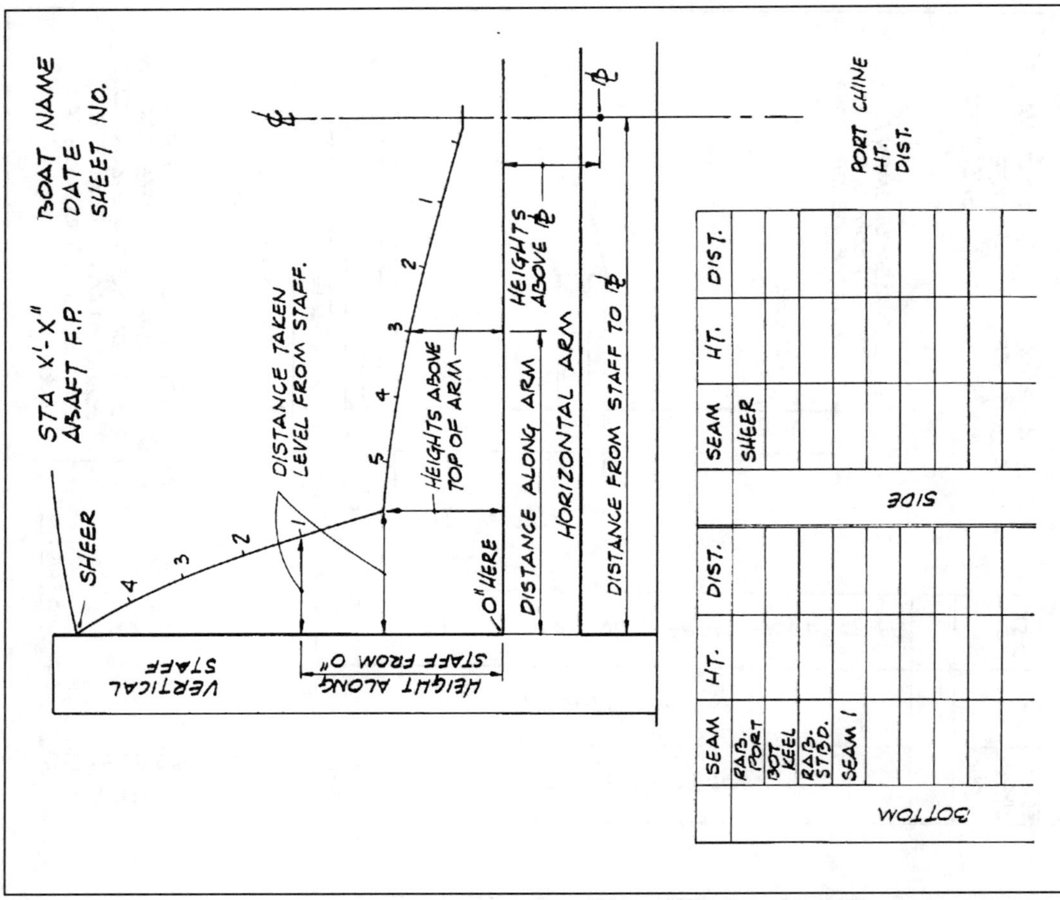

Figure 7.35. Sample note page for method 5, vertical staff with horizontal arm.

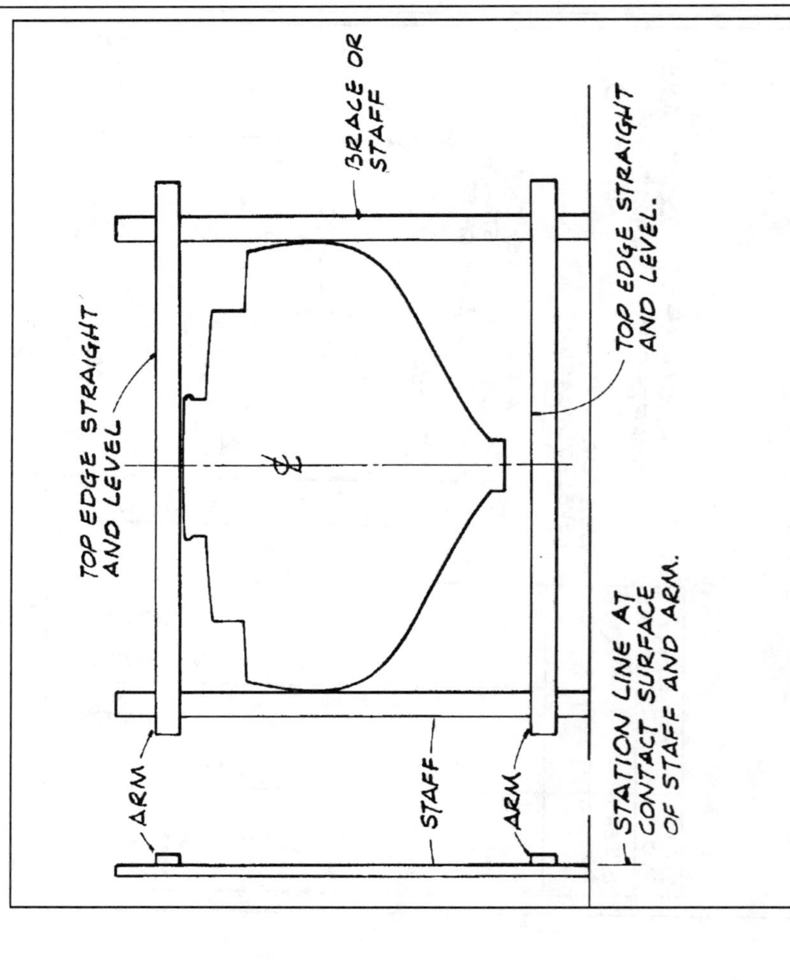

Figure 7.36. A second horizontal arm above the boat to measure deck detail.

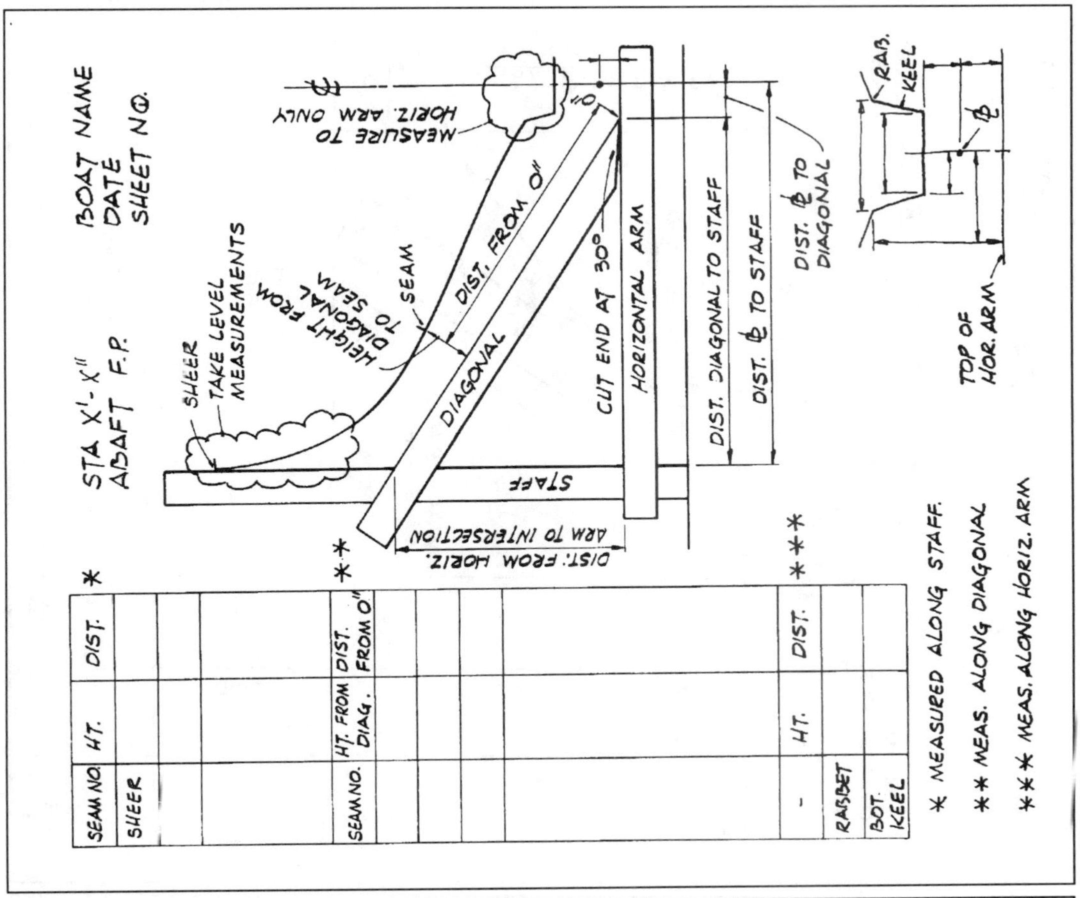

Figure 7.37. Sample note page for method 6, vertical staff with horizontal arm and diagonal.

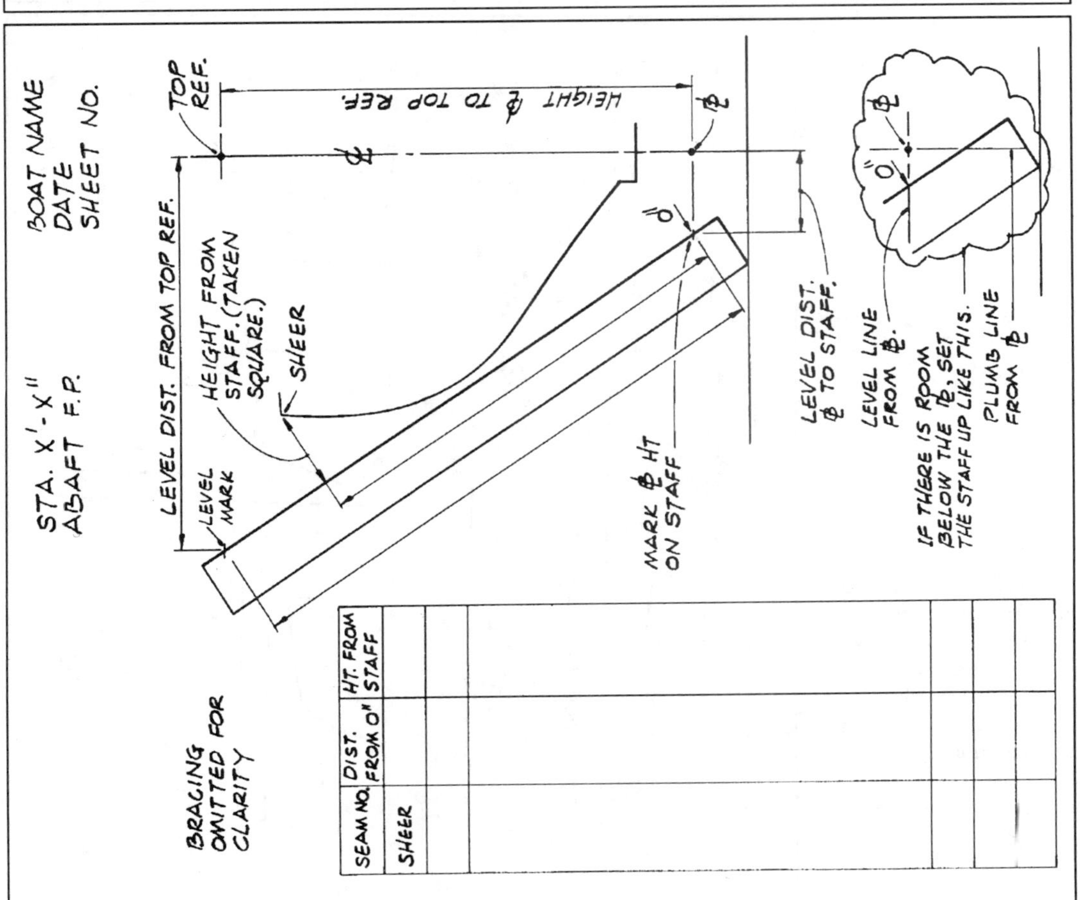

Figure 7.38. Sample note page for method 7, sloped staff with square measurements.

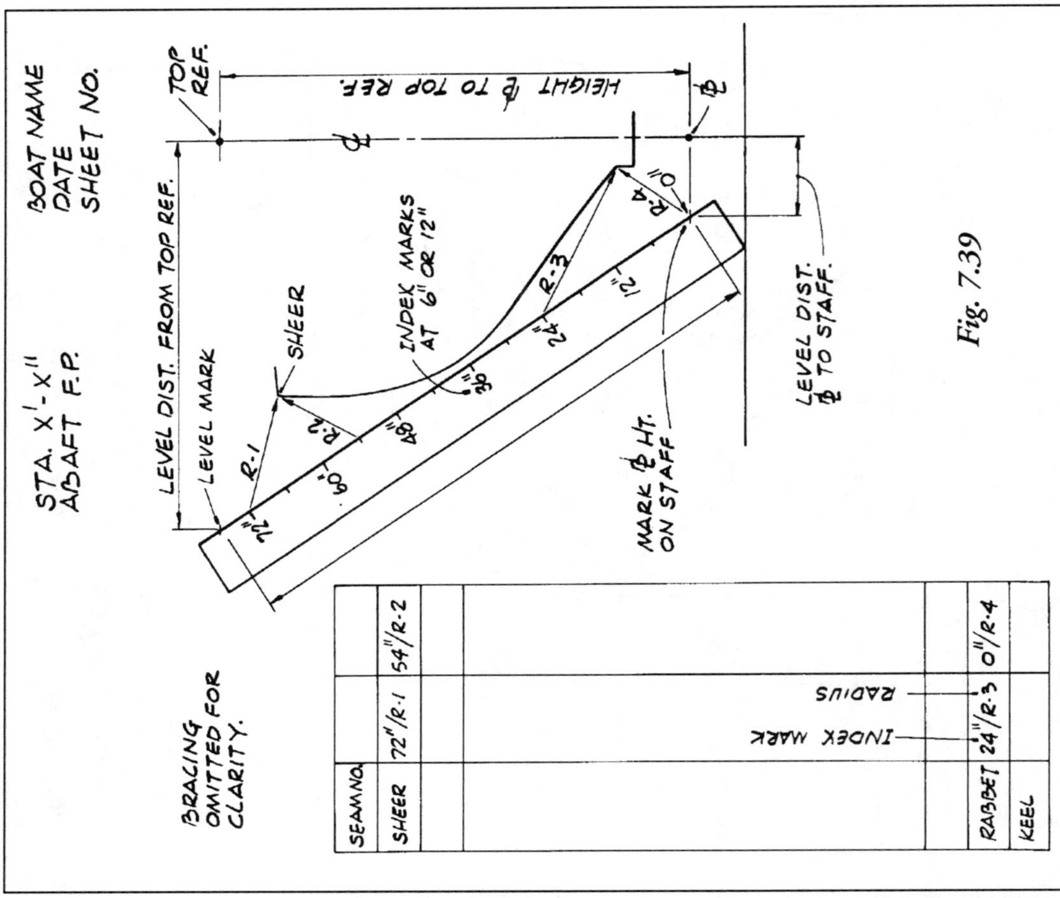

Figure 7.39. Sample note page for method 7, sloped staff with triangulated measurements.

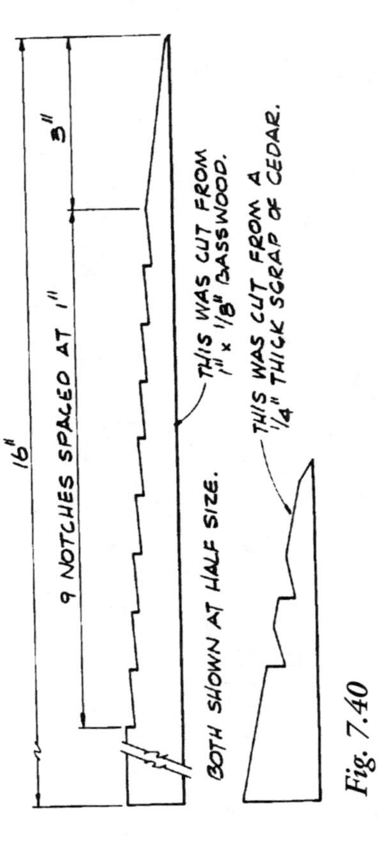

Figure 7.40. Two joggle sticks.

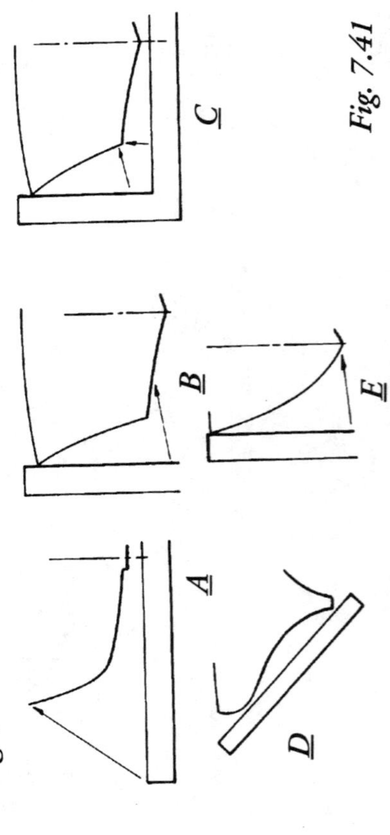

Figure 7.41. Using the joggle stick to take off various hull sections with the scrive boards in different positions. (A) level or leveling board and round hull with the arrow indicating an excessively long distance to measure. Arrow indicates problem area. (B) vertical staff and vee-bottomed hull; the arrow indicates where accurate measuring will be difficult; (C) vertical staff and horizontal arm; the arrows indicate the limits of use for the staff and horizontal arm; (D) sloped staff; (E) vertical staff at forward section of hull.

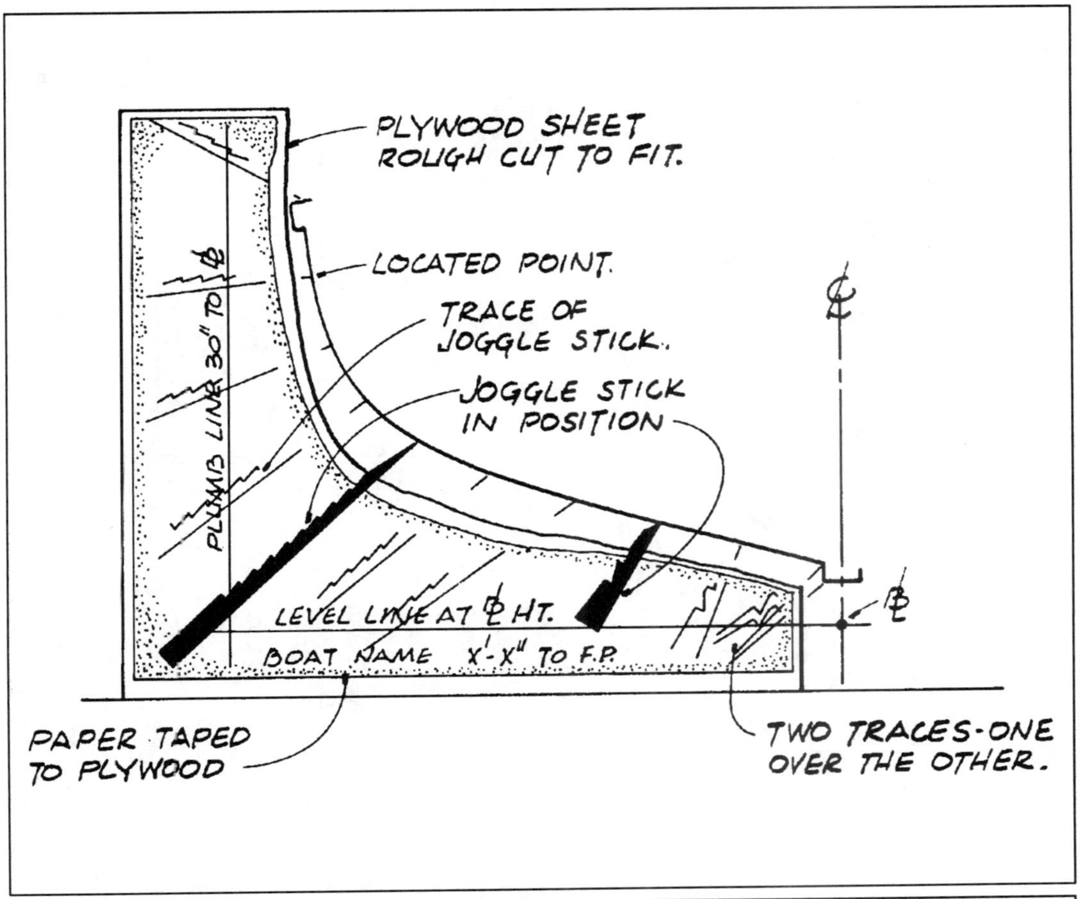

Figure 7.42. Method 8; plywood scrive board with pattern paper in place prior to taking off body section to outside of plank with joggle stick.

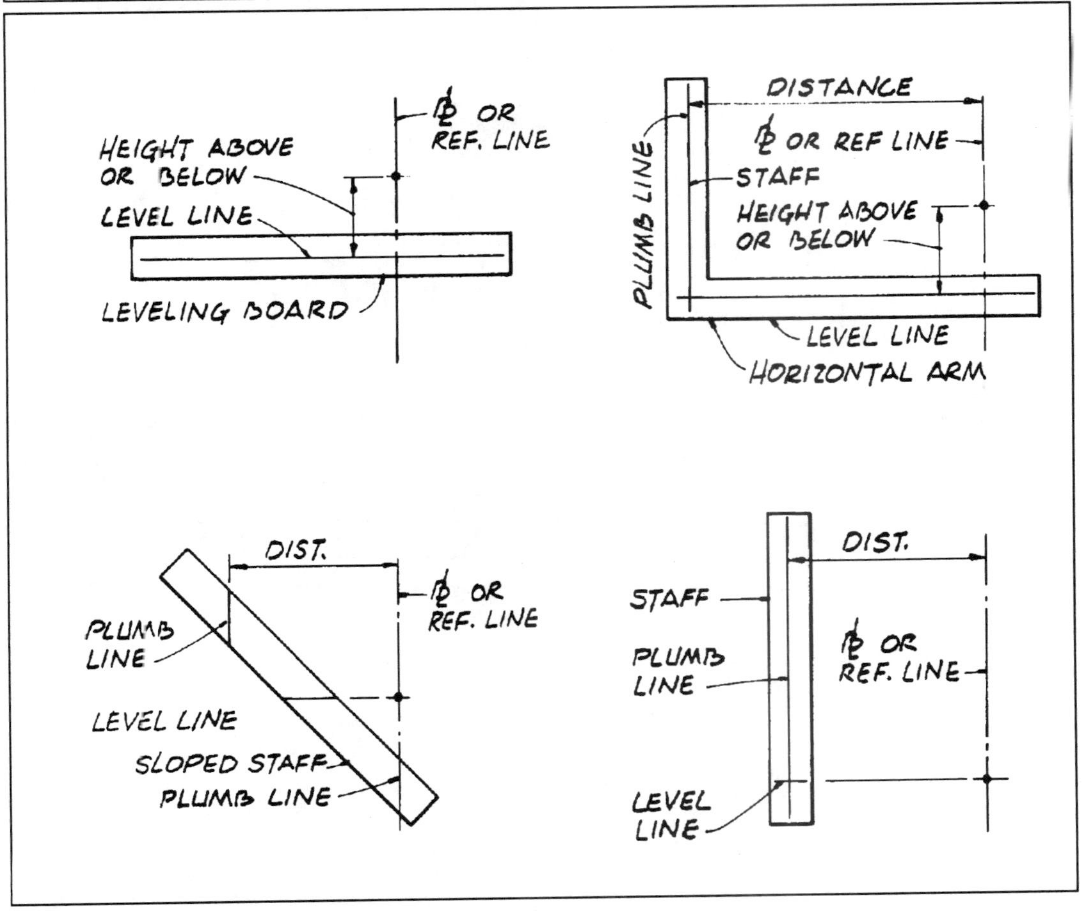

Figure 7.43. The reference marks that should be put on the patterns made with a joggle stick are indicated here.

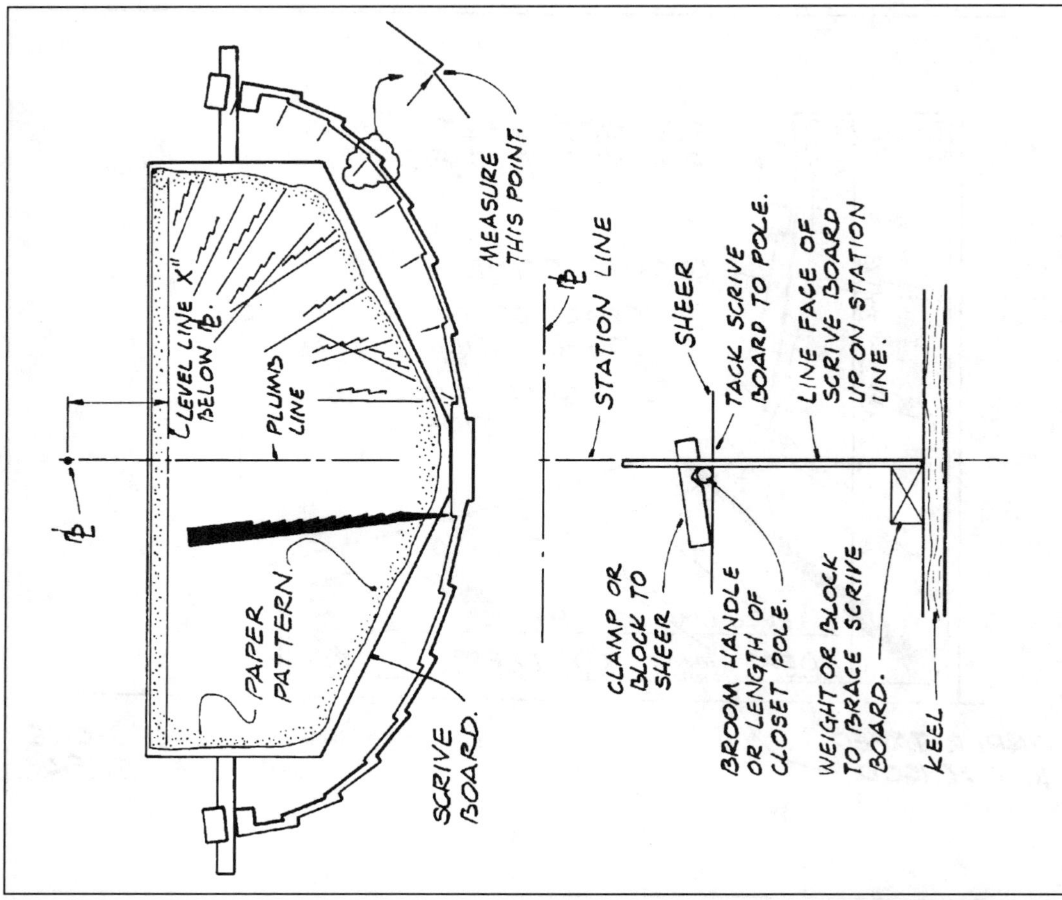

Figure 7.44. Arrangement of scrive board for taking off lines to the inside of plank with a joggle stick.

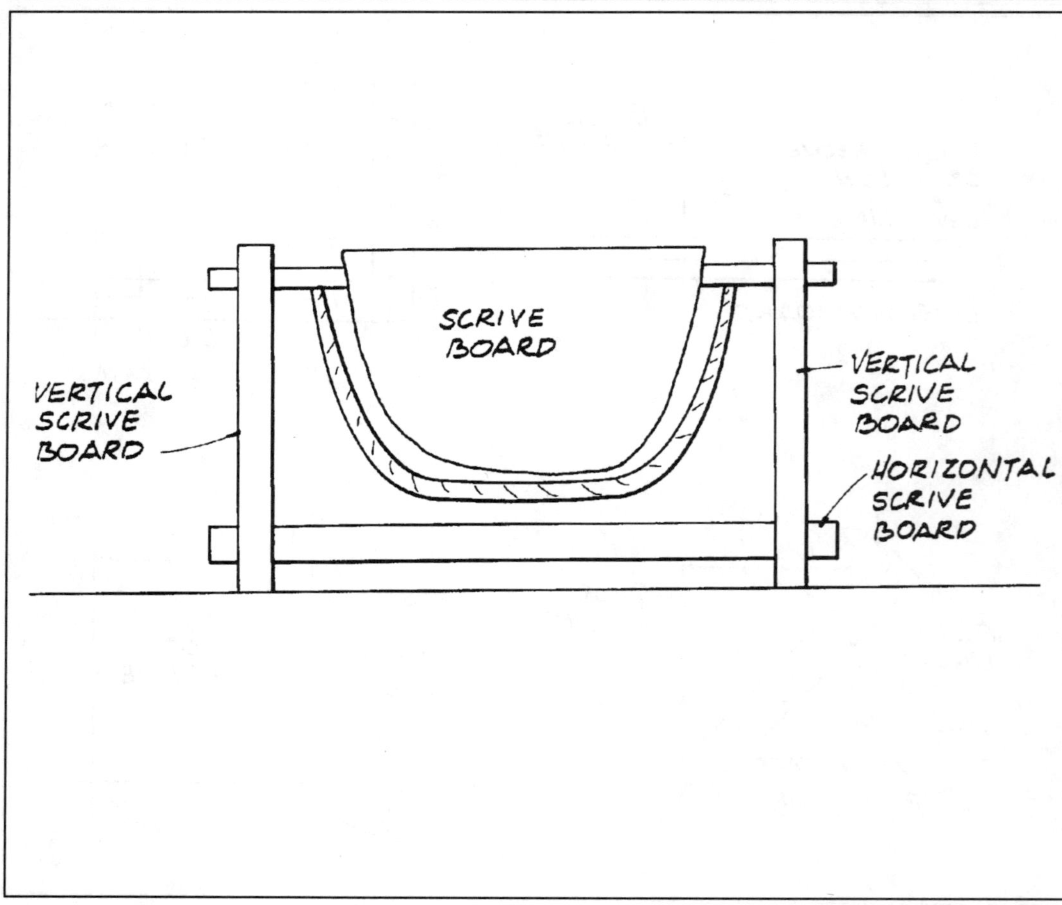

Figure 7.45. An arrangement for taking off inside and outside of hull simultaneously.

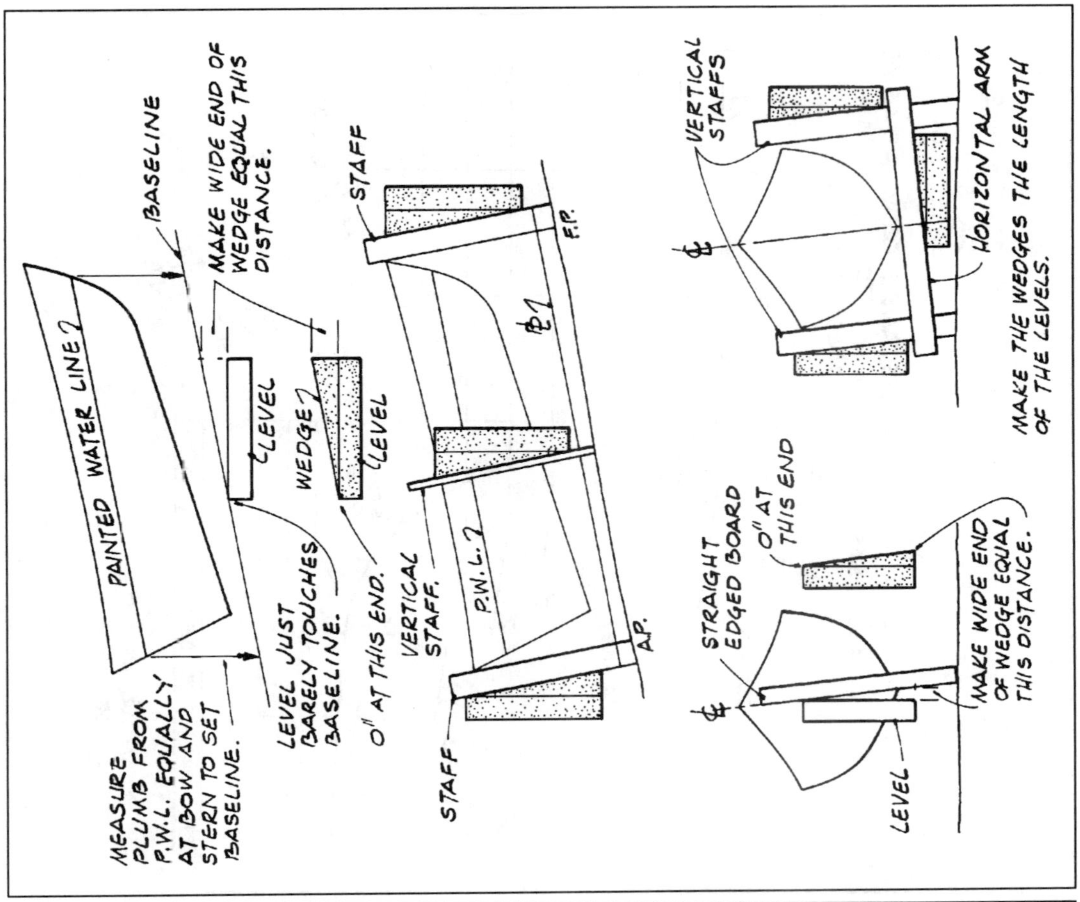

Figure 7.46. The declivity board; compensating for out of plumb and out of level.

Figure 7.47. Taking off the bow profile.
Peter Schmid

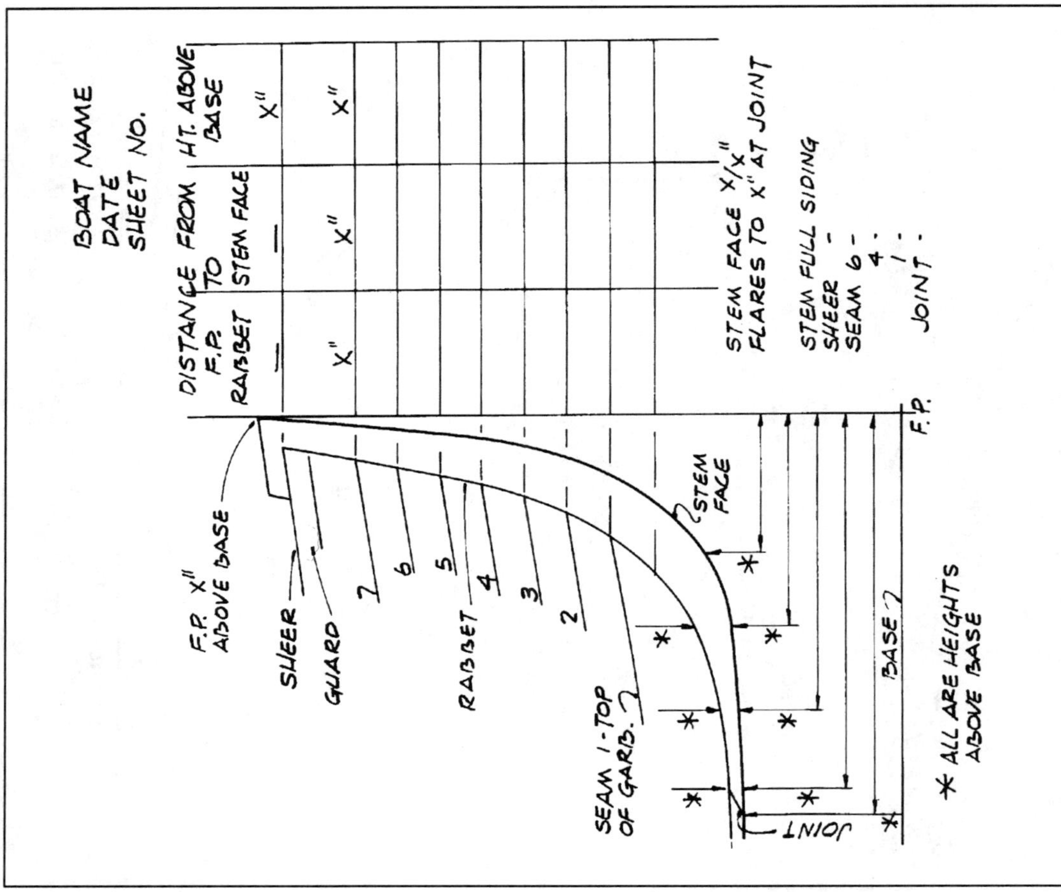

Figure 7.48. Taking off the bow profile with a staff at the F.P.—sample note page.

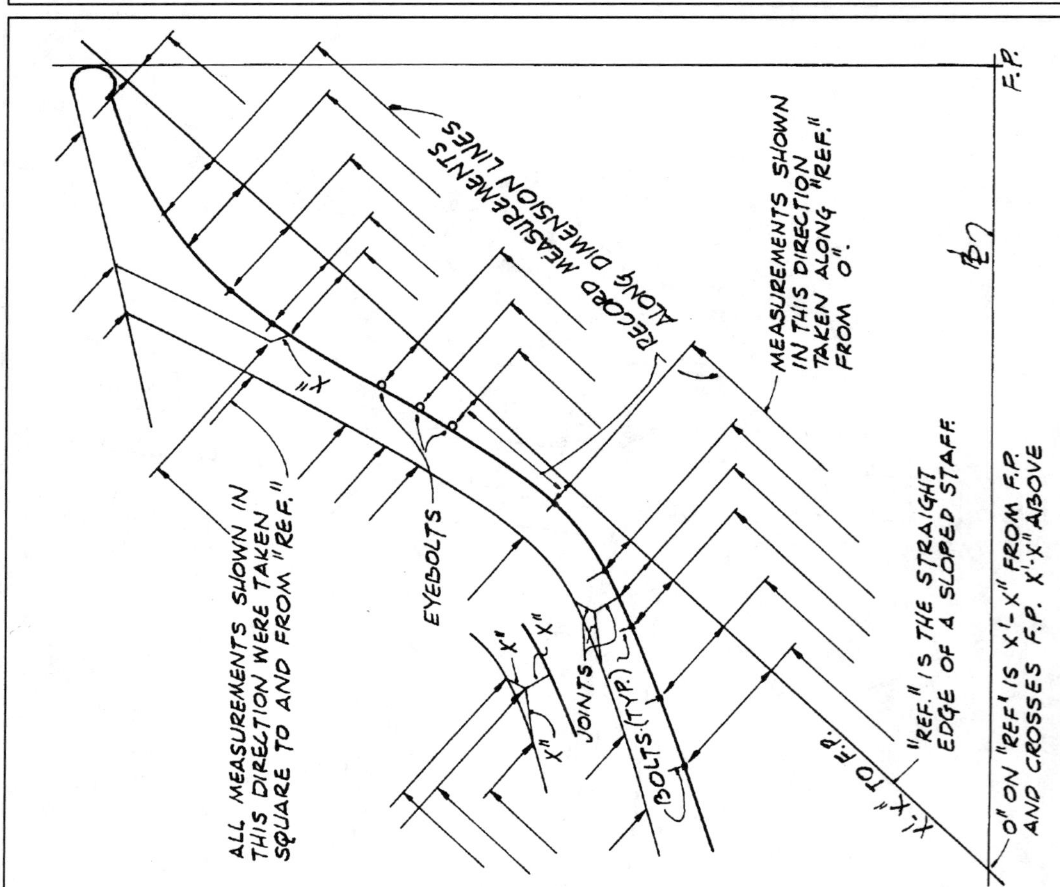

Figure 7.49. Taking off a complex bow profile using a sloping staff. Suggested form of recording measurements.

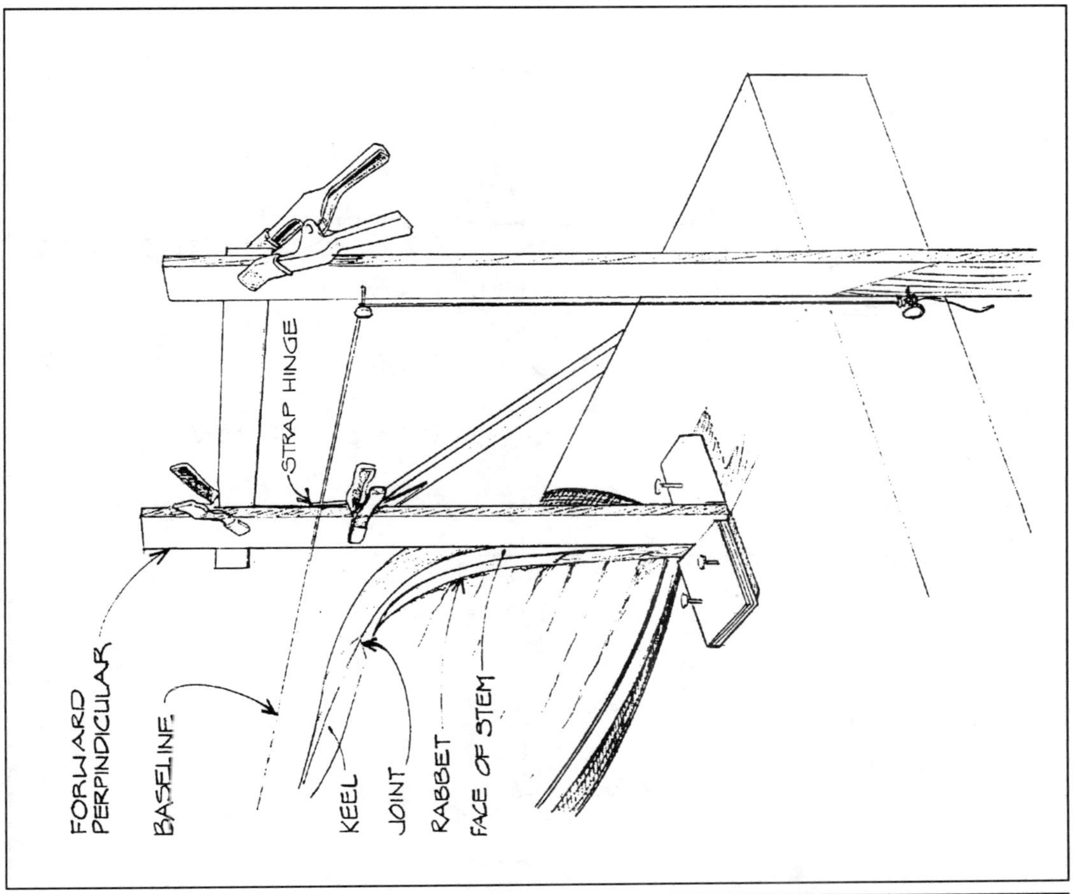

Figure 7.50. Taking off the bow profile with the boat upside down and a staff in place at the F.P. *Peter Schmid*

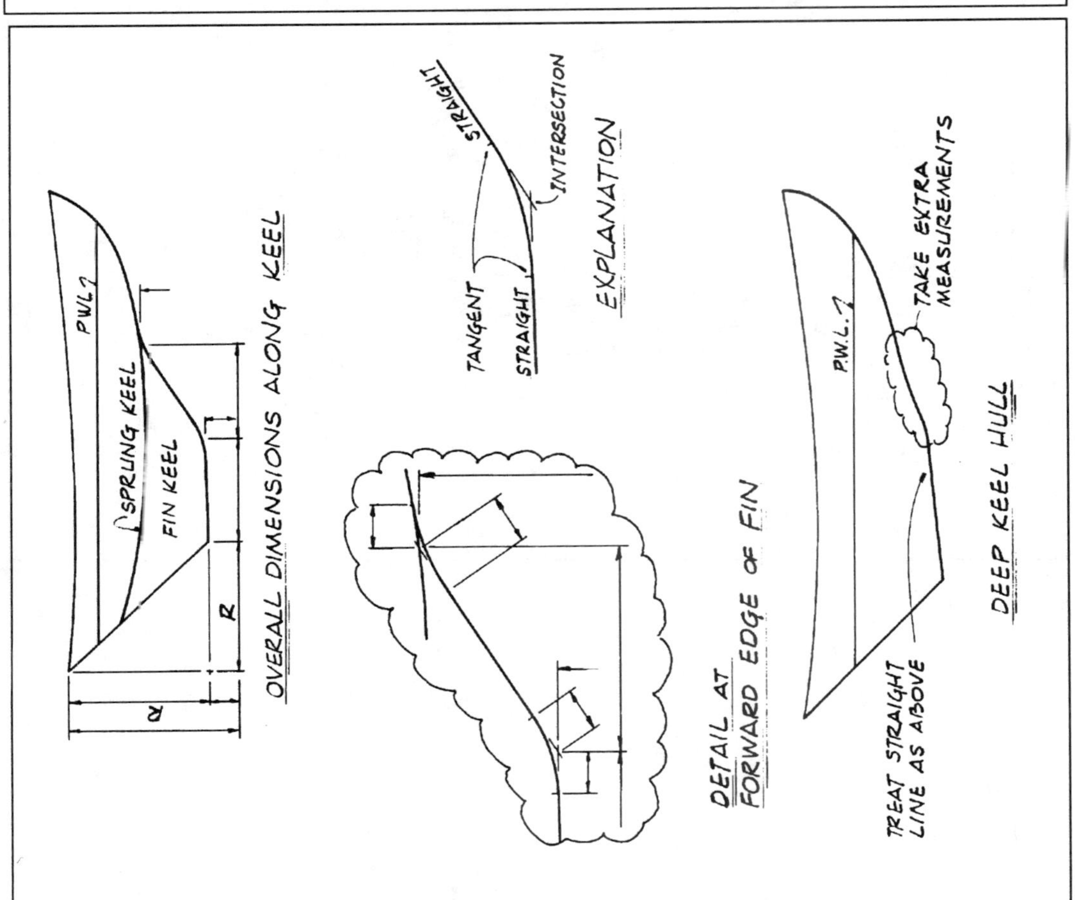

Figure 7.51. Compound shape hull profiles—taking straight-line measurements.

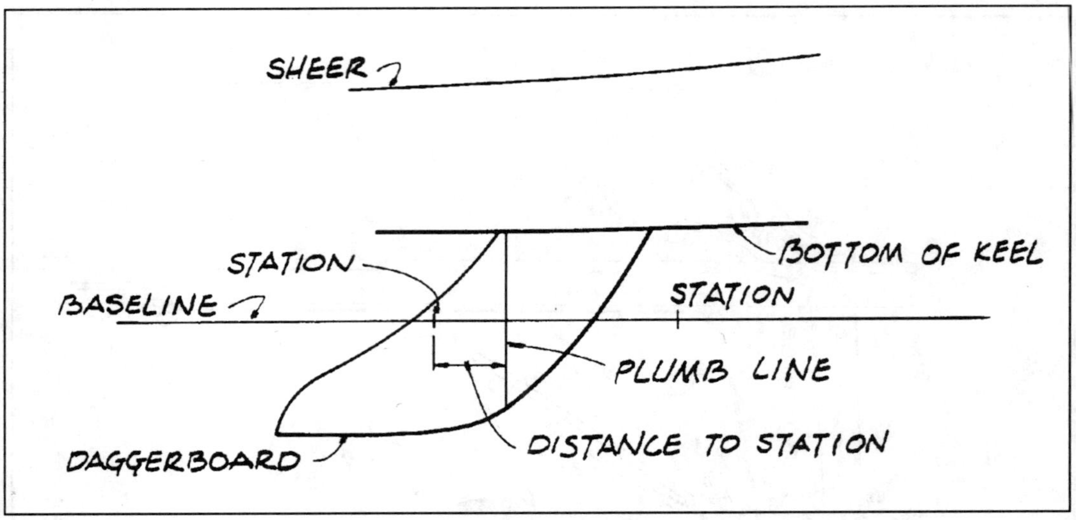

Figure 7.52. Marking a reference line on a daggerboard.

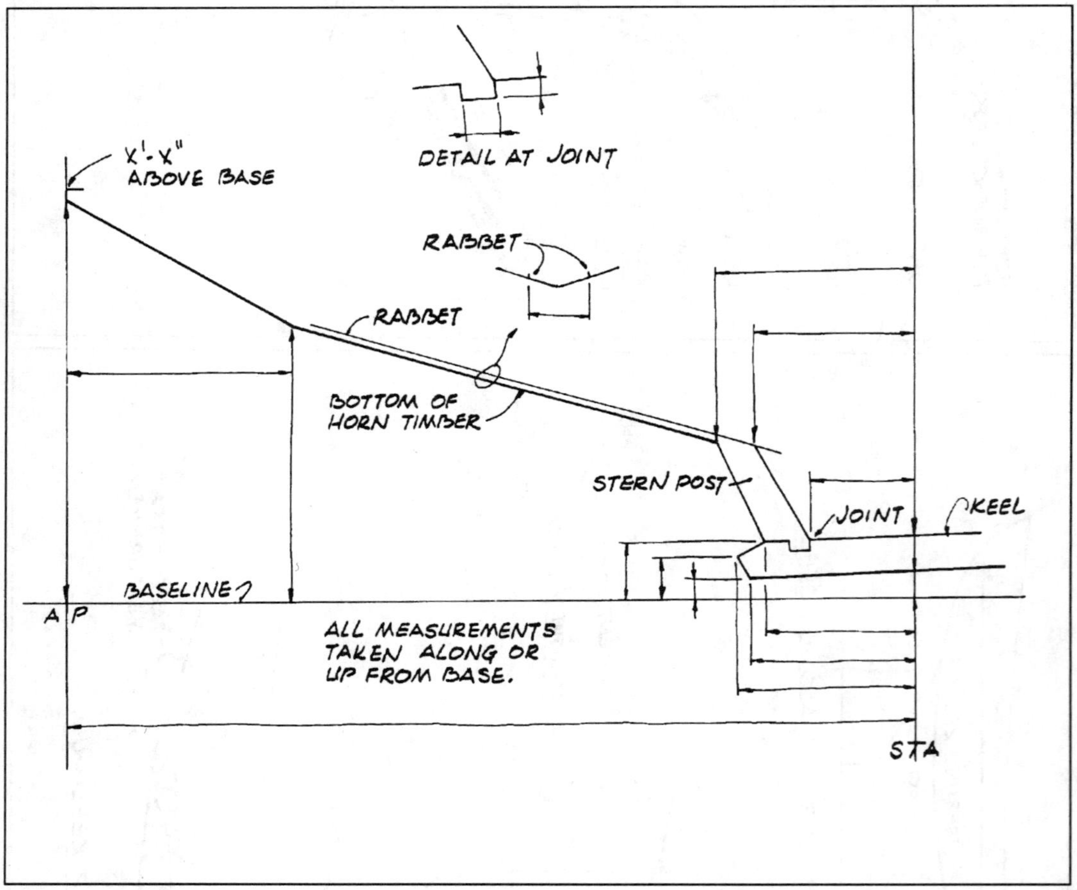

Figure 7.53. Measurements to take from a counter stern.

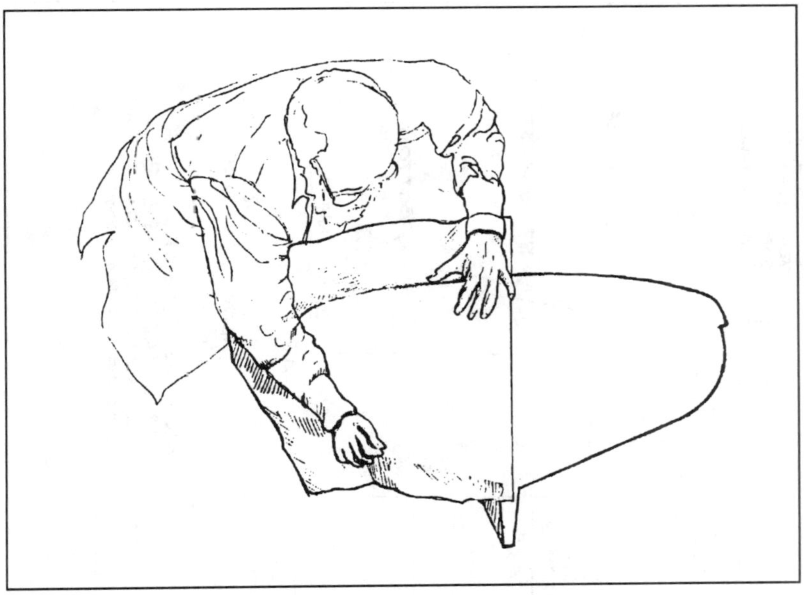

Figure 7.54. Making a transom pattern. *Peter Schmid.*

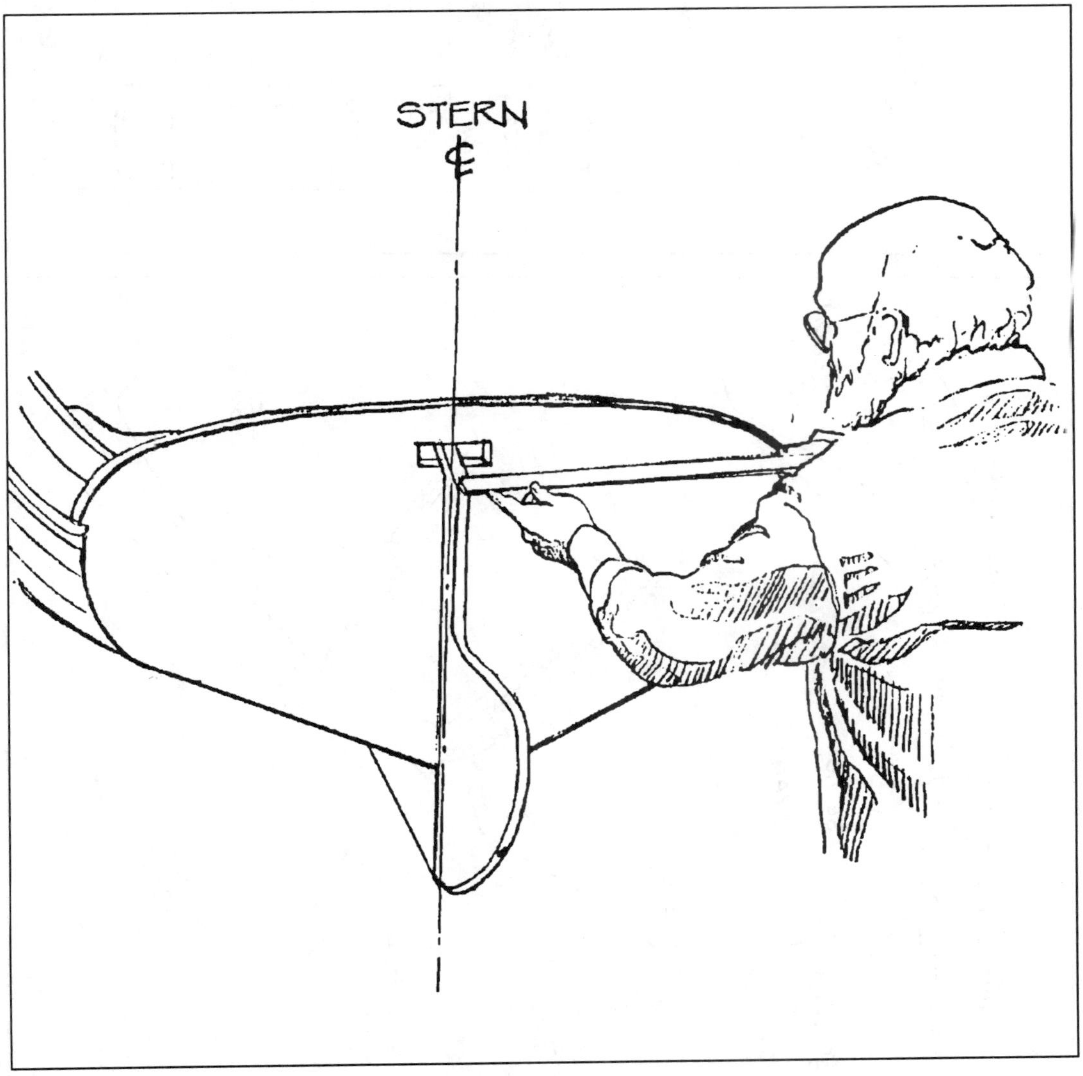

Figure 7.55. Measuring a transom. *Peter Schmid.*

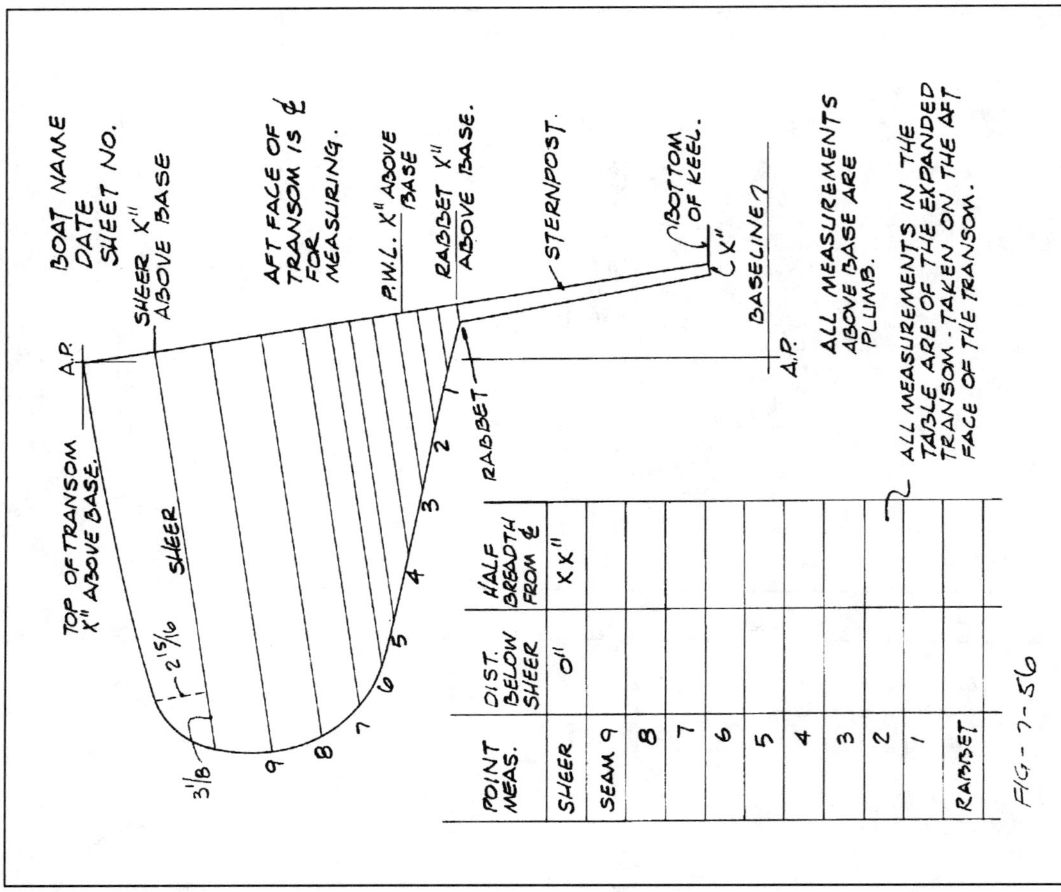

Figure 7.56. Form for recording measurements taken from a flat transom.

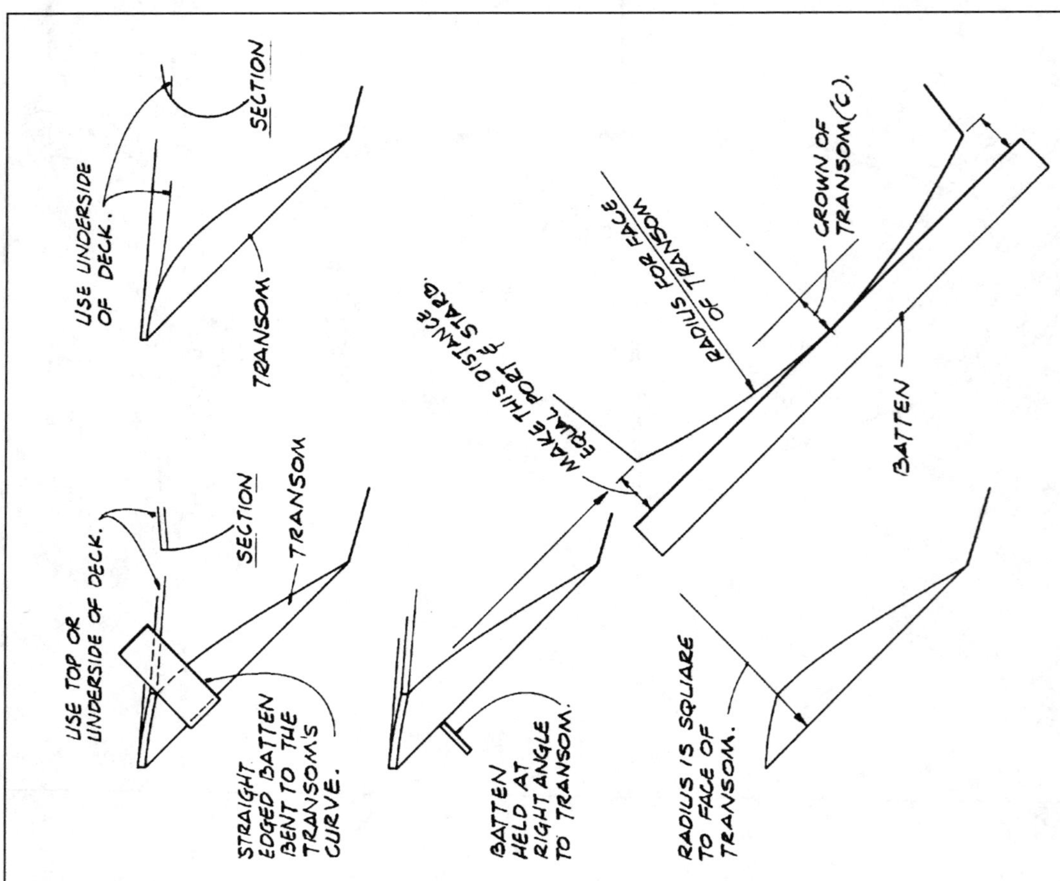

Figure 7.57. Taking off a curved, raked transom.

Chapter 8

Organizing and Plotting Field Notes

Keeping clear, useable notes

D. W. Dillion

The field notes are possibly the most important part of the recording process. They are the record—written notation and sketches—of the measurements and other information taken off the boat: its shape, construction, and joinerwork. If the field notes are cryptic, poorly organized, or sloppy, they will not be a solid foundation for the next step in the recording process—drawing, lofting, or replication.

Waiting until the take-off is complete to prove the accuracy of your work can be costly in time and effort. It is therefore necessary to make sure that your work is accurate as you progress.

The best way to check the accuracy of your work is to plot the measurements as they are taken. To do this, you must understand how to plot—to make a scale drawing from the data taken off the boat. Do not confuse or substitute plotting in the field with the final plotting done on the loft floor or on the drawing board.

Plot the take-off of the hull sections and profile as accurately as you can. It is nearly impossible to work as accurately under field conditions as it is on the drawing board or loft floor, so don't expect hairline accuracy.

The take-off of construction and joinerwork should be plotted accurately enough to show that things "look right" and fit. Plotting your notes in the field will prove or disprove that:

1. the measurements taken give a shape that is visually similar to that measured
2. there are no gross measuring errors
3. the measuring equipment location and relationship to the baseline was properly noted
4. the measurements are properly taken and recorded and are the correct measurement of the correct point.

What skills are needed? The ability to:

1. keep track of what others are doing
2. letter (print) or write legibly
3. use a scale or ruler to plot measurements.

To plot your notes, you will need the following tools and materials:

1. pencil and eraser—a fineline pencil with a separate eraser works well
2. clipboard or other portable writing surface
3. scale or ruler
4. paper: 8 1/2" x 11" will cover most situations. Use either plain or gridded (eight squares to the inch preferred). Lined writing paper is good for general notes, such as the scantling list and the index.

Neither a great deal of equipment nor artistic skill is needed to produce a good set of notes, though some patience and ingenuity are required. A modicum of drawing or drafting knowledge is necessary. The ability to draw fine perspectives is a luxury, not a necessity. If perspective drawing is one of your skills, use it, but don't overuse it. Any object that can be drawn in perspective can be more than adequately shown in a two- three- or four-view sketch, and it will be easier to show measurements on than a perspective drawing.

Figure 8.1 is a copy of a sheet of field notes. It is an unfortunate example of note-taking, both confusing and sloppy. When deciphered however, the measurements shown do plot quite nicely to represent a body section. There is nothing on this sheet or anywhere in the full set of notes, to tell the relationship between the horizontal, vertical, and diagonal references. Nor is there specific mention that the triangle drawn in the upper part of the sheet is related to the horizontal and vertical reference lines. Those things are left for the reader to decipher.

The symbols in fig. 8.2 indicate several possibilities that might develop when plotting the section from these notes. Because height measurements were taken from both the baseline and the horizontal reference, it is quite possible that a plotting error of 4 5/16 inches (the vertical distance from the baseline to the horizontal reference) could be made when plotting the vertical measurements.

If the heights 12, 18, 24, 30, 36, and sheer along the vertical reference were interpreted as heights above the horizontal reference instead of above the baseline, the points indicated by the circled cross marks would result when the distances 30 3/8", 17 1/16", 8 1/8", 3 7/8", 2 5/16", and 1 13/16" are plotted. The plain cross marks are the points as plotted when the vertical reference is related correctly to the baseline.

When the heights above the horizontal reference indicated at 6, 12, 18, 24, and 30 along the horizontal reference are plotted, the locations indicated in triangles result. A comparison between these two plots makes the gross error immediately apparent. The plotted points enclosed in squares were all assumed to have been measured from and square to the diagonal reference.

The notes would be clearer and both the measuring and the plotting easier if all heights were taken from the horizontal reference. Somewhere there should be an indication of what the numbers along the reference lines are. A statement similar to "Arrangement of Reference Lines" or the notations "dist. from c.l.", "dist. from horiz. ref.", and "height above b.l." would do the job. Also, the measurements for the keel and rabbet could have been more clearly sketched. This is the data enclosed in a cloud at the bottom right of fig. 8.1.

A sketch such as "arrangement of reference lines" at the top of fig. 8.2 should be included to illustrate the relationship between the three reference lines.

Figure 8.3 shows the note sheet in fig. 8.1 reworked into a more readable form. Also look at figs. 11.46 and 11.47. The body section shown could have been taken off without the diagonal reference. Measurements taken from the horizontal and vertical references would have been sufficiently accurate.

Figure 8.4 is a copy of an actual note sheet made to record a sail. Even for someone familiar with the type of sail and the form of notation, these notes are barely adequate.

The sail dimensions written in the upper left corner of the sheet note all measurements except that for the leech as "stretched." Was the leech stretched or not? Is this an inconsistency or an error in notetaking? The reefing points, noted in the upper right corner, indicate that there are nine of them 8 inches long, 17 inches up from the foot of the sail. This is fairly clear. But, are the reefing points at the seams or in between? The roping notes in the middle of the sheet are cryptic. The semi-calligraphic lettering is also a weak point. If your natural writing style is calligraphic, either control it so that it is legible, or don't use it. Use simple, block letters instead.

Figure 8.5 is an improvement on fig. 8.4 but still has shortcomings. Missing from both note sheets are the size and shape of the stitching, the sail material, and the sailmaker's name. (See chapter 9 for more information on recording sails and rigs.)

Figure 8.6 shows several ways of arranging, segregating, and identifying information on a note sheet. Each subsequent sheet should include the data shown in the upper right-hand corner of fig. 8.6. At the bottom of the page are two samples of the grids found on gridded paper.

The notes should contain the following information and should be organized in the following manner:

1. Title Sheet (fig. 8.7) with the following information:
    a. Boat name.
    b. Owner's name and address. If owned by a museum, the museum's name and address, and the boat's accession number. If privately owned, the owner's name and address.
    c. Where recorded.
    d. The date recorded.
    e. Names of recording team.
    f. General description of the boat.
2. Index Sheet (fig. 8.8). Keep a working index while taking off measurements, and rewrite it later if you wish to have related bits of data kept together. Don't forget to enter each sheet of notes into the index and indicate the sheet number in the index and also on the note sheet.
3. Scantlings List (fig. 8.9). This list will expand as you find more material to note and cross reference. This sheet is a catch-all and might include random notes and observations, such as the reminder that the plank is screw-fastened or the cross reference at the entry for the stern post. Include the scantlings and hardware for the details with the detail sketches, but note where that data is to be found. The scantling list should include:
    a. Name of piece
    b. Its dimensions
    c. Material it is made of, i.e., what type of wood, metal, etc.
    d. How fastened
    e. Any special detail that can be noted in the space available or the name of

the piece and a reference to where it is shown in detail

f. Small sketches showing the sectional shape of the piece. Figure 8.9 stops at the listing for the riser but would in reality continue on to list all the major items in the boat. The smaller items should be dealt with in detail sketches.

4. Sectional sketch of the hull at approximately amidships (fig. 8.10). Start to make up the scantlings list and sectional sketch before doing any other measuring. Walk around the boat, get under it, and list everything you see. Repeat the process inside the boat. If you don't know the name of a piece, write a letter or number on a length of masking tape, stick the tape on the piece, and record the number or letter on the list. When the recording is completed, you can research books or plans or simply ask a boat builder to determine the name for the piece. As you work throughout the boat, you will find more items to add to the list. All members of the recording team should participate in this work.
5. Sketches to describe the details of hull and deck construction, joinerwork, and hardware.
6. Sketches of every opening in the hull whether hatch, cockpit, or hardware. Refer to figs. 8.11 and 8.12.
7. Sketches showing the measuring equipment, reference lines, and station spacing as shown in figs. 8.13 and 8.14.
8. The take-off for each body section as shown in figs. 8.15, 8.17, 8.24, and 8.25. (Also refer to chapter 7.)
9. The take-off for the bow, as shown in fig. 8.26. (Also refer to chapter 7.)
10. The take-off for the stern, as shown in fig. 8.23.
11. The location of structural members and joinerwork, either on the detail sketches or as a separate list.
12. Details of the sailing rig. (Refer to chapter 9.)
13. Details of any mechanical equipment: engine, fishing gear, winches, windlasses, or hoisting gear.
14. Sketches of the working accessories normally found with the boat, such as oars, special oarlocks, poles, seat backrests, etc.

Figure 8.11 shows a hatch opening and the construction around it as sketched on plain paper. The two details, "batten ends" and "joint—carlin to deck beams" are free-hand sketches made with no attempt to have them either at scale or to proportion. The "hatch opening" was drawn using a 6-inch pocket scale as an aid. No attempt was made to draw the opening accurately; it was "eyeballed" using the end of a pocket scale for the square corners. Measurements taken on the deck showed that the planks at the forward and aft ends of the hatch are 6 inches wide or three planks of equal width across the opening, which is what is shown in the sketch.

The size of a detailed sketch is up to the notetaker. The more complete the detail, the larger the sketch should be.

Figure 8.12 shows the cockpit arrangement of a small, open sailing boat as plotted on paper gridded with eight squares to the inch. To make the plotting easier, each square was given a value of two inches. It was then easier to plan the sketch, make it fit on the sheet, and have it look right. Examine fig. 8.12 carefully. A substantial amount of information is presented in a small space and in a variety of ways.

Figure 8.13 shows the measuring equipment set up to measure a boat and also includes the information that should be given about the set-up in the notes. This semi-perspective drawing is a nice idea, but fig. 8.14 combined with the other note sheets can do the same job. It would be difficult to show the stations and the measurements locating them on fig. 8.13. The only real use for a perspective illustration in the notes is its pictorial value, as it may be more readily understood by more people. On the other hand, anyone who has studied this manual should be able to read any of the sketches presented here.

Figure 8.14 shows and describes the reference lines and planes from which the boat was measured in profile (side view): the baseline, what it was, and what it was attached to; and what the perpendiculars at the ends of the boat were. Refer to figs. 8.20, 8.21, and 8.22 for a guide in setting up a profile sketch.

Figure 8.15 shows the section of a chine-hulled power boat plotted on gridded paper. Each square equals 1 inch. The outline of the L-shaped measuring frame is shown. The frame was made from 3/4-inch plywood with wooden yardsticks tacked to it. The two reference lines are all that is really necessary to show in the notes. The baseline location measurements indicated by the circled letter A can be noted either with written notes or schematically.

Figure 8.16 shows how fig. 8.15 was plotted. Start thinking in terms of the grid, because to use the grid, each square must be given a value. It may be most convenient to make each square equal to 1 inch. A standard 8 1/2" x 11 sheet gridded with 1/8-inch squares, will then equal 68 inches width and 88 inches in height. If a 1-inch margin is used at the left hand edge, and 1/2 inch for the three remaining edges, the useful area is 56 inches by 80 inches.

Start by drawing the horizontal and vertical reference lines on grid lines. The horizontal reference is at 0 inches for all heights or up-and-down measurements. The vertical reference is at 0 inches for all distance or side-to-side measurements. Do not think of distance or side-to-side measurements as "horizontal." Use the word horizontal only in reference to measurements along the horizontal reference. Draw the reference lines near the bottom and right- or left-hand edge of the sheet. This will allow plenty of room for the section to be plotted, or turn the sheet 90 degrees to suit a low, wide section. The table of measurements then goes at the top of the sheet.

Plotting the points is a matter of counting squares and estimating by eye (interpolating) the fractional parts of the square to divide it into halves, quarters, and eighths. The "detail" is an 8 times enlargement of the square 33-34 above horizontal reference and 9-10 distance from vertical reference.

As an example, to plot the sheer given in the table at the bottom left corner of figure 8.15 as 33 3/4 inches height and 9 5/8 inches distance, mark the 33rd and 34th lines above horizontal reference. On them, mark the 9th and 10th lines out from the vertical reference. Then, on the 9th and 10th lines, halfway between the 33rd and 34th lines, make a light mark that will represent 33 1/2 inches above the horizontal reference. Divide the space between 33 1/2 inches and line 34 to get the height of 33 3/4 inches. Draw a light line between the two 33 3/4 inches heights. This is the sheer height above the horizontal reference.

To plot the 9 5/8 inch measurement, make a mark halfway between lines 9 and 10 on the sheer height line at 33 3/4 inches. This is 9 1/2 inches from the vertical reference. Divide the space between 9 1/2 inches and line 10 in half to get a distance 9 3/4 inches from the vertical reference. A mark halfway between 9 1/2 inches and 9 3/4 inches locates the distance of 9 5/8 inches and completes the plotting of the sheer.

A further example is the plotting of the bottom seam 2, which has a height of 11 13/16 inches and a distance of 25 3/8 inches. In this case, the dividing of the grid square into fractions is done in the square outlined by lines 11 and 12 above the horizontal reference and lines 25 and 26 to the right of the vertical reference.

Begin dividing the grid squares with the one that has the coarsest division—as 33 3/4 inches, rather than 9 5/8 inches; or 25 3/8 inches rather than 11 13/16 inches. A bit of practice using fractions in a number of single block spaces will demonstrate that it is quite easy to divide grid squares as needed.

When all the points are plotted, connect them with a series of lightly sketched dots or dashes. For verifying data in the field, this works quite well and has the side benefit of training your eye to see shapes and errors, especially if you look at the hull shape while sketching.

Figure 8.17 shows a round-hulled section plotted on gridded paper. The boat was measured upside-down using two methods of taking off. The point indicated by the arrows could have been located with measurements taken from points 30 and 6 on the reference line to give a clearer crossing of the two arcs. The leveling board, indicated by the heavy outline, had two sections of wooden yardstick tacked to it to give reference points for taking the measurements. This is why the numbers along the reference line repeat, reading from the left 0 to 36, then 0 to 11.

Figure 8.18 shows how fig. 8.17 was plotted. Draw the horizontal and vertical reference lines (center-line) on the paper. In the grid, each square equals 1 inch. Mark the intersection of the reference lines as 12 inches.

Then mark the horizontal reference line at every third or sixth line and number as shown in fig. 8.15.

Refer to the table of measurements and the detail at B.L./Keel in fig. 8.17. Except for those measurements given for seams 5, 6, and 7, the measurements in the table and the detail are plotted by counting and dividing squares. Using seam 3 as an example, the "out" measurement, 25 1/2 inches, falls between lines 25 and 26. Mark them lightly; count down and mark the sixth and seventh lines below the horizontal reference. Divide the square into fractions to locate the point. Repeat the process for seams 1, 2, 4, the sheer, and the rabbet and the bottom of the keel.

Seams 5, 6, and 7 were measured differently. For example, in the table at the bottom left of fig. 8.17, for seam 6 the number 30 in the "out" column is the 30 inch mark on the horizontal reference line; 16 7/8 inches is the distance from the 30 inch mark to seam 6. On line 30, mark 16 7/8 inches down from the horizontal reference. Place the point of a compasses on the 30 inch mark and draw an arc from the 16 7/8 inch mark on line 30 as indicated by the dashed arrow-headed line.

The other measurement locating seam 6 is in the "down" column. On line 36, mark 16 5/8 inches down from the horizontal reference. Place the compasses point on the 36 inch mark and draw an arc from the 16 5/8 inch mark to intersect the previously drawn arc. This is the location of seam 6. Plot the location of seams 5 and 7 in like manner, using the appropriate measurements.

Figure 8.19 represents a stern profile with the circled letters "A" through "K," indicating points to be located by measurement. Figures 8.20, 8.21, and 8.22 show the steps required to plot the finished note sheet, complete with the measurements taken, shown in fig. 8.23.

To plot points "A" through "K" and draw the lines defining the stern, you must build a grid composed of lines at 90 degrees to each other, reproducing the horizontal and vertical lines along which the measurements were taken. The Arabic numerals in quotes are the measurements taken from the boat and will be written along the dimension lines as you progress. The circled Arabic numerals indicate where those measurements are plotted to construct the grid. The circled Roman numerals indicate points from which the measurements—the circled Arabic numerals—are plotted as instructed in the following.

Begin by letting the margin lines in figs. 8.20 through 8.22 represent the edges of a plain sheet of note paper.

Figure 8.20. From the corners of the paper, points I, measure up 1-1/2 inches and draw base. From the edge of the paper at points II, measure 1 inch along the top edge of the paper and base. Draw the aft perpendicular (A.P.). Then along the base and the top edge of the paper, scale off and mark the distance, indicated by the circled numbers 1, from A.P. to the station (STA). Draw in the dimension line and note the A.P.-to-Station distance "1" on it.

Figure 8.21. Scale off and mark the measured height of point "A" along A.P. above base. Write in the measurement as "Top of deck at C.L. —' - —" above base". Repeat the process for point "B" and record the height, indicated by the "2" at A.P. Next, scale off and mark the circled measurements 3 from A.P. along base and the top edge of the paper and draw the line III-III. Then, scale off and mark the circled measurement 4 along line III-III to locate point "C". Record "3" and "4", the measurements from A.P. and the height above base respectively, on the dimension lines. Plotting point "D" is a repeat of plotting point "C". Working from STA aft, scale off and mark the circled measurement 5, along the top edge of the paper and base. Draw the line IV-IV. Scale off and mark the height of "D" above base—the circled measurement 6—on line IV-IV. Draw the two dimension lines and record the measurements "5" and "6".

Up until now all of the plotting has been done in the large area defined by the edge of the paper and base. The remaining points, "E" through "K", fall in a much smaller area and that area should be defined. Scale off and mark the circled measurement 6 above base from point V along the STA-STA line and mark point VI. Draw the line D-VI.

Figure 8.22. To plot point "E", scale off and mark the circled measurements 7 along base and line "D"-"VI" from the STA. Draw line "VII"-"VII". Scale off and mark the height of "E" above base on line "VII"-"VII" to locate "E". Draw the two dimension lines and record the measurements "7" and "8". Plot "F" by scaling off and marking the height above base on the line STA-STA. Draw the dimension line and record the measurement "9".

Figure 8.23. The finished note sheet with all the measurements taken from the boat recorded and shown as the dimensions "1" through "17".

Other examples of notes plotted on plain paper are shown in figs. 8.24 and 8.25. These figures illustrate the same hull sections plotted in figs. 8.15 and 8.17. Figure 8.26 is a bow profile plotted on plain paper.

Figures 8.14, 8.24, and 8.26 are plotted using the same principles demonstrated for figs. 8.20, 8.21, and 8.22. A system of lines is drawn square and parallel to each other and are used to scale off the measurements taken from the boat. The boat illustrated in fig. 8.25 was measured using two methods of taking off.

To plot data as shown in fig. 8.25, draw lines parallel to the centerline at 24, 30, and 36 on the horizontal reference line (top of ref. arm). Then scale off the distance below the horizontal reference along those lines locating the bottom of the keel, the rabbet, seams 1, 2, 3, 4, and the sheer. To plot the locations of seams 5, 6, and 7, follow the steps provided in the discussion of fig. 8.17.

Occasionally it is impossible to show in one sketch all of the measurements taken off part of the boat. To solve the problem, make an outline sketch, as shown in fig. 8.27, which shows a bow profile. Then to add more information, either make photocopies or traced overlays of the basic outline as shown in figs. 8.28 and 8.29.

These three figures illustrate the bow of a large vessel. A total of five sheets were used to record all of the detail in the bow. Figure 8.27 is a tracing of a bow profile showing the baseline, forward perpendicular, stations 1/2, 1, and 1 1/2, and the sheer and stem face. Figure 8.28 is the bow profile with the first layer of detail and measurements shown. Figure 8.29 is a tracing of fig. 8.28 without the measurements but with the additional detail and new measurements shown.

To conclude, how many note sheets are needed to plot the field notes? The only truthful answer is "As many as it takes." Realistically, a 16-foot open boat might need 15 sheets; a 30-foot power boat, 45 or 50 sheets.

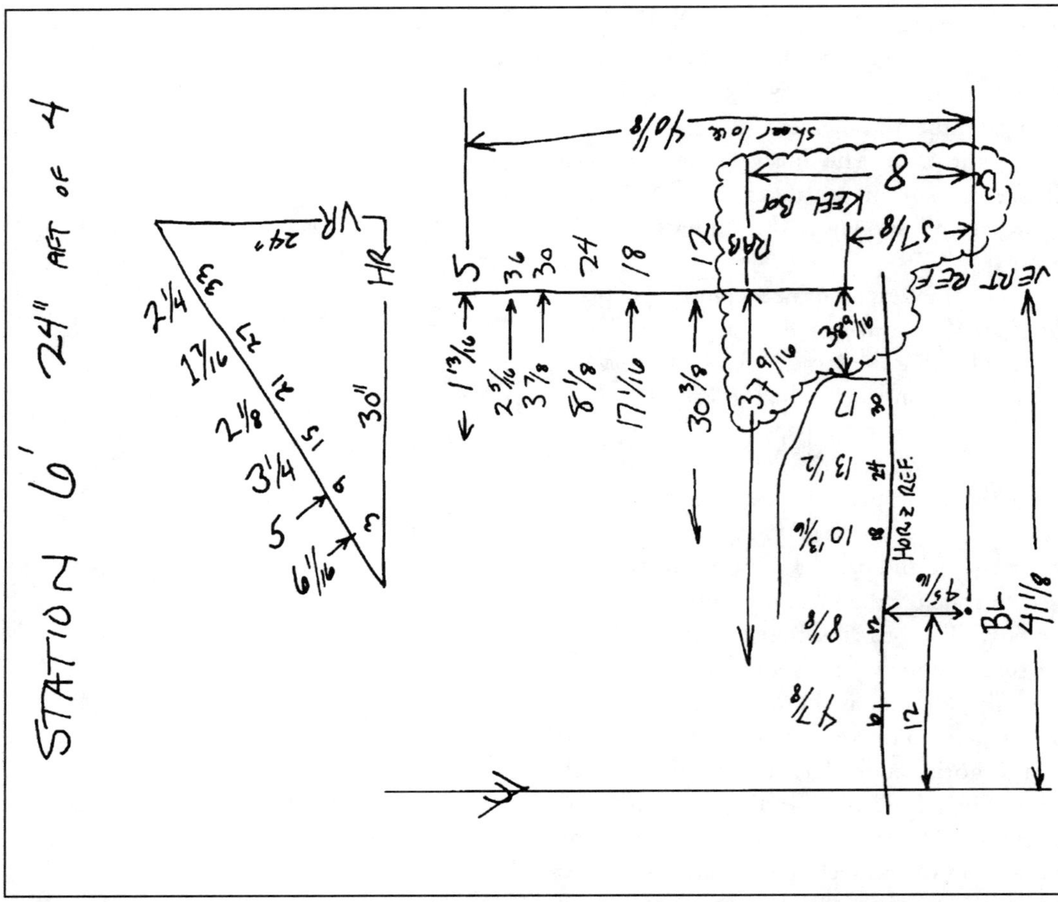

Figure 8.1. Poorly organized notes.

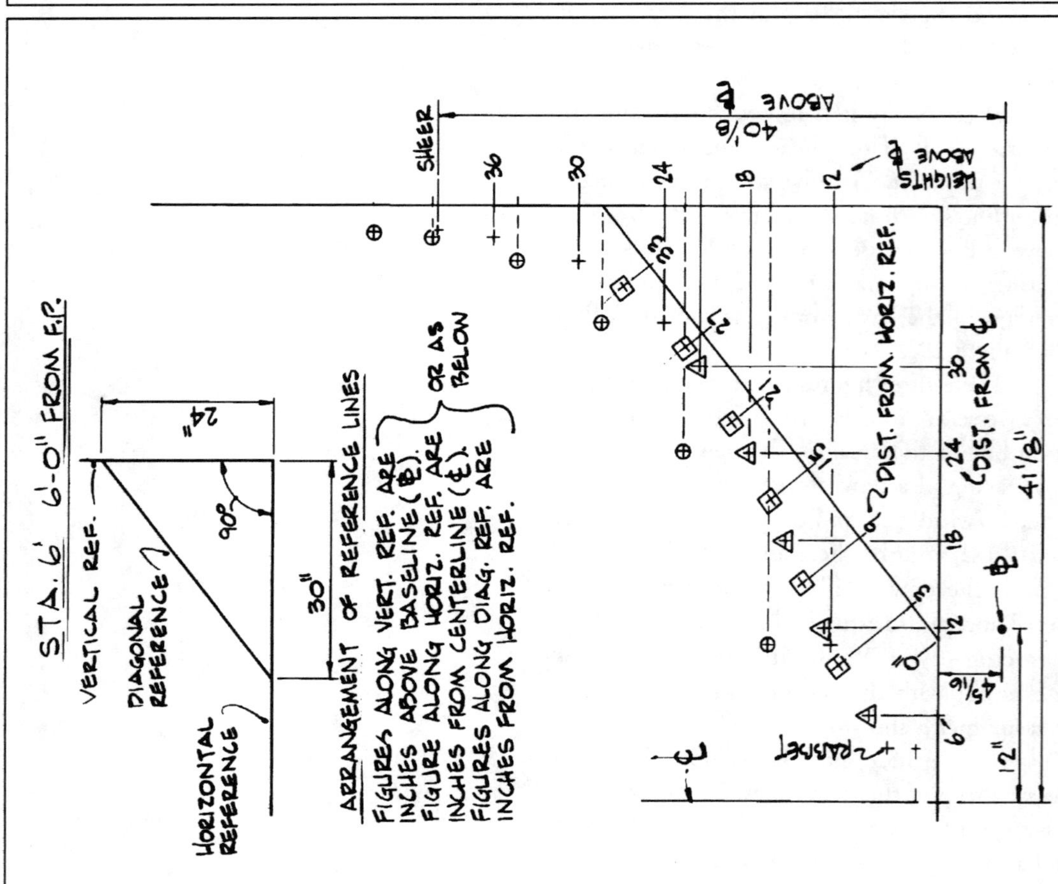

Figure 8.2. Figure 8.1 redrawn to show possible erroneous interpretations.

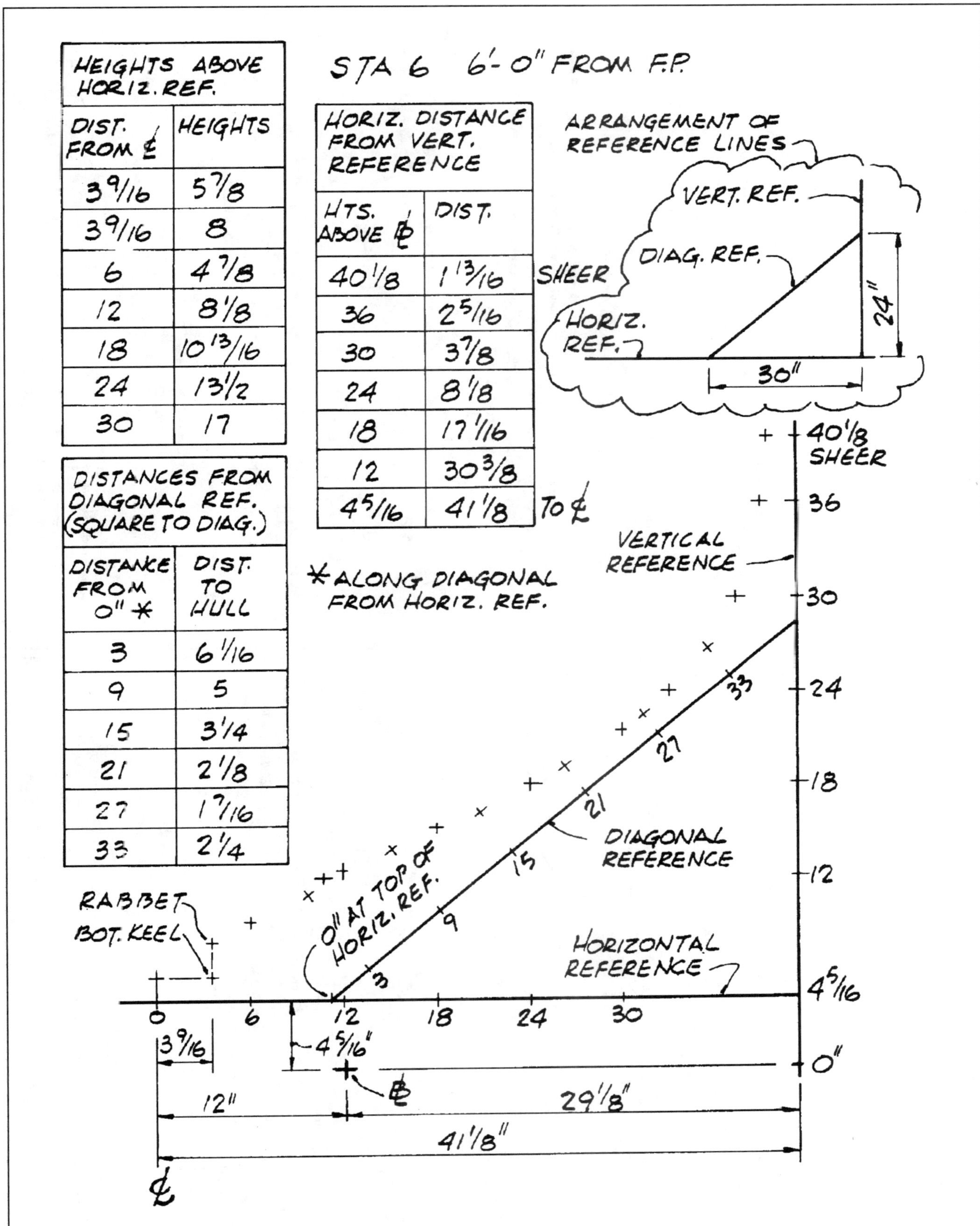

HEIGHTS ABOVE HORIZ. REF.	
DIST. FROM ℄	HEIGHTS
3 9/16	5 7/8
3 9/16	8
6	4 7/8
12	8 1/8
18	10 13/16
24	13 1/2
30	17

DISTANCES FROM DIAGONAL REF. (SQUARE TO DIAG.)	
DISTANCE FROM 0" *	DIST. TO HULL
3	6 1/16
9	5
15	3 1/4
21	2 1/8
27	1 7/16
33	2 1/4

HORIZ. DISTANCE FROM VERT. REFERENCE		
HTS. ABOVE ℄	DIST.	
40 1/8	1 13/16	SHEER
36	2 5/16	
30	3 7/8	
24	8 1/8	
18	17 1/16	
12	30 3/8	
4 5/16	41 1/8	TO ℄

Figure 8.3. Figure 8.1 in better form.

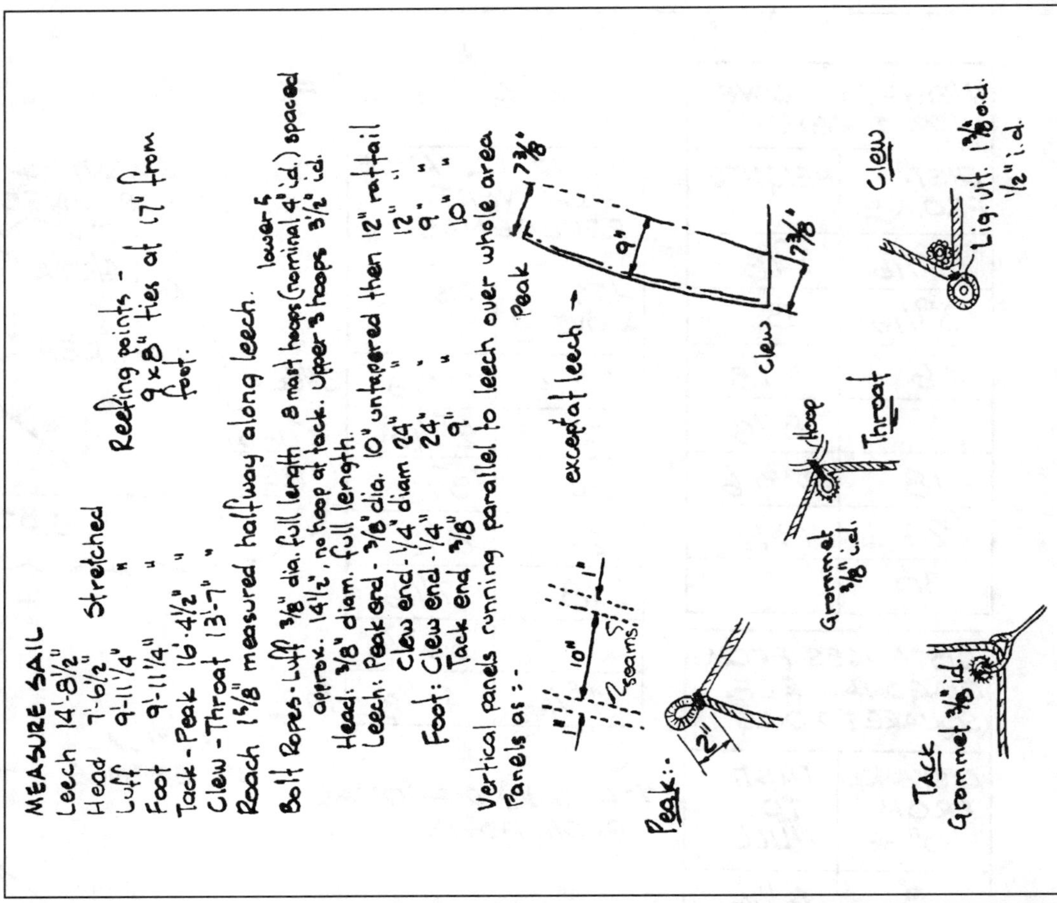

Figure 8.4. A sail recorded in cryptic form.

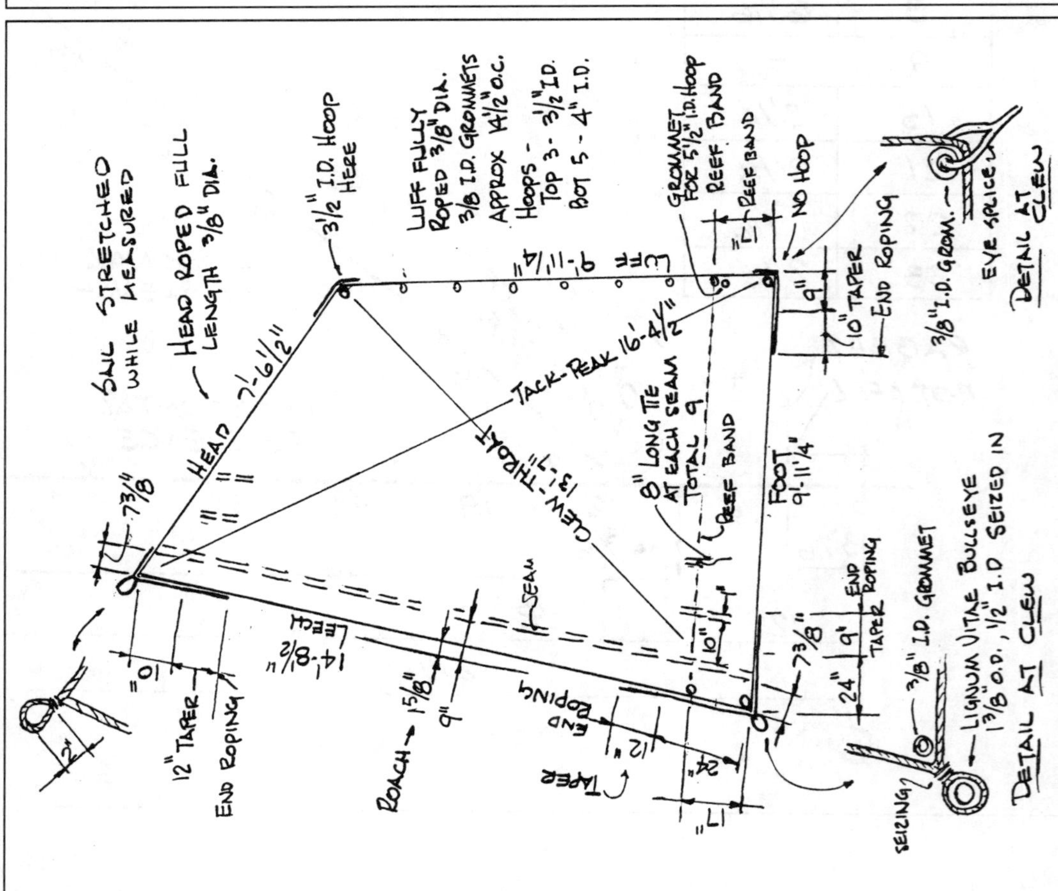

Figure 8.5. An improved version of fig. 8.4.

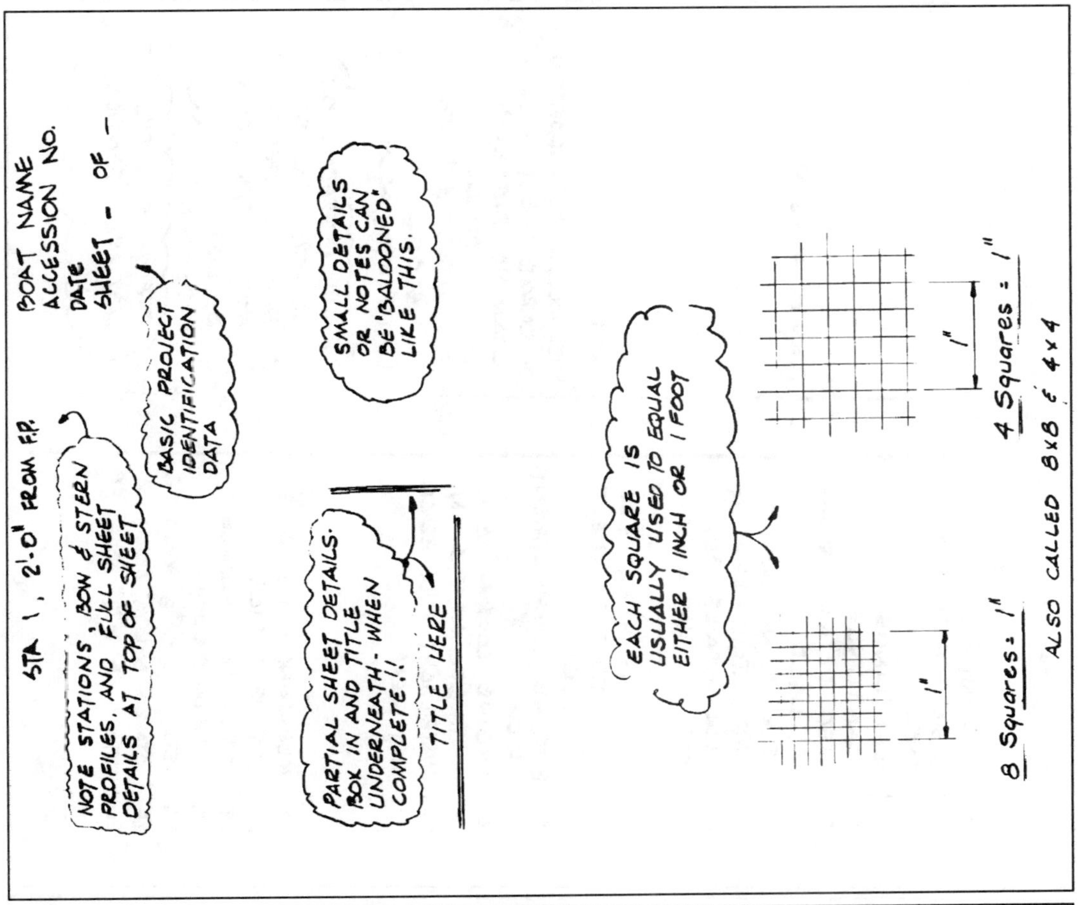

Figure 8.6. Suggestions for arranging a sheet of field notes.

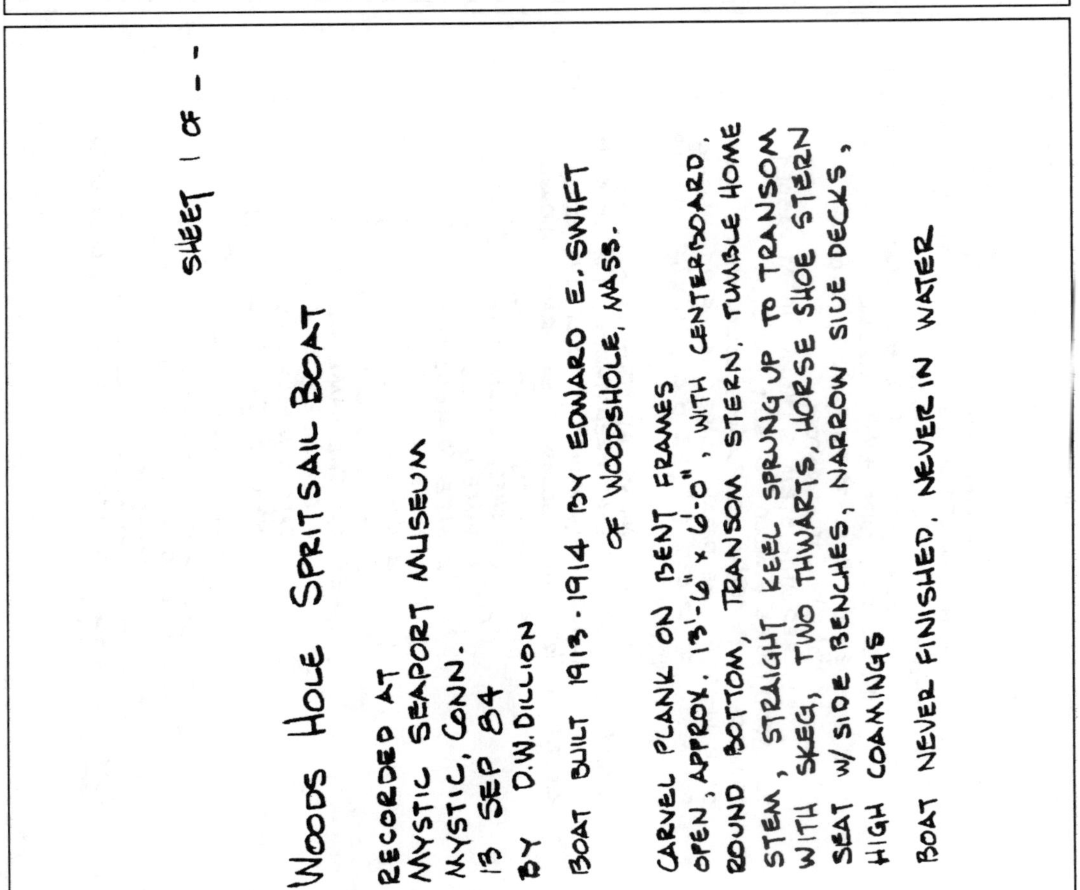

Figure 8.7. First sheet of the field notes.

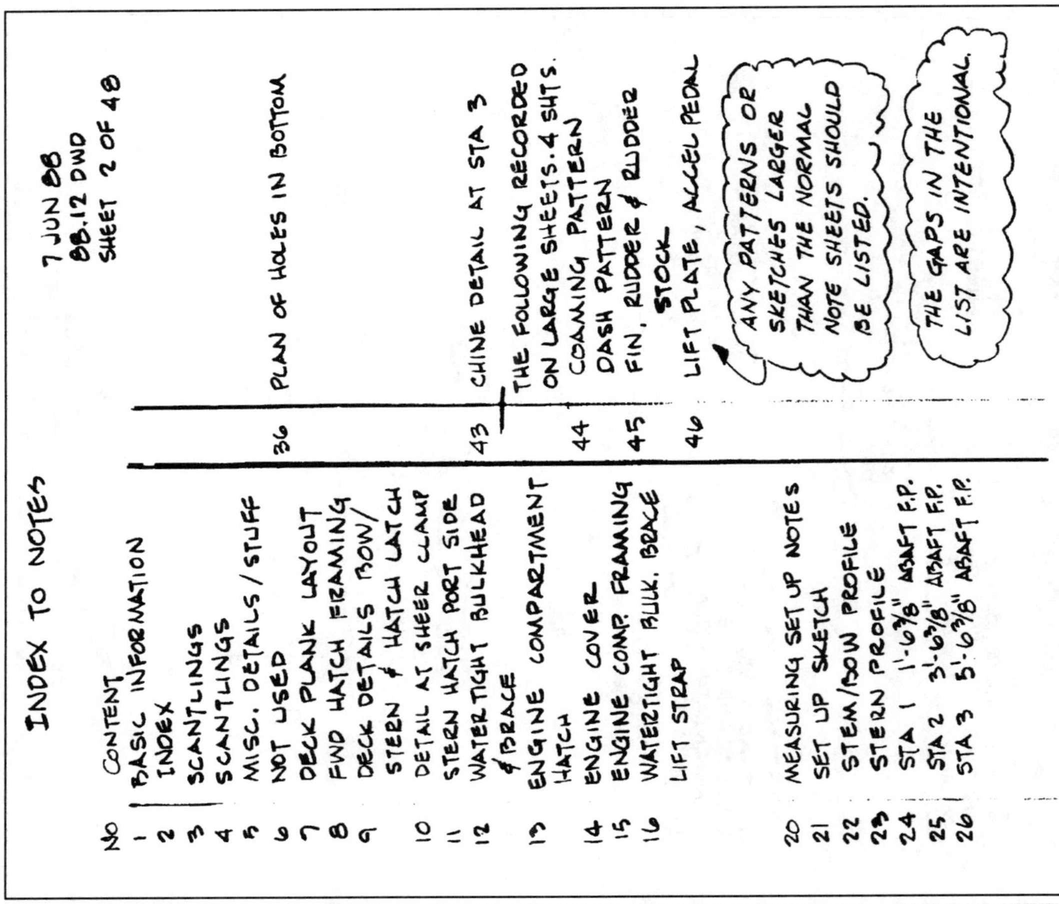

Figure 8.8. Index sheet for the field notes.

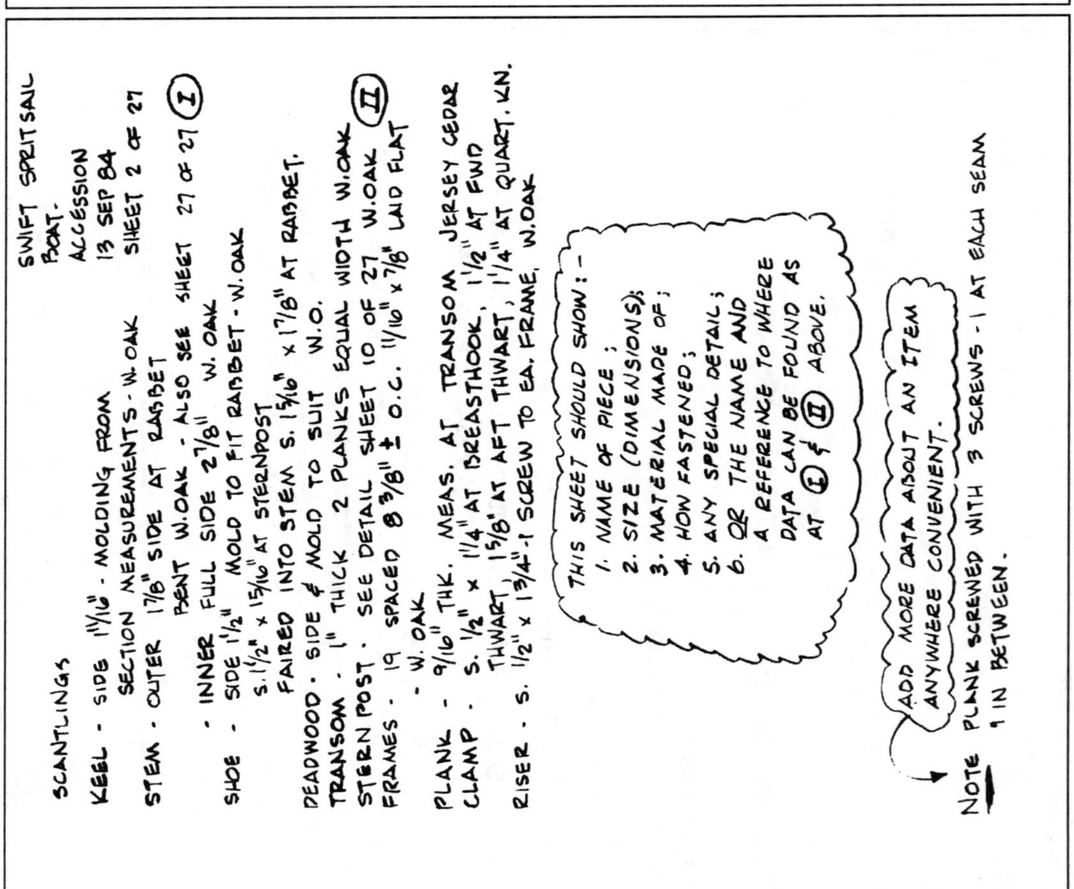

Figure 8.9. Scantlings list.

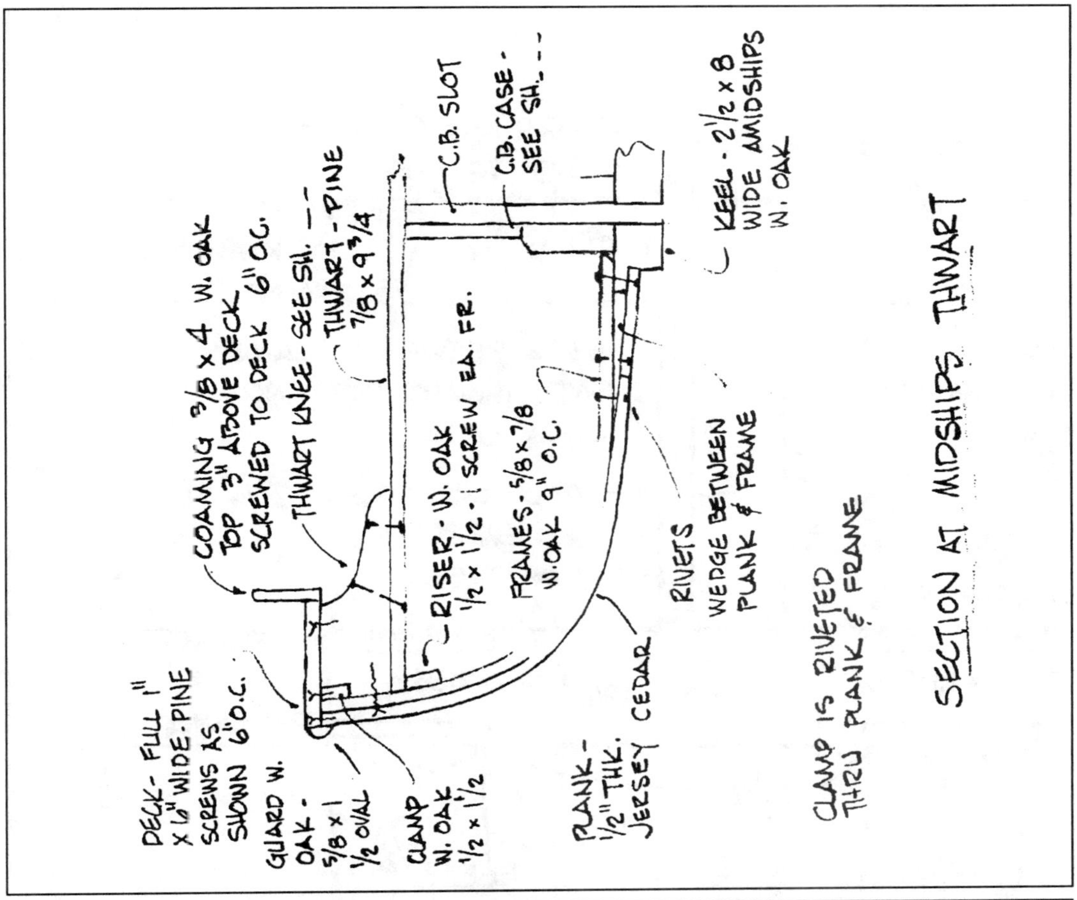

Figure 8.10. Field sketch of *Susie*'s midship section.

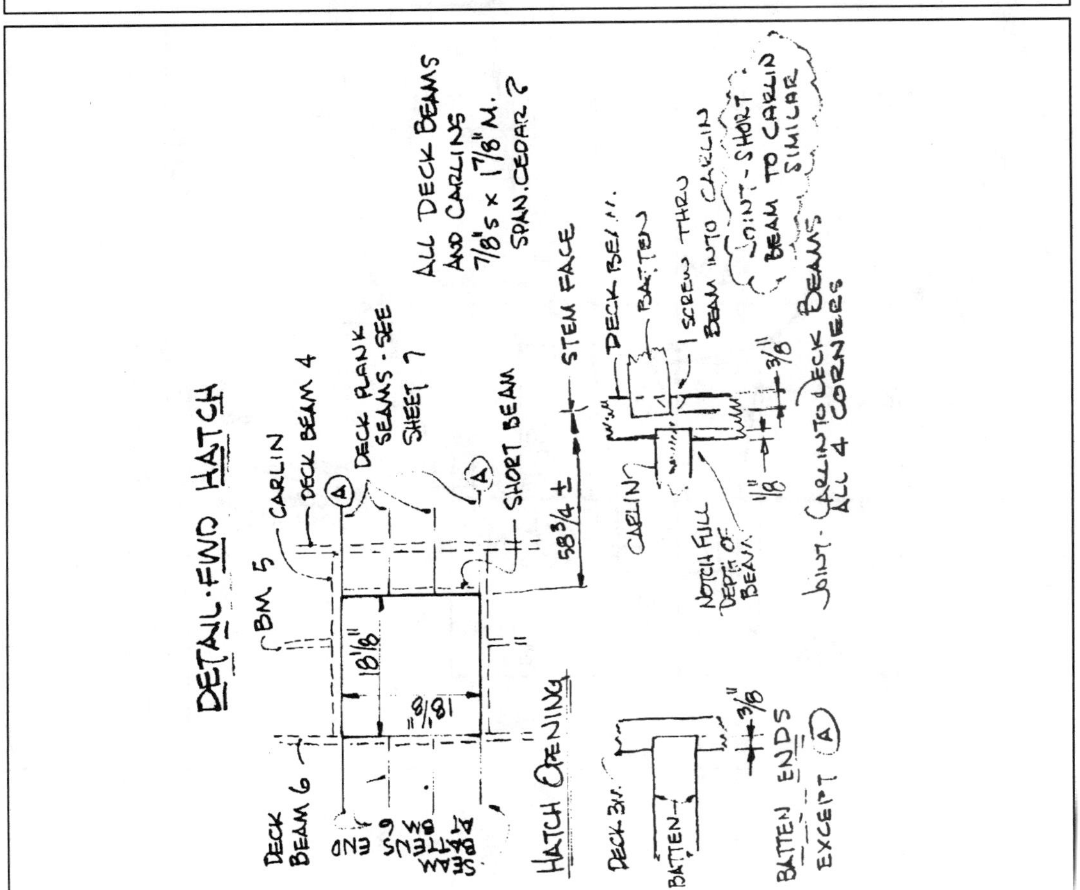

Figure 8.11. Recording a hatch opening and the deck framing around it.

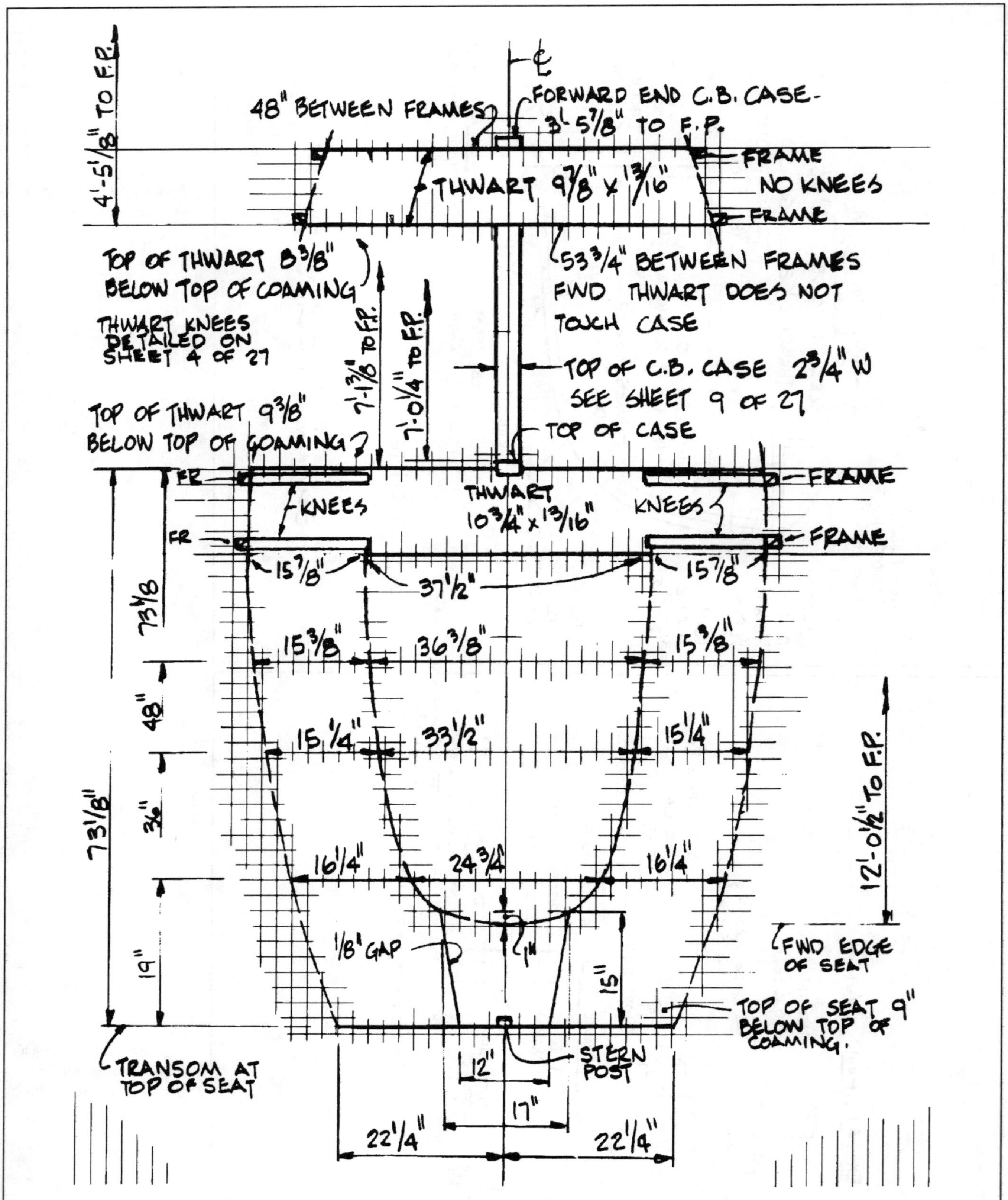

Figure 8.12. Layout of thwarts and stern seat in an open boat as recorded on grid paper.

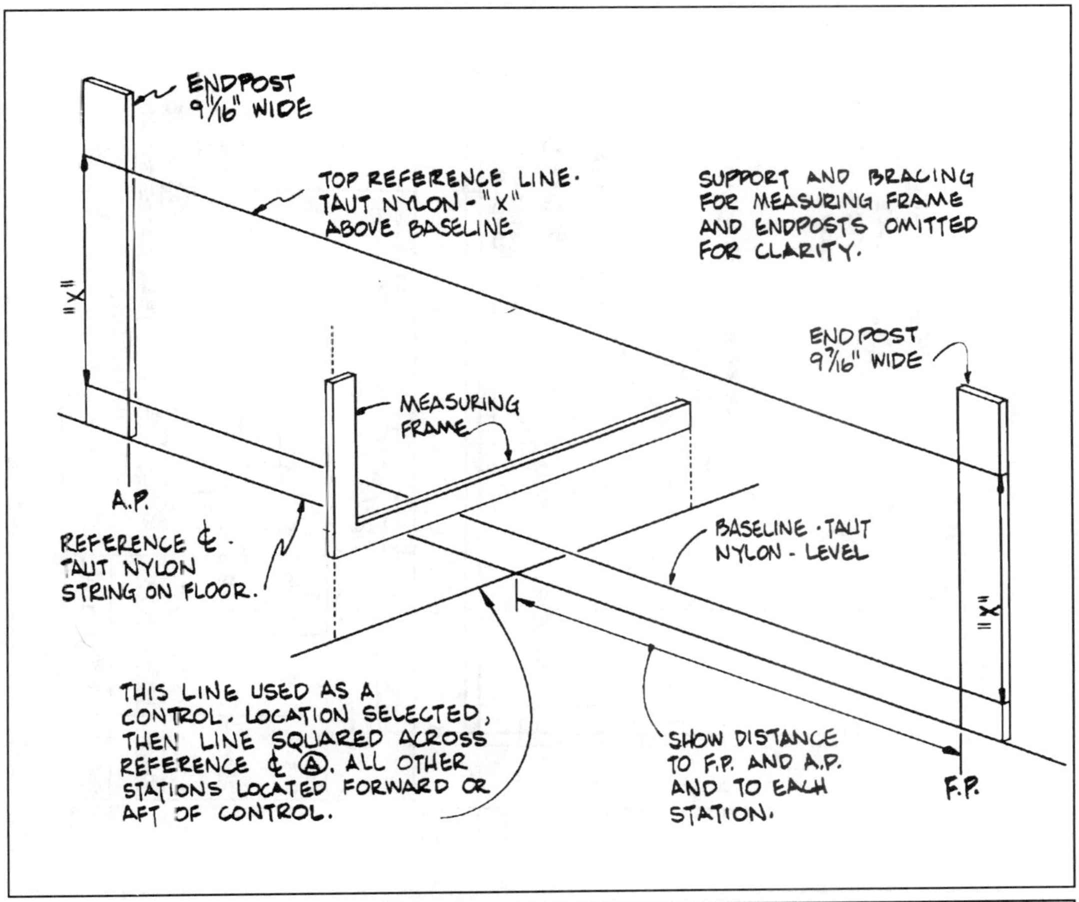

Figure 8.13. A perspective view of measuring equipment set up and ready to be used.

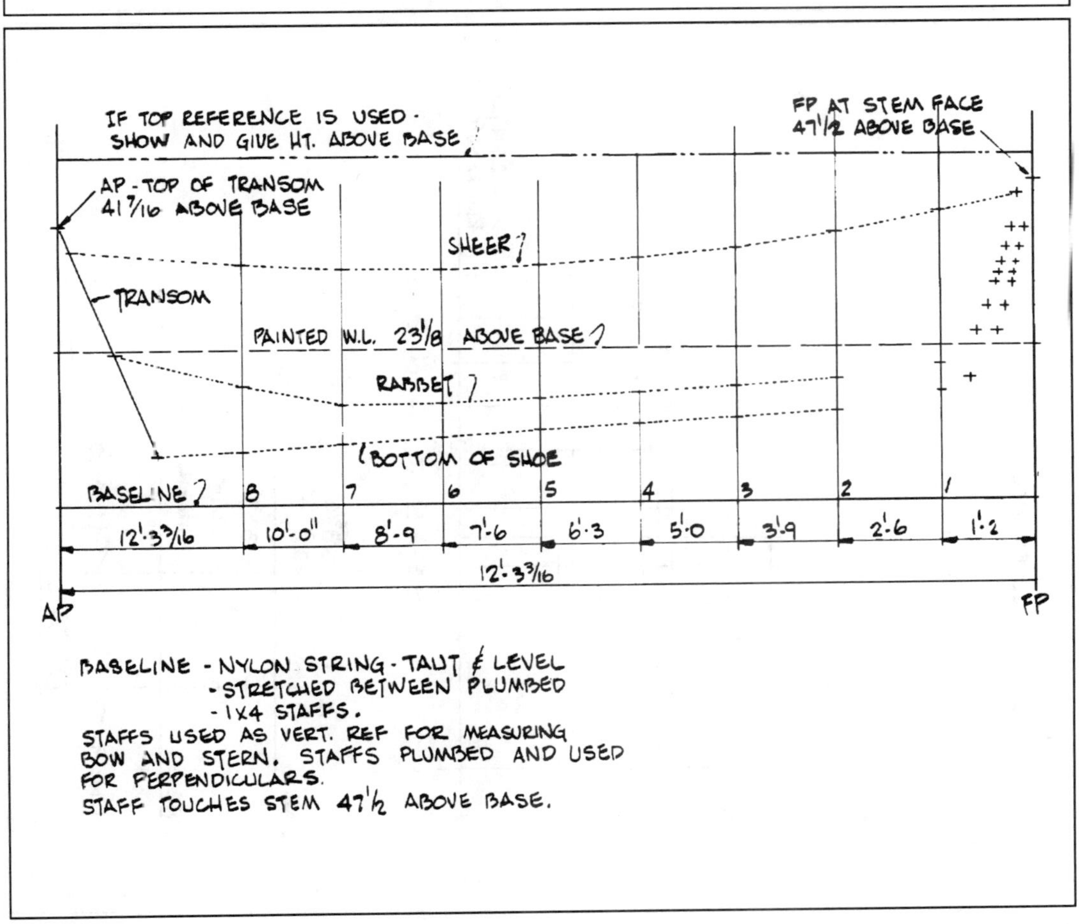

Figure 8.14. Profile view showing the arrangement of the baseline, reference line, perpendiculars, and stations.

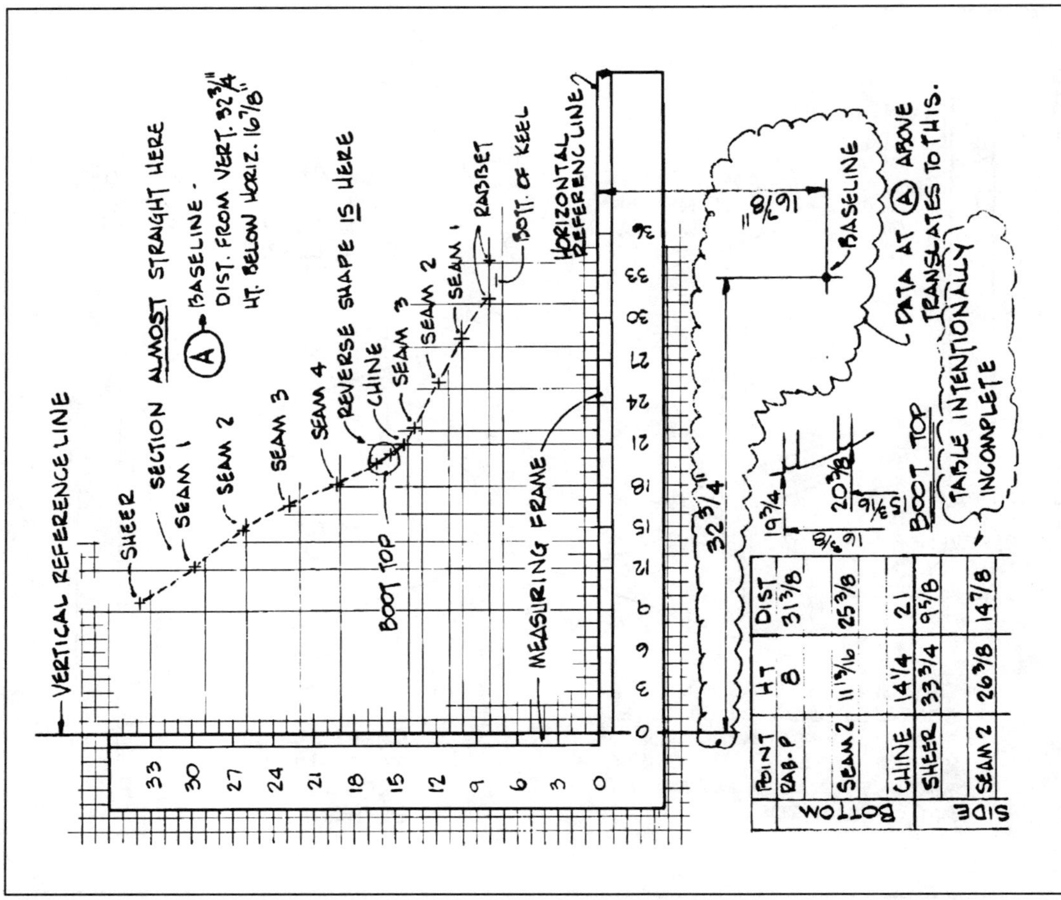

Figure 8.15. Measurements taken off a V-bottomed boat and the resulting plot on grid paper.

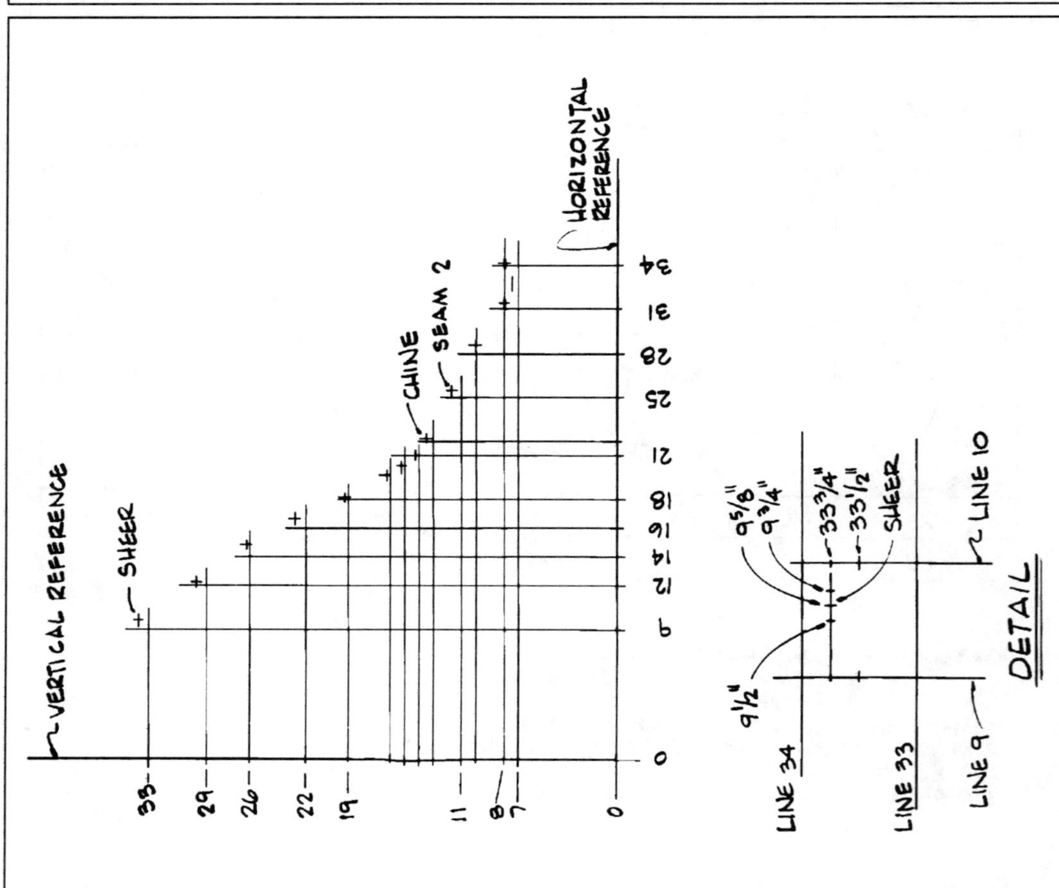

Figure 8.16. The data provided in fig. 8.15 as plotted.

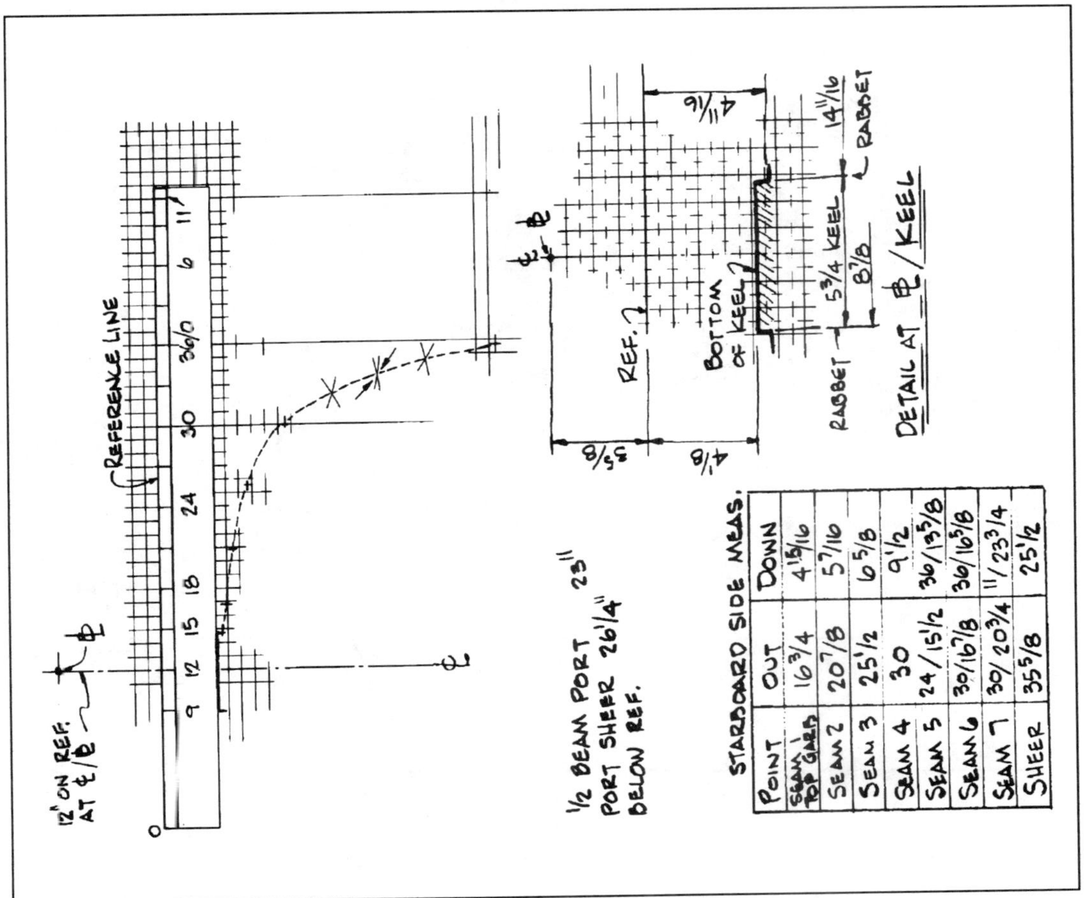

Figure 8.17. Measurements taken off an upside-down, round-bottomed boat and the resulting plot on grid paper.

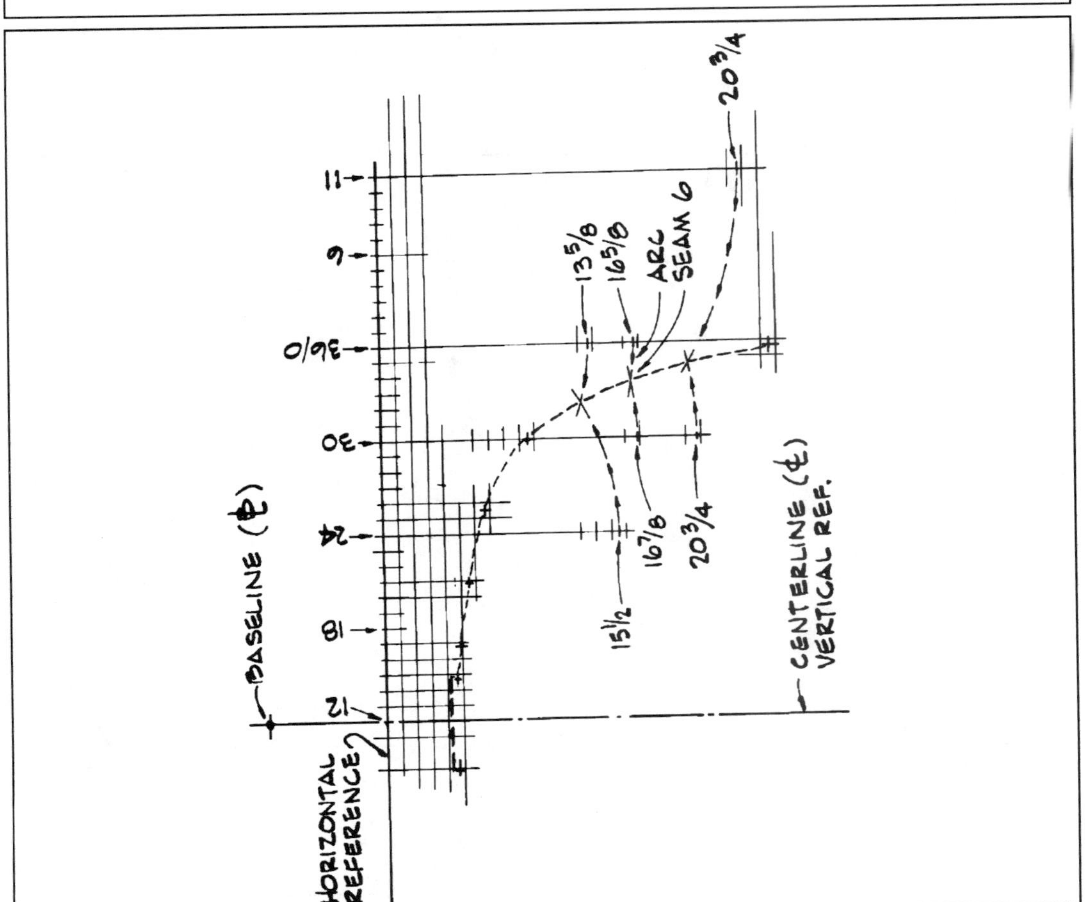

Figure 8.18. The data provided in fig. 8.17 as plotted.

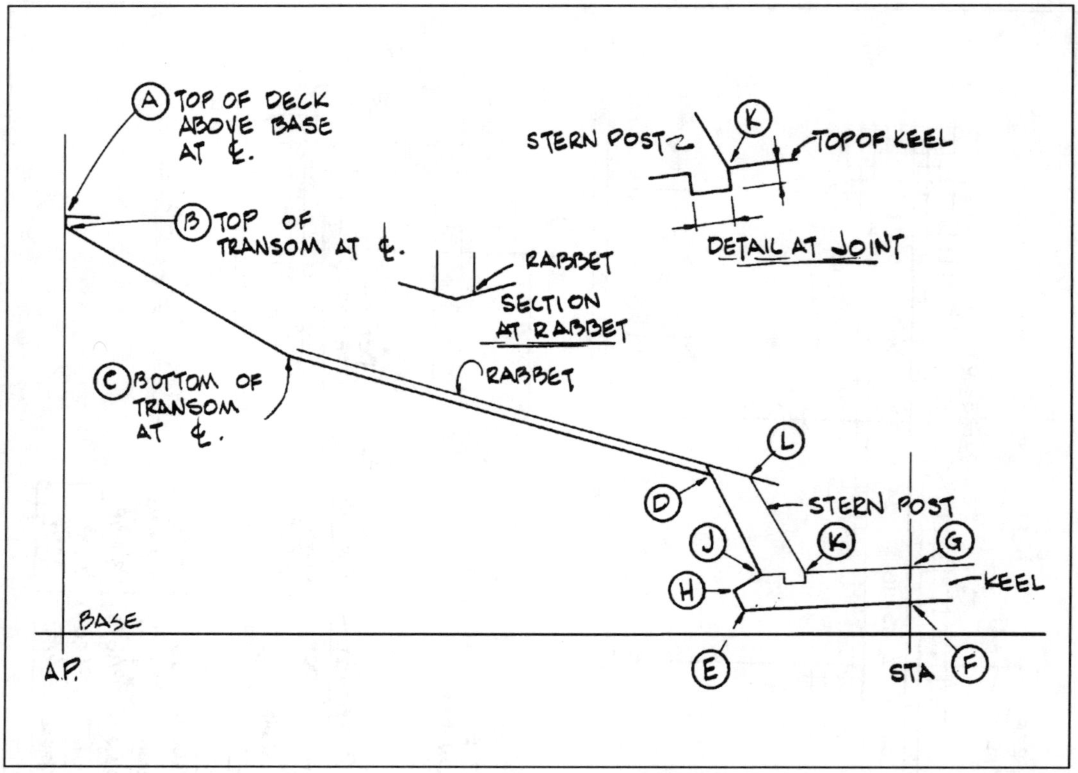

Figure 8.19. Profile of a stern showing the points to be measured.

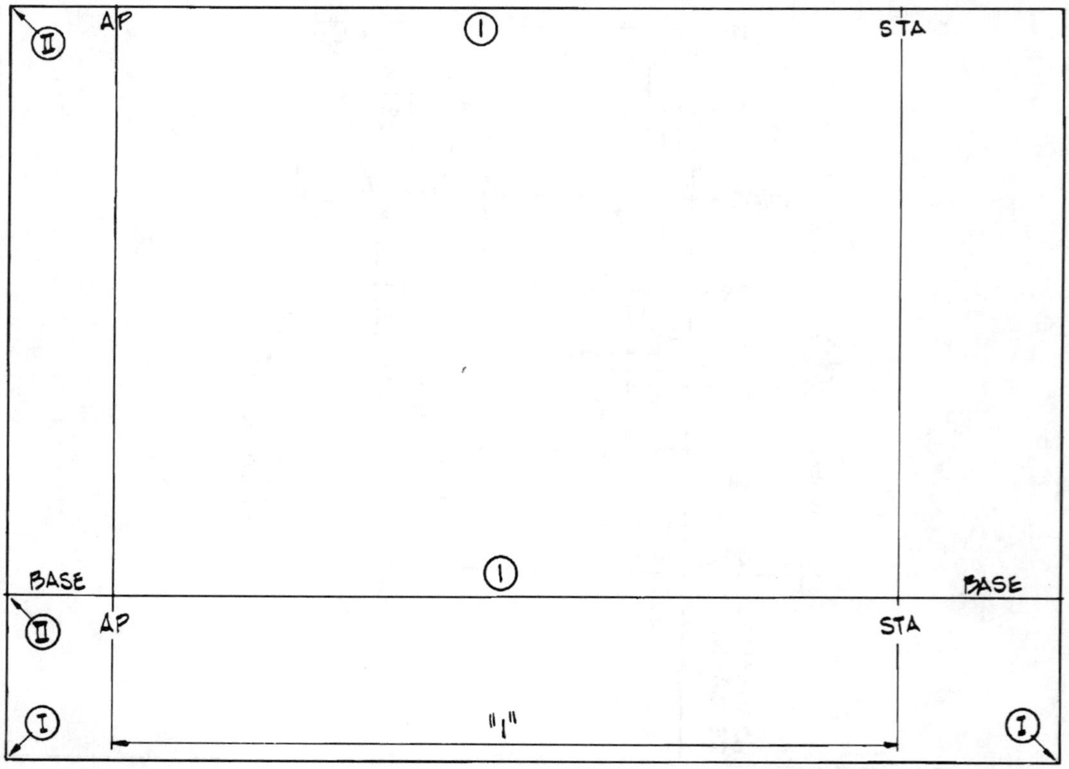

Figure 8.20. Plotting the points measured in fig. 8.19.

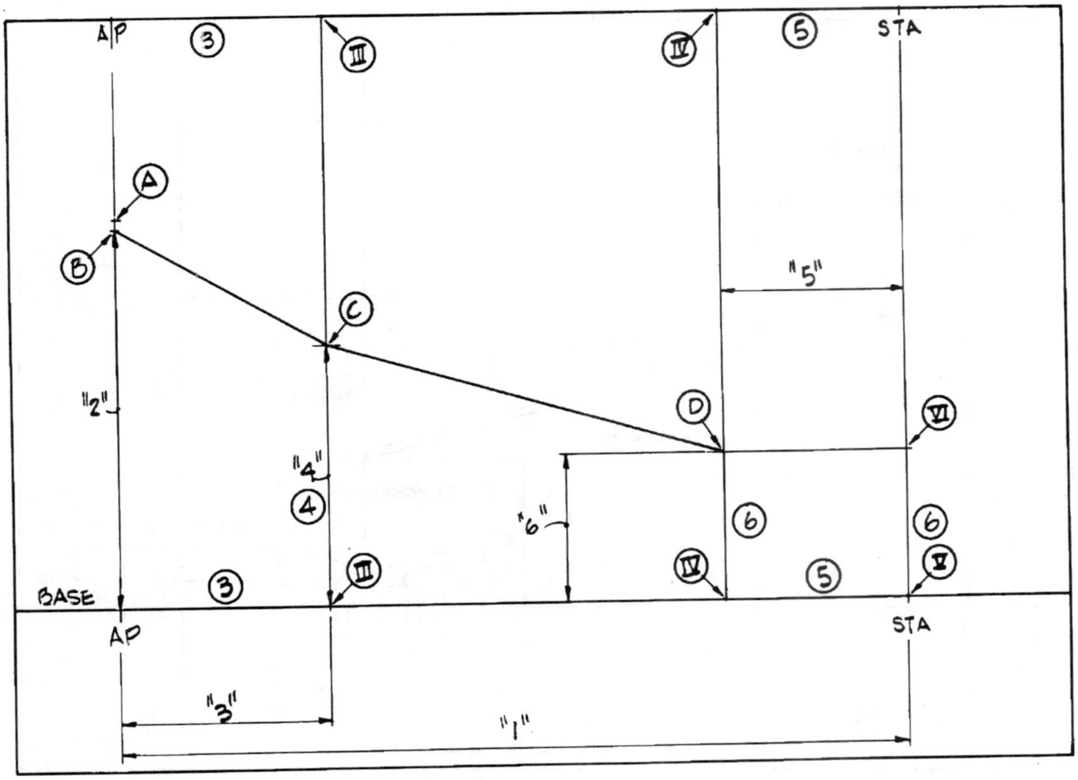

Figure 8.21. Additional points from fig. 8.19 as plotted.

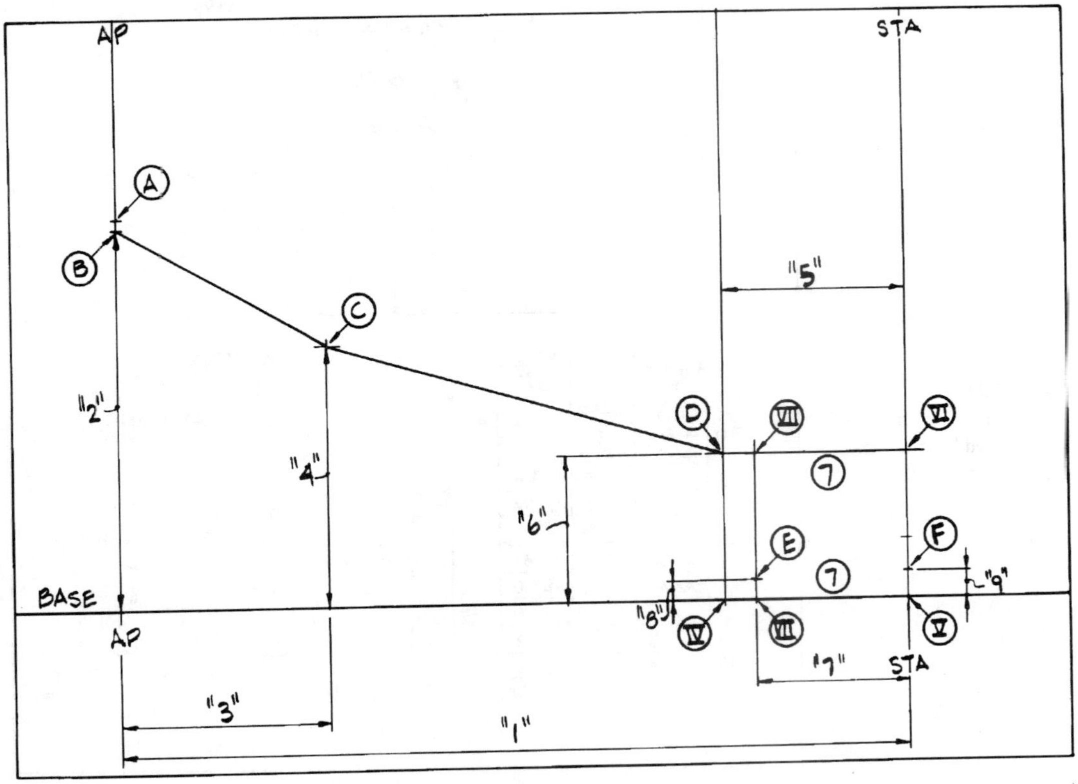

Figure 8.22. Additional points from fig. 8.19 as plotted.

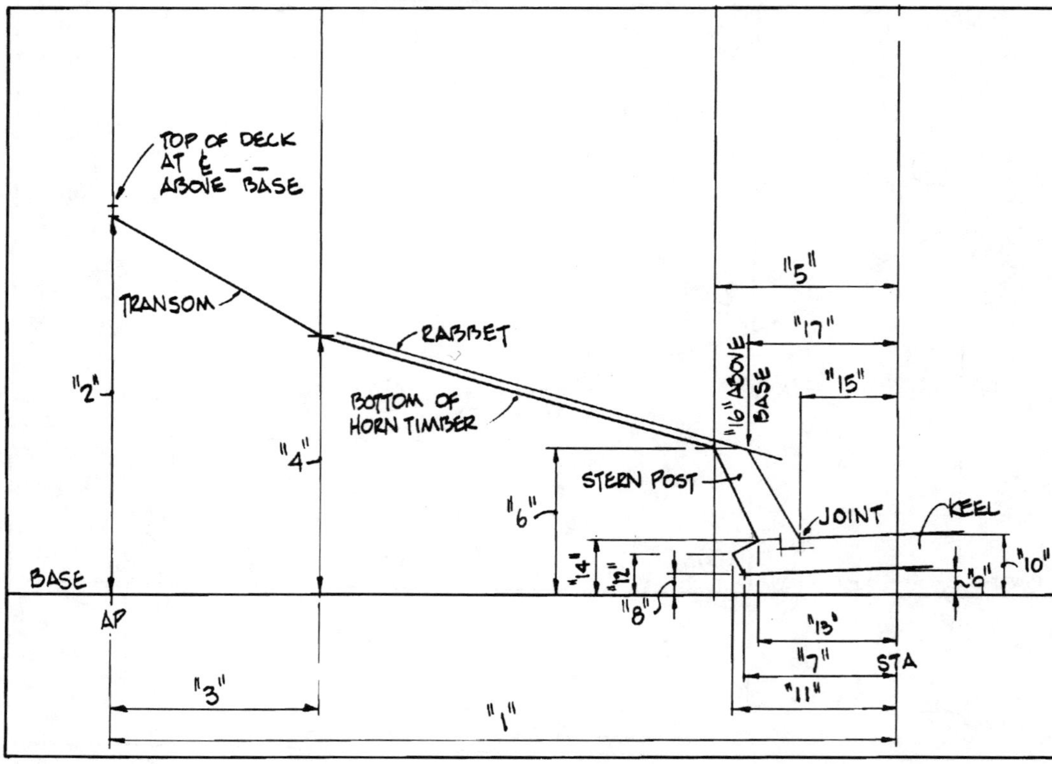

Figure 8.23. The completed plot of all the measurements taken in fig. 8.19.

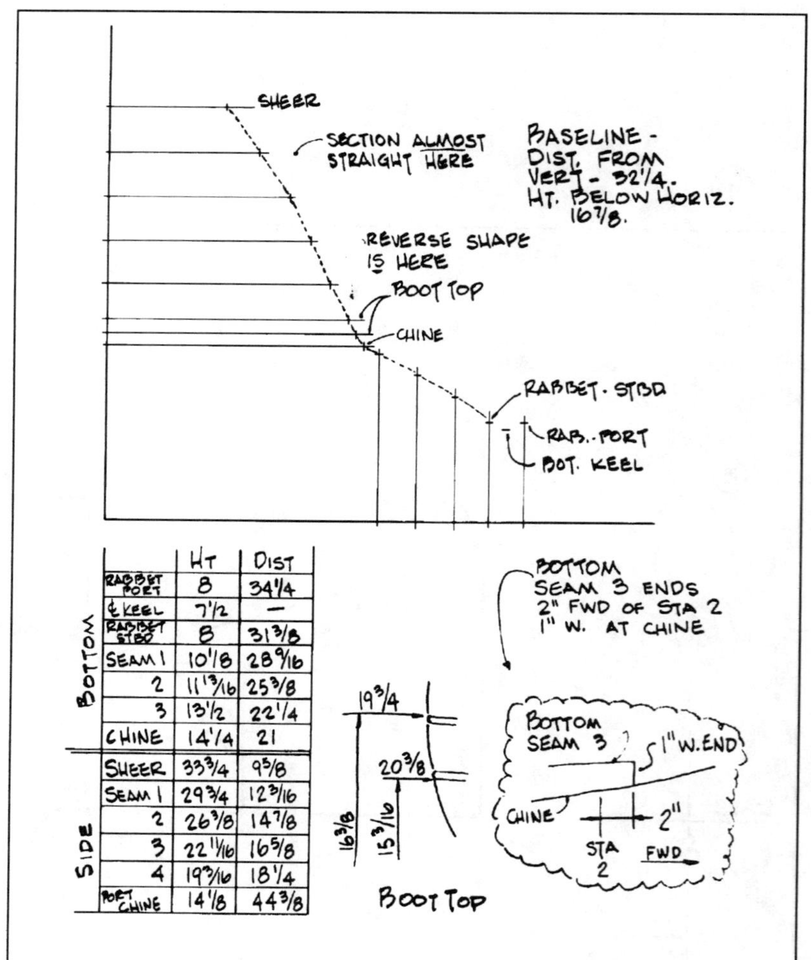

		HT	DIST
BOTTOM	RABBET PORT	8	34¼
	℄ KEEL	7½	—
	RABBET STBD	8	31⅜
	SEAM 1	10⅛	28 9/16
	2	11 13/16	25⅜
	3	13½	22¼
	CHINE	14¼	21
SIDE	SHEER	33¾	9⅝
	SEAM 1	29¾	12 3/16
	2	26⅜	14⅞
	3	22 11/16	16⅝
	4	19 9/16	18¼
	PORT CHINE	14⅛	44⅜

Figure 8.24. Measurements taken off a V-bottomed boat as plotted on plain paper.

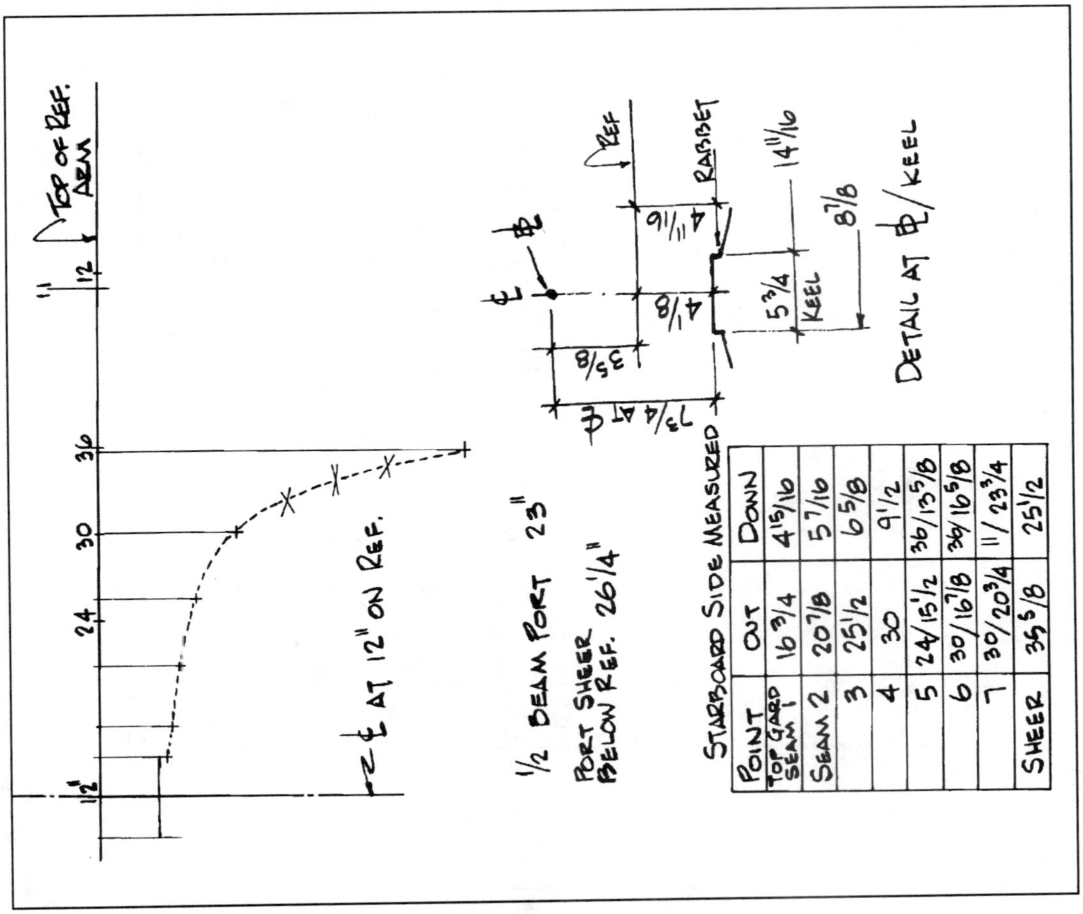

Figure 8.25.
Measurements taken off an upside-down, round-bottomed boat as plotted on plain paper.

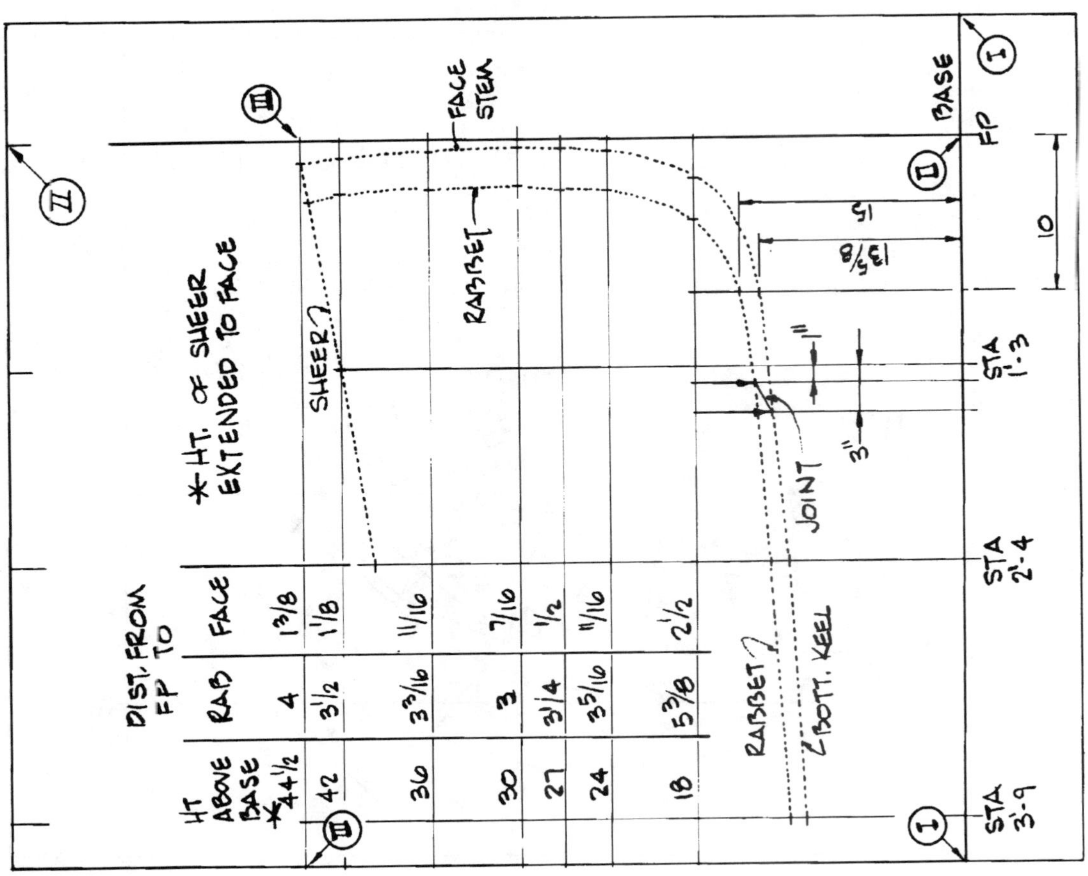

Figure 8.26.
Measurements of a bow profile recorded and plotted on plain paper.

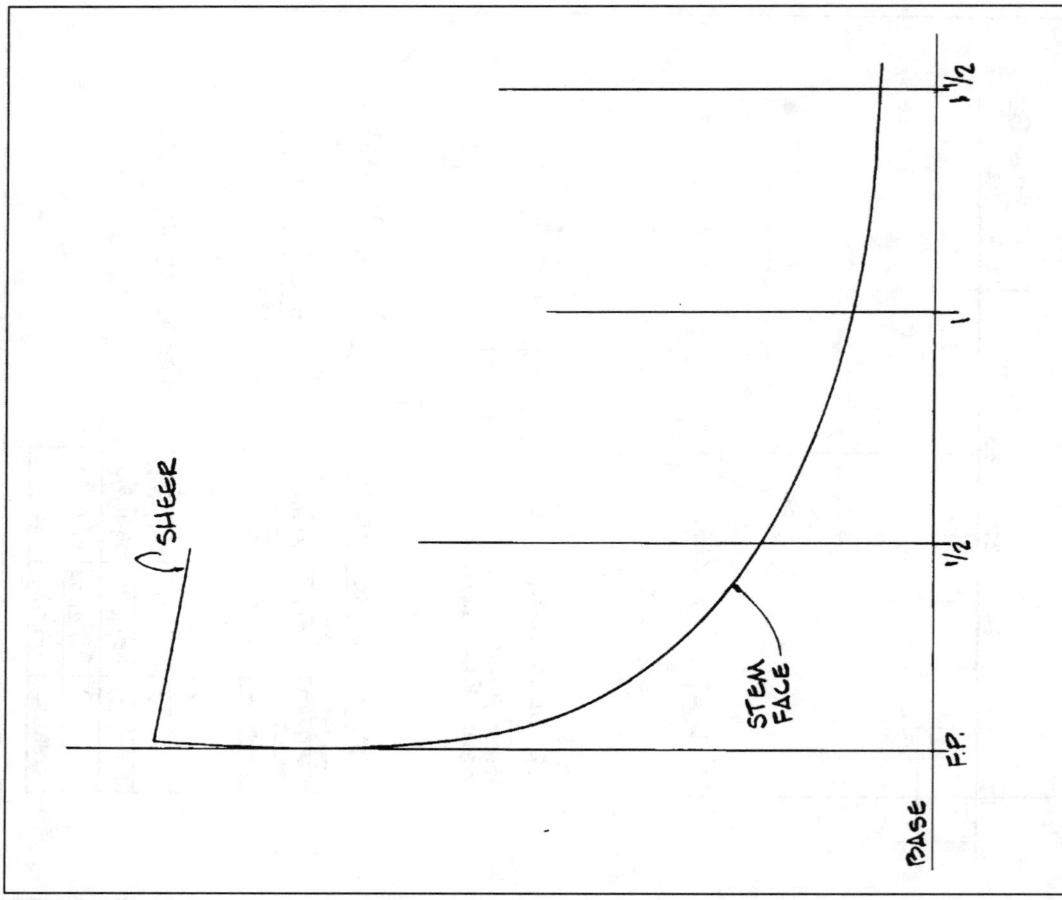

Figure 8.27. The base sheet for recording complex bow construction.

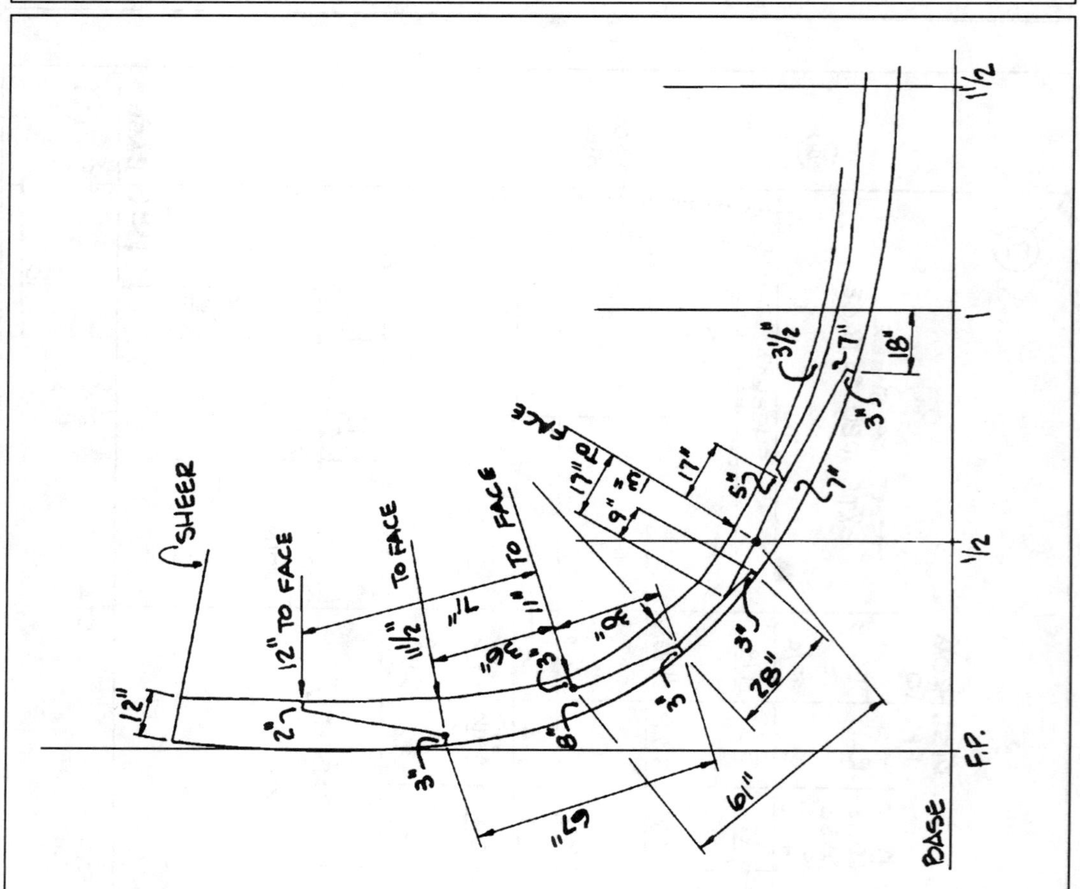

Figure 8.28. The second step in recording complex bow construction.

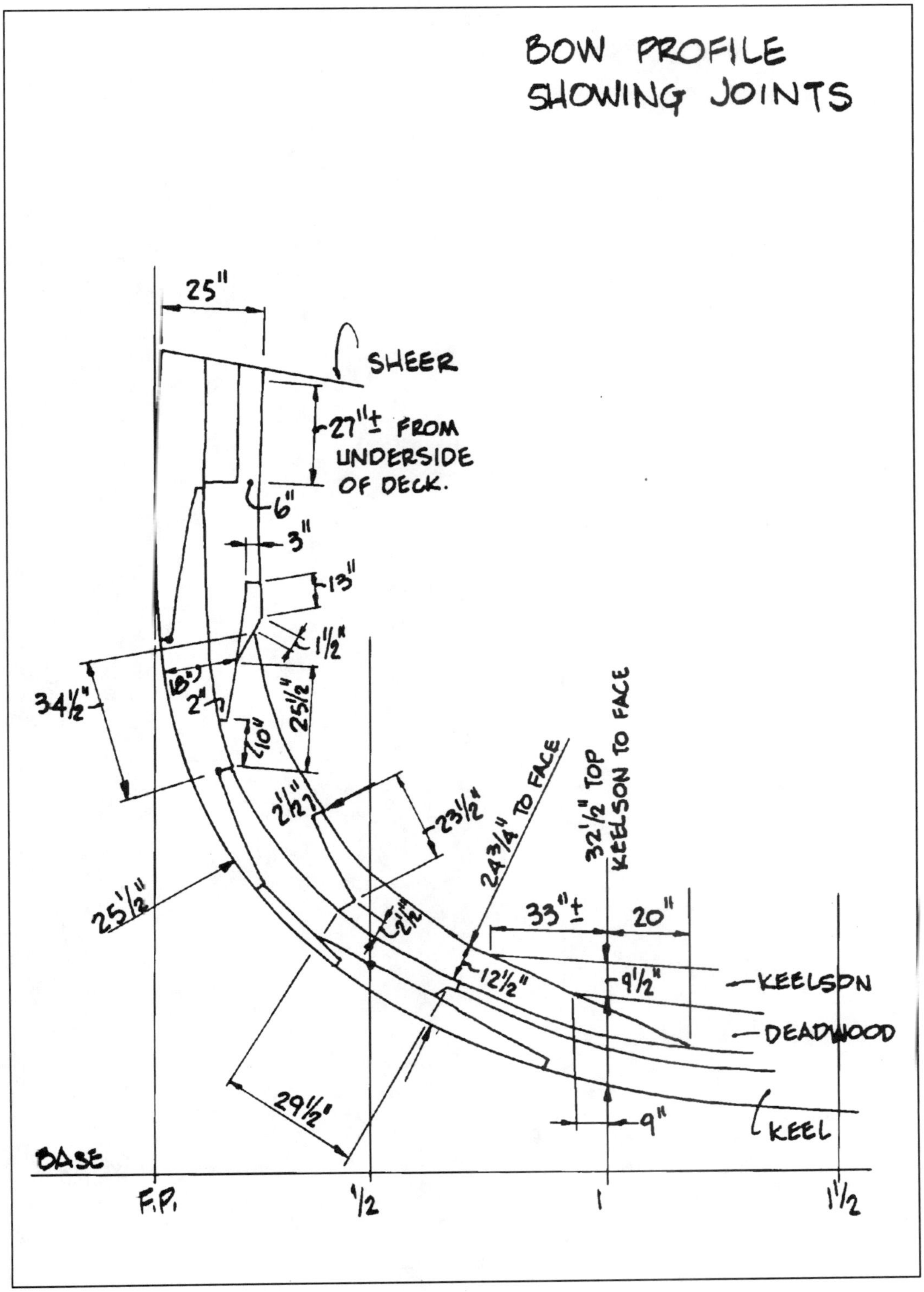

Figure 8.29. The final step in recording complex bow construction.

Chapter 9

Recording Sails, Sailing Rigs, and Rigging

Recording the sailing rigs of watercraft, including illustrations of types and terms

D. W. Dillion

Just as there are many types of hulls, there are many types of sailing rigs: catboat, sloop, ketch, yawl, schooner, etc. And then there are the variations of the main type: gaff headed, jib headed, staysail, etc. To the uninitiated, all of the names, rags, sticks, and strings can be a welter of confusion. Confusing or not, you must take the time to learn the various traditional rigs, because you can't describe something you don't understand.

Only after a rig has been measured and described can it be drawn, replicated, or in dire circumstances, rebuilt or replaced. Therefore, it is necessary to sort out all of those names, rags, sticks, and strings so that they become sails, spars, and rigging.

Your first step, then, is to identify the rig. Identification should be easy if the boat is in a museum collection, because when it was taken into the museum, it likely was logged into the museum's working catalog files. These files are kept either by the curator or the registrar and usually contain other data pertaining to the boat: photographs, oars, sails, spars, etc. Even with the type of rig identified, you may have difficulty determining how the sails, spars, and rigging fit together. Not all parts of a rig are necessarily used at the same time—especially sails—so that photographs alone may not give conclusive identification.

In some cases, especially work boat rigs, whole rigs underwent seasonal changes from a summer rig, with a large sail area, to a winter rig, with a reduced sail area and possibly fewer sails. If the rig is in the boat, the shape of it is obvious and a comparison to the figures in this chapter for identification will be sufficient. In some cases, it may be possible to talk with the owner. He or she can tell you what the rig is. But, as the owner may not talk about the possible combinations of sails, plead ignorance and ask questions. So be careful. The rig that is in the boat may not be the only one that was used.

In what order should you measure and record a rig? That will be dictated by circumstances, especially if the rig or any of its pieces are stored separately from the boat.

Spars are easier to measure when they are set up on saw horses. However, it is easier to describe both standing and running rigging with the rig stepped (set up in the boat) and with the sail or sails bent on. Measuring and describing are little or no problem with small boat rigs because they can be manhandled in and out of boats without difficulty. All measurements can then be easily taken off, and all leads and connections can be determined easily.

Eventually you will have to measure a stepped rig. The sails usually can be unbent for measuring, but the rest of the rig will have to be measured in place. Someone will have to go aloft so that the mast height, standing rigging, and fittings can be measured. Try to do as much as possible in one trip and use the 0-inch end of the tape measure aloft. Beyond this, the solution to the problem, other than an admonition for caution, is beyond the scope of this manual.

The rig may be stored separately, and it may not be possible to step the rig because it is too large for the space available. Then the arrangement of the spars, sails, and rigging will have to be determined by laying them out on a clean floor or by drawing them at scale to develop a sail plan, that is, a side view of the boat showing the general arrangement of sail, spars, and running and standing rigging.

To begin, here are several definitions. Other definitions are given by illustration in the figures for this chapter. Many of the figures are simple line sketches of a form which is adequate for field notes and can be referred to for guidance. Further help identifying rigs and their components will be found in *American Small Sailing Craft* by Howard I. Chapelle, which contains illustrations of rigs and hull types; *Gaff Rig* and *Spritsails and Lugsails* by John Leather, which illustrates the way the rigging is arranged and works; and *The Lore of Ships* by Tre Tryckare, which shows the sail arrangement of almost any rig imaginable.

Mast: the primary support for the rig.
Spars: used to spread the sail to its shape and as an aid in raising or positioning the sail.
Rigging: a word that refers to standing and running rigging. Also the process of setting up the spars, ropes, and lines—both a noun and a verb. Masts are "stepped," bowsprits are "bedded."
Standing rigging: braces the mast and other nonmoving spars; also to support spars semipermanently—generally nonmoving.
Running rigging—controls the sails' relationship to the wind; also to raise and lower the

sails; not fixed in place—moveable.
Sails: the power source in the rig, generally three or four sided.
Bent on: how the sails are attached to the spars
Lead: (pronounced "leed") the routing of the running rigging.
Belaying point: where the running rigging is temporarily tied off after the sails are adjusted.

What information must be taken off to record properly a rig?

1. Location of mast or masts
2. Size, construction, and details of mast or masts and other spars
3. All fittings on and holes in mast or masts and other spars.
4. Attachment points for standing rigging
5. Dimensions of the sail or sails
6. How the sail is built and its material
7. How the sails are bent on
8. Leads and belaying points for running rigging.

Note: Items 1, 2, and 3 above will be taken off as part of recording the hull. Use the figures in this chapter as instruction sheets, samples, guides, and glossary.

Photographing the Rig

Finally, the rig and its details should be photographed. Remember, if the rig is in the boat, the sky will be the background for most of the pictures and will put much of the detail photographed into silhouette. The best light for this sort of photography is in the early morning, the late evening, or during a dark, overcast day. A polarizing filter will offset a bright sky. Take photographs from several vantage points around the boat, so that the relationship of various features in the rig stand out. Do not use a wide-angle lens to take a full profile view of the rigged boat. The distortion at the edge of the frame may prove misleading. A long-range view taken from across the harbor with a telephoto lens will be more accurate. Do not take only eye-level photographs. Take overlapping frames along the mast to get full coverage and, if possible, views looking down onto the deck. Refer to Fig. 9.45 for suggested coverage.

A ruler or scale inserted into close-up views will aid you in recalling proportions and locations, but it is not a substitute for a good job of recording.

Specific references used for this chapter follow. See the bibliography at the end of this manual for full citations.

Ansel, Willits D. *The Whaleboat: A Study of Design, Construction, and Use from 1850 to 1970.*

Bower, R. M., and S. A. Budd. *Make Your Own Sails.*

Bray, Maynard E. *Watercraft.*

Chapelle, Howard I. *American Small Sailing Craft.*

__________. *American Fishing Schooners.*

__________. *Yacht Designing and Planning.*

Crowley, William. *Rushton's Rowboats and Canoes: The 1903 Catalog in Perspective.*

Kerchove, Rene de. *International Maritime Dictionary.*

Leather, John. *Gaff Rig.*

__________. *Spritsails and Lugsails.*

Tre Tryckare. *The Lore of Ships.*

Underhill, Harold A. *Masting and Rigging the Clipper Ship and Ocean Carrier.*

Figures 9.1 through 9.7 show the sail plans most likely to be encountered in small craft.

Figure 9.1. Single-sail sail plans.

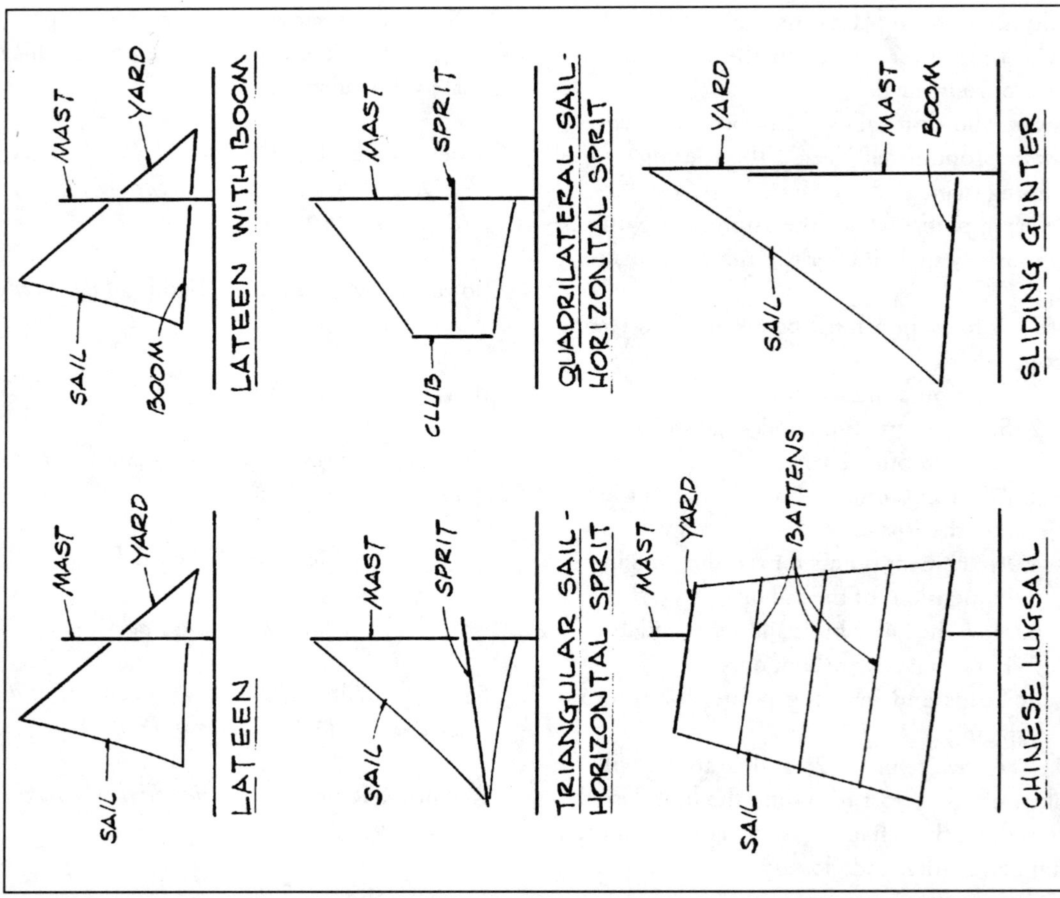

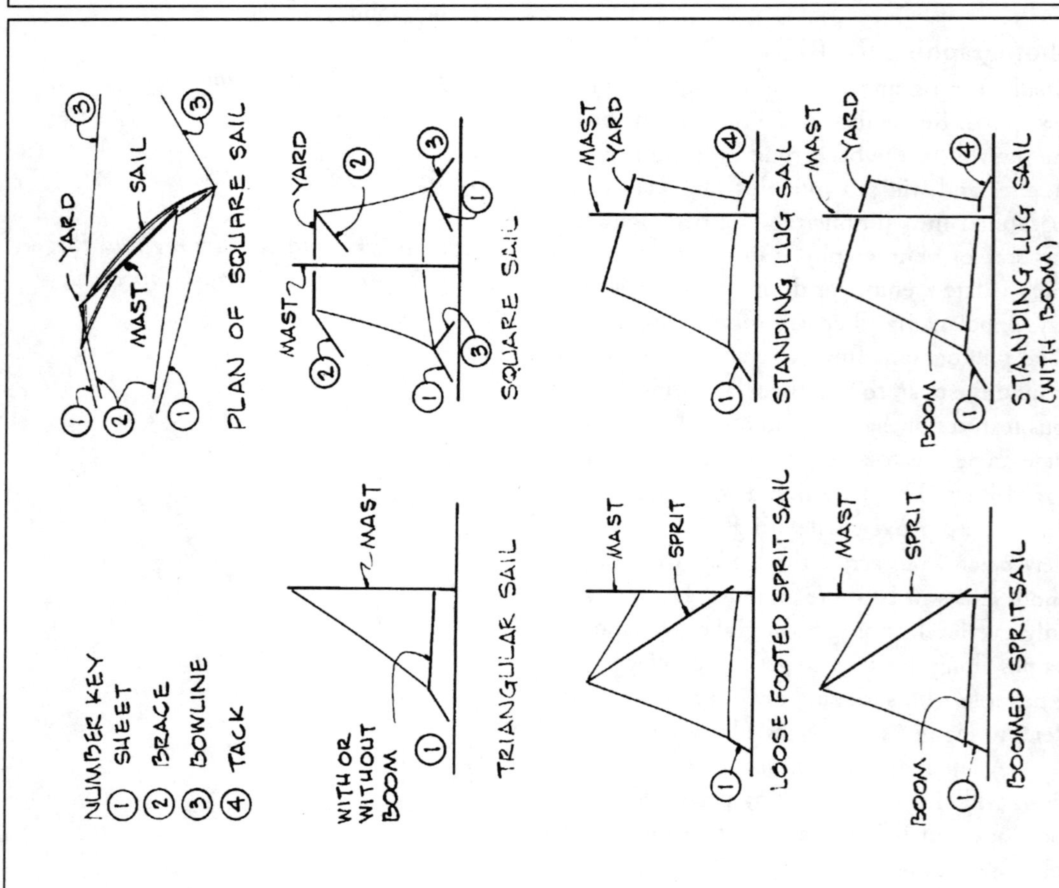

Figure 9.1. Single-sail sail plans.

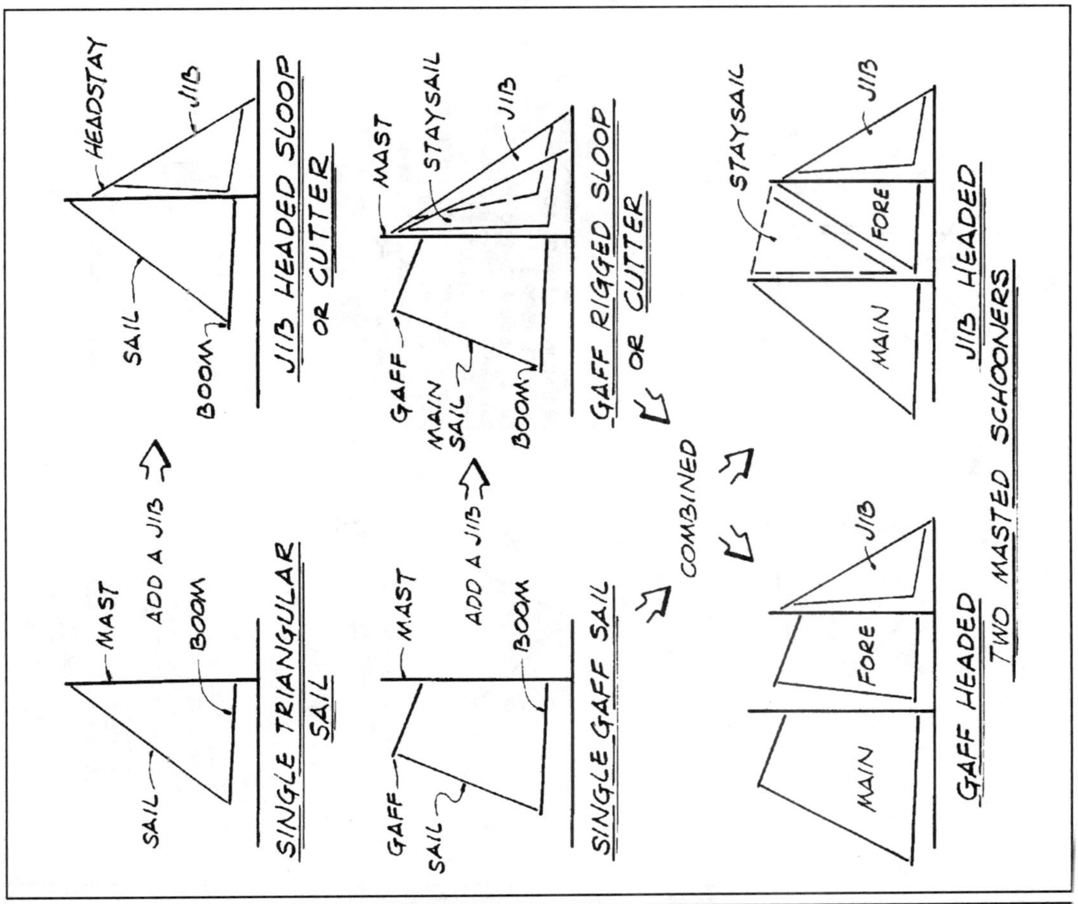

Figure 9.2. The progression from the single triangular and quadrilateral sails to the schooner rig.

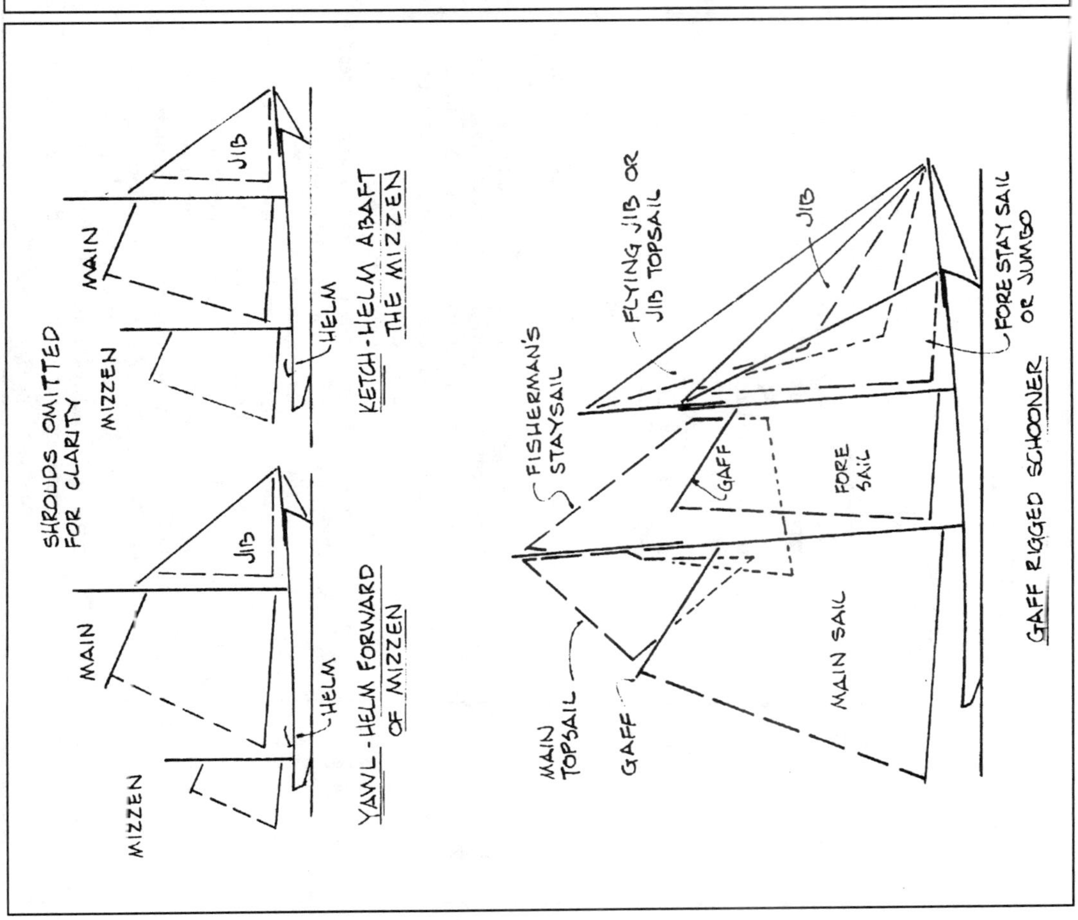

Figure 9.3. The yawl, ketch, and schooner. Each can be jib-headed or gaff-headed.

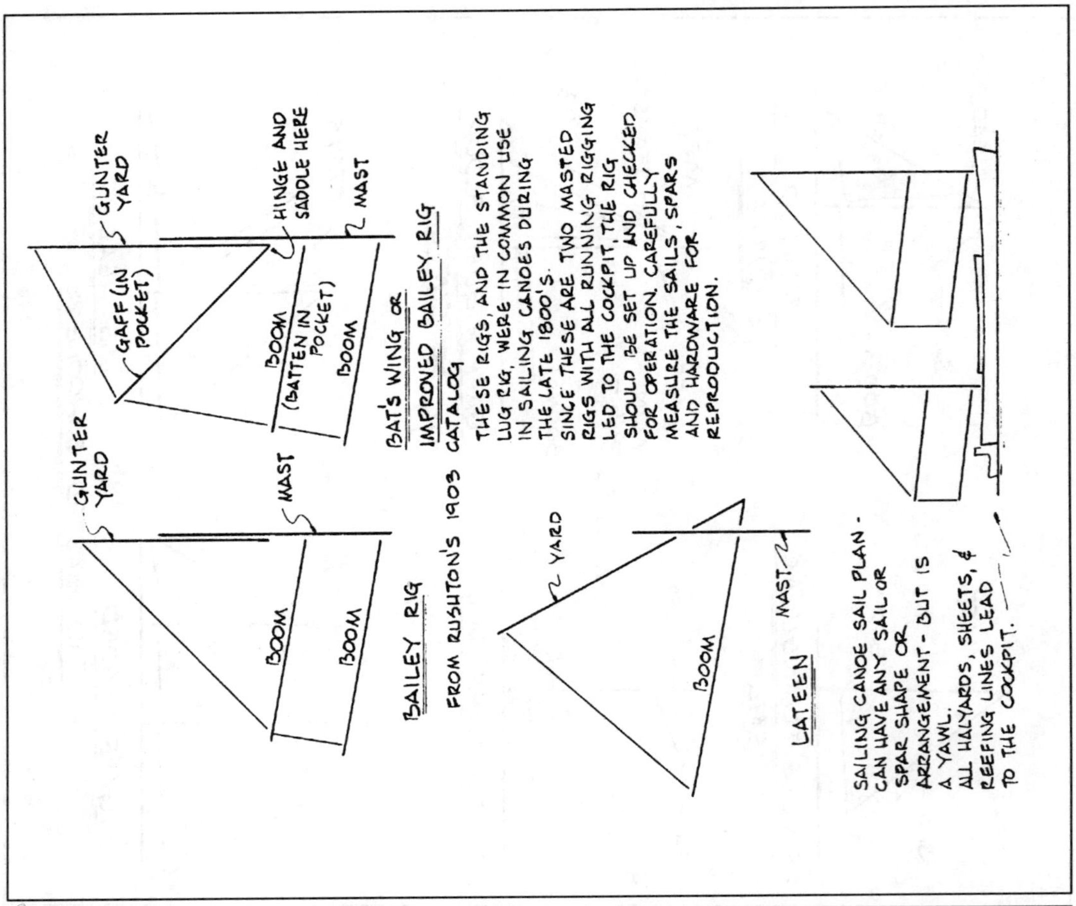

Figure 9.4. Several rigs found in sailing canoes.

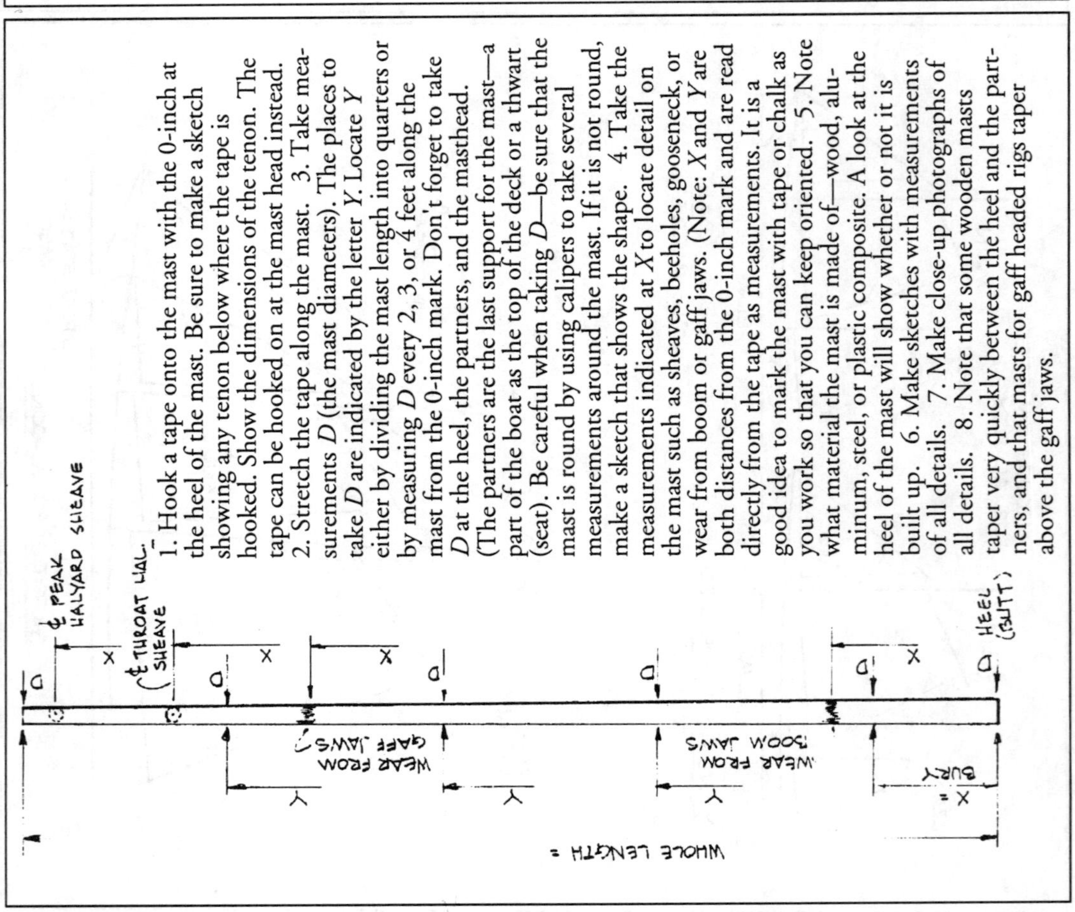

Figure 9.5. Instructions for measuring and recording a mast.

Make up a note sheet for the measurements taken off a spar. Two forms of notation should appear. One should be a list of the distances to the various locations along the mast with a description of what happens at that location. (See list below.) The other should be a simple line sketch—like that at the right—on which the various distances with the measurements noted and the detail found there described. Draw as much as possible to scale. If you use 3/8"=1'0" for the length, use 3/4" or 1 1/2"=1'0" for the diameters so that measuring errors will be apparent.

0-inch on the tape at the heel of mast-tenon projects beyond. See sketch.

Distance	*Detail*
0"	3" dia. heel
2'-7"	5" dia. partners
3'-9"	5" dia. wear from boom jaws taken at c.l. of 3" wide area on aft side of mast
6'-0"	5" dia.
8'-0"	4 15/16" dia.
10'-0"	4 7/8" dia.
12'-0"	4 3/4" dia.
14'-2"	4 5/8" dia. wear from gaff jaws taken on aft side of mast at c.l. of 8" wide area of wear
15'-4"	4 1/2" dia. c.l. sheave for throat halliard
19'-0"	3 3/4 dia. c.l. sheave for peak hall
19'-6"	3 1/2" dia. masthead

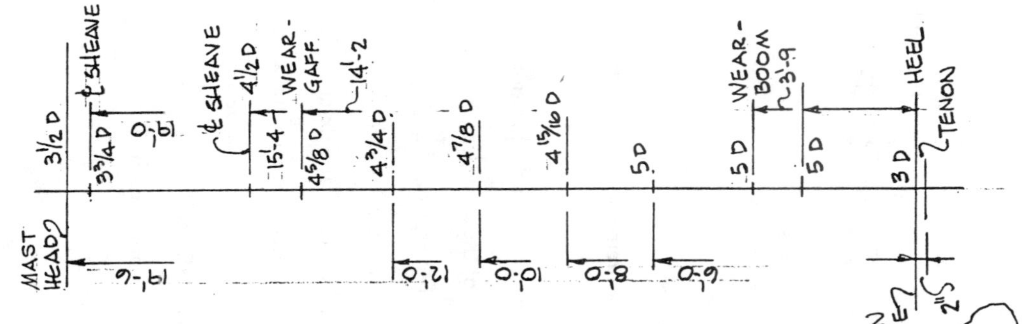

Figure 9.6. A sample note sheet for recording a spar.

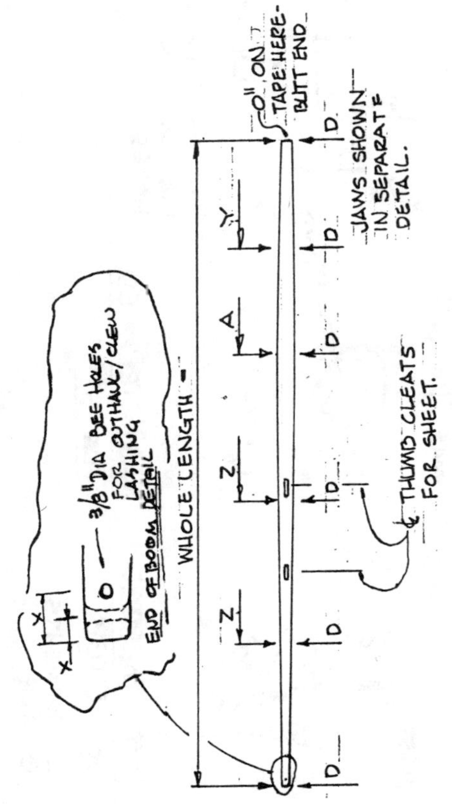

1. Make up a note sheet as shown in fig. 9.6.

2. and 3. See 2. and 3. in fig. 9.5.

4. Find the greatest diameter of the spar, measure and record that, and then measure and record how far it is from the 0-inch mark (distance *A*). The diameter will probably be greatest at one-quarter or one-third of the whole length.

5. Measure and record the diameter at *Y* which should be at about half the distance from the one-inch mark to *A* and *Z* which are at the 1/3 points of the remaining length of the spar.

6. Measure and record the distances from 0" to all hardware, fittings, etc. on the spar.

7. Be sure to note what wood the spar is made of.

8. Photograph all details.

Figure 9.7. Instructions for measuring and recording a boom, gaff, or other spar. (See also fig. 9.5.)

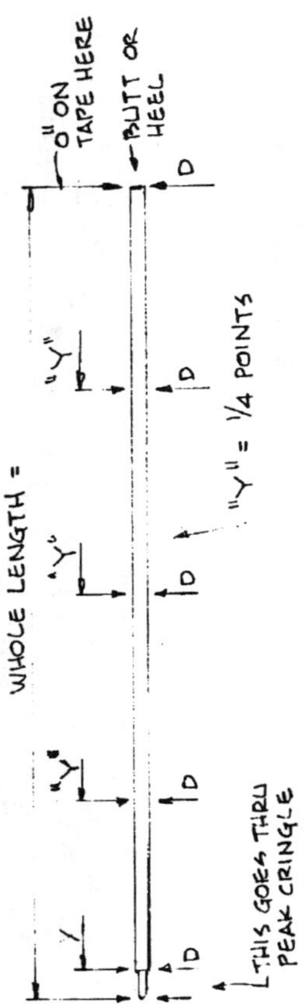

1. Make up a note sheet as shown in in Fig. 9.6

2. Hook a tape on the sprit with 0" at the butt end.

3. Stretch the tape along the sprit. Take all measurements as distances from 0" and read directly from the tape.

4. Take and record the measurements *D* (the diameter) at the locations indicated by the *Y's* which are at the 1/4 points of the sprit (length of the sprit divided by four). Also take *D* at the heel of the sprit and at the shoulder, *X*—and at the peak end. Record the distances *X* and *Y*.

5. Are there any holes or hardware that should be located and detailed? If so, record the distances to them from 0". Make sketches of the details with measurements to show their sizes. Sprits don't normally have any hardware; just a slot or beehole for the snotter at the heel.

6. The greatest diameter of a sprit is usually at its mid-length.

7. Be sure to note what wood the sprit is made of.

8. Photograph all details.

Figure 9.8. Instructions for measuring and recording a sprit. *(See also Fig. 9.5)*

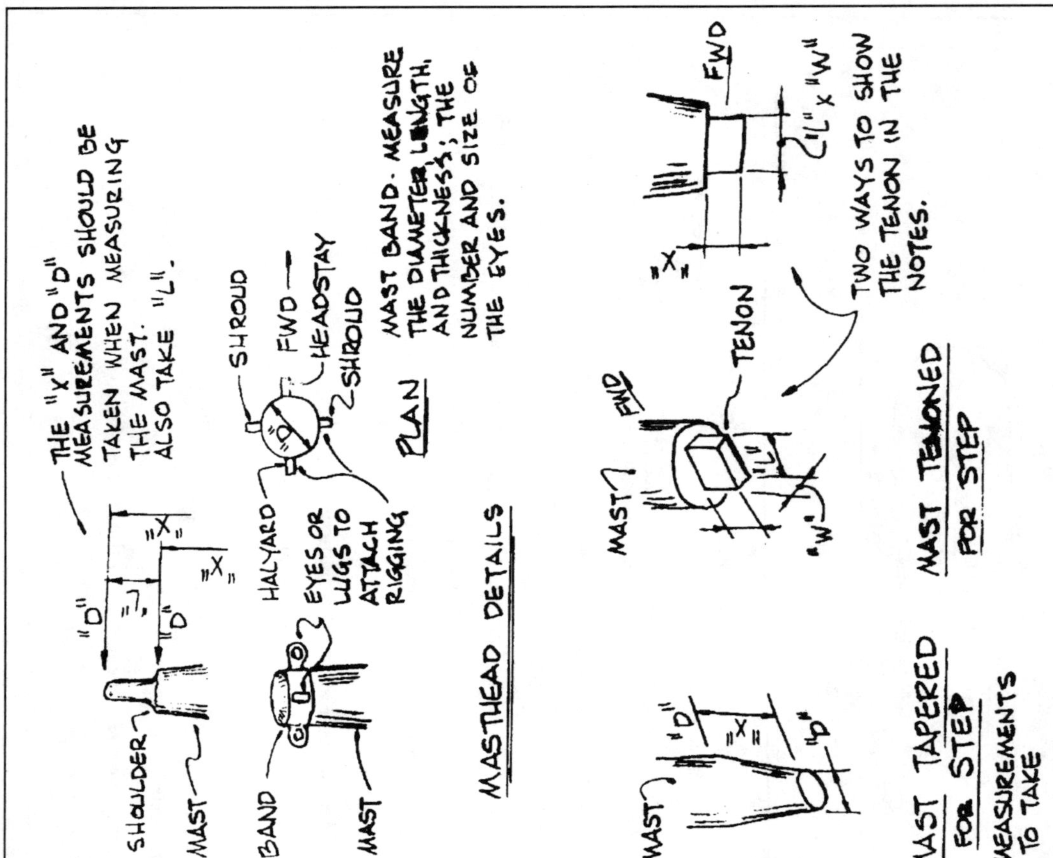

Figure 9.9. Details found on masts and spars.

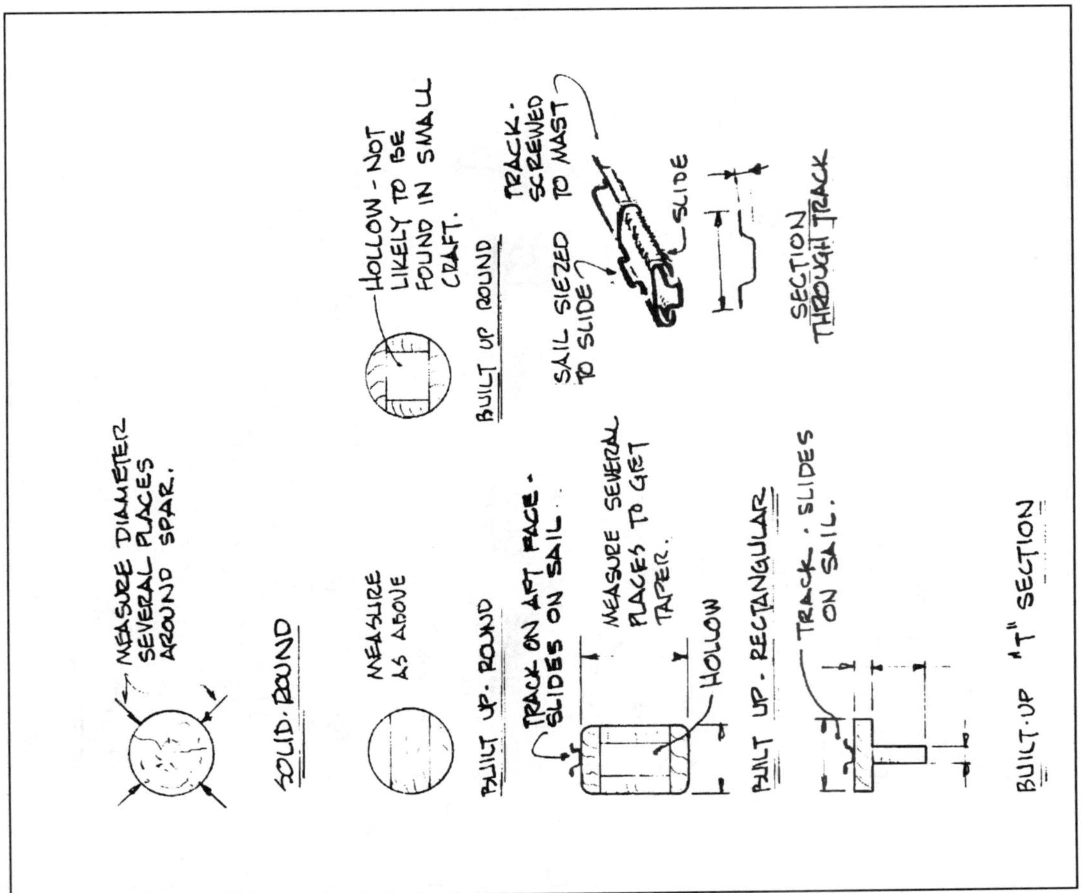

Figure 9.10. Details found on masts and spars.

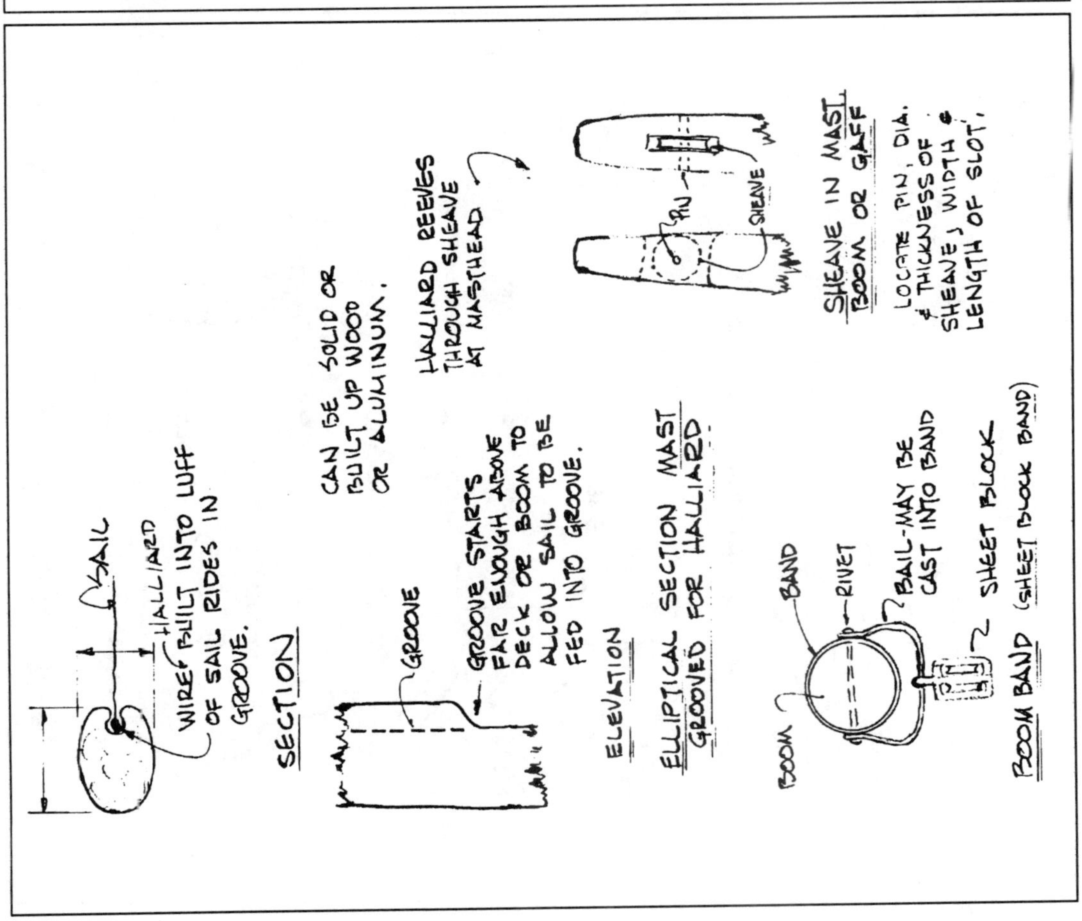

Figure 9.11. Details found on masts and spars.

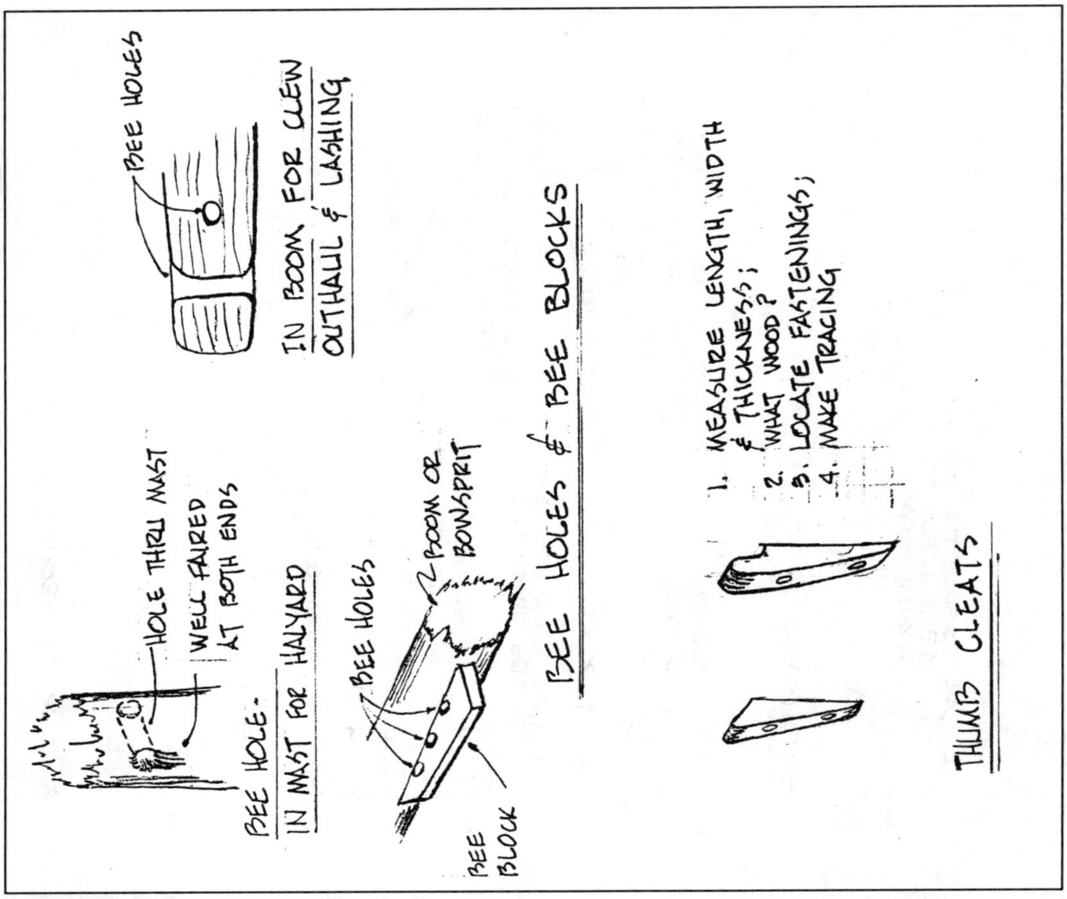

Figure 9.12. Details found on masts and spars.

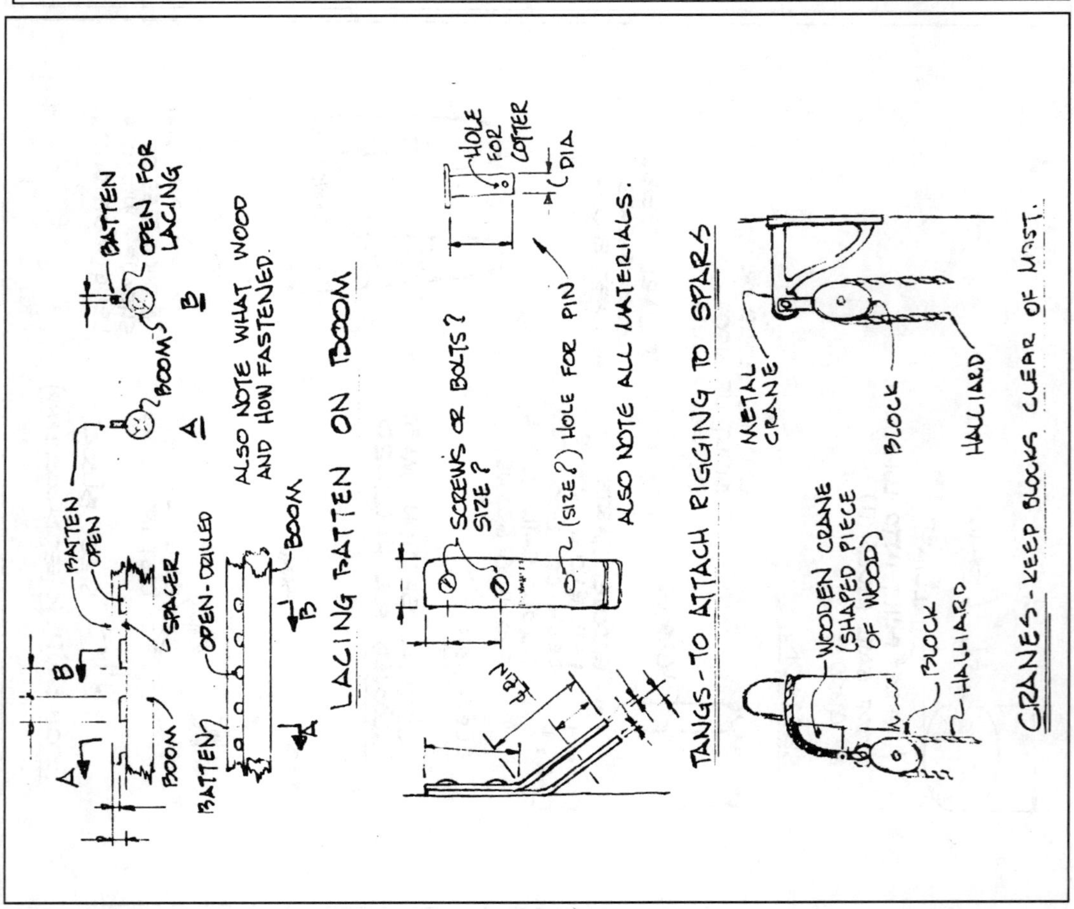

Figure 9.13. Details found on masts and spars.

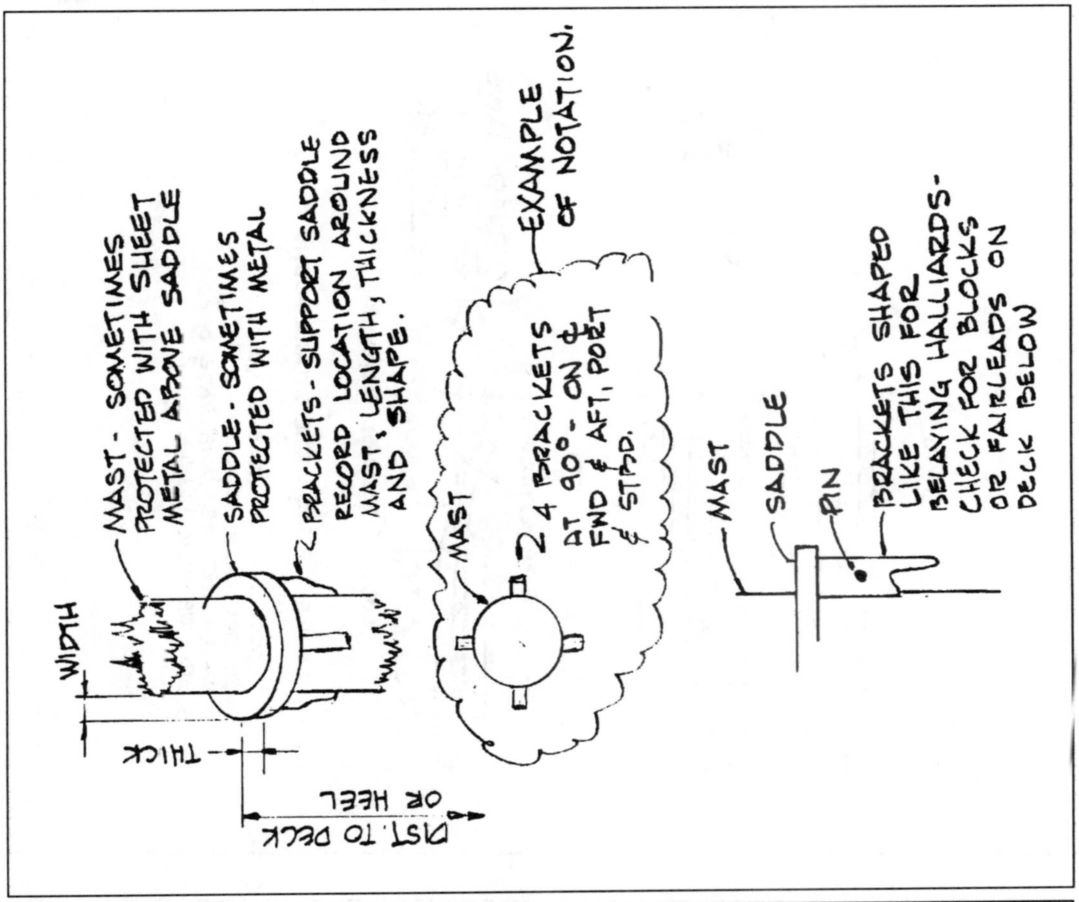

Figure 9.14. Details found on masts and spars.

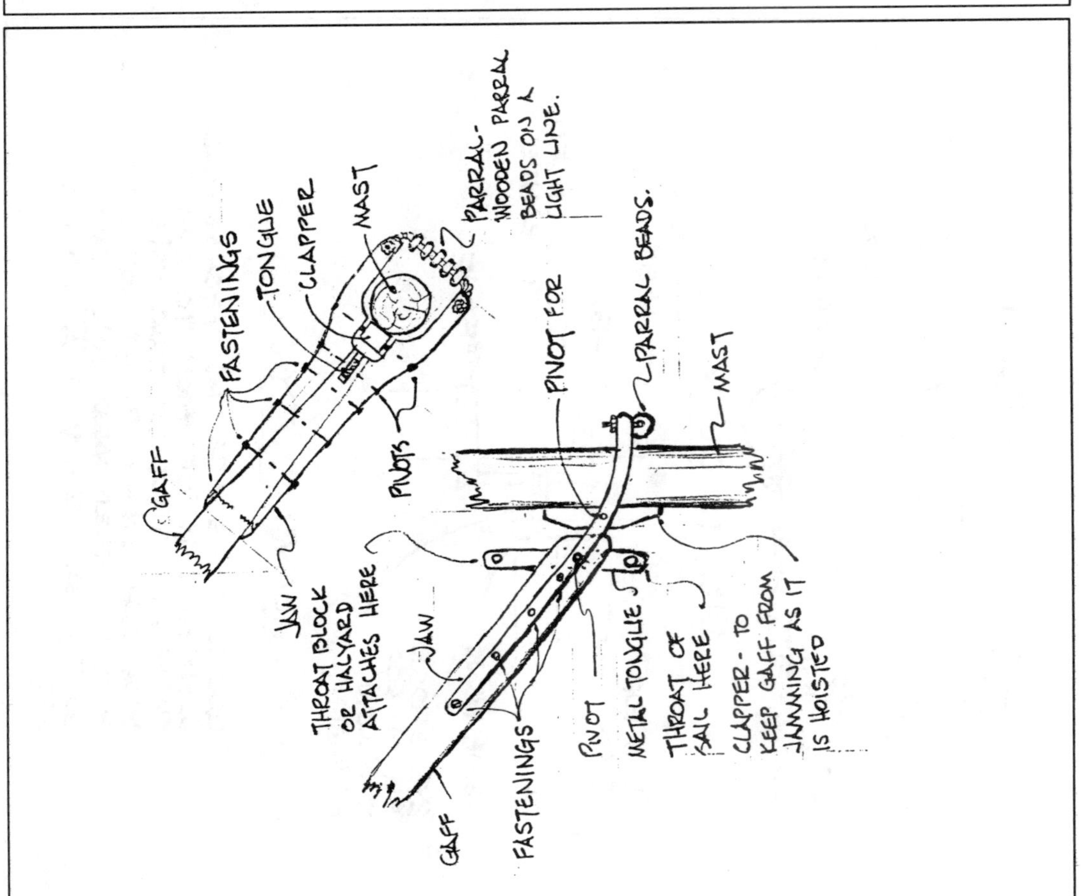

Figure 9.15. Details and nomenclature of gaff jaws.

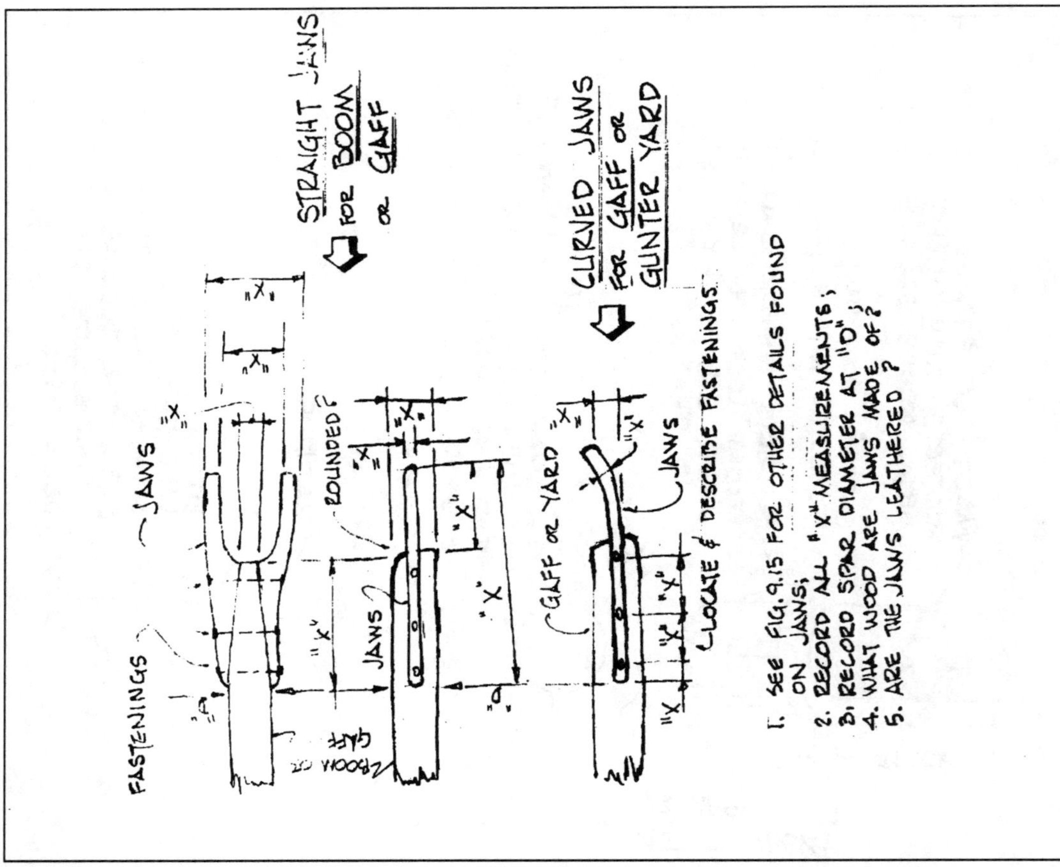

Figure 9.16. Measurements to take from boom, gaff, and gunter yard jaws.

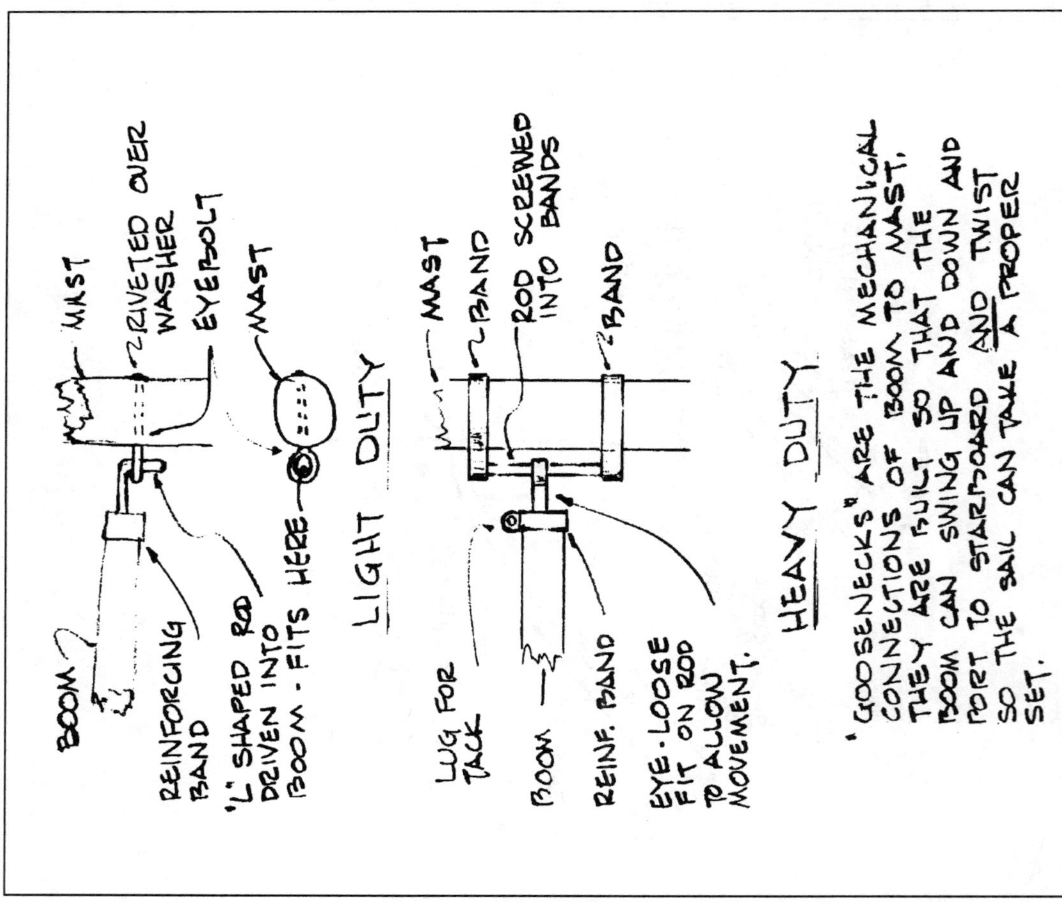

Figure 9.17. Two possibilities for goosenecks. There are many variations.

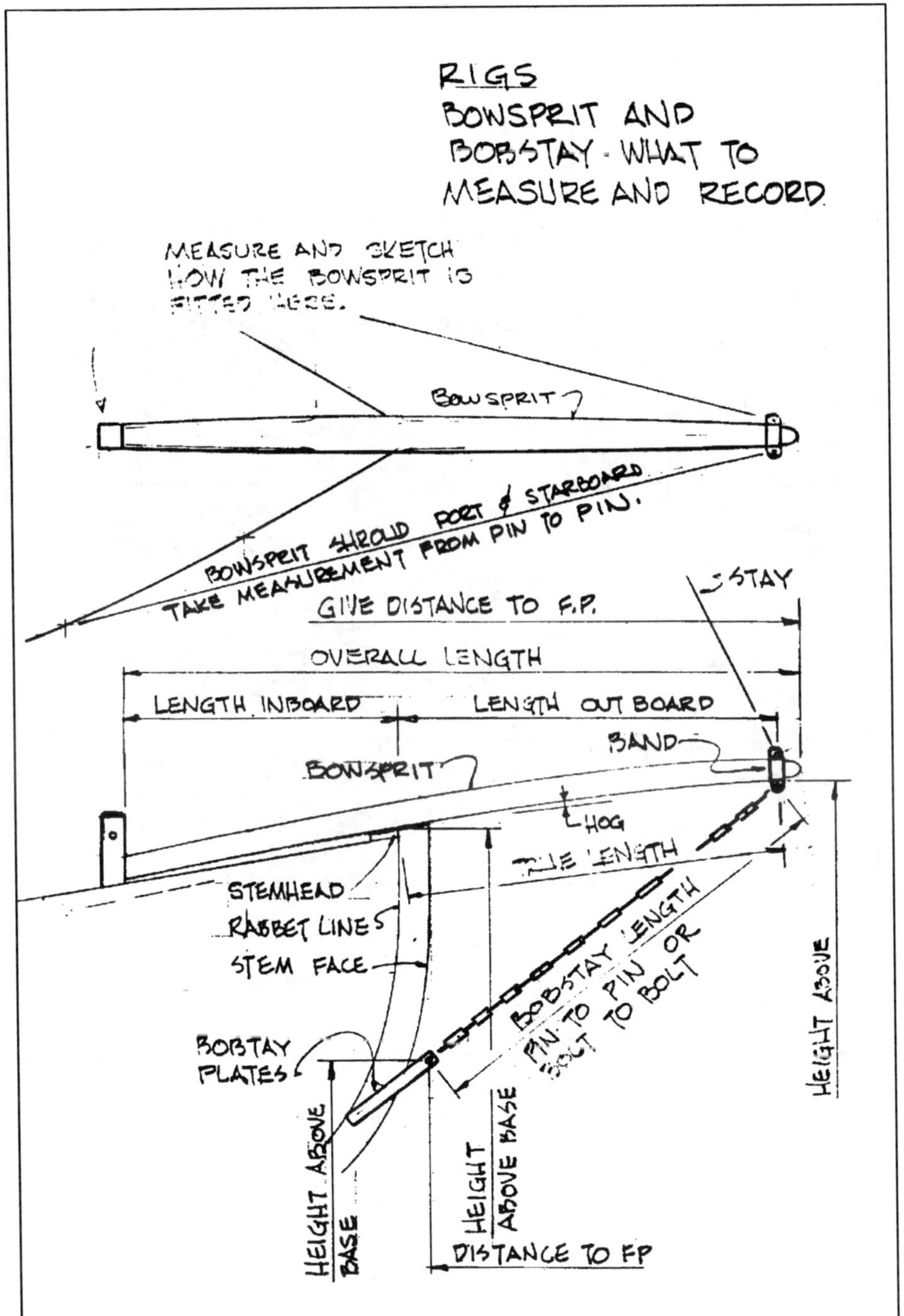

Following the illustration, take off all of the measurements at the bowsprit when the bow/stem profiles are taken off so that the baseline and the forward perpendicular are in place and can be used for references. Other things to observe and record:

1. Bowsprits are not always either round or square. The sectional shape may change from square with chamfered corners (an irregular octagon), to square to round in its length. Frequently they are wider than thick. Bowsprits are rarely straight. A bowsprit that is curved downward in profile is said to be hogged, either sprung (bent) or sawn.
2. Is the bowsprit bolted down snug to the deck? If not, measure how far it is above the deck at the bitts and stemhead.
3. If the bowsprit appears to be snug to the deck, check to see if there are wood shims or canvas underneath it.
4. Does the bowsprit set right down onto the stemhead? Is the stemhead mortised Into the bowsprit? Is the stem head covered with metal, lead, or copper?
5. What is the bobstay made of? An iron hooked in place? A length of chain? A length of wire rope?
6. What wood is the bowsprit made of? Is it wood?
7. How is the heel of the bowsprit held in place?
8. How is the band or other fittings at the outboard end of the bowsprit held in place?
9. Any other details, fittings, or hardware on the bowsprit.

Figure 9.18. The instructions for measuring and recording a bowsprit.

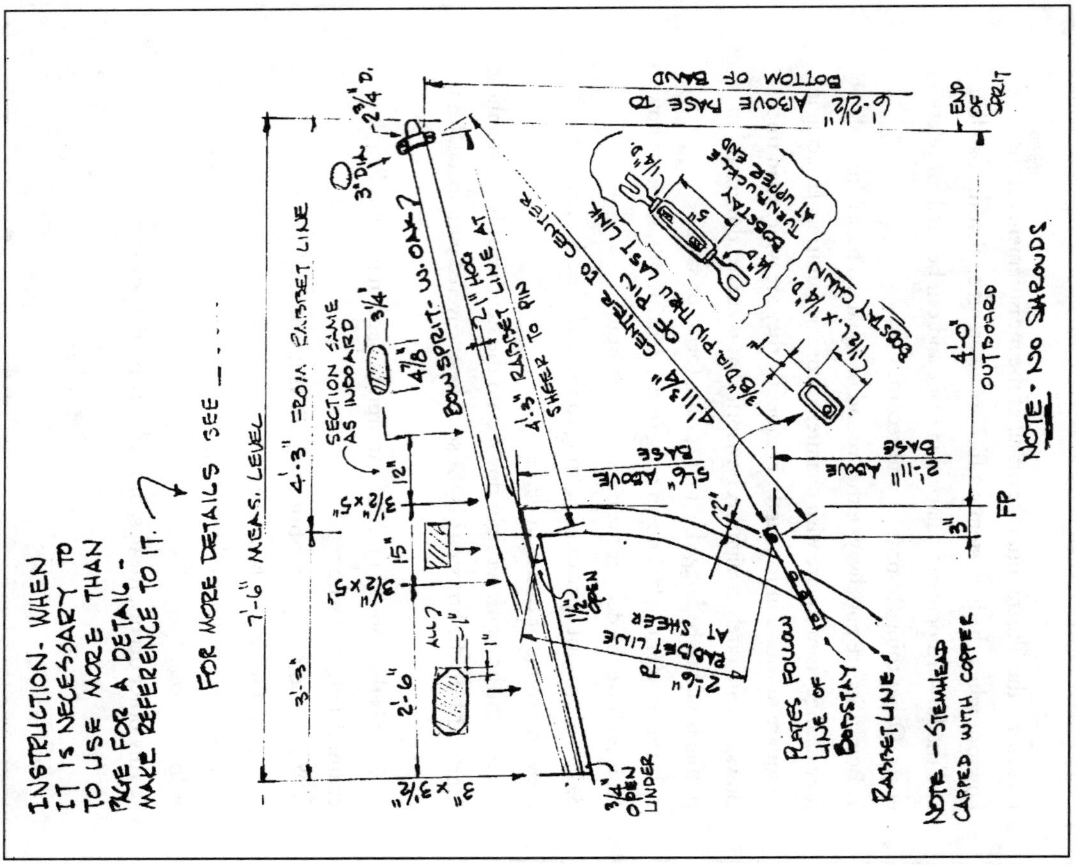

Figure 9.19. Sample note sheet for a bowsprit take-off.

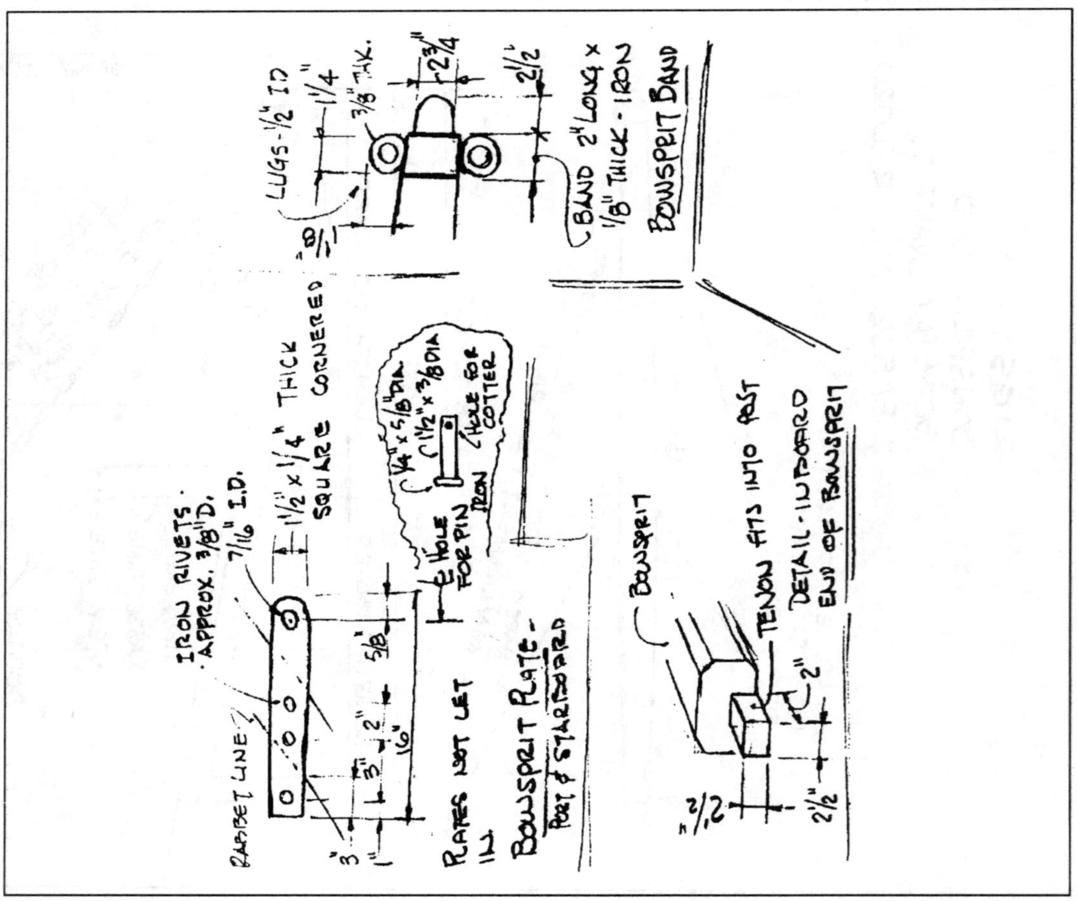

Figure 9.20. Details found on bowsprits.

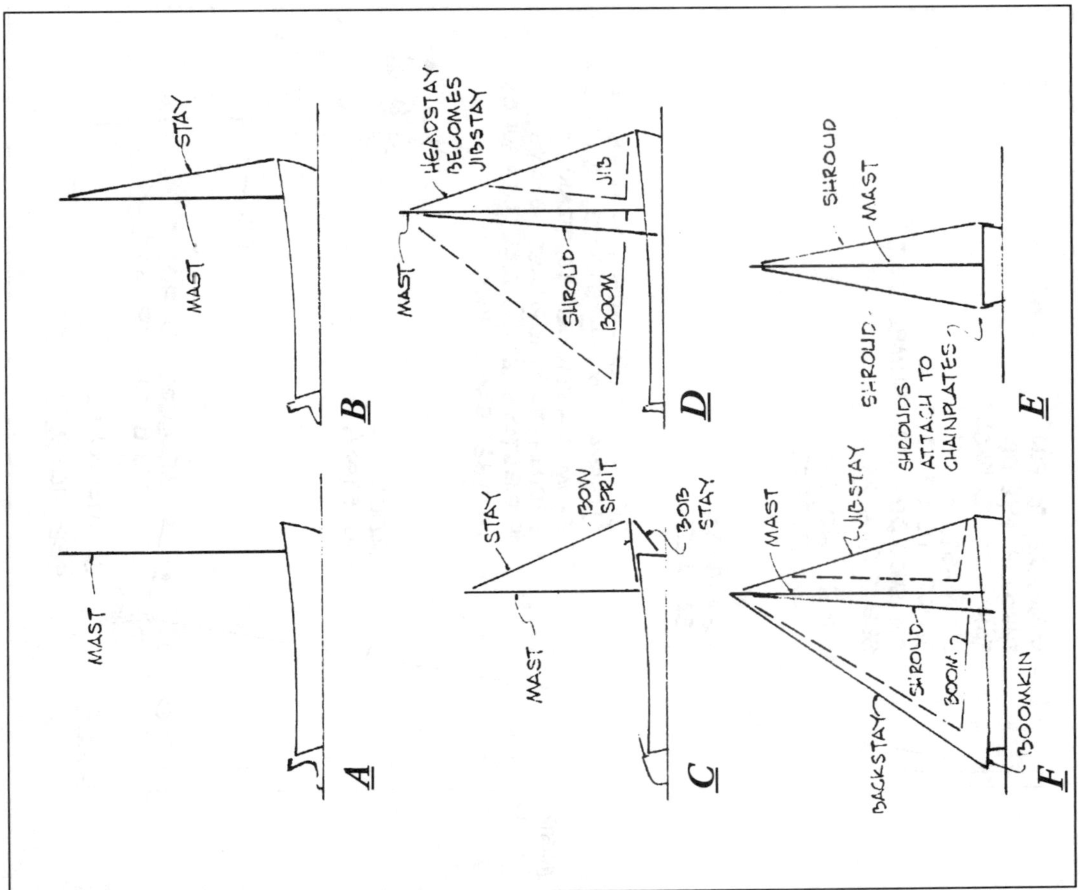

Figure 9.21. Standing rigging: rigging that is permanently set up (fixed in place) and rigging that is only occasionally adjusted. (A) The unstayed mast is common among small, open boats. The rigs shown in figs. 9.1 through 9.3 have been carried on unstayed spars. (B) and (C) Arrangements found in catboats and small sloops. (D) and (F) Shrouds and backstay added for a larger sloop or cutter rig. The difference between the modern cutter and sloop rigs is the location of the mast. The cutter's mast is located at about 40 percent of the waterline length from the bow; the sloop's 25 percent to 35 percent. The difference in "classical" times was that the cutter's topmast and bowsprit could be housed (lowered) while underway; the sloop's could not. (E) A view showing the relationship of the shrouds and mast. The backstay is rigged from the masthead to either the deck or a brace called a boomkin at the centerline of the deck and projecting aft from the hull.

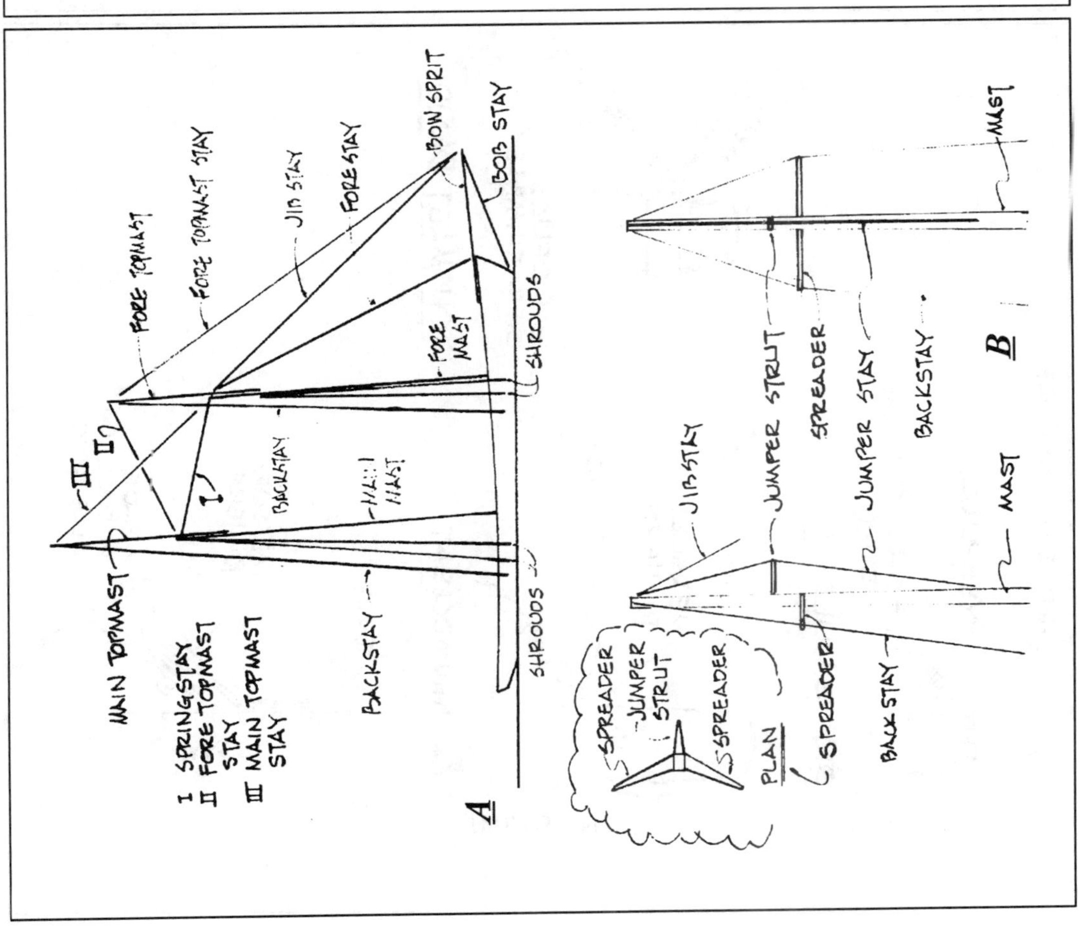

Figure 9.22. The standing rigging for a schooner. (A) Note the addition of the jibstay, forestay, and springstay which is rigged from the fore masthead to the main masthead. Note also that the backstays are rigged to the top mastheads port and starboard. (B) Spreaders and jumper strut.

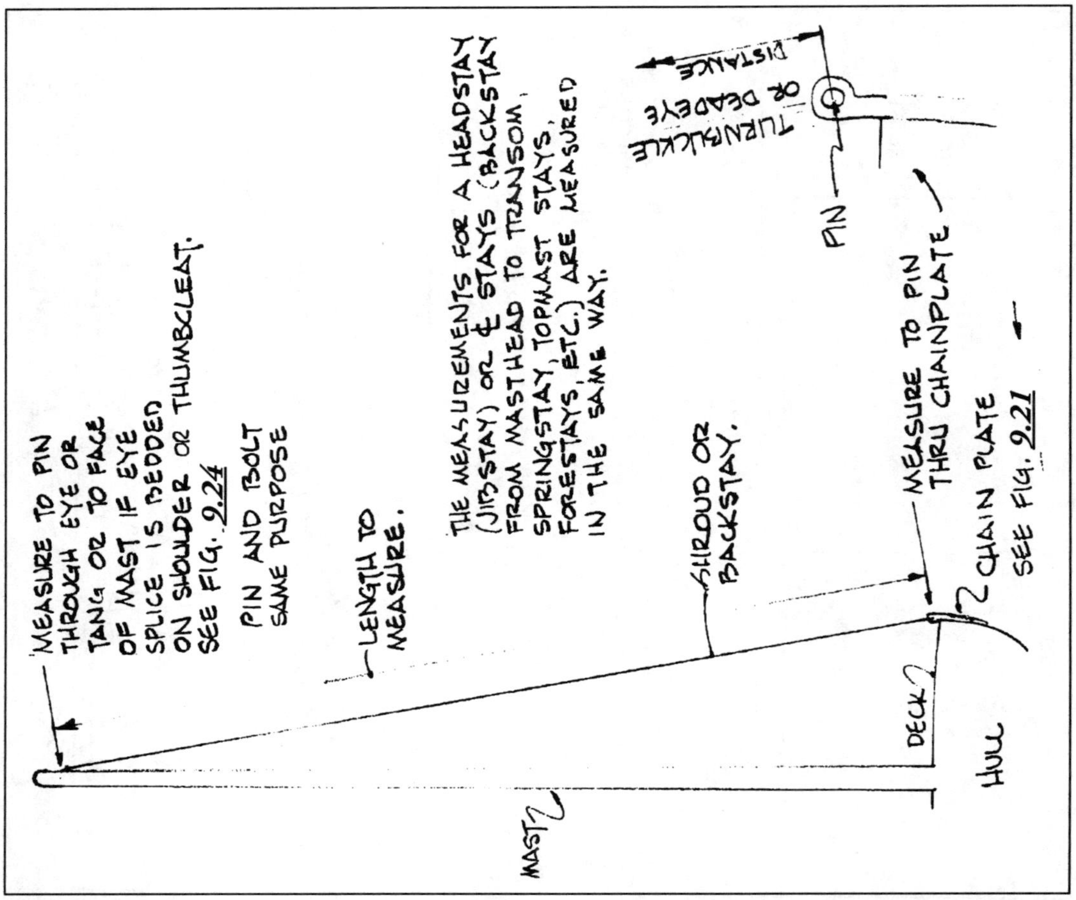

Figure 9.23. Instructions for measuring and recording standing rigging. It is important to record where the rigging was when it was measured because in place on the boat, it will most likely be longer than when it is lying slack on the ground. For example, a jibstay measured on the boat may have a length of 20'-6". Lying on the ground, it may measure 20'-5 1/2". Measure, sketch, and record the following:
1. What part of the standing rigging it is
2. Material—wire rope or natural (manila or hemp)
3. Size of material
4. Describe the ends—deadeye, eye splice, cut splice, swaged on spade end, or turnbuckle with fork, eye or spade end—and record their sizes
5. The length—the examples show the arrangements most likely to be encountered and what measurements to take and the sort of observations to be recorded. Be careful when measuring shrouds and stays that are spliced. Measure not only the splice, but also the spar's diameter where the splice is bedded.
6. The names of the manufacturers of all materials if known.

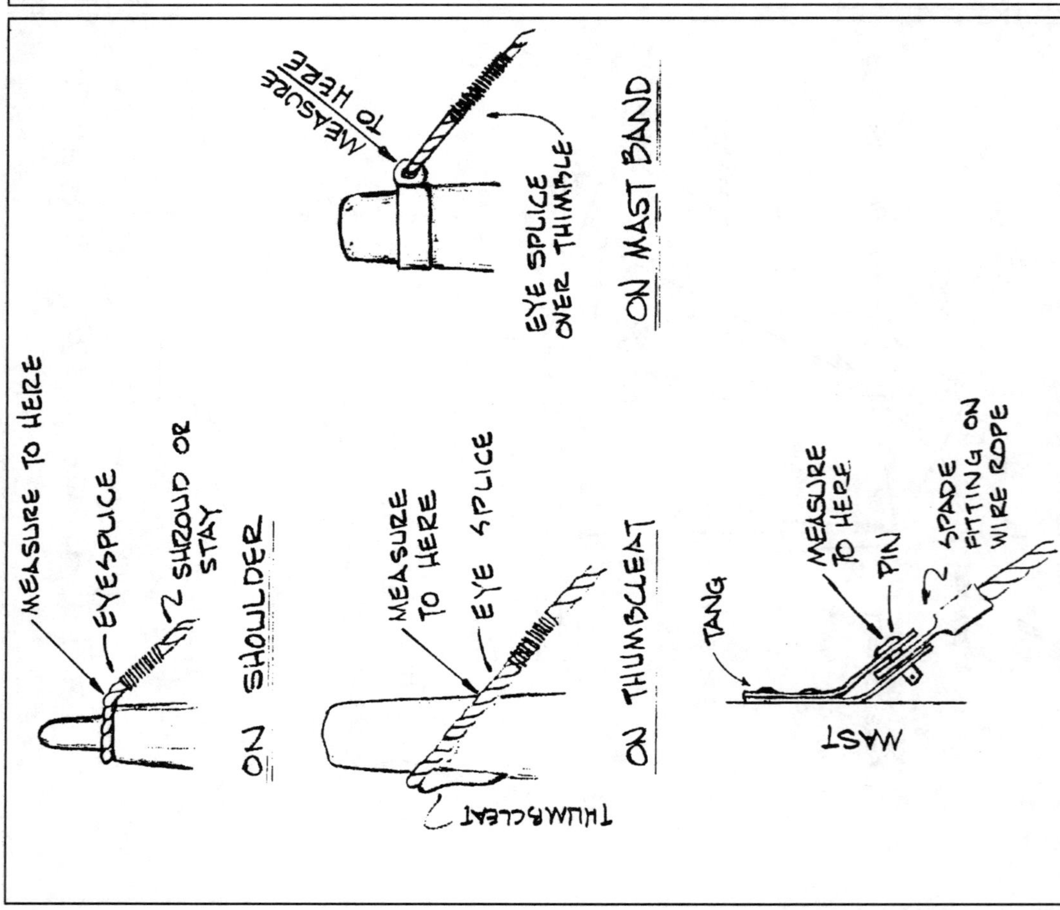

Figure 9.24. Standing rigging: where to take measurements at the upper end of the shrouds and stays.

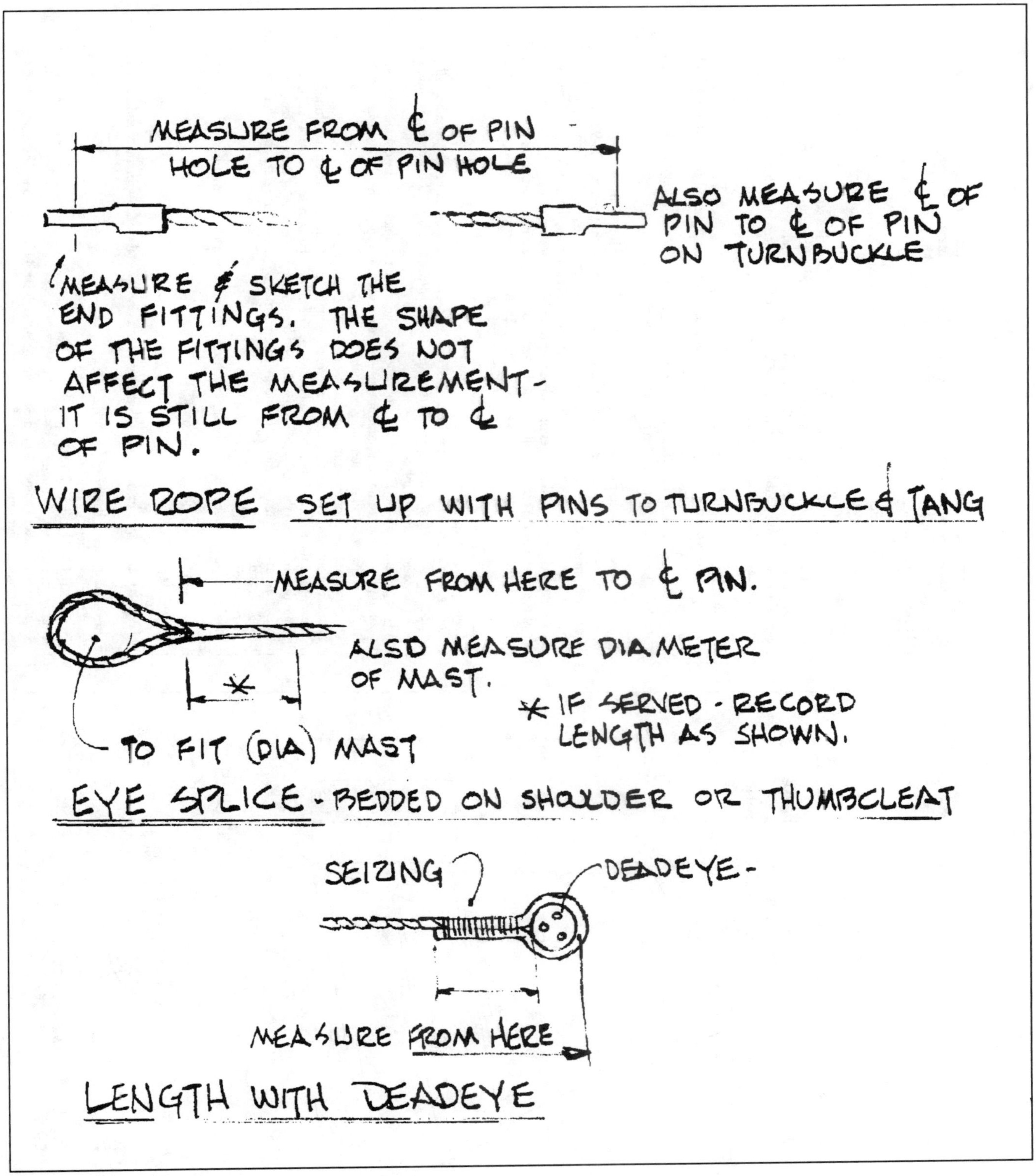

Figure 9.25. Standing rigging--where to take the measurements when the rigging is lying slack on the ground.

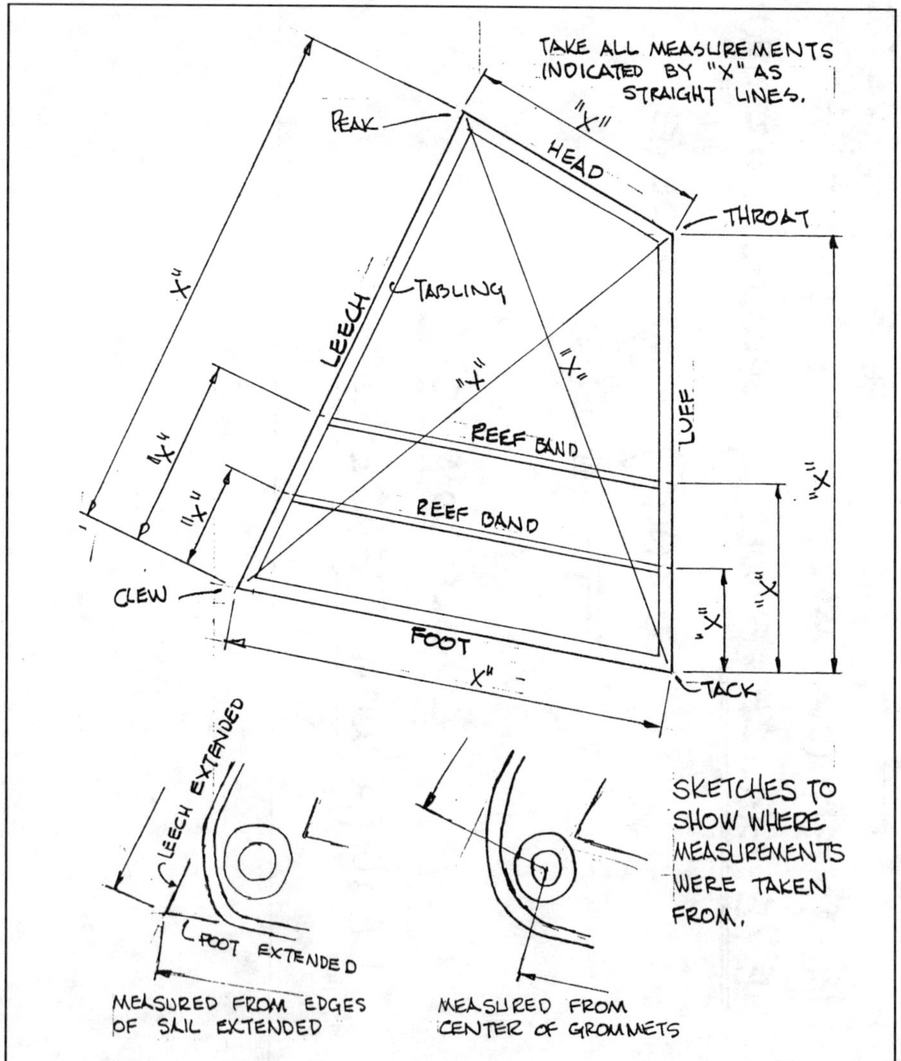

Figure 9.26. (above and right) Instructions for measuring a sail.

The object of measuring and recording a sail is to get data so that the sail can be replaced or accurately drawn and also so that changes in sail design and construction can be analyzed.

A sailmaker's plan of a sail is drawn with the edges represented by straight lines. But a sail is not built along straight lines—as a flat sheet assembled from a number of cloths or panels (narrow strips of fabric). The sailmaker designs and builds in a series of complex shapes by varying the widths of the cloths. These shapes differ with the outline and dimensions of the sail, the use to which it will be put, and the material from which the sail was built. The outline of a sail also has curved shape called roach which may be either concave (positive) or convex (negative).

There are two other factors that affect the way a sail is measured. If a sail has been used, it has stretched so that it is not the same shape as when it was built. Also, the fabrics used for sails all have a different stretchability. The stretching allowance for Egyptian cotton is 6 inches in 20 feet along the length of the fabric and double that if across the fabric. For Dacron (terylene, British), negligible or none in either direction. From this it can be seen that care must by taken when measuring sails. To measure:

1. If possible, lay the sail out on a reasonably flat dry surface.
2. Use the grommets or cringles at the corners of the sail to gently stretch the sail and hold it to its full shape and so that it lies reasonably flat.
3. Measure the outline of the sail. If the sail is triangular, measure the luff, foot, and leech. If the sail is four sided, measure the luff, foot, leech, head, from tack to peak, and clew to throat. See figs. 9.26-9.29.
4. Make a scaled sketch from the measurements showing the "straight line" outline shape of the sail and the major interior items and record all of the dimensions on the sketch. See figs. 9.26-9.29.
5. Note where the measurements were taken from. That is, from the center of the grommets, or the "theoretical" corners of the sail. See fig. 9.26. Note: From this point on, the sketch made in item 4 above will get cluttered. Make copies of it to use for recording the following data.
6. Measure and record the roach on all sides of the sail. Do this as the outline of the sail is measured and save time. If there is no roach, note that too. See fig. 9.28. A simple way of recording roach is shown in fig. 9.28.
7. Locate and sketch:
   a. Tabling: measure its width in several places and note its width.
   b. Roping: Is it the same everywhere? Check the diameter several places. Measure the rattails at all ends. See fig. 9.31. What is the roping? Is it natural fiber, synthetic, or wire? If the roping is wire, is it exposed or enclosed in the tabling?
   c. All cringles and grommets. Are cringles of rope? Do they have thimbles? Are the cringles formed in the bolt rope? Are the grommets pressed metal? Are they sewn in (a rope or metal ring sewn into the sail)? Record the size of all cringles and grommets. See fig. 9.31.
   d. Direction and width of all seams and cloths. See figs. 9.29 and 9.30.
   e. Number of stitches per inch and stitch pattern: straight, zig zag, hand, or machine stitching.
   f. The width of the reef band. Is there one on each side of the sail? How many reef points are there? How are they built into the sail? What is the size and length of the pendants? If there are separate patches, how many? What are their sizes and shapes? Are they on one or both sides of the sail? See figs. 9.26, 9.27, and 9.31.
   g. All reinforcing patches—at corners, reef bands, etc.
   h. Hardware such as slides, clips, or hoops attached to the sail. Show how hardware is attached to the sail.
   i. Battens and batten pockets. See fig. 9.31.
   j. All insignia and emblems.
   k. Repairs.
   l. Manufacturers' names or trademarks (logos) as appear on all metal work—grommets, slides, clips, etc.
   m. Any special lines or hardware attached to the sail
8. Was the sail ever bent to a club?
9. What material is the sail made of?
10. Identify the designer, if known.
11. Record the sailmaker's name and address from the maker's label if there is one. If the sailmaker is identifed from another source, record it.

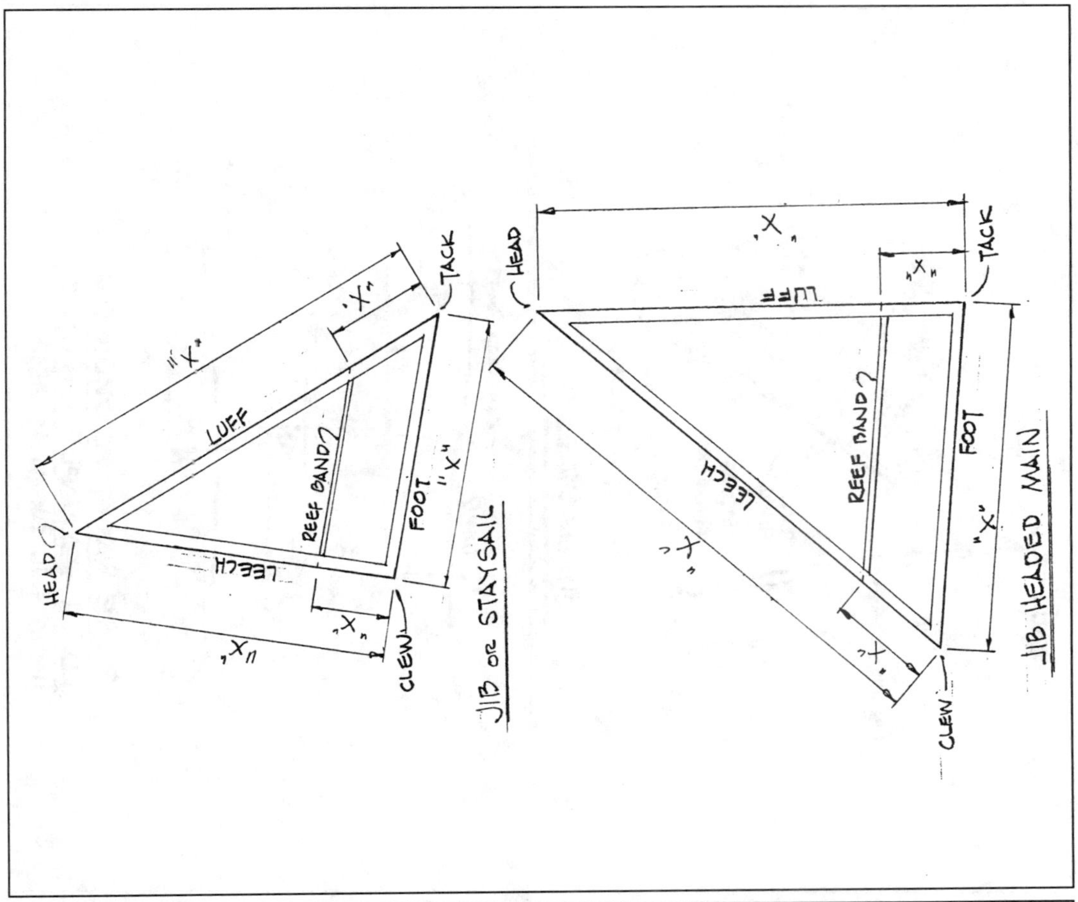

Figure 9.27. The basic measurements for a jib and a jib-headed mainsail.

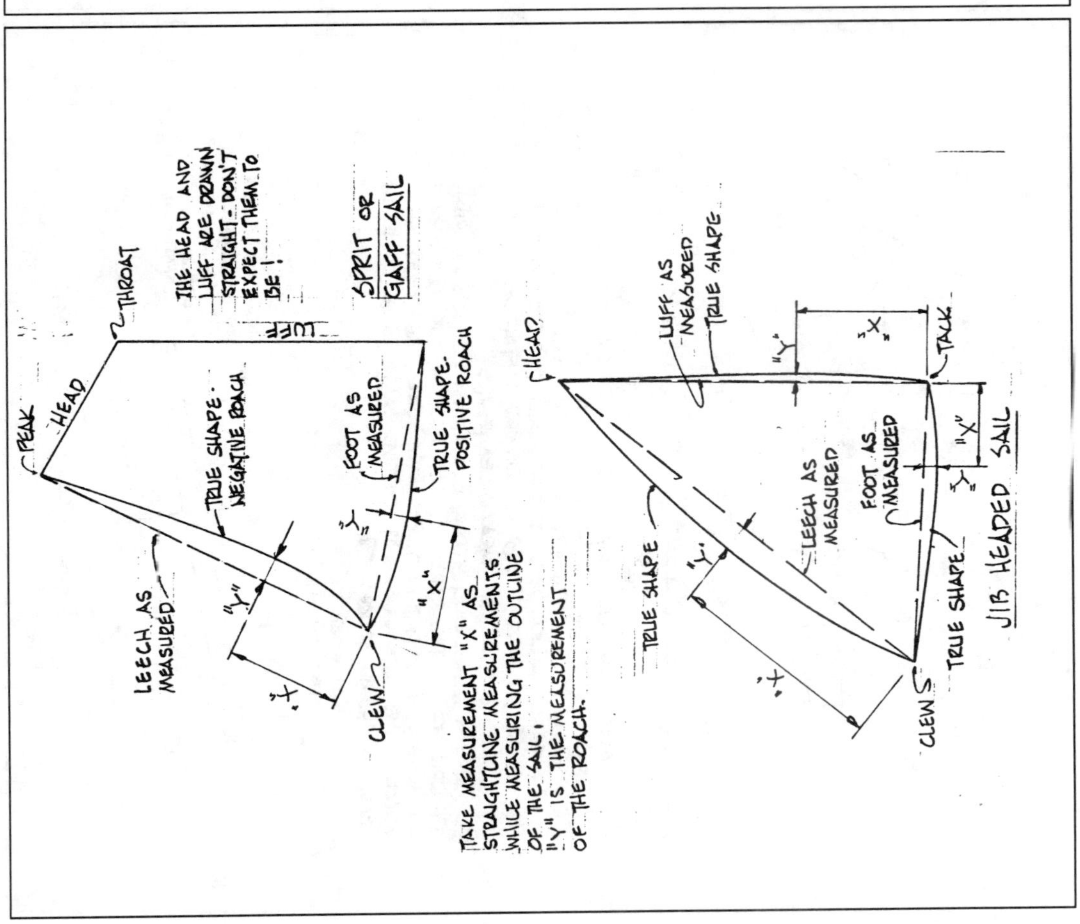

Figure 9.28. Measuring and recording roach.

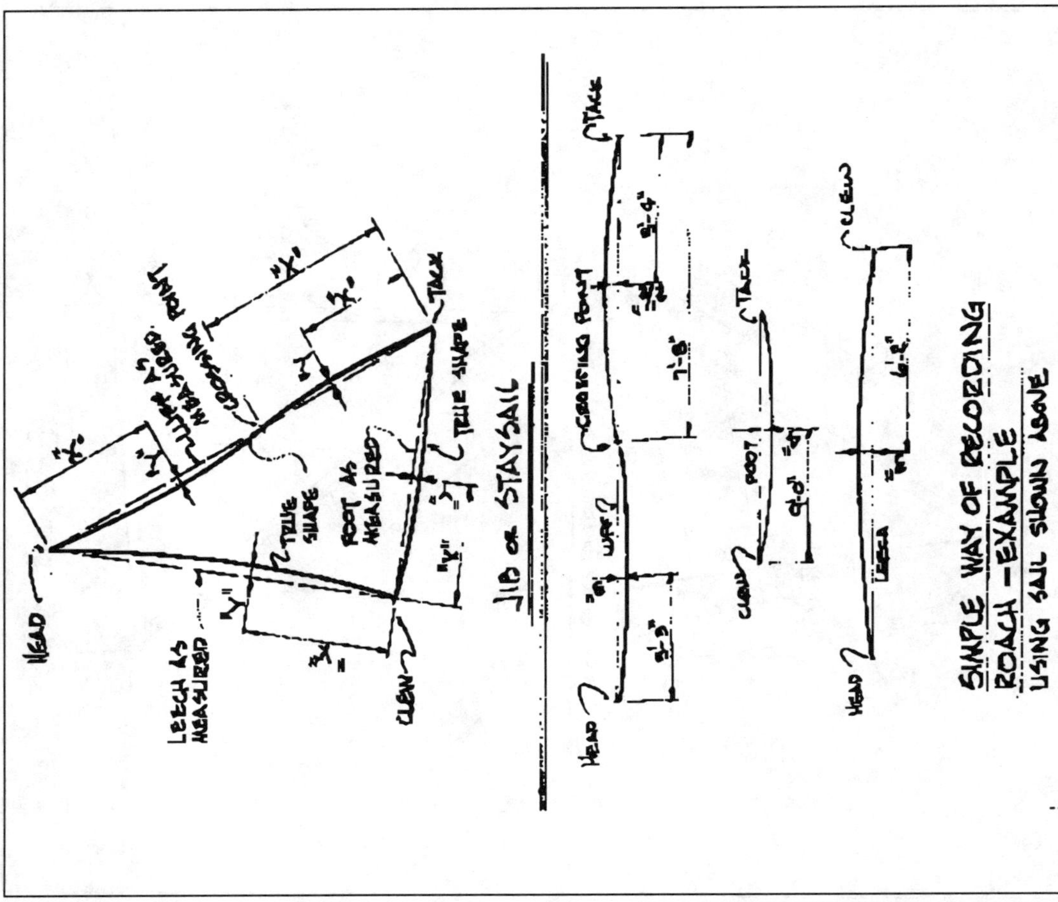

Figure 9.28. Measuring and recording roach.

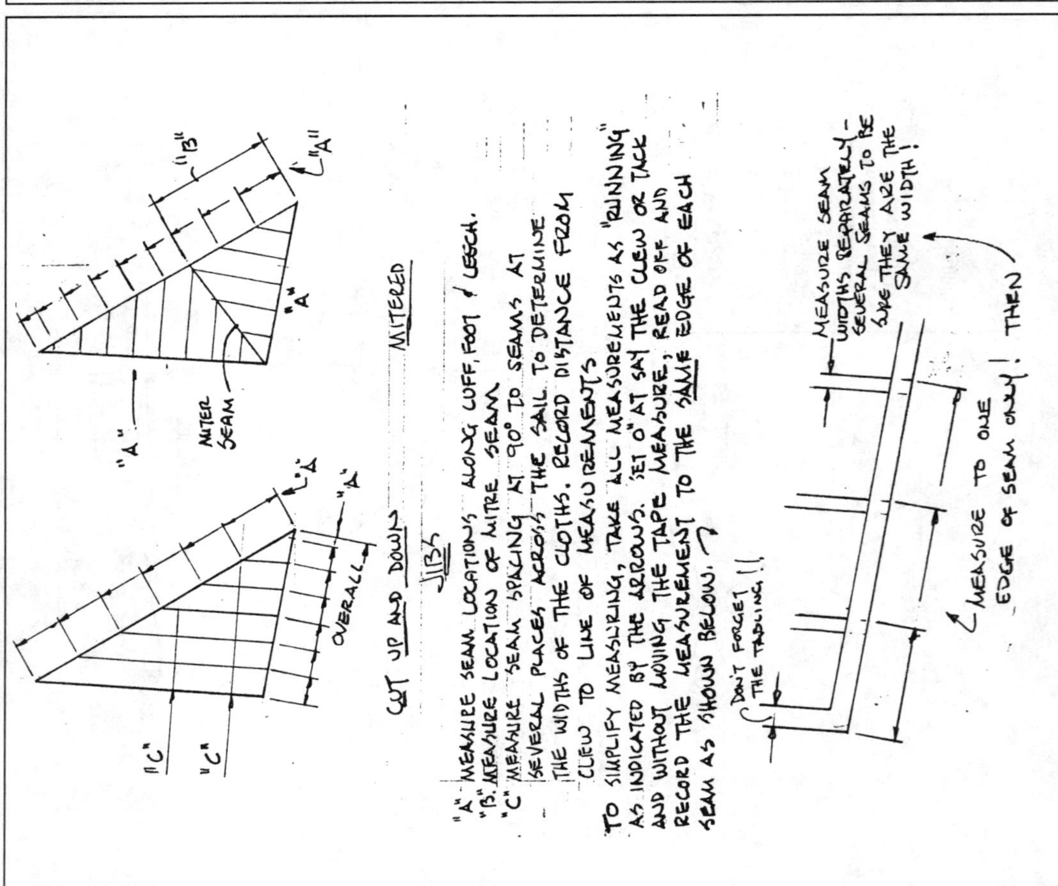

Figure 9.29. Measuring the width of the cloths (panels).

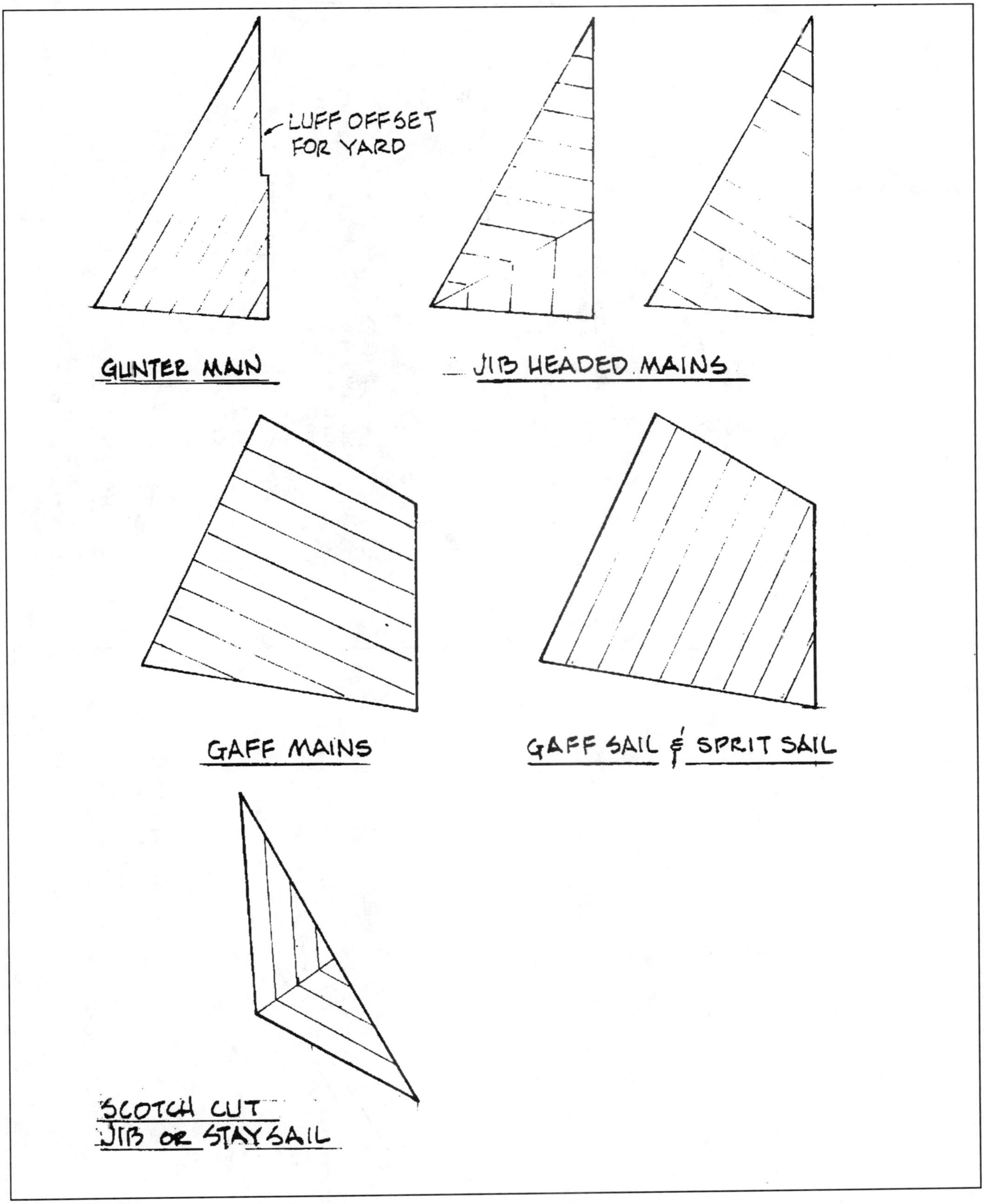

Figure 9.30. Different seam patterns.

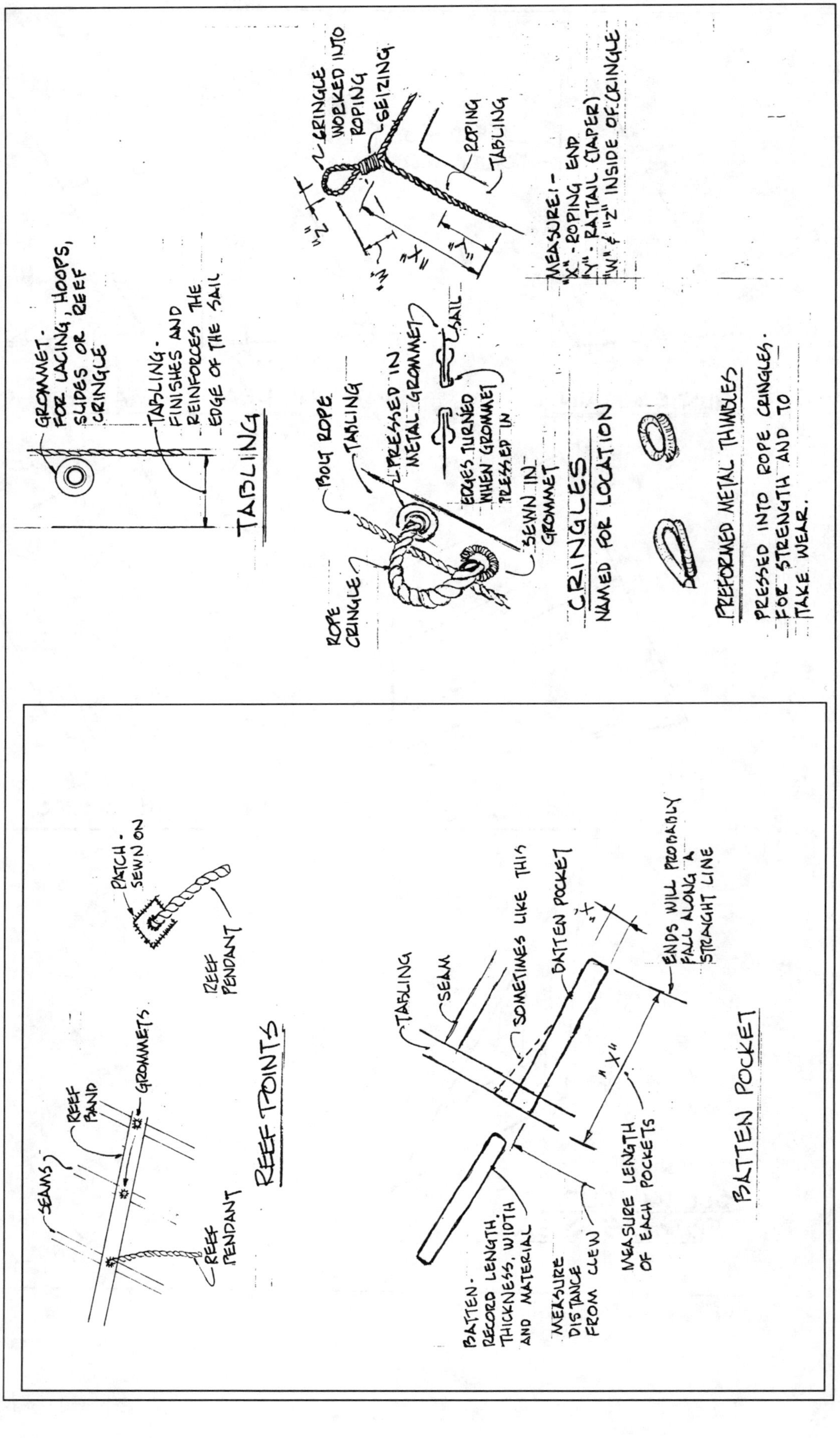

Figure 9.31. Details found on sails.

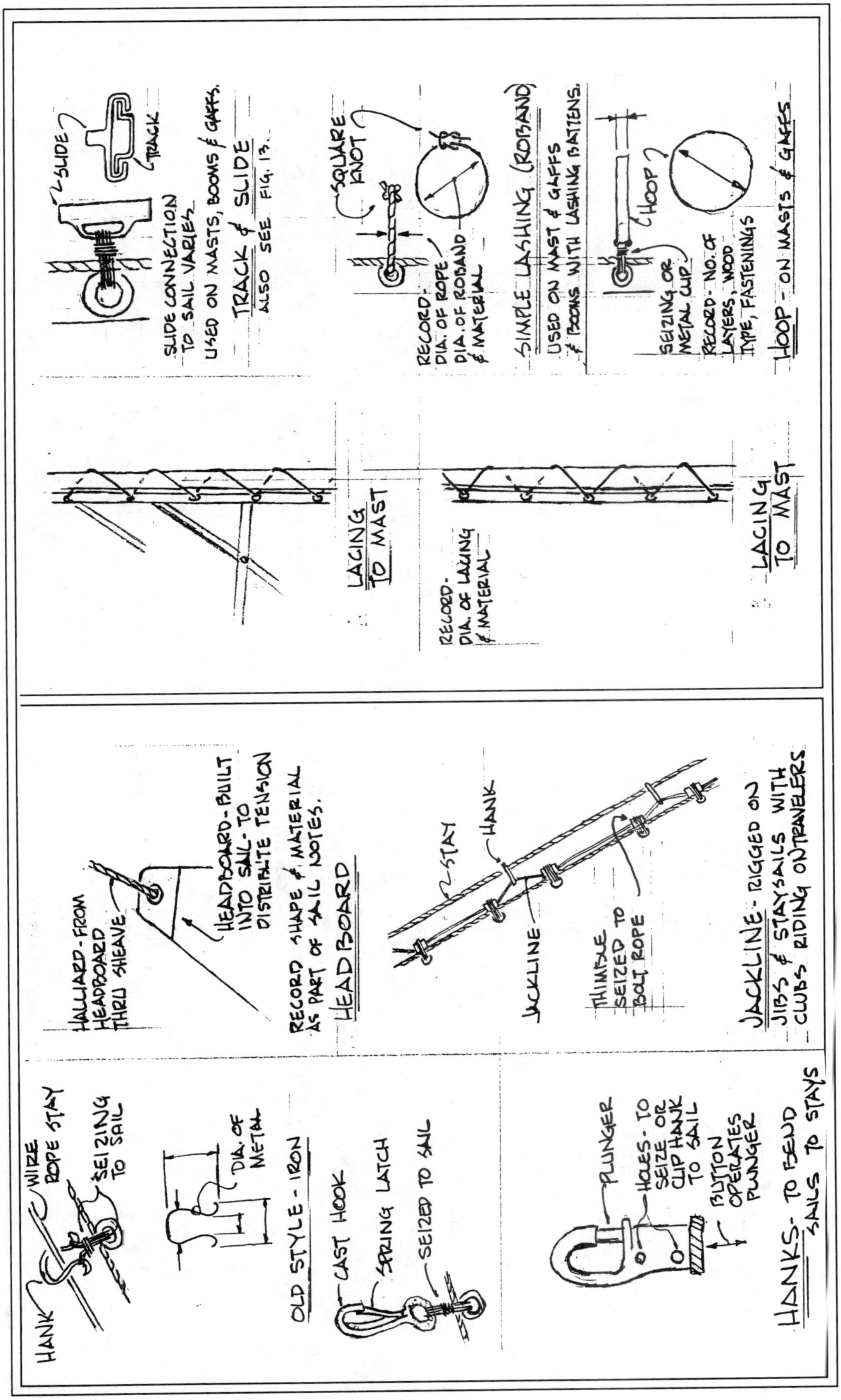

Figure 9.32. Sails—bending on.

Figure 9.33. Measuring and Recording Running Rigs. Some terms you should know:

Main halyard—used to hoist sail and tension luff

Tack—used to keep sail down. Usually made off to cleat on mast.

Outhaul—used to pull foot of sail taut along boom.

Sheet—controls angle of boom to wind and boat. See fig.9.39.

Reef tackle—to pull leech of sail down when reducing sail area or "reefing."

Topping lift—supports boom. One end is fixed in place (standing). The other end is adjustable (hauling end).

Lazy jacks—used to contain sails when reefing or furling.

Recording running rigging is mostly descriptive work that can be dealt with using schematic sketches and supplementary dimensioned sketches. Keep in mind that running rigging is used to move or control moving pieces of gear. Where it attaches to the moving or controlled is the beginning of the description.

For the measuring and recording, make up note sheets using figs. 9.33-9.39 as guides. Show the following:

1. Where and what attaches to: sail, gaff, yard, or club? Or is it attached to hardware on a spar?
2. How attached: eye splice, shackled, or snap hook?
3. Make a sketch in the notes of the attachment.
4. Material: natural, synthetic, wire rope, combination of wire rope and synthetic (natural or synthetic rope spliced onto wire to rope)?
5. Size of material. See figs. 9.44 and 9.45.
6. Make a sketch of the lead of the line.
7. Note and describe each thing or place that the line passes through, over, or under, i.e. sheaves in spars, blocks, fairleads, or beeholes.
8. The belaying point.

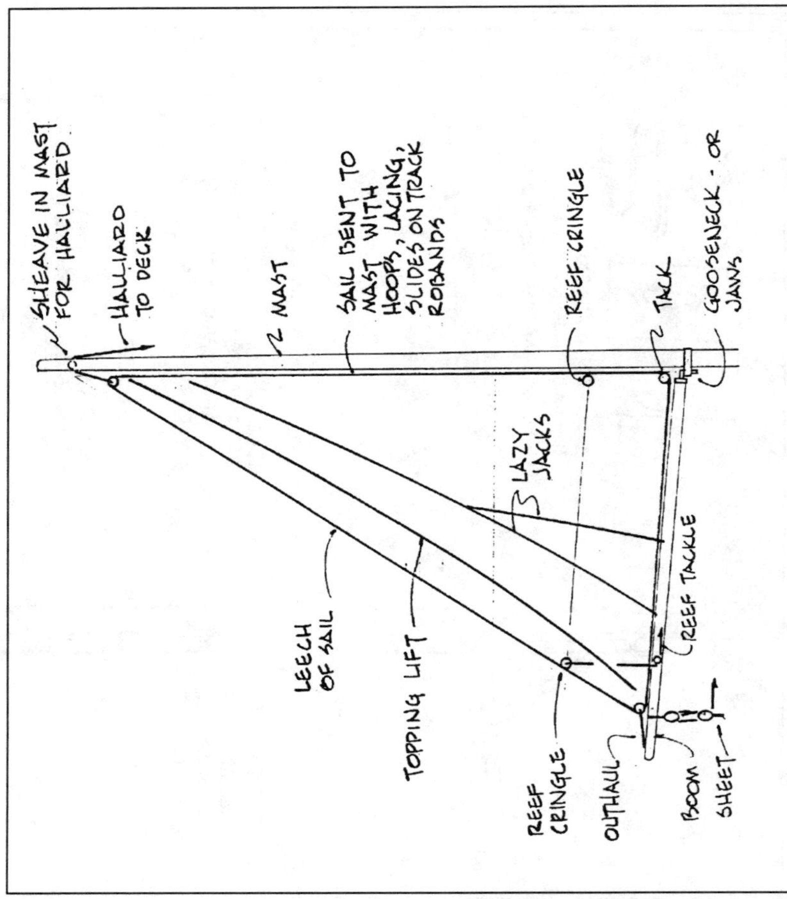

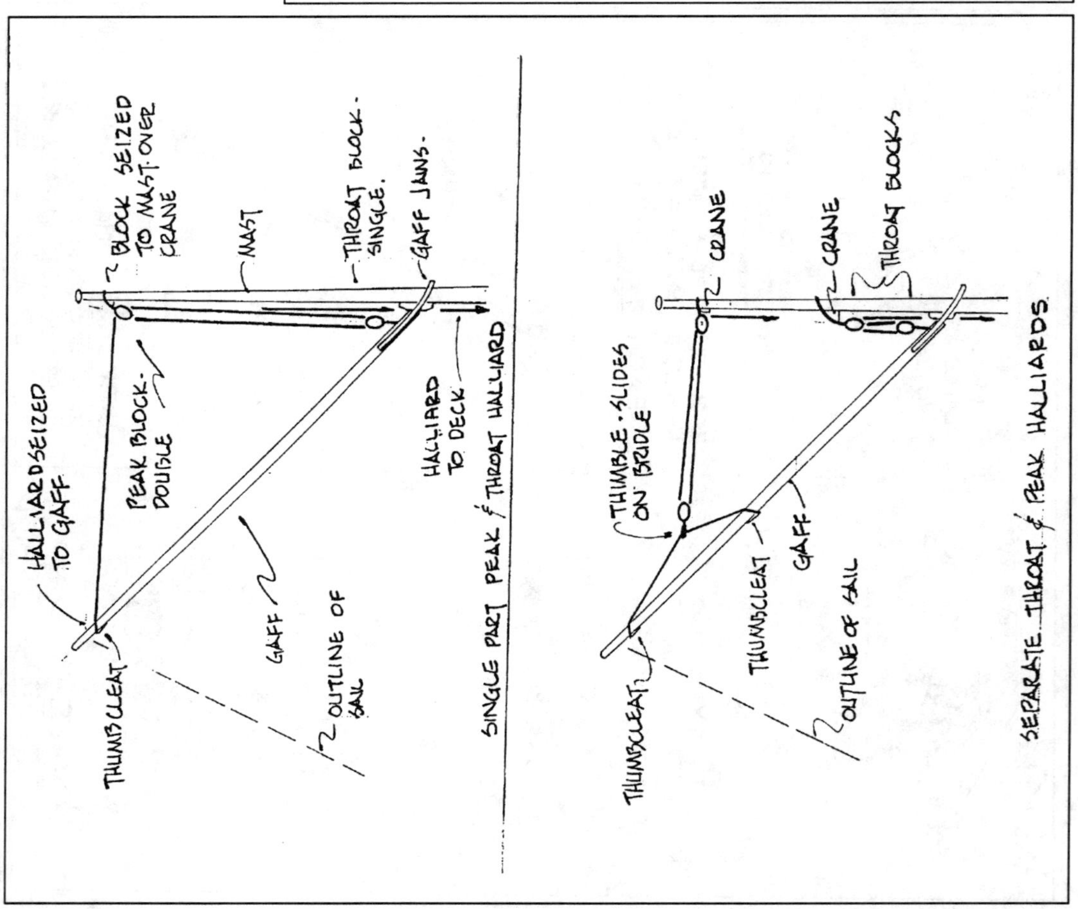

Figure 9.34. The running rigging for a gaff-headed mainsail.

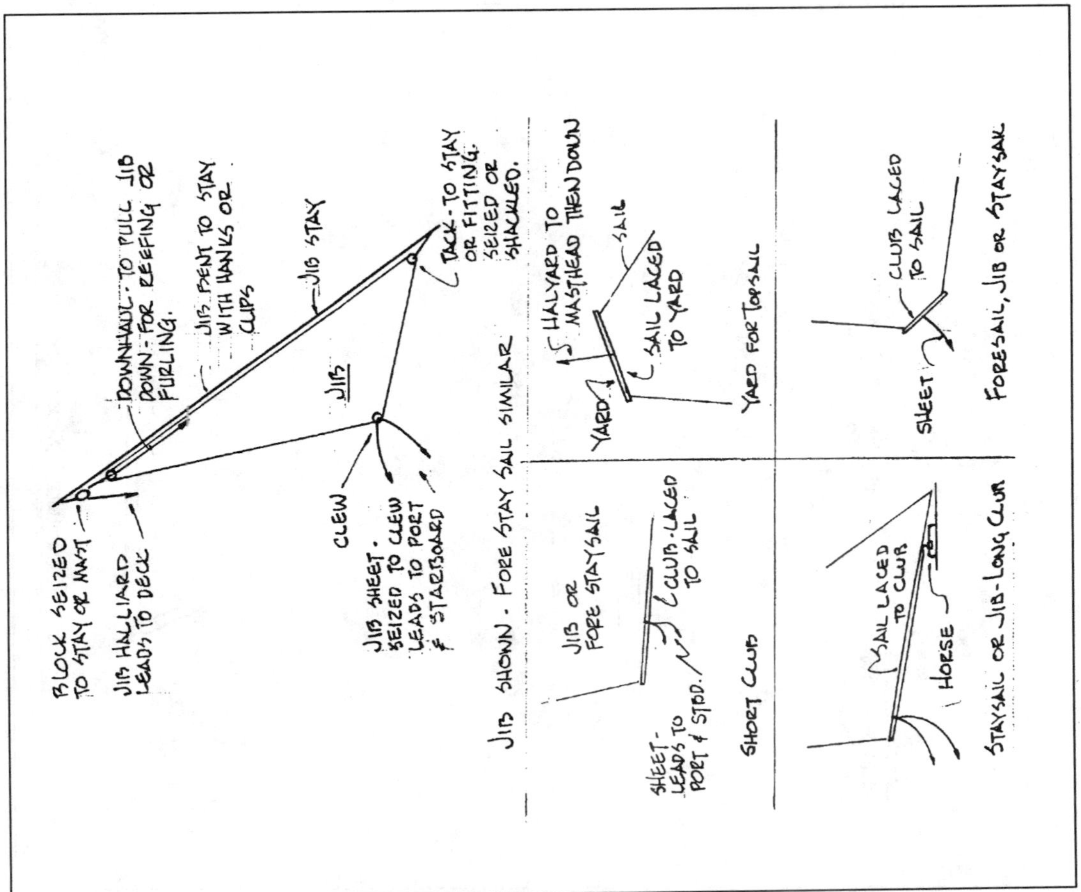

Figure 9.35. The running rigging for jib, fore staysails, clubs, and yards.

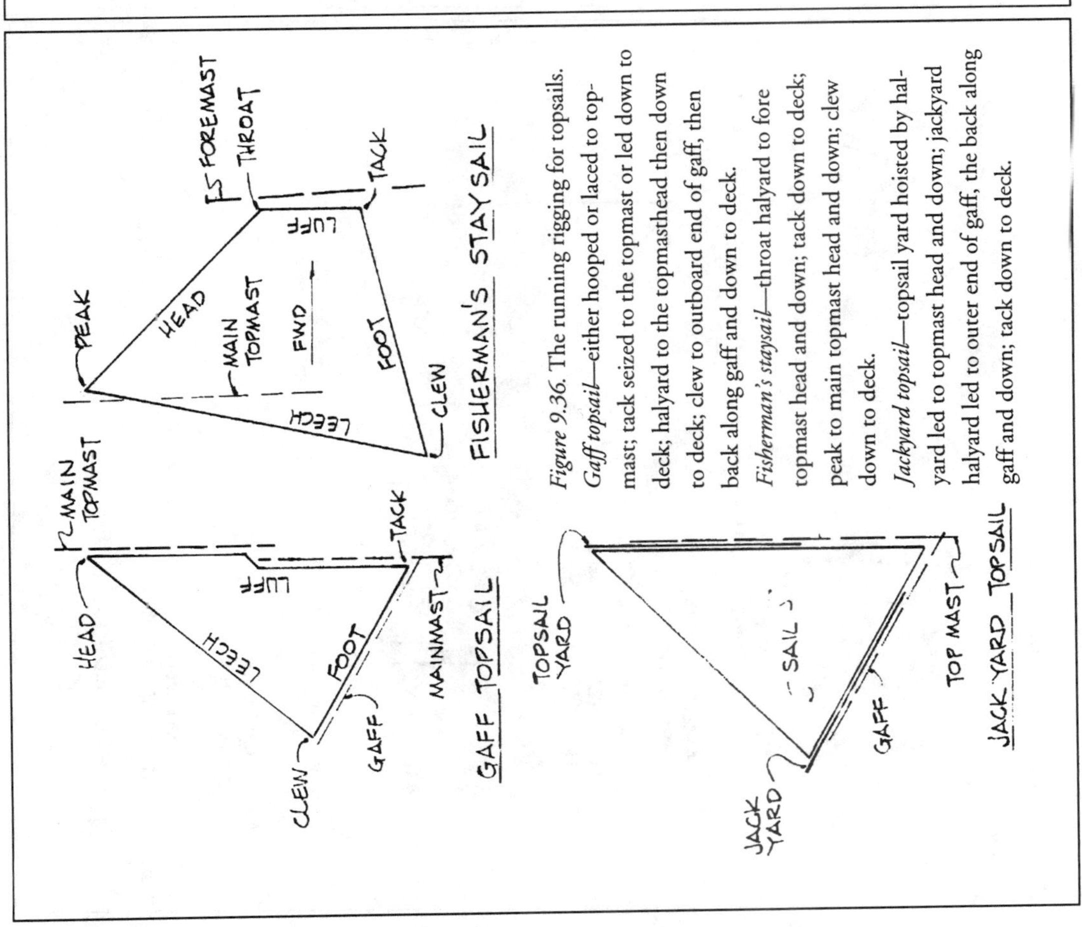

Figure 9.36. The running rigging for topsails.

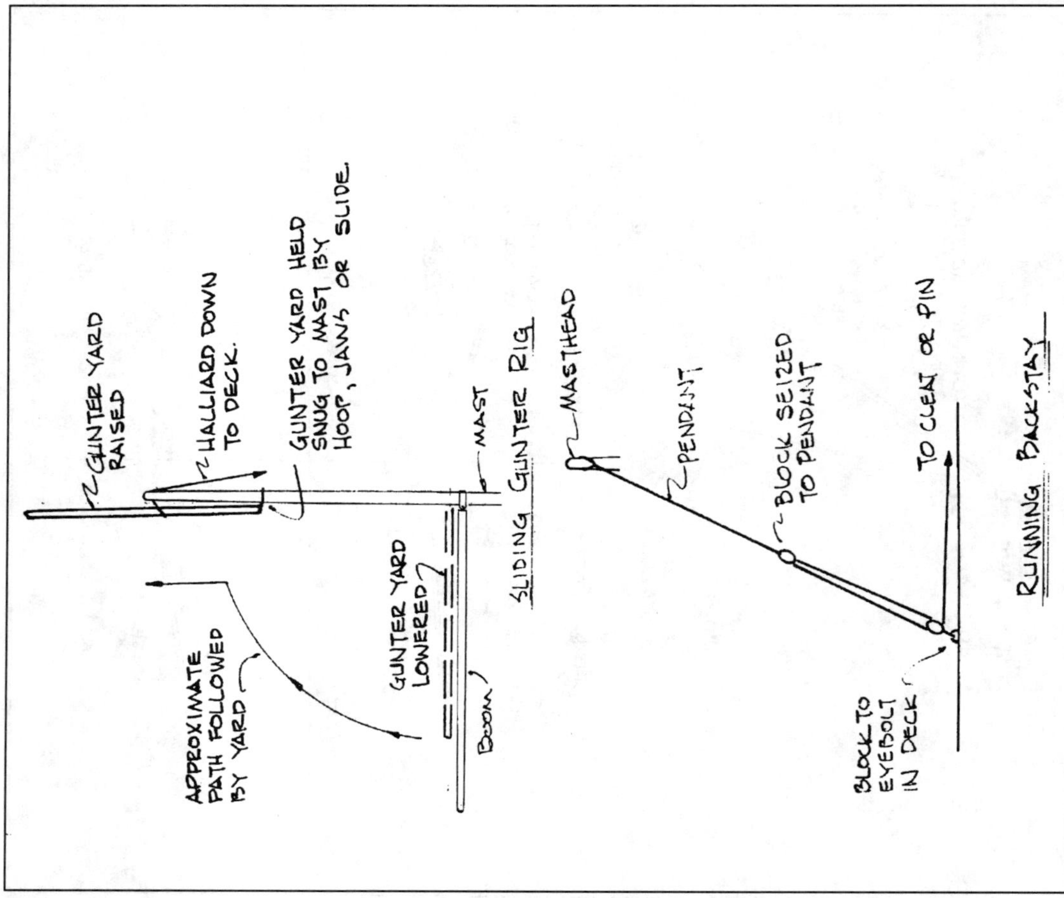

Figure 9.37. Running rigging—the sliding gunter and running backstay.

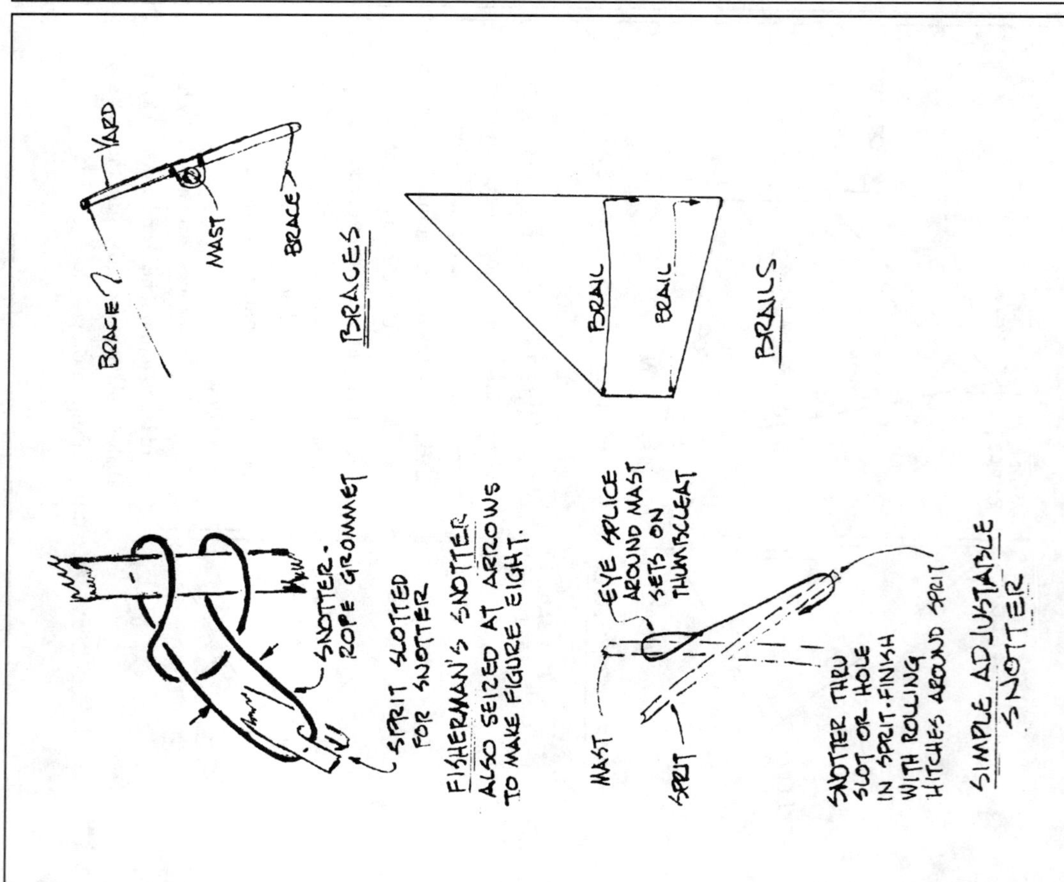

Figure 9.38. Running rigging—miscellany.

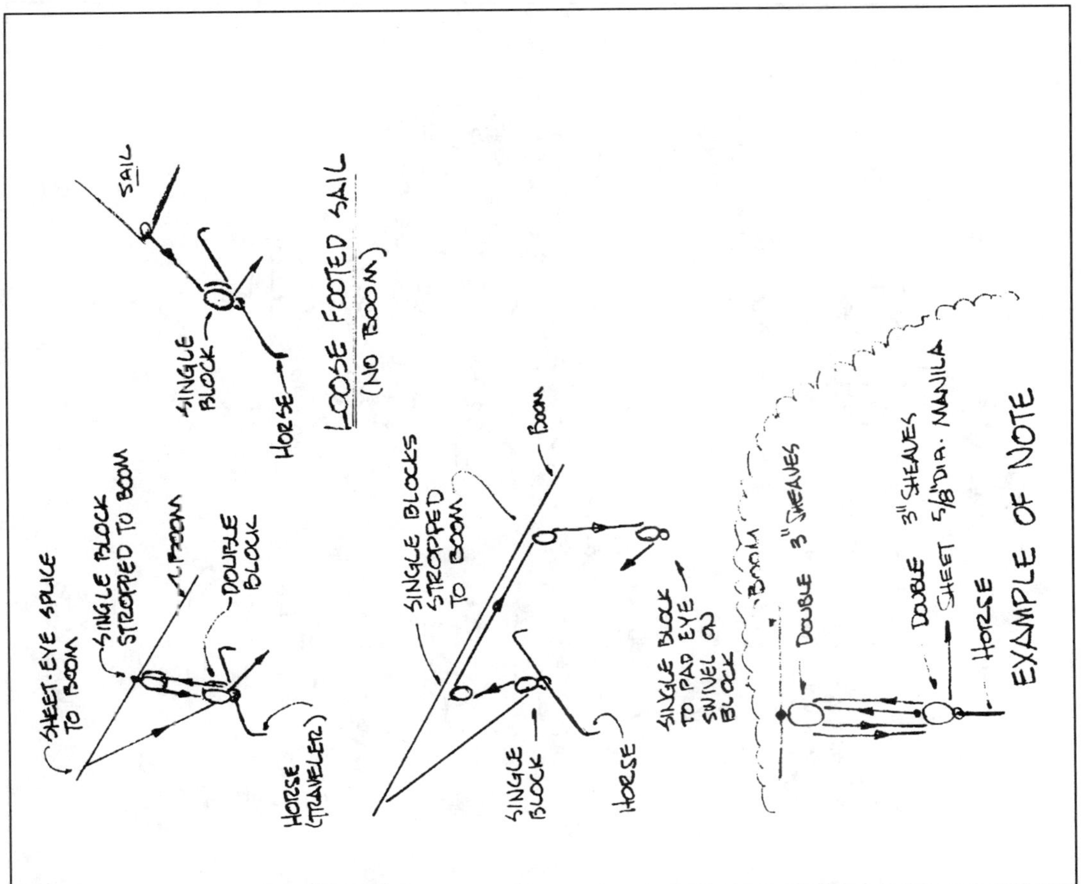

Figure 9.39. Running rigging—sheets.

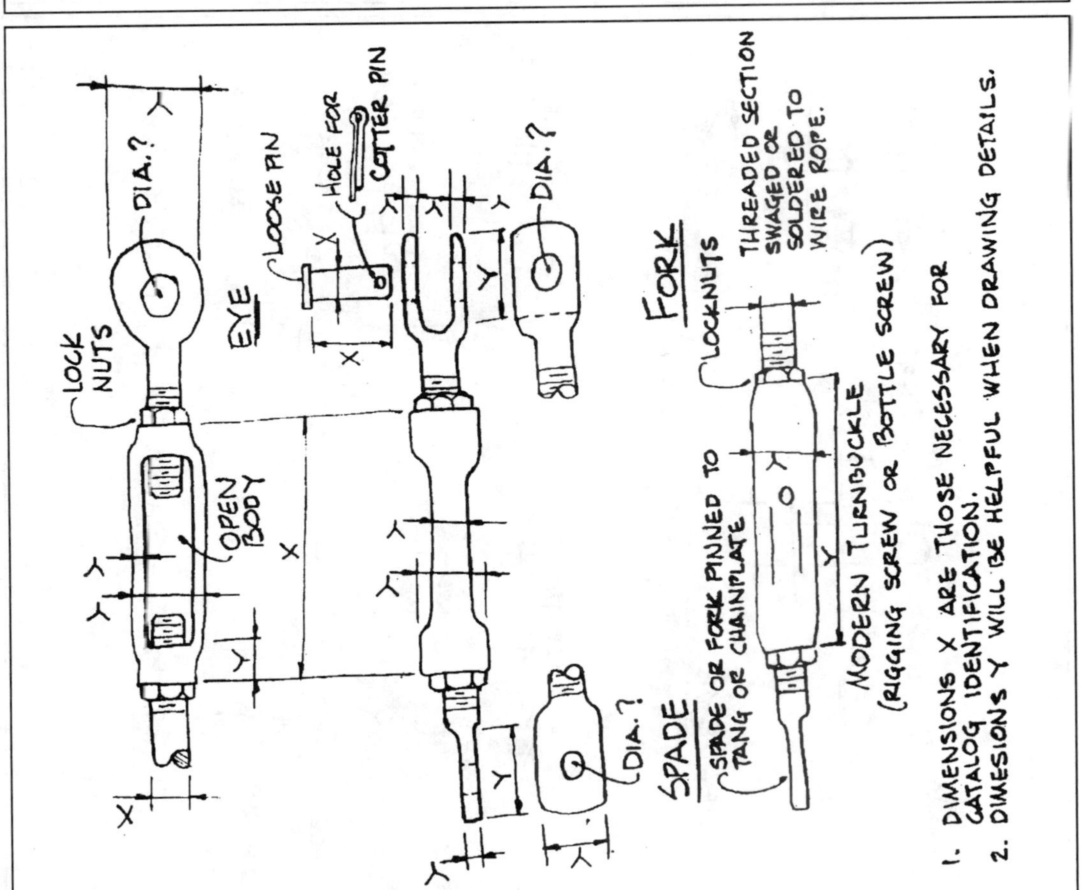

Figure 9.40. Detail measurements of a turnbuckle. Notice the difference between the data needed to identify an item in a catalog and that needed to draw it in detail.

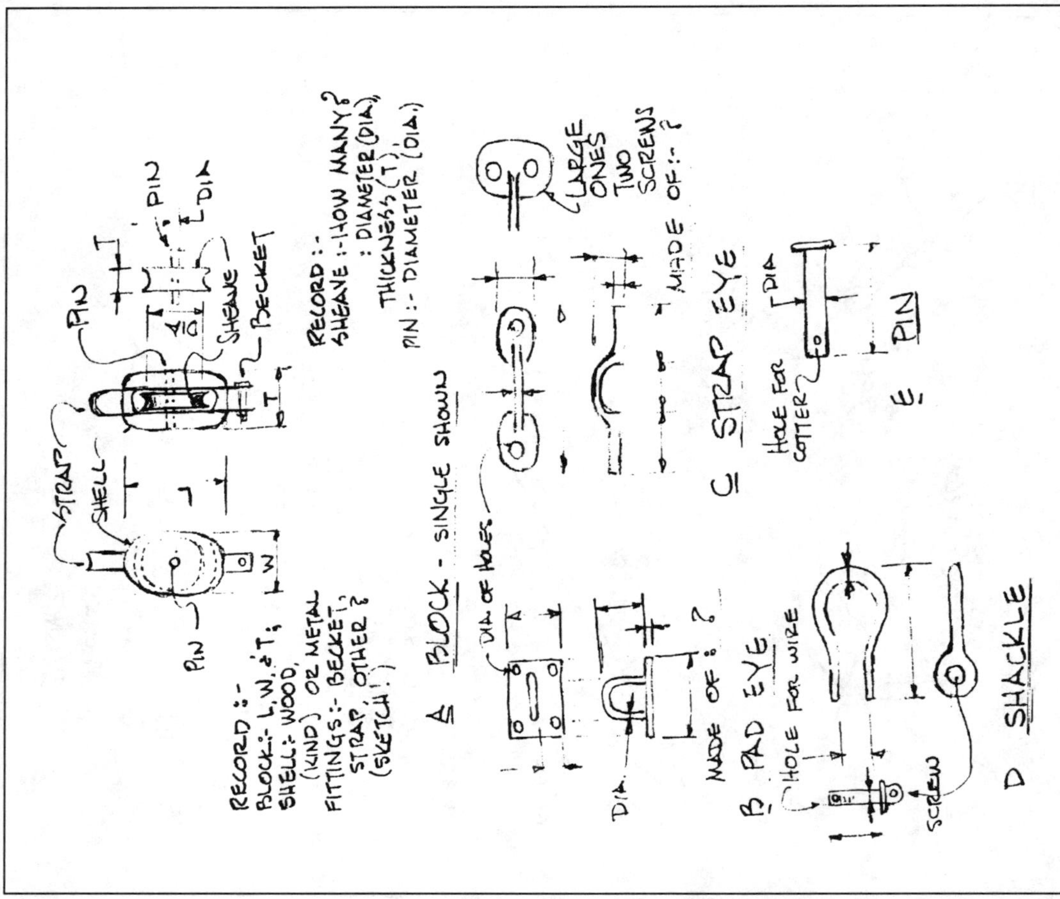

Figure 9.41. Measuring miscellaneous pieces. Make note of maker's name, logo, and catalog number when hardware and "patent" fittings are used on rigging. If all of the blocks in the rig are of the same type, i.e. wood or metal, accurately describe one of them. For all similar blocks, record the location, sheave size, and whether single, double, swivel, becket, or other.

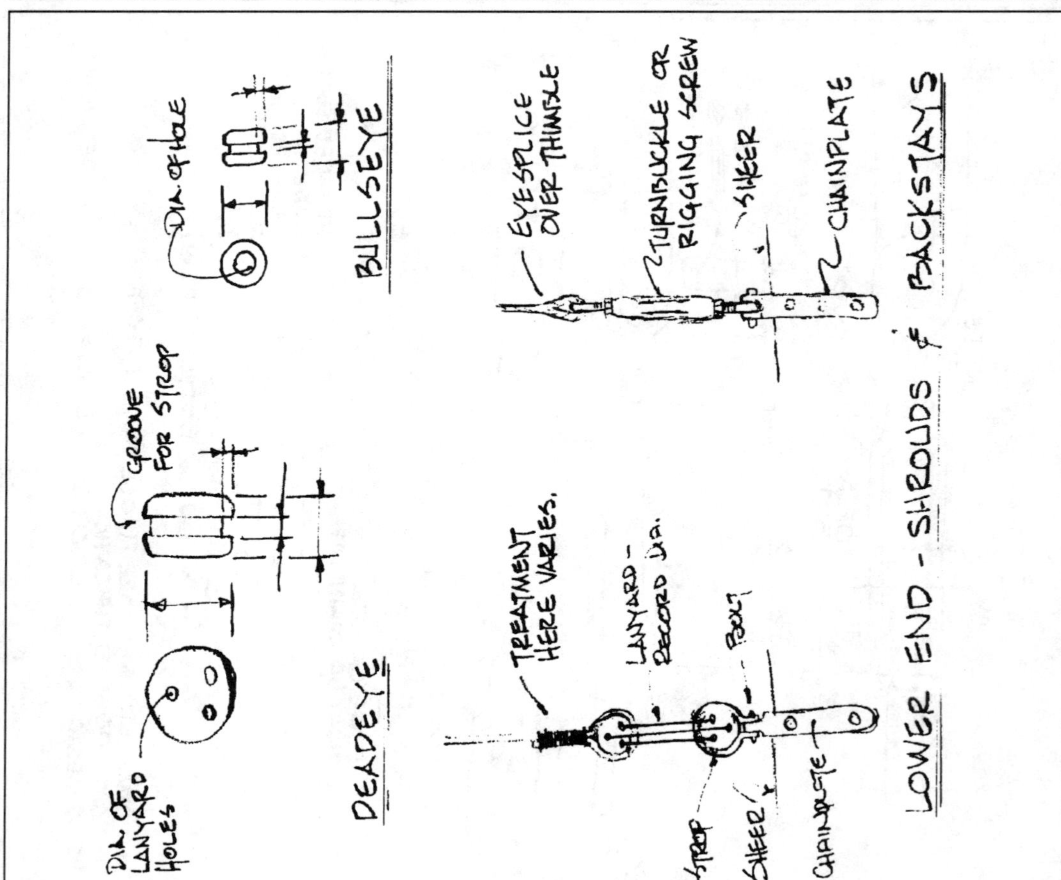

Figure 9.42. Chainplates, shrouds, backstays, bullseyes, deadeyes, and turnbuckles.

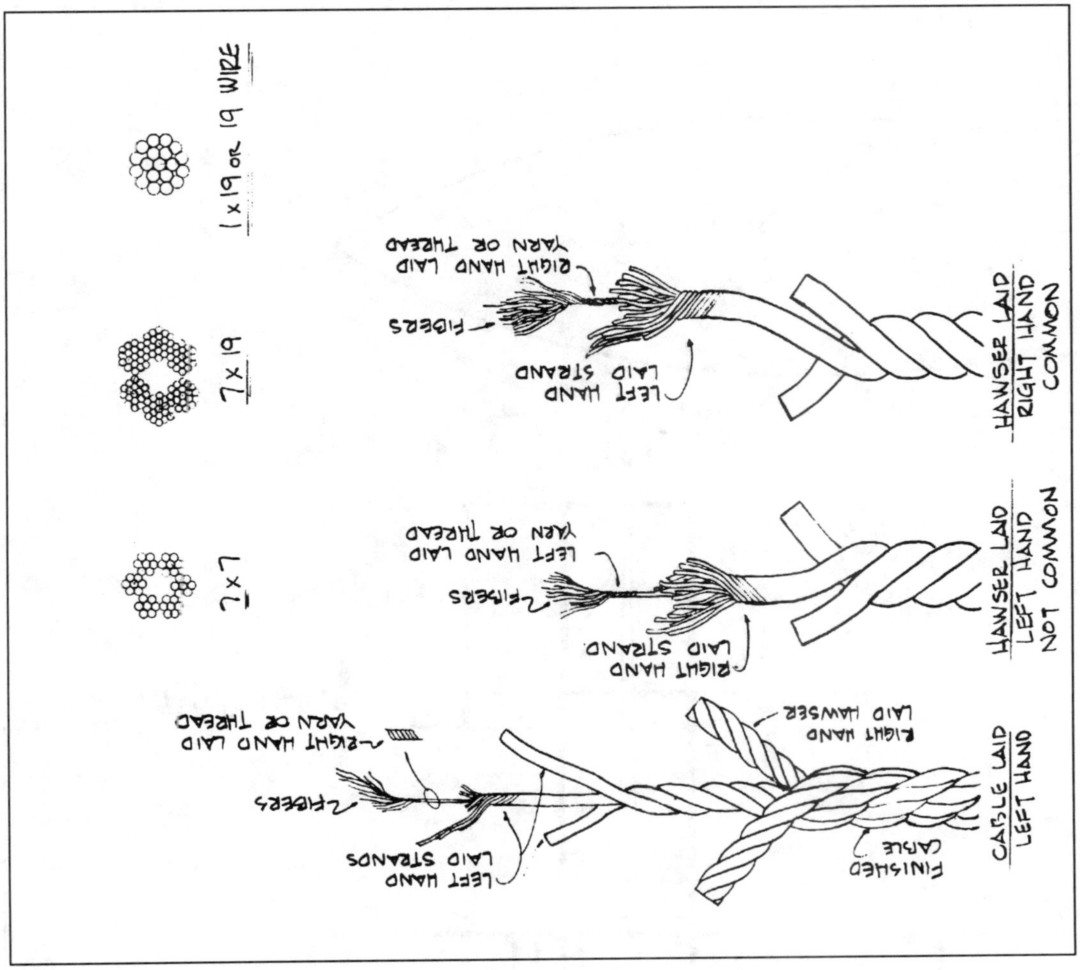

Figure 9.43. (above) Cordage—rope and wire rope: "Cordage: ropes, esp. the ropes in the rigging of a ship," from the *American Heritage Dictionary.* "Cordage: all twisted rope of whatever size or material." Ashley. "Cordage: under this heading comes what is known as the 'small stuff'—all the lines, twines, marlines, etc." Smith. "Rope: a flexible, heavy cord of twisted hemp or other fiber." *American Heritage Dictionary.* "Rope: anything in cordage over one inch in circumference." Ashley.

These definitions make it clear that there is disagreement about what cordage and rope are. Your concern is with recording what you see, so define cordage as all materials used in rigging a boat.

The size of rope over 1/2 inch in diameter is usually given in circumference—slightly more than three times its diameter. That is to say, that rope five-eighths of an inch in diameter will measure slightly less than two inches around. Take the measurement in several places to make certain of the diameter.

When you measure well-used cordage, take the measurements as close to the bitter end as possible. The section near the end whipping or stopper knot is not subject to the stretching or wear that reduces the diameter at its mid-length. A machinist's pocket caliper is handy for measuring. Be sure to rotate the calipers around the cordage to get the greatest diameter.

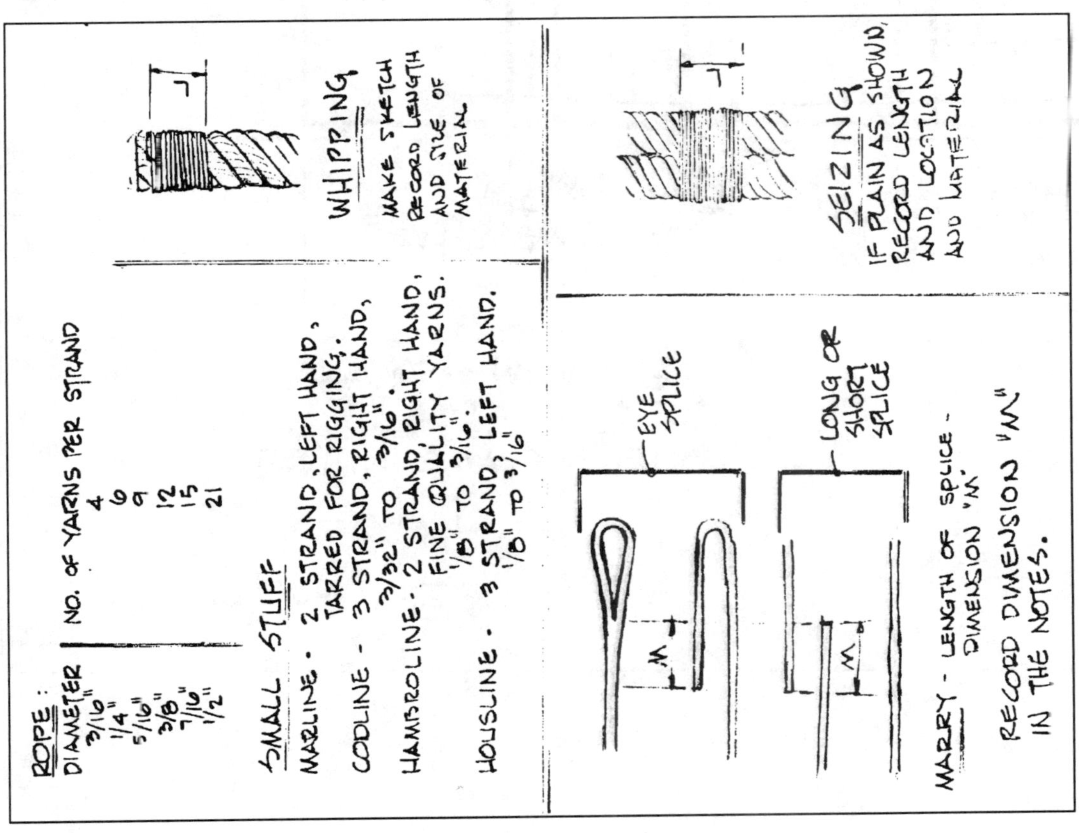

Figure 9.44. (below) Cordage: sizes of rope and small stuff, whipping, seizing, and "marry."

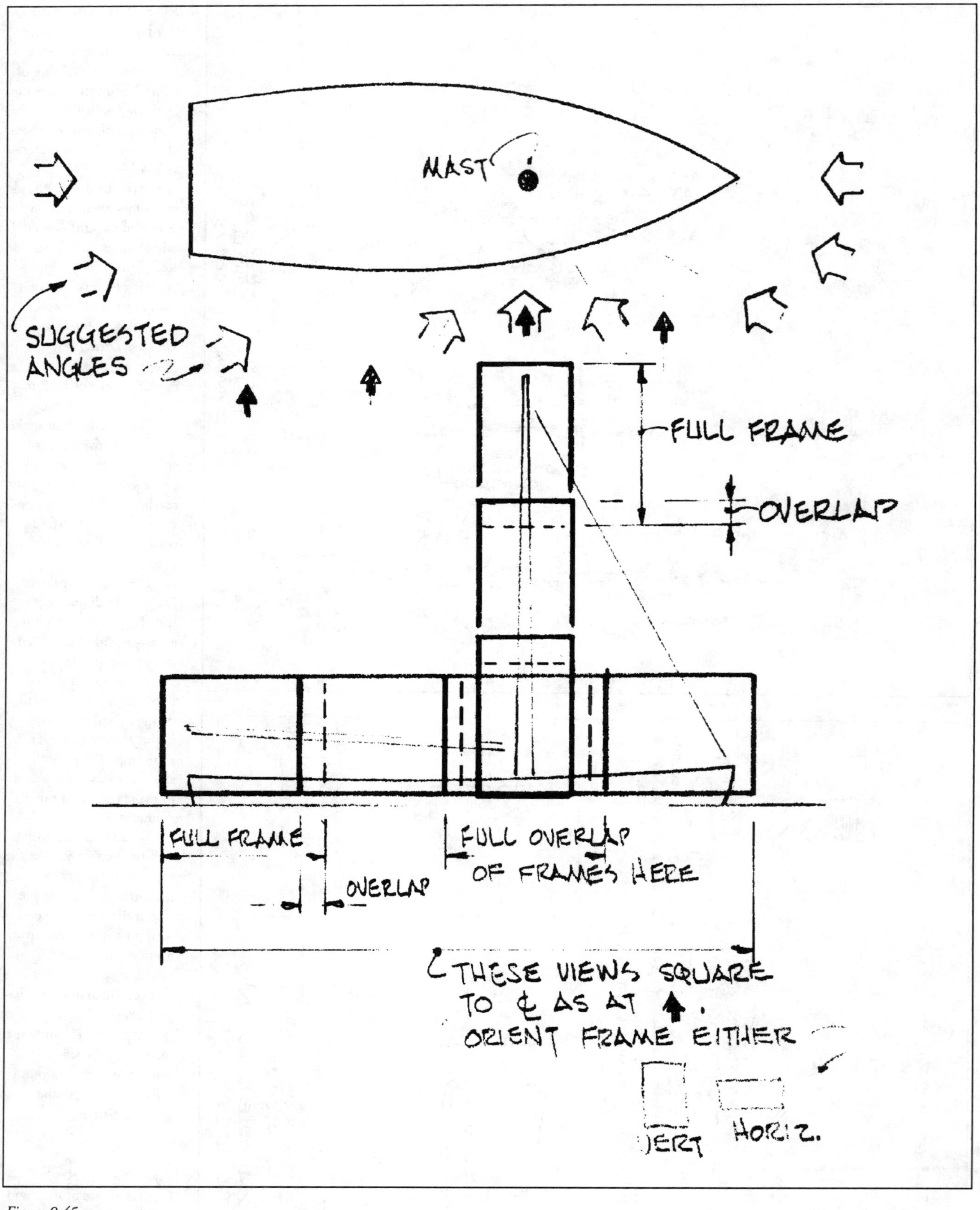

Figure 9.45.
Photographing the rig.

Chapter 10

Selective Documentation: A Guide to Quick Measuring

Recording quickly the information needed to describe a particular boat

Maynard Bray and Paul Lipke

The measuring method provided in this chapter produces a *fast,* evocative "survey drawing" of the boat by roughly measuring and photographing its profile and deck plan. If your goal is to achieve high accuracy and detail, see the more elaborate methods presented elsewhere in this manual. The technique is fast and highly discriminating as opposed to accurate and indiscriminately detailed. It relies on substantial drafting from photographs to fill the voids in measuring caused by restrictions of time and circumstance.

"Selective documentation" is a term the authors struggled to develop in order to describe this unusual approach. The term deserves a brief explanation. As the words imply, a useful archival record is produced. Unfortunately, since the method involves many unspoken, unwritten, snap decisions, the record produced is both narrower in scope and typically less accurate than those usually associated with "documentation" in the field of historical preservation. Hence "selective" was added.

Within these limitations, selective documentation can be useful for: capturing a boat's most important characteristics; surveying for type; distinguishing major differences within a type; as training for learning the relationship between measurements and the completed drawings before using more accurate methods; or as a basis for a new design that retains the flavor of the original. The method has been used to develop a simple profile for planning a painting scheme, to assess differences between Maine lobsterboats, and to record (before it was lost) the original arrangement of the 83-foot Maine sardine carrier *Pauline.*

More than with some other methods, the speed, accuracy, and ease of recording and drafting are largely determined by your expertise. Maynard Bray and Joel White measured the case study boat, the Holmes lifeboat cruiser *Barracuda,* with the former producing the finished drawing. Since *Barracuda's* field notes and finished drawing are presented here as an exemplary case study, it is important the reader be aware that between them Bray and White have many years of experience studying, designing, drawing, building, and rebuilding wooden boats. Their extensive maritime backgrounds give them a feel for what is important to record and what looks right when the drawing is made. Decisions on what measurements to take are made on the spot, minute by minute, on the basis of the recorder's experience. The resulting drawings, although not precise, are a close approximation that helps determine type and character.

With this approach, a 40-foot boat's portrait can be coarsely measured (but not drawn) in as little as two man-hours. Complex deck structures, such as hatches, and details, like winches, consume much more time than hull shape or size. But generally it is just this kind of detail that captures the "look" of a boat. In the example illustrated here, drawing the details of the stern, the rudder and propeller consumed 50 percent of the total drafting time. Use the selective documentation method when you have little time overall. It is especially valuable if you need to minimize time spent in the field. The 34-foot, fairly complex *Barracuda* consumed four man-hours in measuring and photographing and ten man-hours in drafting. The latter broke down into five hours for drawing the basic profile and deck plan and five hours needed to draw the rudder-stern details.

Selective documentation could include a reasonably accurate lines plan as well. Since the sheer, waterline, and rabbet line are measured and drawn in the profile and deck plan, they can be used to provide three fixed points on which to hang the selections. With a few auxiliary measurements and some notations regarding the character of each section, an experienced boat draftsman could develop a full set of lines by filling in and fairing the space between these points. This in turn would make it possible to measure and draw up construction sections and sectional views of the interior arrangement.

Alternatively, field measuring could include data for key body sections—on each forward, amidships, and aft. The data taken from these sections could be inserted into the profile view with another hour or so of drafting, albeit at some sacrifice to the pictorial nature of the finished plans. In the field this would require, first, more care in locating the bow and stern sections than is generally given to the rest of the measuring, and second, using either additional measuring tools or an extra pair of hands during an additional one or two hours in the field. An experienced eye can roughly interpret the sections from the sheer, waterline, and rabbet, but those less familiar will welcome the assistance of direct measurements.

For maximum efficiency, a measurer without drafting skills might wish to consider giving rough but proven plan layouts, photographs, and clearly legible field notes to a skilled draftsman so that he or she can produce finished drawings with appropriate detailing. In this way, the measurer's limited time can be focused on collecting data and then on locating gross measuring and recording errors, rather than on work better left to a faster, more experienced hand. There are hazards, however. With *Barracuda,* the measuring team noted "28 inches" instead of "18 inches" for the rudder aperture. You can see the error in fig. 10.3. The problem was quickly solved because the note taker and draftsman were the same person. A draftsman who had never seen the boat could have wasted a good bit of time trying to reconcile this dimension with the other measurements and with what could be seen in photographs. You must allocate time for cleaning up your notes.

Work is organized around, and compressed into, the allotted time. It proceeds in a sequence that allows you to be cut off at any time—by the owner's need to go fishing, by nightfall, or for any other reason—and still come away with some worthwhile information. The guiding principle is to get enough measurements to complete a specific line or element of the finished drawing. If you won't be drawing a particular line, don't waste the time measuring for it.

An ability to sketch what you see greatly speeds the recording as well. A second set of hands and eyes greatly reduces the time needed and improves accuracy. One person can sketch and note measurements as the other wields the tape measure. Another person, if experienced, can also double check the completeness and accuracy of measurements.

The process can be conducted with the boat in the water or out, usually using the painted waterline as datum in the latter case. Measurements are apt to be coarse, especially if improvised equipment is used. In the case study presented in the figures with this chapter, the measurer used a 36-inch cardboard tube found nearby as a straight edge. Measurements were based on rough approximations of right angles. There was no compensation for sag in the measuring tape. Measurements greater than 3 feet were noted in feet, inches, and quarters of an inch to make drafting easier with a scale rule. Those less than 3 feet were noted in inches and fractions.

Tools

The tools needed for selective documentation follow.

- *Pocket-sized spiral bound note pad.* The pages in a notebook won't blow around as sheets on a clipboard are prone to do. In addition, the small pages prevent you from putting too much information on a single drawing. Pocket-sized pads are more portable, plus you can flip pages and start over when you make a mistake. Save your mistakes, but mark them "void." They may be helpful later. Redraw only the subject or area in which the error was made.
- *Inexpensive lead holding pencil.* These pen cils allow you to maintain line thickness without sharpening.
- *Camera with powerful flash.* (You may need to stop the lens at *f*16 for maximum depth of field). Be sure to use film rated at least ASA 400.
- *Plumb bob*
- *Measuring tape*

Optional equipment:

- *"Sidewinder,"* a flexible steel pattern recorder, timber mold, or straight edge-spiling board for measuring sections
- *Black-and-white scale* ("footbar") to place in photographs to facilitate scaling
- *Tape recorder*
- *Flashlight*

The Process

To increase your understanding of selective documentation, refer to the figures in this chapter as you read the text.

I. Profile

A. Fig. 10.1. Assess the situation and select a convenient reference plane from which vertical measurements can be taken. Most often this plane will be the water surface for boats that are afloat or the painted waterline—assuming it is reasonably straight—for boats that are on land. If your measuring site has a level floor or cradle or railway carriage, it can be used as well. Your objective in this step is to find the most readily available base plane that will appear as a straight line in the profile drawing.

B. Fig. 10.2. Determine five to eight stations at which to measure the sheer heights from the base plane. One station will be at the bow, and another at the stern. The others can be at key points between, such as at the mooring bitt, the cabin front, the mast partner, the cabin back, and the end of the cockpit. You will need the fore and aft locations of all these points for drawing the above-deck part of the boat, so you might as well use them as station locations rather than setting equally spaced stations.

C. Figs. 10.1 and 10.2. Sketch the profile, note where the stations are to be located, and record the sheer height measurements to the base plane. You may need to extend the plane using a straightedge. If the boat is tipped significantly athwartships relative to the base plane you've chosen, take the sheer heights along both sides and average them. *Barracuda,* the boat used in this case study, was bow down in her cradle by 20 inches, so height measurements were taken squared to the painted waterline. A plumb bob was used in places where the resulting error fell within the generous range of accuracy this method assumes.

D. Figs. 10.1 and 10.2. For boats that are hauled out of the water, the underwater profile should usually come next. If the keel is straight for part of its length, a measurement from the base plane to the keel at the two stations nearest the ends of the straight portion are all that are needed. Measure for the rabbet line in the same manner, or measure it up from the keel profile.

E. Fig. 10.2. Estimate and sketch the overhang at the bow as well as the shape of the curve in between the stemhead and the straight part of the keel. Estimate and sketch the rake of the transom or sternpost and the nature of its shape. A photograph of these areas, taken in profile, is a must and will be a valuable backup to the measuring.

F. Fig. 10.3. Take measurements below the waterline aft to enable you to draw the rudder, sternpost, skeg, and propeller, and show them all on a sketch. Again, a photograph of this area in profile is a must, but notice how many measurements were taken for this important area. (So far in the process of measuring this boat, the two-person team spent 25 minutes).

G. Fig. 10.4. Sketch and record measurements of guardrails, coves, and other exterior features or trim. The measuring of the boat's hull exterior is complete for now.

H. Fig. 10.5. Climb aboard and take measurements aft from the stem and along the centerline of all the stations used earlier for sheer heights. These measurements can be cumulative if the tape is long enough and there are not obstructions from stem to stern, or you can start from 0-inches at each station. In drawing the boat's profile, these fore and aft measurements will be the first ones plotted, so don't forget to take them all with reasonable care.

I. Figs. 10.6 and 10.7. Measure, or estimate, and record the deck crown at each station where it is unobstructed rail to rail. Measure the transom radius (its curve in plan view), and record it as well. Take the heights and locations of the deck structures that don't fall on stations. Record the toe rail, taffrail, railcap dimensions, and any other characteristics that should show on the profile drawing of the hull. Later, make sure to take the photographs you'll need to back up the measurements and to show details not measured.

J. Figs. 10.6, 10.7, and 10.8. Measure or estimate the crown in the cabintop in the manner used for measuring the deck. Measure and record the locations of major bulkheads, platforms, and cabin soles.

K. Figs. 10.9, 10.10, and 10.11. Record locations and key measurements for windows, portholes, and trim.

L. Figs. 10.9, 10.10, and 10.11. Record locations and key measurements for fittings such as cleats, anchor windlasses, and hatches, and back them all up with photographs. (Steps K and L, although simply stated, may take more time than all the others put together. Unless the information they represent is somehow available when the drawings are made, the boat drawn may not be recognized as the one measured.)

II. Deck Plan

A. Fig. 10.12. Sketch, measure, and record the boat's beam at each station, taking reasonable care, by eye, to keep the measurements square with the boat's fore and aft centerline. Note the arrow labeled "for'd" that eliminates any possible misunderstanding of the orientation.

B. Fig. 10.12. Do the same with the above-deck structures and cockpit. Sometimes this is best done indirectly by measuring inward from the deck edge.

C. Figs. 10.13 and 10.14. Sketch, measure, and record the sizes and locations of hatches, seats, windows not needed for the profile (such as those across the front of the cabin), and other structure and trim needed for drawing the deck plan.

D. Figs. 10.15 and 10.16. Sketch, measure, and record details, such as mooring chocks, cleats, handholds, fairleads, and bitts, that will show in the deck plan drawing. Photograph these and other areas, especially complex ones, to have as a reference when preparing the drawing. Film is very cheap compared to the time invested in measuring and sketching.

III. Rig

(NOT APPLICABLE IN THIS CASE STUDY)

A. Measure the spars for cross-section and length, and record the location of rigging attachments and fittings. Photograph the details, especially at the ends—and, for masts, near the gooseneck.

B. Measure the sails, if available, especially working sails such as jibs whose size cannot be determined from the spars. Measure or photograph the size and location of the sail insignia, if there is one.

C. Estimate the rake of the mast in the quickest possible way. A photograph of the boat, taken in profile, is one way; obtaining the lay of the line between mast step and mast partner is another.

IV. Photography

The following list includes all the areas of *Barracuda* filmed in 29 frames within 10 minutes of photography. They were later printed as 5" x 7" prints. This print size was chosen because it balances the need for good detail with minimizing cost. (See pages 214 and 215.)

A. Bow profile
B. Cabin trunk profile
C. Stern/rudder mechanism
D. Doghouse viewed from athwartships and end on
E. Foredeck
F. Foredeck windlass
G. Tiller/steering mechanism
H. Cockpit from forward looking aft and from aft looking forward
I. Forward companionway from cockpit
J. Steering station
K. Forward cabin interior looking fore, aft, and athwartships towards both sides
L. Builder's plate
M. Aft companionway from cockpit
N. Engine room in all directions
O. Interior construction details of sternpost, steering cables, and sheer clamp, and framing
P. Bilge looking fore and aft from hatch in cockpit
Q. Boarding ladder and binnacle.

Without clear photographs, preferably black-and-white prints, selective documentation would be so imprecise and require so much more time in the field, it might be no faster than the more precise methods outlined elsewhere. Photographs enable you to determine shapes, such as the transom, and stem and finish details which are very laborious to measure in enough detail to permit accurate drafting. Unusual or complex details, such as fancy finish carpentry and hardware and steering systems, can be roughly captured in photographs. Err on the side of taking too many photographs, you will find them all useful when measuring errors and omissions become evident during draftings.

V. Scantlings

While not essential, a scantling list is often very useful in defining or refining our understanding of a boat. With a tape recorder, it only takes a few minutes on site to measure the following items—far less time than it takes to write the list. You may not be able to get some details due to obstructing sheathing and joinery.

Figure 10.19. Complex areas such as the helm position should be photographed.

A. Skeleton

1. Keel siding or keel batten width
2. Stem and stern timber siding
3. Hull framing dimensions
4. Frame spacing
5. Floor timber siding dimensions and location or spacing
6. Dimensions and location of bilge clamps and sheer clamp (a sketch here can be very useful for getting the arrangement)
7. Deck beam dimensions and spacing
8. Engine bed dimensions and spacing

B. Shell

1. Fastening method
2. Planking method and thickness
3. Anything unusual, such as thicker garboard or sheer strake
4. Unusual widths, stealers, or deck thickness, width, and pattern

VI. Arrangement Profile and Plan

A. Heights and lengths are all that are needed for an arrangement profile; widths don't show up in this view. Thus, depending upon the time available and the relative importance of the profile alone versus the profile and plan views, you may decide to record only the information needed for the profile.

B. Since, by definition, this is an expeditious method for documenting a boat, only the major features of the arrangement need be measured; for example the engine, tanks, berths, bulkheads and partitions, doors, ladders, and platforms. Some details may be important for indicating a boat's character or its use, but the recorder must be selective in this regard since drafting such details can be especially time consuming.

C. In the plan view, athwartship locations will generally have to be measured outward from the boat's fore-and-aft centerline or inward from the sheer.

VII. Lines

If you decide to measure and draw any aspects of hull shape other than those essential to the profile and plan, here are some pointers.

A. You have already taken measurements that will define the endings of the sections—that is, their upper and lower terminal points. The upper terminal is, of course, at the sheer, whose height and width (half-breadth) were measured in sections I. B. and II. A. The lower ending is at the rabbet (where the curving section line usually ends). In step I. D. the height of the rabbet was taken, but not its width.

B. To determine the width of the rabbet, measure the width of keel at each station, and divide by two. This will be the rabbet half-breadth, which, along with its height, will define the lower ending of each section.

C. Next, draw the proper curve between the two end points for each section. By establishing the half-breadths of either the painted waterline (for boats on land) or the load waterline (for boats that are afloat), a third point can be easily plotted as explained in the next step.

D. Fig. 10.17. By suspending a plumb bob from the sheer at each station and measuring inward to the hull at either the painted or load waterlines, the third point of each section can be established. Place the tape measure so that it is approximately square with the boat's centerline and parallel with its base plane. When working on a boat that is afloat, you should take this measurement while in a dinghy in order to avoid altering the boat's trim.

E. With three measured points for each section and an estimate of the line's character between them (whether the section line is straight, hollow, or bulging), a fairly accurate approximation of each section can be drawn. The nature of the curve can be estimated using a straightedge held vertically against the hull, or by crudely "taking off" each section's shape with spiling board or a "shape duplicator," such as timber mold or template guide. For many boats one section taken at or near amidships and two others taken near the bow and stern are adequate to define the overall hull shape roughly.

VIII. Interior Arrangements (not included in this case study)

Though interior arrangement can be hard to determine and time consuming to draw, there may be times when these details are wanted. If possible, reference as many measurements as you can to the fewest possible major units included in the profile and deck plans, such as bulkheads and the cabin sole.

IX. Drafting

As soon as possible, check over your notes for legibility, and process the photographs. It is best to draw the boat as soon as you can after taking off the data, while everything is fresh in your mind. Given the many quick decisions and visual observations you made on site, delay can only hurt the finished product. Before handing the notes over to someone else to draw, thoroughly review them for clarity and accuracy.

Pencil on drafting film is the most forgiving, stable medium readily available. Establish the scale to be used, bearing in mind the most convenient size for the finished drawings. Draft the plan by reversing the measuring process. See fig. 10.18.

A. Lay down the profile baseline and mark off the fore and aft locations of the stations taken at major construction features. As mentioned earlier, the profile baseline selected for survey documentation is usually the painted or load waterline.

B. Set up the deck plan centerline parallel to the waterline and square over the same stations.

C. Fair in the sheer, keel, rabbet, stem, and stern in profile at these locations while referring to the photographs.

D. Lay out the sheer half-breadths and other major features of the deck plan, again using photographs as a check.

E. Drop these features on to the profile, pulling their heights off the field notes.

F. Fill in the details on both plans using notes and photographs as needed.

G. Lay out and fair in the waterline plane in the deck plan.

H. Lay out and fair in the section views (if any) over the profile at the sections longitudinal locations or in a separate body plan. If you have only taken three sections, you may find a separate body plan fails to transmit the shape of the hull adequately.

I. Lay out the title block, clearly indicating the type of plan you have produced with the words "selective documentation." The need for this indication cannot be overemphasized.

You will probably find the drawing process presents you with challenges: measuring errors, measuring omissions, and research questions. The process helps you to learn about the boat and to see details lost in the rush of work on site.

Figure 10.20.
Photograph the stern and rudder details.

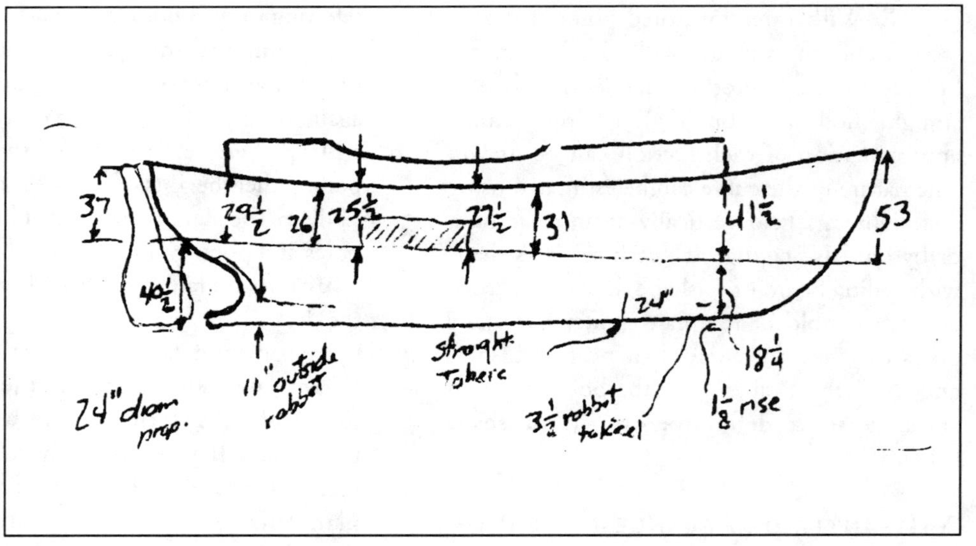

Figure 10.1. Select a reference plane for vertical measurements.

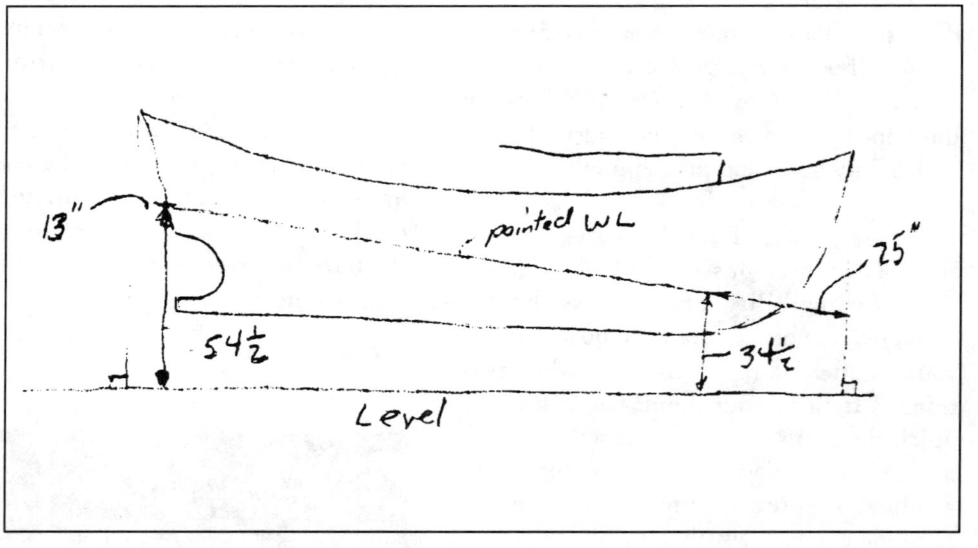

Figure 10.2. Record significant pitch fore and aft.

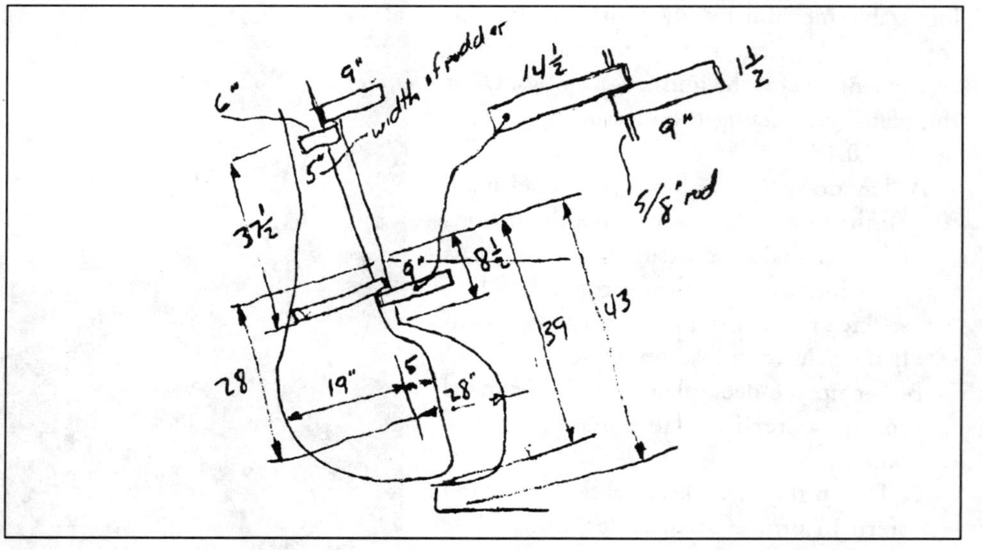

Figure 10.3. Note the numerous measurements taken of the stern below the waterline.

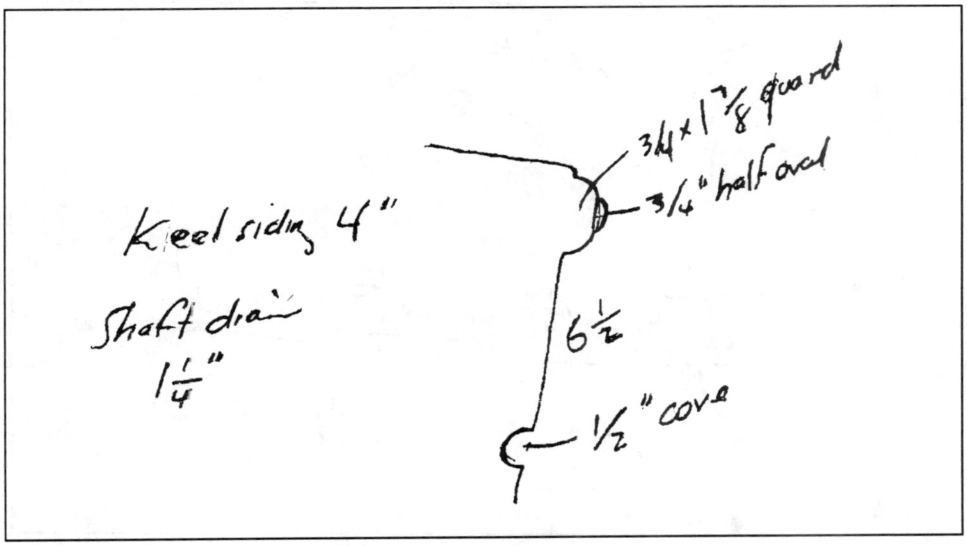

Figure 10.4. Record guard rails, coves, and other exterior trim.

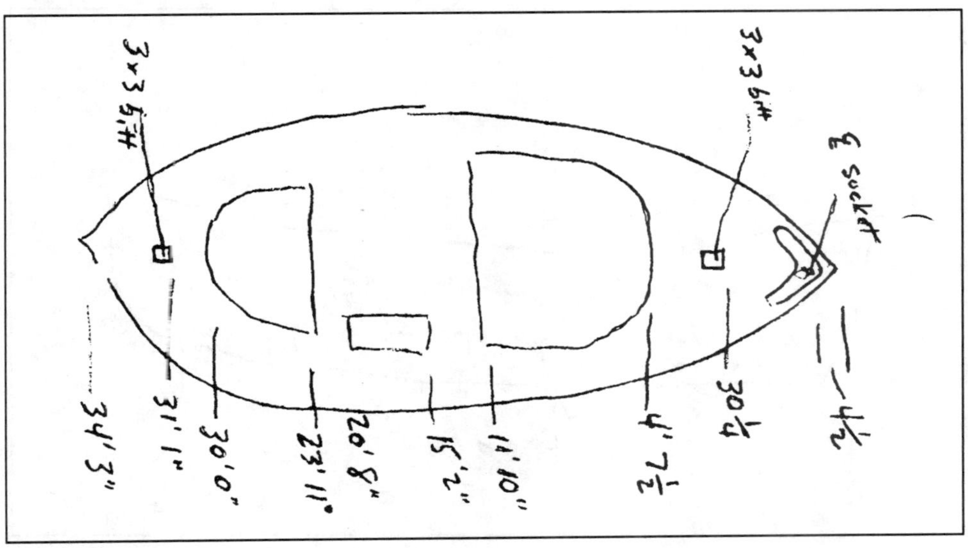

Figure 10.5. Record the fore and aft locations of the stations.

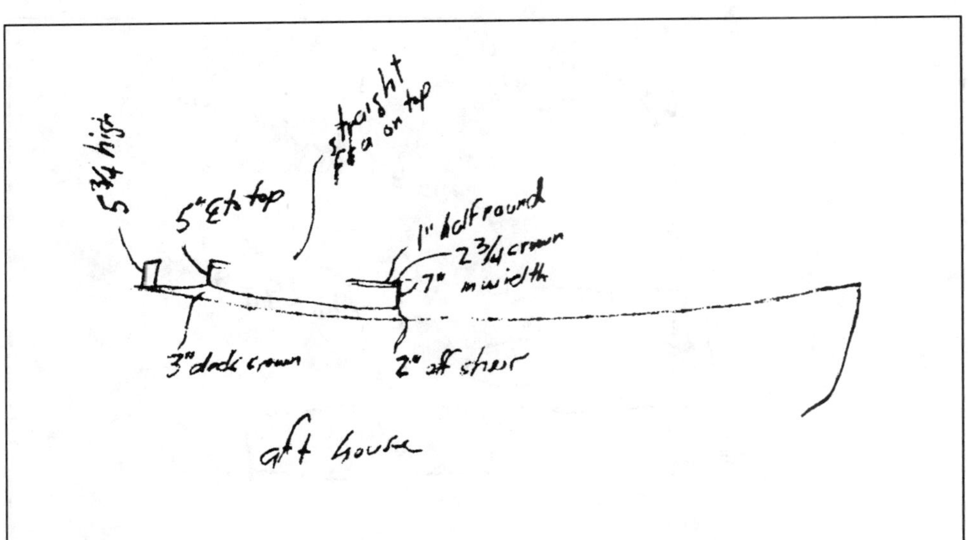

Figures 10.6 and 10.7. (left and top of next page) Record the deck crown, cabin heights, and trim details.

Figures 10.7 and 10.6. (right and bottom of previous page) Record the deck crown, cabin heights, and trim details.

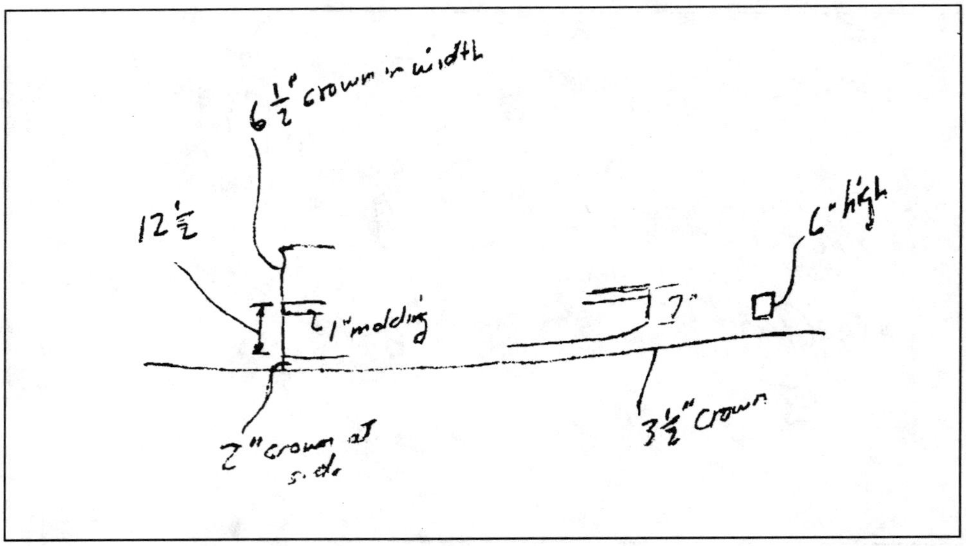

Figure 10.8. Record the locations of major bulkheads and the cockpits.

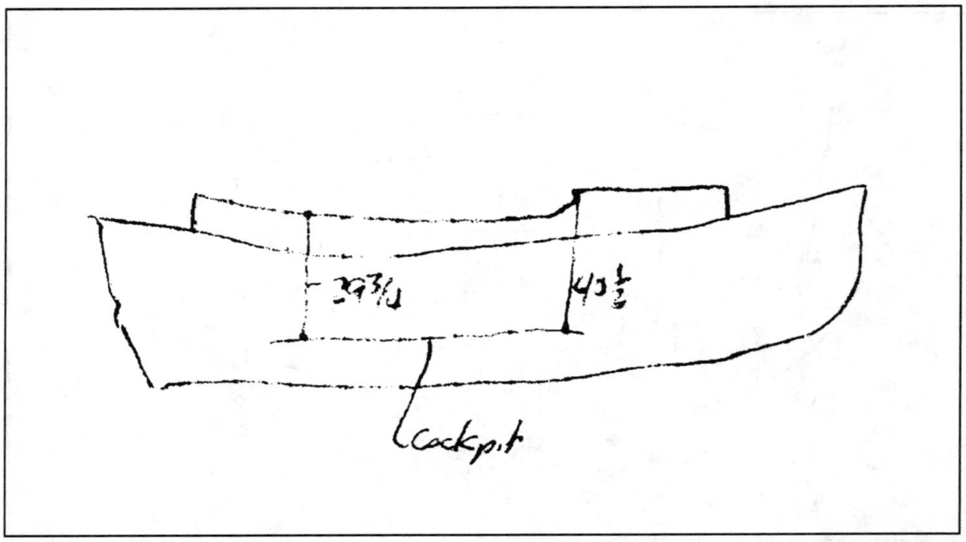

Figures 10.9 and 10.10. (right and top of next page) Record porthole locations and details.

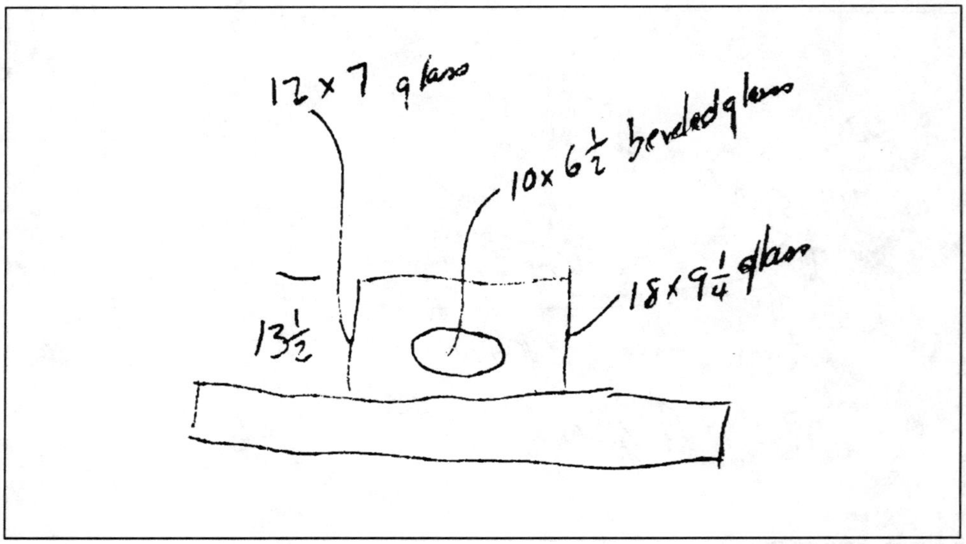

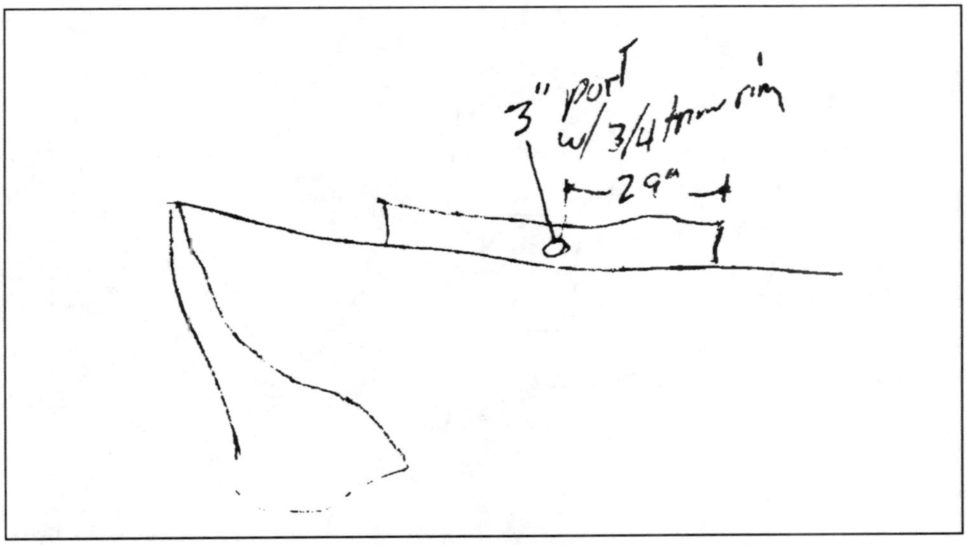

Figures 10.10 and 10.9. *(left and bottom of previous page)* Record porthole locations and details.

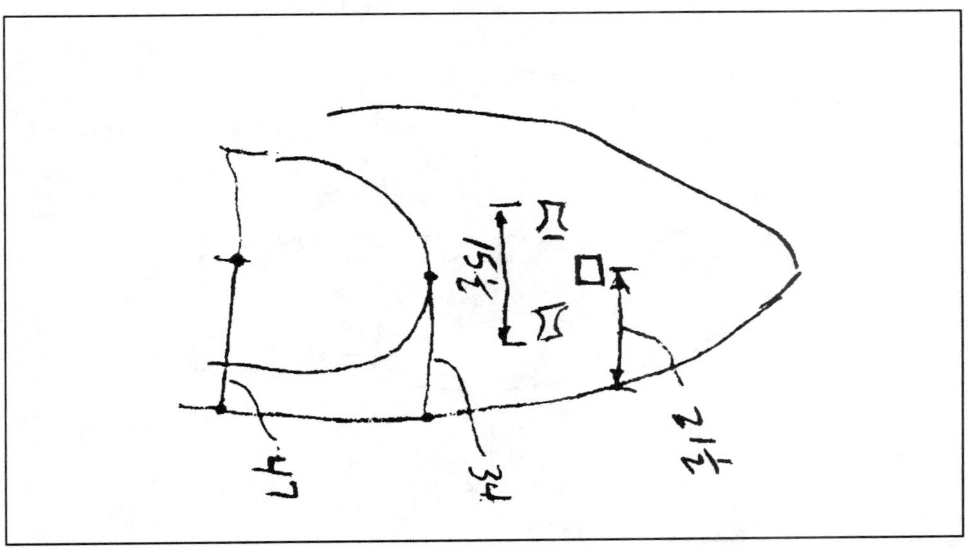

Figure 10.11. Record locations for deck details such as cleats, anchor chocks, and hatches.

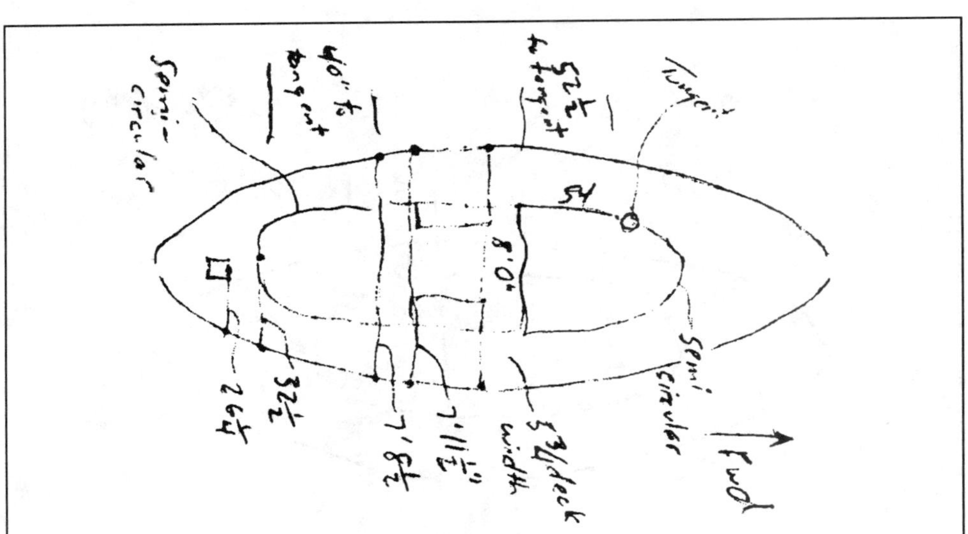

Figure 10.12. Record the beam of the deck and cabins.

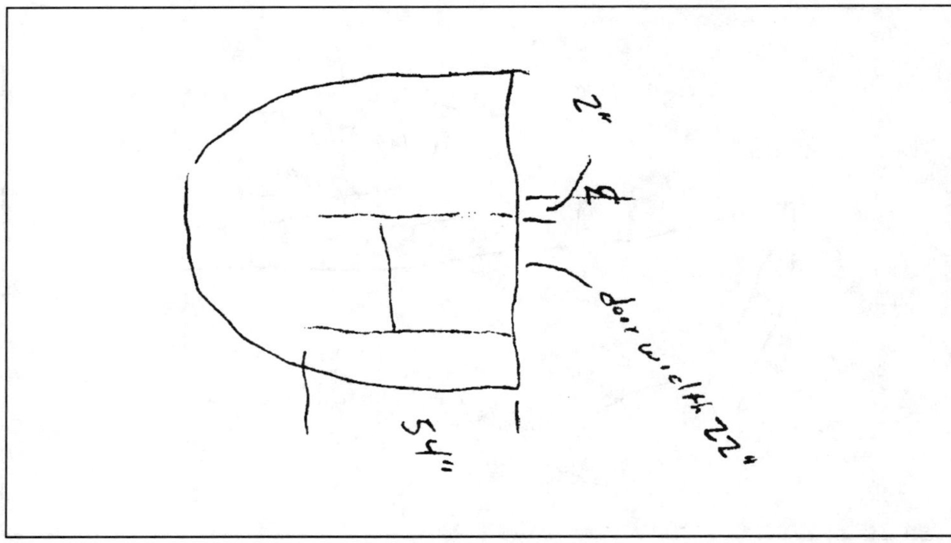

Figures 10.13, 10.14, 10.15, and 10.16 (this page and next) Record hatches, seats, and other details for the deck plan.

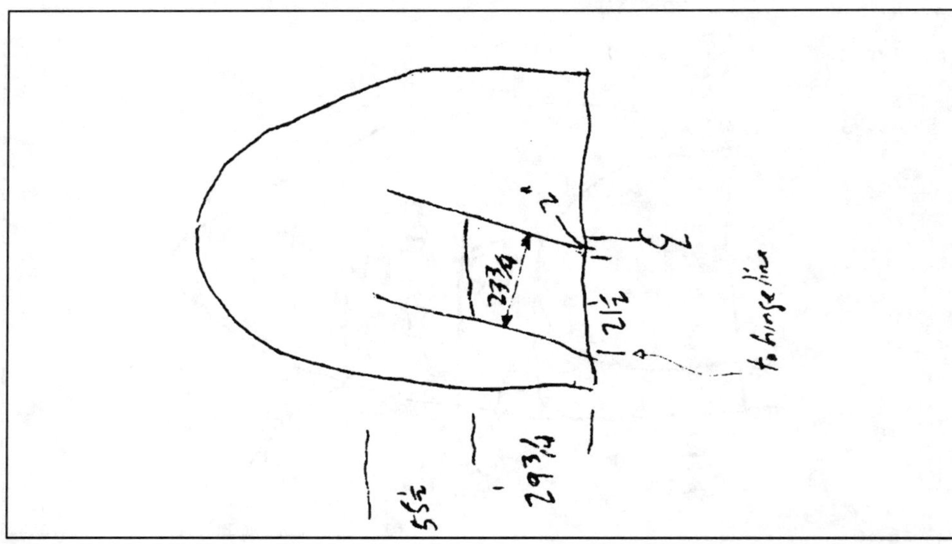

Figure 10.14. (see fig. 10.13 caption)

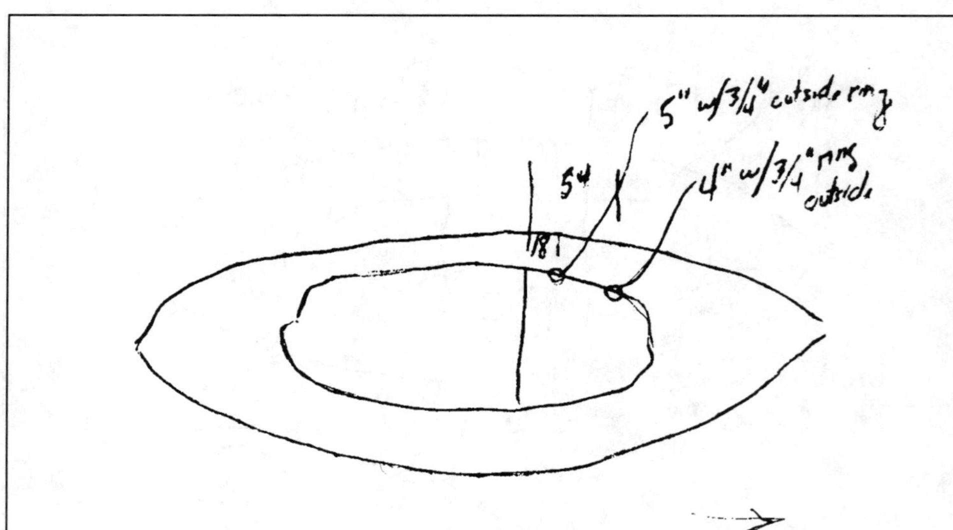

Figure 10.15. (see fig. 10.13 caption)

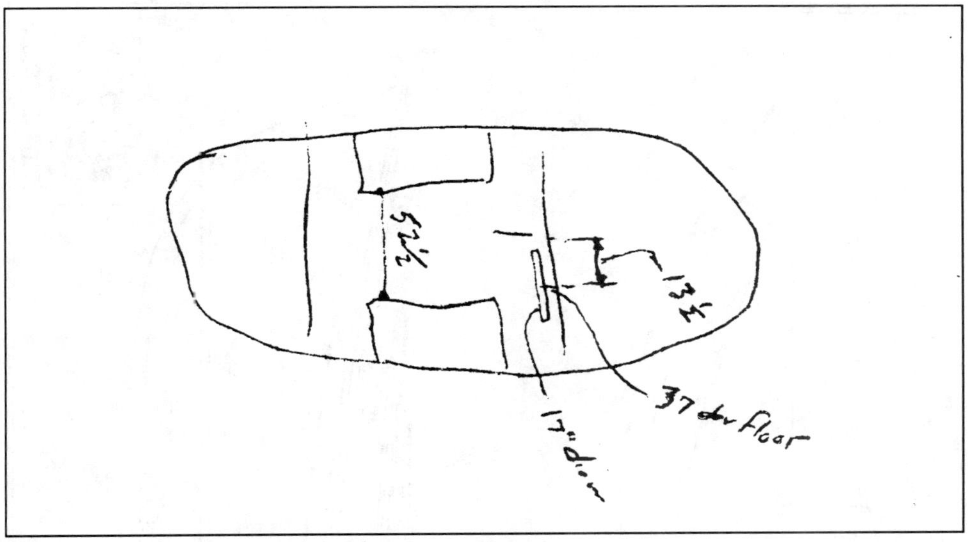

Figure 10.16.
(see fig. 10.13 caption)

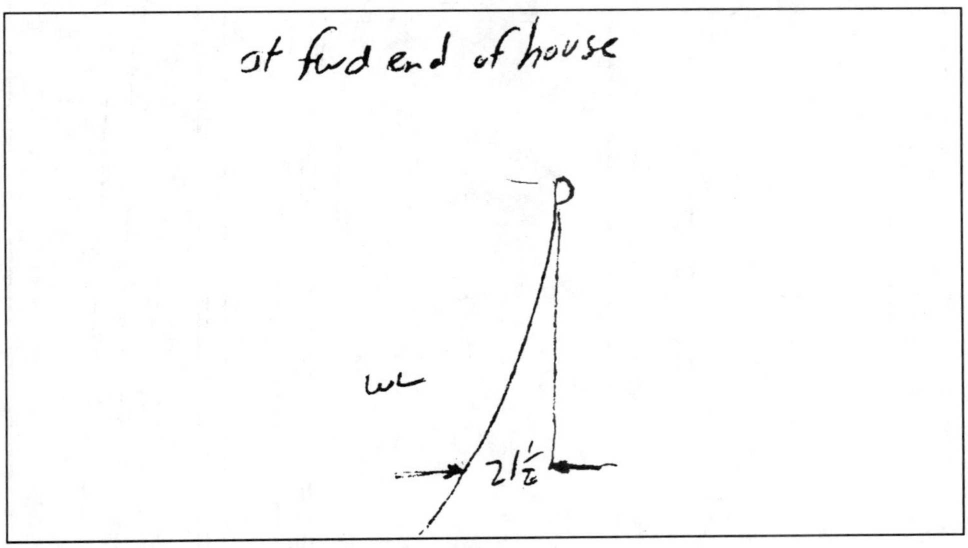

Figure 10.17.
Record the breadth of the waterline

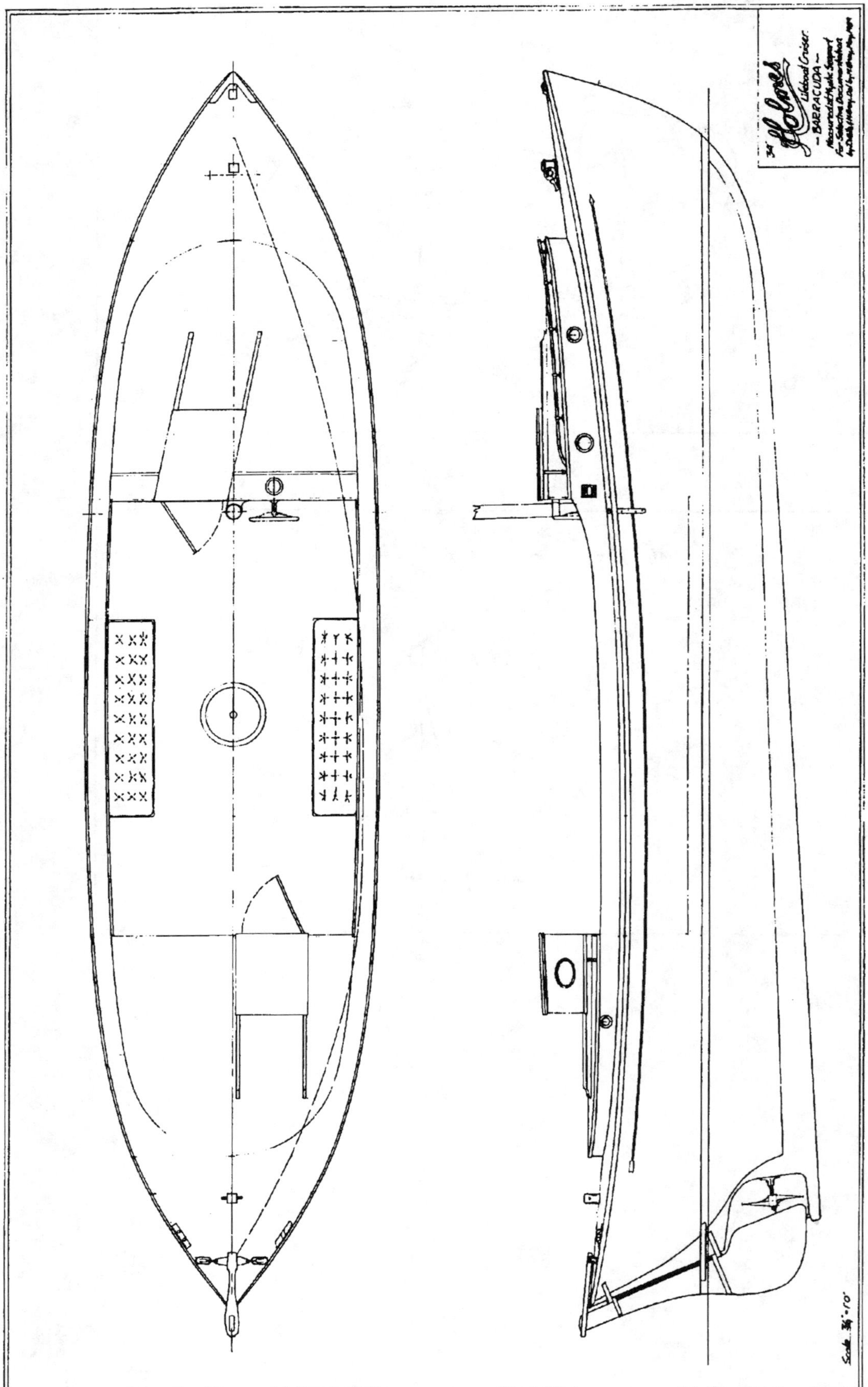

Figure 10.18. The finished drawings of *Barracuda.*

Chapter 11

Drawing and Lofting from Field Notes

The techniques needed to produce finished drawings

D. W. Dillion

With the lines and construction take-off complete, the next step in the documentation process is to develop a set of drawings showing the boat as found or reconstructed. Even if the boat is lofted directly from the take-off, plans must be drawn to satisfy the archival requirements of the project.

There is a good deal of sense in drawing the boat before lofting it, especially if you have no lofting experience. The size of a drawing is physically and visually manageable. Shapes, fairness, and unfairness are easier to see on a drawing—an advantage when laying down a boat that is badly out of shape or heeled or down by the head or stern.

The mechanics of drawing and lofting are basically the same. Drawing is lofting at reduced size; lofting is drawing at full size. The most significant difference is the possible magnitude of error. Drawings are made proportional to full or real size. For example, a scale drawing might be 1/8th, 1/16th, or 1/32nd of full size. Therefore, any inaccuracy in the drawing will be eight, sixteen, or thirty-two times greater when converted to full size.

Many folks look at a set of plans for the first time and throw up their hands in dismay. They see all those lines and become confused. They don't realize that a plan is nothing more than drawn instructions describing the shape of an object and how it is built. A plan is read just as a letter or book is.

If you are new to drafting and lofting, study the books listed in the bibliography or take a brief course in mechanical drafting or lofting to learn the fundamentals. There is a great deal more to both processes than can be mastered during most courses, but, once a number of basics are mastered—especially the principles of projection and the intersection of surfaces—the practices of drawing and lofting are just that, practice.

No matter what your experience and education are, the best way to learn about plans is to draw them, so let's have at it.

A boat's plans have several components. The lines plan is the two-dimensional representation of the three-dimensional shape of a boat. The construction plans show what a boat's structure and joiner work are and can be very (simple such as an expansion of the lines drawing for a flat-bottomed skiff) or quite complex and detailed (as for a forty-foot work or sailing boat). Additionally, a sailing boat would have a sail plan showing the outline of the sails, the spars, and the rigging.

11.1 The Lines Plan

The lines plan is a graphic solution which is fair—that is, accurate—only when the intersections of the curved lines with the straight lines fall exactly the same distance from corresponding reference points in the three different views of the boat: profile, half-breadths plan, and body plan. The intersections can be determined only if all of the lines, ink or pencil, are very fine. A pen size of #0000 (4 x 0, or 0.18 mm.) is good for an inked grid; #00000 (5 x 0 or 0.13 mm.) is better! Pencil lines must be as fine as possible. Try for a line no wider than a knife cut. The lines must be fine so that locations can then be accurately scaled from the drawing when making up the table of offsets.

The flat-bottomed skiff *Amy J.*, as set up for measuring, appears in fig. 11.1, with the body sections shown in fig. 11.4. This boat was designed as an excercise to get you started, to show the measurements needed to make a simple lines plan.

How do you start a drawing? By answering a few questions. At what scale should the boat be drawn? How big should the drawing be? How big can the drawing be? The size of the boat, the scale used to draw it, and how the finished drawings will be stored determine the size of the drawing. If plans are to be rolled, the maximum size is probably limited by width which is usually length of the tube they are rolled into because any length drawing can be rolled. If they are to be stored flat, the size is limited by the size of the storage space, usually a drawer.

"Scale" is the expression used to relate the proportional size of a drawing to the real-life size of the object that the drawing represents. "The scale" is the tool used to define and communicate that relationship. (See fig. 5.8).

"Scale" is written as x/y" = 1'-0". For example:

a. 3/8" = 1-'0", or three-eighths of an inch equals one foot zero inches, or 1/32nd of real-life size.

b. 1/2" = 1-'0", or one-half inch equals one foot zero inches, or 1/24th real-life size.

c. 3/4" = 1'-0", or three-quarters of an inch equals one foot zero inches, or 1/16th of real-life size.

d. 1" = 1'-0", or one inch equals one foot zero inches, or 1/12th of real-life size.

e. 1 1/2" = 1'-0", or one and one-half inches equals one foot zero inches, or 1/8th of real-life size.

f. 3" = 1'-0", or three inches equals one foot zero inches, or 1/4 of real-life size.

Keep in mind that scales are proportional to full size. Study the table in fig. 11.2. The skiff *Amy J.* is 10'-3 1/2" long; read across the line for 10'-3 1/2" until you get to 10 1/4" and 15 7/16", at 1" and 1 1/2" scale respectively. Both are reasonable for a drawing on a 10'-3 1/2" boat. Of the two, use 1 1/2" scale which gives a good size drawing and is workable with a carpenter's rule as shown in fig. 5.8. To figure the drawing size, take the length of the boat rounded up to the next quarter of a foot—round 10'-3 1/2" up to 10 1/2 feet. Now, the size of the drawing, not the material it is on, is the rounded-up length of the boat times the scale of the drawing or 10 1/2 times 1 1/2 which equals 15 3/4".

If you are not used to reading a scale, study fig. 11.3 and use it as a guide to scale off imaginary measurements. Make them up and mark them off along a straight line. Don't stop with those that can be conventiently marked right off within the length of the scale. Using each of the scales shown, mark off the examples as in fig. 11.3; then using the 1 1/2" scale, plot 3'-7", 4'-9 1/2", 6'-5 1/4", 7'-3", 10'-3 1/2", 16'-0", and 18'-9 1/4". Also try several with 1/8" and 1/16" fractions as 7'-3 11/16", and 9'-5 7/8". (Note that "scale" is a generic term as used by architects and engineers and in catalogs. Boatbuilders call it a "scale rule".)

For the rough lines plan, estimate the size of the drawing. Multiply the length of the boat in feet (fig. 11.1)—times the scale of the drawing—10 1/4 times 1 1/2 (10'-3 1/2" is close to 10 1/4') equals 15 3/8", the length of the boat on the drawing. Estimate the width of the drawing by adding 26", the greatest height above base, to the greatest half-breadth, 20 5/8" (Sta. 3 and 4 in fig. 11.4), for a total of 46 5/8", which can be rounded to 4'-0" and then multiplied times 1 1/2" to get 6'. To allow plenty of space for the drawing, round the 15 3/8" length and the 6" height up to 18" and 9" respectively.

Mylar is optional for any of the other "rough" drawings, such as the construction plan and profile, sections, tables of offsets, and sail plans. Don't be misled by the word "rough." In this case, "rough" means a drawing that is rough in appearance only (sometimes downright filthy), containing all of the erasures and changes made to get a faired set of lines. The drawing must be accurate in all possible ways in spite of appearance. When the rough drawings are completed, they are traced to make the finished drawings.

11.1.1. The Grid

Following the Roman numerals in figs. 11.5 and 11.5A, draw the grid in for laying down the skiff's lines at 1 1/2" = 1'-0". Lines drawn over an inked grid with a 6H or 9H lead can be erased without having to repair the grid lines.

I. Draw a straight line, which will be base, 4 1/2" down from the top of the sheet.

II. From the left hand edge of the sheet, scale off along base, 10'-3 1/2" at 1 1/2" = 1' 0", and make a mark (fig. 11.5A). Divide the remaining distance in half, and mark that off along base from the left edge of the sheet to locate F.P.; ("scale off" or sometimes simply "scale" are commonly used as instructional verbs meaning "with scale and pencil mark off").

III. Scale off the station locations and the length to A.P. Do this with as few moves of the scale as possible. Every time the scale is moved to mark off long distances in small increments, small discrepancies sneak in to create inaccuracy.

IV. Erect perpendiculars at F.P., station 3 (6'-0" from F.P.), and A.P., following the steps in fig. 11.6. Draw them the full height of the sheet (from top to bottom).

V. From base, scale up 30" and down 24" along each perpendicular and station 3. Draw lines from F.P. through station 3 to A.P. passing through all three points. If the lines do not pass through each point, check the scaling of the 24" and 30" distances.

VI. To check the grid for squareness, compare

the diagonal (corner to corner) distances of the rectangular spaces within the grid using a beam compass or a trammel. For example, distances indicated by the arrows A-A and A'-A' should be equal. This should also hold true for the diagonals in the smaller spaces as B-B should equal B'-B', and D-D should equal D'-D'. Additionally, the diagonals should intersect at the exact mid-point of the space in which they are drawn. The diagonals B-B and B'-B' intersect on the line at station 3 which is equidistant between station 2 and station 4.

A grid can be laid down using a straightedge and a triangle, but lay down the grids for several lines plans in the above manner before depending on them. Constructing grids in the above manner is a good way to find out what accurate work is and how accurately you can work. Then, when you have determined that your straightedges are straight and your triangles are straight and square, use them for laying down grids.

11.2 Plotting

Plot the take-off notes beginning as shown in fig. 11.7.

1. Scale 26" up from base along F.P. to locate the stem head.
2. Scale 4 5/8" along the base and the top reference line from F.P., and draw a line from mark to mark. Scale 7 3/8" up from base along F.P. and station 1. Draw a line from F.P. aft to intersect with the 4 5/8" line. A line drawn between 26" and intersection "A" is the stem face.

To draw the rabbet line 2" abaft and parallel to the stem face (fig. 11.1, "detail at stem"), follow the steps in fig. 11.8.

1: Align triangle "B" with the stem face; support "B" with "A";

2: Hold "A" firmly and slide "B" away from the stem face;

3: Hold "B" and move "A" to position shown and draw a line along "A". The line along "A" is perpendicular to "B";

4: Scale 2" from the stem face along the line;

5: Reposition "A" to support "B" and slide "B" up to the 2" mark; as you slide "B" up to the mark, check that it still lines up with the stem face; draw the rabbet line; fig. 11.8.

Using two triangles in this manner allows you to draw lines parallel or perpendicular to each other.

A variation is shown in fig. 11.9. Alternately move the triangles "A" & "B" while maintaining their alignment with the line X-Y. All lines drawn along "B" as it is moved along A, will be parallel to X-Y. Where large distances are involved, or on the loft floor, use the method shown in fig. 11.10.

1: Scale up 24 5/8" on F.P. and station 1, and draw a line through the rabbet line. The intersection of the two lines is the sheer height.

2: Scale 24" up on A.P. Then scale 6 5/8" up from base and 4 1/2" forward from A.P. to locate point B. A line drawn from the 24" mark on A.P. to point B defines the raked line (aftermost face) of the transom and skeg. Point C, the forward end of the skeg, 30" forward of A.P. and 4 3/4" above base is plotted and drawn in the same manner.

3: Since the transom face (expanded transom) was measured, the transom measurements must be plotted using the raked line as the centerline (fig. 11.4). The heights of the chine and the top of the transom were taken as plumb heights (fig.11.1). The plumb height of the transom—the difference between the chine height and the top of the transom is—12"; the measurement along the rake is 12 3/8" (fig. 11.11). Because of this difference, plotting the measurements along the wrong line will result in an incorrectly proportioned transom.

Following the steps in fig. 11.12 plot the expanded transom.

1: Mark the location of the chine and sheer.

2: Align triangle A with the rake line and support it with triangle B.

3: Hold B firmly in place and position A; hold A and slide B to align with the chine mark and draw a line; slide B along A until B lines up with the sheer mark and draw a line.

4: Scale the half-breadths off along the lines and draw the bottom and side. Plot the shape of the top of the transom using the data from fig. 11.4.

Using triangles as in figs. 11.8, 9, and 12 is not difficult. It is a matter of proper alignment and keeping one triangle immobile while you slide the other along it.

7. Plot the heights and half-breadths (distances from centerline) of the sheer and chine on the station lines using the data from fig. 11.4. The corrections for height at stations 1, 2, and 4, are dealt with arithmetically, as in fig. 11.7. Do the arithmetic on the field note sheet where the take-off is recorded so any errors can easily be checked. Anyone following your work in the future will have a record of what you did.

With the corrections calculated, scale up from base along the station lines and mark the chine and sheer heights. (Note well: corrections for heights can be made only when measurements have been taken plumb and/or level as described in methods 1, 3, and 5 in chapter 7. Corrections for height cannot be made for methods 2, 4, 6, 7, and 8.)

8. Plot the endings for the sheer and chine at the rabbet line and transom in order to draw the sheer and chine. Begin at the stem. Draw the half-breadth of the rabbet parallel to and 1" out from the centerline. Project the intersections of the sheer and chine with the rabbet line in the profile to the rabbet line in the half-breadth plan as shown by the arrows (fig. 11.7). ("Project" means to move a piece of information from one location to another parallel to the grid lines—the perpendiculars or the base.) Do the same at the transom using the intersections of the sheer and chine with the raked line. Then with a compass—not a scale—pick up and transfer the half-breadths of the sheer and chine from the expanded transom to the lines drawn in the half breadth plan.

Note: Each field measurement is used only once—for the initial plotting, except to check for errors in plotting or scaling.

9. Drawing the line of the sheer is not a matter of "connecting the dots" as in a child's amusement book. Finesse and skill are involved, as are practice and visual experience.

The techniques described in the following quotes involve fairing the sheer. The effect of these techniques should be carefully considered before they are used at the drawing board or on the loft floor.

"When the stern is reached, it is probable that the batten will bend so stiffly that it will not meet the station marks in one or two stations ahead of the stern. Spring up the extending and free end of the batten so that is does meet the marks, and secure. Do the same at the bow. It is very important that the free ends of the batten always be sprung and secured so that it is possible to remove the holding weight or nail at any one of the forward or after stations without the batten's moving." (Chapelle, *Boatbuilding*, p. 80).

"The batten's ends, which project beyond the boat, should be sprung to extend the sheer fairly." (Steward, *Boatbuilding Manual*, p. 77)

"Sometimes it [the sheer] may be a bit flat near the ends due to the ends of the batten straightening out over the end picks [ice picks], in which case you should pull up the ends of the batten gently and pin them there. This should result in a better looking line as well as one more like the plan." (Vaitses, *Lofting*, p. 21)

Using fig. 11.13 as a guide, begin fairing the lines of the skiff. Lay a spline against the plotted points for the sheer in the profile. Use a duck at each station line, F.P., and A.P. to hold the spline in place. Put the ducks down beginning at one end of the spline or the other, or start at the middle and work outward to the ends. Either method allows the spline to run freely and fairly. The spline must not be "captured" or held in place against its natural tendency to run free. If you are lofting the boat, with a wooden batten, do not nail through it. Nails capture the batten, and the holes weaken it.

There is no guarantee that the spline will lay fair when it is against all of the points. Look very carefully at the spline. Move around and eyeball it from a number of vantage points—from the bottom and top of the drawing board, from the left- and right-hand ends, from up high, and from down low to foreshorten it. Does it have any quick rises or sags, humps or bumps? Does it lay in a continuous, eye-pleasing curve? Assuming that it doesn't, let's have a go at fairing it.

At the ends of the boat, leave the ducks where they are. Look at the other points where the spline is not fair. Lift the duck from the spline at an unfair point, and observe the change in the spline's lay. Do this at successive points until the spline lays fair, and the changes made are the smallest possible. This usually means that the spline will weave its way between some of the points and lay right on others. Those points that are way out of line should be checked for plotting accuracy. If they are still way out of line, ignore them unless moving them by an even inch (or foot) horizontally or vertically will bring them closer to where the spline wants to lie.

When the spline (or batten) looks fair and is at or as close as possible to all of the points try:

A. Leave all the ducks where they are.

B. Out beyond F.P. and A.P., at a distance equal to the last station space, place a duck on the spline as indicated by the black triangles in fig. 11.14.

C. Remove the ducks at the perpendiculars. The spline will most likely move away from the marks at F.P. and A.P. as is shown by the dashed lines.

D. Put the ducks back at F.P. and A.P.—alongside the spline, not on it—to hold the spline at the point. Look very closely at the spline directly above the duck. You will see a very slight hump caused by the spline sagging inboard beyond F.P. and A.P. The hump is unfairness (fig. 11.14).

E. Gently move the outboard duck upward to lift the spline away from the duck at the perpendicular. When the spline is no longer supported at the perpendicular, back the duck off until the spline is just supported. Look at the spline again. If the hump is gone, and there is no hump at the first station inboard, and if the spline is fully supported at the perpendicular, the line is fair.

A line is fair when it does not offend your eye with humps, bumps, or flat spots. The best fair line is the one achieved with the smallest possible variation from the data given in a table of offsets or field notes. The process described above works well for fairing the sheer and chine. Fairing the lines of a round hulled boat is more complex and will be dealt with further along in this chapter.

A final point about straight sided hulls: the sheer and the chine may be fair, but the boat may not be. Carefully observe the flare of the sides; it must not waver in and out. For example, the flare may be greatest amidships and gradually straighten up toward the ends of the boat. If the flare wavers in and out, refair either the chine or the sheer, but not both, so the flare changes properly.

10. Draw the chine as described for the sheer in section 9. Note in figs. 11.13 and 11.14 that the effect of fairing up the spline shows up between station 3 and 4 as well as between station 4 and A.P. This is because there is more curvature in the chine compared to the sheer. Forward, where the sheer is quite flat, the difference is neglible or nonexistent.

11. Develop the body plan with station 3 as a centerline. Heights are projected parallel to base within the profile. Half-breadths are picked up from the half breadth plan and transferred to the body plan.

Begin by projecting the heights from station 1 to the centerline. Draw lines extending forward for the sheer and chine (fig. 11.15). Transfer the half-breadths to the appropriate line from the half breadth plan, and draw in the side and the bottom. Plot the remaining sections in similar fashion—1, 2, and 3 forward of the centerline; 4 and the transom aft.

To plot the transom, project the heights from the raked line forward to the centerline. Transfer the half-breadths, T-1 and T-2, to the lines drawn in the body plan. Draw the bottom and side. Develop the top of the transom by first projecting the height of the knuckle to the rake line and then forward to the centerline. Transfer T-3 to this line, and draw the top of the transom (fig. 11.15).

Plotting the shape top of the transom in plan will be described in the section on drawing the rough construction plan.

Draw the half-breadth of the stem in the body plan 1 inch from and parallel to the centerline. Project the sheer and chine heights to it from the profile. With all of the body sections plotted, complete the body plan by fairing the sheer and chine. The stem face, keel, chafing strips, and skeg do not affect the fairing and can be omitted from the lines plan.

If you elected to loft rather than draw the skiff, *Amy J.*, the lines would be arranged as in fig. 11.16 and require an area approximately 4'-0" x 12'-0".

By now it should be obvious that drawing from the field notes is a reversal of the taking-off process. If the field notes are complete and the references used for locating the measuring equipment properly recorded, the drawing will move along quickly.

11.3. The Rough Construction Plans

Draw the rough construction plans in pencil. Do not try to do the whole drawing on one piece of paper or Mylar. Use several pieces as necessary. It is easier to organize the data from the field notes, determine the space needed for each view, and arrange the views to make the finished drawings an attractive piece of work if you are not wrapped up in doing everything right the first time. Even though you are making a rough drawing, the neater and more complete it is, the easier it will be to trace for the final drawing.

Lay a piece of drawing material over the rough lines plan. Trace the grid in ink and the lines in pencil. To trace the other half of the plan, separate the two drawings, turn the lines plan face down, realign them, and trace.

The Construction Profile

With the lines plan traced, draw a line in the profile from the stem head to the top of the transom representing the tape measure. Along the line, scale off the location of the aft edge of the deck and hanging knee, the aft edge of the forward thwart, the centerline of the oarlock, the aft edge of aft thwart, centerline of the oarlock, and the forward face of the frame at the stern seat from F.P. (fig. 6.60 and 11.17). Label the stations, base, tape measure, sheer, and chine.

The tape measure was not parallel to base, but that makes no difference when scaling off the detail locations, because they were taken plumb down—square to the base—from the tape measure. To reproduce the field notes, then, project the locations from the tape measure square to base.

Scale off the 3/4 inch thickness of the bottom plank and the keel at each station and the transom. Draw the straight portion at the bow using two triangles (fig. 11.8). The curved lines will be easier to get parallel if the thickness of the bottom and the keel are scaled off at and in between the stations. Referring to figs. 6.60 and 6.61 for the scantlings, draw the stem, transom, skeg, frames, deck, hanging knees, and oarlock pads. Draw the forward thwart 5 inches, the aft thwart 5 1/2 inches, and the stern seat 4 inches below the sheer. Add the seat beam and apron. Note that the aft edges of the thwarts are located, and that the width is 9 inches wide scaled forward from the locating measurements, and that the thwart edges are beveled on the underside. The frames and hanging knees are normal to the sheer and stop short of touching the bottom.

The Construction Sections

Sections and detail views are required to help explain how *Amy J.* is constructed. Draw the full width of body section 3 and a half-width section taken at the aft edge of the deck and hanging knee. Develop them forward of the profile by projecting the heights and transferring the half-breadths as you did for the sections in the lines plan. Body section 3, the construction section, will be similar to the "Section at Aft Thwart" (fig. 6.61). The half-width section is needed to show the hanging knee (fig. 6.62).

THE CONSTRUCTION PLAN

When the construction profile and sections are as complete as possible, draw the plan (fig. 11.18). Draw lines, port and starboard, parallel to the sheer, 3/4 inches outboard of the sheer to represent the guard; another, 7/8 inches inboard to represent the plank; and a third 3/4 inches inboard from the chine for the intersection of the side plank with the bottom. Erase the line for the outboard chine.

Project the intersection of the inboard face of the stem with the inside of plank at the sheer and the chine from the profile to the plan (arrowed lines A and B). Connect the intersections with line C. Project the aft edge of the deck and the edges of the thwarts—arrowed lines D and E; then project the forward and aft edges of the stern seat (arrowed lines F and G). Project the intersection of the forward face of the transom at the inside of plank with the sheer and chine to the plan. Line H, drawn from the sheer to chine, is the intersection of the side plank and transom at the inside of plank (fig. 11.18).

The lengths, port to starboard, of the thwarts and stern seat are not given in the notes. The lengths must be determined using partial sections. Draw the sections lightly on the profile as indicated by the dashed lines in fig. 11.18 with the cutting planes as centerlines. Transfer the lengths from the sections to the cutting planes in the plan. The ends of the thwarts and stern seat are drawn as curved lines the same shape as, and parallel to, the sheer. Project the aft edge of the stern seat from the profile. The thwarts and stern seat are notched as indicated by the dimension "frame" on the section, which should be picked up and marked off at each frame location (fig. 11.18).

The Top of the Transom. In the plan view, the top of the transom is represented by two curved lines, which must be the same distance apart for their full length when measured along lines parallel to the centerline. Following the steps in fig. 11.19:

1. Project A and B from the plan to the sheer port and starboard; and C and D to intersect the centerline in the plan. Draw the lines E and F.

2. Scale 2 inches up from the sheer along the transom, and draw the knuckle line parallel to the sheer. To establish the knuckle line in the plan, draw lines I parallel to the centerline and 13 1/2 inches out from it, port and starboard. Project G and H from the plan to intersect line I. Spring a spline through G, C, and G, and H, D, and H to draw the long curved lines at the top of the transom. The lines E and F will help guide the spline when shaping the curve. Finish by drawing in the short curved lines A to G and B to H.

This is a simplified version of drawing a transom in plan view. A more accurate job would require taking additional measurements to define the shape of the transom.

The Quarter Knee. To draw the quarter knee in the profile, follow the steps in figure 11.20.

1. In the profile, scale 8 inches forward from A, and draw a short vertical line at B. Scale 1 1/4 inches down from the sheer and draw a line parallel to the sheer to represent the underside of the knee. Point C, at the forward face of the transom, is the bottom corner of the knee. To locate the top inboard end of the knee, scale up vertically from A, 3/4 inches (fig. 6.62), and project that height to the forward face of the transom at D. A point 3/4 inches above C is the bottom inboard end of the knee.

2. Project B from the profile to the plan; project D as the line E. Then scale 8 inches from A onto E to locate D. Draw the line A to D, the line of the knee along the transom. Note that the line A-D rakes aft because the transom rakes aft and the knee rises along it. Draw the curved lines of the knee in plan and profile following the information in the field notes (fig. 6.62) and as shown in the finished drawings (figs. 11.18 and 11.26).

Developing the transom and the quarter knee are relatively simple in the flat-sided skiff. The principles and procedures are identical when drawing round-hulled boat but require more practice and discipline. So again, either take a course or study the books listed in the bibliography. Especially recommended is *French's Engineering Drawing* because of its chapters dealing with the intersection of surfaces. Any course on drafting that you consider taking should deal with that material.

The Table of Offsets

"Offsets" are the draftman's instructions that tell the builder how to develop the boat's shape. With the lines faired, and following fig. 11.21, make up a table of offsets on scrap paper. *Amy J.*'s table of offsets has two components—vertical (heights above base) and horizontal (the half-breadths out from the centerline of the boat, given at the stem rabbet, stations 1, 2, 3, and 4, and the transom). There will be four rows of spaces across the table: one each for the sheer and chine heights; and one each for the sheer and chine half-breadths, each row with six spaces. The table is read horizontally to plot the profile and half-breadth plan, or down the columns to plot the body plan. Additional dimensions, which cannot be tabulated, must be shown in separate details or with dimensions on the lines plan.

Fill in the table by scaling off the sheer and chine heights at the stem rabbet, the transom, and each station.

Offsets are given in the table of offsets in the style of feet-inches-eighths (read as feet, inches, and eighths of an inch), so that 1 7/8 inches is written 0-1-7 and 3'-3 1/4 inches is written 3-3-2. As the offsets are written to eighths of an inch, the problem arises of what to do with sixteenth of an inch fractions.

An additional sixteenth should be noted as a plus sign. The offset 2'-5 9/16" should be noted as 2-5-4+ not as 2-5-5-. Why? Because if 2-5-4+ means 2'-5 9/16" what does 2-5-5- mean? Using a minus only adds confusion, so don't use it in a table of offsets. Where there is no offset to scale off, either put a short dash or an x in the space to show that the blank is intentional.

Before tracing the table onto the finished drawing, check the offsets by scaling them off again, comparing the second effort with the first.

Two tables are shown in fig. 11.21 to illustrate the effect of line weight (pen size) on appearance and readability. A single pen (000 or 0.25mm) was used for the upper table, and three pens (000 or 0.25mm, 0 or 0.35mm, and 2 or 0.5mm) for the lower. The dimensions shown are for guidance only and do not appear on the drawing.

Add notes explaining the table, such as those in fig. 11.21 or 11.28.

The Finished Drawings

Finished drawings should be in ink on Mylar, preferably 4 mil double matt. If you are not experienced at inking, consider making the finished drawings in pencil either on vellum or Mylar. Then have them photographed onto Mylar.

It is a good habit to make a sketch showing how the finished drawings will appear. Figure 11.22, which is based on a drawing size of 24" x 36", shows a layout fitting *Amy J.*'s lines and construction on one sheet. Normally, the views would be aligned with the length of the drawing as in *Susie*'s plans (fig. 11.28).

Refer to the detail in fig. 11.23 for the suggested spacing of the dimension and extension lines and the graphic scale in the lines plan which allows enough room between lines for lettering of proper size.

Figure the space for the construction profile by measuring the length, adding 1 inch at each end, and the height, measuring from the base to the highest point of the boat and then adding 1/2 inch for each dimension line and 5/8 inch for the graphic scale. For the body plan, measure from baseline to the highest point on the boat and the full beam. Measure the length and the half beam for the half-breadth plan. Repeat the process for the pieces of the construction profile, sections, and plan. Roughly sketch out the spaces needed, and write down their dimensions using fig. 11.22 as a guide. Allow 1 inch or 1 1/2 inches between views and at least 1 inch clear all around the edges of the drawing. Add all dimensions in each direction to arrive at the finished drawing size. If you have chosen to use standard cut sheet size, you will have to do some balancing to get satisfactory spacing.

With the layout determined, trace the rough lines into the space allocated for it. If you work in ink, one pen size, possibly #000 (0.25mm), will suffice for all lines; or the drawing can be livened up by using two pen sizes—#00 (0.30mm) or #0 (0.35mm) for the curved lines in the body plan, profile, and half breadth plan, and #000 (0.25mm) for all straight lines. If you draw in pencil, use a fine, dark line that is very consistent in width and density.

The finished lines plan is shown in fig. 11.24. Whether you use ink or pencil, the tracing must be accurate and the curved line and straight line crossings must be maintained.

Livening up the construction plans is more important as it makes them more readable. Using several pen sizes or line weights (widths) makes the plans more readable and attractive. Refer to figs. 11.25 and 26, and observe the visual effect of varying the line weights. Figure 11.25A was drawn with a single line weight and no wood graining to indicate the sectional plane; fig. 11.25B with several line weights (pen sizes) as indicated. In keeping with drafting tradition, the heaviest lines and the wood graining appear where the section is cut through the boat. Further, the heaviest lines of all are on those edges away from the imaginary light source indicated by the large arrow.

Some drafting techniques, especially wood graining and shading, take a lot of practice to have look attractive. If poorly done, they can overwhelm the rest of the drawing. So, study the techniques used by others, and then practice. Judge your work by having it printed. That is how others will see it.

Keeping track of which line weight goes where can be time consuming and frustrating, but the effort is well worthwhile. The job will be easier if as much as possible of one line weight is done at a time. Also, if the work progresses from the upper left portion of the drawing for righthanders or the upper right portion for lefthanders, there is less chance of smearing the ink.

Figure 11.26 shows the appearance of the finished construction plan. The thwart detail—marked "move"—would best be added after the drawing is complete with all lettering added. The pens used were the same as in fig. 11.25B.

Lettering

"Lettering" refers to the words and numbers on a drawing and is done freehand or by hand with a template as an aid. All lettering should be of a size and style that can be read easily both at original and reduced sizes. The minimum height of lettering on the drawing is 5/32 inches. A sample block of lettering is shown in fig. 11.27. Test a sample of the lettering you intend to use by reducing it to approximately 25 percent of the original size. If it is still readable, use that style.

Also refer to *Guidelines for Recording Historic Ships* for commentary about lettering styles and drafting techniques that are acceptable to the Library of Congress and HABS-HAER. Take a special note of the commentary about the various press on and stick on lettering materials.

One of the best aids to lettering is the Ames Lettering Guide shown in fig. 5.32. It can be used to make guidelines for random "call outs" on the drawing or to make up an 8 1/2" x 11" sheet of guidelines that can be slid under the drawing when lettering large blocks of text.

11.4 Round-Hulled Boats: The Planes

The lines that define the shape of straight-sided flat-bottomed skiffs and V-bottomed boats are readily apparent and therefore reasonably easy to visualize. Round hulls, on the other hand, require more visualization because the shape is defined using a number of planes: the centerline plane, body section planes, waterline planes, buttock planes, and diagonals. These appear in *Susie's* lines plans (fig. 11.28).

The waterlines are parallel to each other and the base plane, and at a right angle to the centerline plane, buttocks, and body sections and are the curved lines in the half breadth plan. The buttocks are parallel to each other and the centerline plane, and at a right angle to the waterlines and body sections and are the curved lines inside the outline of the boat in the profile. The body section planes are parallel to each other and at a right angle to the base plane, the centerline plane, the waterlines, and the buttocks, and are the curved lines in the body plan. The diagonals pass outward from the centerline plane at an angle through the buttocks and waterlines and are at a right angle to the body section planes. The diagonals are drawn separately from the remainder of the lines plan and are rarely developed in either the profile or the half-breadth plan.

As an aid to visualizing the way that the planes and shapes relate, build the eggcrate model shown in figs. 11.29 and 11.30.

Drawing the Round Hull

For practice drawing the lines of a round-hulled boat from a table of offsets, refer to figs. 11.31, 11.32, and 11.33, the "Exercise Boat."

Lay down the grid at a scale of 1 1/2" = 1'-0", spacing the stations, waterlines, buttocks, and diagonals as shown.

The Outline of the Boat. Use the offsets and the dimensions on the lines plan to lay down the profile: the bottom of the keel and the rabbet line; the stem profile and rabbet line; the transom rake; and the sheer heights at F.P., each station, and the transom.

In the half-breadth plan, lay down the rabbet line 7/8 inch from the centerline. Project the intersection of the sheer and each waterline with the rabbet line and the transom from the profile to the half breadth plan. Scale off the half-breadths of the sheer at the stations and the transom, and fair the plotted points. Do not forget that the intersection of the bottom of the keel and the stern post, point A, or the rabbet line at the bottom of the transom, point B, in figs. 11.31 and 11.33, are plotted on the rake line of the transom, not A.P.

The Body Plan. With the outline of the boat faired, lay down the body plan on the profile, using station 5 as the centerline. Do not use the offsets for plotting the sheer. Once an offset is plotted, it is not used except to check the accuracy of the plotting. Instead, project the sheer and rabbet heights from the profile, and transfer the sheer and rabbet half-breadths to the body plan for each section and the transom. Use the offsets for plotting the remaining portion of each section. Fair each section as it is plotted. Since you are working from a table of offsets, the transom is plotted on the body plan just as though it is a body section.

When the body sections have been plotted and faired, all of the data needed to lay down a complete set of lines is on the drawing. A fair hull cannot be built from the drawing as it is, because the sections have been faired as isolated lines, not related to each other or to the buttocks and waterlines.

Fairing. A fair hull results from a matrix of curved lines, individually faired, whose intersections with the other planes coincide when compared among views in the lines plan. With the eggcrate model in hand, refer to *Susie's* line plans and look at the points where the body section and the waterline intersect.

The half-breadth must be the same on both the waterline and the section. The same is true of the buttock height—the buttock must be the same height above base on the section and the station line in the profile. The relationship of the waterline-buttock-station intersections is critical to fairing. All intersections must have the correct relationship in the body plan, the profile, and the half-breadth plan. Also, using the intersections is frequently the only way that the end portions of the waterlines and buttocks can be faired, as illustrated in fig. 11.34, the stern of the "Exercise Boat."

Fairing a round-hulled boat involves fairing an irregularly shaped, curved surface. The process begins with plotting the offsets and ends when all lines have been laid down with no lumps, bumps, ripples, hard spots, or flat spots. Fairing is an averaging process—a balancing act that takes time and patience.

It seems to make little difference, either in the time needed or to the lines themselves, whether the next step is developing the waterlines or buttocks. The conventional method is to start with a waterline that is near the top of the turn of the bilge, or the buttock nearest to one quarter of the beam.

The Buttocks. Start with buttock II, the nearest to one quarter of the beam. Project the heights from the body plan to the station lines and the transom in the profile. The buttock ends forward where it passes through the sheer in the half breadth plan at point C (figs. 11.31 and 11.35) and aft at the transom in the profile at point D. Project point C to the sheer in the profile, and project point D onto the half breadth plan as a line for later use. Fair and draw the line from station 3 aft, using a fairly stiff spline. Then, lay a supple spline down from the sheer aft so that it overlaps and fairs into and along the aft portion from station 3 to station 5. Draw the forward part of the buttock line (fig. 11.35). All of the crossings of buttock II with the body sections are strong and should prove quite reliable (fig. 11.34). Forward, in the profile, some adjusting may be necessary when laying down the waterlines to have the intersections project accurately. Also, exercise care when laying down the 13-1/2 inch waterline.

Even though the buttock is a more

complex line than any previously dealt with, the same rules and procedures apply. The points as plotted do not necessarily define a fair line. Check the plotting of any points that are way off a fair line. The line is fair when it is as close as possible to as many points as possible, averages out the others, and has no lumps or bumps.

Since the buttock crosses the section at station 2 in the body plan at a scant angle, the shape of the buttock as it sweeps upward to the sheer is best controlled by the intersection of the buttock with the 13 1/2-inch and 18-inch waterlines and with the sheer in the half breadth plan, which give the stronger crossing angles. "Scant" and "strong" crossings are illustrated in fig. 11.34. The buttock can be fine-tuned after the waterlines are laid down and faired.

All of the points that do not fall on the faired buttock must be corrected in the body plan. Do not proceed until that has been done. It may prove necessary to make slight changes to the body section and the buttock, rather than making the whole change to one of them. Trying to make the corrections after more lines have been laid down can result in a nightmarish situation because a change in the first line laid down and faired is very likely to affect every line that is laid down afterward. Think of it as trying to balance your checkbook after finding an error that was made one hundred checks earlier.

The 13 1/2-inch Waterline. Lay down the 13 1/2-inch waterline in the half breadth plan by transferring the half-breadths from the body plan to the station lines and the line projected from point D. The waterline begins at the rabbet line, ends at the transom, and crosses buttock II in two places.

Fair the plotted points using a stiff spline from the rabbet line forward and aft to station 9 and a supple spline from station 9 to the transom. Forward, project the intersection of the 13 1/2-inch waterline and buttock II from the half-breadth plan to the profile. If necessary, refair the buttock and the waterline to gain an accurately projected intersection. Aft, project the buttock-waterline intersection from the profile to the half-breadth plan. Both lines will have to be treated carefully between station 9 and the transom because there is not much leeway in the shape of the buttock. Allow the spline to run free to more closely approximate the run of the planking. Also, because the section shape between buttocks I and II at station 9 and the transom is nearly a straight line, the waterline will be quite straight. Make corrections as necessary in the body plan to buttock II and the waterline before going any further.

The Lines Plan and Plank Lines. Fairing lines that approximate the run of the planks requires special attention during the take-off and when developing the lines. The plank ends at the stem, transom, or sternpost must be straight, unless the planks were sawn to shape, so the fastenings will not have to be used to haul the plank ends into place. At the stern, *Susie*'s waterlines curve very quickly toward the centerline, while the buttocks, which approximate the run of the planks, are nearly straight (fig. 11.36). *Spy*, illustrated in fig. 11.37, has a garboard plank whose shape could be misleading. The garboard has such extreme shape that when viewed end on, as in the body plan, the garboard seems to present much of its surface bluntly to the water. Instead of knifing its way through the water, it seems as if it would be pushing. If you study the waterlines and section lines through the garboard in the body plan, you will see that that is not so.

The 18-inch Waterline. Lay down and fair the 18-inch waterline as described for the 13 1/2-inch waterline; use a stiff spline for fairing the full length of the line. The end of the waterline at the transom must be projected as a line to the half-breadth plan, and the half-breadth transferred to it from the body plan. Also project the intersection of the 18-inch waterline with buttock II at the bow (fig. 11.31). This intersection is perhaps the only place where difficulty might be encountered. Make the necessary corrections.

The 9-inch Waterline. Lay down the 9-inch waterline, paying particular attention to the hollows (reversed curvature) forward and aft. Take care to prevent hard spots developing where the curves reverse. The tricky spots in the 9-inch waterline will show up when developing buttock I. Any corrections necessary that result from laying down the waterline should be made.

Buttock I and III. Buttock I should lay down quite nicely. Except for where the buttock crosses the body section at station 1 in the body plan, the crossings are strong and the waterlines have all been laid down giving useful intersections to project into the profile.

Of all the lines in the exercise boat, buttock III has the greatest number of difficult crossings to work with: at stations 3 and 9, and at two places on the 9-inch waterline. The intersections of the buttock with the 9-inch waterline are especially difficult and must be treated delicately, with the most control coming from the body plan.

The Diagonals. After checking all the heights and half-breadths, lay down the diagonals, so called because they radiate diagonally from the centerline in the body plan with no apparent relationship to the body sections, waterlines, or buttocks. The diagonals must pass through as many of the body sections nearly normal to the section as shown in fig. 11.38. This is to compensate for the many scant crossings in the lines and also to allow the diagonals to pass through places where only the body section defines the shape, as through the bilge. The lines of the diagonals shown in the body plan represent the planes of the diagonals that cut the hull lengthwise from bow to stern, beginning at the rabbet line forward and ending at the aft face of the transom aft, thereby describing a line on the skin of the hull. So, when thinking of diagonals, think of the skin of the hull.

The locations of the diagonals in the "Exercise Boat" are where there is a lack of other fairing information. When the diagonals are fair and any corrections that are necessary have been made, then the lines are fair.

Except for the ends, laying down and fairing the diagonals is straightforward. So, following fig. 11.38, let's deal with the ends first, beginning with diagonal A. Project the intersection of diagonal A and the rabbet line in the body plan to the rabbet line at the bow and from there across the base, as indicated by the arrows marked AR. Following the arrows marked AT, do the same at the stern. Then, transfer the dimensions marked AR and AT to the projected lines below base. Use base as the centerline for laying down the diagonals. The remaining points are plotted as shown by A1. Transfer the distance shown to the station line, repeating the process for each station. It is a good idea to lay down one line at a time, as occasionally the diagonals cross as shown at station 1.

11.5 The Round-Hulled Boat from the Field Notes

Figures 11.39, and 11.41 through 11.50 are field notes made during the take-off of *Susie's* lines adapted for the exercise of developing lines from the field notes. Note that figs. 11.47 and 11.48 are presented in dimensional form to provide the exercise of plotting from notes made in this manner. Chapter 6, "Taking Off Construction," contains data which will allow drawing *Susie's* construction plan once the lines plan is completed.

The steps in developing *Susie's* lines are similar to those used for *Amy J.* and the "Exercise Boat":

1. Draw the grid, base, forward and aft perpendiculars, top reference line, and the stations as in fig. 11.40, using the information in fig. 11.39. For convenience, draw the centerline forty scale inches (40") below base to allow room for the waterlines and diagonals. Do not locate and draw the waterline, buttock, and diagonal planes at this time.
2. Using the measurements in figs. 11.41 and 11.42, plot the bow and stern profiles.
3. Using station 6 in the profile as the centerline for the body plan, plot each body section and fair it as it is plotted. Plot sections 1 through 5 forward and 6, 7, 8, and the transom aft of the centerline. Erase all unnecessary lines after each section is faired.

Station 3. As shown in fig. 11.43. Draw a line parallel to base, 5 1/2 inches above it and forward of station 6 to represent the top of the level used when taking-off. Scale off and mark the distances out from the centerline along the line. At each mark on the level line, draw a perpendicular line. Scale off and mark the heights to plot the points measured.

Station 4. As shown in fig. 11.44. Draw a line representing the top of the leveling board 3 3/4 inches above and parallel to the baseline forward from station 6. Using station 6 as 0 inches, mark off distances as shown from 6 inches through 48 inches along the leveling board line. With a compasses swing intersecting arcs with radii equal to the measurements given to plot each point measured.

Station 1. As shown in fig. 11.45. Draw a line representing the edge of the vertical staff 10 inches forward (to the right) of station 6. Scale off and mark the heights above baseline along the line. Draw lines parallel to base at each mark and scale off the appropriate distances from the vertical staff to locate the points measured.

Station 2. As shown in fig. 11.46. Draw a vertical line 21 inches forward of station 6 to represent the edge of the vertical staff, and mark heights at 12, 18, 24, 30, 36, and 42 inches above baseline on it. Following the example shown for seam 6, use a compasses to swing arcs equal to the measurements taken to plot the points.

Station 6. As shown in fig. 11.47. Draw a line representing the horizontal arm parallel to base and 5 1/2 inches above it aft from station 6. Draw the edge of the vertical arm 37 1/8 inches out from the centerline (station 6), and note that the intersection of the two lines is 0 inches for vertical and horizontal distances. Mark off the distances from 0 inches along both lines. At each mark on the reference lines, draw a line at a right angle and scale off the distance to the plank seam.

Station 5. As shown in fig. 11.48. Draw a line 6 1/4 inches above and parallel to base forward from station 6 for the top of the horizontal arm. Draw a vertical line from the top of the horizontal arm 36 13/16 inches forward of station 6 to represent the edge of the vertical staff. Then scale off 20 inches along both the horizontal and vertical lines and draw the diagonal reference. Mark the distance up from 0 inches on the vertical staff, along the horizontal arm from 0 inches, and along the diagonal from the horizontal arm. Note that the distances marked off along the diagonal are from the top of the horizontal arm. Plot the seam locations by scaling off the distances to the plank seams along lines drawn square to the reference lines at the distance marks.

Station 8. As shown in fig. 11.49. To locate the edge of the sloped staff, scale 21 inches along base aft from the centerline (station 6) and 42 inches horizontally aft from the top reference line. Draw the edge of the sloped staff between the two marks. 0 inches on the staff is at base with reference marks every 6 inches along the edge of the staff. To check the location of the staff, scale its length which should be very close to the 52 3/8 inches shown in fig. 11.49. Note that 21- and 42-inch measurements should be exact as they are horizontal while the 52 3/8 inches is on a diagonal between them and might scale as much as 1/8 inch out. Plot the seam locations following the instructions for station 4 (fig. 11.44).

Station 7. As shown in fig. 11.49. Following the dimensions given, draw the lines representing the sloped staff. Scale out from the centerline (station 6) 12 inches aft along base (point A). Draw a line from the top reference aft from the centerline and parallel to base. Scale out along it from the top reference line 43 1/2 inches and draw a vertical line and scale down along the vertical line 15 7/8 inches (point B). Draw a line between A and B for the edge of the staff. Locate the seams using the instructions for station 5, fig. 11.48—scale off the distances along the staff from 0 inches; at each mark square off the distance from the staff to the seam.

4. Develop the transom in the body plan. This process is very similar to developing the transom in the skiff *Amy J.* The difference is that the skiff had only the sheer and the bottom of the transom to work with. *Susie's* transom has the sheer, the rabbet, nine plank seams, and the top of the transom to work with. The measurements needed are given in fig. 11.42 with the procedure for developing the transom illustrated in fig. 11.51:

a. Scale the sheer height up along the

centerline (station 6) and draw a line representing the sheer height aft to halfway between stations 7 and 8.

b. Project the top of the transom, the sheer, the rabbet, and the bottom of the keel forward to the centerline. Draw lines of suitable length aft from the centerline.

c. Scale the distances out from the centerline, aft, 22 3/4 inches at the sheer, one inch at the rabbet, and 11/16 of an inch at the bottom of the keel. Draw a line from the rabbet to the bottom of the keel.

d. Using seam 8 as an example, scale the height 5 inches from the sheer down along the rake line of the transom and make a mark. Project the mark forward to the centerline and draw a line extending aft beyond the sheer. Scale off 22 7/8 inches—the distance out from the centerline—along the line. The remaining seams, 1 through 7 and 9, are plotted in exactly the same manner. When fairing the transom, do not lose the hollow between seam 3 and the rabbet.

e. To plot the top of the transom, scale off from the centerline along the sheer line 12, 17, 19 7/8, and 23 inches. Then, square to the sheer line scale up from it 4 11/16, 4 5/16, 3 7/8, and 1 inch respectively. Draw the top of the transom.

5. The next step is to plot and fair the outline of the boat: the sheer, the rabbet line, and the bottom of the keel in both the profile and the half breadth plan. Develop the profile by projecting the sheer, rabbet, and bottom of keel heights from the body plan to the station lines. Draw fair lines passing through or near as many of the points as possible. Remember that the bottom of the keel in the profile is a straight line from station 4 to the stern post.

6. Develop the half breadth plan by transferring the half-breadths of the sheer, rabbet, and bottom of the keel from the body plan; the half-breadths of the stem rabbet and face from the bow measurements; and the half-breadth of the bottom of the sternpost (keel) from the stern measurements (see fig. 6.36). After the lines are faired, it may prove necessary to go back to the body plan and make minor corrections. Develop the waterlines, buttocks, and diagonals as was done for the "Exercise Boat." Draw the waterlines at 12, 14, 16, 20, 24, and 28 inches above base; the buttocks at 6, 12, 18, and 24 inches out from the centerline; and diagonal A up 28 inches on the centerline and out 12 inches on a line 4 inches above the baseline, diagonal B up 40 inches on the centerline and out 24 inches on the 20-inch waterline, and diagonal C up 36 inches on the centerline and up 32 inches on the 24-inch buttock.

The procedure for lofting from the field notes is exactly as described above except that all measurements are reproduced full size.

11.6 Working from Patterns

If you took patterns off the boat, reconstruct them at full size, either on the loft or on a large sheet of paper. Make the reference lines that control the locations of the measuring equipment just as they appear in the field notes. Align the pattern with the references, and, using the method by which the shape was taken off, work backwards from the pattern to arrive at the shape.

At this point, move to the loft to complete the take-off represented by the pattern. To go from the patterns to the drawing board, all of the data taken by pattern will have to be plotted at full size, measured, and then plotted to scale on the drawing. Repeat this procedure for each section to complete the body plan, and then the bow and transom and stern. Decide where the waterlines, buttocks, and diagonals are best located, and lay them down. Then work up the lines just as you did for the "Exercise Boat" and *Susie.*

11.7 The Curved, Raked Transom

You have, no doubt, heard various rumors of boat builders and loftsmen going starkers trying to lay down a curved, raked transom. As you proceed through the work, get a boiler-plated friend to standby with a tape recorder. What you will utter should be preserved as a warning to the next soul who attempts to derive a curved, raked transom from your field notes.

Keep in mind that in all probability the face of the transom was built to the arc of a circle—on the surface of a cylinder—and that the pattern or measurements taken off are the expanded, or true, shape of the transom's face. The measurements necessary for plotting the transom, beginning with determining the radius of the face, are the full beam at the sheer and the crown.

Following fig. 11.52, the geometric solution for the radius, draw line Z - Z'. At V, erect the perpendicular W-V-W'. From V, measure off the distance C, the crown (fig. 7.58). On W-V-W' measure off B/2 (one half of the full beam measured at the transom). Using W, U, and W' as center points, swing the arcs X with a radius equal to the distance W-U. Through the intersections at X, draw the lines Y-Y'. Their intersections with Z-Z' is the centerpoint for the arc and the distance centerpoint-U is the radius of the transom face. To check the accuracy, swing the arc W-U-W' from the centerpoint. The arc should intersect all three points. If not, adjust the radius slightly by increasing it until all three points are intersected.

The algebraic formula for determining the radius of the face of a curved, raked transom is shown in fig. 11.53. Quite straightforward, the formula is quicker to use than the geometric solution. However, the results you get from the formula should be checked using the geometric solution.

It is unlikely that either solution will give a nice, even dimension for the radius because the transom may not have been built to an even dimension, it may have changed over the years, wear and tear may have eliminated the sharp corners needed for accurate measurement, or the measurements were not sufficiently accurate. Consequently, the solution may result in an odd dimension that should be rounded to the nearest even inch. It is probably really necessary to furnish the radius only if a boat is likely to be built from the plans.

The step to convert the field notes or pattern of a transom to the drawing or lofting are:

1. Draw the profile, except for the sheer but including the raked line of the transom. The sheer in the profile cannot be completed until the sheer at the transom is plotted from the take-off.
2. Determine the radius of the arc of the transom face (as shown in fig. 11.51 or 11.52).
3. At point "Y" the bottom of the raked line of the transom in the profile, draw line Y-Y' square to the raked line of the transom. Mark off the radius of the transom face along line Y-Y' and draw the arc representing the face of the transom (fig. 11.54).
4. Working from point Y, lay down the expanded transom using the raked line as the centerline. Reproduce all of the reference lines used for taking off. Then lay down the measurements taken at either full-size or to scale. If a pattern was taken, fold the pattern in half along the centerline and align it with the raked line of the transom. Draw reference lines on the pattern parallel to the centerline or the sheer to sheer line to aid in plotting the transom's shape. Draw them at even-inch locations to make scaling them accurately on the drawing easy.
5. Transfer the distances A, B, C, D, and E from the expanded transom to the arc of the face of the transom, individually, not as a total A through E, to locate points 1, 2, 3, 4, and the sheer.
6. At points 1, 2, 3, 4, and the sheer, draw lines parallel to the radius, resulting in the spaces A', B', C', D', and E'. Also, from each of the points 1 through 4, and the sheer, draw lines parallel to the raked line in the profile as indicated by the arrows.
7. From the centerline of the body plan, working aft, draw lines parallel to the centerline and spaced at the distances A', B', C', D', and E', and number them 1, 2, 3, and 4.
8. Project the intersections of lines 1, 2, 3, 4, and the sheer with the outline of the expanded transom and parallel to base with lines 1, 2, 3, and 4 in the profile.
9. Fair the shape of the transom. In each case, the direction of projection is indicated by the arrows.
10. With the outline of the transom faired

in the body plan, erase lines 1, 2, 3, and 4 in all locations. Do not erase the radius line Y-Y'.

To develop the transom in the profile and the half-breadths plan, use a standard form of grid with waterlines, buttocks, and diagonals as shown in fig. 11.55:

1. Draw the lines grid; waterlines 1, 2, and 3 in the profile; buttocks I, II, III, IV, and V in the body and half breadth plans; and the diagonal in the body plan.
2. Draw the buttocks parallel to the radius line and intersecting the face of the transom. Project the buttock-arc intersections into the profile parallel to the raked line (shown in fig. 11.54 as arrowed lines).
3. Project the buttock-transom outline intersections from the body plan to intersect the buttock lines in the profile.
4. From the profile, project the intersections into the half-breadth plan.
5. The waterlines and diagonal can also be used to develop the shape of the transom. Pick up the short distances WB and DB and draw them as short buttock lines and project them as shown in fig. 11.54.

11.8 The Lines of the Lapstrake Boat

The lines of measured (existing, traditional, or old) lapstrake planked boats have frequently been drawn as though they were carvel planked but to the inside of the plank (see *American Small Sailing Craft,* Chapelle). This assumes that when the planks were lined off that the laps fell where the plank touched the mold. That is not always the case.

Many older boats were built by eye and experience in a made-to-work manner. In these boats, the frames may touch the planks and be fastened to them between the laps. In some cases, the frames may even be knuckled as shown in chapter 6, fig. 6.29. Because of this, the shapes of lapstrake planked boats are best represented by drawing the plank lines and body sections rather than waterlines, buttocks, and body sections. To lay down the rough lines of the lines plan:

1.a. Construct the grid.

b. Plot the bow and stern profiles.

c. Plot the body sections at their respective station lines using the recorded measurements. If you took patterns from the boat, use measurements taken from the patterns.

d. Using the sheer height and half-breadth, and the rabbet heights and half-breadths from the sections, plot and fair the out one of the boat. Note: if the boat is hogged or twisted, refer to section 11.10.

e. Project the plank line heights from each body section to its station line and transfer the plank line halfbreadths to the half-breadths plan. Note that the body sections were plotted at their station lines to reduce the distance that the plank line heights are projected to the station line, thereby reducing inaccuracies when projecting.

f. Working very carefully, fair the plank lines in the profile and half-breadths plan. The spline or batten must be used very gently. Any variation from the projected points in the profile must be applied to the body section and any resulting change must be transferred to the half-breadths plan. The reverse must also be attended to—variations from the transferred points in the half-breadth plan must be applied to the body plan and any effect projected into the profile. In all cases, keep the variations from the projected and transferred measurements to the smallest possible.

g. With the plank lines in the profile and halfbreadths plan faired, plot the body plan. The plot should give fair lines there also. If not, check every point that you have transferred from the profile or the halfbreadths plan for accuracy. An example of a rough lines plan appears in fig. 11.56.

h. When the plank lines are fair in the profile, halfbreadths plan, and the body plan, construct sections at each station showing the planks, the lap, and the line that the frame followed in the boat measured (fig. 11.57). These sections should appear on the construction plans.

i. The components of the finished lines plan—profile, half-breadths plan, and body plan.

The Construction Plan

Construction plans for a lapstrake planked boat should include the plank lines at least as an outboard profile and a half breadths plan (fig. 11.59). An end-on view similar to a body plan is also nice for its pictorial effect. Details showing the points that the offsets are given to, the shape and dimensions of the lap, and sections through the rabbet should all be shown.

11.9 Developing the Joggle Stick Patterns

Patterns taken off with a joggle stick, as described in chapter 7, figs. 7.43 and 7.45, can be plotted on the loft floor for full size use or on paper to be scaled down to drawing size. The procedure is straightforward and is shown in figs. 11.61 and 11.62.

11.10 The Out-of-Level and Out-of-Plumb or Out-of-Shape Boat

If you have taken off an out-of-plumb or out-of-level boat with a declivity board as described in chapter 7, drawing from the notes is straightforward. The declivity board uses wedges to create artificially a squared grid system within which to measure the boat. The result is that the boat is measured as though it is plumb and level. The notes can be plotted as they were for the "Exercise Boat" and Susie. If the declivity board or boards were correctly made and used, the results should be good.

Before drawing on out-of-shape boat, or indeed even before taking-off, review chapter 6 (Taking Off Construction) and chapter 7 (Taking Off—The Measuring Process, section 2, Choose Which Side to Take Off, and 3.1, Plumbing the Boat).

Laying down the lines of an out-of-shape boat presents several problems. You recorded the boat and know the what and where of its out-of-shape condition, and you have developed a feel for the boat. This puts in your hands the ability to judge what adjustments should be made to the shape as taken off. The following steps will help you make the necessary adjustments.

1. On a sheet of Mylar, lay down the measuring grid as recorded in the field notes.

2. On a separate sheet, the section sheet, plot the station sections exactly as recorded in the notes, but individually, with their own centerlines (fig. 11.63). Be sure to plot the sheer height and breadth, port and starboard, so that any twist can be seen and corrected. With the sections on a separate sheet you always have your original plot, and you can also see what changes have been made as the drawing progresses. In addition, you can superimpose the lines plan over the section sheet to align the section shape with the sheer and rabbet line.

Do not plot the station sections as shown in fig. 11.31. By doing so, you will lose your original plot and create an indecipherable mess of lines and erasures. The arrangement shown in fig. 11.31 is fine when the boat has no problems to straighten out or when working from a table of offsets.

Do not draw the waterline and buttock grid on the section sheet. If the sections have to be rotated or otherwise moved out of the squared grid system, those waterline and buttock lines may prove a hindrance. Rely instead on the waterlines, buttocks, and diagonals grid drawn in step 1 above.

3. Using the heights and the breadths from the station sections plotted in step 2 above, plot and fair the outline of the boat; the sheer, the chine, or the rabbet line and the bottom of the keel in the profile; and the sheer and the chine or the rabbet line in the half breadth plan exactly as done for *Amy J.*, the "Exercise Boat," and *Susie.*

Concentrate on getting those lines to look as you feel they should. Make your observations work for you; if the sheer, or the keel have hogged, remove the hog; if the hull has opened up, or closed up, deal with those in the outline first. Unless the boat is severely out of shape or damaged, keep the lines as close as you can to the take-off measurements. Do not start by fairing the station sections. The outline of the boat must be dealt with first so that the sections can be fitted into it. Trying to start with the sections first can be frustrating and lead to starting over again—a waste of time.

4. Locate and draw the straight lines on the grid for the waterlines, buttocks, and diagonals in the profile, the half breadth plan, and the body plan. Don't worry about ideal locations at this stage. The grid lines can be relocated when the lines are faired.

Remember that the faired body plan should be laid down on the profile using a station line at about amidships for the centerline. Draw the vertical lines for the buttocks on either side of the centerline (fig. 11.31).

5. Develop the lines plan. First slide the

section sheet under the lines plan and, using the sections or parts of sections that appear to be good shapes, mark off heights on the profile and half-breadths on the half-breadths plan.

Move the section sheet up or down to align with the faired sheer and rabbet line when marking off the waterline, half-breadths, and buttock heights. If there is a difference in the distance between the faired rabbet line and sheer from the distance in the notes, mark off the heights and half-breadths close to the rabbet line and the sheer with the plotted section at those locations. Then locate the plotted section halfway between, and mark off the desired points. The section sheet can also be rotated to adjust for twist.

Start by fairing the quarter beam buttock and a waterline at the top of the bilge (fig. 11.64). A diagonal cutting through the bilge might prove helpful in the middle third of the boat. Do not start the body plan yet.

To use sections that are obviously out of shape, lightly sketch on the section sheet what you feel is the correct shape. Use the sketched line to mark points to fair the long lines.

Using the points marked off, plot and fair as much of each of the "long lines," waterlines, buttocks, and diagonals, even if the result is several pieces of the long lines. Do not use the end points of the pieces and only the middle two of four points. Do not use any line for fairing if it has less than four points (fig. 11.65).

As you begin to get fair long lines, develop the body plan.

When there are enough points in the body plan to fair a section in it, no less than four, do so. Correct the long lines to agree with this section, if possible, or revise the section. Continue the process until you have a complete, faired lines plan.

11.11 Hints

Remember that the rabbet line and the transom are reliable for the ends of lines.

Expect the floors of small, open boats to be straight lines rather than convex.

When fitting a plotted section into the grid, remember that the girth, the total of the width of the planks, or the measured distance from rabbet line to sheer, cannot change. The girth can be flexed to give a faired section that fits into the faired lines plan.

A boat with gentle, sweeping long lines might be faired using sections at the quarter length of the boat, the painted waterline, the quarter-beam buttock, and a diagonal through the bilge, if the sections not used initially are fitted into the long line matrix and then all lines are faired.

11.12 Summary

To summarize, a set of plans might contain:

1. The lines plans (sometimes more than one sheet)
   a. Profile (sheer plan)
   b. Half breadth plan
   c. Body plan
   d. Diagonals
   e. Table of offsets
   f. Detail dimensioning for those things not covered by the table of offsets, such as bow-stem profile, stern post rake, and rudder

2. The construction plans
   a. Construction profile: a section taken lengthwise through the boat along the keel on or parallel to the centerline to show the backbone structure—stem, keel, sternpost, knees, centerboard and case, and all frames, floors, and deckbeams where they cross the centerline
   b. Hull sections showing the structural members and joinerwork
   c. Plan view of framing for deck and house top
   d. In a complex boat, an accommodation plan, profile, and sections to show sleeping, eating, and storage facilities
   e. Detail drawings to show dimensions of cupboards, hatch covers and slides, ladders, and bunks

3. Mechanical plans
   a. Engine installation with fuel and exhaust systems
   b. Separate electrical generating system

4. Sail, spar, and rigging plan

The preceding list shows what a designer might include in a set of plans for a boat that is to be built. The record plans of an existing boat should include any and all details listed above that are in the boat.

Should the plans be "measured drawings" or "dimensioned drawings"? "Measured drawings" are done accurately to scale, with only the overall dimensions and a graphic scale given. Generally, measured drawings are made of vessels considered unlikely candidates for replication, such as very large ones. Several examples of measured drawings appear in *Guidelines for Recording Historic Ships* (Anderson, 1988).

"Dimensioned drawings," on the other hand, should show sufficient dimensional information to aid the builder in locating the major components of the boat. The best way to illustrate this is to consider examples.

In figure 11.26, *Amy J.*'s construction plan, locations are given for the aft edge of the deck and the hanging knee; the aft edge of both thwarts and—because the thwarts are 9 inches wide, thereby automatically locating the forward edge—the locations of both frames since they are at the thwart edges; the forward edge of the stern seat, but not the aft edge, which is at the transom; the centerlines of the oarlock pads; and the distances to the top of the thwarts and stern seat below the sheer. No dimensions across the boat are given. These should be taken from the boat being built from the plans so as to allow for any slight variation from the offsets.

Susie's lines (fig. 11.28) and construction plans (fig. 11.68) leave too many dimensions to be scaled, which may be appropriate for an experienced builder, but not for the beginner. The centerboard slot is located, but the width must be deciphered from the scantlings, which say:

"7. Case head ledges — 1" x 2 1/4" w.o."

This cryptic notation indicates that the slot is 1 inch wide. The thwarts and stern seat should be located forward and aft and below the sheer, as should the floor boards. And, most importantly, there are several dimensions that could be added to help in laying out the centerboard case (fig. 11.66). The mast location and rake are given by notation in the space below the tables:

"C.L. of mast crosses sheer 12 3/8" aft of F. P.
Rake forward 4 1/2" in 8'-0"."

Dimensions on the plans would have done a better job. Additionally, a graphic scale should be shown on the lines and sail plans, and for the construction plan, profile, and sections. The detail sketches, if well-dimensioned, do not need graphic scales, but the undimensioned "full-size" details should have individual graphic scales (fig. 11.67).

The dimensions should be given in the system by which the boat was built; that is, if the English system was used to build the boat, give feet and inches; if the metric system was used, give meters and decimals. Graphic scales should be shown at least in the system used for the dimensions. No harm will be done if both English and metric graphic scales appear (fig. 11.67).

The next step beyond the fully dimensioned plans are the archaeological plans, showing the damage, deterioration, repair, and rebuilding that has taken place during the boat's lifetime.

The best way to know if plans are adequate is to get feedback. This may consist of a review by a knowledgeable person or commentary from someone who has tried to build from the plans. Note the "tried," because if a proper job has not been done, the plans may prove useless.

Figures 11.70-11.81 are included to illustrate different styles of work and different styles of presentation. *Dog Star* (figs. 11.70-11.75) was designed by Philip H. Rhodes, Sr., of New York City, as design No. 1600, in 1930.

11.13 Reconstructions

A reconstruction is a plausible representation of missing or incomplete data and should be supported by a clear explanation of what was done. Reconstructions are acceptable when a boat is damaged, out of shape, or has been extensively modified or repaired and when the goal of the project warrants the reconstruction. The reasons for the reconstruction should be stated in the records or on the plans.

11.14 Suggestions, Hints, and Tricks

When fairing—if scaled, projected, or transferred points don't agree or fair, check the projection, scaling, or transferring before making any change. Also, a line laid down with a spline is not necessarily fair, especially a soft or sweeping line that becomes a quickly curving (hard) one. In this case, the spline must be worked or forced, strenuously controlled to give the desired shape. But the line must still be fair! When fairing a curved line into a straight line, extend the straight line beyond where you think the curve ends to avoid having the curve sag. Try the same technique when fairing a tight curve into a flat curve.

The sheer in the body plan can be difficult to fair. For this reason, it must be carefully observed during the take-off to give a good visual check on the boat's appearance.

To practice with ship's curves, make a series of dots on a sheet of paper and make several photocopies. Then, on one, sketch what you think is a fair line through the dots; on another, do the same thing using ship's curves; on a third, fair a spline through the same points. The results may amaze you.

Using curves: If a ship's curve almost wants to follow a line, but not quite, flip it end for end and try again; or, if the curve is tightening up, flip the ship's curve end for end and let it ease coming out of the shape (fig. 11.76). When working on the rough lines plan, alongside the line with it, write the number of the ship's curve used alongside the line, or trace the holes in the curve onto the drawing and enough of the curve to readily identify it.

When it is necessary to move a line during the fairing process, try to make the change away from grid intersections where no other line is affected (fig. 11.77).

When plotting the notes at the drawing board or loft, you are trying to reproduce a shape, not create one. Make the spline lay along as many points as possible with as few ducks as possible. Shift ducks around until this is achieved. Where the spline does not want to fair through several points, weave between them, averaging the "errors."

As you check the rough lines plans, make a small red check mark at points that are okay.

It is easier to work accurately to a cross mark at the plotted point, as shown at the plank seams in fig. 11.42 than to a small dot. The cross mark draws the eye to the intersection, which is the point wanted.

Use colored pencil when making changes on the drawings. Leave the original pencil work until all changes are made, then clean up the drawing.

Occasionally a missing measurement can be compensated for. If you have the bottom of the stem measured and drawn, as in fig. 11.78, but not the height of the bottom, and you know the depth of keel farther aft, swing the arc as shown, and then draw a line between the points.

When drawing the plans, don't waste time drawing nuts and bolts in full detail. Instead, indicate their centerlines and state sizes.

When using dashed lines to show hidden edges, the nearer should predominate, with those farther from the viewer broken around them. In some cases, the nearer lines can also be heavier (fig. 11.79).

There are a number of rules given in text books for taking sections (cutting through an object to show a view at an angle to the drawing surface). The most important rule is to take the section where it shows what you want to show.

Tracing can be a difficult process when done on a regular drafting table, yet a light table (glassed top with lights below to backlight the work) can be beastly to work over. As an alternative, try sliding a piece of clean white paper under the work; it will highlight what you are tracing.

As you trace the "rough" drawing, use the same ship's curve for as many pieces of line as possible. It is faster.

When drawing any object, block in its outline or maximum size and shape to aid in accurately drawing it (fig. 11.80), especially when inking the finished drawing. This same technique is helpful when using templates to compensate for any misalignment between the cutouts and the printed indexing marks.

Keep drawings clean by covering the unused areas with scrap paper. To erase ink or plastic lead from Mylar, moisten a vinyl eraser, and go to it. Don't forget to use an erasing shield. If ink skips on Mylar, wipe the surface briskly with a clean, soft cloth. Don't use solvents to clean Mylar unless the manufacturer recommends it.

If you draw in pencil *do not* roll the pencil. Since the point is probably not a true cone, the line will waver. Also, rolling the pencil causes the hand to wander, and this will contribute to more extreme wavering. This means that a straight line will not be straight and a curved line will not be drawn as you want it to be.

The general rule for drawing details is that they be drawn at double, triple, or quadruple the basic drawing scale. For example, if the basic drawing is at 3/4-inch scale, the detail should be at 1 1/2-inch or 3-inch scale or even full size.

Don't overlook using full-size details. If you have made a pattern of a piece of hardware, trace the pattern full size onto the plans to save time. Make a half-size reduction by photo reducing the pattern, then tracing it onto the plan.

If a drawing must be small for publication, make it large and then have it photographically reduced to suit. Adjust line weights, so that the drawing will reproduce properly at the reduced size.

Many lines drawn to the same intersection—as radials or diagonals—soon obliterate the intersection, which will then be impossible to locate accurately. Do not draw the lines all the way into the intersection (fig.11.81).

Sharpen compass leads on the side away from the metal center point to make a chisel point as shown in fig. 11.82. This is easier to see, line up on marks, and sharpen, and permits drawing smaller diameter circles.

Use compass or a bow pencil instead of dividers for transferring measurements as the pencil makes the necessary marks as you work.

Drafting compass points slide on mylar, so put a piece of drafting tape over the point and draw the intersection on the tape. Or use easy-release mending tape.

Splines are made of relatively soft vinyl and dent and scar easily in use. Smooth them up with a fine file while holding the spline on a straight surface such as a saw table.

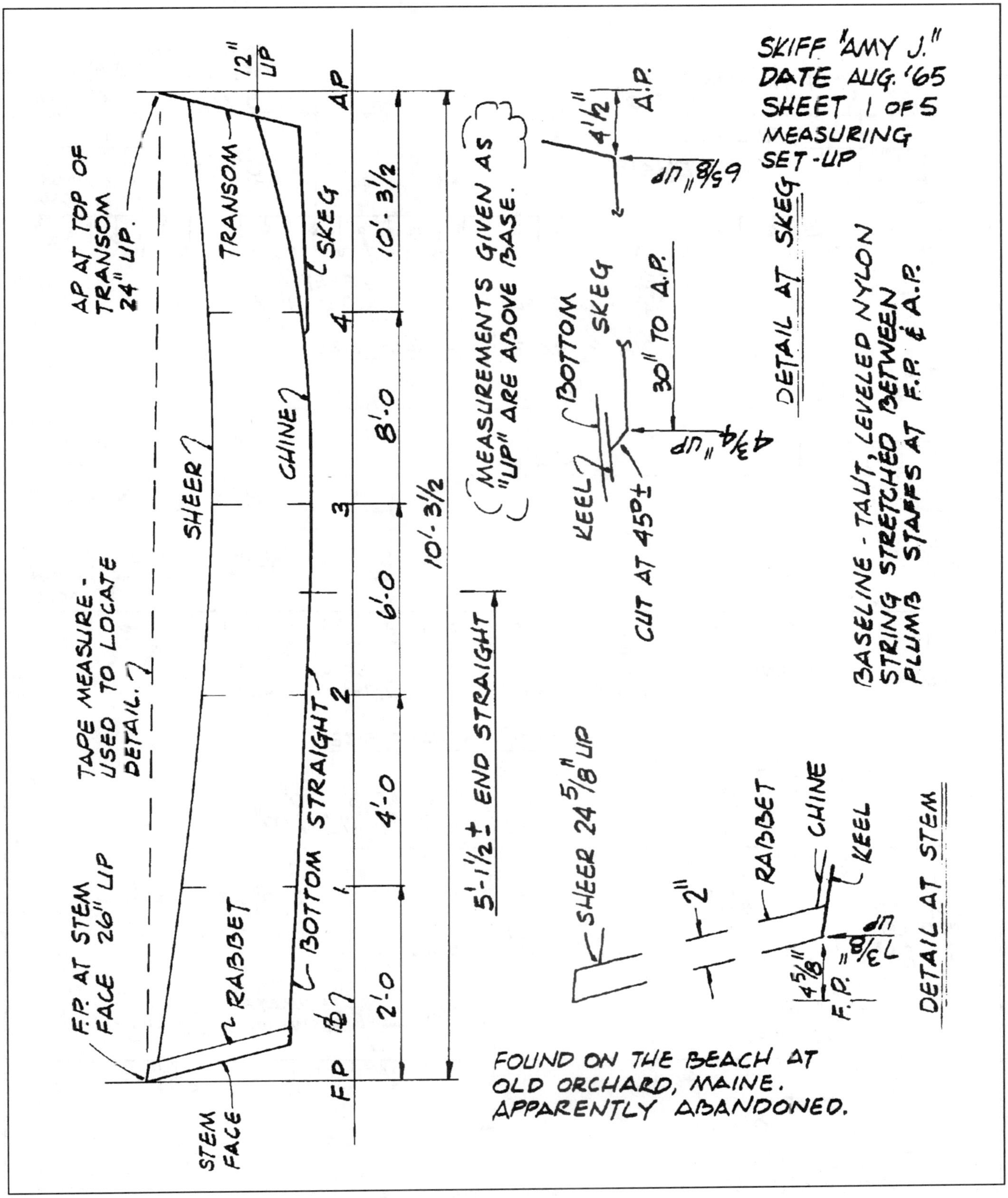

Figure 11.1. Amy J.—set up to be measured.

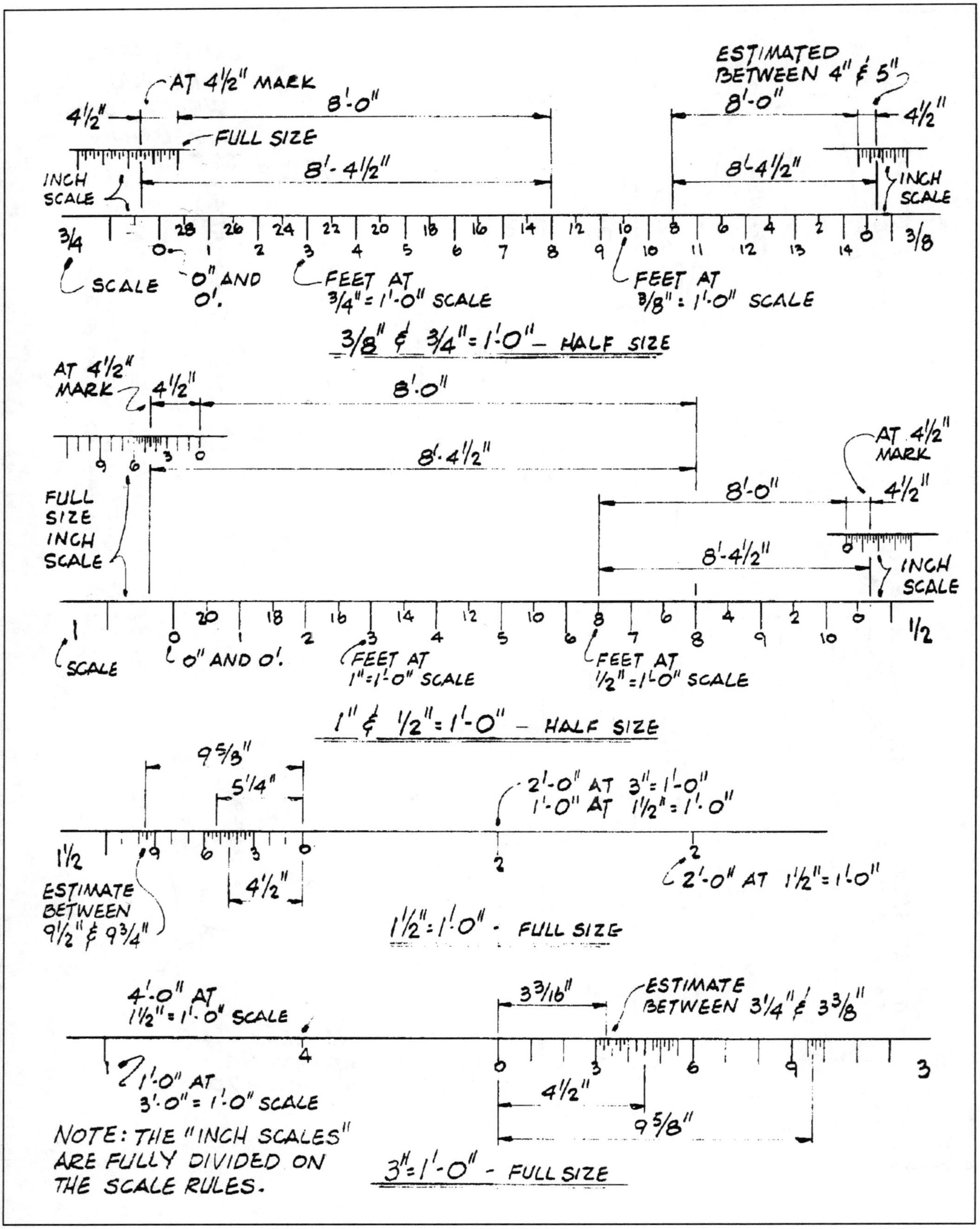

Figure 11.3. Reading commonly used scales.

		(A)	(A)	(B)	(C)	(C)	(D)	(D)
SCALE OF DRAWING	FULL SIZE	1/8"=1'-0"	1/4"=1'-0"	3/8"=1'-0"	1/2"=1'-0"	3/4"=1'-0"	1"=1'-0"	1 1/2"=1'-0"
LENGTH OF BOAT AND SIZE AT SCALE.	45'-0"	5 5/8"	11 1/4"	16 7/8"	22 1/2"	33 1/4"	45"	67 1/2"
	30'-0"	3 3/4"	7 1/2"	11 1/4"	15"	22 1/2"	30"	45"
	15'-0"	1 7/8"	3 3/4"	5 5/8"	7 1/2"	11 1/4"	15"	22 1/2"
	0'-3 1/2"	1 9/32"	2 9/16"	3 7/8"	5 1/2"	7 5/8"	10 1/4"	15 7/16"
							1"=1'-0"	1 1/2"=1'-0"

(A) [shaded] DO NOT USE.

(B) — — MARGINAL - PROBABLY ONLY SAIL PLANS.

(C) —— GOOD FOR ALL

(D) TOO BIG

Figure 11.2. Table showing effect of scale on drawing size.

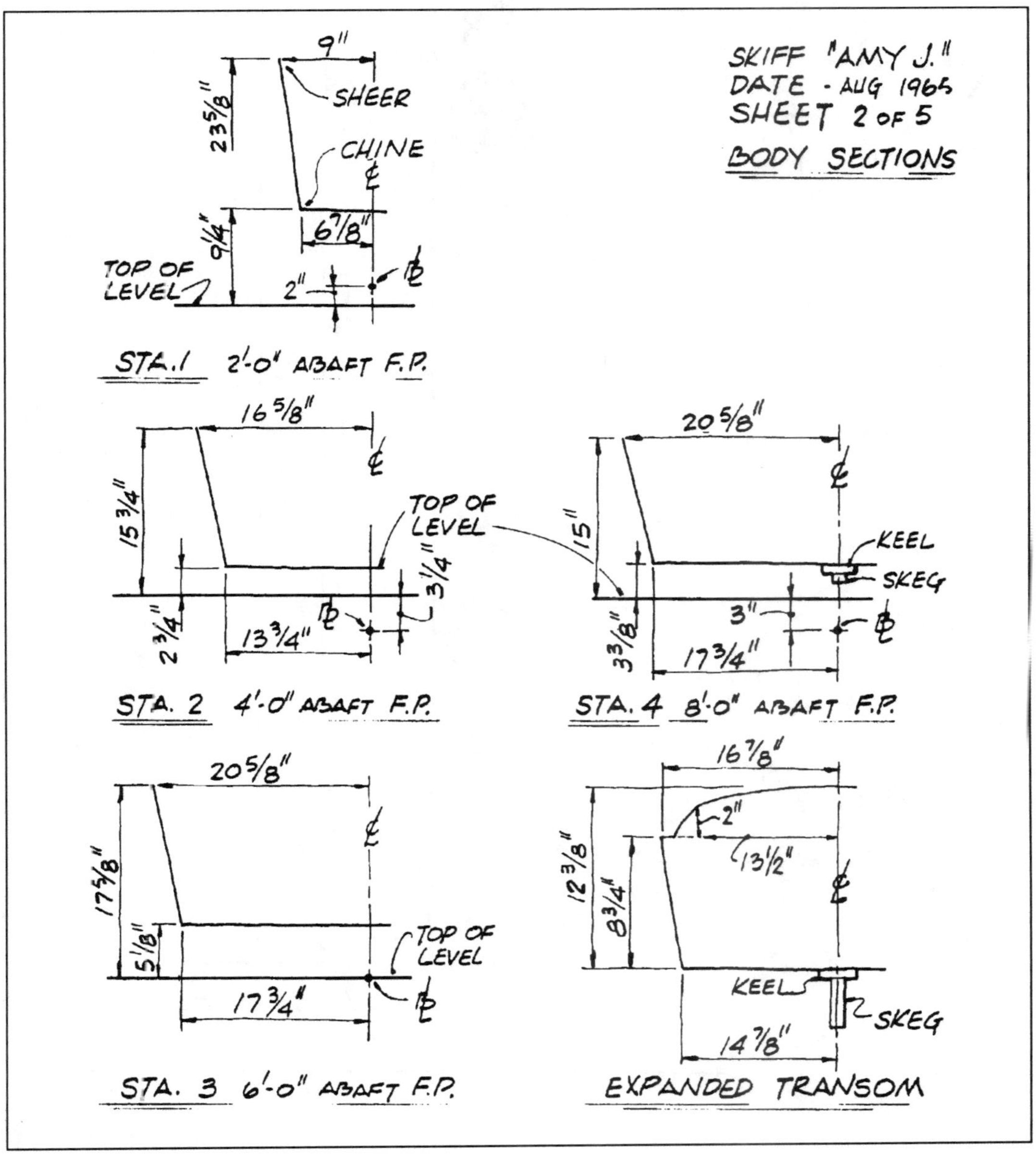

Figure 11.4. Amy J.—body section take-off.

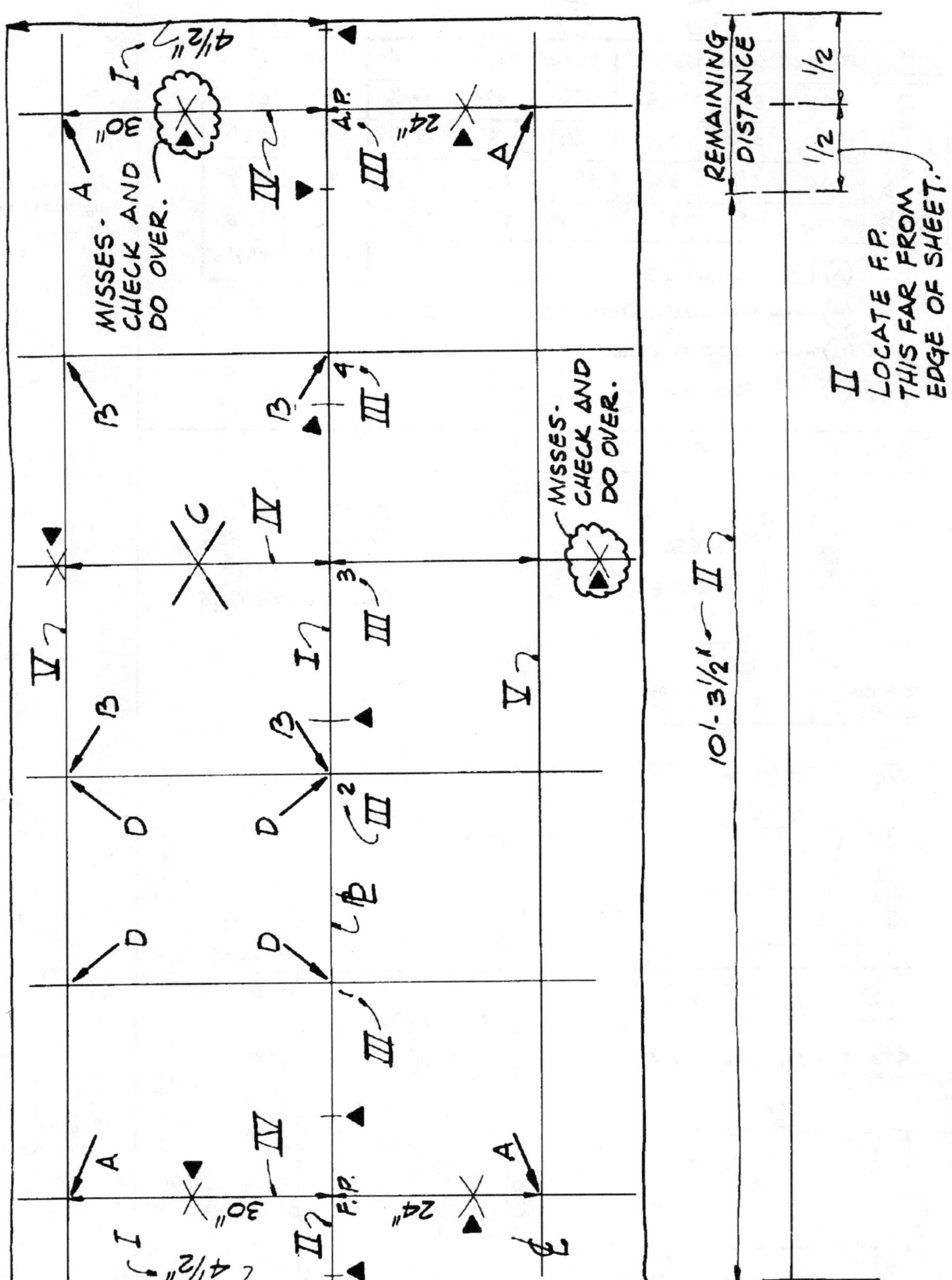

Fig. 11.5

Fig. 11.5A

Figure 11.5. and 11.5A
The grid for the rough lines plan.

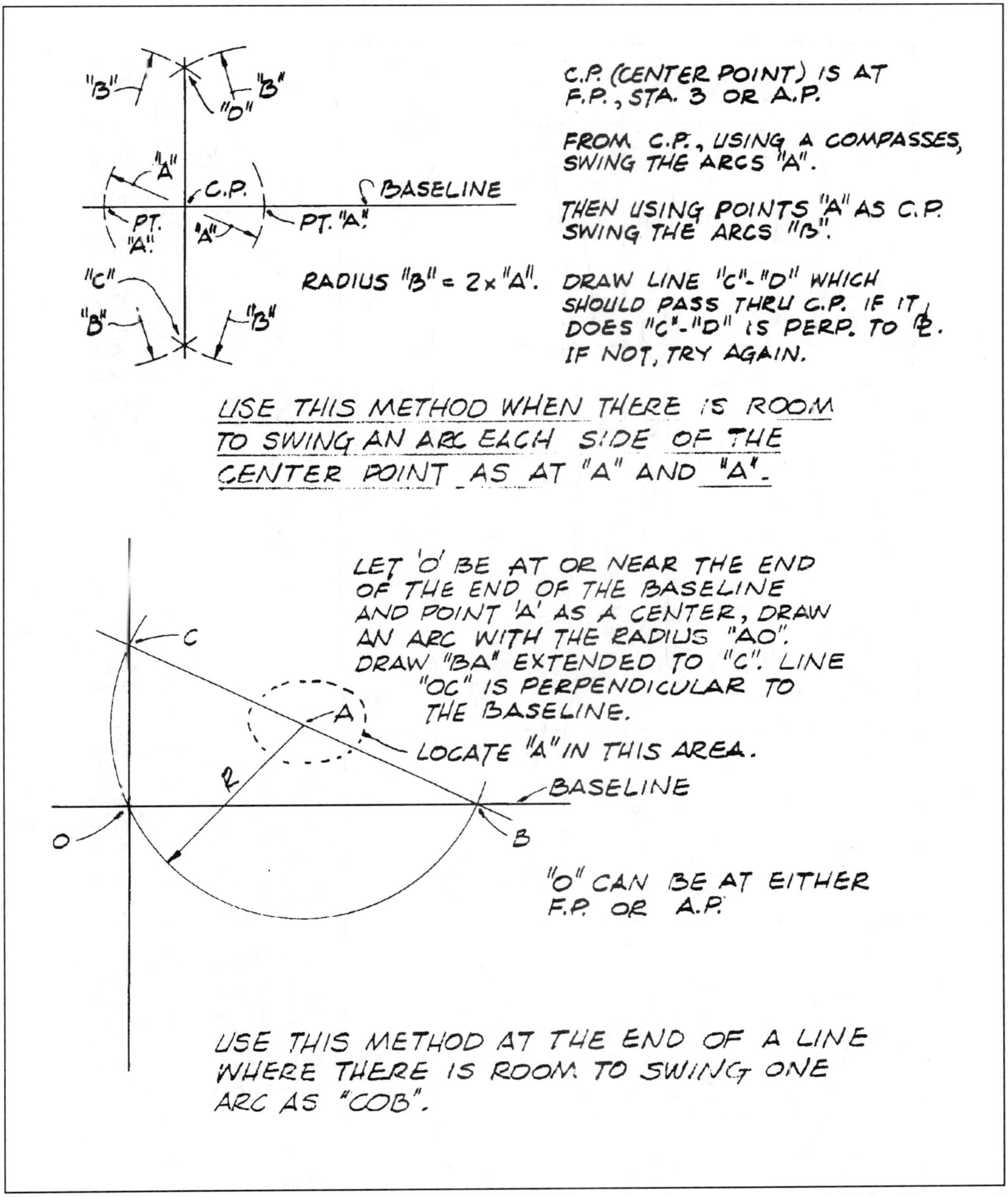

Figure 11.6. Two ways of erecting a perpendicular.

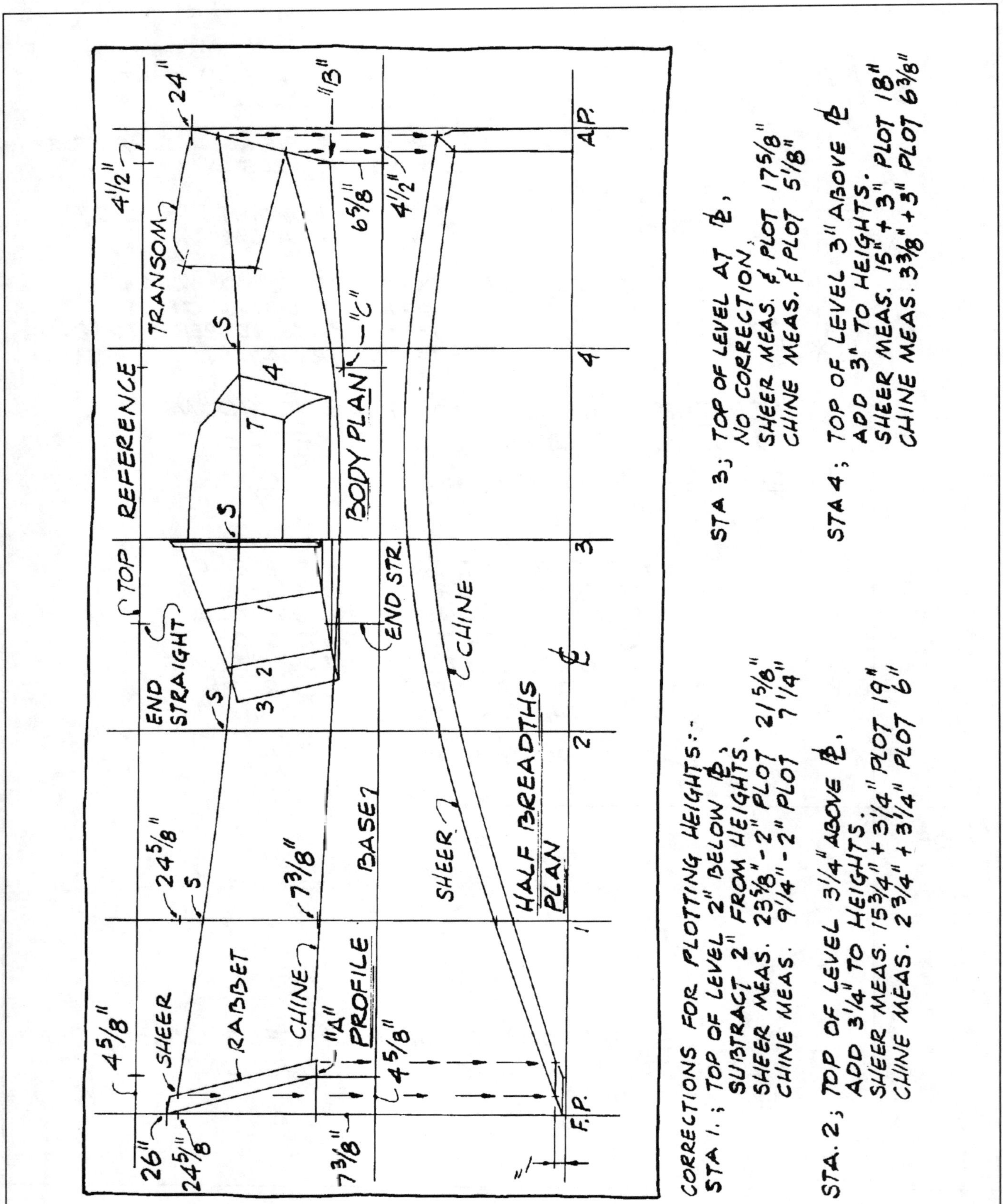

Figure 11.7. Amy J.—the rough lines plan.

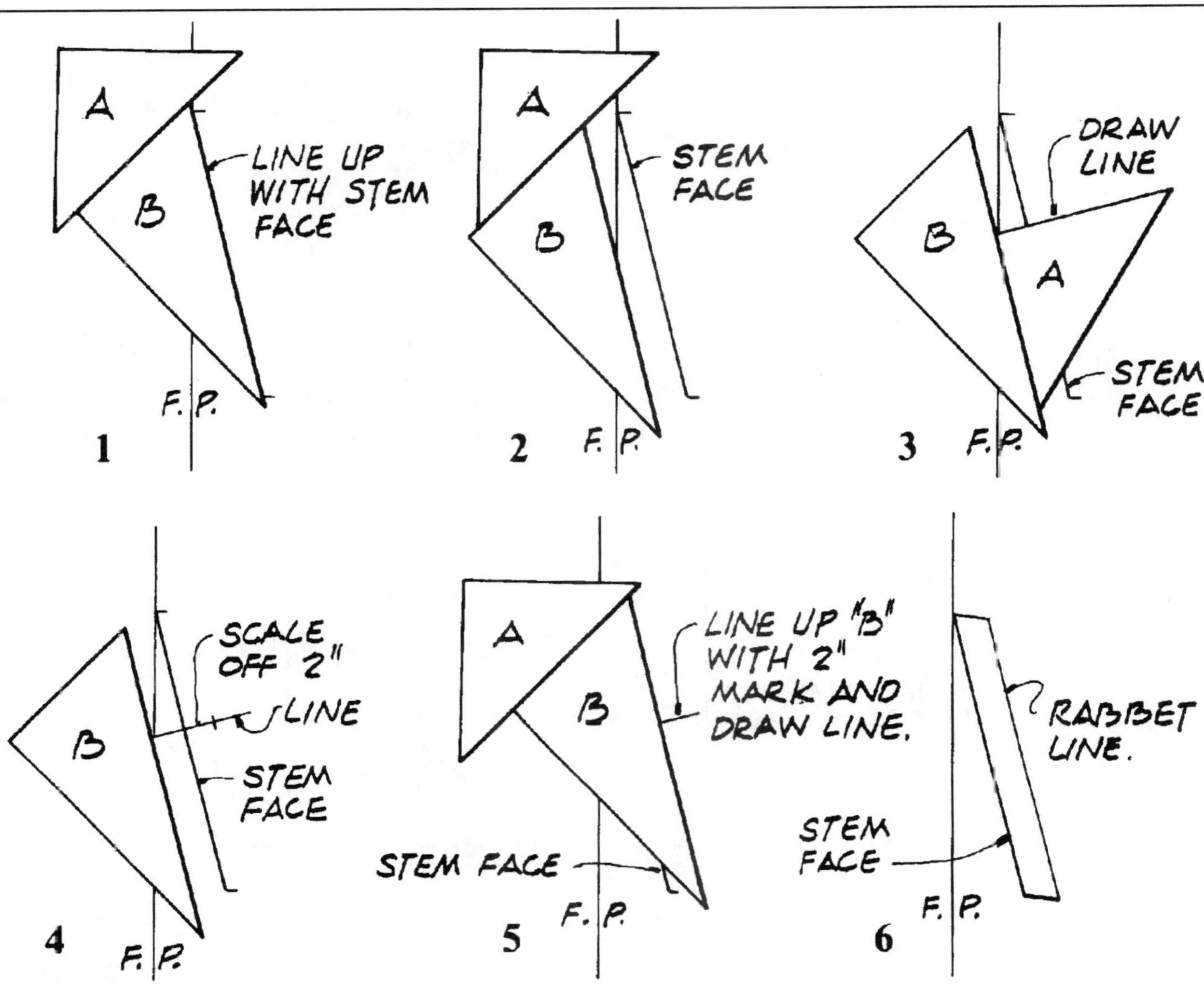

Fig. 11.8

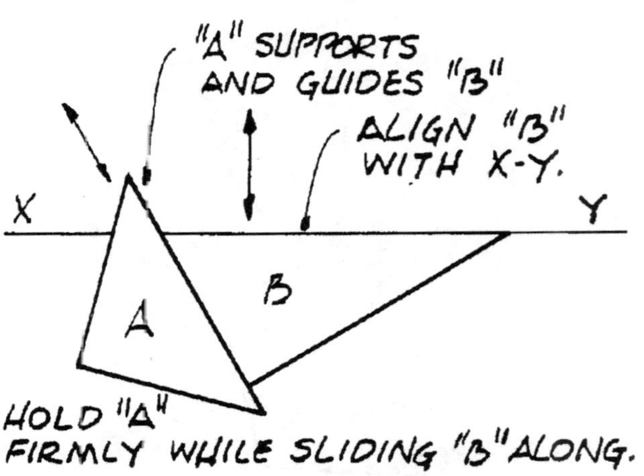

Fig. 11.9

Figure 11.8. Sliding triangles.

Figure 11.9. Sliding triangles.

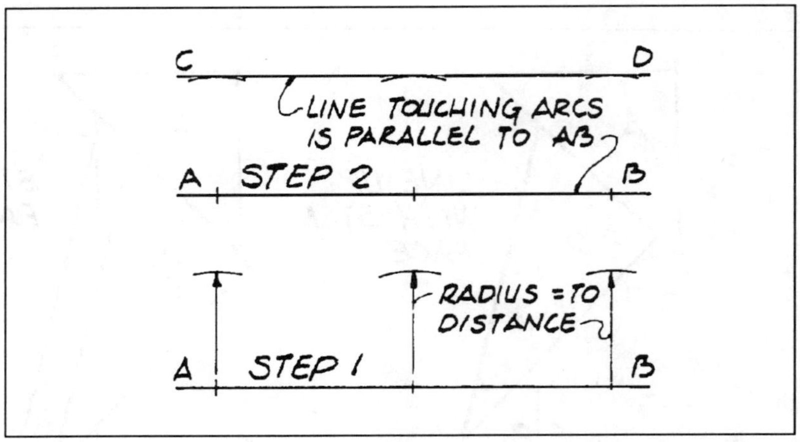

Figure 11.10. Using equal arcs to draw parallel lines.

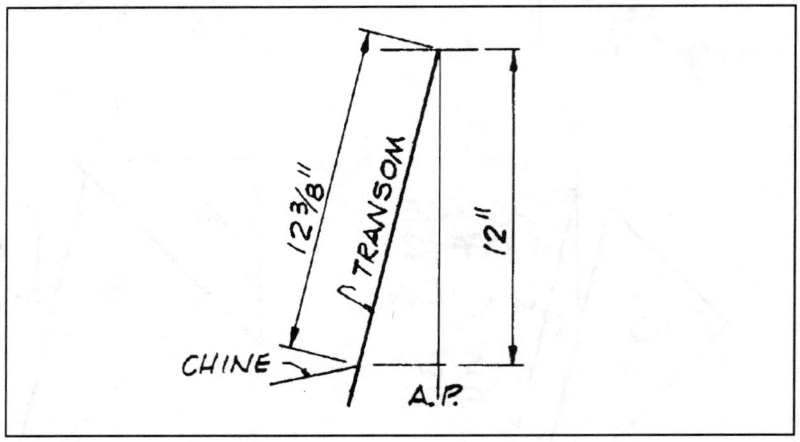

Figure 11.11. Amy J.—plotting the true height of the transom face.

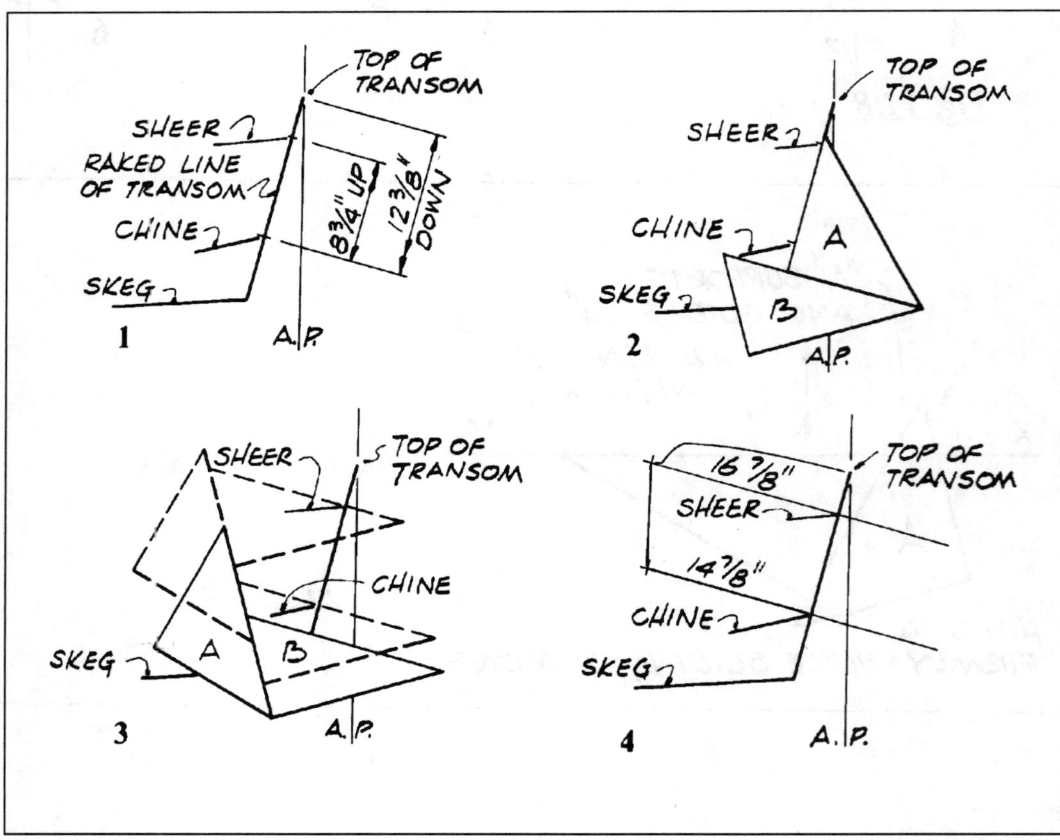

Figure 11.12. Amy J.—plotting the expanded transom.

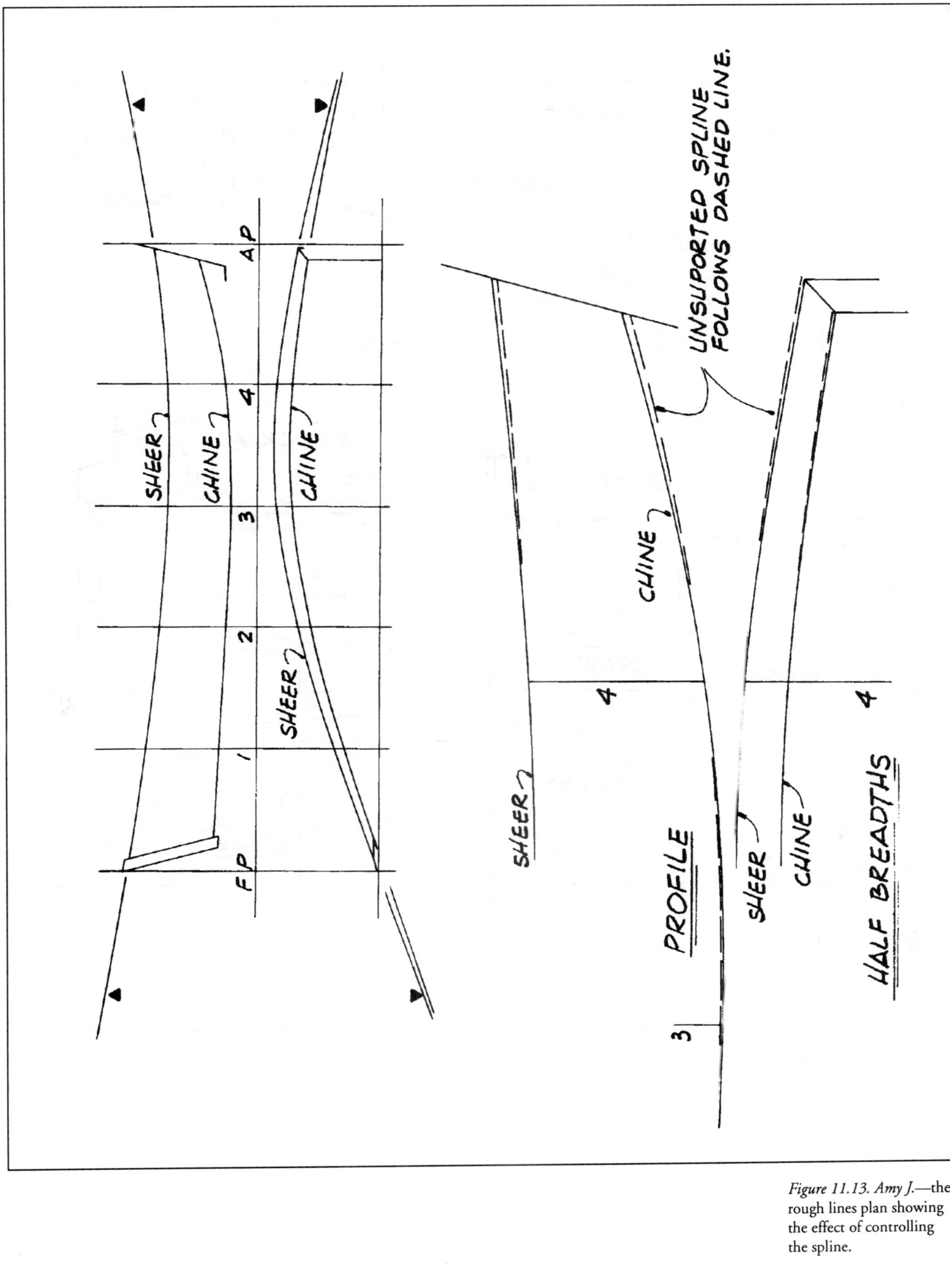

Figure 11.13. Amy J.—the rough lines plan showing the effect of controlling the spline.

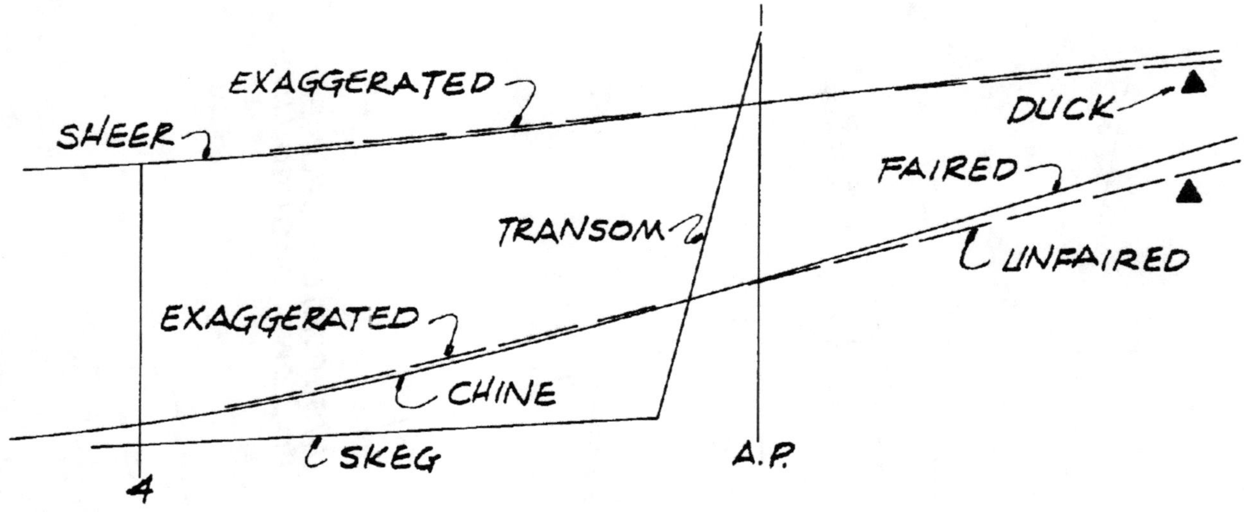

Fig. 11.14

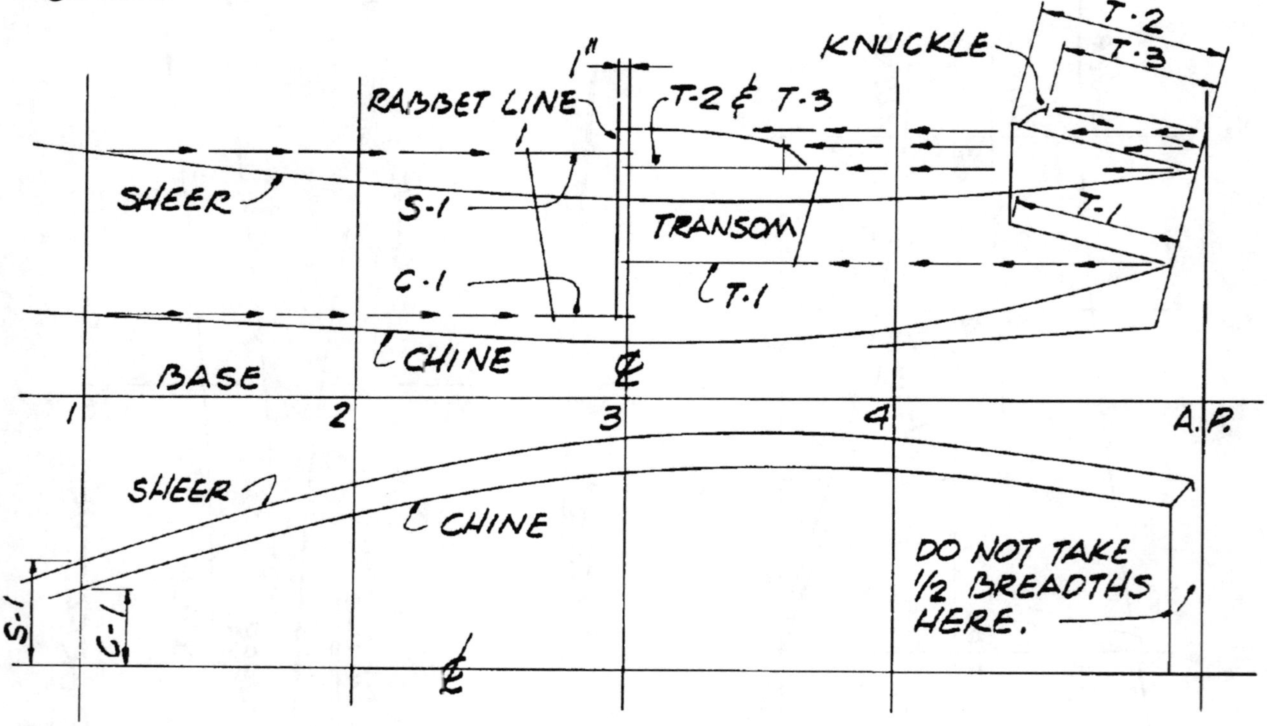

Fig. 11.15

Figure 11.14. Detail of the effects of controlling the spline.

Figure 11.15. Amy J.—developing the body plan.

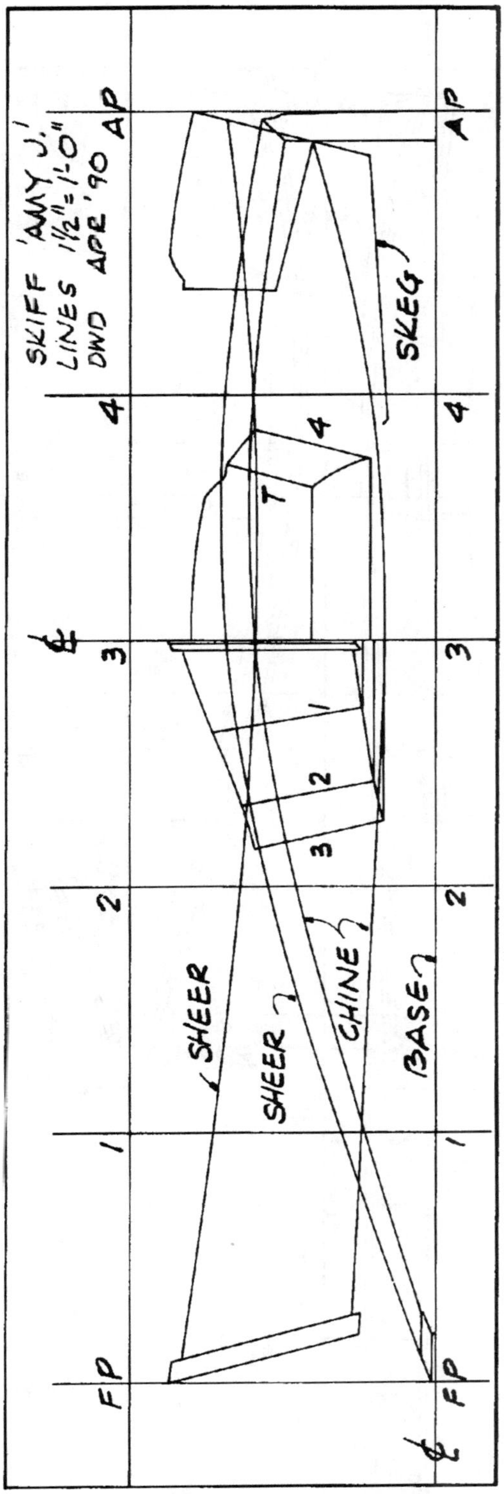

Fig. 11.16

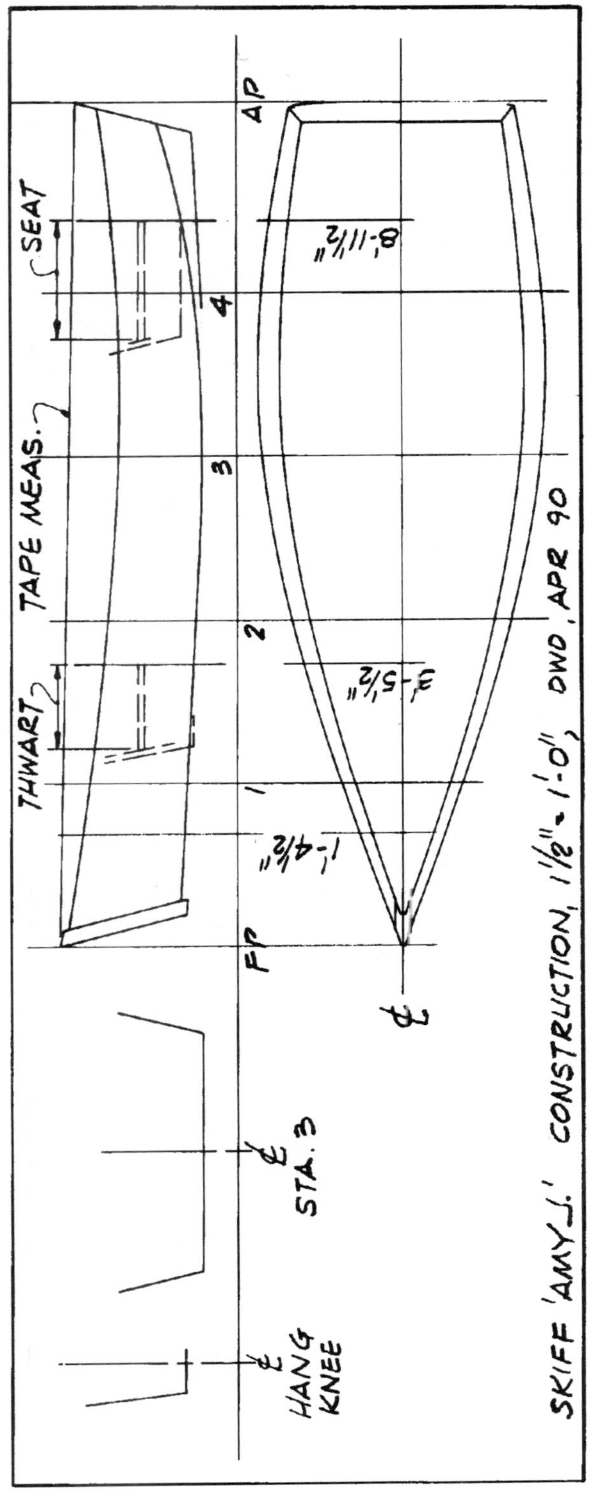

Fig. 11.17

Figure 11.16. Amy J.—lofted.

Figure 11.17. Amy J.—beginning the rough construction drawing.

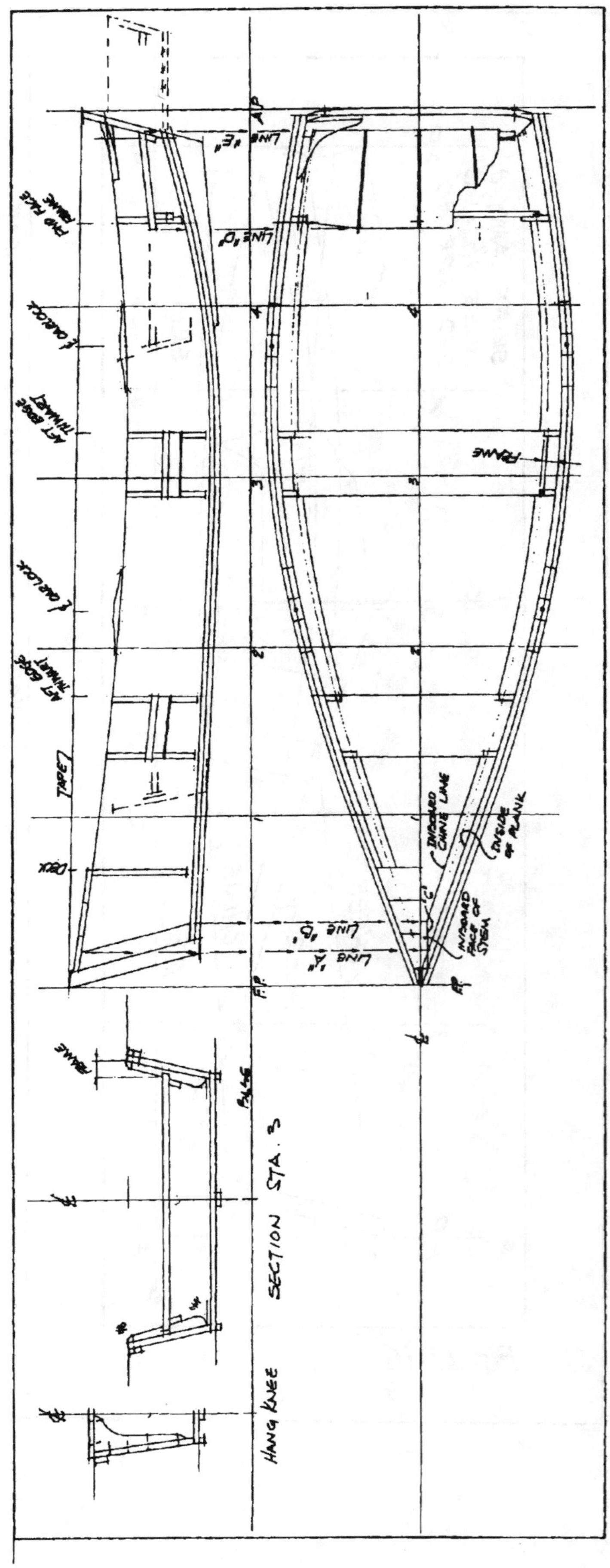

Figure 11.18. Amy J.—the completed rough construction drawing.

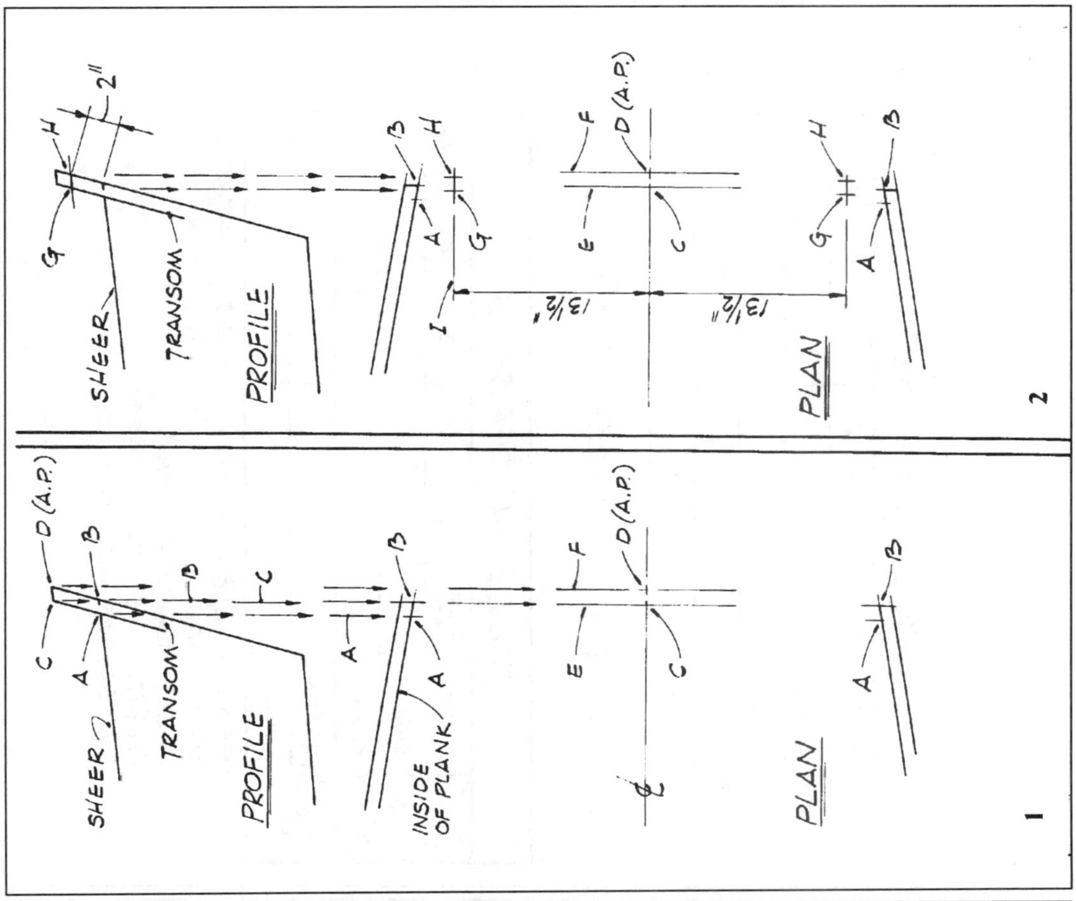

Figure 11.19. Amy J.—developing the transom in plan.

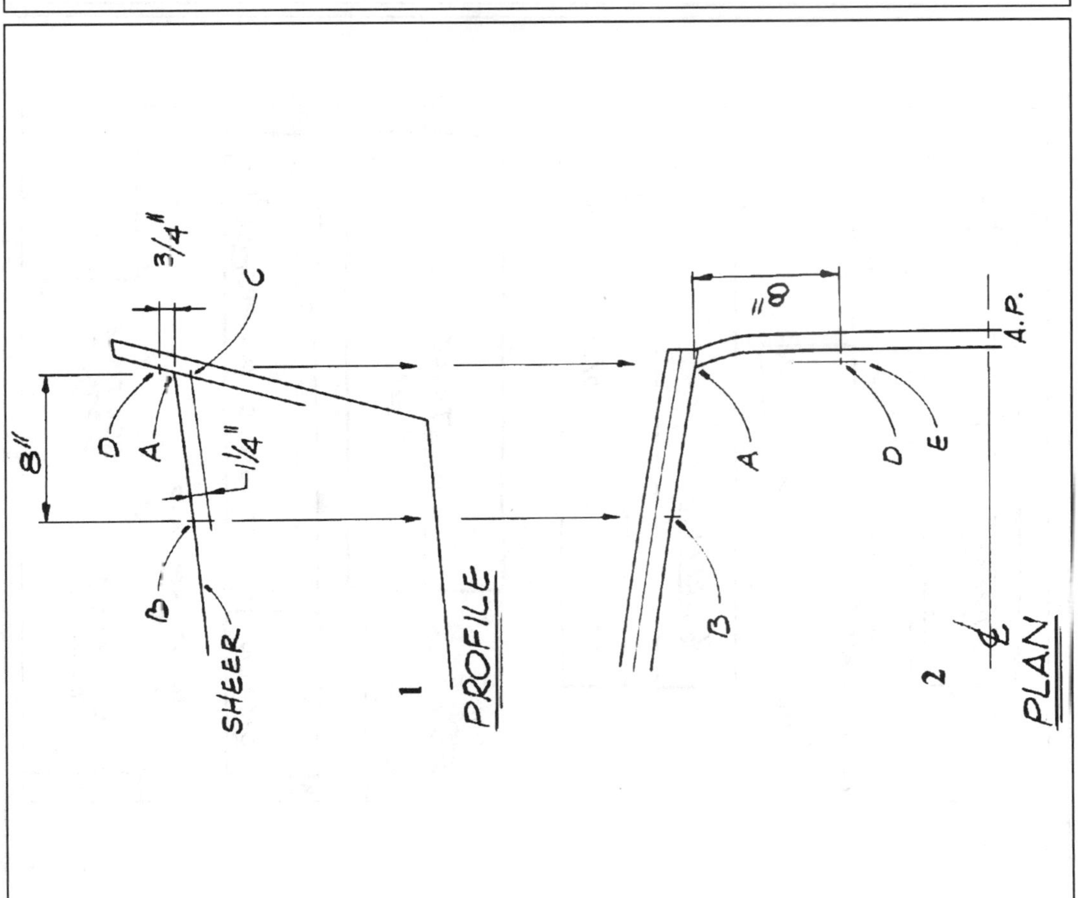

Figure 11.20. Amy J.—developing the quarter knee.

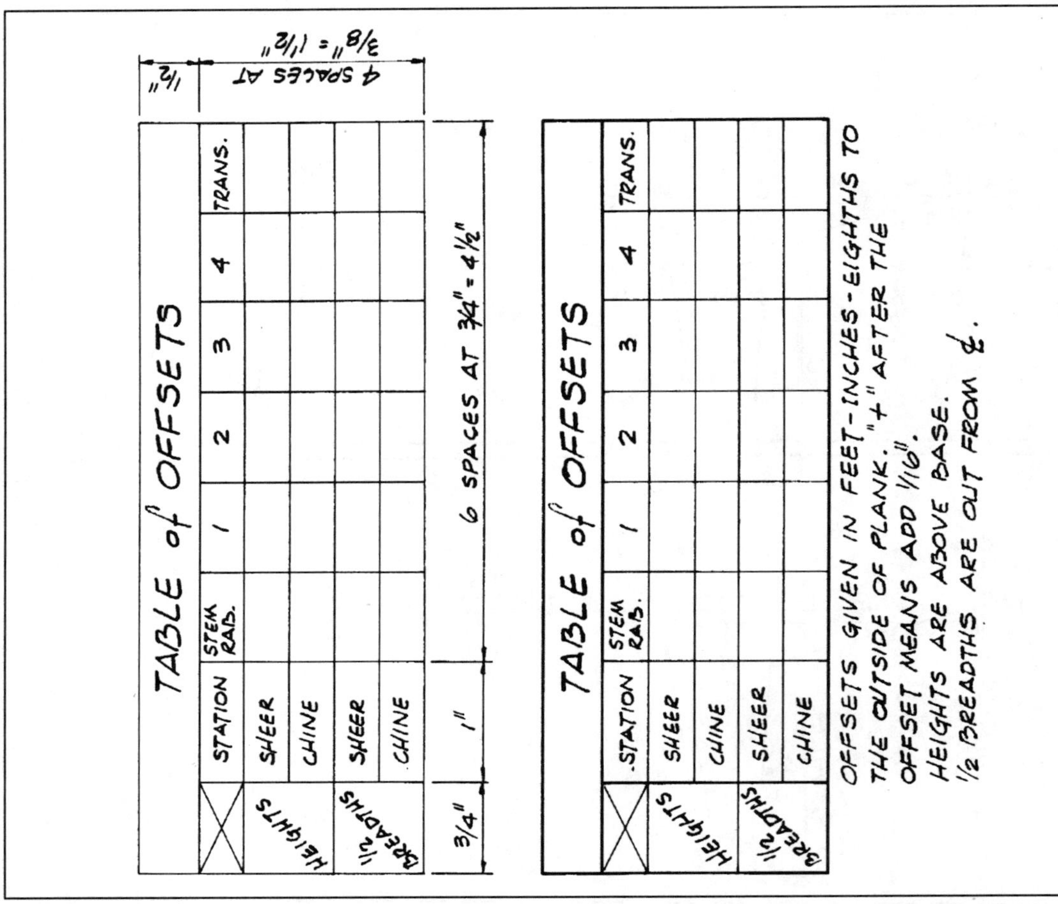

Figure 11.21. Amy J.—the table of offsets.

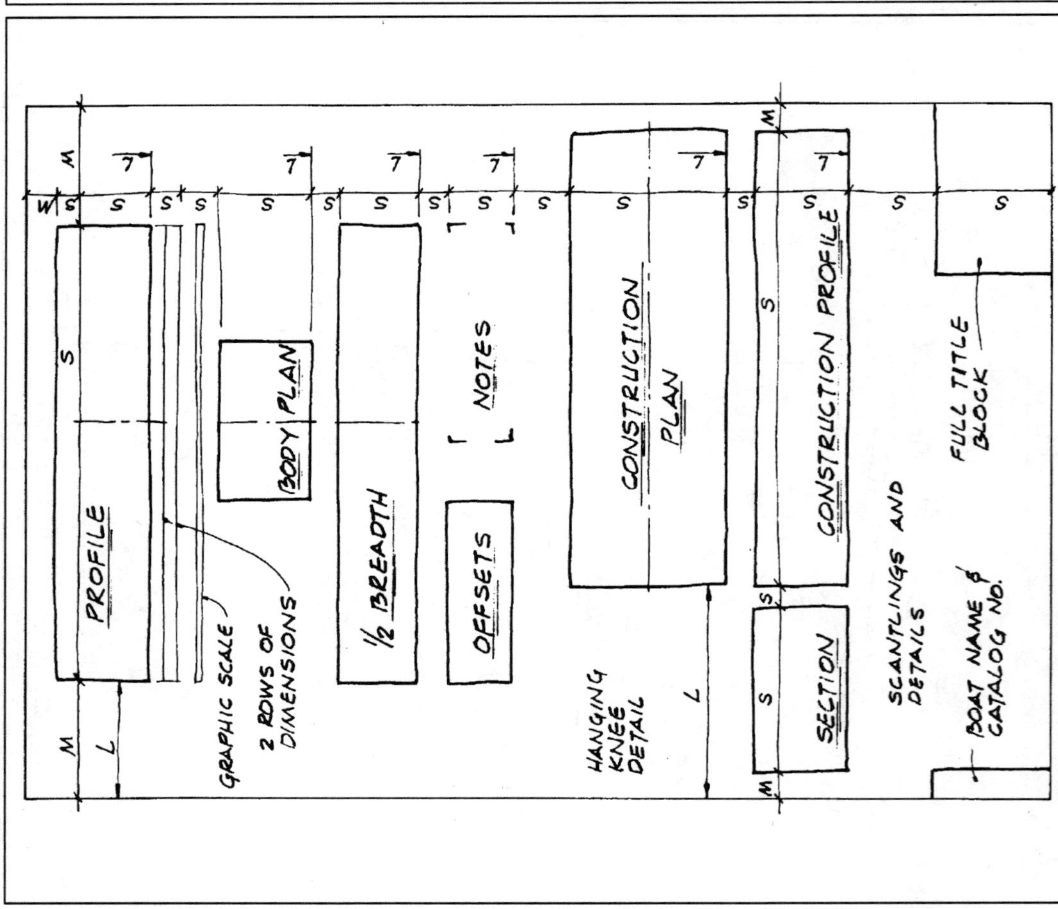

Figure 11.22. Amy J.—the drawing layout.

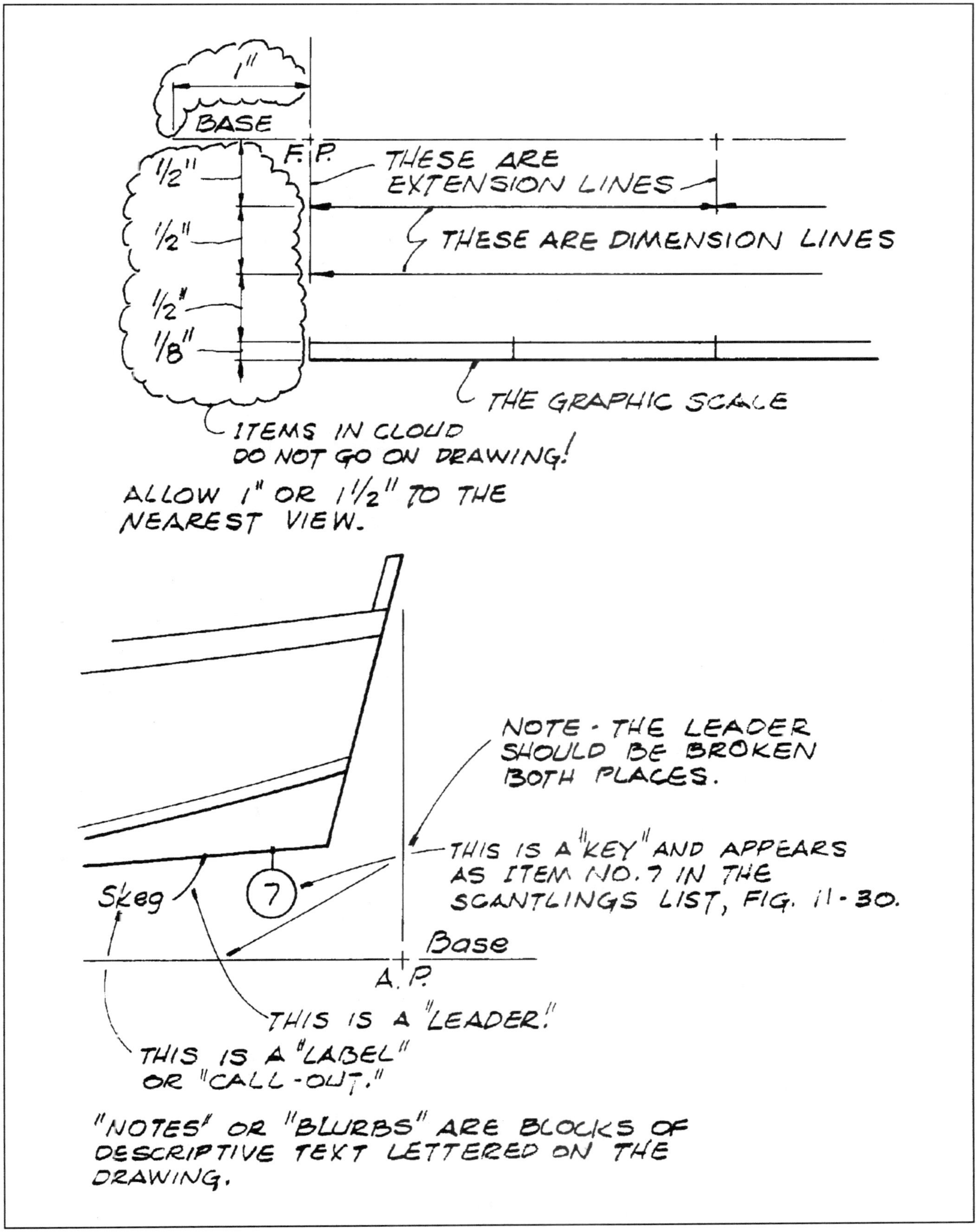

Figure 11.23. Amy J.—basic components of the drawing.

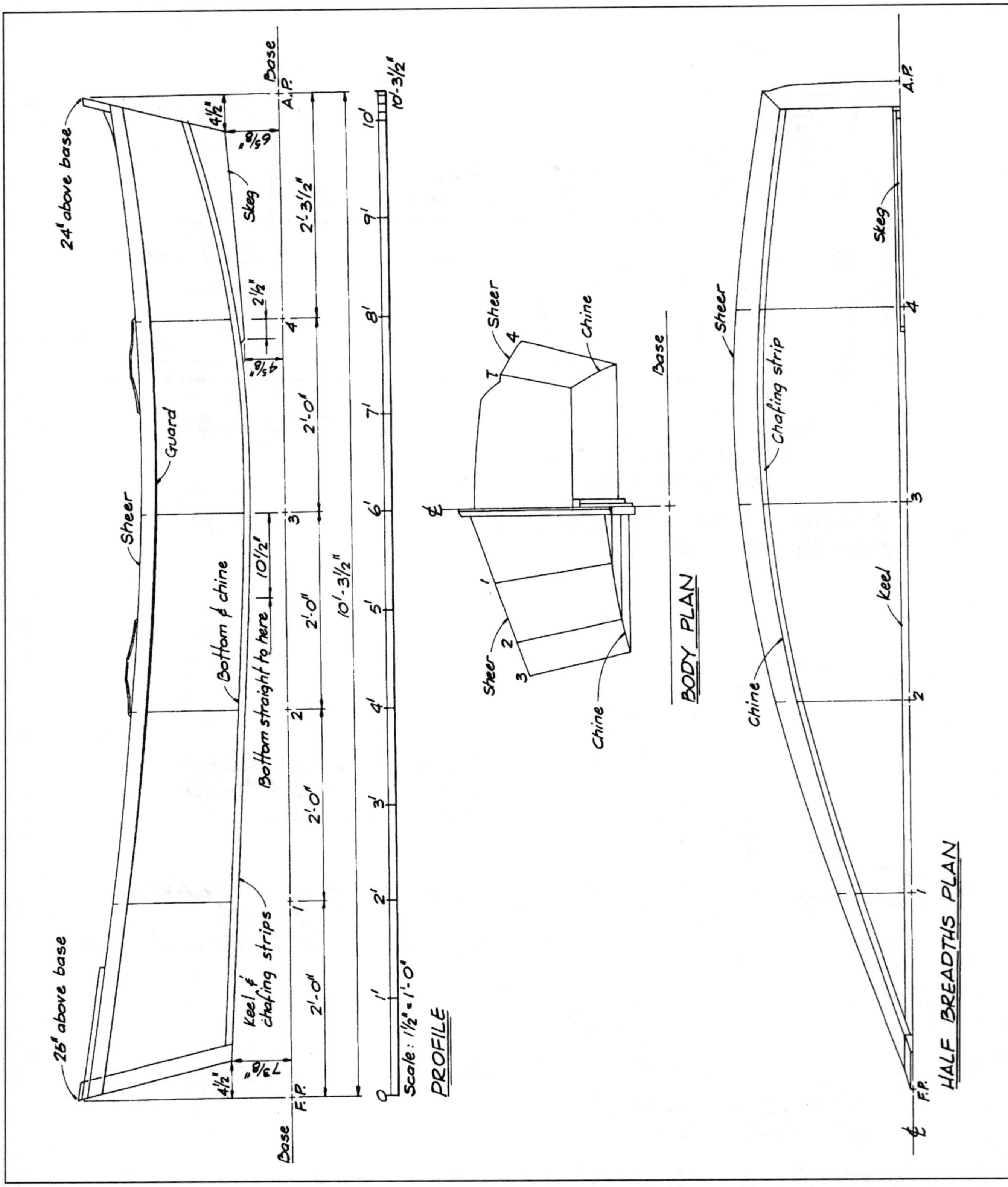

Figure 11.24. Amy J.—the finished lines plan.

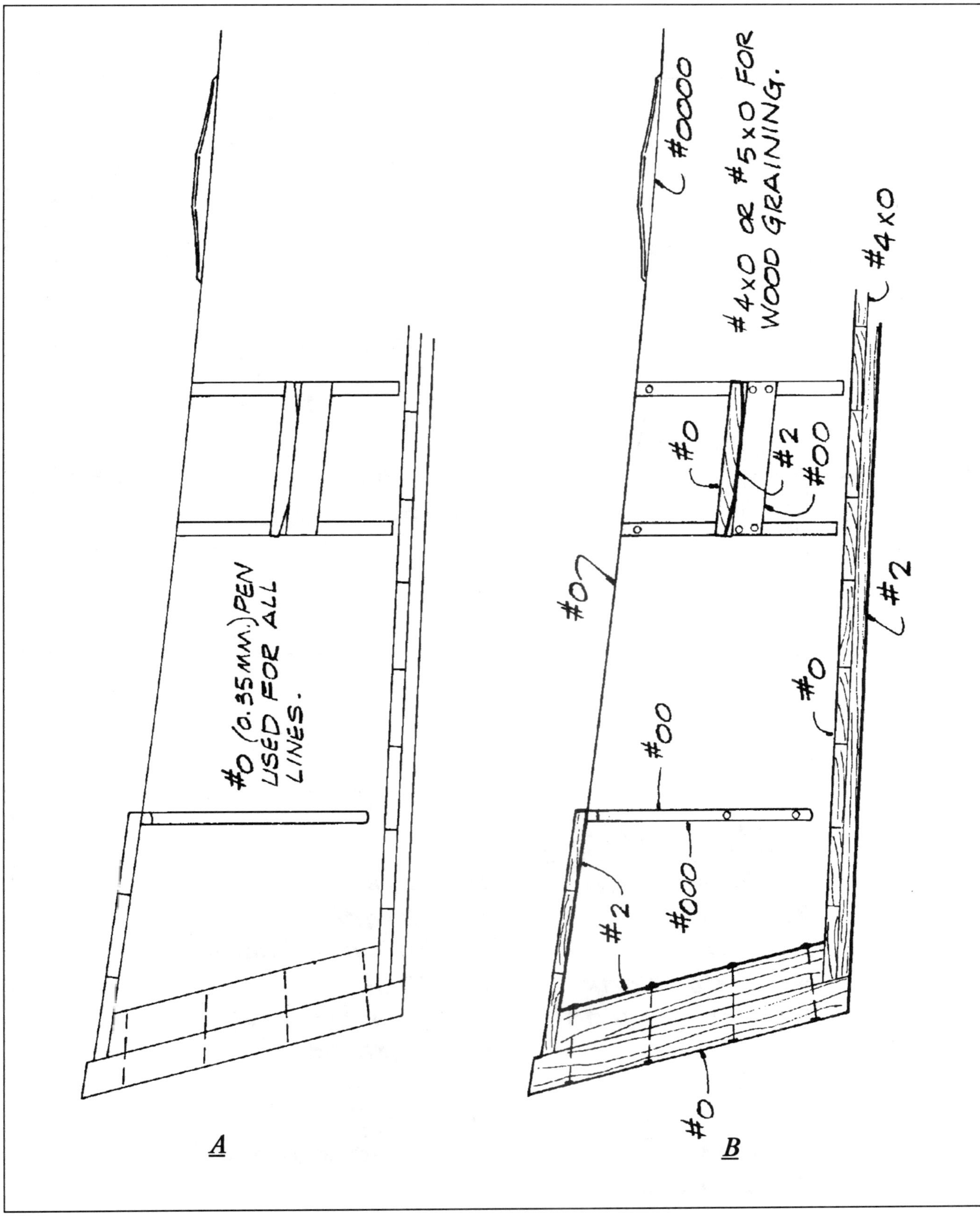

Figure 11.25. Amy J.—visual effect of line weight.

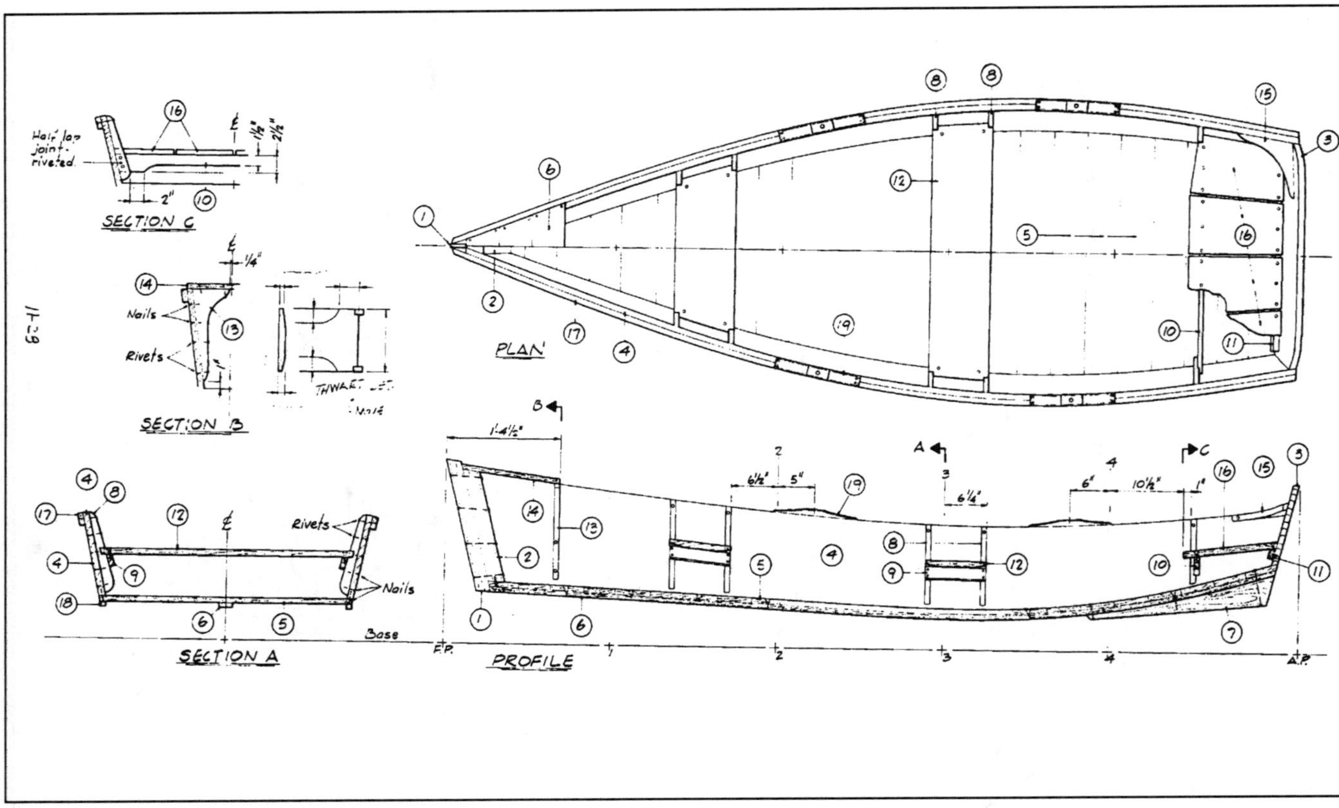

Figure 11.26. Amy J.—the finished construction plan.

SCANTLINGS

1. False stem
2. Stem
3. Transom
4. Side plank
5. Bottom plank
6. Keel
7. Skeg
8. Frames
9. Risers
10. Seat beam
11. Apron
12. Thwarts
13. Hanging knees
14. Deck
15. Quarter knees
16. Stern seat
17. Guard
18. Chafing strips
19. Oarlock pads

Figure 11.27. Amy J.—the scantlings list. A complete scantlings list appears in fig. 11.68.

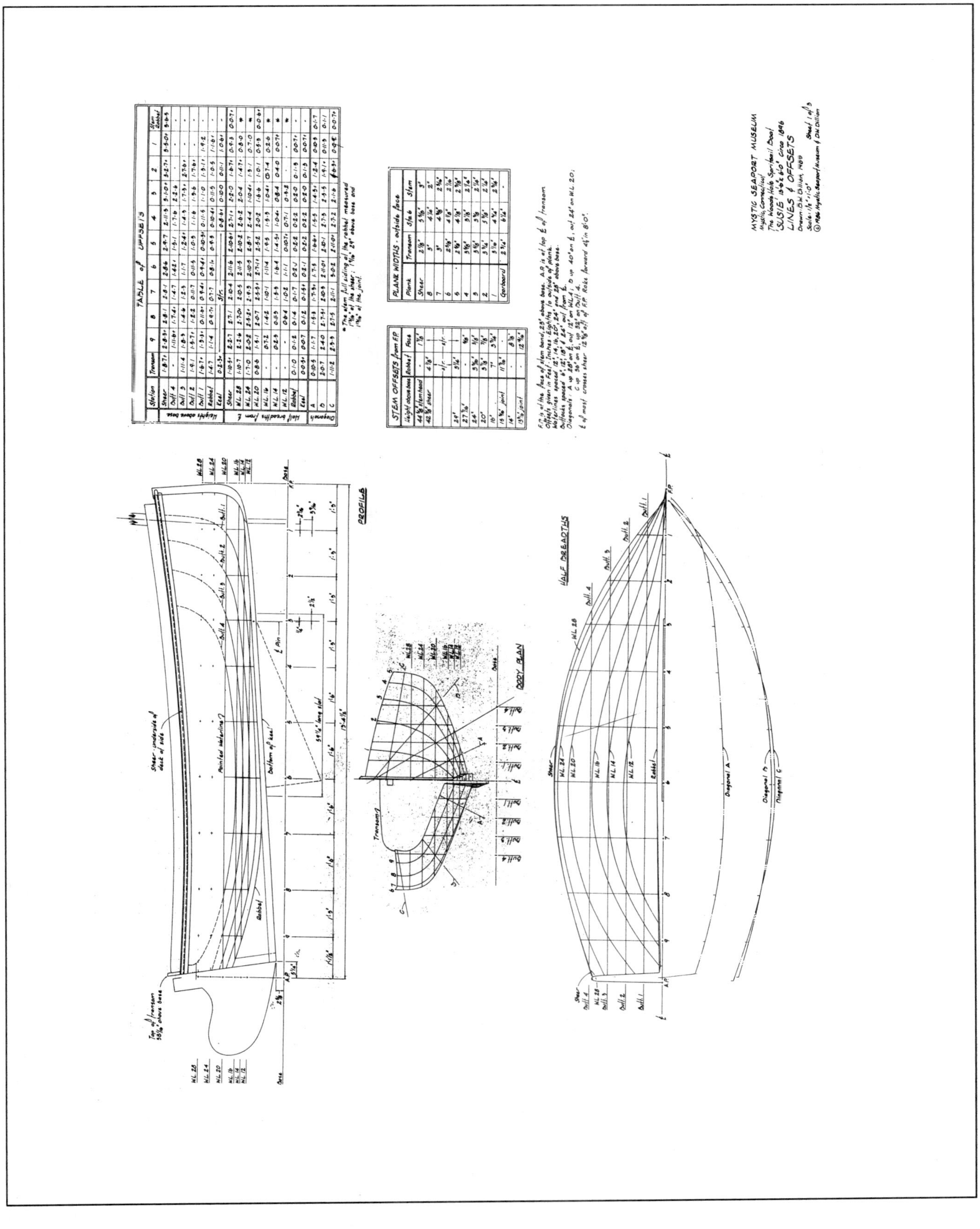

Figure 11.28. Susie's lines plan. *Mystic Seaport Museum.*

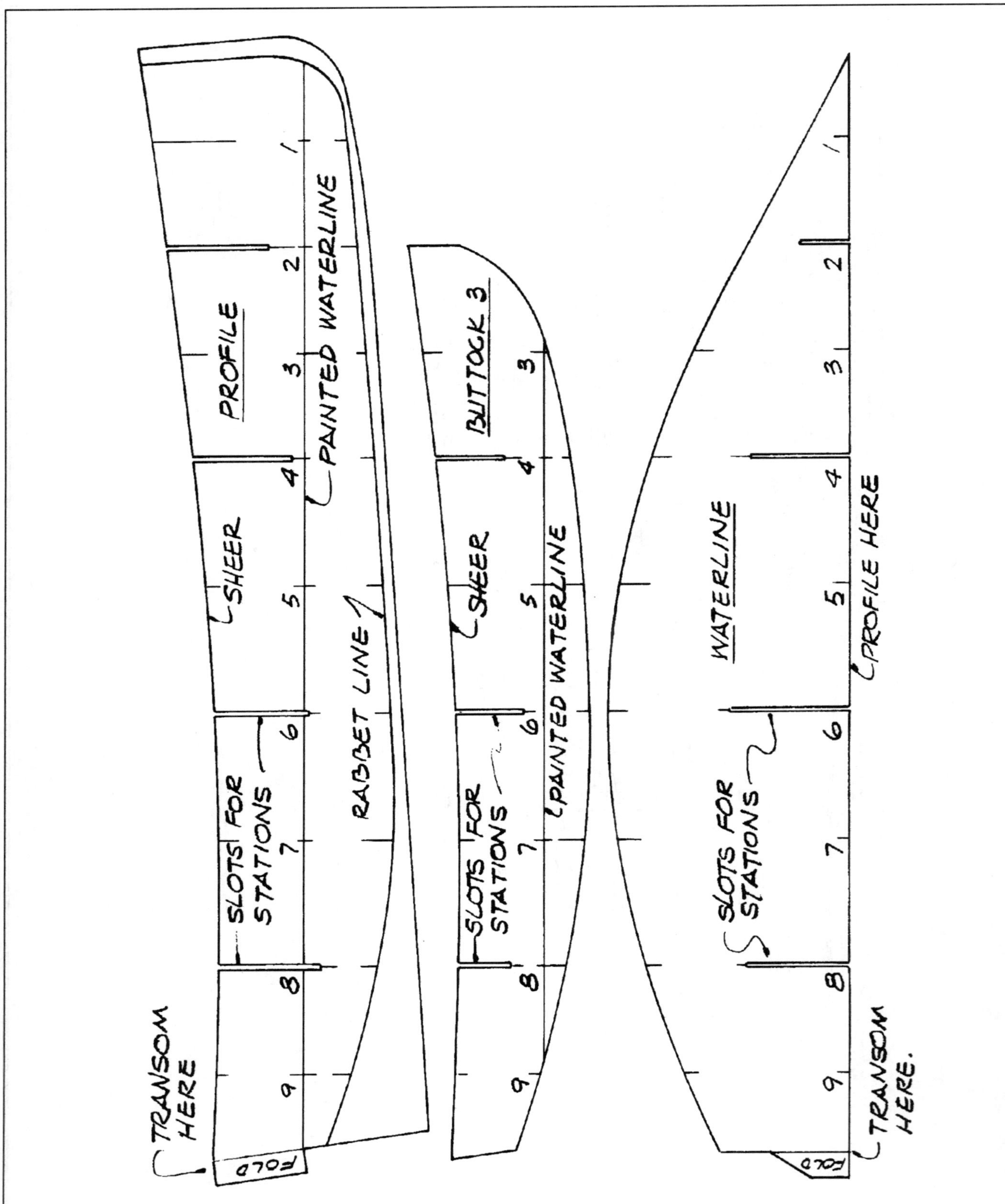

Figure 11.29. Susie—the eggcrate model. Assembly instructions: photocopy both figs. 11.29 and 11.30, and glue the photocopies to light cardboard such as a manila file folder. Using scissors, a single-edge razor blade, or a sharp hobby knife, cut out all the pieces. Fit each body section into the appropriate slot, and either tape or glue in place with white glue. Slide the buttock plane into the slots marked for it and fix it in place. Repeat the process with the waterline and diagonal. Finish the model by fitting the transom in place. Fold the tabs on the profile and waterline over and glue the transom to them.

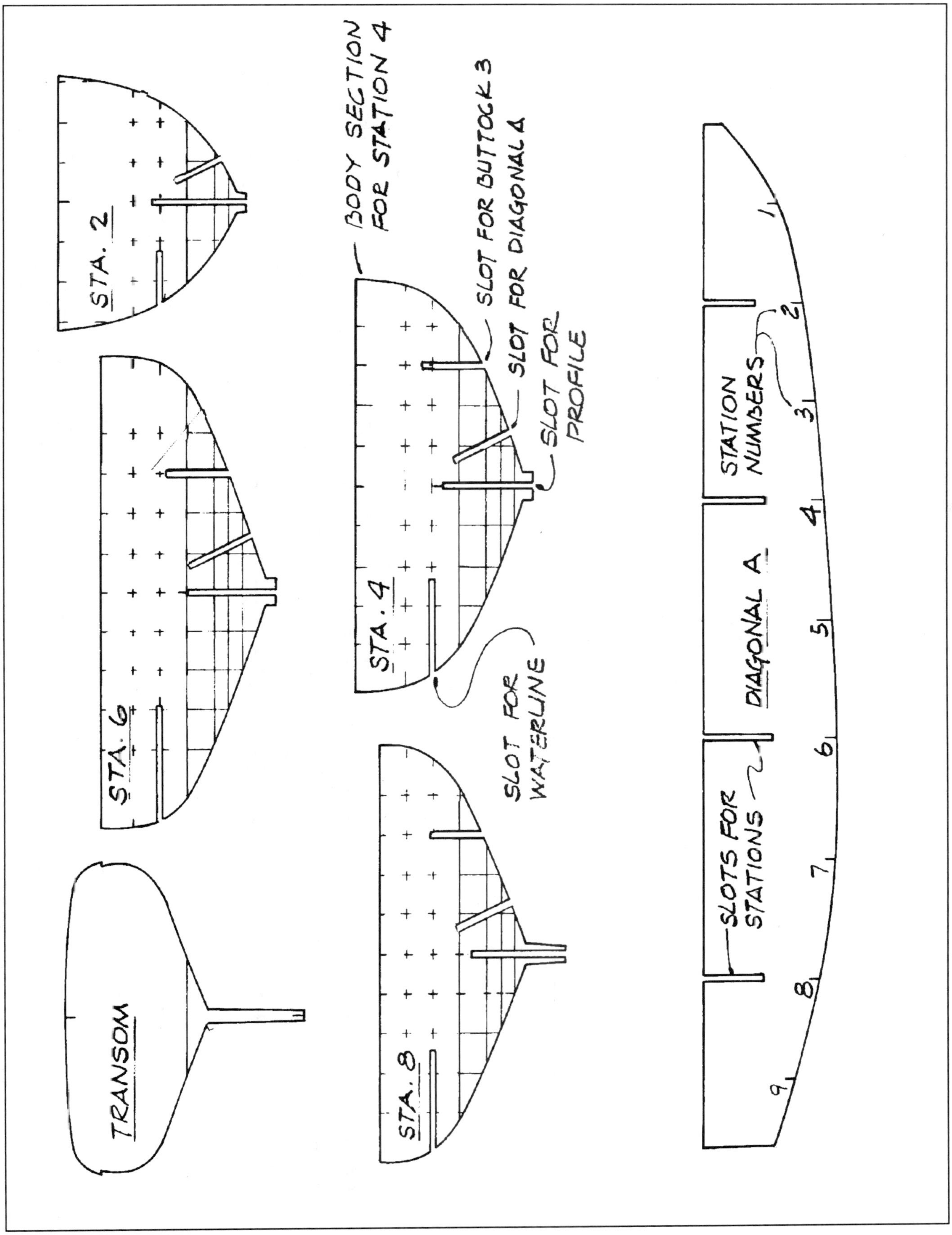

Figure 11.30. Susie—the eggcrate model.

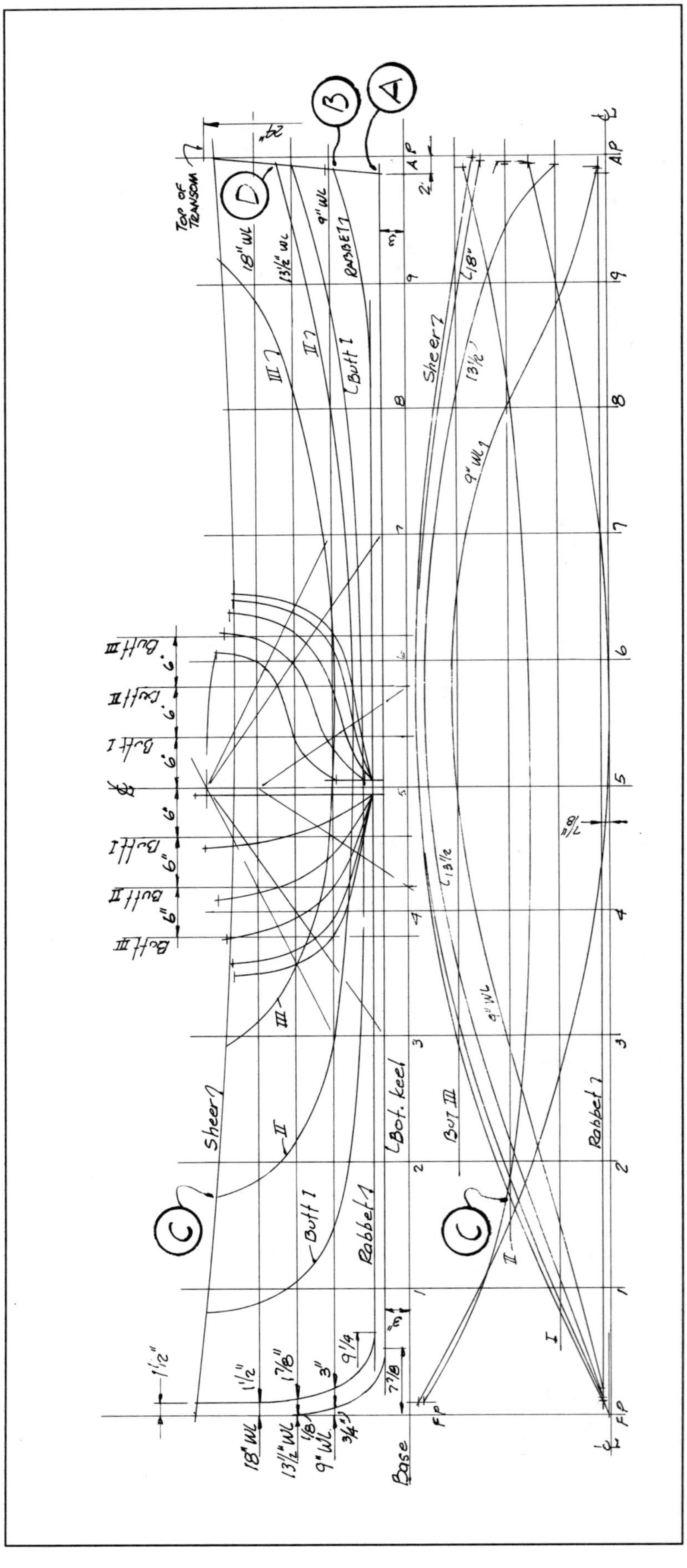

Figure 11.31. The "Exercise Boat"—the lines plan; an example of a "rough drawing."

TABLE of OFFSETS IN FEET-INCHES-EIGHTS TO THE OUTSIDE OF CARVEL PLANK.											
	STATION	1	2	3	4	5	6	7	8	9	TRAN.
HEIGHTS ABOVE BASE	SHEER	2-0-0	1-10-5	1-9-6	1-9-1	1-8-6	1-8-5	1-8-6	1-9-1	1-9-6+	1-10-6
	BUTT. I	1-2-5	0-7-1	0-5-6+	0-5-3	0-5-2+	0-5-3+	0-6-1	0-6-2	0-9-5	1-1-4
	BUTT. II	-	1-3-2	0-8-6	0-7-2	0-6-5	0-6-4	0-6-6	0-9-0	0-11-6	1-3-3
	BUTT. III	-	-	1-8-0	0-10-6	0-8-6+	0-8-5	0-9-7	1-1-0	1-7-4	-
	RABBET	0-4-1	←			STRAIGHT		→	0-4-1	0-5-1	0-8-4
HALF BREADTHS	SHEER	0-7-3	1-1-5	1-6-3	1-9-2	1-10-6	1-11-0	1-10-1+	1-8-5	1-6-3	1-3-6+
	18" W.L.	0-6-6+	1-0-6	1-5-5	1-8-7	1-10-4	1-10-6	1-10-0	1-8-1	1-5-5	1-3-0
	13½" W.L.	0-5-5+	0-11-1	1-3-7	1-7-5	1-9-3	1-9-6	1-8-5+	1-6-3	1-2-4	0-6-0
	9" W.L.	0-4-0	0-8-2	1-0-3	1-4-0	1-6-2	1-6-5	1-4-6	1-0-0	0-4-6	0-1-0+
DIAGONALS	A	0-8-5+	1-0-2+	1-1-7	1-2-4	1-2-5+	1-2-3	1-1-5	1-0-3	0-10-0	0-6-6+
	B	0-8-5	1-2-6	1-7-1	1-10-2	1-11-7	2-0-0	1-10-7+	1-8-6	1-5-7	1-2-7
	C	0-7-0	1-2-2	1-6-6	1-10-1	1-11-7	2-0-1	1-11-1	1-9-1+	1-6-3	1-3-6

LENGTH F.P. TO A.P. 12'-6"; STATIONS SPACED 1'-3".
WATERLINES SPACED 9", 13½" & 18" ABOVE BASE.
BUTTOCKS SPACED 6", 12" & 18" OUT FROM ℄.
DIAGONALS - A 18" UP ON ℄, 12" OUT ON BASE
B 24" UP ON ℄, 3" UP 30" OUT
C 24" UP ON ℄, 12" UP 30" OUT

Figure 11.32. The "Exercise Boat"—table of offsets.

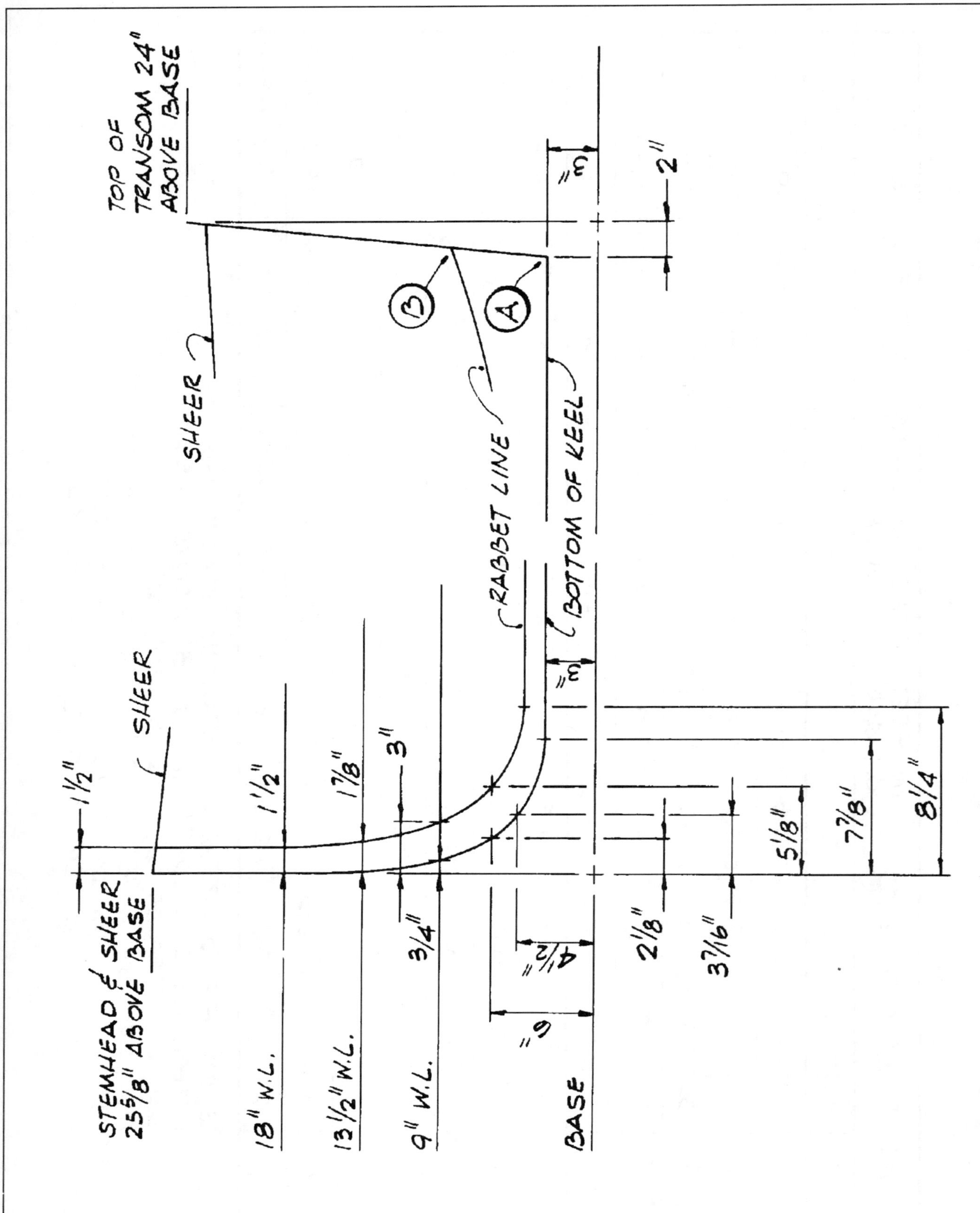

Figure 11.33. The "Exercise Boat"—detail of bow and stern layout.

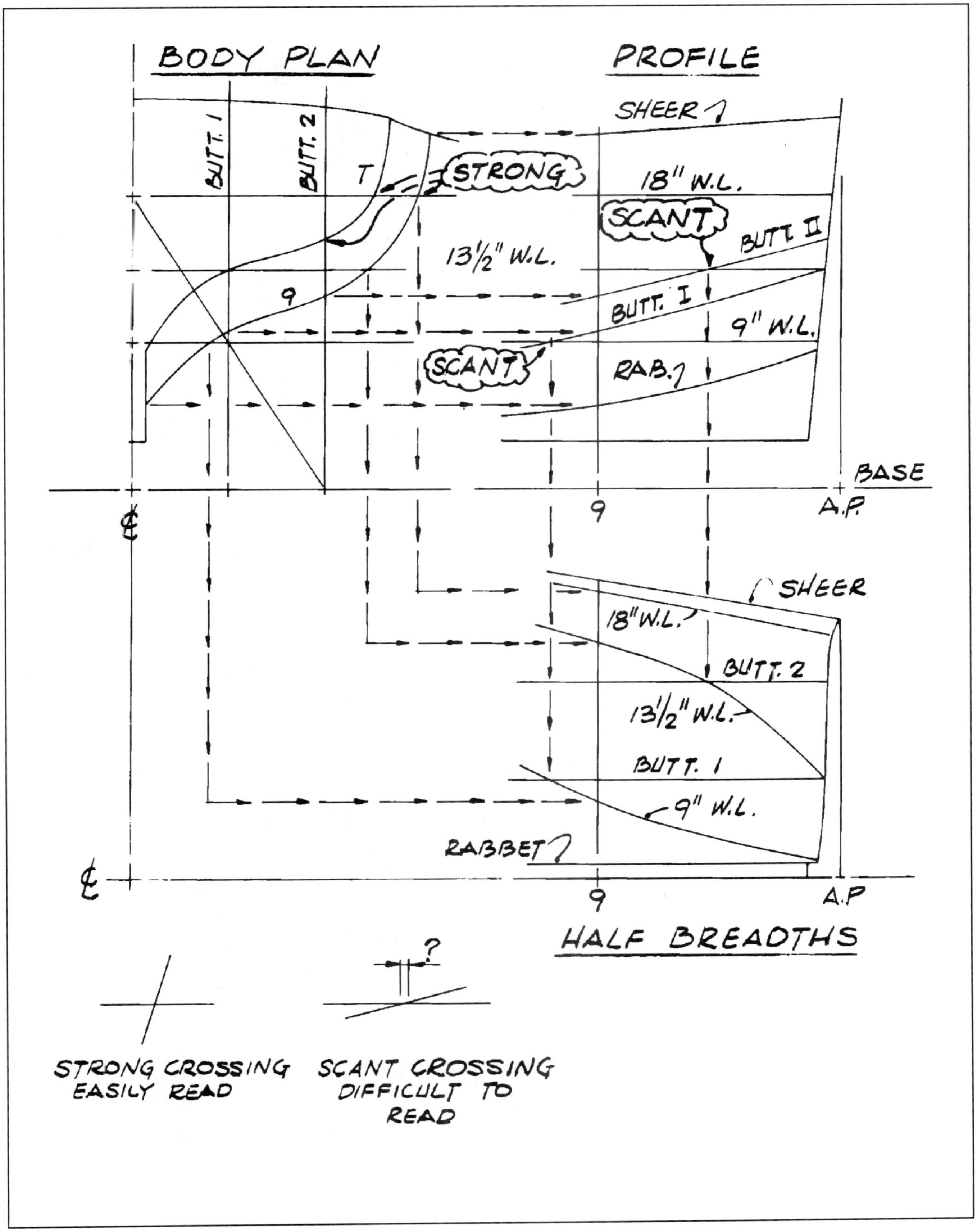

Figure 11.34. The "Exercise Boat"—accurately projected intersections and scant and strong angles.

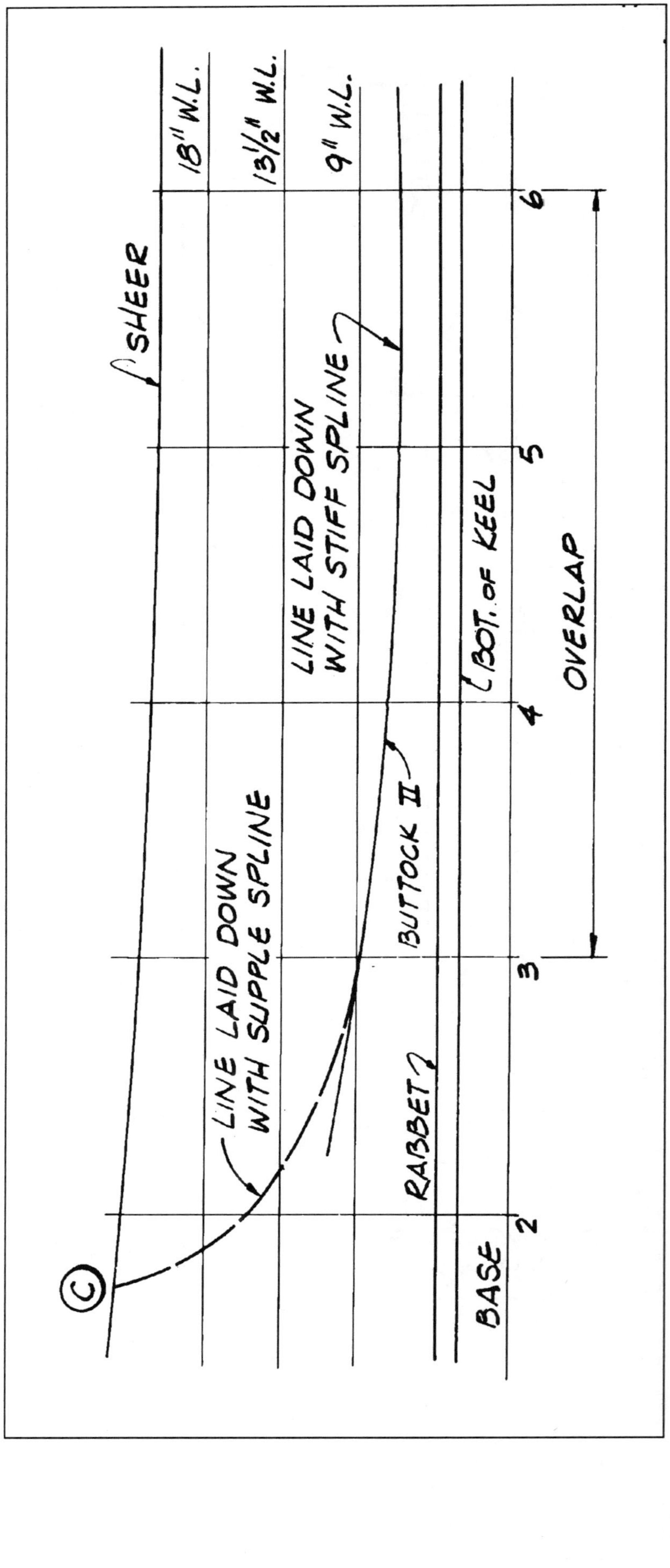

Figure 11.35. Using supple and stiff splines to fair a compound curve.

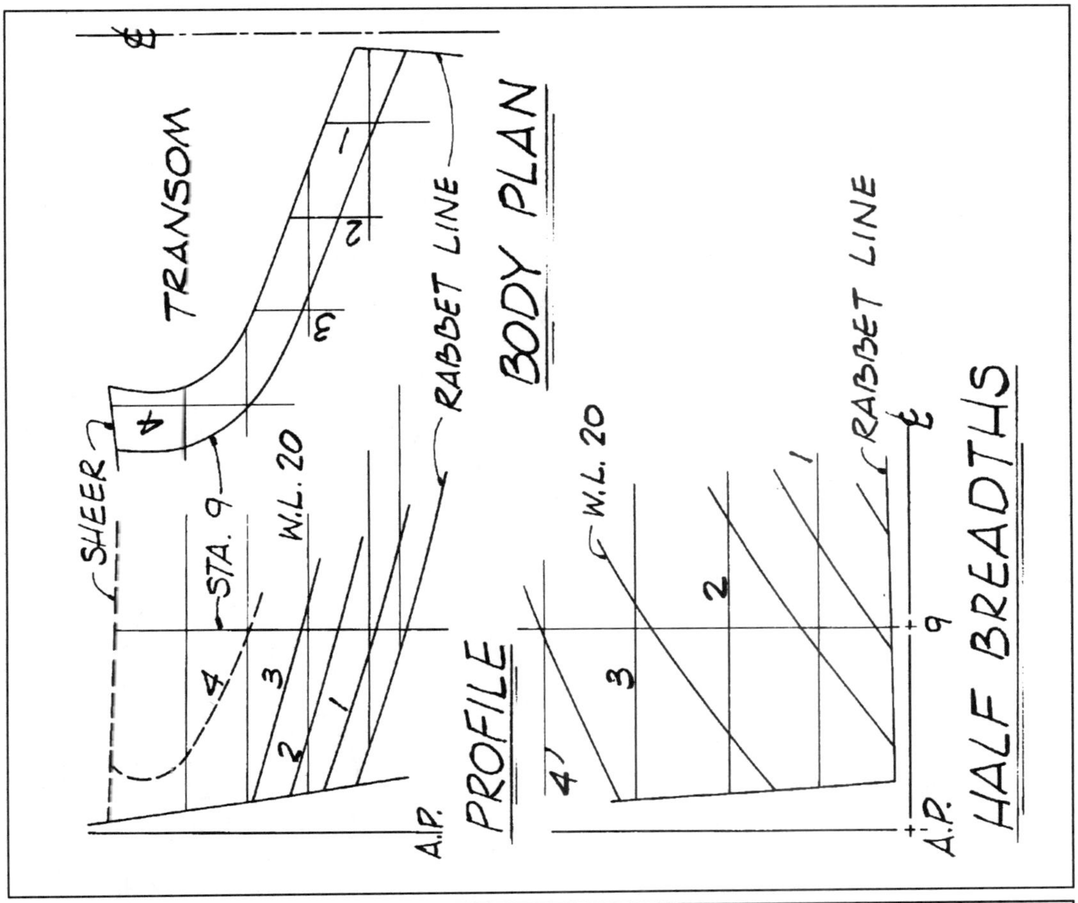

Figure 11.36. Susie—lines at the stern showing the run of the buttocks.

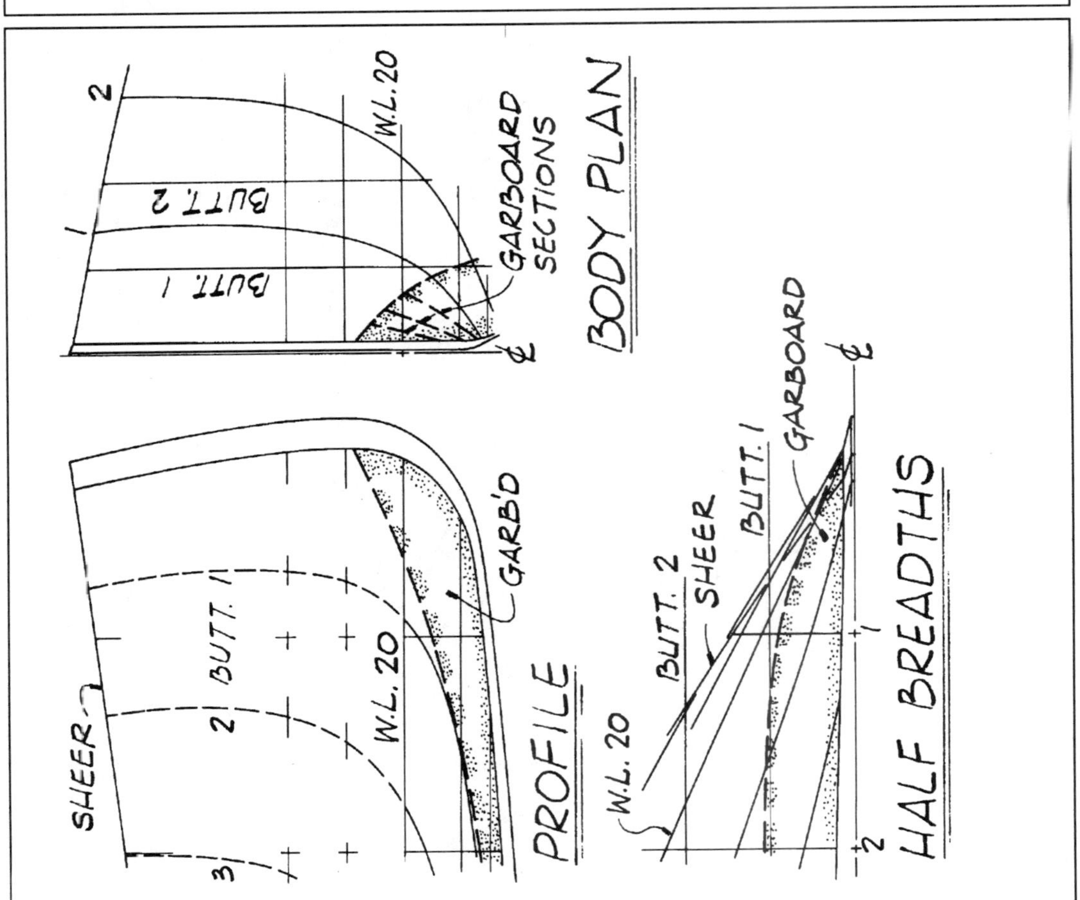

Figure 11.37. Spy—lines at the bow.

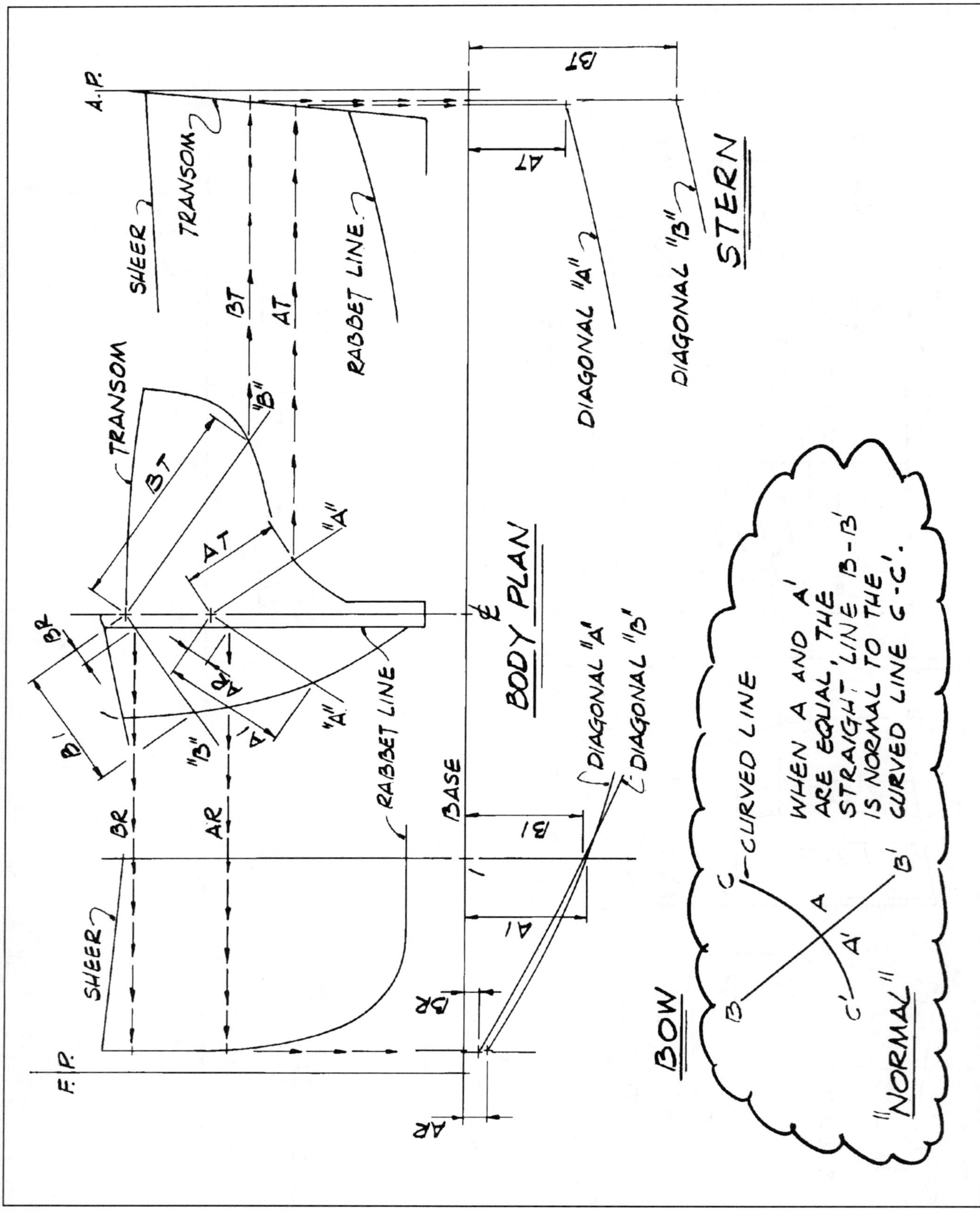

Figure 11.38. Diagonals—determining the endings.

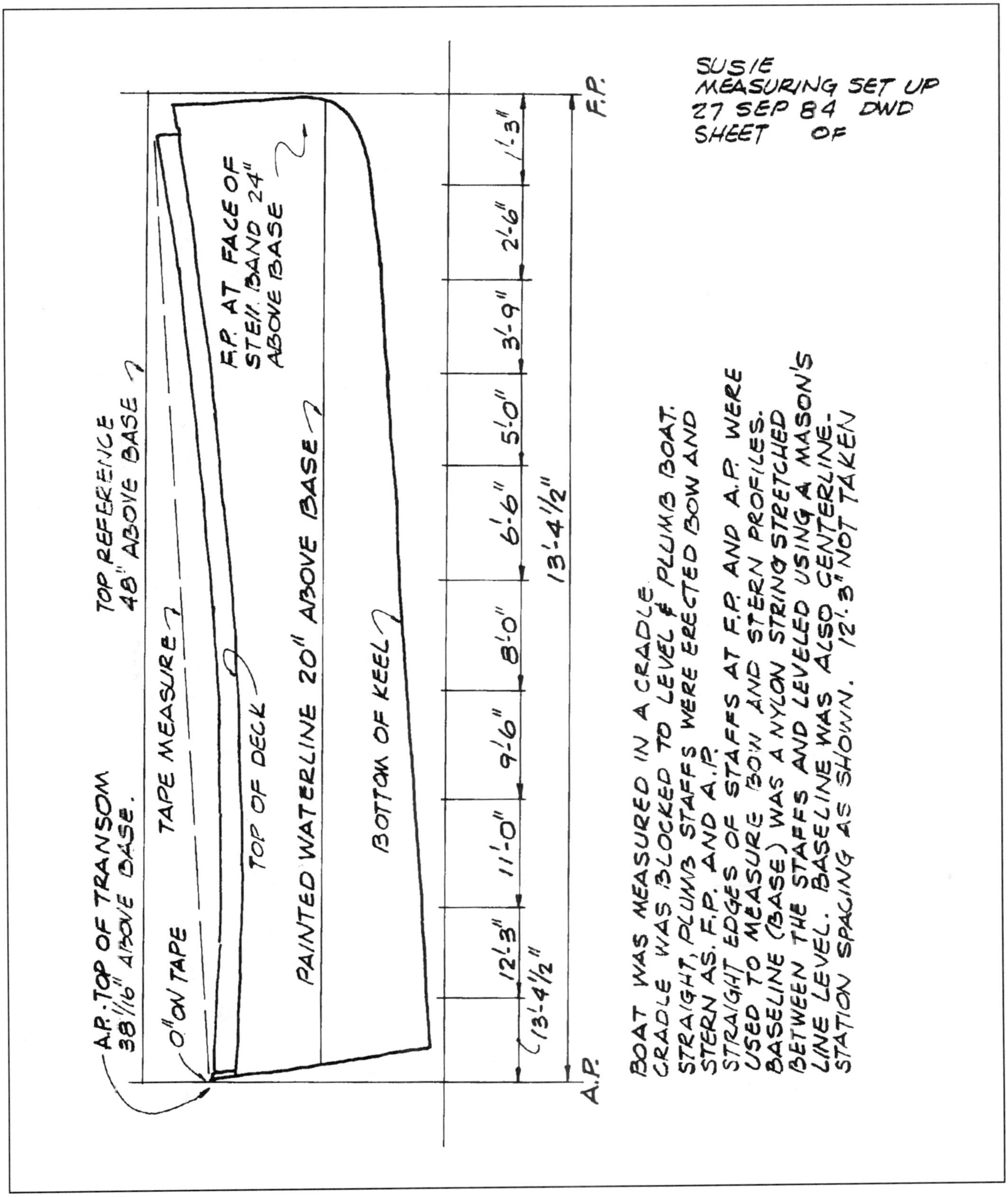

Figure 11.39. Susie—the field notes showing the measuring set-up.

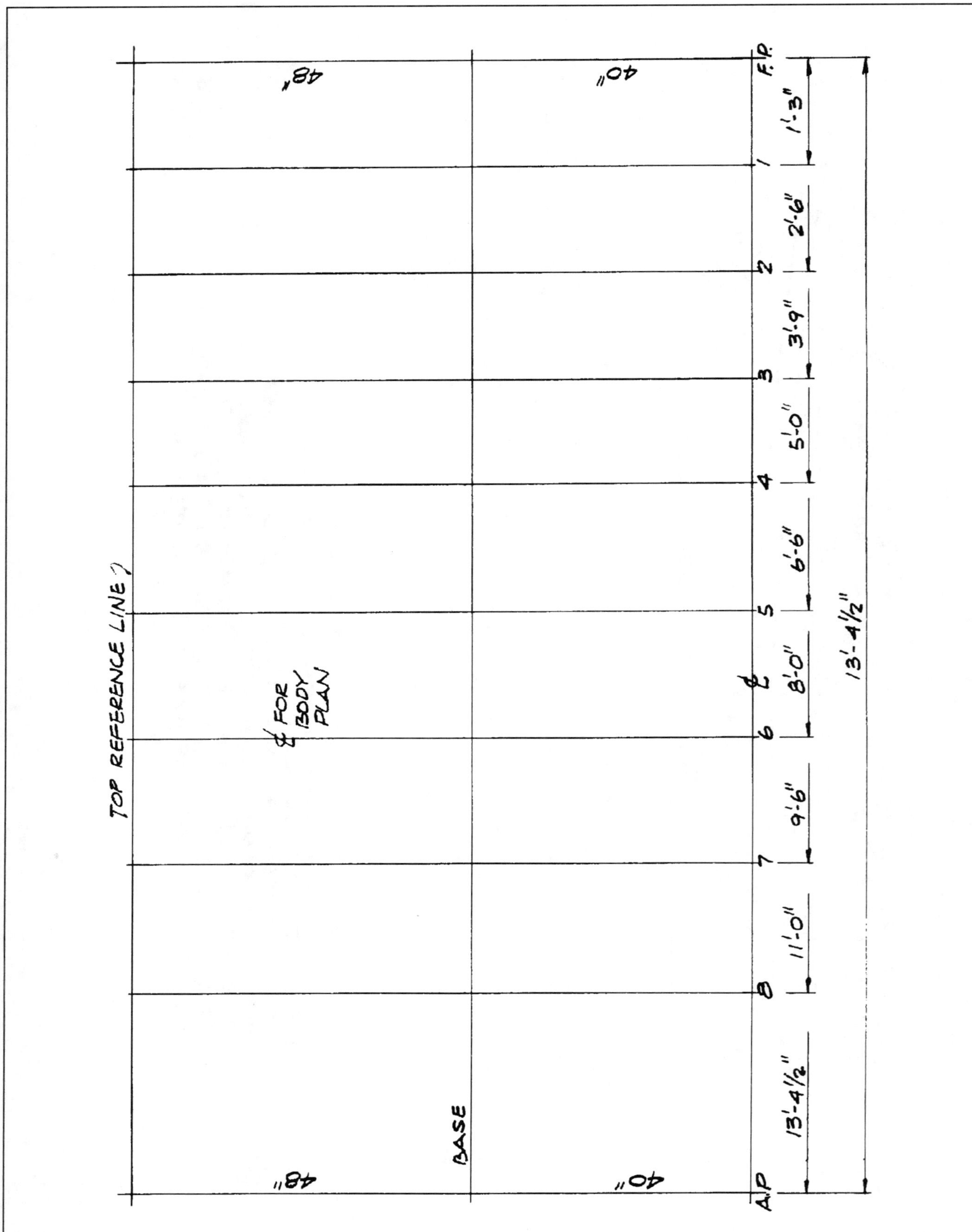

Figure 11.40. Susie—plotting the field notes—the grid.

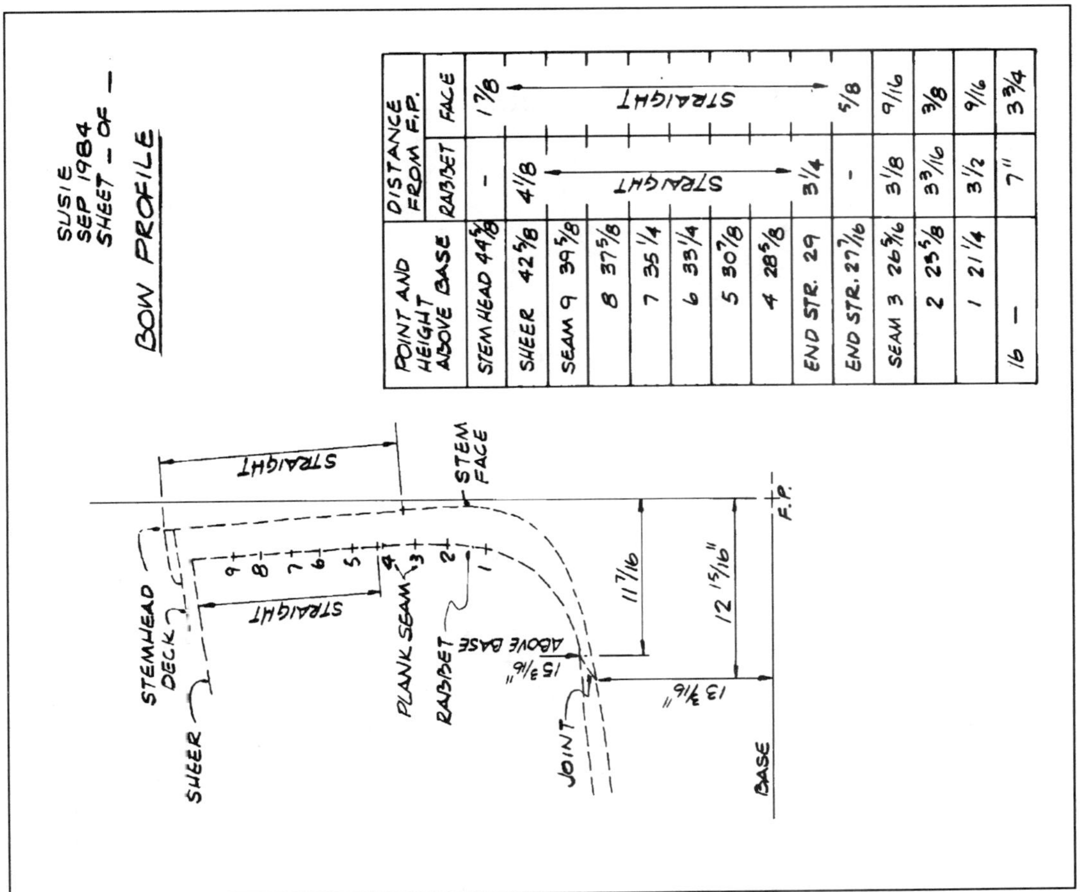

Figure 11.41. Susie—the field notes for bow profile.

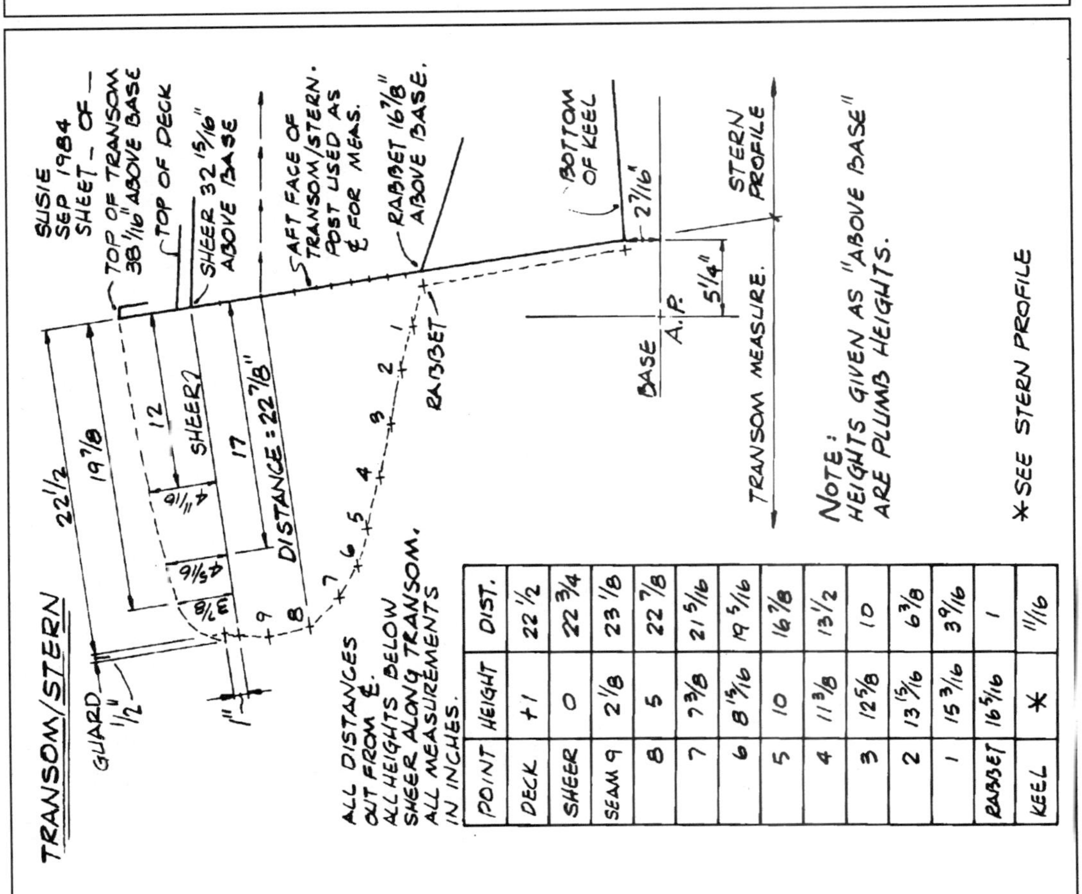

Figure 11.42. Susie—the field notes for stern-transom measurements.

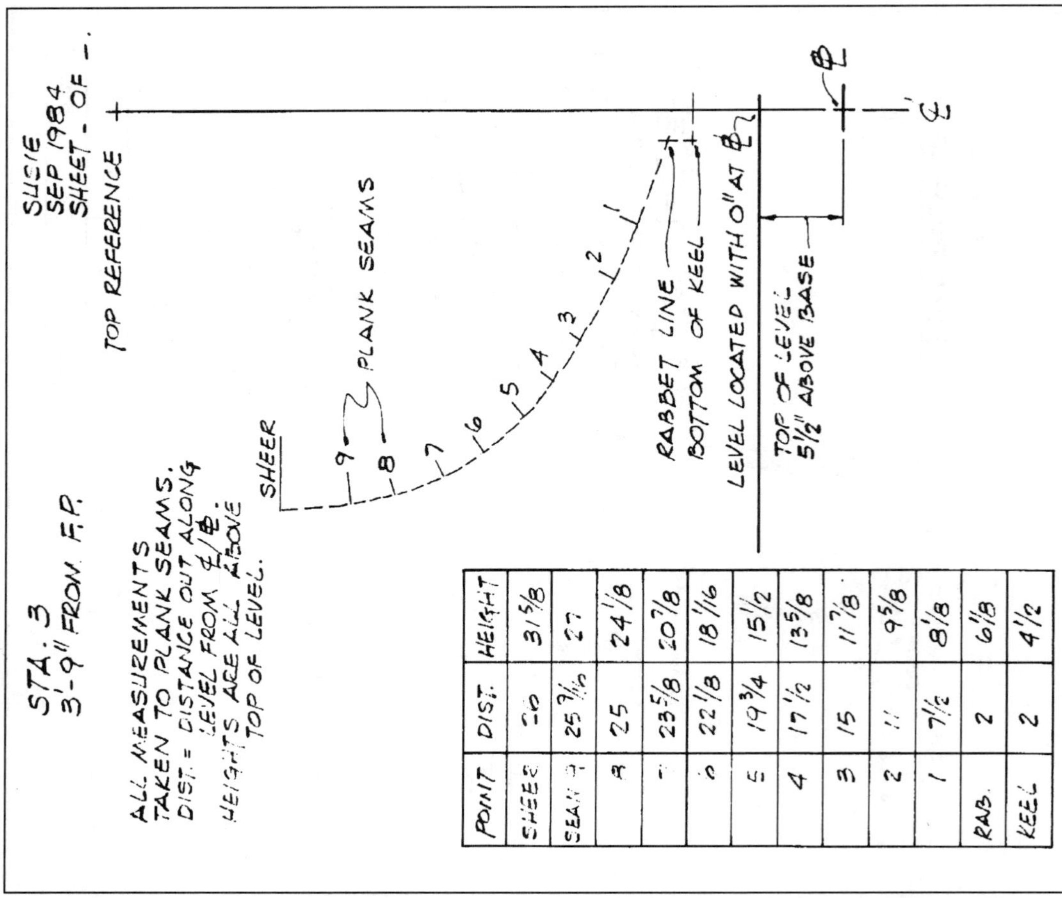

Figure 11.43. Susie—station 3, method 1, level and plumb bob.

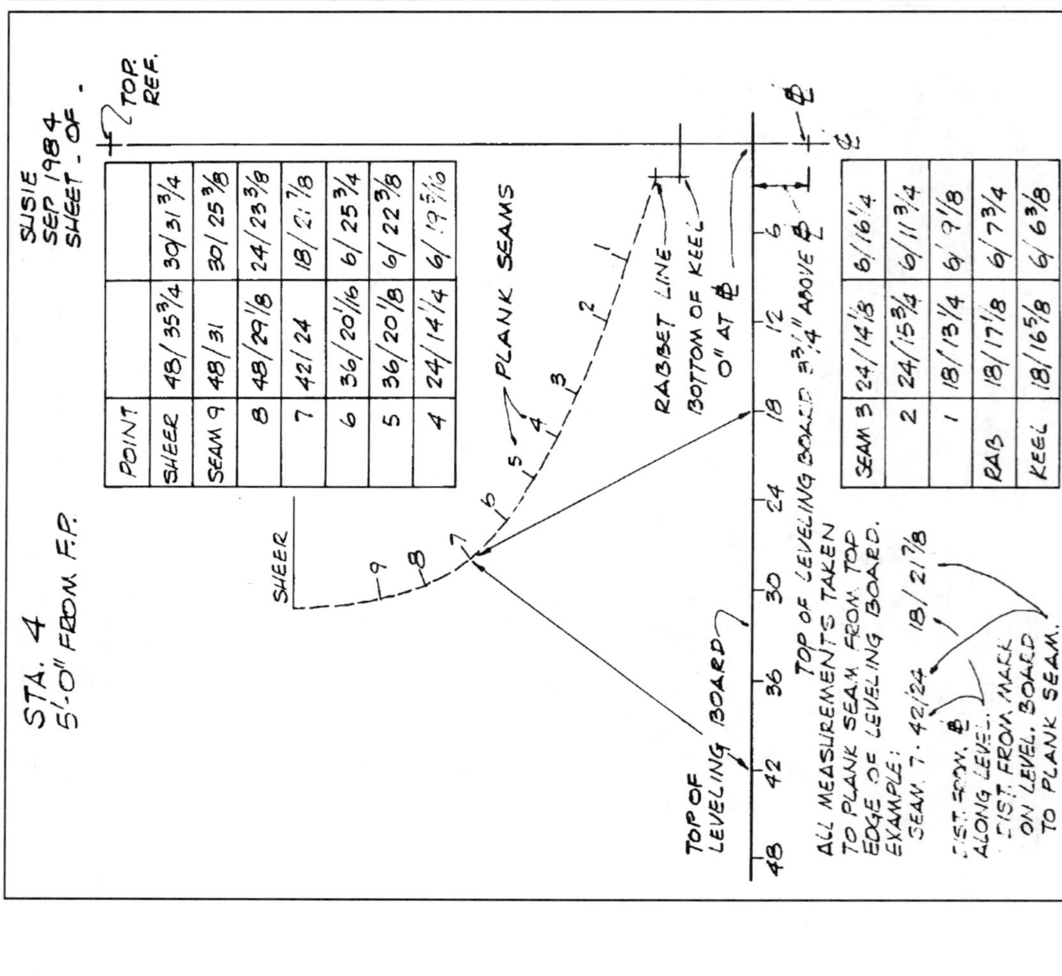

Figure 11.44. Susie—station 4, method 2, leveling board with triangulated measurements.

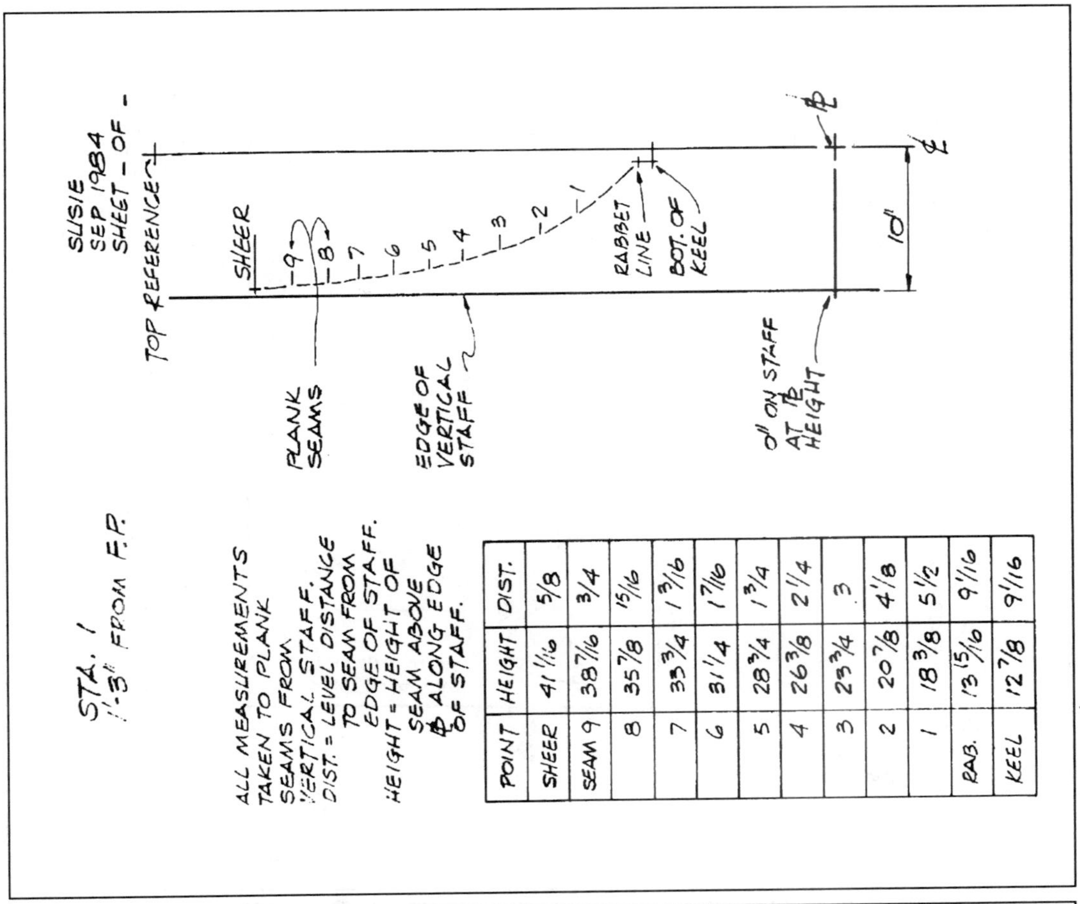

Figure 11.45. Susie—station 1, method 3, vertical staff with level measurements.

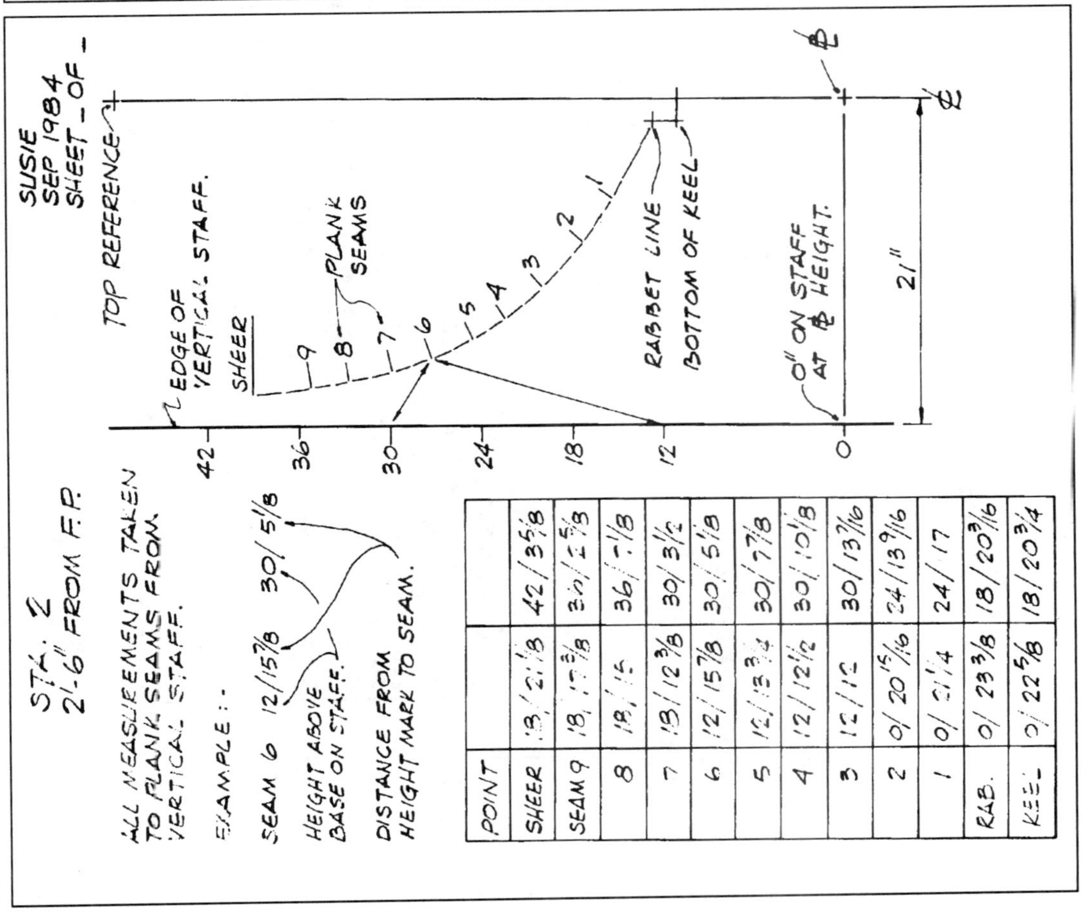

Figure 11.46. Susie—station 2, method 4, vertical staff with triangulation.

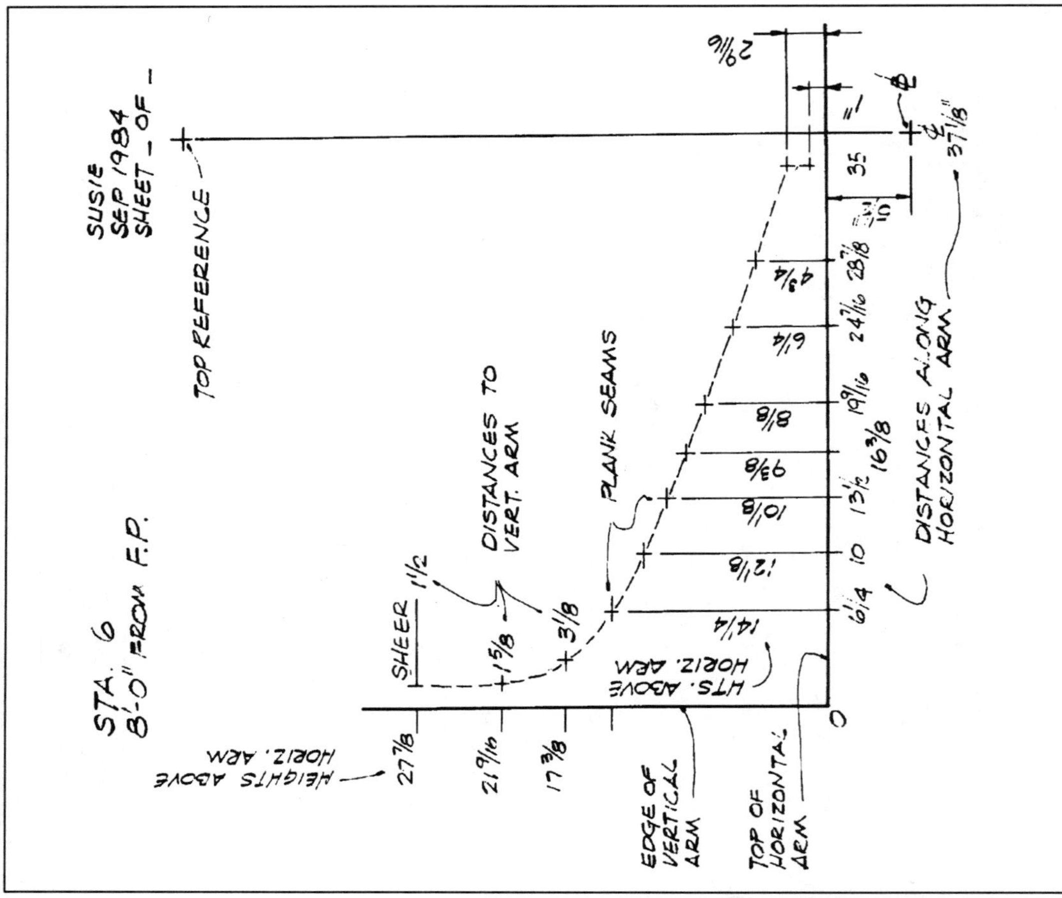

Figure 11.47. Susie—station 6, method 5, vertical staff with horizontal arm.

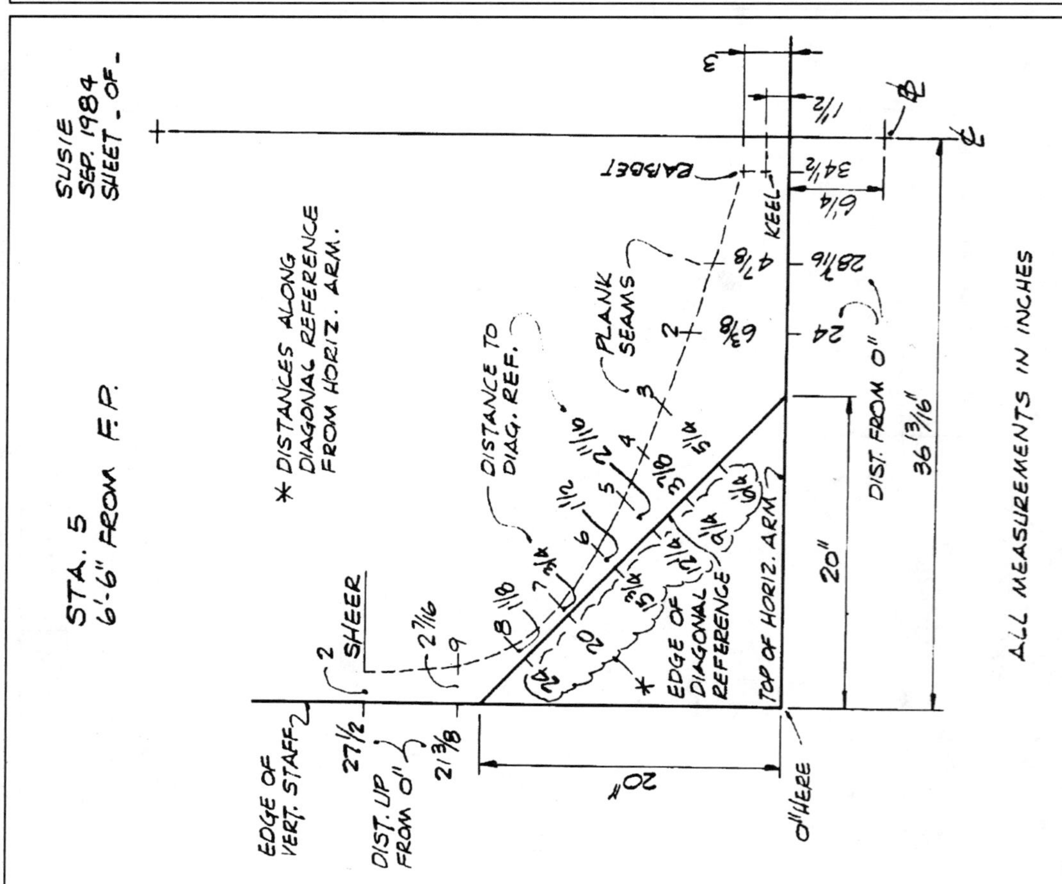

Figure 11.48. Susie—station 5, method 6, vertical staff, horizontal arm and diagonal.

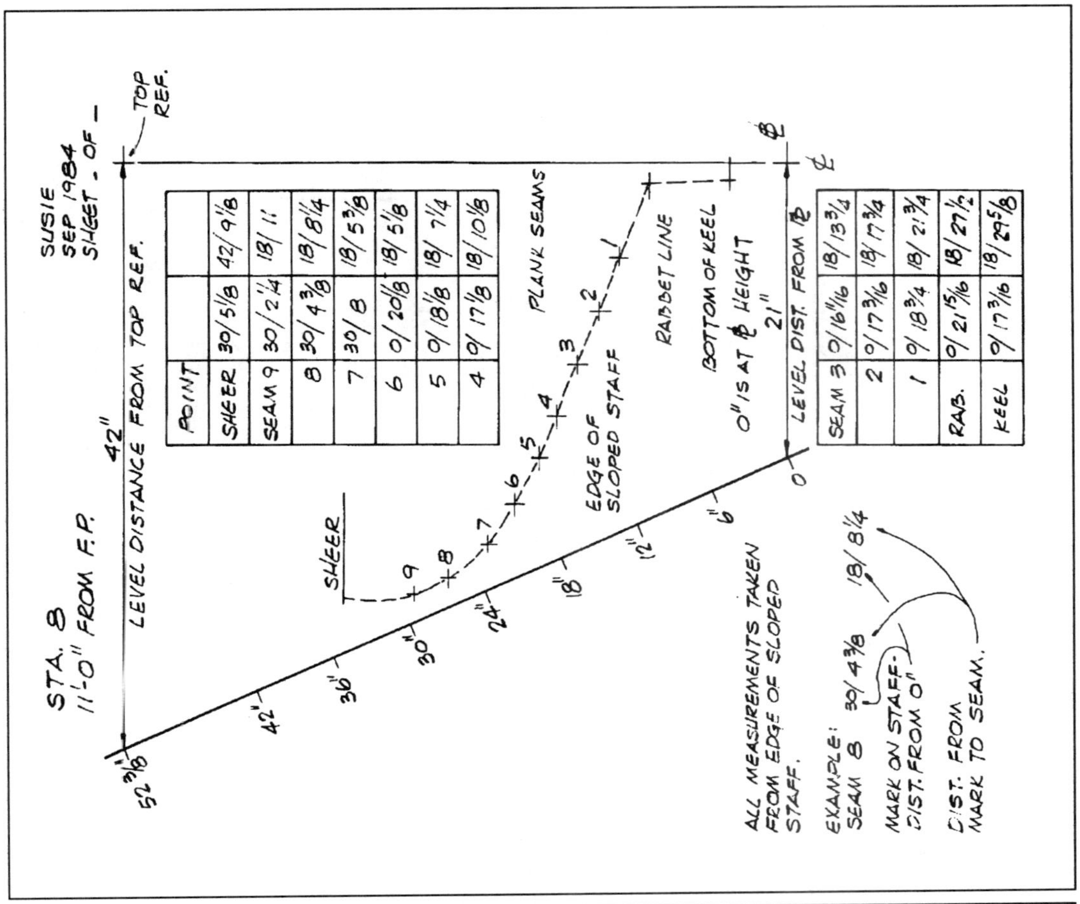

Figure 11.49. Susie—station 8, method 7, sloped staff with triangulated measurements.

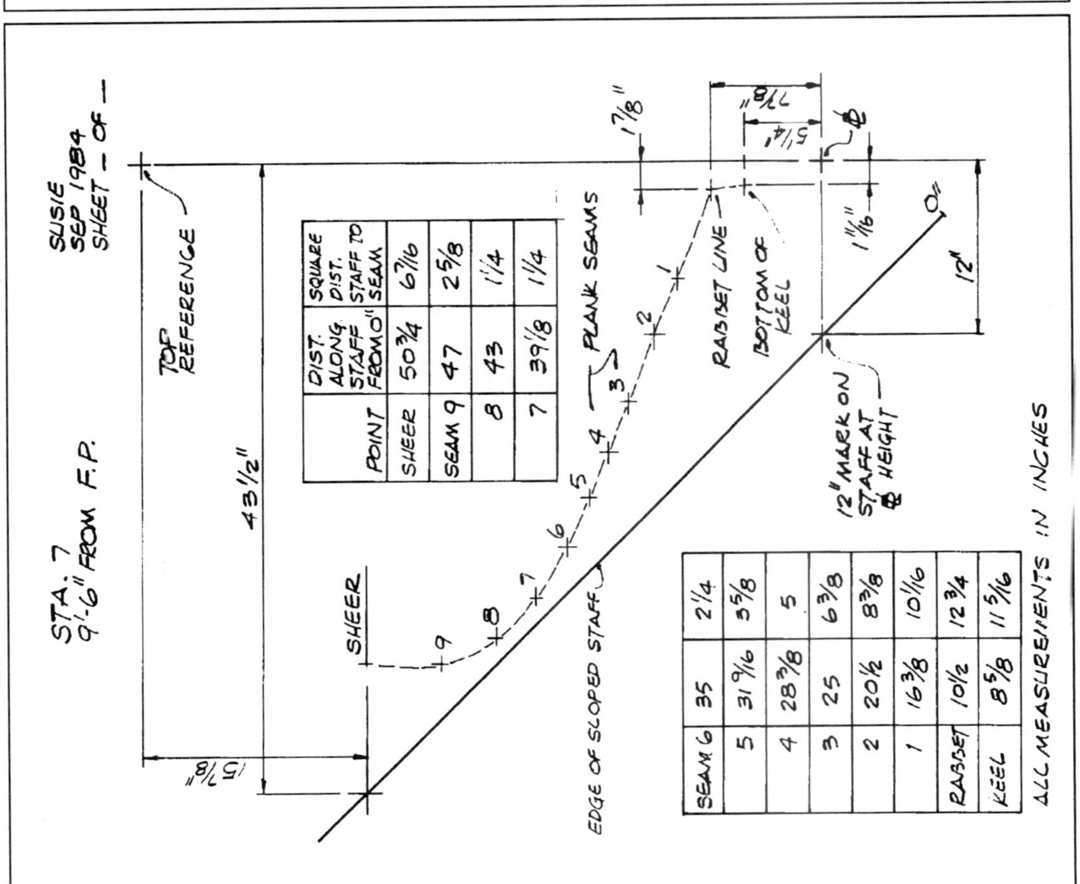

Figure 11.50. Susie—station 7, method 7, sloped staff with square measurements.

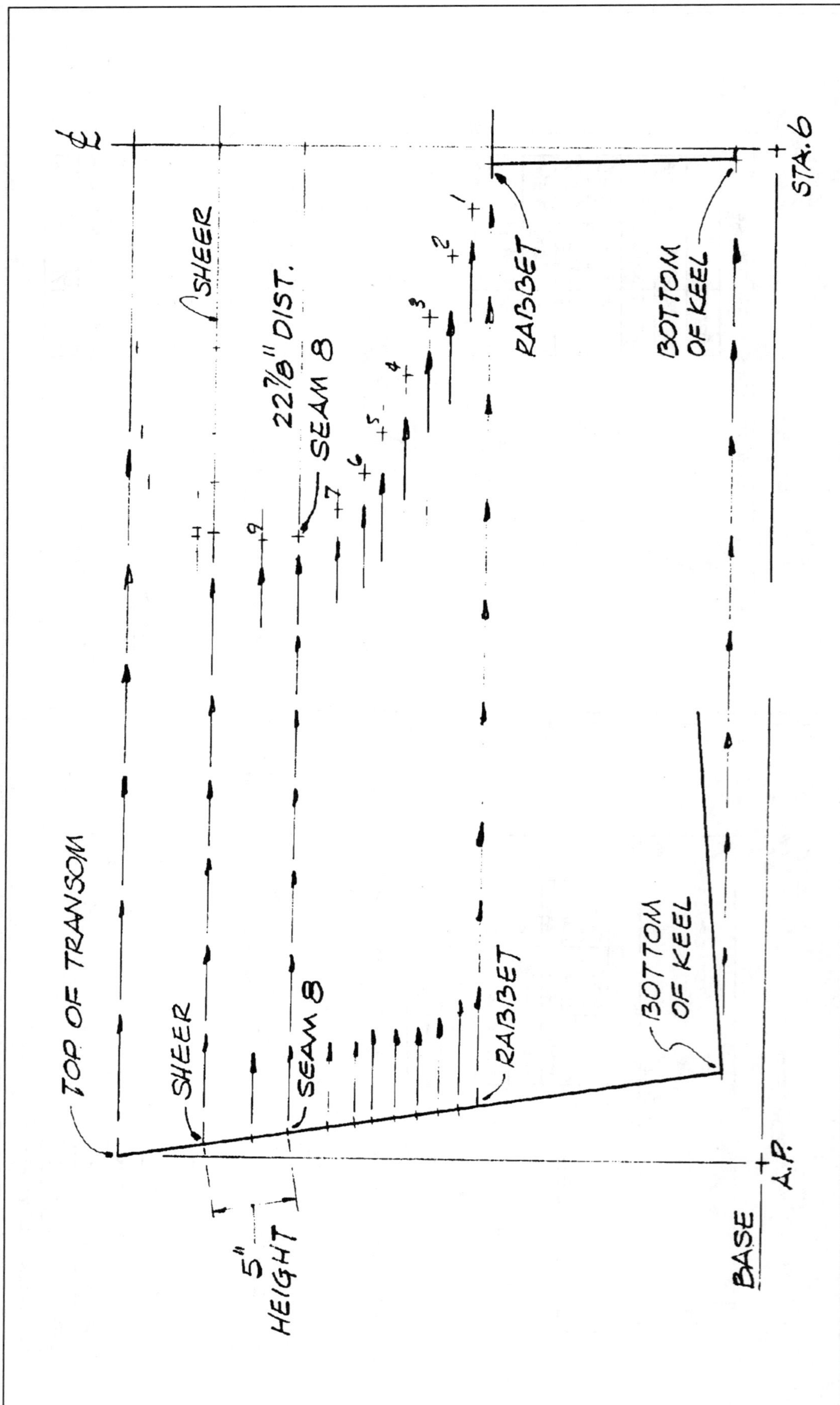

Figure 11.51. Susie—projecting the transom to the body plan.

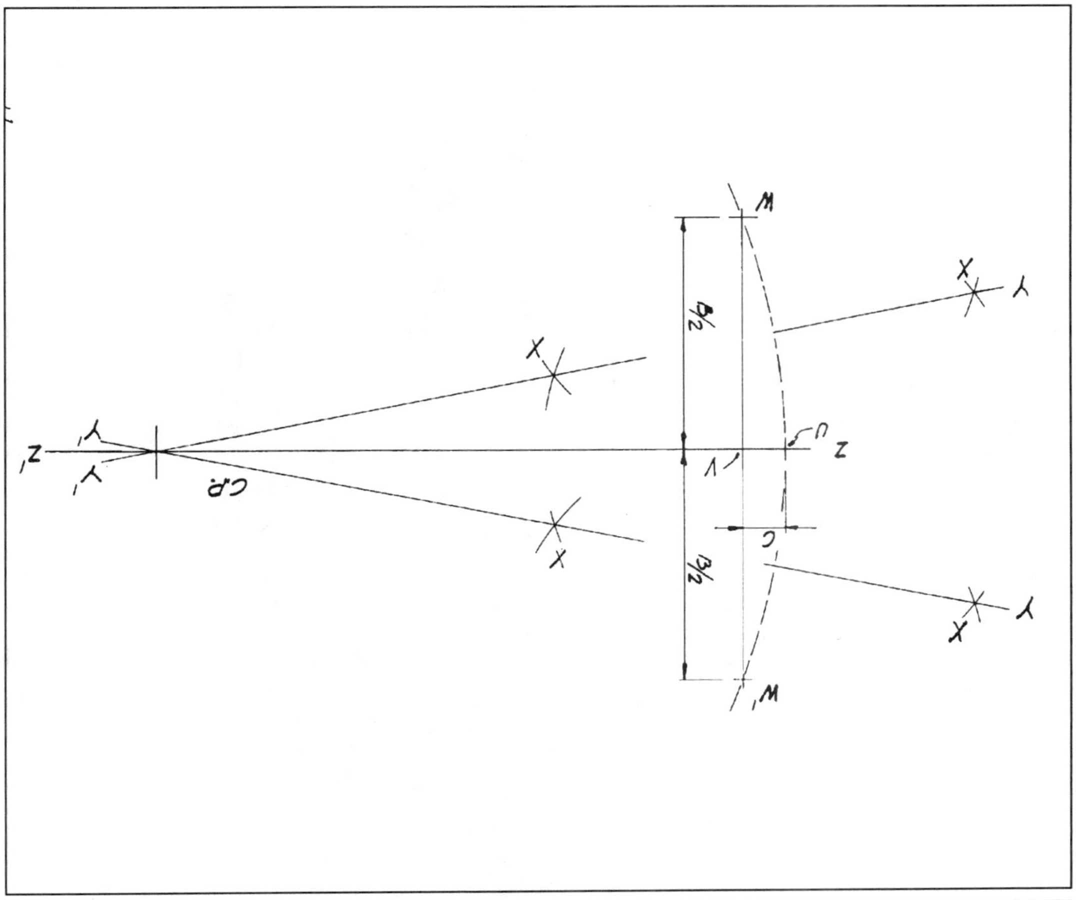

Figure 11.52. The curved, raked transom: the geometric solution for radius of the arc of the transom's face.

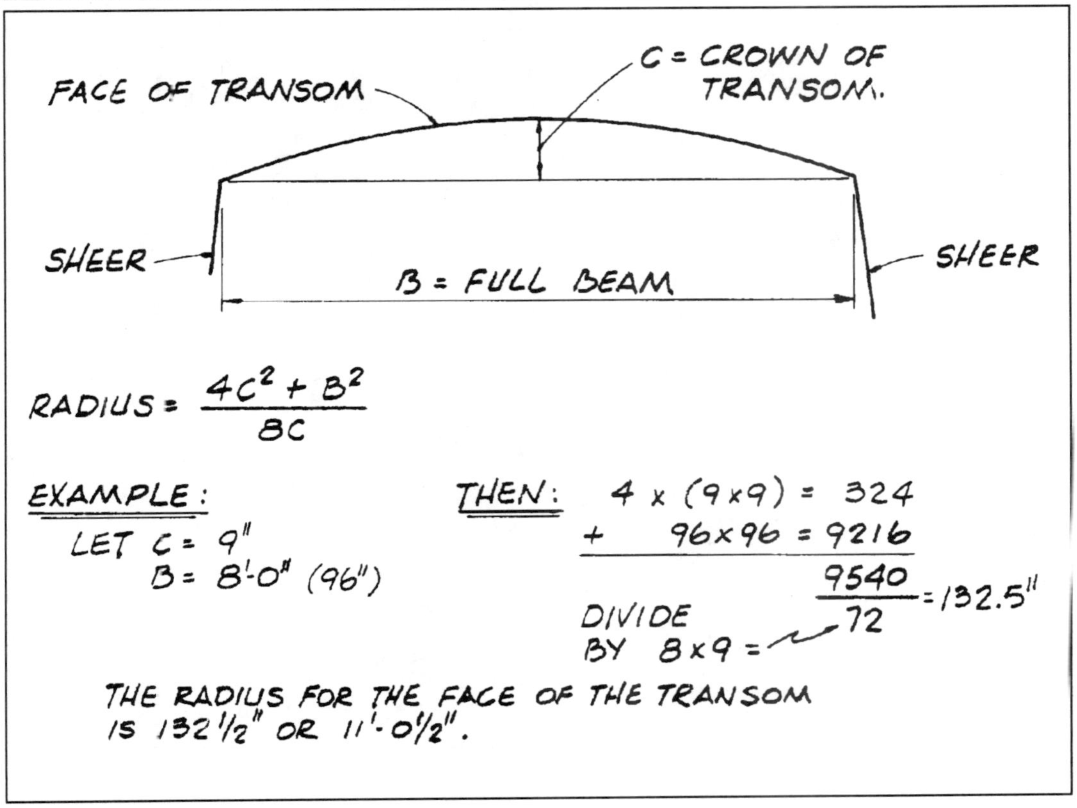

Figure 11.53. The curved, raked transom: the algebraic formula for the radius of the arc of the transom's face.

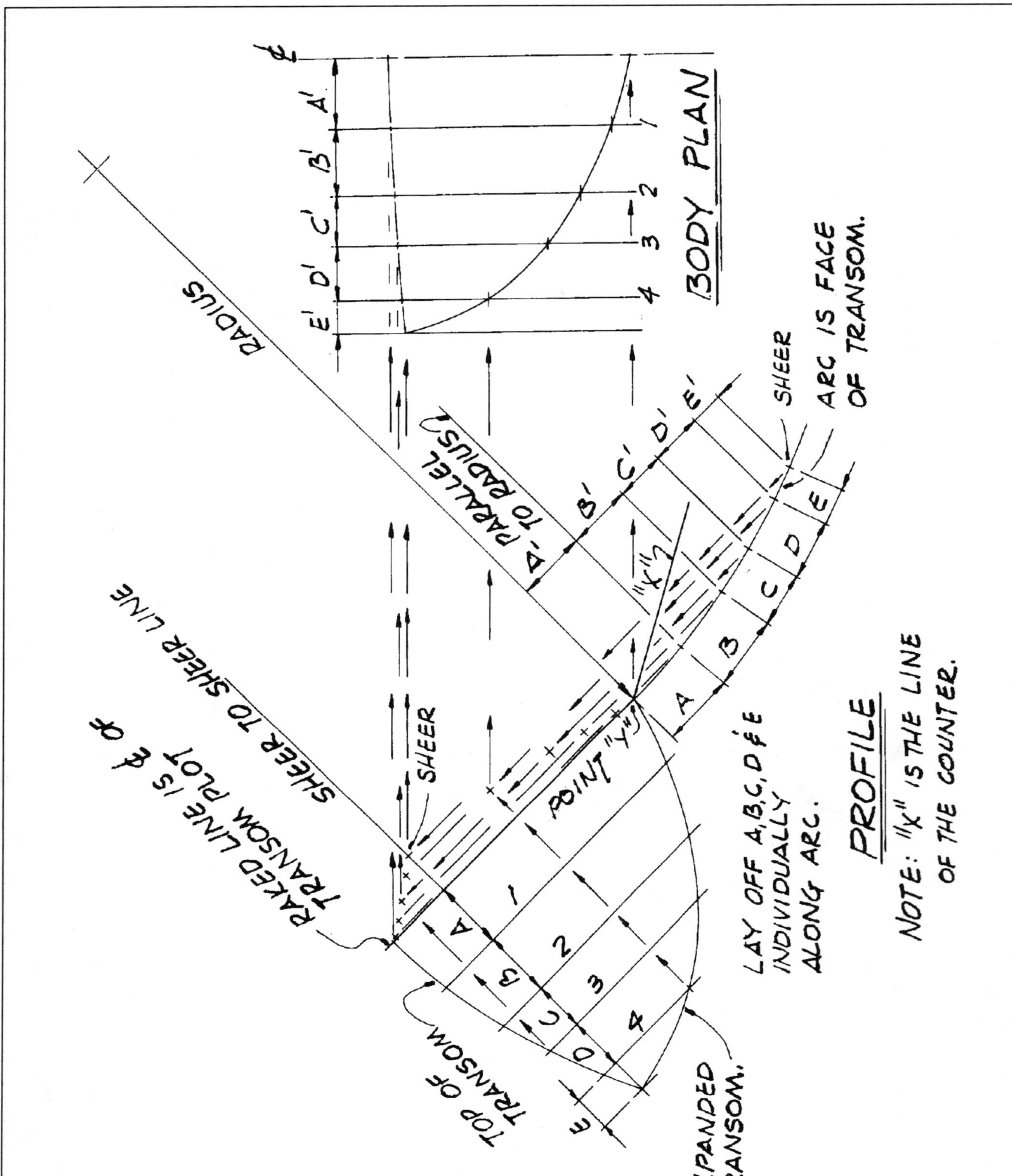

Figure 11.54. The curved, raked transom: converting the field notes to the body plan shape.

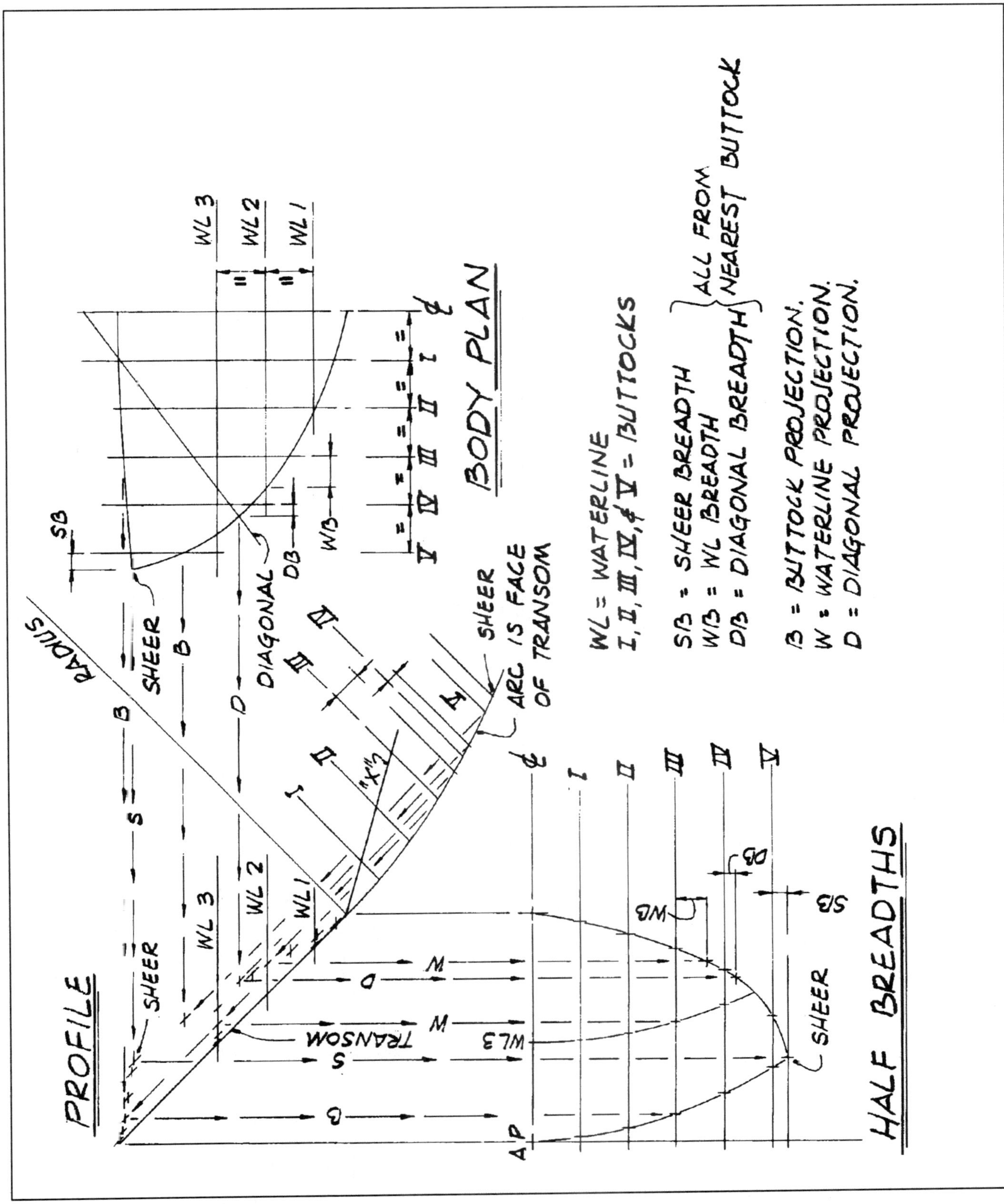

Figure 11.55. The curved, raked transom: developing the transom in the profile and half breadth plan.

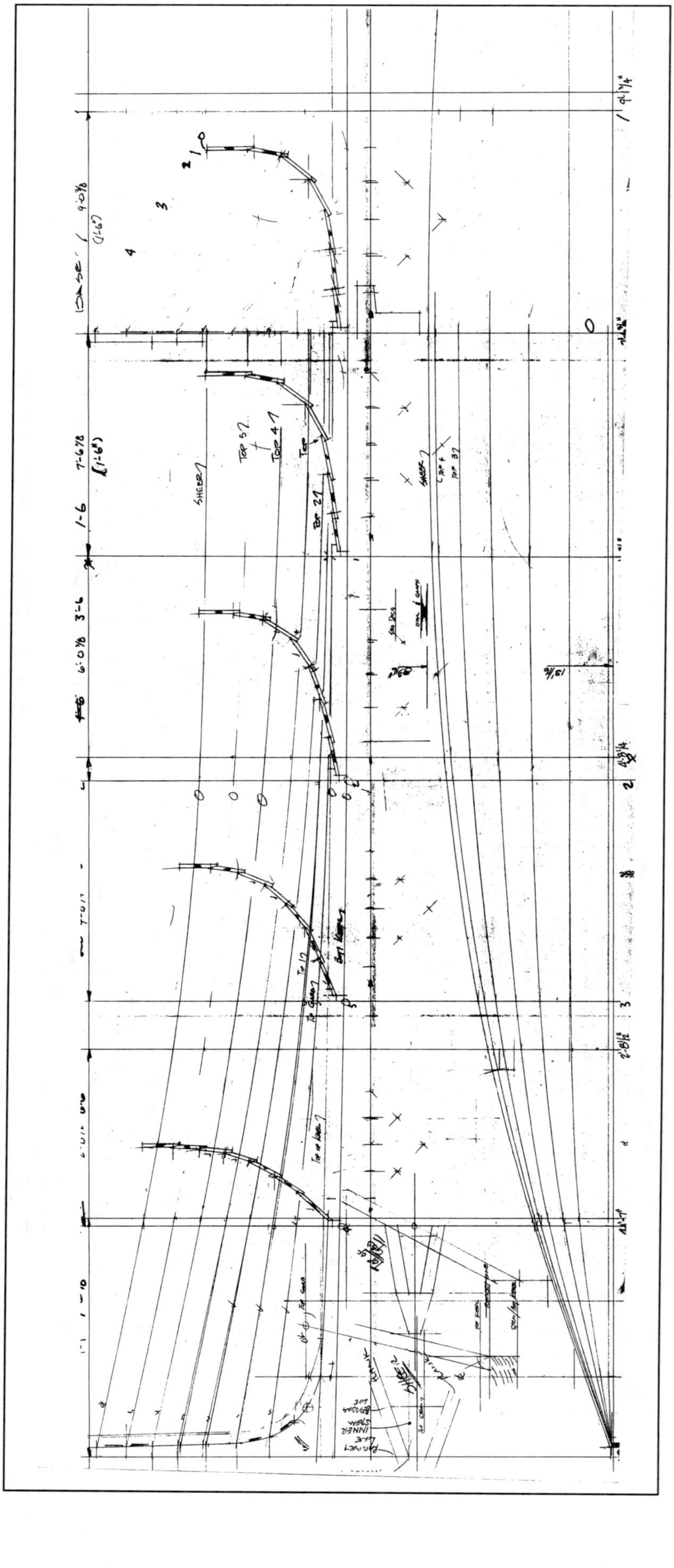

Figure 11.56. The rough lines plan for a lapstrake planked boat.

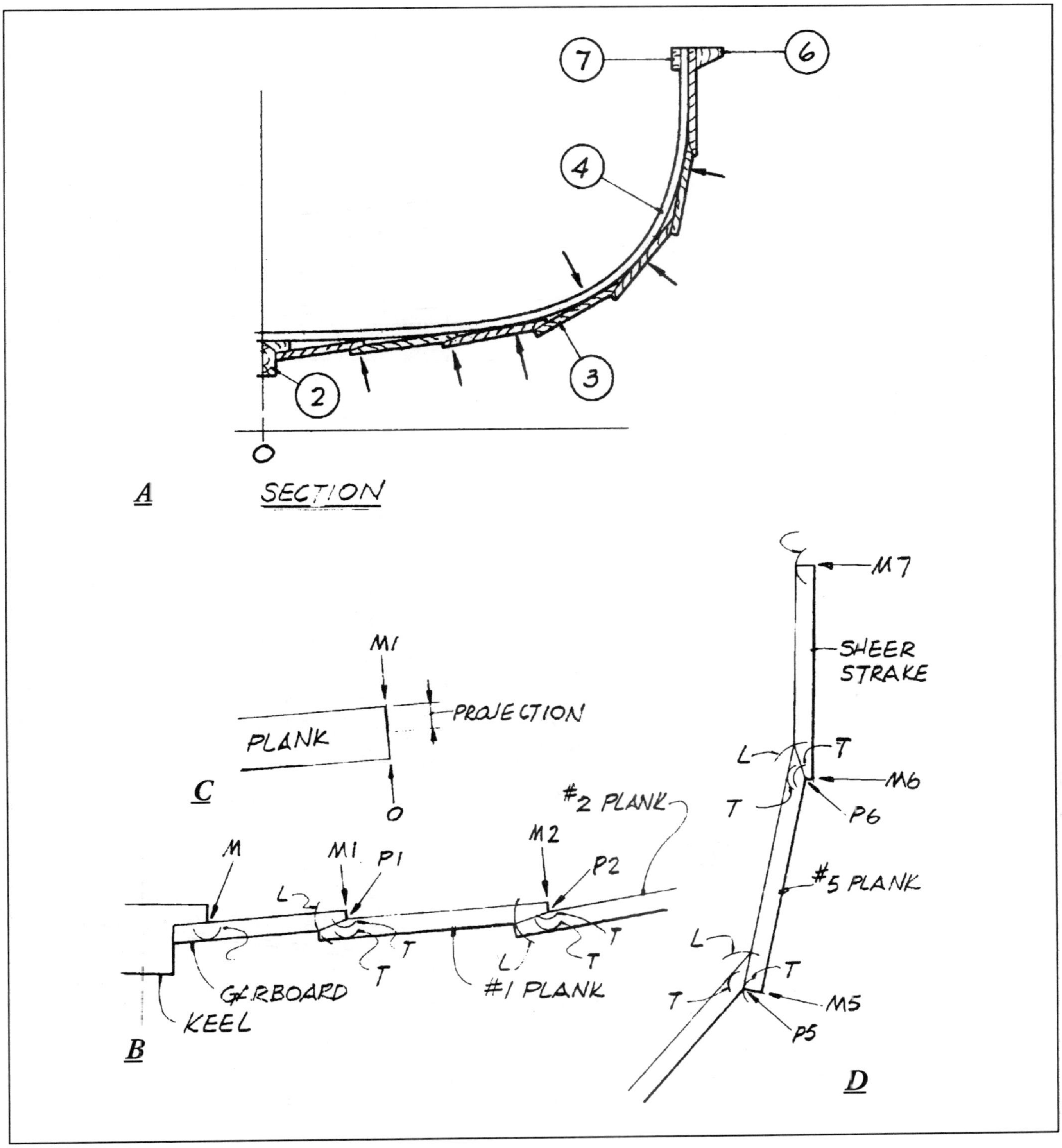

Figure 11.57. Developing the hull sections of a lapstrake planked boat. (A) is the section as it is shown on the finished construction plans. The arrows indicate the locations of the plank to frame fastenings. The principle controlling the development of the planks in a hull section is that demonstrated in fig. 11.10. From measurements taken to the inside of plank proceed as in (B) plotting the location of the points measured on the planks and indicated as M and M1, and M2. Draw the line M-M1 for the inside of the garboard. Using points M and M1 as centers, swing the arcs T equal to the plank thickness. A line drawn touching the arcs is the outside of the garboard. At M1 draw the line M1-0 perpendicular to M-M1, see C. Mark off the projection (shown in C.), and fig. 7.24,) along M1-0 to establish P1. Using P1 as a center swing the arc L equal to the lap width. Where arc L intersects the outside of the garboard is the edge of the #1 plank. The #1 plank is laid out in similar fashion. A line drawn from P1 to M2 is the inside of the #1 plank. Swing the arcs T using P1 and M2 as centers. Draw a line touching the lines for the outside of the plank. Using P2 as a center, swing the arc L to locate the edge of the #2 plank. Drawing the remaining planks up to the sheer is a repetition of the process.

If you measured the outside of the plank as in D, the process is identical except that you will work from the outside in b plotting the measured points.

At M7 and M6 swing the arcs T. Draw a line touching the arcs for the inside of the sheer strake. At M6 draw the edge of the sheer strake square to the line M7-M6 and mark off the projection, point P6. At P6 and M5 swing the arcs T and draw a line touching the arcs for the inside of the #5 plank. At P6 swing arc L to establish the edge of the #5 plank. Repeat the process until you reach the garboard.

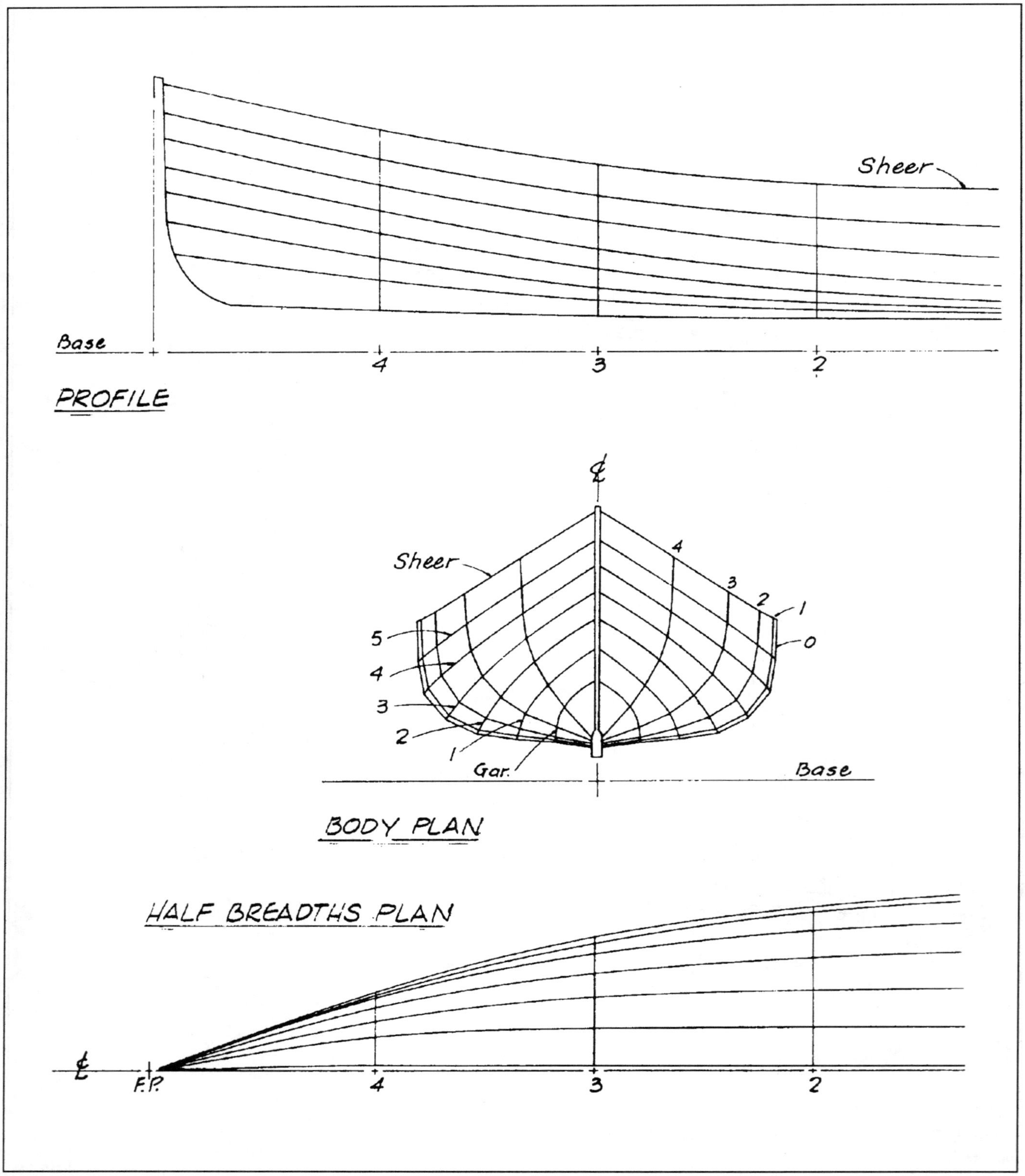

Figure 11.58. Components of the finished lines plan.

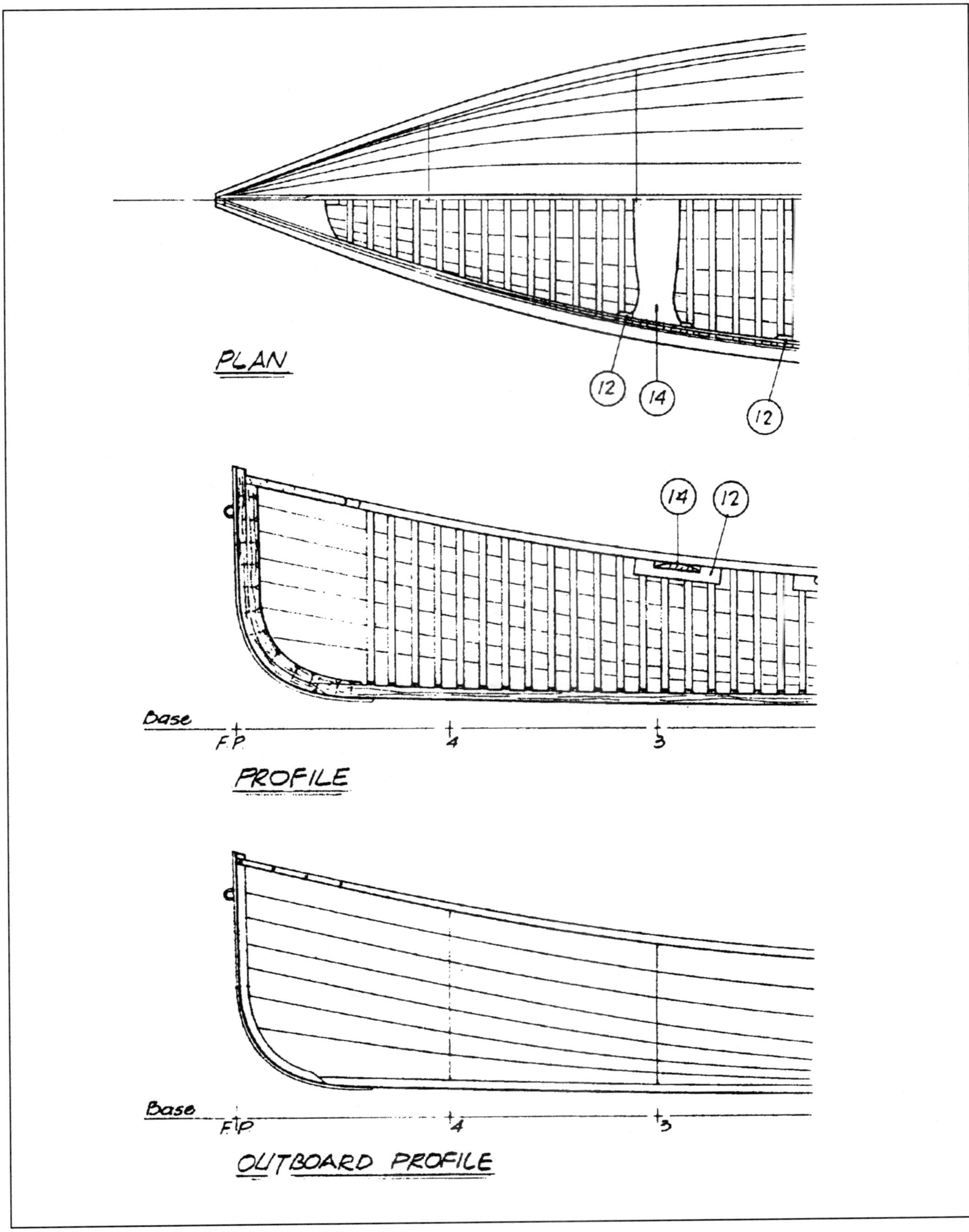

Figure 11.59. The lapstrake planked boat, plan, profile, and outboard profile.

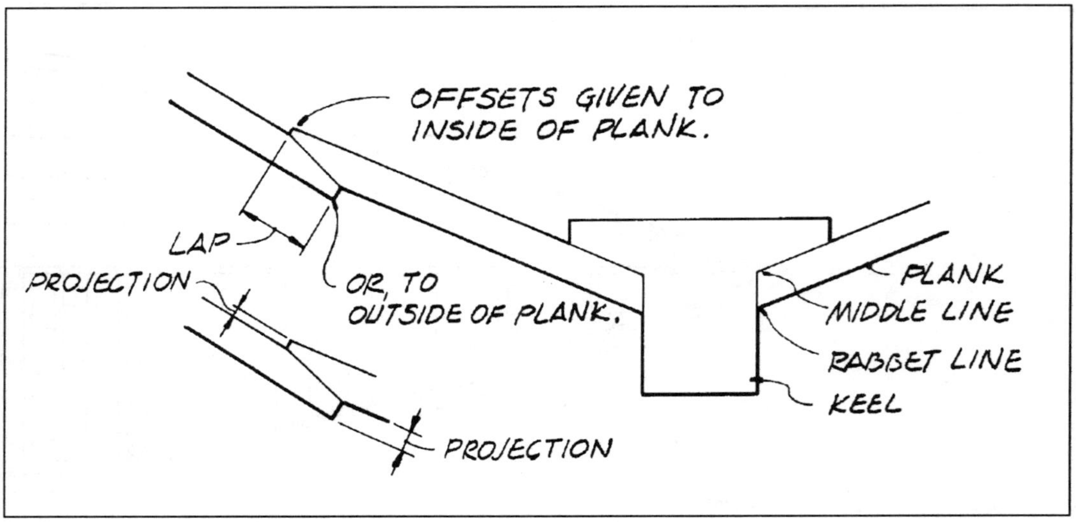

Figure 11.60. Indicating the point that the offsets are given to, the lap width and the projection.

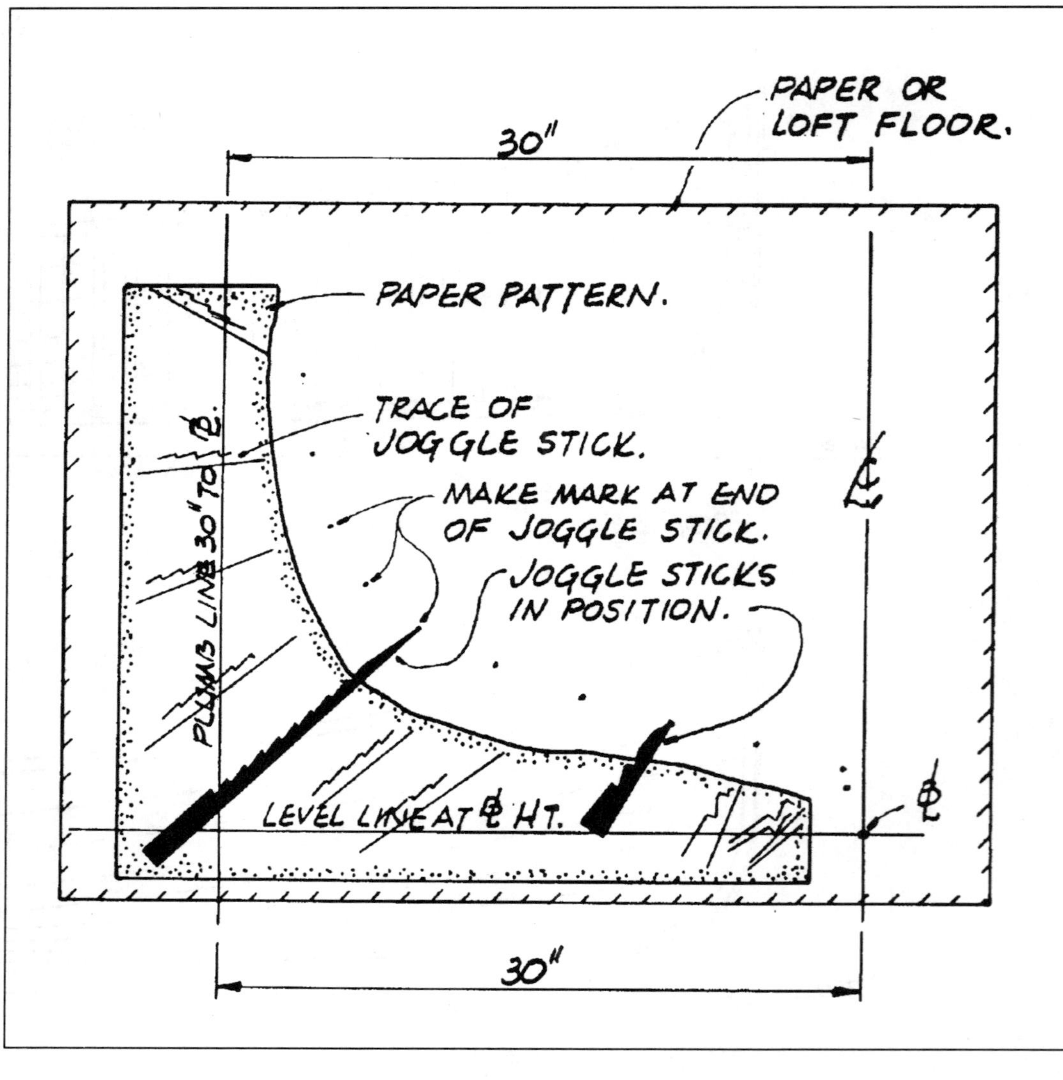

Figure 11.61. Plotting sections taken to the outside of plank with a joggle stick. a) On the loft floor use a station line for the centerline and base for the level line. b) On paper, draw the centerline, and the level line square to it. c) On both the loft and the paper, draw a line 30 inches from and parallel to the centerline. d) Lay the pattern down so that the level line and the 30-inch line on the pattern align with those on the loft floor or the plotting paper. e) position the joggle stick on the traces and make marks at the ends. f) Fair a line through the marks for the section of a carvel planked boat or develop the planks of a lapstrake planked boat.

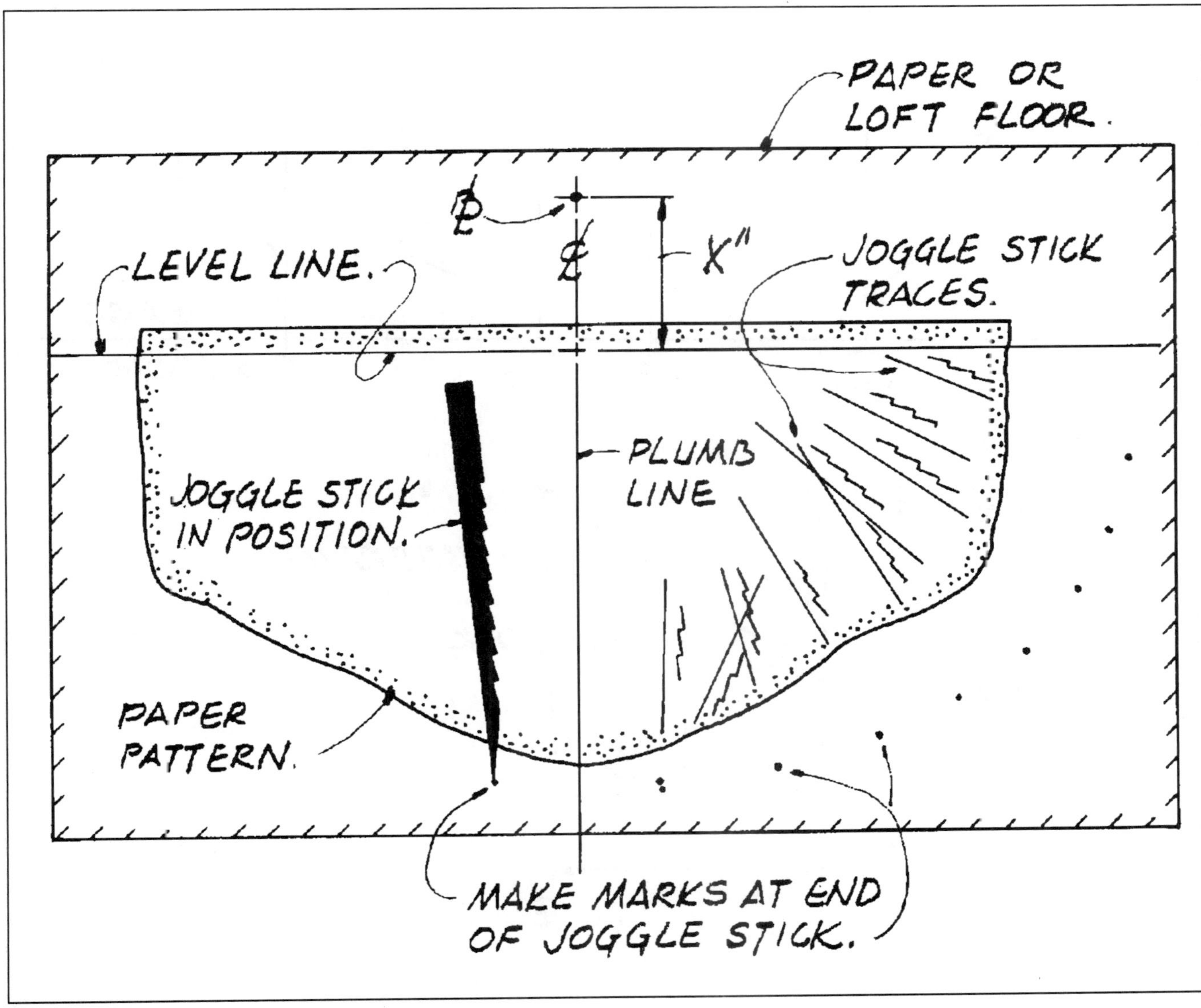

Figure 11.62. Plotting sections taken to the inside of plank with a joggle stick. a) On the loft floor use a station line for the centerline. b) On paper, draw the centerline and mark the baseline on it. c) On both, draw the level line at the recorded distance below the baseline (dimension X in fig. 11.62) and square to the centerline. d) Lay the pattern down so that the centerline and the level line on it align with those on the loft floor or the plotting paper. e) Position the joggle stick on the traces and make marks at the ends. f) Fair a line through the marks for the section of a carvel planked boat or develop the planks of a lapstrake planked boat.

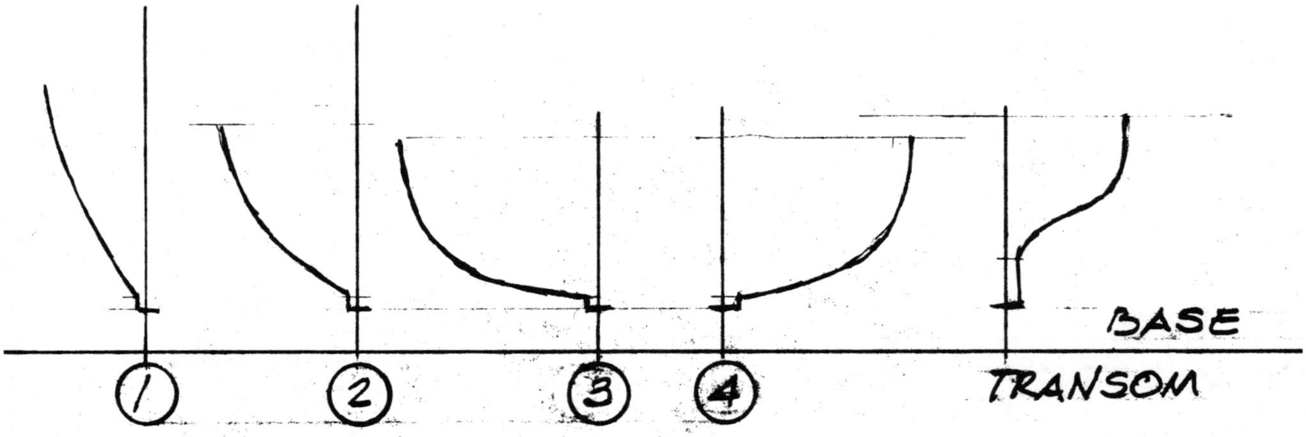

Fig. 11.63

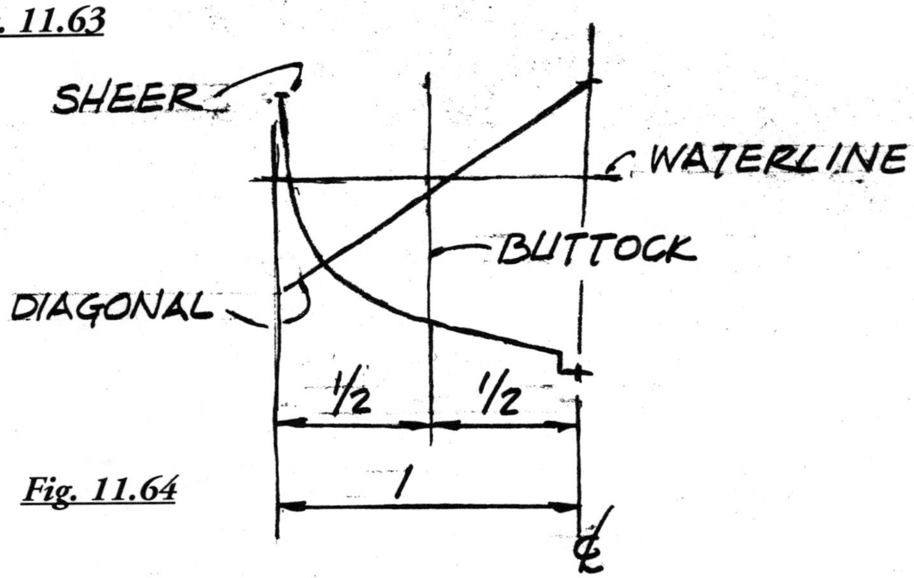

Fig. 11.64

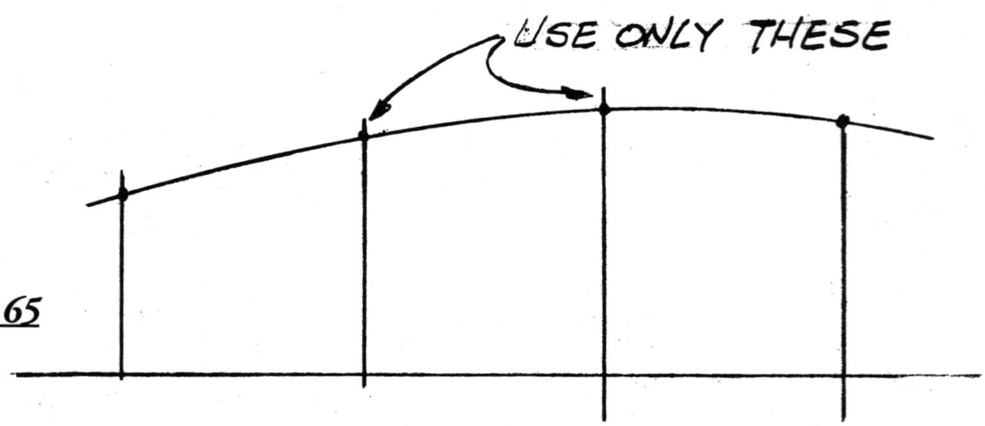

Fig. 11.65

Figure 11.63. The station sections of a twisted boat plotted separately.

Figure 11.64. Waterline, buttock, and diagonal locations.

Figure 11.65. Four points on a line.

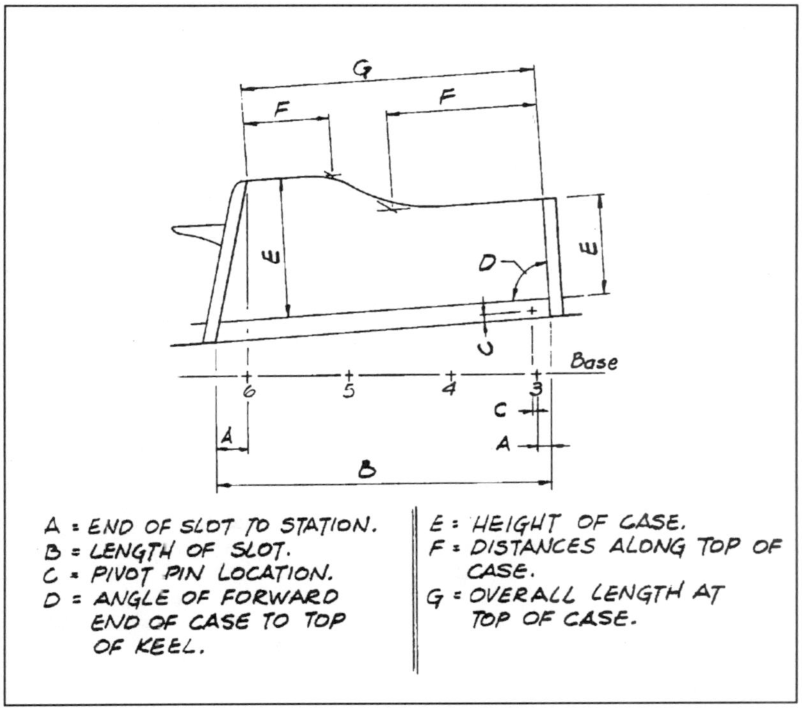

Figure 11.66. Dimensioning the centerboard case.

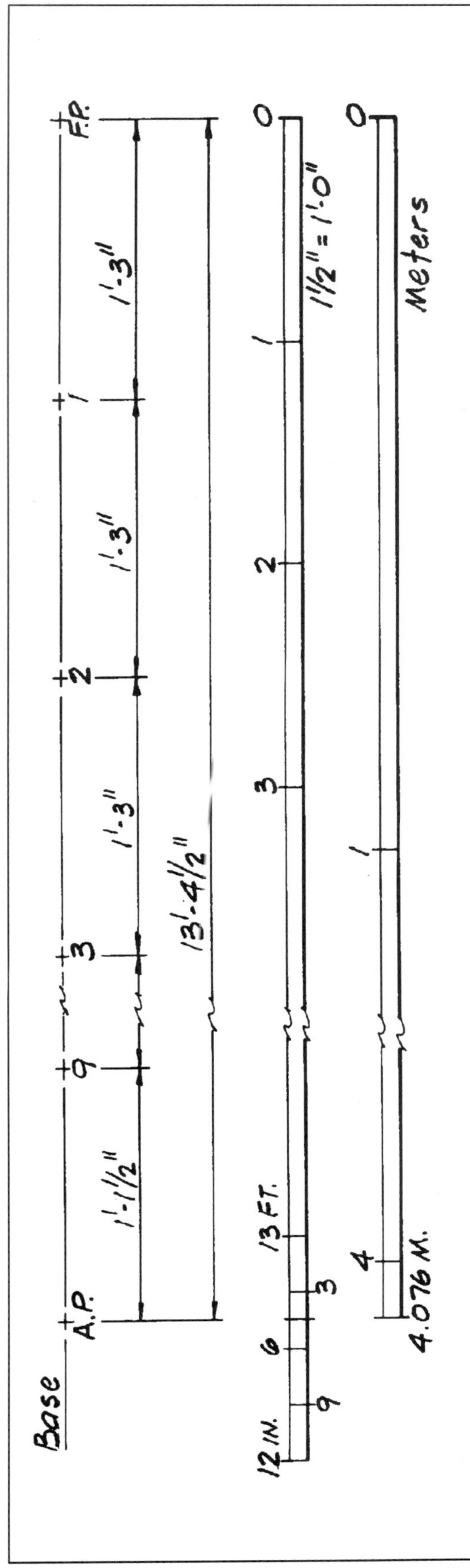

Figure 11.67. The graphic scales.

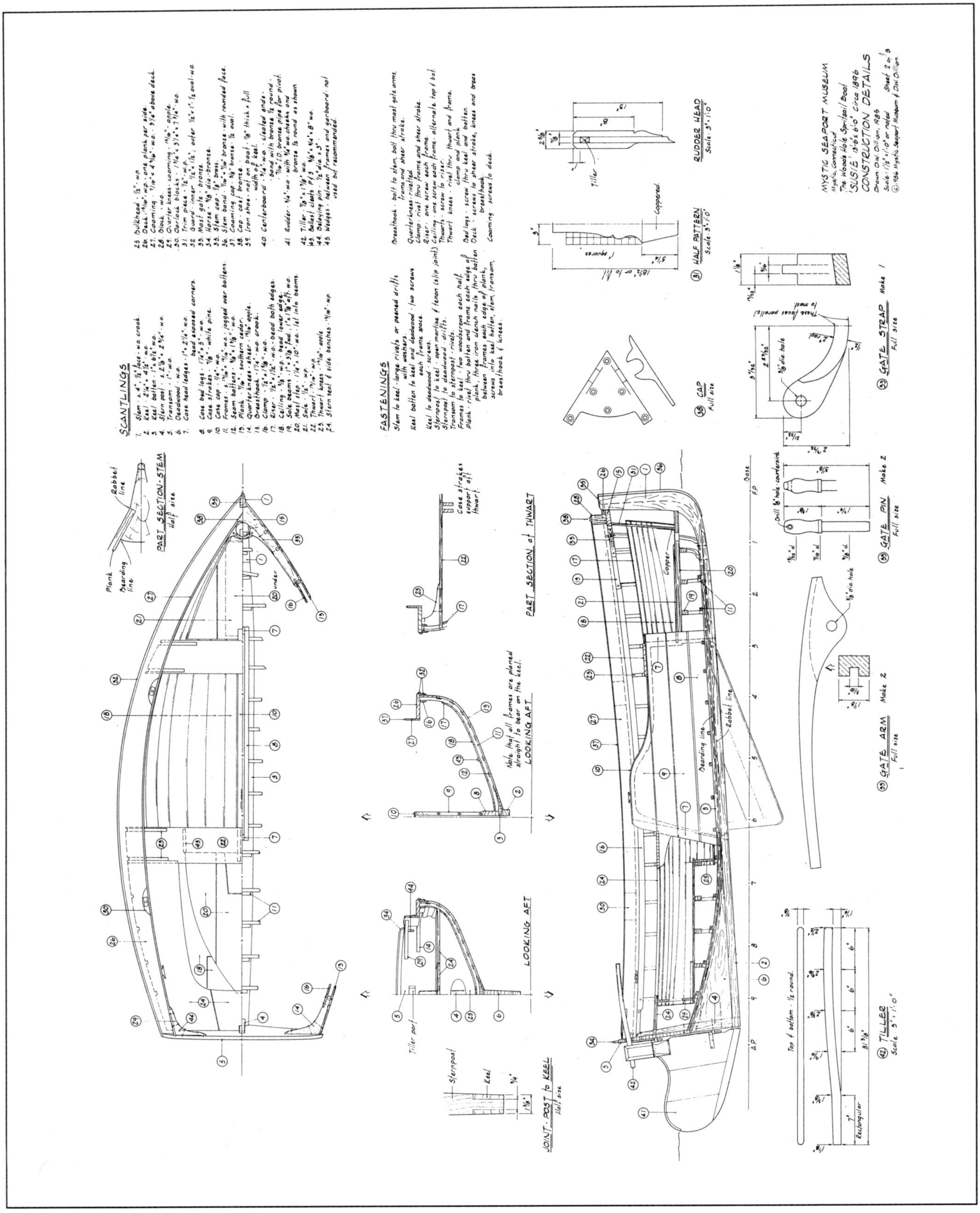

Figure 11.68. Susie's construction plan. *Mystic Seaport Museum.*

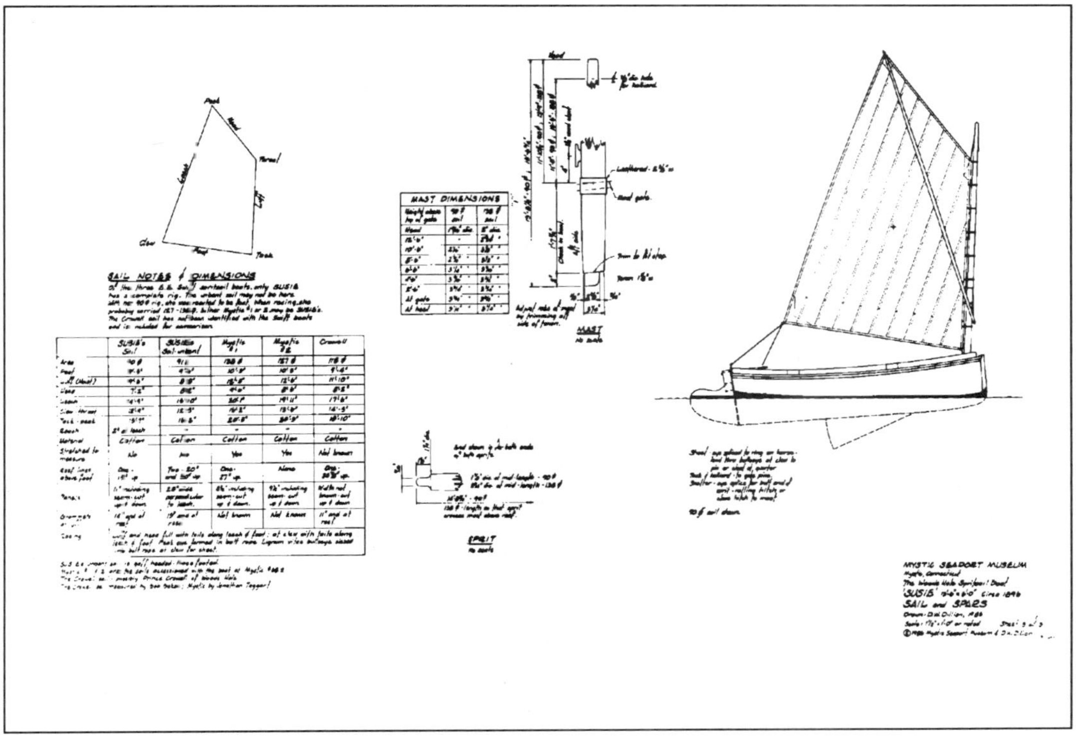

Figure 11.69. Susie's sail plan. *Mystic Seaport Museum.*

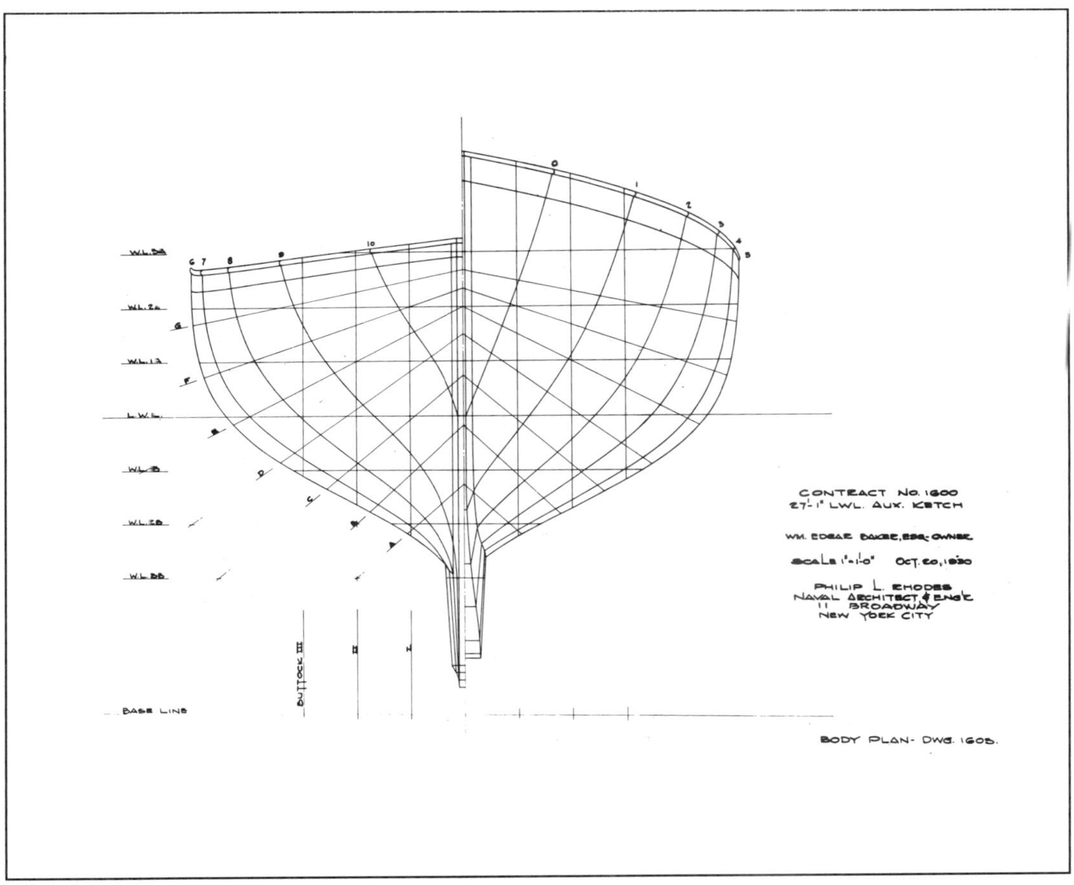

Figure 11.70. Dog Star's body plan, by Philip H. Rhodes. *Mystic Seaport Museum and Mr. Philip H. Rhodes, Jr.*

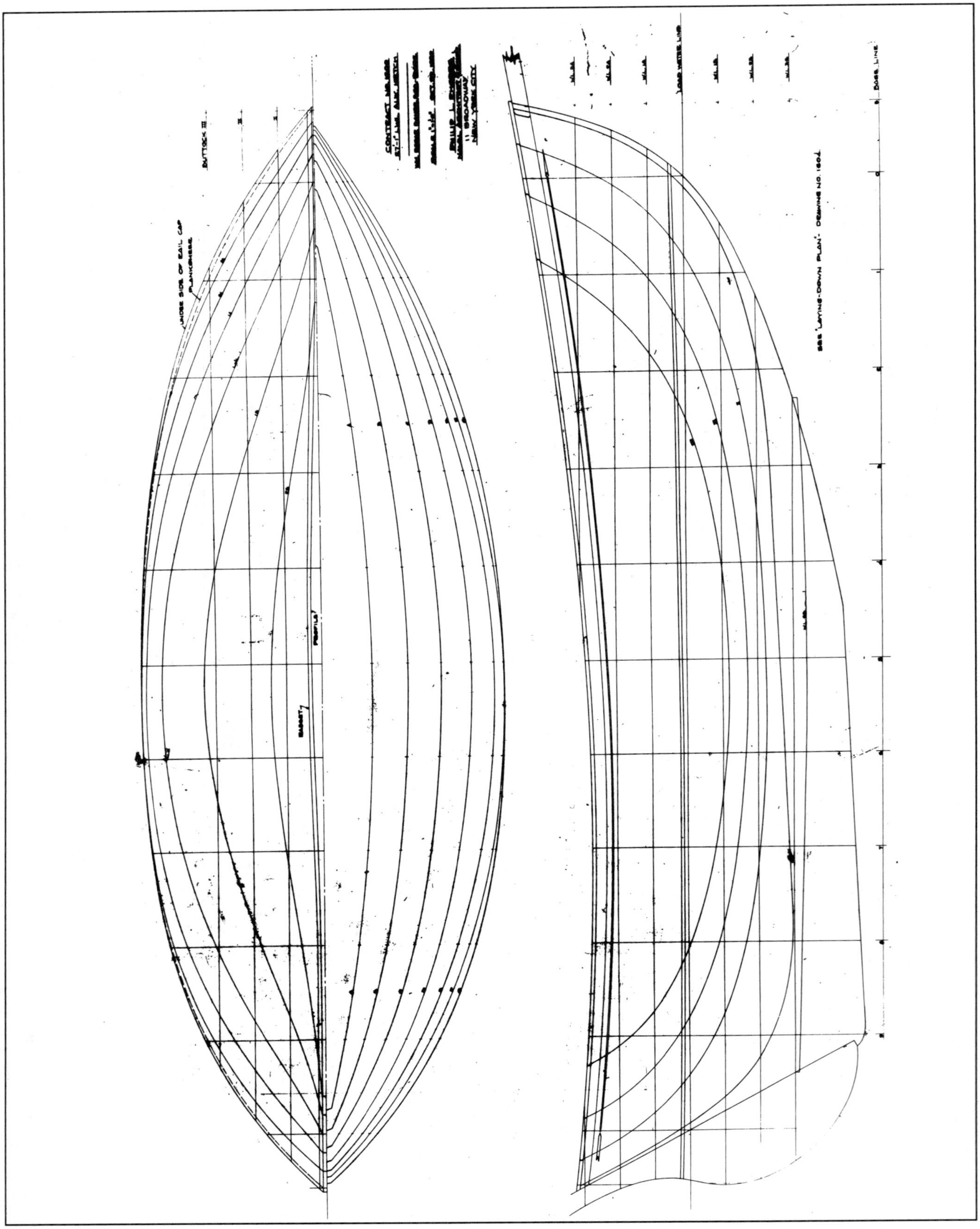

Figure 11.71. Dog Star's sheer and half breadth plan, by Philip H. Rhodes. *Mystic Seaport Museum and Mr. Philip H. Rhodes, Jr.*

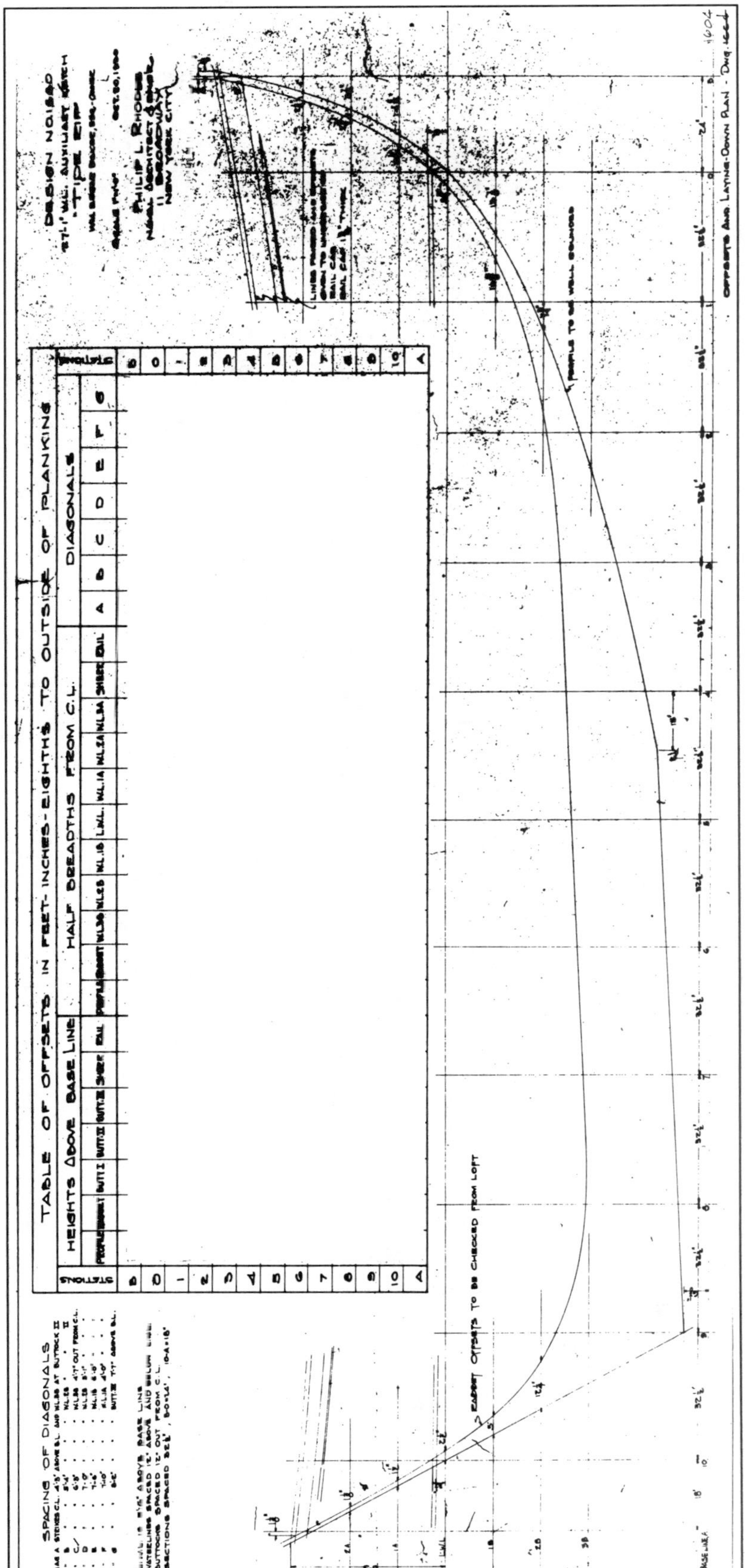

Figure 11.72. Dog Star's offsets and laying down plan, by Philip H. Rhodes. *Mystic Seaport Museum and Mr. Philip H. Rhodes, Jr.* Note the arrangement of the table.

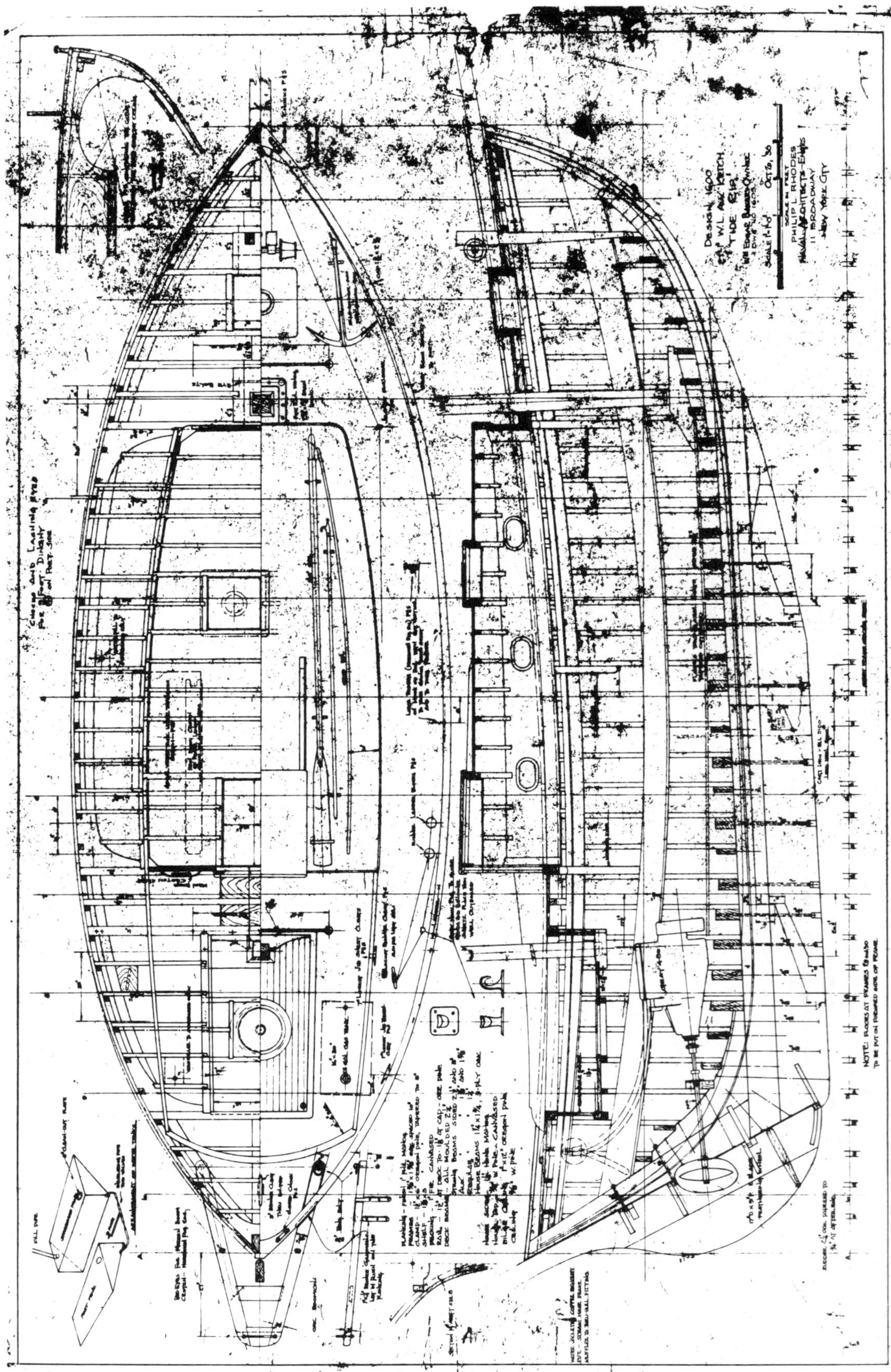

Figure 11.73. Dog Star's construction plans, by Philip H. Rhodes.
Mystic Seaport Museum and Mr. Philip H. Rhodes, Jr.

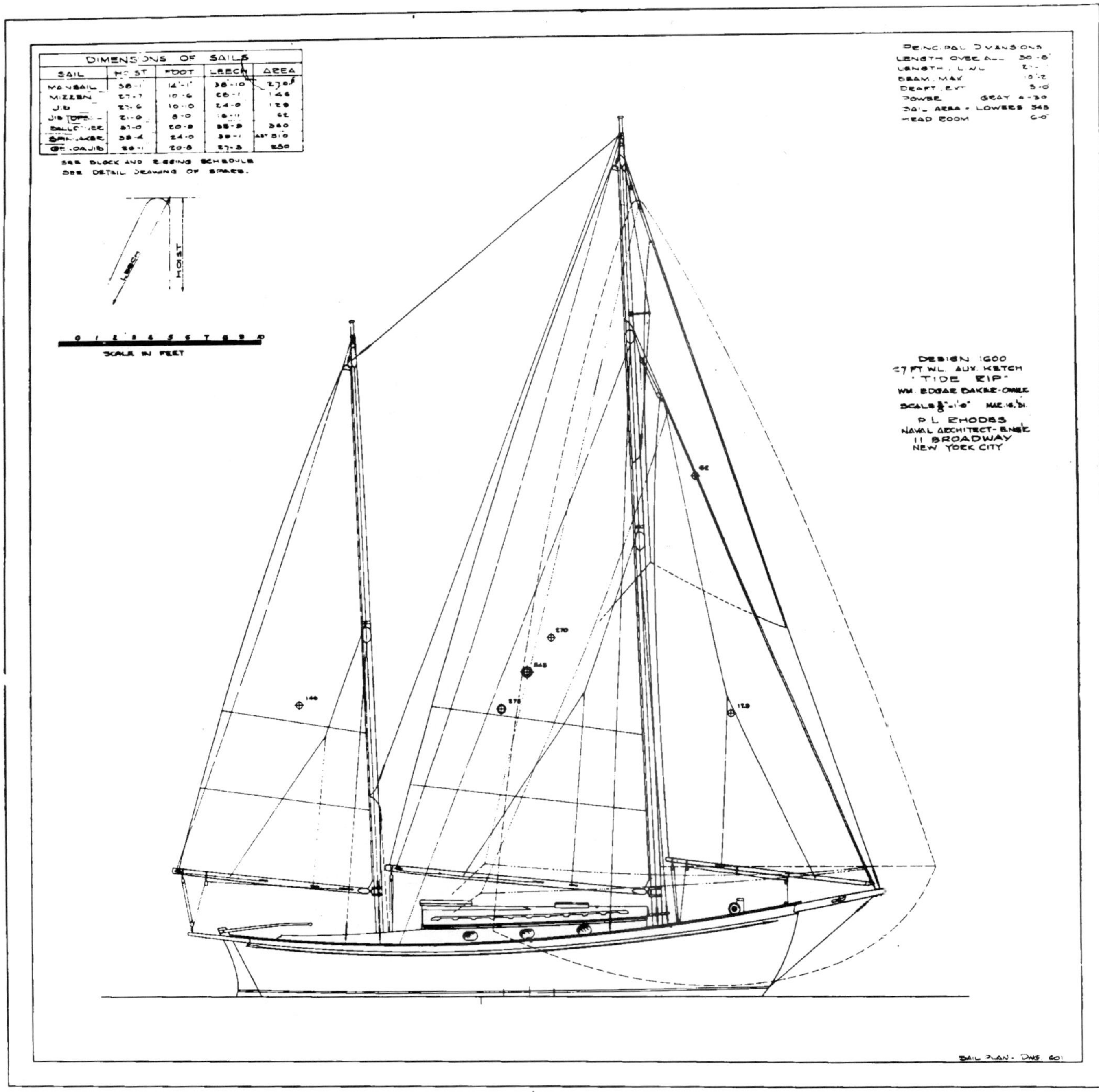

Figure 11.74. Dog Star's sail plan, by Philip H. Rhodes. *Mystic Seaport Museum and Mr. Philip H. Rhodes, Jr.*

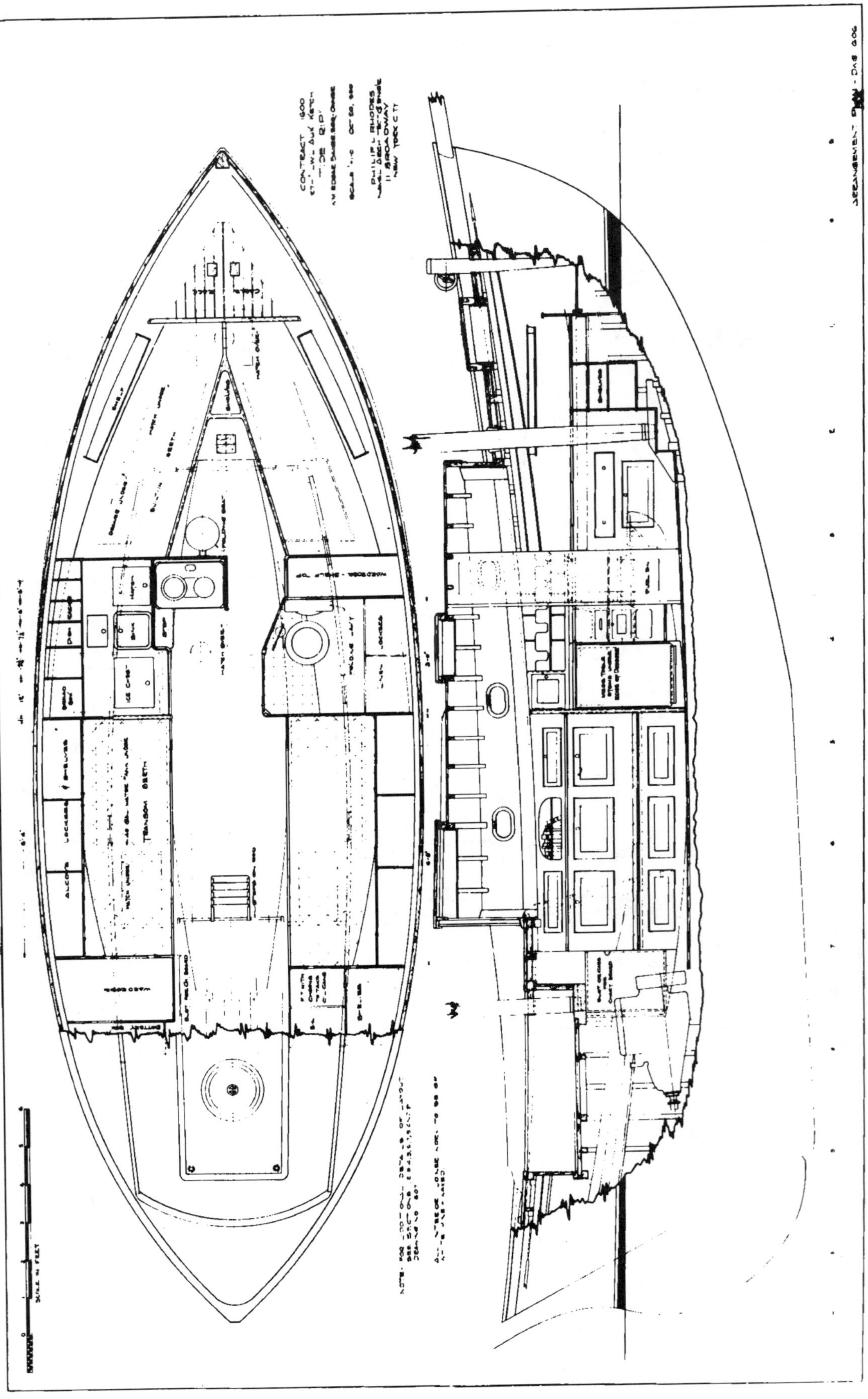

Figure 11.75. Dog Star's arrangement plans, by Philip H. Rhodes. *Mystic Seaport Museum and Mr. Philip H. Rhodes, Jr.*

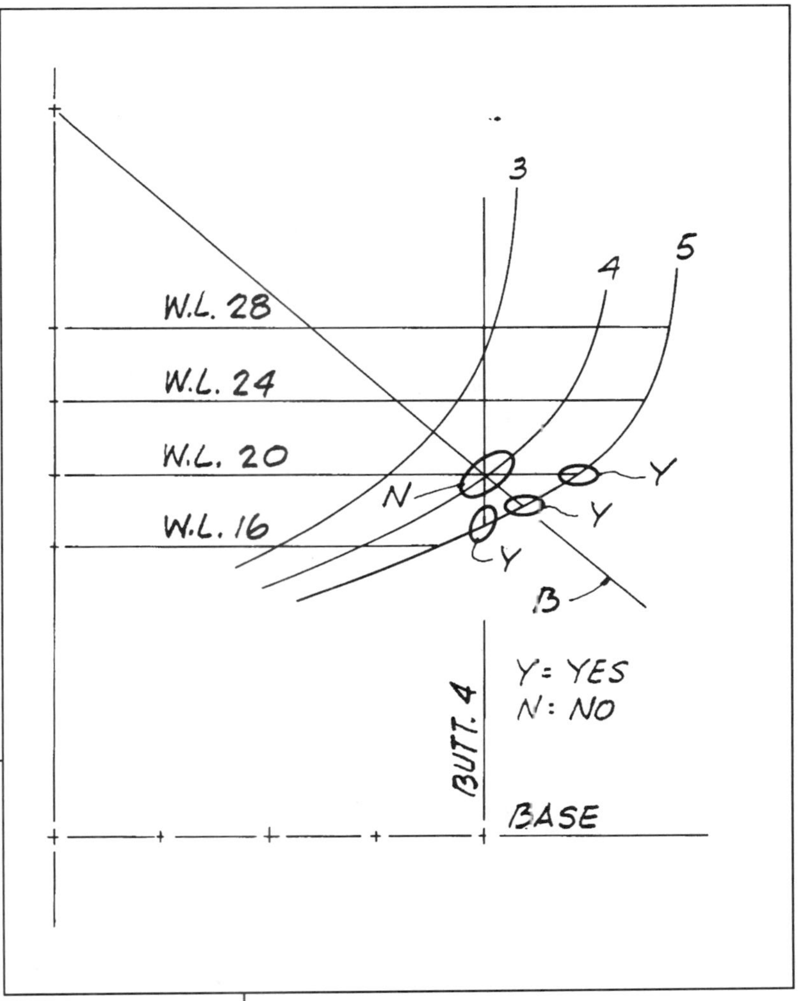

Figure 11.77. Partial view of *Susie*'s body plan showing likely places to move line for fairing. Compare the intersections shown as likely locations to move lines with the profile and half breadth plan in fig. 11.28.

Figure 11.76. Flipping a curve end for end.

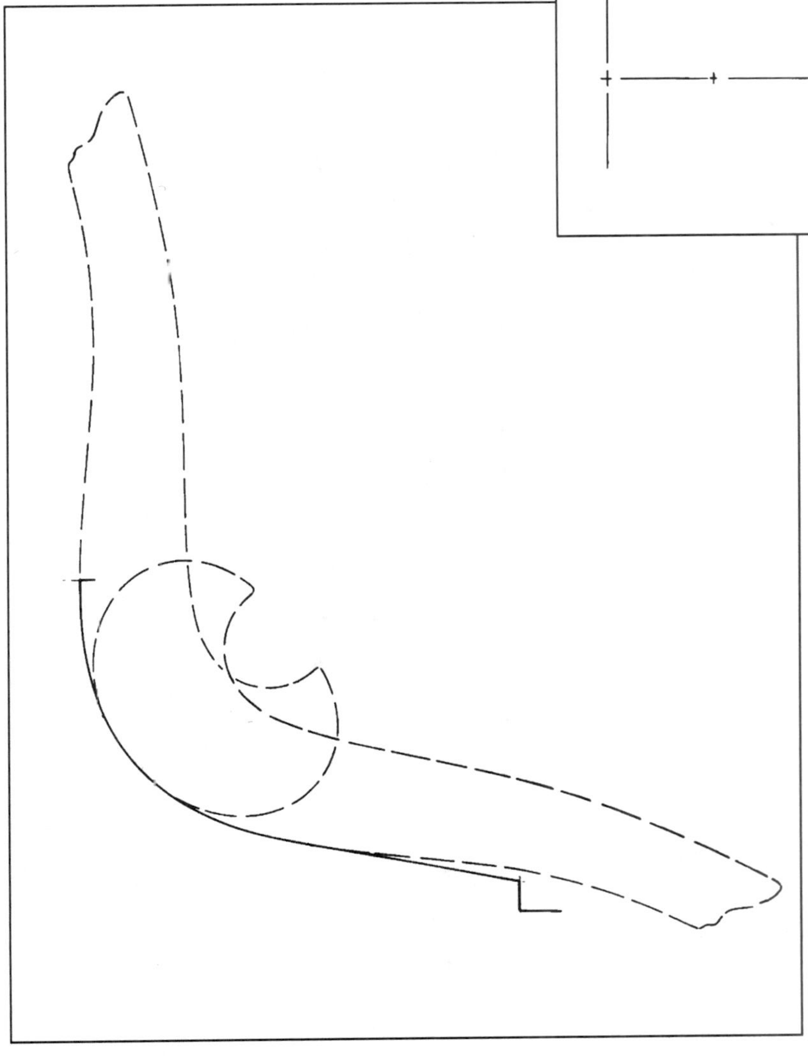

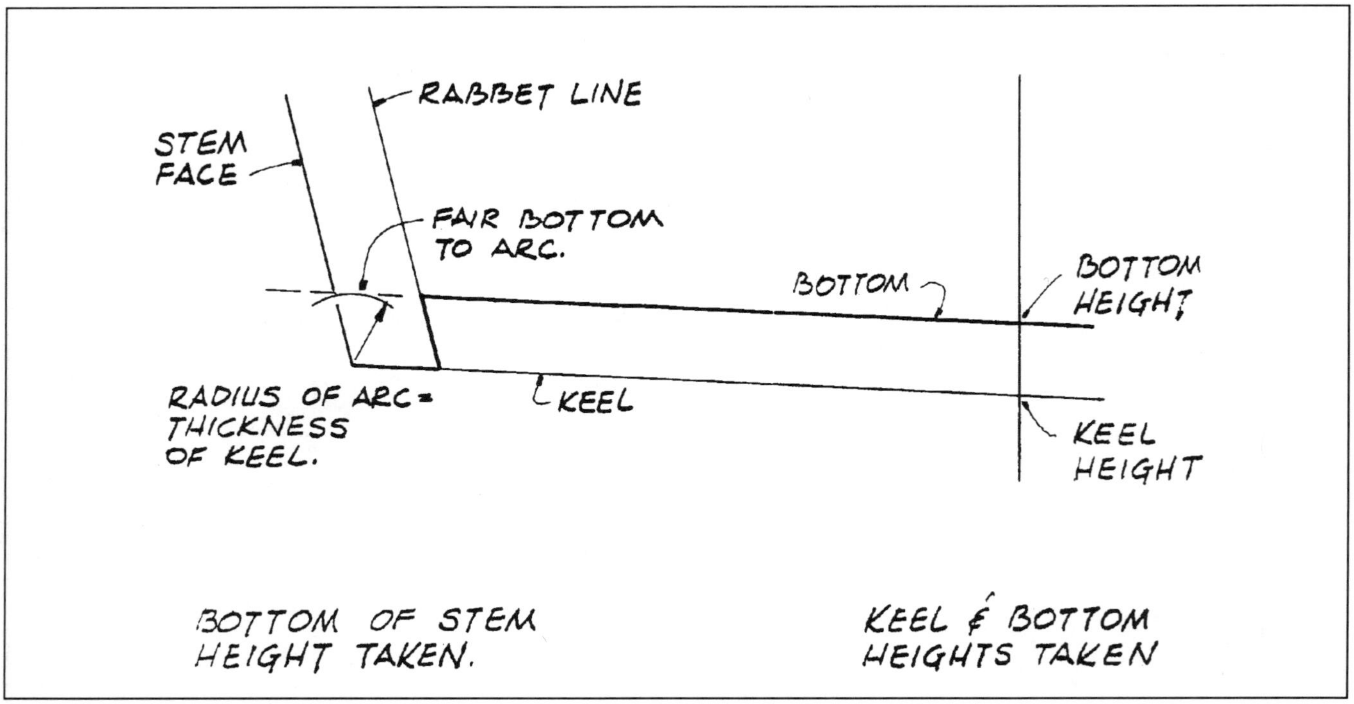

Figure 11.78. Using recorded data to compensate for missing data.

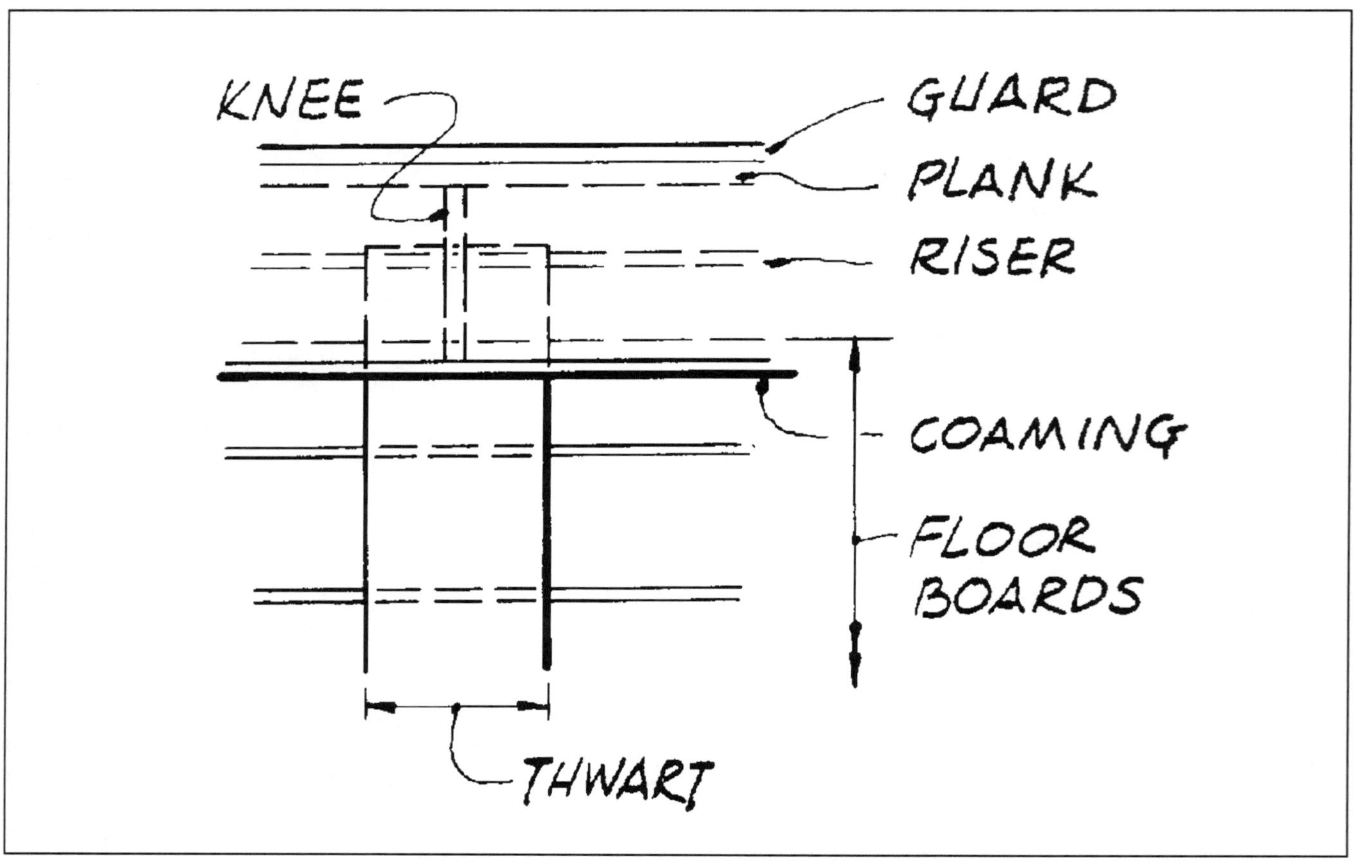

Figure 11.79. Dashed lines as hidden lines.

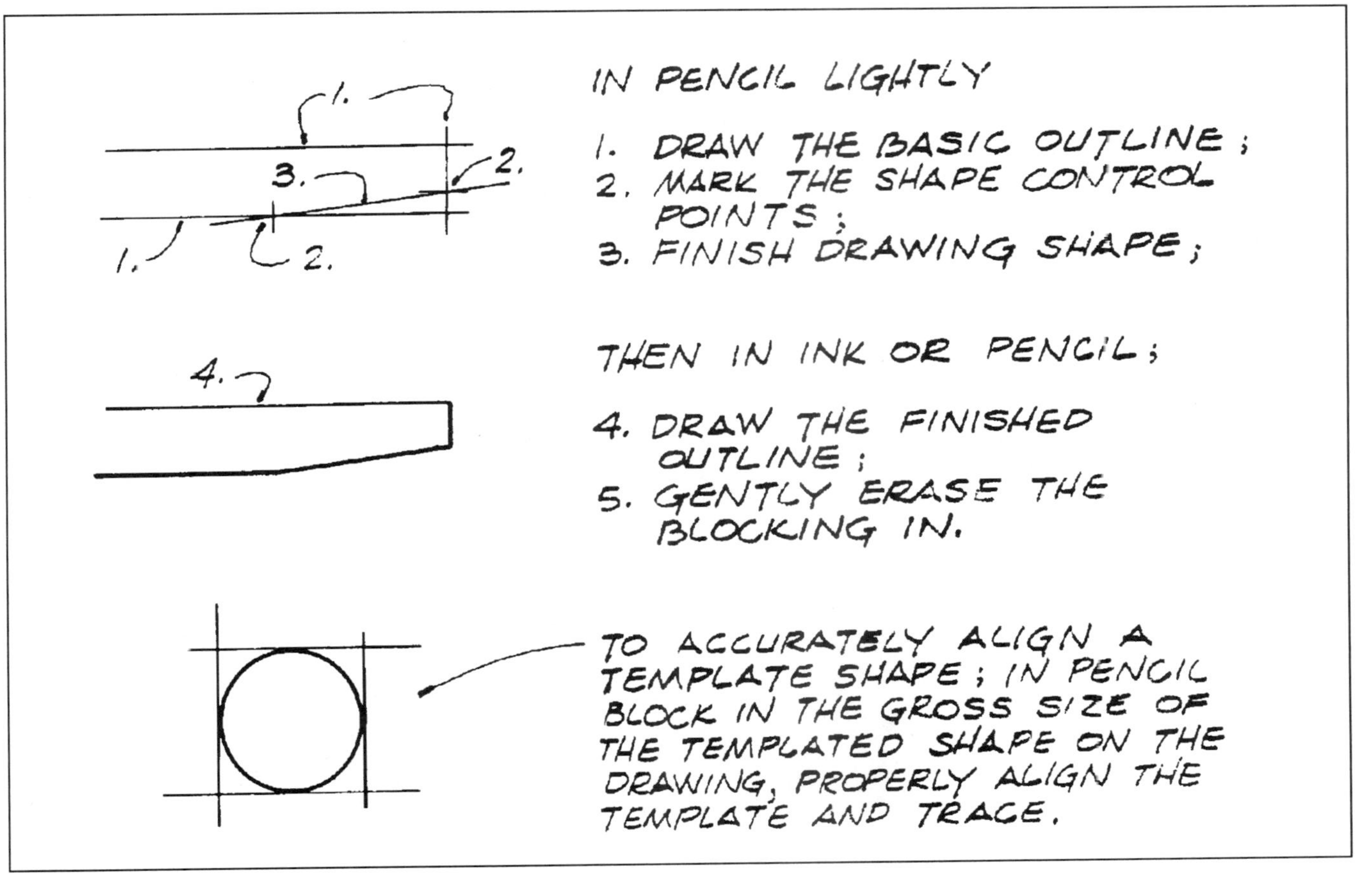

Figure 11.80. Blocking in a detail.

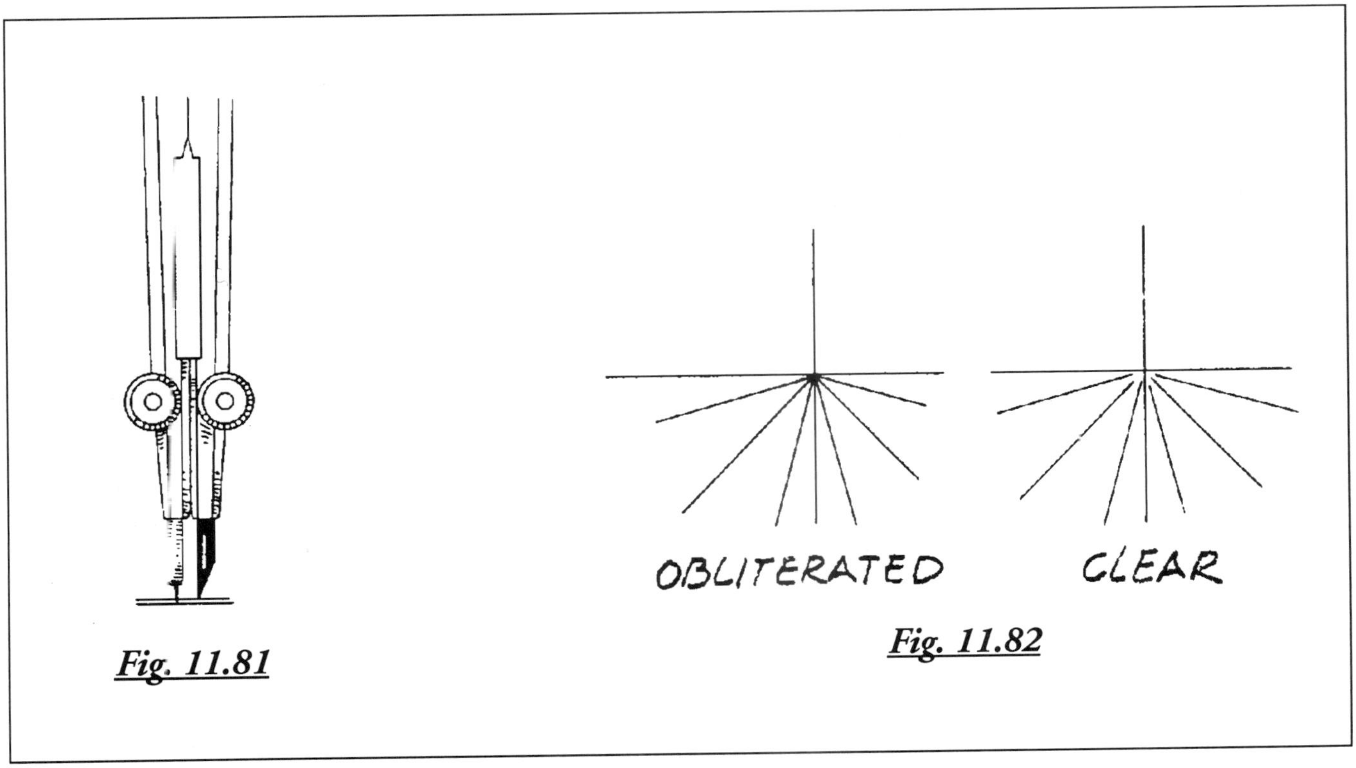

Figure 11.81. Sharpening compass points.

Figure 11.82. Many lines drawn to the same intersection.

Chapter 12

An Archaeological Approach

Recording boats from their remains and learning from their parts

Kevin J. Crisman

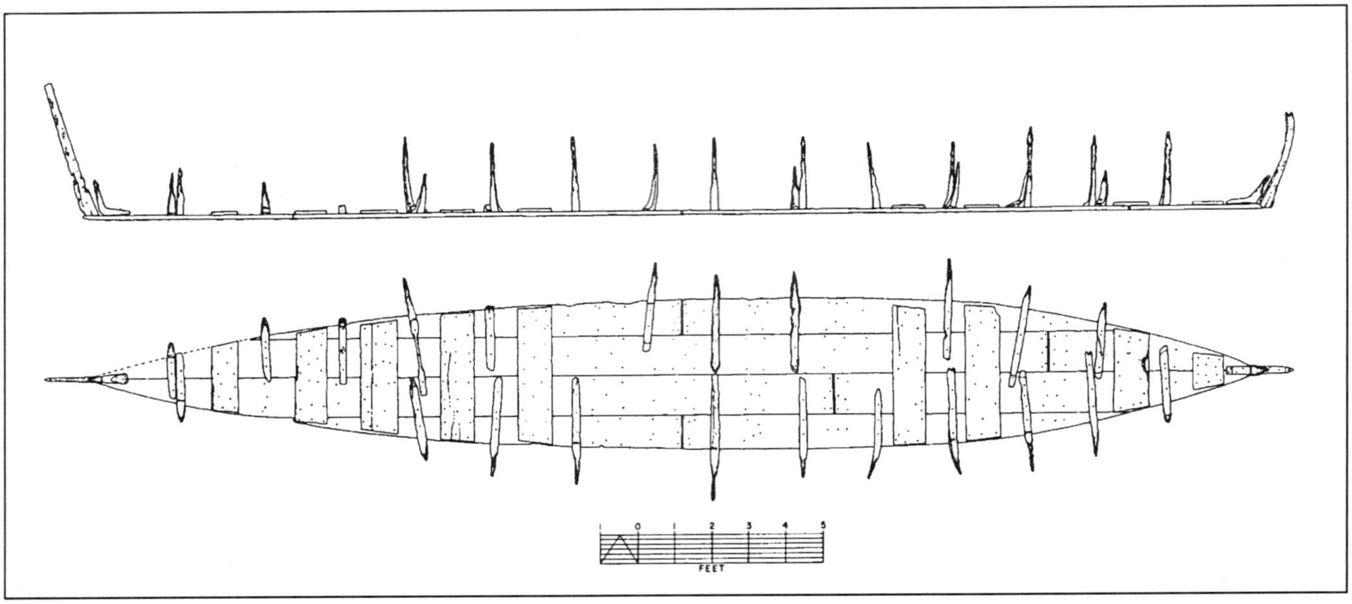

Figure 12.1. The remaining timbers of an eighteenth-century Lake George Bateau, profile and plan.

Figure 12.2. Reconstructed midsection of a Lake George Bateau.

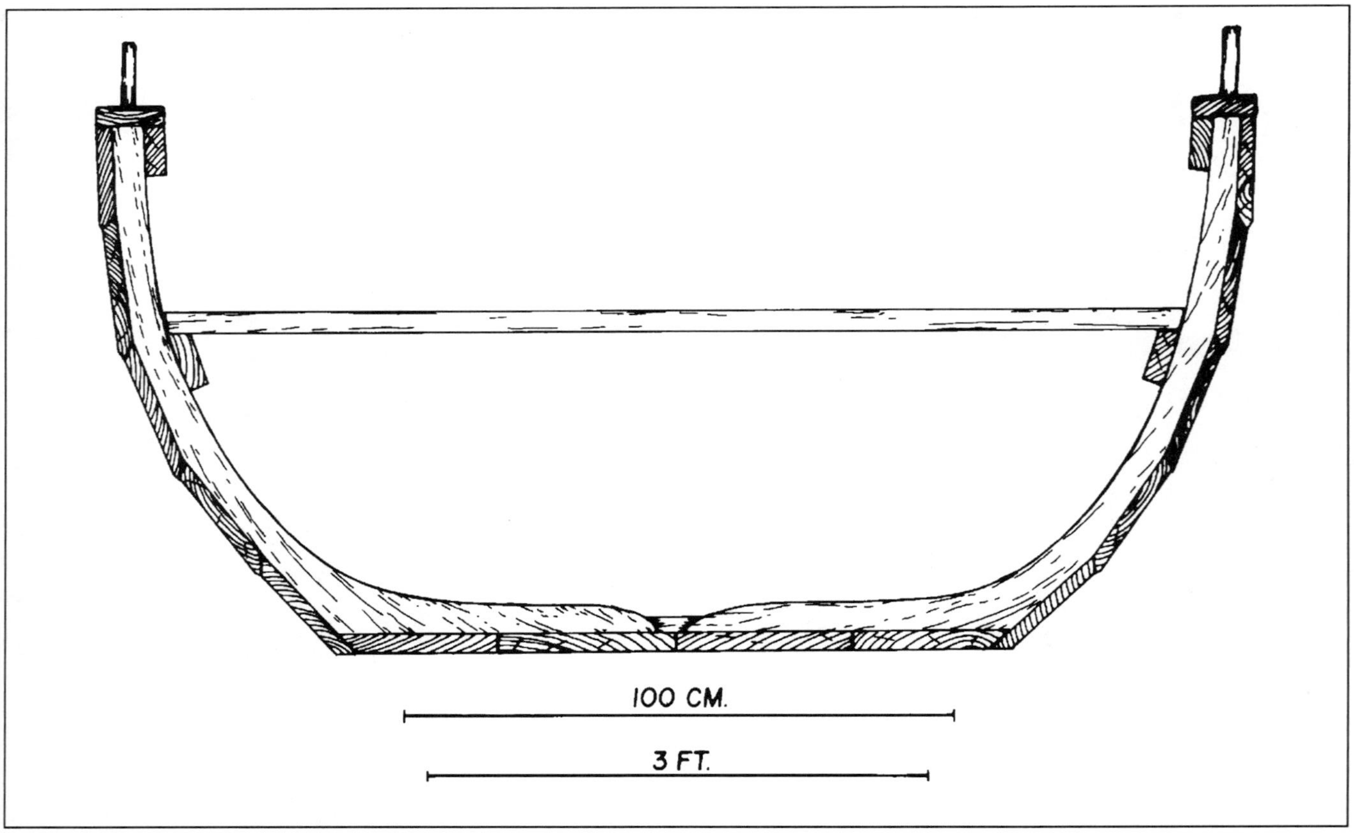

Every small craft recorder, both professional and part-time, will eventually encounter a hull that is rotted, dismembered, or represented by only a few fragments of planking and frames. Deteriorated boats and detached hull timbers too often are assumed to contain no useful information and are dismissed as unworthy of study or preservation. This is hardly the case. In fact, such remains generally contain a wealth of significant data on the design, assembly, and career of a boat or a class of boats. The amount of data that can be gained from them is limited only by the observational skills and tenacity of the small craft recorder.

Techniques of archaeological study have been applied to ship and boat finds, particularly in recent decades, and the result has been a dramatic increase in our knowledge of the hull types and construction methods used throughout human history. In the Mediterranean, the Institute of Nautical Archaeology and other organizations have examined ancient wrecks and from them traced the development of ship building over two-and-one-half millennia. The Kas wreck (fourteenth century B.C.), the Kyrenia wreck (fourth century B.C.), The Yassi Ada wreck (seventh century A.D.), and the Serce Limani wreck (eleventh century A.D.) have documented the gradual shift from material-and-labor-intensive, edge-joined plank construction to plank-on-frame methods of construction.

Ship and boat finds in northern Europe have similarly helped define the evolution of building practices in that part of the world. Examples include early sewn-plank boats (the Ferriby I hull of mid-second millennium B.C. and the Hjortspring Boat of circa 500 to 300 B.C.), clinker-built craft of the medieval era (the Sutton Hoo ship of 600 A.D. and the Skuldelev Wrecks of 1000 A.D.), and transitional carvel-planked hullsof the early Renaissance period (the Mary Rose of 1545 A.D.).

Hulls investigated by archaeological methods in the Western Hemisphere likewise have provided vital clues to the water craft used by Native Americans and to the boats used by Europeans to explore and settle North and South America. Excavations sponsored by the Canadian Government at Red Bay, Labrador, have uncovered examples of chalupas, small, light-weight whaling boats used by Basque whalers in the sixteenth century. One popular boat type employed in North America, the flat-bottomed, double-ended bateaux was built in the thousands during the seventeenth, eighteenth, and nineteenth centuries, but detailed contemporary plans or descriptions of them are scarce. Discoveries of bateaux hulls in Lake George in the early 1960s and at Quebec City in 1985 have greatly increased our knowledge of these significant craft by showing how they were built and what they looked like. Archaeological measurements of one of the Lake George hulls were used by boat builders at the Lake Champlain Maritime Museum to assemble a replica for testing under sail and oars in 1987.

These archaeologically studied hulls shared a common condition: all were in a badly deteriorated or fragmentary form when discovered. In spite of the hulls' incomplete states, archaeologists and marine historians documented and analyzed them in detail and, in the process, contributed to our knowledge of seafaring through the ages. These studies of battered wooden craft remains often demanded a leap of faith—the potential for finding new information was not always apparent when they were first recorded. In many instances a continual, lengthy process of collecting and sorting small pieces of evidence was required before a vessel's full significance could be understood.

If this chapter has a central point, it is this: a historically or architecturally significant wooden boat should never be considered worthless or undeserving of careful documentation on account of its condition. The hull may be beyond preservation and restoration, but a wealth of information is always there, waiting to be recorded and shared. Using archaeological techniques of recording and analysis, it is often possible to determine original hull form and construction.

Reconstruction can take any number of forms, including scale plans, models, plank diagrams, or even full-scale replication. Even when a vessel is so fragmentary or decayed that the possibilities of precisely reconstructing it are limited, an astounding range of information can be gleaned from the surviving pieces. Close inspection of boat timbers can reveal types of wood and fastening materials available to a builder, techniques of shaping and assembly, as well as the builder's attitudes about craftsmanship and the product he is creating. The fragments of a hull can even answer broader questions about the society in which the builder lived, by demonstrating levels of technology and the speed at which that technology evolved. More specialized efforts such as archaeo-botanical analysis of cargo have been used in reconstructing a vessel's ports of call.

This chapter discusses some of the tools and techniques of archaeological documentation and is designed to give the recorder of historical small craft some ideas about what can be recorded from an incomplete hull, how that information can be recorded, and how to put the resulting data to good use. It is not intended to provide a comprehensive answer to every situation, nor is it a substitute for archaeological training and experience. Contact a local maritime museum, state archaeologist, or other knowledgeable professional if you have questions about how to proceed.

Planning Considerations

The boat recorder, presented with one or more battered hulls in need of documentation, may wish to refer to chapter 1 and its discussions concerning levels of documentation and allocation of resources. The considerations reviewed are in most cases as equally applicable to deteriorated hulls as to intact examples.

Among the first tasks confronting the archaeological recorder is the determination of a hull's age and its relative historical and architectural importance. The context of the find or artifacts associated with it may provide evidence of age and use. When such artifacts are absent, the vessel's construction details can often be used to determine age. Hull fastenings are especially good for identifying approximate period of construction; in North America exclusive use of treenails or hand-forged spikes and bolts generally means that the boat predates the mid-nineteenth century, while the machine-made fasteners point to a later date. Building techniques can also serve as a means for dating hulls, although some caution must be used since older practices occasionally continued in some areas long after they had been abandoned in others.

Determining the historical, cultural, and architectural importance of a wooden boat is likely to be more complicated than determining its age. Historical importance might be defined as a vessel's relationship to specific periods, events and places, while cultural importance is a broader category, referring to a vessel's significance in understanding the society that designed and created it. Most old wooden boats are of local interest at a minimum, while a general lack of photographic and construction information lends a particularly high level of significance to small craft dating prior to the middle of the nineteenth century. Architectural importance might be defined as the degree of uniqueness to be found in the form and assembly of a hull and the availability of information on similar boat types. When in doubt about a boat's significance, seek advice and comments from people and organizations such as the Museum Small Craft Association, maritime museums, present-day wooden boat builders, nautical archaeologists, and maritime historians. Many of these are listed in the appendices. Libraries and historical societies can also provide helpful information.

The planning process must also include an assessment of immediate threats to the boat remains and a determination of the level of documentation that is practical under the circumstances. Ideally, the vessel will be found in a location where it can be studied with minimal disturbance over the longest possible time. It is axiomatic that moving a decayed wooden hull or pieces of a hull will result in the loss of information. Therefore, avoid moving the hull whenever possible, or at least delay moving it until after the completion of the recording process.

Small craft of great historical interest do occasionally turn up in situations that demand a crash program of study. Good examples of this are the occasional discoveries of boats and ships during the excavation of building foundations. In such circumstances, it is best to get in as much in situ documentation as possible, concentrating on noting the location and relationship of individual pieces. The details of the individual timbers can be recorded once they have been moved. Photographs should be taken of hull components in their original, undisturbed positions, prior to their removal. Be sure to keep a photographic log and to include reference points in each shot such as a north arrow or a sign board that explains the subject; without some sort of reference you may find it difficult to interpret the photographs at a later date.

One final consideration in planning the study of boat remains is the ultimate disposition of the information that is being recorded. At an absolute minimum, the data that you have gathered should be organized into a legible record of measurements, drawings, and photographs, then copied and filed at public institutions (libraries, museums, historical societies, and state archives), in order to make the results of the study beneficial to future generations of small craft researchers. While depositing raw data in a file drawer or archive shelf does ensure its preservation, it does little to help other researchers or educate the public. The results of your work should also be presented in other forms, including books, monographs, magazine or newsletter articles, a full-sized replica or scale model of the original craft, or museum exhibits. Analysis of hull remains and the preparation of reports or exhibits invariably requires more time and effort than the original recording process, but without them the job must be regarded as incomplete.

When initiating a project to document a deteriorated small craft, the recorder will do well to ponder the following questions:

- What is the age of this hull?
- What is its historical and cultural significance?
- What is its architectural significance?
- Is the hull threatened by further decay or dismemberment?
- Based on the above factors and the available resources, how detailed should the documentation be?
- How can the information generated by the study be shared with others?

Procedures for Recording

The procedures used to record the archaeological remains of decayed small boats are in many instances similar to those used to document intact hulls. The equipment used in both cases is for the most part identical. The difference is primarily in the level of information needed to finish the job. When a boat is intact the task is fairly straightforward: measuring and describing form and construction. Documenting a deteriorated or partially dismembered hull initially calls for a similar procedure, but the recorder must go on to become a "naval architecture detective," and search for clues to hull shapes and timbers that are no longer present. Small details take on infinitely greater significance. The detective work further intensifies when the time comes to render the original appearance of a boat in the form of a written description, lines plans, construction plans, a model, or a full-sized replica. The reconstructor must carefully analyze what currently exists and from this extrapolate what the boat looked like when it was a floating, functioning craft.

Three levels of field data should be recorded during a hull documentation project; in descending order these levels are context, structure, and components. "Context" might be described as the location, conditions (both environmental and cultural), and the materials (both natural and man-made) surrounding the remains of a boat hull. "Structure" refers to the spatial relationships of timbers; in other words, the assembly and shape of a hull. The "components" level is just as its name implies—the study of individual pieces of timber and other materials that make up a hull.

The importance of context can vary greatly from boat to boat. In some cases, it might be almost irrelevant to the study of a hull, and in others it can be crucial to the successful analysis and reconstruction of a boat. For an example of basic contextual information, we might examine the case of the bateaux found in Lake George, New York, in the early 1960s. When found, these boats lay in fifteen feet of water, slightly offshore from English fortifications built in the mid-1750s. The hulls were filled with rocks and aligned in neat rows on the bottom of the lake. Some of them still contained cargoes of thirteen-inch mortar bombs, as well as other eighteenth-century debris. The upperworks of the bateaux were badly deteriorated, but the bottoms of the hulls, covered with several inches of mud, were in excellent condition.

The position of the Lake George bateaux and artifacts within them together indicated that they were probably built in the English colonies sometime after 1755, that they were

Figure 12.3. The Basin Harbor Maritime Museum's full scale replica of a Lake George Bateaux after launching into Lake Champlain, 1987.

used to transport military stores, and that they were filled with stones and intentionally scuttled. The fact that they were sunk in rows rather than in a haphazard manner suggests that the scuttlers intended to retrieve the hulls at a future date. Mud from the lake bottom clearly had a preservative effect upon the timbers, which otherwise decayed in the shallow, fresh water. The context of the bateaux at Lake George thus provided much of the "who, when, and why" information that was essential to the analysis of both the construction and history of these craft (figs. 12.1, 12.2, and 12.3).

Start recording contextual information by preparing a site description, both verbal and pictorial, for the hull you intend to document. Questions to answer in your field notes include:

- Where was the hull found?
- Who found the hull and how long have people been aware of its existence?
- Does the hull appear to have been lost accidentally or was it purposely abandoned?
- Does the hull appear to have been moved since it was lost or abandoned? If it has been moved, does the structure seem to have been damaged in the process?
- What are the environmental conditions around the hull, and how have they acted to deteriorate or preserve the hull structure?
- What bodies of water and geological features are located near the wreck (swamps, rivers, lakes, oceans, reefs, mudflats)?
- What, if any, man-made structures are located near the hull (other boat or ship hulls, buildings, wharves, landings, bridges, fortifications)?
- What artifacts, if any, were found inside the hull? Outside the hull? Are these artifacts related to the operation of the boat, or were they part of its cargo?

If the hull has obviously been moved from the site where it was originally lost or abandoned, most of these contextual questions will have to be answered by interviews or research or simply left unanswered. When the hull does appear to be in its original location, then it is up to you to collect as much contextual information as possible. Use existing maps or charts of the area or, if those are not available, create your own maps to record locational data. Use nearby structures or topographical features as reference points to plot the precise location of the boat. Take photographs of the site and its surroundings to show the position of the hull relative to natural or man-made features. If artifacts are present, accurately record their locations relative to the hull structure before they are moved or disturbed. The hull and any detached hull timbers that are present should not be moved or rearranged during this stage of the recording process, since important clues may be lost by changes in location or orientation of pieces.

The documentation of a deteriorated boat's structure and components requires the same sort of measuring and recording equipment that is used for intact boats. If possible, use the units of measurements (metric, feet and inches, etc.) employed by the boat's builder. Your selection of measuring implements should include a half-dozen rules and "yardsticks" of varying lengths, two or more medium-length (50-foot) tapes, and if you intend to create a site map or plot the locations of widely scattered hull timbers, a long (100- or 200-foot) tape. Fiberglass tapes are less likely to break and will remain legible longer than steel tapes, but they can stretch over long distances. Short (6- to 12-foot) spring-retractable steel tape measures are excellent for working in dry environments, but wet or dirty conditions quickly destroy their internal mechanisms. Get an open reel tape at a surveyor's supply store or good drafting supply house; the reels, being open,

Figure 12.4. Two of the eighteenth-century bateaux under excavation in Quebec City. Working from the staging, the timbers were cleaned of mud and labeled prior to measuring and photographing. *Kevin Crisman, courtesy of Daniel LaRoche.*

are easy to clean and allow the tape to dry. Your tool kit should also include bevel gauges, plumb bobs, line levels, plenty of stout string, a heavy-duty stapler and thumb tacks, and prenumbered or lettered plastic tags (I use squares of plastic drafting film or white vinyl kitchen shelf paper, marked with indelible black "Sharpie" pens). Plastic zip-lock bags and film canisters come in handy for holding samples of wood, fastenings, or other small finds. Survey stakes, nails, and a mallet can also come in handy when it is necessary to create datum points and baselines for mapping.

If the hull you are recording is under cover, or if the weather is good, stenographers notepads can be used for sketches and notes. For underwater or other adverse-environment work, I suggest plastic drafting film (the thinner, .003 base, one-sided matte finish film is cheaper by the roll). These rolls of film can be precut into sheets and flat-clipped to masonite clipboards that have been spray-painted glossy white on both surfaces. To provide ample writing and sketching surfaces, attach a sheet on each side of the clipboard. Standard wood-and-lead pencils can be used under most recording conditions, but mechanical pencils are a better bet, since they do not require a pencil sharpener. With drafting film, use leads especially designed for film. Whatever writing implement you use, it should be attached to the clipboard with a lanyard to prevent loss.

Start your documentation of a boat's structure with a preliminary survey of the hull. This includes jotting down a brief written description (type of boat, prominent features, condition of fastenings and timbers, and orientation of hull) taking a few broad dimensions (overall length and maximum beam), and sketching rough plan- and profile-views that identify important construction features such as the stempost, sternpost, keel, and keelson. If the hull is broken up and scattered, try to determine how much of the original structure is present and which parts may be missing or damaged. This is also the time to tack numbered or lettered reference tags to frames and planking. The frames, for instance, might be numbered in sequence from stem to stern, and the outer planking lettered from the keel to the gunwales. The tags should be prominently placed to ensure that they will appear in photographs but should not cover potentially significant features such as joints, fastenings, or tool marks. Finally, if some parts of the boat cannot be reached except by walking or kneeling on the hull, you may want to consider suspending some type of platform over it to increase accessibility and prevent damage (figs. 12.4 and 12.5).

Use the preliminary descriptions and sketches to outline a work plan and checklist. A little strategy now can spare you the anguish of later discovering you neglected to record an important timber. Create a list of simple, defined measuring tasks—for example, "Keel: record molded and sided dimensions and length; locations and dimensions of scarfs, rabbets, and fastenings" or "Frames: record location of forward edge of each frame on a centerline tape; molded and sided dimensions of each frame; space between each frame; method of fastening to keel." Select the locations where you will want to record the form of the hull or the curvature of endposts with offset measurements. If your recording time is limited, determine which hull features are worthy of concentrated study and which may be assigned lower priority. Preparing a work plan beforehand will make your field time much more productive and will ensure that you have all the data necessary to analyze the hull.

Several different mapping techniques can be used to record the structure of a boat. The condition of the hull, its location and accessibility, and time considerations will determine which techniques you select. Throughout the process, your goal is to obtain accurate data on methods of assembly, the location of individual timbers relative to one another, and three-dimensional form. Any measuring technique or combination of techniques that permits you to achieve this goal is satisfactory. Obviously, the more time and care you devote to measuring, and the more redundancy you build into your measurements, the more accurate your final results will be.

An essential part of any small boat documentation project is the recording of hull curvatures for the preparation of lines plans. One major difference between documenting well-preserved boats and deteriorated ones is how horizontal baselines are established for taking offset measurements. A small boat that is intact can be shifted around until it is level, but attempting to do this with a hull that is in an advanced state of decay will result in damage and the loss of structural information. Instead, you must devise a system that will permit you to record frame and endpost curvatures in situ.

Figure 12.5. Archaeologist William Bayreuther cleaning frames and bottom planking of bateaux 8B4 found in Quebec City. *Kevin Crisman, courtesy of Daniel LaRoche.*

The simplest means of recording structure is to use the structure itself as a baseline. The keel or keelson, if present, can be used as a centerline, over which you may affix a baseline tape parallel to the longitudinal axis of the hull; if these timbers are no longer present, you can extend a baseline tape over the existing structure, taking care to place the baseline at a right angle to the axis of the frames. Measure the spaces between the frames, as well as the sided dimensions (or width) of the frames themselves, from stem to stern. Then measure the spacing of the frames relative to the baseline by noting where the forward edge of each frame is located along the baseline. The locations of other key hull components, such as the stempost, the sternpost, and mast steps (if present) must also be tied in to the baseline. At selected intervals along the baseline, measure the width of the existing structure with a horizontally oriented tape held at a right angle to the baseline. These dimensions will provide the basic information necessary to prepare a plan view of the hull structure.

Triangulation can be used to quickly map a hull for preparation of a plan view of the remains; it is particularly effective for mapping the horizontal distribution of broken and scattered timbers. In this system, three or more datum points are placed around and slightly above a wreck, their positions relative to one another are established, and then distances from each datum to selected points on the wreck are measured. A scale plan of the site is created by establishing the datum points on paper and using a pair of pencil-pointed dividers to relocate the points with intersecting arcs (figs. 12.6, 12.7, and 12.8). If executed with care, triangulation can produce remarkably accurate results.

Another way to map the structure and horizontal distribution of a deteriorated hull is to construct a grid over it (fig. 12.9). Many different materials can be used for a grid, including rigid plastic pipe, steel channel, or wooden stakes and string. Using a grid and a plumb bob, record the relative positions of timbers and fasteners precisely. This is a time-consuming way to document a hull, and on dry land a large grid can be awkward to work around. However, grids constructed from steel channel are ideal for excavating and documenting wrecks sunk in murky waters.

If the boat you are studying has collapsed, or was buried in the ground, chances are that you will have to take your offset measurements from the interior of the hull, using a horizontal datum level (fig. 12.10). Long flexible curves can also be used to obtain the curvature directly from the frames or planking. Whatever method you employ to record hull sections, be sure that you determine the vessel's angle of list so that the proper, "upright" orientation can be determined during the reconstruction phase of the project. Remember that even if an important structural timber (such as a frame or endpost) has been removed or is hopelessly deteriorated, adjacent timbers will often hold that piece's original shape. Careful, thorough mapping of undisturbed timbers will preserve these clues and ensure that they are available to you when it is time to analyze and reconstruct the hull.

Your measurements and sketches of a hull's structure should be augmented with photographs taken from many different angles. A series of overlapping photos taken from the sides and from above can later be assembled into a rough but very useful photomosaic of the wreck. At a minimum, the photographs will provide the reconstructor with a handy reference in cases where the written data is sparse or confusing. In some instances, you may find that, due to oversights in field recording, the photos will provide the only information available on a particular detail of assembly.

The third and most detailed phase of archaeologically documenting a small boat concentrates on a hull's component pieces. Obviously, recording a boat's structure also involves recording the dimensions and shapes of many individual timbers, but going back over these timbers with an eye for details is well worth the expenditure of time. The small details—materials, tool marks, fastening patterns, wear marks, and paint stains—will fill in the blanks when it comes to understanding a boat and the people who built and used it. Your attention to details becomes even more important when a boat is severely decayed or in pieces, for the observations will provide clues that are crucial to the successful analysis and reconstruction of the hull.

Measuring the components can be organized according to the approximate construction sequence; that is, beginning with the keel and proceeding on to the stem, stern, frames, keelson, deck structure (if any), and external and internal planking. If several individuals are involved in the recording work, they can each be assigned a group of similar or related timbers to examine—for instance, all of the keel timbers or all of the frames. The methods used to record the individual pieces will vary according to the complexity of the hull member, the amount of time available, and the skill of the recorder, but ideally there should be sufficient documentation to prepare plan, profile, and section views of every timber.

Measured sketches of each timber's shape and dimensions will provide the foundation of data on the boat's construction, but such sketches are only the beginning of your investigation. Try to identify all of the materials that went into the construction of the hull. Extract samples of nails, spikes, and other fasteners from timbers, and identify them with tags that state exactly where they were found. Collect samples of other building materials, wood, paint, pitch, caulking, and sheathing or other protective coatings for later analysis.

The identification of woods is a particularly important objective, since they have a direct influence on assembly techniques and hull strength and longevity. Woods can also reveal much about the sources and quality of materials used by the boat builder. Wood identification experts can often be found at universities that offer a degree in forestry or in state forestry departments. If there is a possibility that the hull might someday be restored, bring the expert to the wreck for a direct, minimal-damage analysis of the timbers; however, if there is little possibility of restoration, samples can be sawn or chiseled from each timber for later study. The samples should be large enough to allow inspection of the cross-grain structure (about 1-inch square is adequate) and should be securely labeled. If the timbers were found in a waterlogged condition, do not allow samples to dry before they are identified, because drying of waterlogged wood can obscure characteristic features.

You can learn a great deal about the tools, techniques, and craftsmanship of a boat's builder by systematically examining each timber's surfaces for marks left during construction. A series of regularly spaced cut marks across the grain of the wood may indicate squaring with a broadaxe; dimpled wood surfaces with an occasional cut mark may be evidence of shaping with an adze; and closely spaced parallel scores are likely to be signs of sawing. Be alert, too, for unusual lines, symbols, or numbers scored into the wood surface, as these may be marks used by the builder to guide the construction process. Evidence of now-missing hull pieces can frequently be found on timber surfaces in the form of mortises, empty fasteners holes, stains in the wood, or places of uneven wearing.

The surfaces of timbers may carry clues that indicate how a boat was used, how long it was used, and what happened to it after abandonment. Such clues will often be very subtle and may include areas of wear or damage on the interior surfaces of frames and planking, as well as patterns of scratches, cuts, notches, or stains. Unusual concentrations of fasteners, patches, and timbers with wear patterns that do not match neighboring pieces may be signs of repairs and thus indicate length of service. Try to record all of these features. No matter how insignificant they may seem, they will almost invariably contribute to your analysis and reconstruction of the boat.

Dismantling a wooden hull in order to record it is a lot like pushing Humpty-Dumpty off the wall: dismantling provides a much better idea of what is inside, but the structural integrity is ruined. This destructive step should not be taken on a casual basis, for it will irrevocably alter the hull's condition and may prevent you or others from re-examining assembly details at a later date. However, there are times when disassembly is not only acceptable but desirable. If the boat you are studying is certifiably doomed to imminent destruction by natural forces or human activity, or if it is going to be taken apart anyway for conservation treatments, then this action is probably warranted.

Reducing a hull to its component pieces invariably reveals a wealth of information that was previously obscured by surrounding timbers. Dismantling also greatly improves the accuracy of the recording work: curvatures of frame endposts, as well as the angles of scarfs, bevels, and other construction features can be closely examined and precisely measured. How you take the boat apart will of course depend on how it was put together and on the information that you hope to obtain. Reversing the original sequence of construction and gently sawing or prying off timbers will produce the best results; avoid using excessive force whenever possible, for it can destroy subtle clues on timber surfaces or inadvertently create a set of new (and misleading) tool marks.

At a minimum, draw plan and profile views of each timber to show dimensions and the locations of fasteners, marks, and stains. If possible, trace the curvatures of frames and endposts directly. The highest level of accuracy may be attained by tracing every piece at full scale. To do so, place each timber on a flat surface and suspend a sheet of glass or plexiglas just above it; then spread clear acetate over the glass and outline the timber, trace fastener holes, tool and wear marks, and other features with color-coded indelible pens. The process is difficult and time-consuming, but it will ensure that you have collected most of the data that a hull has to offer (fig. 12.11).

Organizing your field notes, drawings, wood and fastenings samples, and photographs is the all-important final step in the recording process, for it will profoundly affect the ability of you or anyone else to use the data at a later time. Archaeological study demands close attention to detail, which in turn generates a tremendous amount of paper. Illegible writing, untitled lists of measurements, sloppy sketches, and haphazard filing will waste hours of time and cause great frustration during analysis. It may even prevent the successful completion of the project. If the field notes are a mess, recopy them, and then order them according to subject (three-ring binders are useful here). Be sure that every sheet has the project name or site number, the name of the recorder, the date it was recorded, and a title that fully explains the subject. Samples of hull materials and photographs and slides should get the same treatment.

Analysis of the Hull

Accurately recording construction is only the first stage in your archaeological analysis of a deteriorated wooden boat. The next stages—analysis of the timber remains, cultural context research, reconstruction of hull assembly and lines, and preparation of a final report—will require considerably more time, effort, and persistence. Expect to spend six months on follow-up work for every week spent in the field. Many archaeological projects dealing with boats fail to advance beyond the initial documentation work because of the time required for follow-up work. Measuring and sketching a deteriorated hull merely preserves information; analysis and publication will ensure that the information is available to present and future researchers.

How you go about analyzing, researching, and reconstructing the boat will depend upon how well the hull is preserved, how much is already known about this vessel and similar vessels, and how you conducted your investigation. For example, if the boat is relatively well-preserved but little is known about its cultural context, more time might be spent researching the latter subject. A work plan is just as important here as it was in the field-recording stage of the project. Start by outlining what you already know about the hull, what questions you have about its construction and career, and what products you hope to generate from your work.

Begin your analysis of a small craft by drafting site and wreck plans. Preparation of these plans allow you to examine, organize, and interpret the data collected and creates a permanent graphic record of what you have found. Site plans consist of one or more maps that show the geographic and cultural context of the boat find. The maps should include the location and orientation of the hull relative to natural features (such as bodies of water) and human structures (such as docks, buildings, and canals); detailed maps might be drafted also to illustrate the distribution of artifacts or other site-related materials around the hull. Of course, preparation of a site plan may be impossible if the hull was moved prior to the documentation work and no records were kept on the original context.

Detailed scale plans of the hull in its "as found" condition are extremely useful for archaeological analysis. They condense the vast amount of information contained in your field notes into a form that is easily examined, interpreted, and shared with others. The work that goes into drafting of a set of wreck plans invariably has beneficial side effects: patterns in the hull's construction and repairs are identified, clues to missing timbers and hull form are recognized, and errors made while recording the hull are found and corrected. Keep a logbook handy during the drafting to jot down observations, major discoveries, and questions. The plans should consist of top, profile, and section views of the hull to portray its three-dimensional shape and general condition when studied.

Supplement the plans of the site and the hull with a written summary organized in the approximate order of construction. This report should describe where the boat was found, its condition when examined, timber dimensions and wood types, methods of assembly, types of fasteners, caulking and coatings, and any artifacts found in its vicinity. Include as well any thoughts or opinions you may have on the vessel's age, origins, and significance.

The preliminary analysis of the wreck is simply a process of summarizing what you have learned about the boat during field study. Converting your sketches and notes into plans and written descriptions is an excellent way to become more familiar with the evidence you have collected and to create a document that can be sent out to small craft experts for comments. The plans and report also provide a foundation for the subsequent stages of the project: cultural context research and reconstruction.

To understand a boat's design, materials, and construction, you must first understand the people and society that produced it; for this reason you should be prepared to spend a substantial portion of your time interviewing informants and examining the collections of libraries, museums, and archives. Research of this kind is particularly important if the boat you are studying is deteriorated and incomplete, for in many cases the dimensions and appearances of missing pieces can only be reconstructed from photographs, plans, or descriptions of similar craft. See chapter 15 for instructions on how to conduct ethnographic research.

The penultimate stage in the archaeological analysis of a deteriorated wooden boat is the reconstruction of its original, "as built" design, construction, and appearance. The reconstruction can take one or more forms: scale construction and lines plans, conjectural drawings, plank diagrams, scale models, or full-scale replication. The forms you choose, and how detailed you make them, will depend in some measure upon the extent and condition of the hull you studied, the quality of your field notes, your time and material resources, and how you intend to disseminate the results of the project.

Gather together the evidence that is the basis of your work—field notes, photographs, wreck plans, and the preliminary report summarizing the hull's construction. Your archival research and informant interviews should provide the other body of evidence needed to make informed judgments about the boat's appearance and career.

Reconstructing hull lines (its three-dimensional shape) is the first order of business, since lines represent the framework that defines a boat's maximum dimensions and form. Reconstruction can be done in two ways, either graphically or by scale modeling.

The graphic method consists of an on-paper, piece-by-piece and shape-by-shape rebuilding that generally proceeds in the original order of hull construction. Begin by creating half-breadth, body, and sheer plans of the intact, undistorted portions of the hull, extending the curves of frames and endposts as far as their original shapes and surfaces will permit you. The techniques described in chapter 11 to mold and loft from notes and measurements are for the most part applicable to archaeological construction. Be sure to use plastic drafting film for this work, since it holds up well under constant erasing and redrawing.

Once the "solid" part of the hull has been established, you will begin to use the hundreds of clues that you so patiently collected during the field work, preliminary analysis, and cultural context research. The importance of good recording in the field should become evident at this time. Locations, dimensions, and shapes of heavily deteriorated or missing timbers may be determined by inspecting similar pieces that survived, fastening patterns, and marks on adjacent timbers. The curvature of missing or damaged parts of frames and endposts can be arrived at by fairing lines up from intact portions. In areas where the archaeological evidence is sparse or entirely absent, it will be necessary to extrapolate from plans, drawings, or photographs of similar craft. The process of reconstruction involves reviewing and selecting the bits of evidence that best fit what is known about the hull, and then incorporating them into the lines plans. Be sure to indicate on the lines plans (and in an accompanying report) exactly where the facts ended and where the conjecture began.

Most projects start with graphic reconstruction to get information on paper, to ease research by different team members, and to speed publication. However, if the skills, time, and resources are available, modeling is a particularly effective method of reconstructing a small craft (and easier in some respects). Building a research model allows you to replicate the boat builder's work in miniature and to determine the form of a broken-up wooden boat. It is also extremely useful for experimenting with different hull configurations and for identifying possible errors in the reconstruction. A relatively small selection of tools is required for wooden scale modeling: a band or jig saw, small-blade hand saws, chisels, planes, spokeshave, files, and sandpaper should be adequate for most of the work. The best materials are fine-grained, relatively soft woods such as bass wood, pear wood, or fine-grained, non-splintery pine.

Build your reconstruction model in as large a scale as possible; larger wood pieces will give you a better feel for how the original timbers behaved during the construction of the boat. If the model is intended to be a meaningful representation of your boat, make exact scale copies of each hull timber as it was found, including the angles of bevels, the location of nail holes, the scars of tool marks, and areas of wear, damage, and decay. Make a detailed reduced drawing of each timber and attach it with a few drops of glue to one side of the replica piece to guide the initial cutting and shaping. The research model is a tool that permits understanding of a hull's design and assembly; accuracy of replication takes precedence over aesthetic qualities (fig. 12.12). A museum-quality display model can be built later.

Begin the construction of the model by assembling those portions of the hull structure that were more or less intact when you measured the boat. Let the partial structure and wreck plans serve as guides to positioning the detached timbers where they best fit on the hull. Line up nail holes, paint lines, mortises, and other features with corresponding features on other timbers. If the real boat was crushed flat, reverse this process on the model by gradually raising up the sides until fastener holes and other marks on the wood align with one another. Thin wooden batten strips can be used to hold the sides temporarily in position during the reconstruction work; once you have arrived at a hull form that agrees with the evidence, add more battens to "ghost in" hull parts that are entirely missing. When the reconstruction has been completed, take offsets from the model and draft a set of hull lines.

The reconstruction of the boat's lines can and should be augmented with other plans that convey information about the form and assembly of the hull. Timber dimensions and building methods can be illustrated with interior profiles and section views, and construction techniques can be shown in detailed plans or with "exploded" drawings. The shape of the boat's hull can also be explained and compared with other small craft finds by preparing strake diagrams that show the curved shape of hull planks in the flat. By aligning all of the planks from one side of the boat amidships and ordering them from garboard to sheer strake, a strake diagram indicates the hull's stem-to-stern curvatures. Plank edges that remain equidistant indicate continuity of form, edges that separate indicate narrowing of the sides, and closing edges show hollowing in the sides.

Models are a particularly effective way to educate the public or inform small craft researchers about the boat you found and its significance. Models can take a variety of forms, such as wreck models, hull-section models, cut-away models that reveal construction techniques, and, of course, complete reconstruction models that show the boats as they may have looked when they first came out of the builder's shed. Full-sized replicas of the boat will allow you to learn firsthand what went into building it and to test its seaworthiness. If you decide to produce a full-sized replica for testing, every effort should be made to duplicate the original design, materials, and carpentry in order to ensure that your copy behaves as much like the real thing as possible. Study the best of previous efforts in this area, particularly the replicas of the Kyrenia ship and Skuldelev III.

Finishing Up

The publication of some type of report on your boat will guarantee that the knowledge you have painstakingly acquired will be available to the widest possible audience, both now and in the future. The report should include a discussion of the craft's historic and cultural context, a technical description of its assembly, the site maps, wreck plans, reconstructed lines drawings and construction plans, and clear statements of the rationale behind major decisions in recording, research, and analysis.

Archaeological techniques for studying small craft remains are similar in many respects to those used in conventional recording of intact boats. In both cases, you are attempting to extract as much information as possible from the hull in order to create an accurate rendering of its design and construction. Archaeological study is, however, a case of making do with less, and making what you have count for more. Maritime archaeologists around the world have demonstrated that battered, squashed, decayed boat remains can still be made to yield a wealth of information about the seafaring past of the human race.

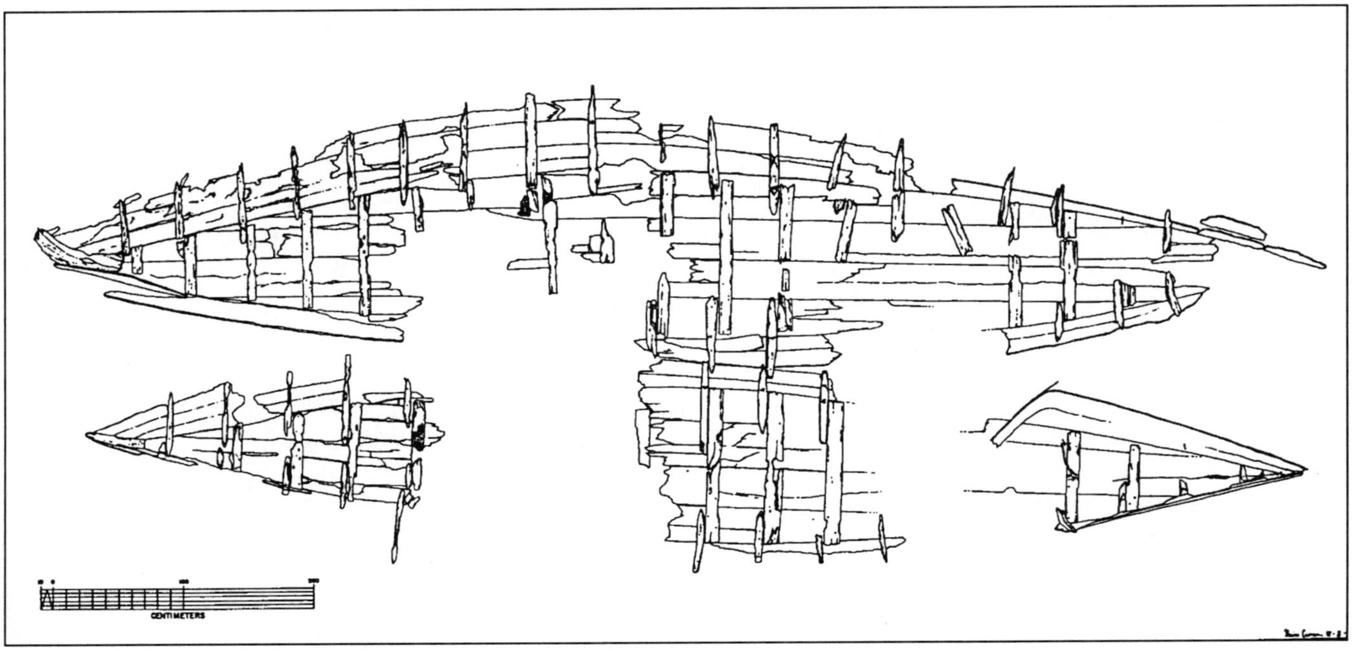

Figure 12.6. Plan of the Quebec City Bateaux 8B3 and 8B4.

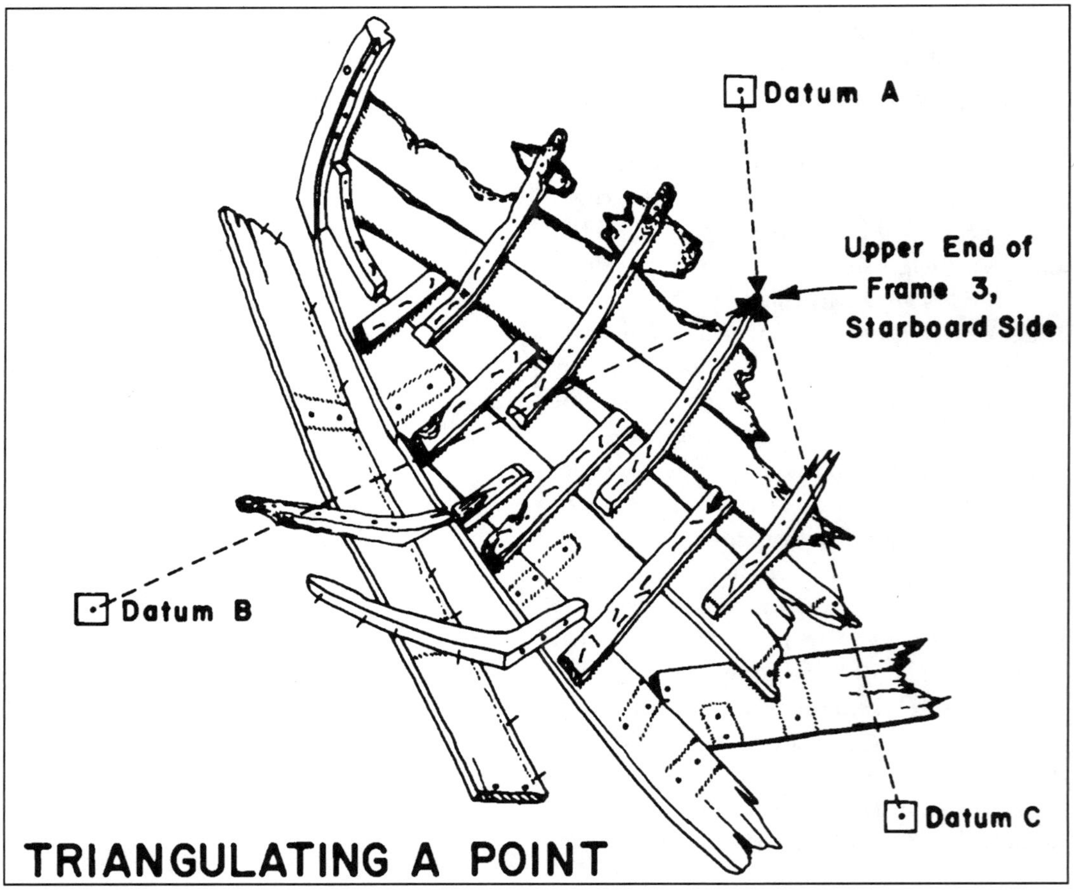

Figure 12.7. Triangulation of broken and scattered timbers.

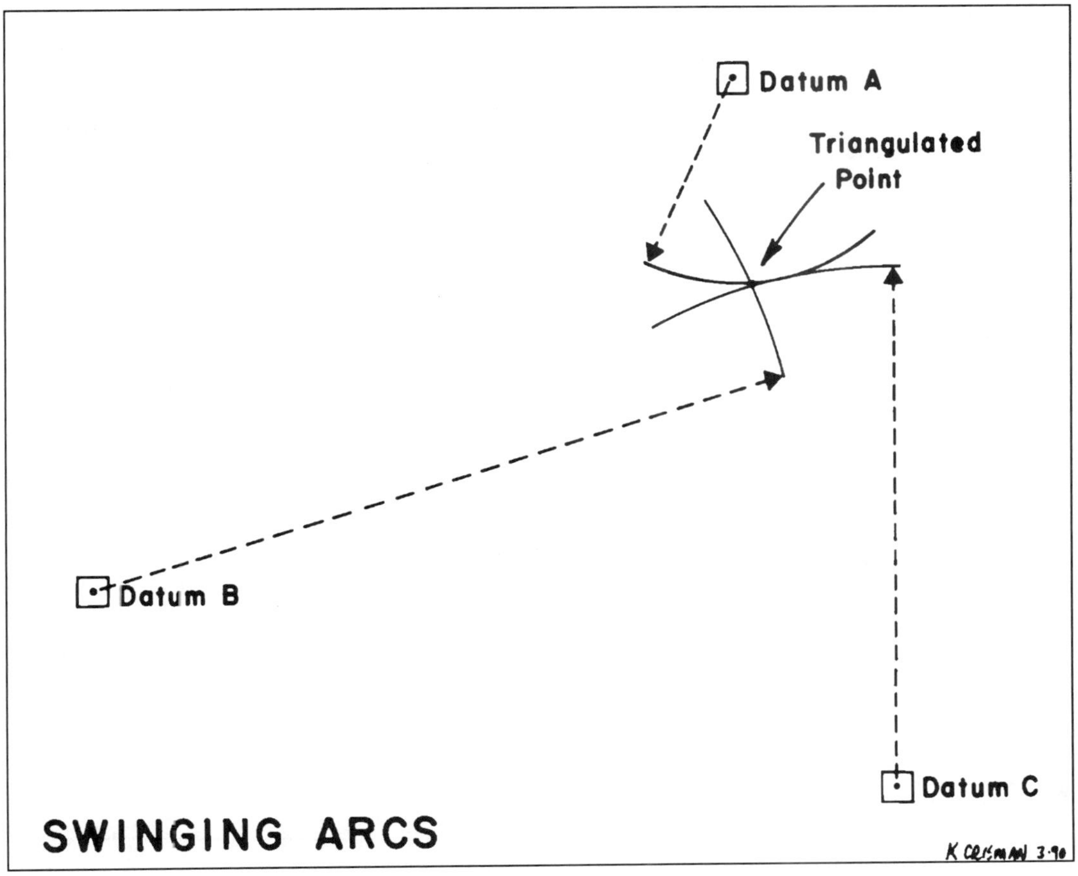

Figure 12.8. Sweeping arcs to locate a triangulated datum point.

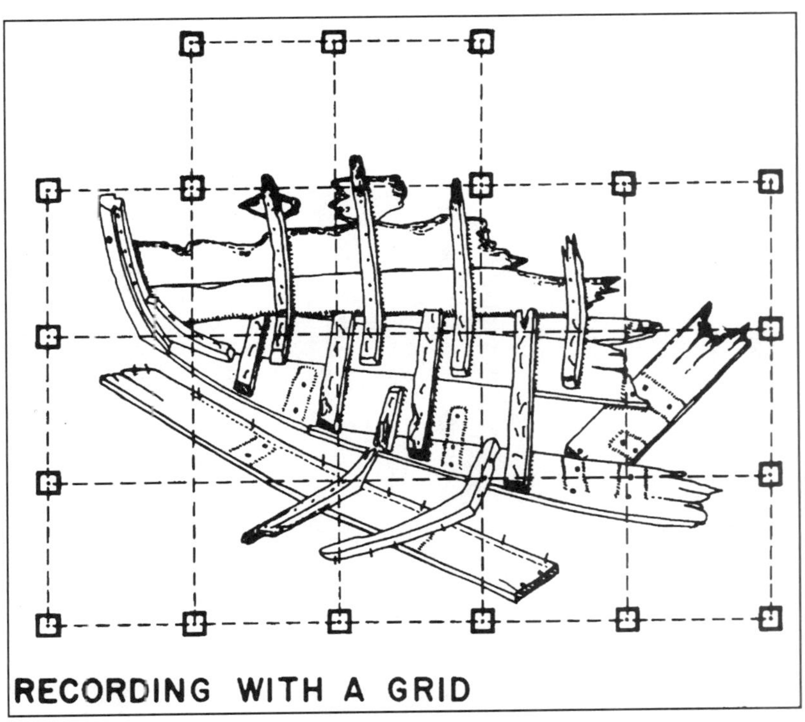

Figure 12.9. Using a grid to record a wreck site.

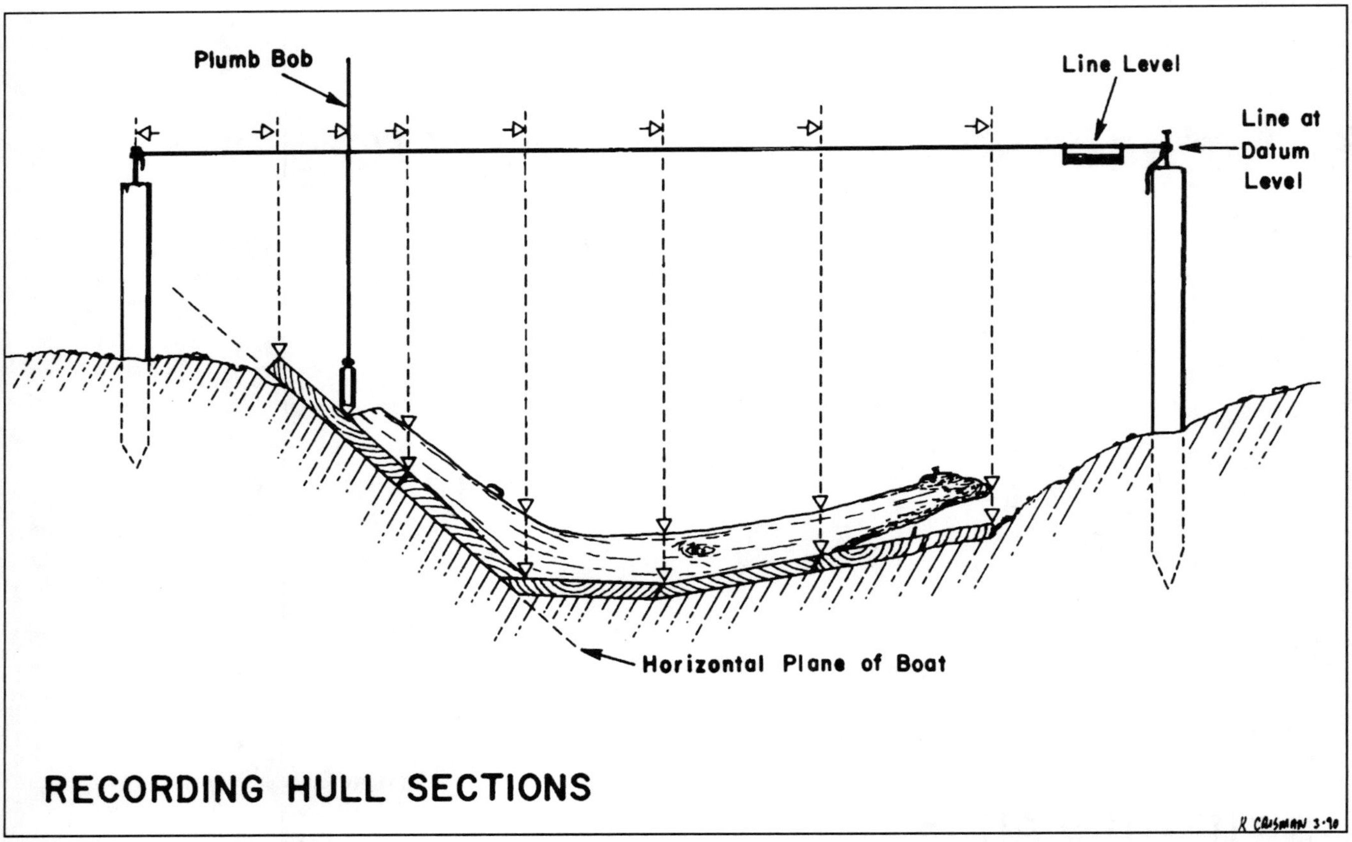

Figure 12.10. Using horizontal datum lines to record a collapsed, buried hull.

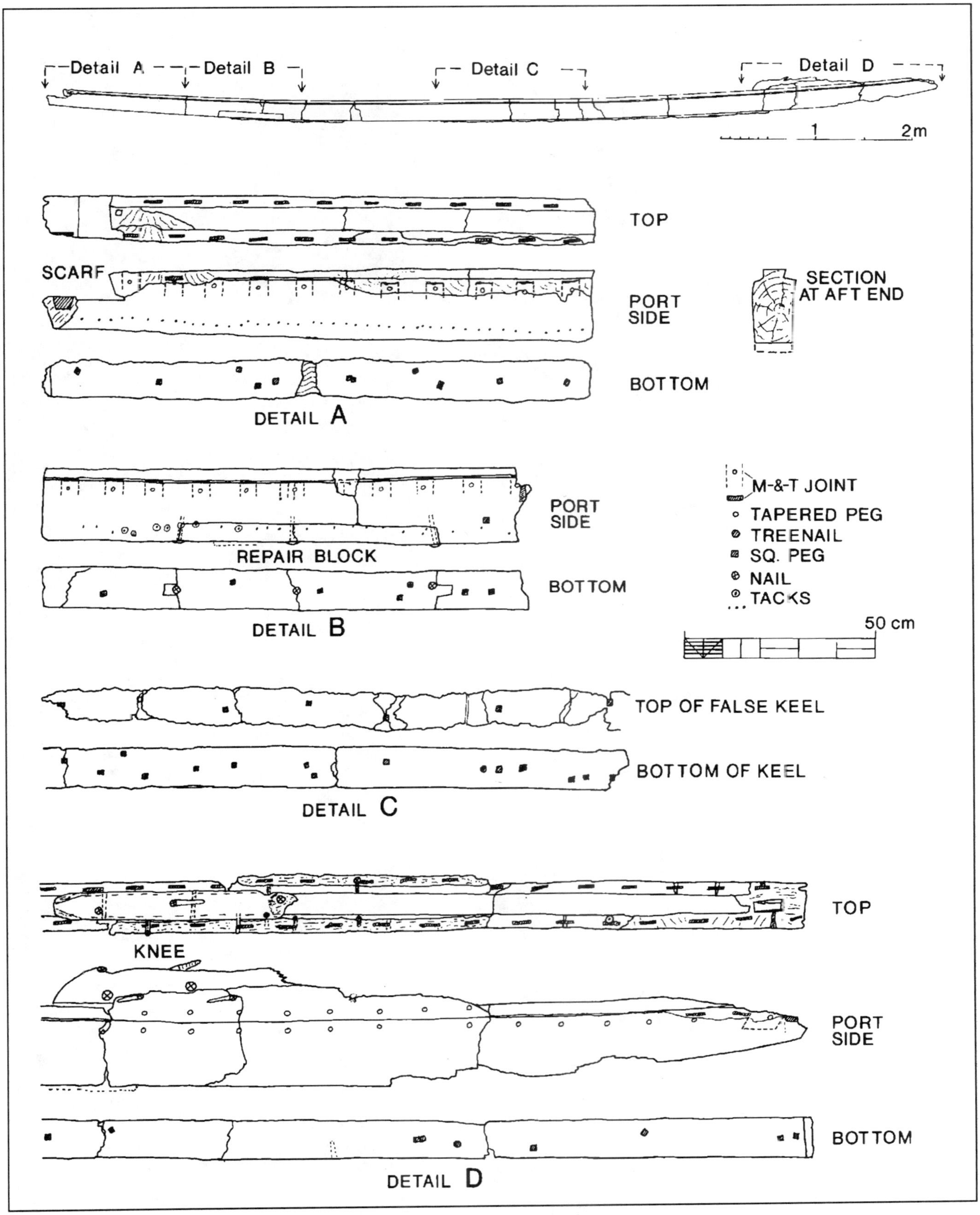

Figure 12.11. A few of the hundreds of detailed drawings of the fourth-century B.C. Kyrenia wreck.
J. Richard Steffy.

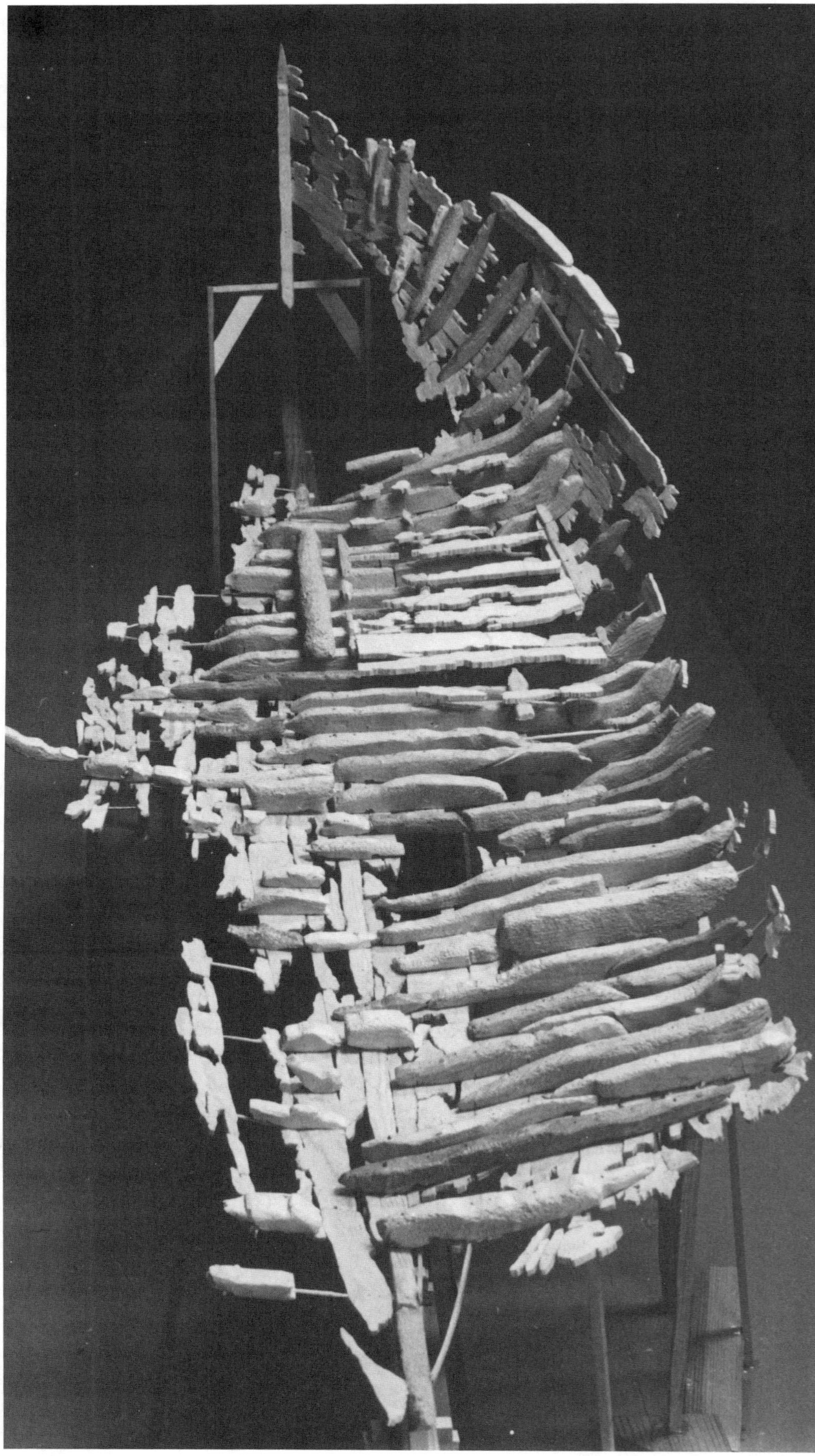

Figure 12.12. J. Richard Steffy's working fragment model of the eleventh-century Serce Limani wreck. It details erosion, patterns, fastenings, and joinery. *Cemal Pulak.*

Chapter 13

Sketching and Photographing Boats

The photographs and sketches that should be part of the recording process

Peter Schmid and Paul Lipke

Describing the complex, curved shapes found in boats is especially difficult. For this reason, illustrations—sketches, perspective drawings, and photographs—play an important role in documentation.

Sketching

A sketch is an illustration done on site. A perspective drawing is an illustration showing three dimensions; it is often of a quality suitable for presentation or publication. The purpose of both the sketch and the perspective drawing is to illustrate a point or points not best described in words, conventional construction or lines plans, or photographs.

Figure 13.1 shows a tender cut in section to illustrate its planking, frames, and other construction features at an optimum angle. It took just twenty minutes to draw, but to translate in words the total information it reveals would take pages of text. This illustration also shows the advantages and opportunities of the sketch or perspective drawing. In the drawing, the boat is dissected, a technique that allows the subject matter to be portrayed at any stage of dismantling without damage to the artifact. There are some drafting conventions used in this sketch to aid our understanding of the construction. First, the dissection is partly accomplished by drawing the boat in section. Second, a cut-away technique is used to represent material as if it had been sliced through, instead of as a complete drawing of an incomplete boat.

The sketch shows the planking cut in section with a pattern that is symbolic of wood: wood grain. Other materials are designated by less obvious symbols when they are shown in section; refer to a book on technical drawing for conventional symbols of other materials. Note the rear thwart is outlined in dashed lines, in order to show its edges while allowing other details underneath the thwart to be viewed. The relationship of the oarlock to its pad and the pad to the fastenings that secure it to the boat are shown in an exploded diagram, which allows an assemblage of parts to be viewed in isolation. In this illustration, the width of the keel is given, but many other dimensions could easily have been added. Dimensions, labeling, and notes help clarify the illustration, but leaving out noncritical information is equally important. A minimum of style is quicker to execute and easier to read. Finally, there is the freedom of sketching. It allows views that may be unachievable except in the imagination. The sketch can be judged successful if it illuminates the subject matter quickly, accurately, and in a way we comprehend with little or no additional commentary.

As in any illustration, the information to be transmitted determines the medium for drawing, its style, its point of view, and its level of detail.

The material on which to draw should be a good quality white sketching paper. A looseleaf format permits accessible storage in getting the paper to and from the site, ease in filing with other documentation, and ability to reproduce on copying machines with standard-sized papers. However, there may be times when looseleaf notebook paper is too small for the drawing or to accommodate the style of the illustrator. Paper with preprinted perspective lines helps to sort out problems in perspective and to avoid errors in the field. Available in only a few angles, preprinted perspective paper forces the illustrator to draw the subject from a point of view that may not be the most effective. Avoid kraft (brown) paper, cardboard, and handy scraps; they are not suitable for records.

The medium of illustration can be pencil, fiber pens, technical pens, watercolors, or markers. Black is the preferred color since it reproduces best. Pencil is the best medium to use. Inexpensive and erasable, pencils are available in wide ranges of hardness, permitting sharp lines and graded shading. Other media, such as watercolors, should be used only by someone with exceptional artistic skills.

Line quality and lettering style should be clear, crisp, and free of embellishments, and the drawing style should be subservient to the transmission of information. There is, however, room for self-expression. In producing a drawing for a museum or government agency, a highly finished, strictly technical approach, with inked lines and a minimum of freehand pen work, is desirable. Such illustrations are developed in several stages, never on-the-spot in the field. An illustration for a magazine article might justify a more stylistic or graphic approach—a freehand pencil or pen-and-ink drawing.

The point of view must be guided by what information is being transmitted, not which point of view is easiest to sketch. A bird's eye, three-quarter view from above is usually the most instructive point of view for capturing an entire boat; a two-dimensioned line drawing conveys the deck layout best; a three-dimensioned line drawing of plank seams with shading clarifies the compound curves of a hull. The level of detail to include is best determined by asking the question, "What will best help me convey the illustration's intended purpose?" Consider the following questions in deciding how much detail to include:

- Do you plan to show different types of materials?
- Is the apparent condition (rot, rust, or damage to frames) of the material important, or will notations suffice?
- Will you illustrate the functioning of movable parts (for example, the sweep to an oar or boom) with directional arrows?
- Do you need to show labels, descriptions, dimensions, or annotations for any part of the subject?
- Do you need to show methods of attachment?
- Do you need to show quality of workmanship?

Depending on what kinds and how much information you need to show, you may decide to make several sketches of the same object or feature rather than just one illustration. Attempting to cram too much information onto a single view is a very common mistake and can be self-defeating. Remember that if the sketches are prepared as part of the field notes compiled during documentation, each sketch should be labeled with the project title, date, time, and subject title.

The Method

Because boats are composed almost entirely of compound curves, they often lack the straight lines you need to determine the vanishing points of a perspective drawing. You should impose or imagine a working grid, in the form of a box subdivided by a system of perfect or imperfect cubes, around the subject. The outside edges of the box and cubes become the vanishing lines (fig. 13.2). The box becomes a placement system for the important lines that define a boat's length, beam, depth and shape: namely, the stem, stern, sheer, keel, and bilge. The cubes help define the bend or sweep of each line and its correct placement in relationship to the others.

This system of establishing a perspective grid is best appreciated when sketching on site. After you have determined the point of view and set the grid for a particular perspective, some lines—especially those viewed obliquely—are distorted and therefore difficult to define. You can leave that point of view and observe the line in pure profile, plan, or other less distorted angle. Obviously, viewing the artifact as though it were orthogonally projected (such as in pure profile) changes the perspective of the cube into a square without any apparent depth. Figure 13.4 shows an example. From the rear quarter, it is hard to determine the exact location and height of the sheer's lowest point. However, if you move to a profile view, you can clearly see that the lowest point falls just slightly forward of the aft face of the cube marked "X," and its height can be placed three-fifths from the bottom.

If you have only one opportunity to gather information in the field, you should make extensive sketches, photographs, and measurements. Because a field sketch most often is informational in purpose, a polished look is unnecessary. However, should the sketch be developed into a perspective drawing suitable for publication, the drawing can be verified only against the documentation you have on hand after leaving the site.

The Procedure

Look at the boat, and imagine a superimposed grid of cubes around it as shown in fig. 13.4. To help develop those first basic lines, set up a fore-and-aft centerline and plumb lines to determine the relationship of the boat to the grid. Draw this grid on paper and reverse the process, imagining the boat superimposed within it.

Look at fig. 13.2. Draw a fore-and-aft centerline at the keel. Draw one more line parallel to centerline and tangent to the hull's highest point. Drop verticals from the upper baseline to the lower baseline to complete a plane located along the boat's centerline.

Using this plane as a reference, draw the sides of the box containing the boat as parallel planes. Look again at fig. 13.2. You can see the athwartships locations of the sides are determined by the boat's maximum beam, and the top and bottom lines of the sides are at the same heights as the centerlines. Extend the top and bottom of the sides to create the right vanishing point at their intersection. Close in both ends of the box and locate the left vanishing point by connecting the ends of the box and then extending these lines.

Look at fig. 13.3. Break the box down into cubes. To determine the first cube's size, you need to know the measurement of half the maximum beam. Note the centerline plane divides the end faces of your box into two equal squares. These equal squares provide the basic unit to be developed proportionally in the remaining cubes. Estimate the correct length of the first square by proportionally projecting the half-beam width to unit "Y."

Once the longitudinal face of the first cube is determined, draw a line (A) diagonally through its opposite corners. A line (B) parallel to the first diagonal will run through the adjacent cube. To determine the proportional width of the second half-beam, draw another diagonal line (C) and a second parallel line (D). Repeat the process until the entire box if fully divided, keeping in mind that the last unit may be a rectangle. If the boat is significantly deeper than it is wide, you may need to divide the height into several cubes. To position superstructures, you might develop a separate series of cubes above the hull's primary box. A beamy catboat might be drawn in as little as four cubes or rectangles, but a canoe's proportions might require as many as sixteen.

The bow, stern, cabin top, sheer, and turn of the bilge constitute the major defining lines of the boat. The rectilinear grid will help you assemble them. Mentally or on paper, subdivide the cubes vertically and horizontally to position the sweep of each line. In fig. 13.4, the illustrator subdivided the port stern cube to clarify the sweep of the line that constitutes the port side of the transom. Remember that as each line is added to the grid, it becomes reference for those that follow.

In drawing an entire boat, you may find that the far sheer is difficult to represent. The line you draw may look strangely contorted, because the sweep of the sheer simultaneously falls away from the top of the box and swells away from the box's center. By carefully applying the grid while shaping the line, you can develop the correct proportion. Trust the grid, and add more lines until either awkwardness of the original line disappears or the source of the awkwardness becomes evident. In any event, the grid makes it easier to locate problems. If the point of view you have selected creates lines that *are* awkward, the grid method will give you confidence that what you have drawn is correct.

In your illustration, you must create a curve, or what is needed to imply a curved surface, so the viewer can interpret the surface shape. Use plank lines, deck seams, other boat parts, dashed station lines, or shading to clarify or emphasize difficult, compound curved surface areas or lines.

Smaller objects in the boat can be placed quite easily using the gridded envelope. If needed, develop a smaller set of cubes within the primary ones for a fine set of reference points. Or you may decide to make exploded drawings for components such as deck hardware. (See the oarlock socket in fig. 13.1).

Good working styles are best learned by analyzing successful examples, such as those included in the books listed in the bibliography. For the artist unfamiliar with marine illustration, a few minutes with any of these works will be very instructive. For superb hull depictions using the fewest possible lines, see Eric McKee's *Working Boats of Britain.*

Photography

In photographing boats for documentation, be prepared to make an extra effort in order to overcome significant obstacles in lighting and composition.

The 35mm camera loaded with black-and-white print film is acceptable for most documentation purposes. If you have access to larger formats, so much the better. In either case, the publications of the American Association for State and Local History and the Historic American Engineering Record, cited in the bibliography, are recommended. Bear in mind, however, the standards advocated in these works have architectural orientations. For instance, they recommend all photographs be taken plumb and square to the subject, while capturing the detail of a boat's shape and layout sometimes requires taking photographs at varying heights from the quarter or forward of a beam. If you follow the checklist provided at the end of this chapter, you will capture not only the curves of the hull but also the location and construction of the many components above and below deck.

Two lenses of approximately 50-58mm (standard) and 28mm (wide angle) offer compositional advantages over the convenience of a pocket camera with a single fixed lens. The wide-angle lens creates a good deal of distortion, but there will be situations where the boat cannot be moved, where overlapping pan shots do not satisfy your needs, and where nothing else will enable you to fit the entire boat in a single frame. A zoom lens will suffice; however, zoom lenses need more light and may not provide resolution quite as sharp as either of the single lenses.

The lighting inside boats and storage facilities is often poor. If true archival quality photography is warranted, take the time and trouble to move or illuminate the boat so that slow, high-resolution film can be used. If this is impossible or inappropriate, use faster ratings and then "push" the film. With standard film at normal speeds, the wide lens openings (*f*5.6 or less) and slow shutter speeds (1/30 of a second or slower) allow both camera movement and shallow depth of field to affect picture quality. For black-and-white prints, Tri-X can be pushed to ASA 1600 and T-Max to ASA 3200 with special processing. Both Tri-X and T-Max show high resolution when enlarged to 8 by 10 inch prints. Contact your local custom processor for availability and limitations, and be sure to tell your processor whenever film has been "pushed."

When ordering contact sheets and prints, be aware that the high-contrast, resin-coated papers used by many processing labs are not considered stable enough for some archives.

Order 8 by 10 inch contact proof sheets for use at the drawing board, and use a loupe or magnifying glass (4X to 8X magnification) to read them. Using a loupe and two or three contact sheets is easier than shuffling through thirty or forty large prints on a drafting board already crowded with drawings, field notes, and tools.

If you want color slides or prints for publication or presentation, take these presentation shots with color film, but do not rely on the slides or prints for the long-term record. They simply are not durable enough.

Whenever possible, insert a black-and-white graduated scale (sometimes called a "footbar") within the image, being sure to place it in (or up against) the plane of the subject. Your illustrator and future researchers will be eternally grateful. If you know the size of something in the image (from your field notes), you can use that to make a scale; but if the viewing angle isn't close to 90 degrees, then the graduated scale can be especially helpful.

If at all possible keep a photographic log, recording film type, film speed, shutter speed, f-stop, subject, and point of view. The last two items in the list are especially important, since pictures of collapsed hulls or close-ups of components, such as the underside of a deck, can be hard to orient and understand, particularly if they are viewed by anyone other than the photographer.

There seems to be some perverse natural law that requires boats to be almost as big as the available storage space. Capturing a complete 90 degree profile is especially difficult. To do so, try flattening your tripod-mounted camera against the restricting wall, and take overlapping shots. Make sure you have the same major construction feature on opposite sides of each adjoining frame so that you can align the images. Removing the thickness of your body from between the wall and the camera can make a big difference.

The best use of tripods is probably not for holding the camera but rather for positioning additional lighting. If you highlight shape and detail by strategically placing lights, you will be rewarded at the drawing board. "Balanced" lights for photography are a wonderful luxury, but you can easily use a couple of cheap, clip-on reflector lamps with good results. By augmenting the camera's flash unit, you will be better able to highlight shape and reduce glare. Below deck, you will probably need both the reflector lights and the flash unit. Having said this, tripods are essential for low-light conditions where separate lights or flash attachments are inappropriate, inadequate, or unavailable. If you cannot make the tripod stable inside the boat at the position you want, try using just one or two of its legs in combination with your hands, while bracing your elbows on some part of the boat.

Be willing to compose the picture by moving the boat if you find that you cannot otherwise achieve clear views of the subject. If the boat is afloat, docks or other boats are certain to be in the way, so plan to move the boat to a better location. Be creative in your use of other boats as shooting platforms; dinghies will do, but boats with large towers, such as sportfishermen, offer a useful variety of heights from which to shoot.

Even pictures taken with inadequate equipment and unstable film (such as "instant" camera film) are far better than no pictures at all. You can always make stable copy prints at a later date.

Organize and label your negatives, contact sheets, transparencies, and prints. There is no point in taking pictures if the material is assembled the way the average family assembles a photo album. A total stranger who has never seen the boat or other supporting documentation should be able to look at any one image and know what it is. Close-ups of construction details and interiors can be especially hard to identify if labels are not provided. If there is any room for doubt, provide some indication of orientation; what appears to be down may in fact be up.

The following is a list of the most useful views to be captured by photography:

- 90-degree profile view of entire boat
- 90-degree plan (deck) view of entire boat (turn small boats on their sides, climb the mast)
- 90-degree view of bow
- 90-degree view of stern
- end view of bow from waterline height
- end view of stern from waterline height
- end view of bow and stern from eye level, or well above sheer
- end view of stern perpendicular to transom face
- 3/4 profile view towards bow from quarter
- 3/4 profile view towards stern from broad on bow
- midsections viewed from the end immediately beside the bow
- midsections viewed from the end immediately beside the stern
- detail of rabbet at stern if not clear in the 90-degree view of bow
- detail of rabbet at stern if not clear in the 90-degree view of stern
- exterior plan view of foredeck area, including mast partner, winches, bulwarks, and rails
- cockpit-companionway exterior
- cabin trunks with skylights, handrails, and vent stacks
- steering position-mechanism
- keel, keelson, floor timbers, and garboard from interior amidships
- keel, keelson, floor timbers, and garboard at bow and stern.

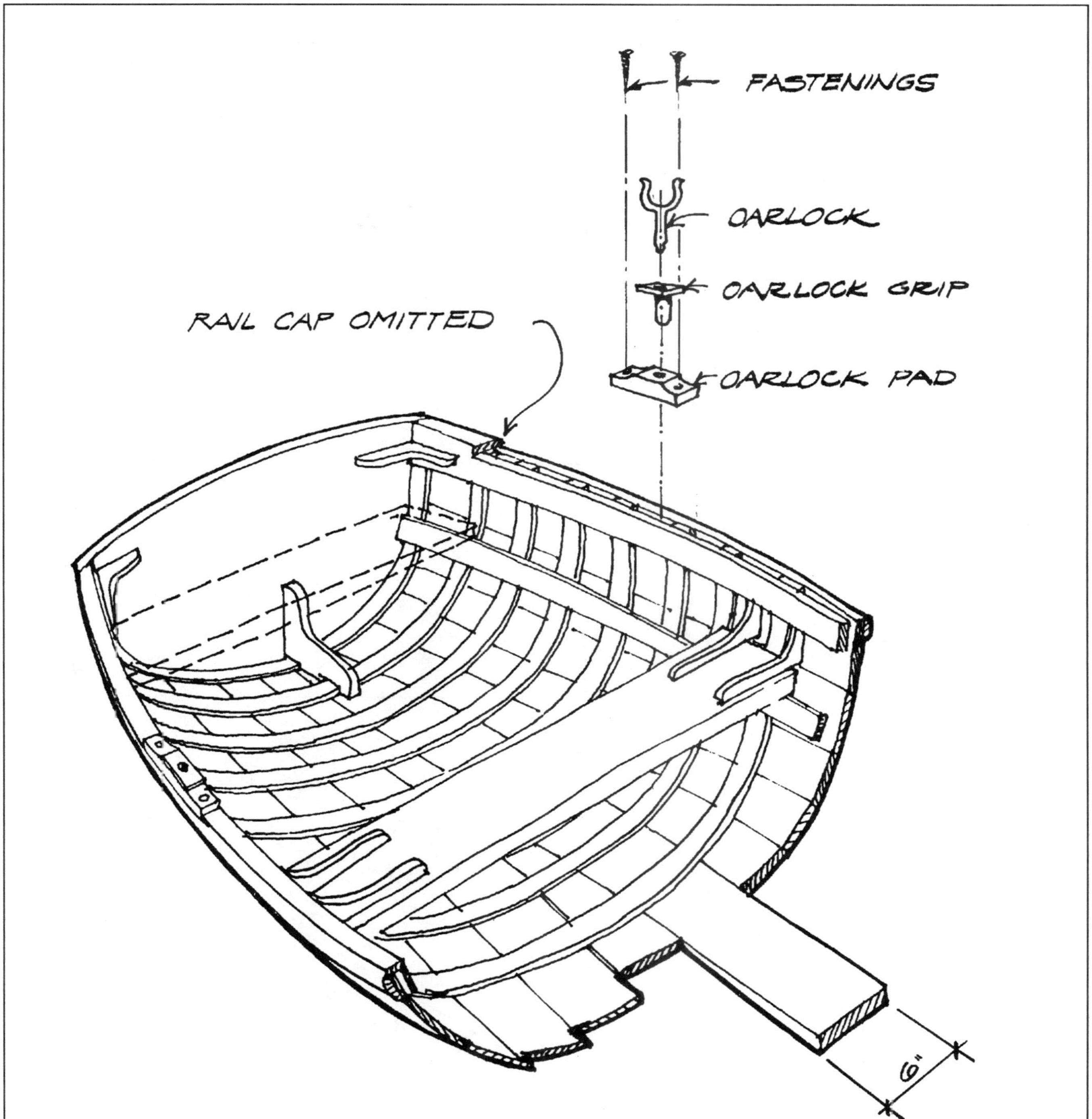

Figure 13.1 Perspective drawing can emphasize important details and convey information quickly. *Peter Schmid.*

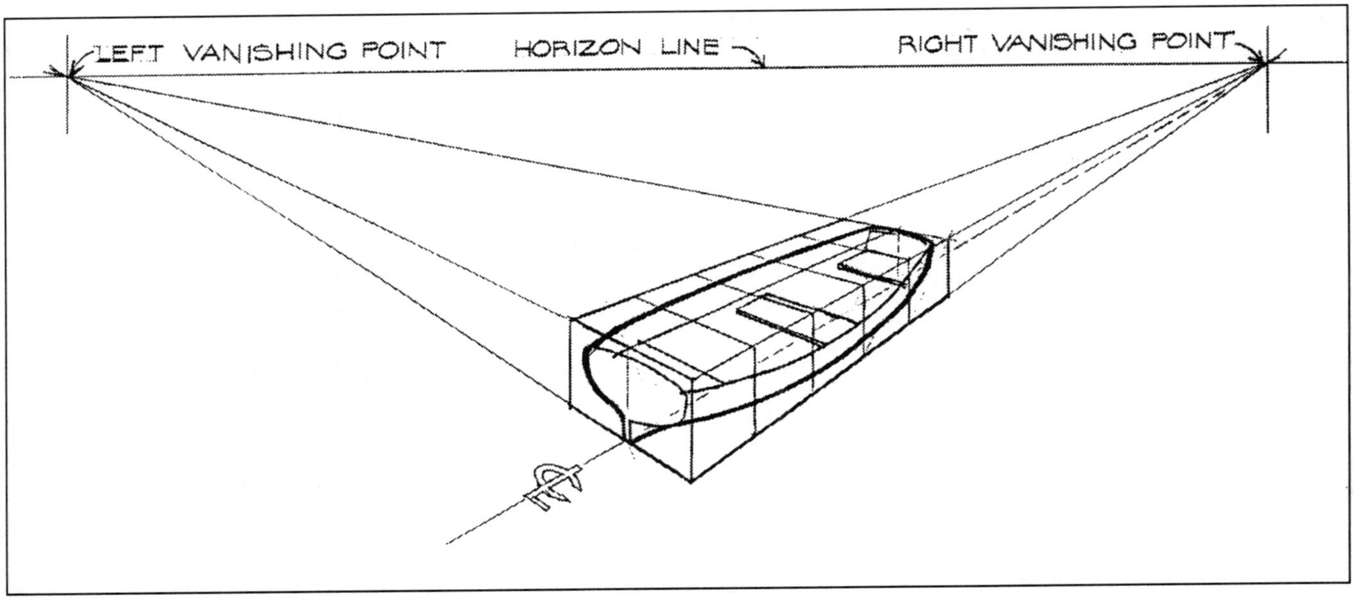

Figure. 13.2 Using the principles of two-point perspective, the boat is "placed" within a box. *Peter Schmid.*

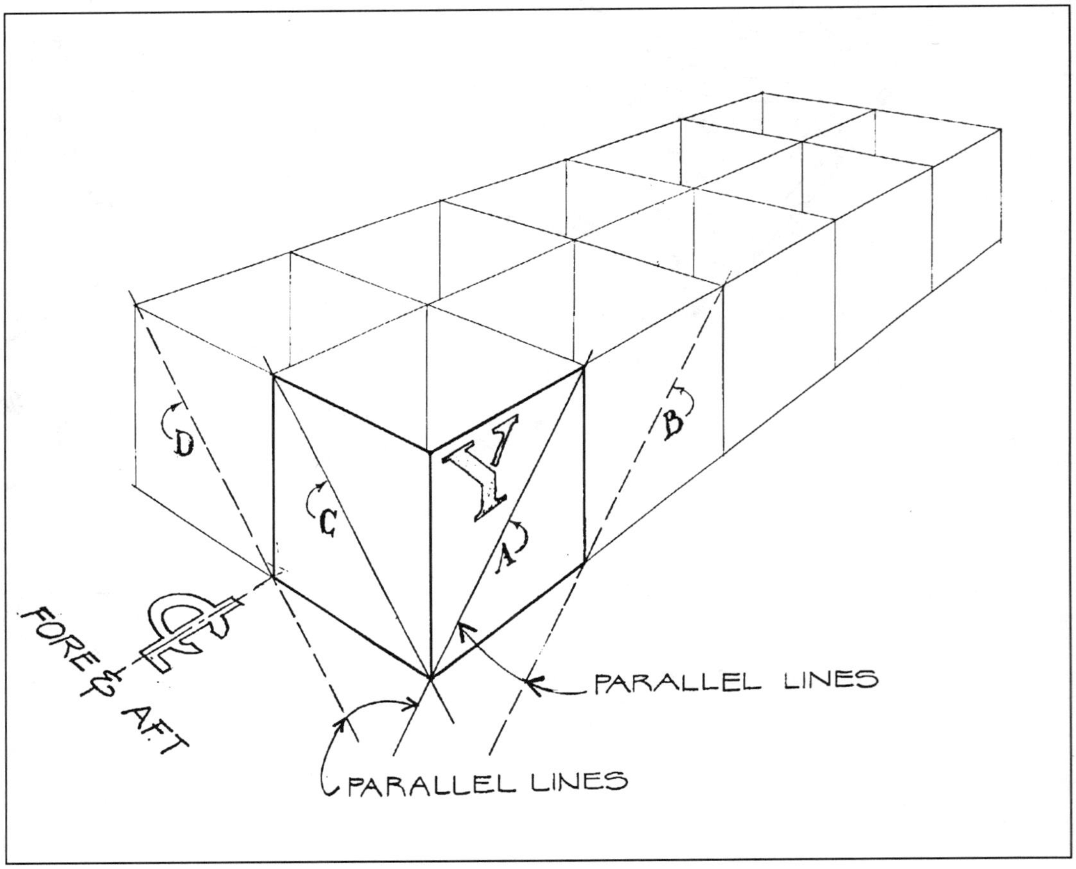

Figure 13.3 Developing a perspective scale within the box makes it easier to keep the boat's many components and lines in proportion. *Peter Schmid.*

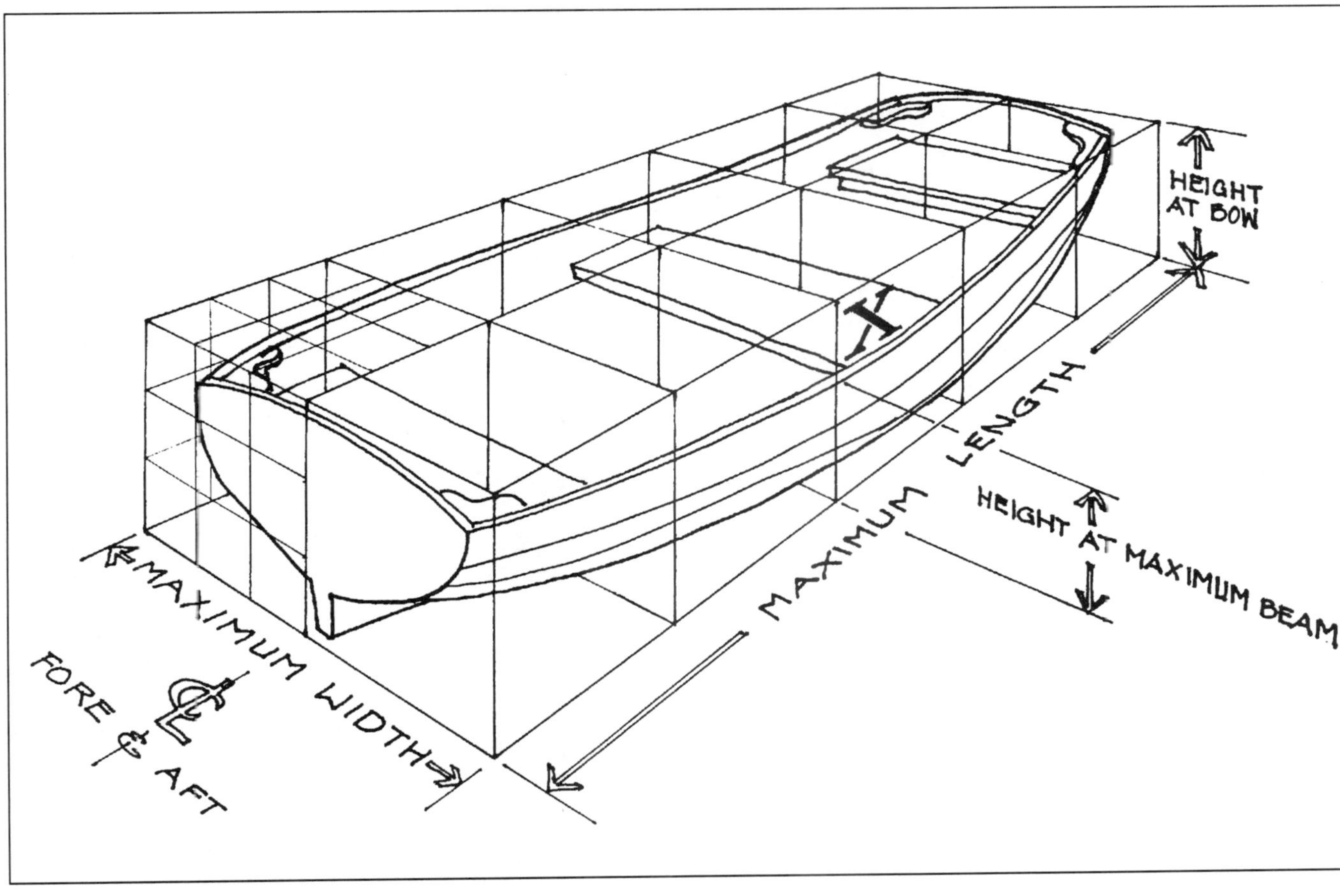

Figure 13.4 Working through the details, it is sometimes helpful to subdivide the squares, as has been done on the port side of the transom. *Peter Schmid.*

Chapter 14

Historical Photographs

Finding historical photographs of boats and learning from them

William N. Peterson

Seeking documentary photographs for any given subject can be a major undertaking. Historical photographs are particularly difficult to find, especially those taken before 1890. Their discovery, even when they do exist, can be complicated when the subject is as arcane as "American small craft." Compounding the problem of scarcity is the lack of a "union list" indexed by subject matter in many of the public repositories that collect photographs. In other words, even when a photograph exists, it may take a Holmesian sleuth to track it down. In addition, most photographs in public collections tend to be catalogued by the photographer's name rather than by subject matter. Be prepared to find your quarry in unlikely places.

Before the late 1880s, the mysteries of the photograph were the domain of the professional picture-taker. High costs associated with the technology—overhead, bulky camera equipment, delicate glass plate negatives, and bottles of developing chemicals—prohibited amateurs. In fact, prior to the early 1880s, if the photographer wanted to take views outside the studio, he or she had to take the dark room along because the photographic process required the glass negative to be exposed while the light sensitive collodion emulsion was still wet. The somewhat more practical "dry-plate" glass negative did not come into general use until 1882.

Wet-plate (collodion) negatives can usually be distinguished by their greenish yellow color as opposed to a darker green hue found on dry plates. Also, dry-plate negatives tend to be uniform in size and thickness, while wet plates are often irregular. Before the discovery of even these cumbersome systems, photographers (at least in the United States) relied almost exclusively on processes such as the daguerreotype, ambrotype, and ferrotype (tintype)—processes that did not yield a negative but were one-of-a-kind (direct positive) images that could not be reproduced.

It is probably fruitless to seek photographic images to document small craft prior to 1859, which is the year the wet-plate process came into general use by photographers. Articles and books that predate this period and which deal with American small craft, particularly working small craft, are seldom illustrated with photographs. Even professional photographers whose work began in the era of the glass-plate negative and who specialized in maritime subjects more often than not concentrated on taking views of larger yachts and commercial sailing vessels. The names of Edwin Levick, Nathaniel Stebbins, James Burton, Charles Bolles, and, slightly later, Morris Rosenfeld come to mind in this regard.

Another reason for the lack of small craft and ship coverage is that early plates were incredibly insensitive, requiring exposures of many seconds. Floating vessels would show up as a blur—not worth attempting to photograph.

To a large degree, the impracticality of the photographic medium for the common man and woman came to an end in the late 1880s with the development of flexible roll film and the introduction of George Eastman's hand-held Kodak camera. At this juncture, the amateur photographer stepped into the picture, so to speak, and began to take images of various topics of individual interest. Images of small craft appeared, particularly those depicting small pleasure boats and yachts. Indeed, small craft went from one of the least common maritime photographic subjects to one of the most ubiquitous. Specific attention to small working craft, however, continued largely to be neglected by both amateur and professional until relatively recently.

Having made the case for the scarcity of early photographic images, there is one once-popular type of photograph that is worth the attention of small craft enthusiasts. The development of the paper stereograph in the late 1850s established for the American consumer a vast window to his or her nineteenth-century world. Stereographic photographs were taken of every conceivable subject. Stereographic views and hand-held viewers were indispensable features of most middle- and upper-class parlors in the nineteenth century. Subjects that especially fascinated the public were images of the seashore and waterfront. Crowded wharfs, steamboats, and shipping scenes abounded. It is by paying close attention to these photographs that the researcher can discover in the background, often partially hidden, the elusive image of the small working boat (fig. 14.3).

Stereographic views can be found in the collections of almost every state and local his-

Figure 14.1. Historical knowledge would aid the researcher in identifying this albumen photograph of an outing on the Kennebunk River as one taken after 1888 since it is a view of one of the annual meets of the Kennebunk River Club, which was not organized until 1889. Judging from the ladies' wide hats and blouses with puffy shoulders, it is probably the mid-1890s. *Mystic Seaport Museum, Inc.*

torical society as well as most history museums. There is even a National Stereoscopic Association, which publishes a quarterly journal *Stereo World.* Queries to this organization (National Stereoscopic Association, Oliver Wendell Holmes Stereoscopic Research Library, Eastern College, Saint Davids, Pennsylvania 19087) may prove helpful in locating obscure collections and in dating or otherwise identifying images. In this regard, there are many useful tricks of the trade in dating stereographs. Often the name of the photographer or publisher is embossed or stamped on the verso (left-hand page) of these images. The active dates for most photographers are known or can be easily discovered. Even when the active period spans thirty or more years, you can analyze clothing, shop signs, structures, and other identifiable features shown in the view to narrow down the period.

Studying the photograph as an object itself can also be useful. Variations in the style of lettering and logo on the reverse or ends of the stereo frequently changed and can, with study, be chronologically identified. Over the years, the card stock upon which the images were mounted also changed in style, shape, and size. By familiarizing yourself with these features, you may be able to date a stereograph broadly. For example, the photograph in fig. 14.2 can be dated to about 1873 by noting its original 3 1/2" x 6 15/16" size, its rounded corners, and its card stock color--yellow with a blue verso--common to that time. Square corners with yellow or pale yellow card stock were commonly used in the early and mid-1860s. Dark gray and curved mounts were typical of the late 1890s and early twentieth century.

The lengthy caption with fig. 14.3 gives some indication of how much information can be gleaned from a single image when backed up by research, yet the image's potential has hardly been exhausted. Additional lines of research leading to more information about the boat (or even its possible location) might include:

- Identities of the men in the boat, particularly the young man in the bow who might still be alive.
- Identification of the flags painted on the second strake near the bow. Since these

Figure 14.2. The union boat house at "boathouse row" was operated by "Capt. W. F. Cline" and is shown tied up along the east bank of the Schuylkill River at Fairmount Park, Philadelphia. A typical stereograph such as this on taken around 1873 can be identified by the card mount which is yellow with blue verso. *Mystic Seaport Museum, Inc.*

Figur 14.3 This view of a fourteen-foot lapstrake skiff was taken in 1914 at the George Lawley & Son Boatyard at Neponset, Massachusetts. The date comes from the photograph's original envelope, which was not discarded until all the information was copied. Information in this particular case also noted that this was a "stock skiff." The time period can be approximated by an examination of the clothing worn by some of the men in the boat and by noting the small two-stacker torpedo boat in the distance. Small craft enthusiasts will also note the fine little carvel-planked yacht tender with the name *Zenda* on the transom, which is a typical example of how small craft show up as incidental features in photographs. Even lacking the convenient date on the old negative envelope, a quick check of the *Lloyd's Register of American Yachts* would indicate that the yacht *Zenda* was built by the Lawleys in 1912 and that in 1914 it underwent an engine change. Sometimes the researcher can save some legwork by looking for small craft where they ought to appear as a matter of geographical or historical happenstance. A study of yawl boats, for example, might lead one to maritime collections noted for their holdings of multi-masted coastal schooners, or the study of "Cowhorns" would logically send the researcher to find views of Block Island. Common sense is always the rule. *Mystic Seaport Museum, Inc.*

look like yacht club burgees, you might start with the yacht clubs near Neponset, or at least on Massachusetts Bay. Sending a letter including a sketch of both burgees to members of the Massachusetts Bay Yacht Clubs Association would be a good place to start. If the flags can be identified, the yacht club's records might well provide a wealth of data.

- Library research on the George Lawley & Son Boatyard would lead you to such readily accessible sources as *WoodenBoat* and *National Fisherman,* both of which have had articles on Lawley and his tenders. You may find contact with the authors of those articles fruitful, since writers often have a lot of information they don't use in print. The authors might even be able to tell you if the boatyard's records still exist.
- Digging deeper might take you to Neponset-area historical societies and contemporary periodicals such as *Rudder* wherein you might see advertisements for this model.
- Ask yourself what other vessels or elements in the image could be informative? Why was the picture taken? Was it to show load capacity? Was this the tender for the larger vessel shown in the view?
- Using the men's bodies to help give a sense of scale, can you estimate the boat's length. If the boat's interior is visible, as in fig. 14.2, it is possible to use regularly spaced features such as framing to scale off lengths or to get a sense of scantling dimensions.

Among the organizations and museums that are known to have extensive collections of maritime photographs are the following:

The Adirondack Museum, Blue Mountain Lake, New York
The Mariner's Museum, Newport News, Virginia
National Archives and Records Administration, Washington, D.C., and its eleven field branches in Boston, New York, Philadelphia, Atlanta, Chicago, Kansas City, Fort Worth, Denver, Los Angeles, San Francisco, and Seattle
Calvert Marine Museum, Solomons, Maryland
Mystic Seaport Museum, Mystic, Connecticut
South Street Seaport Museum, New York, New York
The Whaling Museum, New Bedford, Massachusetts
Society for the Preservation of New England Antiquities, Boston, Massachusetts
San Francisco National Maritime Historical Park, San Francisco, California
Boston Public Library, Boston, Massachusetts
Marine Museum at Fall River, Fall River, Massachusetts
Cape Ann Historical Association, Gloucester, Massachusetts
North Carolina Maritime Museum, Beaufort, North Carolina
Chesapeake Bay Maritime Museum, St. Michaels, Maryland
Connecticut River Museum, Essex, Connecticut
Philadelphia Maritime Museum, Philadelphia, Pennsylvania

Steamship Historical Society, University of Baltimore Library, Baltimore, Maryland
The Antique Boat Museum, Clayton, New York
Great Lakes Historical Society Museum, Vermillion, Ohio
Suffolk Marine Museum, West Sayville, New York
New Haven Colony Historical Society, New Haven, Connecticut
Peale Museum, Baltimore, Maryland
New York State Historical Association, Cooperstown, New York
Historical Society of Talbot County, Easton, Maryland
The Peabody Museum, Salem, Massachusetts

Interest in historic photographs of small craft is likely to take the researcher to many of these institutions. No single archive has a comprehensive collection of small craft photographs. Small craft historian John Gardner attributes this lack to the fact that until the Adirondack Museum was founded in the 1950s, the notion that specific attention be given to the subject was unheard of for public institutions. Nevertheless, many maritime museums have in the meantime rectified, to some degree, this deficiency.

An important note when using any repository: be prepared to pay production costs and in some cases user fees for any images you may want. Increasingly, repositories require user fees and credit lines when your research ends in a published form. Finding, administering, and preserving historical photographs is a costly business, and agencies that provide images for public use have a right to expect some compensation for their stewardship. Some organizations may allow you to copy their photographs on site, but many more prefer to provide you with the images themselves in order to maintain control of the negative. Be prepared to allow two to six weeks for your order.

The days of the "freebies" are largely over because it is too costly to provide free copies of photographs. In fact, in the last decade, requests for photographic reproductions have tripled at most institutions that specialize in historical images. Your order is probably only one of many. In addition, museums and libraries usually make prints to order and do not, in most cases, carry an inventory of images.

Another important source for early maritime images are maritime collectors and scholars, who can often be found at the various maritime symposiums and small craft workshops held each year in such places as the Maine Maritime Museum, the Naval War College, and the Mystic Seaport Museum. In addition, organizations such as the National Maritime Historical Society, the North Atlantic Society for Oceanic History (NASOH), the Gulf Coast History and Humanities Conference, and other groups hold annual symposia at changing locations. The amateurs and professionals who attend these functions can be very helpful, and even if they don't share your particular interest, they can often steer you to someone that does.

Caring for Your Historical Photographs

As a primary research source, vintage photographs, in spite of their relative scarcity, are perhaps the most commonly found historical documents. Most every family has tucked away in the closet or attic a box that contains old photographs. Their familiar appearance in most homes has often led to an indifferent attitude toward their care. Valuable photographs have been thrown away or unintentionally damaged simply because their most current owner had no appreciation for their historical worth.

The historical photographs that remain must be cared for properly. Rule number one is never attempt to clean an old photograph yourself! Photographs taken in the nineteenth and early twentieth centuries were made by many different processes. There are albumen, gelatin, gum bichromate, carbon, platinum, and a vast array of other types. If you can't tell the difference among the types, leave them alone. A cleaning process that may appear to work for one type of photograph can be disastrous to another. There are a number of qualified photograph conservators in the country. Ask at any major history or art museum for the one nearest you. In any event, most images will not need restoration, but there are a few fundamental precautions for their basic preservation.

A cool storage area is preferable for photographs. In fact a temperature under 70 degrees F. maintained all year long is best. A dry storage area is also preferable to a damp one. Again a constant humidity of less than 60 percent is preferable. Attics are not good storage places. In the summer, temperatures rise very high in attics or any closed room and often drop rapidly at night. When this happens, minute condensation forms that eventually will crack and damage emulsions and print images. Humidity is also a breeding ground for mold, which can badly damage an old photograph. If ideal temperature and humidity cannot be maintained, at least try to achieve a stable environment. Take care that the storage site cannot be infested by insects. Gelatin emulsions common in many early photographs are tasty treats for mice, silverfish, and cockroaches.

Never write on the back of a photographic image. When bearing down on the back of a photograph, you can cause the imprint to come through on the image side. In humid areas, ink on the back of a photograph will frequently bleed through to the image side. Or ink on the reverse of one print will transfer to the image side of another photograph stacked underneath. If you absolutely must write on the back of a photograph, do so very lightly with a soft pencil. Do not use grease pencils for marking on original prints or negatives; when you remove the marks, abrasions can occur. Confine your use of the grease pencil to copy prints only.

Never use pressure sensitive tape to attach information or make repairs to photographs.

Never put paper over a photograph and write on top of it. For that matter, it is not wise to place anything on top of unique historical images. The chance that the image will be scratched is very great. In this regard, it is better to use hand-held magnifiers than lupes, which must be placed directly on the image for best magnification.

Never eat or drink while working with historical images. Smoking is also hazardous to the health of photographs. In addition to accelerating the yellowing process, hydrogen sulfide in the smoke will actually hasten the breakdown of the chemicals that produced the image.

When storing photographs, place them in chemically inert acid-free boxes. Place photographs of similar size together, so that wide, heavy images are not sitting on top of smaller ones. Each image, in fact, should also be stored in its own individual acid-free envelope or, even better, in triacetate or polyester sleeves. Both of these products are inert to photographs and allow the viewer to see the image without actually handling it.

When handling original prints and negatives, wear clean white cotton gloves. They are available at most pharmacies for less than a dollar a pair. No matter how dry or clean your fingers seem to you, they invariably will transfer dirt and salts to the image. Fingerprints put on an image today can eat into the chemically sensitive developing emulsion and show up years later as smudges.

Never display or store photographs in rooms that receive direct sunlight. Rooms with fluorescent lights are not recommended either, since this light will also quickly fade photographic images. Even incandescent light, over a long enough period of time, will fade photographs. A completely dark storage area is best.

Cataloguing

Cataloguing photographs can be a highly personal procedure. In my own system, I file the images according to the vessel type or rig and then arrange each group according to the name of the vessel. Unidentified craft are filed behind the last identified image in its category. This system works best when the original photograph is not used at all but is replaced by a contact print from a negative taken from the original image. With a numbering system in place, you can always retrieve the original photograph (which ideally remains in a proper storage box) if it is needed. If you use your photographs often, this system can save a lot of wear and tear on your valuable collection.

It is worth pointing out that the final disposition of your collection needs your attention. Consider how often important collections amassed over a life time by a meticulous collector or researcher have been lost to posterity because no provision for their disposal was made. In some cases, important collections of photographs have been unceremoniously thrown away or sold to antique dealers by unappreciative survivors. In short, make some provision for your collection in a legal will. Your favorite maritime museum or historical society is always a good choice.

Chapter 15

Beyond the Boat: Documenting the Cultural Context

Recording the cultural, social, and economic environment of the boat; ethnographic fieldwork methods

Paula J. Johnson and David A. Taylor [1]

When is the documentation of a vessel complete? After the lines have been rendered on archivally safe Mylar and deposited in the collections of a public institution? When the vessel has been photographed from stem to stern? After a data record has been entered in a computer database?

For those who view small craft as artifacts and, therefore, as expressions of material culture, full documentation demands looking beyond the boat. It demands that attention be directed to maritime resources as significant as the boats themselves—namely, the builders and the diverse cultural contexts within which vessels are conceived, produced, and used.

This is hardly a new concept. Anthropologists, archaeologists, folklorists, historians, and some museum professionals have long asserted that understanding an artifact's context is crucial to understanding its meaning. Recently, in the field of maritime history, Peter H. Spectre called for increased research into the contexts of our maritime artifacts, predicting that, if this is not done, "we will eventually reach the point where we will be surrounded by historical debris and nobody will have a notion of what it all means or even know where to look for an answer."[2]

Vessels, like all artifacts, have many contexts, and it is useful to distinguish among them here. The work of Howard I. Chapelle, for example, offers a typological context that attempts to explain the commonalities and differences among boat types, identifying a boat's place within a given time period, geographical area, or fishery. Social historians also offer contexts; for example, when researching the history of a late nineteenth-century trawler, they might provide a description of the cod fishery, highlighting the contributions of the various ethnic groups who worked in it.

Anthropologists, folklorists, and others have contributed contextual descriptions of a similar kind in their ethnographic approach[3] to documentation. Building upon the foundations established by scholars of artifacts and history, these researchers offer descriptive accounts of community life and the artifact's place within it. Their work characterizes the living, human context in which boats and other cultural expressions—physical and nonphysical—are created and used, emphasizing the knowledge, values, traditions, and perspectives of community members. Among the best examples are Janet C. Gilmore's *The World of the Oregon Fishboat: A Study in Maritime Folklife*, Michael Orbach's *Hunters, Seamen, and Entrepreneurs: The Tuna Seinermen of San Diego*, and David A. Taylor's *Boat Building in Winterton, Trinity Bay, Newfoundland.*

Not all ethnographers choose to use the same methods and strategies to discover and describe the meanings and distinctions used by people of other cultures. An ethnographic study may be broad, describing, for example, the whole of the New Bedford, Massachusetts, from the factory to the church to the fishery. Or, it may focus more narrowly on the context of an occupational community, documenting, for example, the maritime industries of the Chesapeake, including watermen, boats and boat builders, packing houses and workers, but not documenting those aspects of Chesapeake culture outside the marine focus. (See, for example, Paula J. Johnson's *Working the Water: The Commercial Fisheries of Maryland's Patuxent River*). Folklorists have often analyzed a community's expressive culture[4] in order to view the cultural and natural environments through "native" eyes.[5] (See, for example, George Carey's books entitled *Maryland Folklore and Folklife* and *A Faraway Time and Place: Lore of the Eastern Shore*).

An ethnographic approach will not give much insight into the cultural context of *historical* small craft, as it relies on a researcher's ability to document a living, dynamic community. However, the time is ripe for examining the occupational descendants of those builders who created our historical vessels—the traditional builders who are building small craft today.

By *traditional* builders we mean those people who are members of a folk group[6] that builds boats according to local practices. Typically, they learned their craft informally, by imitation of or instruction from other members of the community. And they build boats reflecting regional and occupational preferences, often using highly localized design and construction techniques. Documentation of the skills, knowledge, and lifestyles of these builders is particularly important because few, if any, records of them or their work exists. In addition, traditional builders are becoming fewer and farther between. With the increasing disintegration of maritime communities along America's waterways and coasts, the knowledge and skills of those bearers of maritime tradition are not being passed along to younger generations.

Sections of this chapter offer suggestions for conducting ethnographic research for the purpose of documenting the cultural context of traditional small craft. This type of activity should be an integral part of all small craft documentation efforts, but, in terms of approach, techniques, and goals, ethnographic research is quite different from other types of documentation described in this manual.

One goal of this manual is to help individuals who wish to record small craft do so in a way that will generate comparable data for boats across the nation. This manual, therefore, works toward standardizing procedures, much in the same way the U.S. Department of the Interior and state historic preservation offices follow standard procedures for documenting historical structures and, quite recently, large vessels.[7] These programs separate documentation according to levels of thoroughness, ranging from a quick, elementary survey of a property to a detailed study that makes use of a number of perspectives and techniques to illuminate all of a property's features.

Compared to standardizing the documentation of an artifact, standardization of procedures for documenting cultural context is infinitely more difficult, because both the sources for and collectors of ethnographic data are human beings, with attitudes, personalities, and histories that influence all manner of interaction. Approaches that work well for one researcher may not work for another. In addition, documentation approaches vary from project to project because of differing project goals. A museum may set out to document a community for the historical record as well as for background material for an exhibition. An individual may document maritime culture to fulfill requirements for a graduate degree. A state agency may sponsor a project for locating maritime cultural resources in order to put on a folklife festival. In each case, researchers will approach documentation according to the specific project goals. Still, while taking these variables into account, attempts to standardize procedures have been made and field-tested with some success. (See, for example, Thomas Carter and Carl Fleischhauer's *The Grouse Creek Cultural Survey: Integrating Folklife and Historic Preservation Field Research* and David A. Taylor's *Documenting Maritime Folklife: An Introductory Guide*). In the remainder of this chapter, a number of procedures are presented that can be used to document the cultural contexts of small craft.

Figure 15.1. Jimmy Trossbach, right, learns how to build boats the traditional way from his father, Bill. The Trossbachs, joined by another of Bill's sons, Buddy, built a forty-six-foot workboat for Jimmy in the summer of 1983. Researchers from the Calvert Marine Museum visited the boat building barn at least twice a week to record the process. *Photograph by Paula Johnson, courtesy of the Calvert Marine Museum.*

Formulating a Research Plan

Just as it is advisable to have a clear plan in mind before a vessel is constructed or measured, success in documenting the cultural context of a vessel will be greatly enhanced if the researcher begins with a carefully considered plan of action. Such a plan should clearly articulate: the goals of the research, the specific tasks to be performed, the order of tasks, the approximate time required to accomplish tasks, the personnel required, the documentation equipment to be used, the organization and final disposition of collected data, and estimated project expenses. Having said this, however, once the project is set in motion, the researcher should be prepared to be flexible in his or her approach in response to unanticipated findings. Indeed, flexibility is an asset that permits the researcher to take full advantage of the serendipitous discoveries that frequently occur during the course of research.

Institutional Involvement

During the planning phase of the project, consider whether cultural institutions, such as museums, historical societies, folk cultural agencies, libraries, and archives, should be involved in the project. If you require technical assistance in conducting fieldwork or organizing field data, many cultural institutions will be prepared to supply it. In some cases, institutions may be willing to contribute fieldworkers to a project in exchange for cosponsorship and, perhaps, the field data generated by the project. Similarly, if you need documentation equipment, such as tape recorders, 35mm still cameras, and video cameras, institutions may be willing to lend it. If you wish to place a project's field collection in a repository where it will be preserved and made available to other researchers, state and local cultural institutions may be willing to accept these data.

It is especially important to make a decision about the final disposition of field data as soon as possible, since repositories generally have specific requirements concerning the cataloguing and processing of such data. For example, if a repository assigns accession numbers for photographs according to their chronological order, you should record these numbers as soon as the films have been processed. Or, an institution may have a particular "release" form that it prefers informants to sign to indicate that they have consented to their tape-recorded interviews being kept in the repository. Similarly, some repositories have special catalogue forms for writing down the details of photographs or recorded interviews. In any case, if you have decided to deposit data in an archive, a museum, or some other responsible repository, confer with appropriate repository staff about recommended procedures for processing tape recordings, photographs, drawings, field notes, and other data. Through these discussions, fieldworkers can acquaint themselves with the basic information that should accompany all field data, thus preventing any duplication of effort.

Reaching an early decision about where field data will eventually reside is also important because it will enable researchers to tell others what will happen to this information. (Researchers should provide this information whether or not informants ask.) In many instances, if a researcher is able to tell an informant that collected data will be preserved for posterity at a reputable repository, the project will gain credibility in the informant's eyes and lead to greater cooperation. Many informants take pride in the fact that the information they have passed on is of such value that "they're saving it at the museum." Placement of field data in a responsible repository, beyond ensuring preservation and availability to future researchers,[8] shows, implicitly, respect for those from whom data was collected.

During the planning phase, researchers should also consider whether dissemination of collected data is to be a goal of the project. That is, will they seek ways to inform others about the results of the project through publications, exhibits, films, videotapes, lectures, and conferences? If these are the goals, additional planning will be required to ascertain content, technical requirements, personnel needs, timetables, and costs. Once again, cultural institutions, many of which are directly involved with a variety of such dissemination activities, are often good sources of information.

If a project will require outside funding for such things as the hiring of professional fieldworkers, the processing of photographic film, expenses for fieldworkers, and the creation of a product, it is essential to explore ways to secure such funding before the project is launched. To say the least, it is embarrassing to cancel a project, after it has been announced, because of lack of funds. In fact, it is irresponsible to announce that a specific product—a book or an exhibit—will result unless there is a reasonable chance of this coming to fruition. There are a number of ways to secure funding for projects that seek to document local culture, ranging from relatively simple activities, such as benefit suppers, to the submission of formal applications (to private, state, and federal foundations and granting institutions) for financial support. For information about grant support, contact local and state arts councils, humanities councils, folk cultural agencies, and municipal offices of cultural affairs.

Preliminary Research

It is crucial to start research activities with a thorough review of previous research relevant to the project's topic. Locate and analyze all published works that pertain to the individual craft and boat type, or types, under investigation, and compile a bibliography. Doing so allows you to trace the steps of previous investigators and to determine what conclusions have been reached, what source materials have been used, and what questions remain to be answered. In some instances, published works will be voluminous, as in the case of a famous yacht developed by a designer of renown. In others, as in the case of a localized fishing craft, the published record will be small or nonexistent.

It is also important to determine whether significant unpublished data exist that pertain to the project's topic. Have interviews been recorded with boat builders, boat users, and other knowledgeable individuals? Do business records exist from the pertinent boat shops or yards? Are there relevant photographs in cultural institutions, newspaper files, or private collections? Are lines plans available? Can pertinent half-models, full-rigged models, molds, boat-building tools, and other artifacts be found in museums and private collections?

A wide range of publications that do not focus specifically on the project topic can, nevertheless, help to delineate the cultural context. Local histories, for example, can help researchers better understand the significance of boat building and boat use vis-à-vis other activities, as well as gain a sense of the texture of life in a given community. (A recent community history that is particularly good in this respect is Ben Green's *Finest Kind: A Celebration of a Florida Fishing Community.*) While they seldom detail small working craft, local histories often include data on famous vessels and their builders and skippers. Frequently, the most valuable boat-related data in local histories are historical photographs of harbor scenes, vessels under construction, and vessels in use. Biographies and autobiographies of residents of the area may also contain useful information. For example, anyone researching inshore fishing craft used in Maine during the early years of the twentieth century would benefit from Harold B. Clifford's biography, *Charlie York, Maine Coast Fisherman.* Local newspapers contain a wealth of data about nearly every aspect of life in maritime communities. Of special interest are articles on boat launchings and sinkings, yacht regattas, fish landings, and community celebrations such as blessings of the fleet; obituaries of noted boat builders, ship captains, and fishermen; notices of the arrivals and departures of vessels; and classified advertisements offering the sale of new and used vessels and gear.

Publications that deal with local fisheries and fishery industries often contain considerable data about fishing vessels, including design and construction, crew size and specialization, the use of fishing gear, species pursued, seasonal activities, and innovations in gear and vessel design. One of the most important sources of information about commercial fishing in the United States during the late nineteenth century is the magnificent seven-volume work *The Fisheries and Fishery Industries of the United States,* edited by George Brown Goode. A less well-known, but equally significant, work is the eleven-part *The Fish and Fisheries of Colonial America: A Documentary History of the Fishery Resources of the United States and Canada,* edited by John C. Pearson. Trade publications, such as *Commercial Fisheries News* and *National Fisherman,* can also be mined for profiles of boat builders and fishermen, changes in state and federal fishing regulations, and information on recent trends in the design and construction of vessels and gear. Recent works that are helpful guides to scholarly publications concerning contemporary commercial fishing include James M. Acheson's "Anthropology of Fishing," and Leif Landberg's *A Bibliography for the Anthropological Study of Fishing Industries and Maritime Communities* and its supplement.

Many other forms of published and unpublished documentary evidence can help place boat building and boat use within the context of a community's economic and social activities, past and present. Census records and city directories usually contain valuable information, such as the number of individuals involved in maritime-related occupations, and the names and addresses of boat yards, changleries, fish plants, sail lofts, and other maritime businesses. The tax records of maritime communities sometime include data on the sizes and valuations of locally owned craft. Boat registration files maintained by state agencies usually contain fairly specific information about commercial and recreational vessels, including date and place of construction, names of the builder and owner, principal dimensions, and type of propulsion.

In order to understand how boat builders attempt to build boats that successfully fit the conditions of the local environment, it is necessary to investigate such environmental features as climate, water temperature, high water and low water fluctuations, depth, and condition of waters where boats operate, topography and composition of coastal land, and the degree of protection afforded in harbor areas. Similarly, it is important to know the extent to which builders of wooden boats use local trees for building material. And, in the case of fishing craft, it is especially helpful to learn about the life cycles and seasonal activities of the species fishermen pursue since, in many cases, these factors will directly influence fishing boat design and construction. Perhaps the best study of the relationship between environmental factors and traditional boat design is Eric McKee's *Working Boats of Britain: Their Shape and Purpose.*

Useful information about a study area's natural environment—topography, climate, flora, and fauna—can often be found in publciations of state departments of marine resources, the National Oceanographic and Atmospheric Administration (NOAA), the National Marine Fisheries Service, the Sea Grant College extension system, the U.S. Coast Guard, and the U.S. Geological Service (USGS). Charts of local waters and topographical maps, issued by NOAA and USGS, respectively, are particularly useful for finding out about the lay of the land under water and on shore. Researchers should use maps and charts, as well as direct observation, to familiarize themselves with the key features of the local landscape—features that will, undoubtedly, be referred to by informants.

In order to understand the legal constraints placed on boat builders and boat users, researchers should also become acquainted with the local, state, and federal laws that affect the design, construction, and use of vessels. Do regulations prescribe construction standards, such as specified building materials and minimum scantlings based on vessel size and use? What is the basis for vessel taxation? What safety features are required? Do commercial fishing regulations limit participation in certain fisheries to vessels under a certain length, with a particular kind of gear, or with a certain method of propulsion? Are fisheries subject to open and closed seasons? Do regulations restrict fishing activities to certain fishing zones? Is there a "limited entry" system that holds participation in a commercial fishery to a specific number of vessels? Do regulations prescribe the number of passengers a vessel may carry or the weight of cargo it may transport?

The scope of preliminary research will vary according to the requirements of each project. Given the range of potential sources of data discussed above, researchers will have to decide which ones to explore, as well as the level of effort that should be applied to this phase of research.

Making Contacts

Regardless of whether or not useful data can be found in published and unpublished documents, a large body of knowledge related to boats is not recorded in any fashion. Such knowledge resides in the thoughts and memories of residents of the study area. Sometimes this knowledge is highly specialized and is the intellectual property of a single individual. Sometimes the knowledge is less specialized and, individual variation excepted, is shared by members of specific groups. Sometimes certain types of knowledge are shared by community members and passed along over time as a sort of collective memory, a nonphysical, but very real, aspect of local heritage. Investigating the cultural contexts of boats involves making contact with people who possess these kinds of knowledge: the boat builder who is the only one who can describe how he invented a revolutionary construction process; the group of crab fishermen who can discuss the qualities they desire most in their boats; natives of the community who can relate the stories, passed

Figure 15.2. Information about potential informants can often be garnered at marine supply stores. At the start of a 1986 survey of the maritime traditions of Apalachicola, Florida, a project sponsored by the American Folklife Center and the Bureau of Florida Folklife Programs, folklorist Nancy Nusz, right, asked a store clerk and an engine mechanic at Wefing's Marine Supply about community residents who build boats and make fishing gear. *Photograph by David A. Taylor, courtesy of the Bureau of Florida Folklife Programs.*

down to them, about local Prohibition-era rumrunners and the vessels they used to elude the Coast Guard. Obtaining these oral data is a central part of ethnographic fieldwork.

One of the first questions to confront the fieldworker seeking such information is: How do I find the people who possess the knowledge I'm after? In most instances, the categories of informants will be obvious. For example, a project focusing on contemporary Florida oyster skiffs specific to a certain community would, in all likelihood, involve interviews with those who build them as well as those who use them. In this circumstance, unless the researcher is already familiar with the community, it will be necessary to determine the specific builders and fishermen to approach about assisting the project. Often, it is extremely beneficial to seek the advice of local residents with a broad knowledge of community affairs, especially maritime-related activities. Harbor masters, clerks at marine supply stores, newspaper editors and reporters, town officials, marine extension agents, and yacht brokers are frequently among those who can suggest not only those individuals who have the knowledge that is sought, but also those among them who would be inclined to share that knowledge.

In some cases, it may be helpful to issue a press release about the project to local newspapers so that a large number of area residents will be informed. Likewise, making presentations at meetings of local organizations, such as service clubs, historical societies, yacht clubs, and fishermen's associations, can help spread the word. Sometimes placing a classified advertisement under the heading "information wanted" can result in useful contacts. Disseminating information about the project in these ways can often result in useful suggestions from community residents on potential informants and other sources of information.

Assuming that the topic of a project embraces knowledge that is still to be found in the study area, the researcher's initial queries about individuals who possess it will probably yield a list of possible informants. At this stage, it is useful to take stock of the individuals on the list and decide if there is any merit in contacting some before others. Does it make sense to interview the oldest, most experienced boat builder first? Should the president of the fishermen's union be interviewed before other fishermen out of deference to his status in the organization? In the final analysis, there may be no valid reason for conducting interviews in a specific order, but it is a point worth considering. Also bear in mind that the list of potential informants is likely to grow since those contacted will probably suggest others.

Ethics

"Ethics," writes folklorist Bruce Jackson, ". . . have to do with the moral implications of the role you play when you're doing fieldwork and with the moral consequences of your decisions and actions after you're done with the fieldwork."[9] In the context of fieldwork, ethical questions arise in connection with researchers' explanations of who they are, what they're doing, and why they are doing it. There are also ethical implications to many decisions, including what data should be made public and what should not, how data is presented to the public (in the form of a book, exhibit, video, or other product), what artifacts are borrowed from informants, and how the assistance of informants will be acknowledged.

Whether you make known the details of a project in the ways discussed earlier, or in one-on-one encounters with residents of the study area, it is imperative to portray accurately the scope and purpose of the project. Potential informants have a right to know the goals of the project, the individuals and institutions involved, how collected information will be used, and what will happen to taped interviews, photographs, artifacts, and other material when field research has been completed. Anything less than total honesty about any of these matters is a sure way to torpedo local residents' willingness to cooperate.

If interviews are tape recorded, it is advisable to ask informants to sign a printed "consent form" indicating they have given permission for the recordings to be made, and also, if applicable, to be deposited in a repository. If an informant wishes to place any restrictions on the use of the recorded interview, these restrictions (which can be written on the form) should be respected. The form, a copy of which is given to the informant, should briefly, but clearly, state the purpose of the project and how collected information will be used. (For models of interview consent forms, see Edward D. Ives's *The Tape-Recorded Interview: A Manual for Fieldworkers in Folklore and Oral History* and David A. Taylor's *Documenting Maritime Folklife: An Introductory Guide*).

If photographs, models, or other artifacts are borrowed, lenders should be told how their possessions will be used and when they will be returned. It is recommended that a printed "artifact loan form" be used to confirm that an informant has granted permission to a researcher to borrow a specific item, and that the researcher has received it. The form should clearly specify the purpose for which an item is borrowed and the date by which it will be returned. The form should be signed by both informant and researcher, with a copy going to each. If the researcher cannot, in good conscience, guarantee that an artifact will be returned undamaged and when promised, then he or she should not borrow it.

Acknowledgments

It is essential that researchers acknowlege those who provide assistance to a project. This can be done by thanking people in person, sending thank-you letters, and presenting certificates of appreciation. If a product, such as a book, exhibit, or film, results from the project, the names of the people who helped should be included.

Because the circumstances of fieldwork are so variable, it is impossible to prescribe a set of ethical do's and don't's that will apply to every case. And, in deciding, for example, what information about a boat builder's business practices should be made public and what should not, the researcher will have to use his or her best judgment. However, Bruce Jackson's "golden rule" provides a moral compass:

> When you're in doubt about whether an action on your part is ethical or not, a good starting place is to put yourself in the subject's postition and consider how you would feel if you learned what that friendly person was really up to. If you'd be annoyed and offended that you were made a sample in a study you wouldn't want to be part of or were recorded or photographed without your knowledge or permission, don't do those things to others. If you'd feel betrayed because things you said in confidence were made part of a public report, then don't betray confidences—or at least tell people who think they can trust you that you can't keep secrets. Hemingway once defined the good as "what you feel good after." Think

about how you'll feel later—perhaps how you'll feel if you ever see that person again. If the answer is "not so good," then don't do it.[10]

Keeping Field Notes

A practice all researchers should follow when conducting fieldwork is keeping a field notebook. Such a notebook, which should always be within easy reach, is essential for jotting down the various kinds of data that can turn up during the course of fieldwork, including names and addresses of potential informants, names of photographic subjects, and sketches of boats and boat parts. In the field notebook, or in a separate notebook, researchers should keep a daily journal of their activities. Especially useful when interpreting data later on, such a journal should contain summaries of each day's activities, including observations on successes achieved, problems encountered, and impressions of the progress of the research. In recording daily entries, try to keep the researcher of the future in mind; that is, record information in the field notebook that will enable others not present at interviews and other research events to make better sense of tape recordings, photographs, and other items generated by the project. Describe the settings of interviews, the type of tape-recording equipment used and how it was arranged, sources of extraneous noise during a recording session, the lighting conditions for photography, and the weather conditions when boat measurements were recorded. Describe interactions with informants before, during, and after an interview is recorded. Recording daily field notes is a practice indicative of competent fieldwork. Ideally, it yields a chronological record of each researcher's activities, plus a plethora of background information necessary for more seamless interpretation of other data.

Interviewing

Conducting interviews is an extremely valuable means of collecting information and one that is commonly used by cultural investigators. There are several types of interviews, ranging from unscheduled, informal interviews conducted on a wharf or in a pilothouse to scheduled and very structured sessions involving the use of a prepared list of questions. While most interviews are conducted with a single informant, occasionally group interviews are useful. Sometimes interviews are recorded on audio or video tape; sometimes they are not. Researchers will have to decide what types of interviews are appropriate, how many informants should be involved simultaneously, and whether or not it is appropriate to make recordings. In part, these decisions will be based on the goals of the project, the expertise of fieldworkers, and the willingness of informants to permit their words to be recorded on audio or videotape. Choices will also be influenced by the exigencies of field situations. For example, in the case of a busy boat builder who can spare no more than fifteen minutes, the fieldworker may only be able to scribble down in a notebook the answers the builder gives to his or her questions. In general, however, tape recording interviews is recommended, as it frees the fieldworker from having to write notes furiously as the informant speaks. The recording also captures the subtleties of speech and turns of phrase that often identify an individual as a member of an occupational group or a community. In addition, the recording provides a document that can be analyzed by the fieldworker and, if preserved in a repository, studied by future researchers.

What kind of equipment should be used to make sound recordings of interviews? At present, recordings of the highest quality are made with professional reel-to-reel tape recorders and digital audiotape (DAT) recorders. Most projects, however, will be adequately served by portable cassette recorders in the $200-$400 price range, especially if augmented with external microphones. In making decisions about recording equipment, researchers should carefully consider what level of sound quality is most appropriate for their projects. For example, if a sound recording will be an integral part of a product to be generated by a project (such as the sound track for a slide-tape program), then it is worthwhile to use equipment that will produce recordings of high quality. (Of course, as desirable as it may be to use professional-quality recording equipment, it may not be possible, for financial or other reasons, to purchase, rent, or borrow it.) For detailed advice about the selection of recording equipment and audio tape, consult Bruce Jackson's *Fieldwork* and Edward D. Ives's *The Tape-Recorded Interview.*

Figure 15.3. Folklorist Paula Johnson records the comments of waterman Roy King and the sounds of gill net fishing on the Patuxent River in 1981. In addition to making this recording, researchers also interviewed King and his wife, Clara, in their home. *Photograph by Peter Kurtze, courtesy of the Calvert Marine Museum.*

Addressing Informants' Concerns

As noted above, a fieldworker should take time to conduct background research focusing on the region, occupation, or community in question before setting up an interview. This is especially true for female fieldworkers studying traditionally male-oriented occupations such as fishing and boat building. In these typically male occupations, a female fieldworker should expect some skepticism from the men she interviews; therefore, it is important that she familiarize herself with an occupation's history, tools, and terms.

Folklorist Janet Gilmore, who studied fish boats of the Oregon coast, encountered some resistance when she first approached fishermen in the Charleston-Coos Bay area. She writes:

> No matter how I presented myself to fishermen initially—on the docks or over the phone—I sensed some discomfort and suspicion from almost all of them. During the course of interviews I could actually see the relief spread over their faces once they had realized that my interest was sincere and my questions relatively harmless.[11]

Gilmore found that male researchers in the area had their own difficulties gaining the trust of the fishermen. Fishermen shunned one young man continually because he looked like "a classic case of someone seeking work on a fishing boat." She concluded that, despite initial problems in obtaining interviews, being a woman worked to her advantage. She "did not run as great a risk of appearing a prime candidate for a job, provoking avoidance or enticement behaviors; [she] was less likely to become a fisherman in the future, thus less worrisome as a potential gossip in the wrong contexts, and more trustworthy as a confidant; and [she] was not expected to know much about fishing or mechanics, so fishermen were more likely to take pains to explain things."[12]

During the Patuxent River Folklife and Oral History Project, conducted by the Calvert Marine Museum of Solomons, Maryland, from 1981 to 1982, one female fieldworker found that her lack of knowledge and experience concerning diesel engines and hydraulic oystering gear did not hinder, but actually helped, in the interview situation. Oystermen made an extra effort to explain to her the workings and relative merits of various engines and rigs, while they assumed that her male coworker already knew about such things. The tables were turned, however, in the women's domain, when watermen's wives assumed she already knew how to make crab cakes. When she asked for one woman's recipe, the reply was, "You take your crab meat, mix it up with our seasonings, and fry them up." When a male fieldworker was present several weeks later, the crab cake cook, who assumed he knew very little about cooking, provided a lengthy, richly detailed account of how crab cakes are made.[13]

A related issue is the degree to which fieldworkers are regarded as insiders or outsiders by the people they are interviewing. At first glance, there are some obvious advantages to being an insider. For example, an insider is already familiar with the members of the group being studied and the cultural traditions they possess. Accordingly, an insider hardly has to worry about breaking the ice (with his uncle, the boat builder) at the beginning of an interview. Yet, because the insider already knows so much, it is likely he or she will not ask any number of elementary questions, and that the informant, recognizing the interviewer's comprehension of certain basic issues, will probably not bring them up independently. Edward D. Ives refers to "stranger value" in a fieldwork situation, explaining that "sometimes people will say things to strangers that they would feel awkward or silly saying to members of the family."[14]

As fieldworkers associated with the Patuxent River project discovered, there were some advantages to being outsiders. Residents of the study area did not expect the interviewers to know local terms, place names, families, or events, so, when asked about them, informants generally took great care in their explanations. Also, as outsiders, the fieldworkers were not associated with any of the factions involved in long-standing disagreements. Still, there were at least two occasions when local people refused to be interviewed by researchers they did not know. One was the case of a fish market owner, who, it was learned later, understood the question, "And where did these perch come from?" as a means of investigating him for illegal fish marketing. Actually, all the researcher wanted to know was whether Patuxent perch were available locally. Another researcher encountered suspicion as well when a boat builder mistook his innocent query, "So, how many boats have you built here?" for an Internal Revenue Service ploy to discover unreported income.

Provided the researcher is aware of the advantages and disadvantages of being an insider or an outsider, he or she should not feel that either status is appreciably better than the other when it comes to eliciting useful data from informants. Insiders, who begin ethnographic research with an intimate understanding of an area and its people, should be wary of assuming that they know everything. Such an assumption will tend to cause researchers to skip over a good deal of important basic data. In addition, if too much is taken for granted, the likelihood is great that insiders will neglect to ask about variations in the ways things are done. As Ives notes, it can be awkward, on occasion, to pose a question when the researcher knows that the informant knows he or she already has the answer.[15] However, this awkwardness can usually be overcome if the interviewer provides (perhaps before the interview session begins) an explanation along these lines:

> Because I've lived here all my life, I know the answers to some of the questions I'll be asking you. I'll be asking them because we want to record as much information as we can on tape—the simple stuff and the complicated stuff, too—so that, in the future, people will have a complete picture of how boats were built and used around here.

Outsiders, for their part, while recognizing that their status may prompt richly detailed explanations from informants who understand their ignorance of local activities, should, nevertheless, not let this be an excuse for failing to do background research. Outsiders should also be aware of the fact that their "strangeness" may result in misunderstandings of their motives and subsequent suspicion or outright rejection. The possibility of misunderstandings in this regard underscores the necessity of providing accurate and consistent explanations of every research project.

Interviewing Technique and Strategy

Interviewing is a large topic and one that cannot be easily summarized here. Accordingly, those interested in learning more about interviewing strategies, the cataloguing and transcribing of recorded interviews, and the selection and use of recording equipment are advised to consult some of the many publications that deal in depth with these subjects. Four valuable guides are: *From Memory to History: Using Oral Sources in Local Historical Research* by Barbara Allen and Lynwood Montell, *Interviewing: Strategy, Techniques, and Tactics* by Raymond L. Gorden, *The Tape-Recorded Interview: A Guide for Fieldworkers in Folklore and Oral History* by Edward D. Ives, and *Fieldwork* by Bruce Jackson.

In interviewing, there are a number of central lines of inquiry that should be pursued. Because it is important to know the informant's place in the community—his or her status, experience, and relationships to other residents—ask questions about background. (This is often a good way to begin an interview.)

- When and where were you born?
- What are the names of your parents?
- What are (were) your parents' occupations?
- Do you have brothers and sisters?
- What was the extent of your formal schooling?
- Have you always lived in the same community?
- What types of jobs have you held over the years?

If an informant is being interviewed mainly to elicit information about a particular skill, such as boat building, it is important to determine how knowledge of that skill was aquired. For example, after asking "How did you learn about boat building?" and listening to the informant's reply, follow up with questions that address when the informant began to learn, and *from whom* information was acquired. To what extent did the informant learn about boat building in the community in an informal way, based mainly on observation and imitation? And to what extent was knowledge acquired more formally, through boat-building courses or published manuals? Also ask about the informant's current sources of information about new designs and construction practices; the answer to this question will help determine how, when, why, and by what means innovations are introduced.

Other questions can generate information about the history of boat building in the community, particularly during the informant's lifetime. For example, the question "What were the principal boat types that were in use when you were a youngster?" can yield useful descriptions of boat types as well as the local names used to categorize them. Follow-up

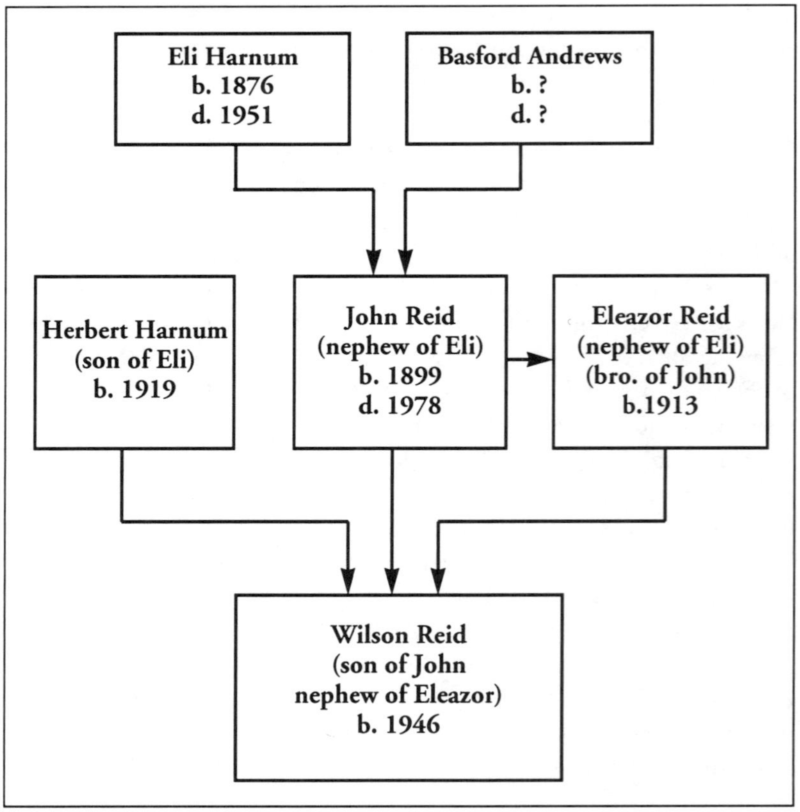

Figure 15.4. Data from interviews and documents can enable researchers to formulate "trees" illustrating the transmission of boat building knowledge over time. This illustration traces the transmission of boat building knowledge through three generations in a Newfoundland fishing community. *David A. Taylor.*

questions about the continuity, modification, or abandonment of these types over time can reveal the predominant trends in boat development and their underlying causes. Questions about past and present boat builders will help the researcher chart a "tree" of builders that reveals the transmission of knowledge over time and will shed light on those individuals whose contributions have been especially influential.

Understanding the Boat

In addition to the transmission of knowledge, cultural investigators are also interested in understanding all other steps within the process of the creation of a boat. Two major segments of the process are design and construction. Questions about design may address such topics as: the sequence of design activities; the use of molds, half-models, and plans; the borrowing of design-related devices (molds, half-models, plans) by local builders; the involvement of clients in the design process; design requirements prescribed by state and federal regulations; design changes introduced by a builder over time; and testing of designs through use of vessels. Questions about construction may include: the procurement of building materials; the sequence of construction activities; the details of major operations such as the assembly of the backbone, timbering, planking caulking, and finishing; tools used; local terms used for different processes, materials, and boat parts; and specialized skills of employees. Detailed studies of traditional boat design and construction within their cultural context include C. Richard K. Lunt's doctoral dissertation "Lobsterboat Building on the Eastern Coast of Maine: A Comparative Study" and David A. Taylor's *Boat Building in Winterton, Trinity Bay, Newfoundland* and his doctoral dissertation "A Survey of Traditional Systems of Boat Design Used in the Vicinity of Trinity Bay, Newfoundland, and Hardangerfjord, Norway."

Another important area of consideration is the boat builder's business practices. How is the market for his products determined? Is advertising used? Are boats ever built on "spec"? How much of the work is "stock" and how much is done on a "custom" basis? To what extent are clients involved with the shaping of boat designs? What payment schedules are clients asked to follow? How is a boat-building project "costed out" with regard to labor and material costs and estimated profit? Who are the builder's main competitors? How much work is contracted out to specialists such as electronics technicians, engine mechanics, spar makers, riggers, and foundry men?

A study of local boat-building traditions should also include interviews with individuals who own the vessels. It is important to find out from them why they chose a particular boat; how the boat performs compared to others they have owned; what they would change (or have changed). Vessel owners are likely to be frank about what they like and dislike about their boats—and everyone else's. Interviews with boat owners often reveal the standards, in terms of both function and aesthetics, for boats in their communities.

If researchers interview several boat builders in the study area and elicit responses to a standard body of questions, they can analyze the responses for evidence of the principal patterns of boat-related behavior shared by the members of this group. This analysis will also expose the subtle variations that exist from builder to builder, as well as the practices that are unique to individuals. This research model can also be applied to boat owners or any other group that possesses knowledge germane to a particular project. Determining patterns of behavior, and the rules underlying these patterns, is an over-arching concern for those who seek to understand boats within their cultural contexts. Key works to consult for a better understanding of the fundamentals of cultural processes are Henry Glassie's *Pattern in the Material Folk Culture of the Eastern United States* and Barre Toelken's *The Dynamics of Folklore.*

Documenting Scenes

Fieldwork that seeks to document scenes—culture in context—is just as important as fieldwork involving interviews. For example, Patuxent River project researchers documented (with sound recordings, still photographs, and field notes) a number of scenes characteristic of the study area: watermen at work aboard their boats, boat builders constructing boats in their backyards, oyster shuckers and company owners at work in local packing houses, watermen repairing and maintaining their boats and equipment at piers and marine railways, watermen talking to each other at oyster buying stations and in other informal settings, community members participating in local seafood festivals and church suppers, and watermen attending public hearings concerning fisheries regulations. In the course of this sort of documentation, researchers did not attempt to record interviews. Instead, they attempted to capture the sights, sounds, and rhythms of the actions they observed. They found that cultural values and attitudes were most eloquently expressed in scenes such as these.

Participant Observation

Another approach that researchers often use to understand cultural processes better is known as participant observation. Briefly stated, to conduct participant observation a researcher involves himself or herself in the activity under study. The assumption underlying this approach is that participation in an activitiy will yield information and insights that cannot be obtained (or obtained as easily) in any other fashion. For example, in connection with the American Folklife Center's Pinelands Folklife Project, folklorist Gerald Parsons went railbird hunting with a group of hunters and guides on a freshwater marsh in New Jersey in order to better understand the structure of the hunt, including how double-ended railbird skiffs are used.[16] Even a very short-term experience, such as a morning spent helping a fisherman apply bottom paint, can yield useful data as well as build rapport. Experiences of these kinds can be an exciting and fruitful part of field research, and they have the added benefit of demonstrating to informants, and to the community at large, your commitment to the research.

Researchers should be mindful of the fact that, however desirable, it may not always be possible to arrange an opportunity for participant observation. A busy boat builder may see absolutely no merit in taking on an apprentice, especially if the prospective apprentice has little boat-building experience and no commitment to the profession. Similarly, a fisherman may be unwilling to take on a helper who has not been steeped in fishing culture since birth. Nevertheless, opportunities for participant observation do occasionally arise, and each researcher will have to decide whether it is worthwhile to pursue them.

Photography

As another contributor to this volume has shown, photography is virtually indispensable for documenting the physical properties of vessels. Cultural specialists generally do not limit their photography to boats but also record: portraits of informants, images of many kinds of human activities, and photographs in family albums and other collections. Standard equipment used is usually one or more 35mm, single lens reflex (SLR) camera bodies and an assortment of interchangeable lenses. If copies are to be made of photographs in an informant's collection, it is advisable to use lenses designed for close-up work, as well as a copy stand that holds the camera securely in place above the image to be copied.

When using photography to document the creation and use of boats, researchers should attempt to capture the essential processes involved, the people involved in the processes and their interactions with each other and with their tools and equipment. Apart from skills associated with the technical aspects of taking photographs, this requires attention to the critical chronological steps that make up any activity, followed by comprehensive photographic coverage of each. Obtaining adequate coverage of human activities usually requires taking shots from different distances, and with different lenses, while walking completely around the scene (or, at least, walking around as much of the scene as possible). For example, adequate coverage of a boat launching would include: a shot that encompasses the scene of the event (boat yard, boat, launchway, harbor, and assembled crowd); shots of the boat situated on the ways; close-ups of blocks and wedges that hold the boat in place; group shots of the builder, owner, owner's family, and dignitaries; shots of the christening as the bottle smashes on the stem; shots of yard workers hammering out wedges; shots of the boat sliding down the ways; shots of the boat hitting the water; shots of members of the crowd reacting to the launch; and shots of post-launch celebrations in the boat shop. In other words, when

Figure 15.5. Documenting how watermen use a vessel is an important part of cultural investigation. Claude Mister, a Patuxent River (Maryland) shaft tonger, needs—and has found—a stable platform for oystering. *Photograph by Terry Eiler, courtesy of the Calvert Marine Museum.*

documenting a cultural scene, attempt to record images that help answer the investigator's fundmental queries: Who? What? When? Where? Why? How?

In order to capture the essential elements of an event, such as the hypothetical launching discussed above, it is important that fieldworkers determine as much as possible about the likely sequence of activities and the lay-out of the site *in advance.* Ascertaining this information will help fieldworkers decide where they should be and when they should be there, as well as the type of equipment and film to use, in order to photograph critical activities successfully. If more than one photographer will be documenting an event, advance planning is required in order to judge how best to coordinate coverage. Any planning for photography should also consider whether permission must be obtained for access to an event or to an especially good vantage point. For example, if it is apparent that the best location from which to document a public blessing of the fleet ceremony is the deck of the vessel carrying the members of the local clergy who will bless the vessels, it will undoubtedly be necessary to secure permission to be on the vessel from the organizers of the event well in advance. Do not expect to make such an arrangement at the last minute.

Field photographs are useful aids for eliciting information during interviews. For example, they can be used to generate detailed descriptions of boat-building activities, local terms for boat parts and gear, and the names of individuals portrayed. Showing historical photographs to informants, including images from their own photo albums, can also elicit long narratives about people and events from the past.[17]

Photography, like interviewing, is a more complicated subject than can be dealt with in this summary. Depending upon your level of photographic skill, you may wish to consult some of the vast number of guides that pertain to the technical side of photography. For useful suggestions about documenting artifacts and landscapes, consult *Architectural Photography: Techniques for Architects, Preservationists, Historians, Photographers, and Urban Planners*

Figure 15.6. Part of the landscape of Broomes Island, Maryland, including waterfront businesses and locally built skiffs. *Photograph by Peter Kurtze, courtesy of the Calvert Marine Museum.*

Figure 15.7. Members of the Broomes Island community worship together on the shore on Easter morning. Researchers associated with the Patuxent River Folklife and Oral History Project recorded the service, which included many references to the river and its importance to the community. *Photograph by Paula Johnson, courtesy of the Calvert Marine Museum.*

by Jeff Dean. The standard work on using photography to document human culture is *Visual Anthropology: Photography as a Research Method* by John Collier, Jr., and Malcolm Collier. Fleischhauer's "Sound Recording and Still Photography in the Field" (in *Handbook of American Folklore*) and Jackson's *Fieldwork* also contain helpful advice about cultural documentation through photography.

Since an important part of documentation is recording explanations of processes and occupational techniques, it is worth considering videotaping the process. While this requires specialized equipment and skills, as well as the permission and patience of the informants, the resulting tapes can provide valuable documentary material that is difficult to capture in any other way. A nonmaritime, but applicable, example comes from research conducted in a northern Nevada ranching community by the American Folklife Center. During the course of the project, two fieldworkers videotaped a Nevada rancher explaining the philosophy behind and process of "parting" cattle prior to branding. What would have been difficult to comprehend with words was made clear when the rancher drew a diagram of the parting activity. Videotape footage of the rancher's illustrated explanation, as well as the activity underway, are included on Carl Fleischhauer's videodisc, *The Ninety-Six: A Cattle Ranch in Northern Nevada.* Useful information on videography is contained in *Videotaping Local History* by Brad Jolly, *Fieldwork* by Bruce Jackson, and "An Introduction to Media Documentation of Louisiana Folklife" by Nicholas R. Spitzer.

Post-Fieldwork Activities

Even after the last interview has been recorded, the final frame of film exposed, and the last page of field notes written, more work remains to be done before the project is completed. For example, if researchers have not already done so, they should catalogue all field data; that is, identify all photographs according to date, place, subject, and photographer, and identify each tape-recorded interview according to date, place, subject, interviewer, and recording configuration.[18] In the case of tape-recorded interviews, interviewers should prepare chronological summaries of the contents of each of their recordings. Clearly, given the fallibility of memory, it is important to attend to cataloguing chores as soon as possible after data are collected.

If data are to be placed in a repository, it is necessary to verify that all materials are properly identified, catalogued, and packaged in accordance with the repository's standard procedures. If duplicate copies of field data are to be made for researchers, it is important to work out how this will be accomplished prior to turning data over to the repository.

The end of the fieldwork phase of the project is also a good time to determine whether all informants and others who have assisted the project have been properly thanked

Figure 15.8. Friends and relatives participate in the launching of a new boat. About seventy-five people turned out to help celebrate the launching of the skipjack *Connie Francis* at Piney Point, Maryland, in October 1983. *Photograph by Paula Johnson, courtesy of the Calvert Marine Museum.*

Figure 15.9.
Photographer Terry Eiler perched in the rafters of Bill Trossbach's barn to shoot this bird's-eye view of the building process.
Photograph by Terry Eiler, courtesy of the Calvert Marine Museum.

Figure 15.10. The beamy deck of the skipjack *Dee of St. Mary's* provides space for dredge winders and dredges, as well as ample work space for five crew members and the day's harvest of oysters. *Photograph by Carl Fleischhauer, courtesy of the Calvert Marine Museum.*

for their help. If they have not already done so, researchers may wish to thank people in person or, perhaps, with letters of gratitude or certificates of appreciation. It may also be appropriate to present informants with copies of taped interviews conducted with them or copies of photographs taken of them, their boats, or other meaningful subjects.

The next major phase of the project will be the analysis of the field data which will, in turn, lead to the formulation of conclusions. If a goal of the project is to produce a book, an exhibit, a film, or some other product, then decisions will have to be made about how this will be accomplished. If some product will result, informants will, naturally, want to know when it will be available. It is the responsibility of the researcher to keep in touch with informants about this, and, when the time comes, determine how the product or products can be shared with them.

How Much Should be Documented?

As we noted at the beginning of this chapter, boats do not exist in a vacuum. As products of human thought and action, they are inexorably linked to the countless strands that make up the complex and ever-changing tapestry of culture. As Basil Greenhill has cogently argued:

> But a boat should be judged only, and I repeat only, in the light of the requirements for which she was built and the resources of the society which built her. She should never be judged by comparison with other boats built for different purposes of different materials in different circumstances. The basic question is one of fitness of purpose in relation to broad local circumstance. To appreciate a boat one must be aware of the factors that gave rise to her building, the timber available, the general environment, the building traditions of the society which produced her and, above all, the purpose for which she was built.[19]

How much should be documented to ascertain the factors that comprise a boat's context? For all intents and purposes, it is impossible to document everything that influenced how a craft has come to be as it is. Each researcher will have to decide the depth and breadth of research that is appropriate for his or her project. For examples of fresh approaches taken in recent years in presenting boats "in relation to broad local circumstance," see *Atchafalaya Swamp Life: Settlement and Folk Occupations* by Malcolm Comeaux, *The World of the Oregon Fishboat: A Study in Maritime*

Folklife by Janet C. Gilmore, *Working Boats of Britain: Their Shape and Purpose* by Eric McKee, and *Working the Water: The Commercial Fisheries of Maryland's Patuxent River* edited by Paula J. Johnson.

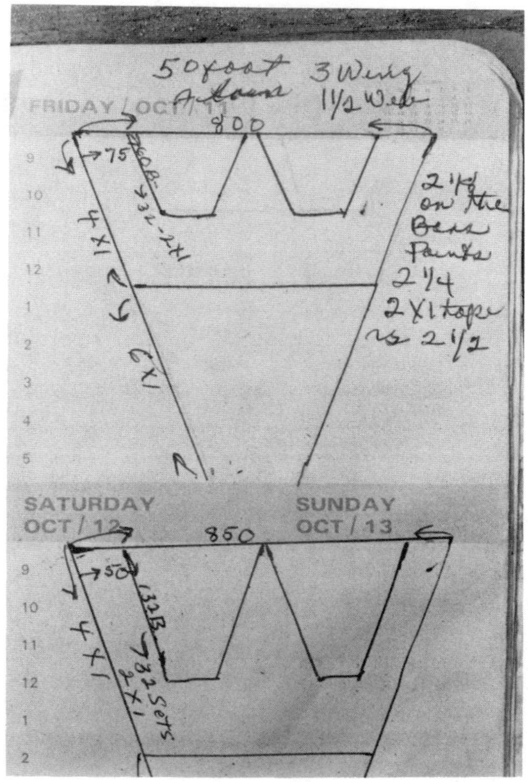

Figure 15.11. Sketches of the pattern for a fifty-foot shrimp net in the notebook of netmaker Costa Buzier of Apalachicola, Florida. These sketches and others were used by a fieldworker to elicit detailed information about net design during a tape-recorded interview with Buzier. *Photograph by David A. Taylor, courtesy of the Bureau of Florida Folklife Programs.*

Figure 15.12. A section of the exhibit *Built to Work: Building Deadrise Workboats in Southern Maryland* at the Calvert Marine Museum in Solomons, Maryland. The exhibit, funded by the National Endowment for the Arts' Folk Arts Program and the Maryland State Arts Council, features workboat designs and construction techniques of four local boat builders. *Photograph by Paula Johnson, courtesy of the Calvert Marine Museum.*

Notes

1. The authors gratefully acknowledge the assistance of Carl Fleischhauer, Mary Hufford, and Michael Taft, as well as the support of the Calvert Marine Museum and the American Folklife Center, and the Smithsonian Institition.

2. Peter H. Spectre, Editorial, *WoodenBoat* 84 (Sept./Oct. 1988): 2.

3. Ethnography literally means "a portrait of a people."

4. *Expressive culture*, also referred to as *folklife expressions*, consists of a wide range of creative and symbolic forms, including custom, belief, technical skills, language, art, architecture, music, play, dance, drama, ritual, pageantry, and handicraft.

5. Mary Hufford, *One Space, Many Places: Folklife and Land Use on New Jersey's Pinelands National Reserve*, Publications of the American Folklife Center, No. 15 (Washington, D.C.: Library of Congress, 1986), 16-19.

6. As Barre Toelken defines the term in *The Dynamics of Folklore* (Boston: Houghton-Mifflin Co., 1979), 51, a folk group is "any group of people who share informal communal contacts that become the basis for expressive, culture-based communications."

7. These standards are available in the U. S. Department of the Interior's "Archeology and Historic Preservation: Secretary of the Interior's Standards and Guidelines," in *Federal Register* (Sept. 29, 1983: 44716-42) and in Richard K. Anderson's *Guidelines for Recording Historic Ships* (Washington, D. C.: National Park Service, U. S. Department of the Interior, Historic American Engineering Record, 1988).

8. Access to some field data (tape-recorded interviews, for example) should be governed by the wishes of the informants who provided it. The use of "consent forms" permits researchers to maintain a record of informants' preferences in this regard.

9. Bruce Jackson, *Fieldwork* (Urbana and Chicago: University of Illinois Press, 1987), 259.

10. Jackson, *Fieldwork*, 278-279.

11. Janet C. Gilmore, *The World of the Oregon Fishboat: A Study in Maritime Folklife* (Ann Arbor: UMI Research Press, 1986), 9.

12. Ibid., 10.

13. In order to balance the differences in the quality and quantity of elicited oral information that stem from informants' gender-based assumptions about what a researcher knows, a team approach to fieldwork is often advisable. In this case, the team would include male and female fieldworkers, and each would interview male and female informants.

14. Edward D. Ives, *The Tape-Recorded Interview: A Manual for Fieldworkers in Folklore and Oral History* (Knoxville: University of Tennessee Press, 1980), 38.

15. Ibid.

16. Hufford, *One Space, Many Places*, 47.

17. Ives, *The Tape-Recorded Interview*, 74-79.

18. "Configuration" refers to: tape format (cassette or reel-to-reel) and length, recording speed, and recording mode (mono or stereo).

19. Basil Greenhill, *Archaeology of the Boat: A New Introductory Study* (Middletown, Conn.: Wesleyan University Press, 1976), 25-26.

Chapter 16

Computer Applications for Documentation and Analysis

Promises and problems in using a computer to help record, draw, and analyze a boat's shape

Garth S. Wilson

The use of advanced computer technology in maritime heritage documentation is still in its infancy. Computers are quickly gaining in popularity in maritime museum collections management, but their use in the measurement, depiction, and analysis of maritime artifacts has yet to be fully explored. The available systems are often prohibitively expensive, and only a few museums have collections large enough to justify the high initial cost of the hardware and software. Nevertheless, computers are used more and more in the design and construction of ships and boats, and the cost and operational complexity of such systems are falling rapidly. As museological tools, computer systems possess great potential and, if carefully adapted, can serve in the documentation, study, and reconstruction of historical vessels.

This chapter will review three computer approaches to small craft documentation (the United States Yacht Racing Union [USYRU] hull measuring method; the Great Lakes Historic Ships Research Project [GLHSRP] system; and the Canadian Parks Service Stereo-Photogrammetry system) in order to provide the reader with a survey of existing systems and an impression of the potential of computer-aided approaches to hull measurement, the production of lines plans, and hydrostatic analysis.

From the outset, however, it must be conceded that much of what follows reflects the experience gained by the author in the course of developing the second of these systems: GLHSRP. Furthermore, the descriptions of the systems provided here will undoubtedly soon be outdated, though there is some compensation in the knowledge that the rate of change will, in all probability, be to the benefit of the prospective user.

Historical small craft often reflect centuries-old traditions of craftsmanship and intuitive skills. A better understanding of traditional small craft design will, therefore, ultimately enrich our knowledge of how our ancestors responded to the demands and opportunities of their marine environment. Indeed, the rapid hydrostatic analysis made possible by today's CAD (computer assisted design) systems offers a valuable quantitative supplement to the subjective descriptions of vessel stability and performance so characteristic of small craft studies.[1] With a more thorough knowledge of boat design we will be better able to reproduce the working and pleasure craft of our forefathers and, in doing so, share something of their experience of the seas, lakes, and rivers around us.

Computer technology promises to increase and facilitate our knowledge and appreciation of small craft history and design, while at the same time greatly improving our capacity to catalogue and share technical information pertaining to historical vessels. Using computers, standardized compact catalogues of small craft design—including complete lines plans, offsets, isometric drawings, and hydrostatic data—can be compiled, copied, and distributed to form libraries of national and international small craft design. With this capacity, further impetus will be given to the important areas of historical research and the preservation or revitalization of traditional boat-building skills through the construction of accurate replicas.

In choosing a system, a number of basic issues must be borne in mind, not the least of which is price. Beyond cost, however, you must think very carefully about the desired application and level of sophistication needed. For example, you ought to consider:

- The number of hulls to be documented and their accessibility.
- The availability and cost of alternative, traditional methods of hull measurement and documentation.
- The availability and cost of local computer expertise.
- The compatibility of available systems with existing or readily available hardware and software.
- The ultimate purpose or long-term intentions of the documentation program. Is the focus to be on data recording, storage and retrieval or hydrostatic analysis?
- The importance of documenting construction features.

Having considered these issues, if a particular approach is appropriate or desirable, next consider which of the systems available on the market is best suited to the budget and capacity of the documentation project. However, in view of the rapid pace of development and obsolescence in the computer industry, even six-month-old information about available systems can be out of date. Ideally, then, you should conduct your own survey by formulating a list of questions regarding the intended application and by sending these to the appropriate authorities or, in the case of the analytical software, directly to the design firms.[2] Loans or purchases of slightly out-of-date software and used equipment may make the use of computers a more viable prospect.

The USYRU is the least expensive and most widely used of the three measuring methods examined here. Essentially, the USYRU system employs a computer to measure and record offsets taken manually from a hull using a special electric device, sometimes referred to as the "wand." The stored data is then translated into a table of offsets or body plan using either the service offered by the USYRU or, alternatively, various personal computer systems operating on customized software. The operation of the USYRU measuring device requires special training, and takes eight to twelve hours for both sides of a forty-foot long hull. Smaller vessels would require proportionally less time. The time frame and expense of further analysis depends on the analytical system chosen.

The GLHSRP Photo-Measuring System was specially developed for the study of historic hull forms as part of a research project undertaken at the Marine Museum of the Great Lakes at Kingston, Ontario, between 1987 and 1989. As the name implies, the system employs a camera, which is used in conjunction with a specially built mechanical device. Although the GLHSRP device was built on a scale appropriate to the measurement of half models, it is adaptable to the measurement of small craft up to a length of thirty feet.

As a rule, a one to two meter-long half model could be measured by a skilled operator in about one and a quarter hours, including about half an hour for the assembly and set up of the equipment. Under favorable circumstances, historical small craft could very likely be measured at the same rate. Basic analysis—translating the photographs into a body plan at the computer work-station—usually takes about forty-five minutes per hull using the "Fast Yacht" system, while surface creation and the production of accurate drawings, a process requiring a high degree of operator training and skill, can take from ten to forty hours.

Stereo-photogrammetry is another photographic approach to the measurement of shape. The method involves simple triangulation of the light rays emanating from an object on overlapping photographic images taken from two distinct vantage points. This approach is most commonly employed in aerial photography for the production of topographical maps. The method described here has been refined and used in the measurement of vessels, large and small, by the Architectural and Engineering Service of the Canadian Parks Service.[3] The photographic process remains essentially a mechanical procedure, though the interpretation of the stereo-photographs can now be performed using a computer. Although the computer-aided systems are not difficult to operate, the cost is high, ranging from $60,000 to $150,000 (Canadian currency) in 1989. Subcontracting may be an option in certain situations.

The USYRU Hull Measurement System

The United States Yacht Racing Union (USYRU) has developed an electronic Hull Measuring Instrument (HMI), which is used in measuring and handicapping yachts. The HMI devices are custom built by the USYRU and cannot be purchased. The HMI works in conjunction with a small portable computer, and this technology could be easily employed to lift lines from historical small craft. The USYRU measuring device operates by means of an electronic wand attached to a cord and works on the same principles as the manual angle-distance method. Essentially the device consists of a box containing shaft encoders, string take-up drum with string, and electronics with RS-232 interface. As described by the USYRU:

> A telescope is mounted on the top of the box for alignment purposes and the box itself is mounted on a four-screw levelling base. Other components include a wand that houses four push buttons and a panel meter and a surveyors tripod onto which the HMI is mounted. The HMI can interface with a variety of computers as long as they support the BASIC programming language, have a display, an RS-242 port, and suitable memory for data storage.[4]

Once the device is placed on a base line at a known distance and height from the hull and the number and location of the stations are selected, the operator simply extends the wand to the hull and electronically records the offsets for each station through a series of wand-to-hull contacts. It is not necessary to mark the hull; a meter on the wand indicates when the measurer is deviating from the station line, and with each contact, the angle and distance are measured and stored. The HMI functions in conjunction with the Tandy TRS-80 Model 102 or any portable computer equipped "with RS-232 interface and sufficient memory to run a 7K-byte BASIC program and store 20K-bytes of hull data."[5] The use of an on-site disc drive is optional. Most measures simply take the recorded data home and copy it onto a floppy disk.

Conversion from raw data to offsets is conducted by the measurer through USYRU. However, the USYRU is solely concerned with handicapping yachts, and while a number of organizations have approached the USYRU in pursuit of other applications—including the measurement of historical ships—as of June 1989, none of these proposals has come to fruition. The problem is more one of means rather than will: the USYRU staff is kept busy by the demand for yacht handicapping. Moreover, the USYRU is also rightly concerned about protecting the interest of naval architects.

Thus, the USYRU system is a viable choice only when there is no possible copyright infringement and when circumstances justify a cost of approximately $800 for a basic table of offsets. The willingness of the USYRU to provide data in a form that would permit ready hydrostatic or performance prediction will depend on the individual project. However, anyone with basic programming skills could write a program to allow the data to be analyzed using other CAD yacht-design and analysis software packages.[6] For heritage conservation purposes, then, the main advantages of the USYRU device are the speed of the measuring process and the high degree of accuracy it provides. If trained measurers are available, individuals or institutions wishing to lift the lines from a small number of vessels might find it appropriate to hire the services of a USYRU measurer on a contract basis.[7]

The GLHSRP Photo Documentation System

The Great Lakes Historic Ships Research Project (GLHSRP) represents the most extensive application yet undertaken of computer technology to the recording of historical vessels' shape and hydrostatics. With funding from the Provincial Government of Ontario and the Federal Government of Canada, GLHSRP recorded one hundred hulls between the springs of 1987 and 1989.[8]

The salient feature of the project was the use of the advanced CAD "Fast Yacht" computer program in the study and depiction of historical hull design. The main focus of GLHSRP was on large nineteenth-century, wooden-hulled commercial carriers, but a sample of nine noteworthy Great Lakes small craft was also included in the data base, the lines being derived from half models or plans. Moreover, a certain amount of time was specifically allocated to the exploration of the system as a tool for small craft documentation (measurement and depiction) and analysis.

When measuring, the GLHSRP device is set up next to the half model or, if the models are mounted and cannot be moved, against a wall. The device functions by defining stations along the length of the hull-artifact with a well-focused shadow (fig. 16.1). Each station is photographed adjacent to a reference "flag" of known dimensions defined by four clearly marked datum points. A 35mm SLR camera employing a standard lens and high speed slide film is used. The resulting transparencies are later projected onto a digitizing tablet, traced with a digitizing pen, and interpreted using special software. This software is complementary to the "3-D Digitizing" routine of the "Fast Yacht" CAD program and uses the known distances between the datum points on the flag to compensate for any perspective distortion. Briefly, the device is aligned with the subject, and stations are chosen. At each measured station, as well as at the bow and stern, a photograph is taken with the film plane kept as closely aligned with the flag as possible.

Since GLHSRP was aimed primarily at the study of larger commercial vessels, the device was developed for the measurement of half models only. Nevertheless, the prospect of adapting GLHSRP to the measurement of historical small craft was seriously discussed. The mathematics of the process are proven, and GLHSRP consultant Steve Killing drew up a set of guidelines for the enlargement of the device for this purpose.

Stereo-Photogrammetry

Yet another computer-aided measurement method is stereo-photogrammetry. Though essentially similar to the stereo-photographic techniques and technology developed for aerial mapping, this approach has been specifically adapted for heritage conservation purposes, including the documentation of historical vessels, on land and underwater, by the Heritage Recording Services Section of the Canadian Parks Service.

The stereo-photogrammetry system involves photographing a given object with stereographic camera equipment. The number of stereopairs (a pair of photographs taken from a parallel orientation) required varies depending on the size and shape of the subject. For example, fifteen stereopairs were required for the exterior documentation of the forty-nine meter Klondike River sternwheeler *S.S. Keno*. Once processed, these photographs are then plotted, either using the traditional analogue machine or using a computer-assisted method.

With the analogue method, the shape is plotted along continuous lines, and a very high level of accuracy is possible. However, a high degree of training and experience is required to operate analogue plotting equipment. With the computer, one of two approaches can be taken: the stereograms can be digitized using a standard digitizing tablet or read using a sophisticated, computer-assisted analytical plotter. With the computer approach, individual point measurements are taken from the images, with the level of accuracy dependent upon the number of points measured. In general, using the simple digitizing tablet, the level of accuracy is within 10 percent of the traditional method, with the final product reflecting what is more an ideal shape than a precise representation of the in-situ form (damage, dents, and all). On the other hand, the system is relatively less expensive than the traditional analogue equipment (ranging from \$60,000 to \$150,000 as opposed to \$100,000 to \$300,000—in Canadian currency—for the latter) and requires less operator training and skill. Given a greater degree of investment, an analytical plotter can also be used to interpret the stereograms. The analytical plotter provides a very dense array of individual point measurements producing what is virtually a continuous line. The result, therefore, is computer efficiency and a level of accuracy comparable to the analogue method.

For heritage purposes, the Canadian Parks Service system has been oriented towards a very high level of accuracy, sufficient even for such purposes as the recording of small architectural decorative elements. However, the Canadian Parks Service has also applied its stereo-photogrammetry technology to the documentation of historical vessels with great success and has produced very fine and accurate lines plans. In the last decade, the Canadian Parks Service has introduced increasingly sophisticated computer technology to what was originally a largely mechanical process, and this has added a high degree of efficiency to the process. The system employed by the Canadian Parks Service consists of a Wild BC II analytical plotter, a Data General 30 central processing unit, and a Wild plotter table. With this equipment, the stereo photographs can be interpreted and lines plans generated more quickly and efficiently. Whether or not the slight decrease in accuracy inherent in simple tablet interpretation is acceptable to the user will depend, of course, on the demands of the project. However, the ease of operation, increased efficiency, and the decrease in unit cost imbue this computer-tablet approach with many advantages, while use of a analytical plotter offers the best of both worlds, though at a higher unit price.

3-D Digitizing and Surface Creation Using the GLHSRP "Fast Yacht" CAD System

The measuring of hull shape through any method, computer-aided or manual, is only the first step in the process of documentation. Once the lines are lifted, a wide range of graphic documentation and hydrostatic analysis is possible using available CAD programs. A brief description of the methods used in digitizing a body plan and generating a surface using the GLHSRP "Fast Yacht" CAD system[9] will provide the reader with an impression of the manner in which computer technology is applied to the documentation and analysis of hull form after either computer-aided or traditional measuring systems have been used to lift lines.

Although a number of other software packages are available for conducting hydrostatic analysis and generating lines plans, the "Fast Yacht" system is among the most sophisticated available. A closer look at the operation of this system will provide the reader with a good sense of the museological potential of CAD systems.

The vessel used in this description is the St. Lawrence River skiff, *Annie*, built in 1885 by Bain & Co. of Clayton, New York. The lines for this vessel were traced directly from an existing set of plans drawn by Andrew Steever, a well-known veteran enthusiast of the St. Lawrence skiff.[10] *Annie* is 17' 8-3/4" long, has a beam of 38 7/8", a depth of 13 1/16" (midsection) and a displacement of about 140 pounds, empty and dry. The monitor images shown in figs. 16.2 to 16.9 were produced directly from the computer using a laser printer; therefore the lines and images are devoid of color and are rather less smooth than they actually appear to the operator.

Figure 16.2 shows the digitized stations of *Annie* in body plan. These lines were traced or digitized from the body plan of the original lines plan using the electronic pen and digitizing tablet. The body plan was placed on the tablet, and the digitized lines were created simply by depressing the pen, which has a small switch in its tip, at various points along each line or station; the more acute the curve, the more depressions, or points, are required. The same set of stations can also be created by entering offsets directly using the computer keyboard, or by tracing the stations from a series of slide images created using the GLH-SRP photo-measuring device and projected onto the digitizing tablet.

Figure 16.3 shows the same set of digitized stations in profile view, with the digitized profile of the stem and stern. Once entered into the computer and stored in a file, these stations and the stem and stern profiles are sufficient to conduct hydrostatic analysis or to serve as guideposts in the surface design process. The speed of the digitizing process depends on the method by which the basic data is entered; that is, whether it is traced from a lines plan, traced from a series of slides, or typed in as offsets, though on average the time required for a single hull is thirty minutes.

Surface Design

Surface design using the "Fast Yacht" software is conducted by means of a net, a grid consisting of vertical "columns" (shown in red on the monitor) and horizontal "rows" (shown in green). The density of the net required depends on the complexity of the shape and the judgment of the operator. At points where the columns and rows cross, the operator can, using the pen and tablet, "pick up the net" and move it in three dimensions. By alternating between profile, body plan and plan view, the net can be changed, sculpted, and ultimately made to conform to the digitized stations in all dimensions.

Alterations in the net correspond directly to alterations in the actual surface (shown as a finer grid of blue lines on the monitor). While it is the net that is manipulated in the hull creation process and which determines the shape of the surface, it is the perfect alignment of the faired surface grid with the digitized stations that is sought by the operator and provides the desired finished form, shown in figs. 16.4, 16.5, and 16.6.

Thus, surface design involves a careful process of net adjustment, a regular switching between net and surface to check developments, and a constant cycle of changes in perspective to ensure precise alignment of the surface with the digitized stations and centerline profile in all three main views. To assist the operator, the software has a number of options, including control of heel, pitch, and yaw to provide the operator an almost infinite number of perspectives on the hull. The net can also be adjusted from any perspective, while a "zoom" function permits the viewer to magnify any segment of the hull for fine examination or adjustment. Furthermore, to eliminate some of the confusion of looking at too many net lines, the software contains a "patch" function permitting the operator to work on small, select sections of the hull in isolation. When desired, net, surface, and digitized data can all be brought up on the monitor at once. Finally, a more natural view of the hull can be generated by means of the "mirror" and "solid" options. Figure 16.7 shows the hull displayed with the "mirror" on, while figs. 16.8 and 16.9 show the hull from different perspective in "solid."

Once the surface has been satisfactorily generated, the results an be presented as a conventional or parametric (fig. 16.10) lines plan or as a "quickplot" isometric drawing using the computer plotter. Full-scale plans can also be produced on mylar for use in lofting.

Why Use Computers?

Aside from their inherent speed and efficiency in documentation, computer systems used to analyze hull form can greatly improve our knowledge and appreciation of the design qualities of historical small craft. From the hydrostatic data, it is possible to test the reputed or relative stability of boat types.[11] Also, a careful examination of the coefficients of form and other hydrostatic factors can shed new light on the hydrodynamic efficiency of working and pleasure boats. In this way, trends in the evolution of hull forms produced by a specific designer or within a given type or class can be defined and analyzed.[12] Information of this sort, if placed within a broader economic and social context, may allow us to appreciate better some of the underlying causes of the regional variation and popularity of certain designs and the nuances in performance that made the product of one builder preferred above others. The latter is an area of especially rich potential using the performance prediction packages now available with the more advanced CAD programs.

The systems and applications described above reflect a new approach to the recording, depiction, and analysis of historical hulls. The systems apply the technology of modern yacht measurement and design to the interests and goals of marine heritage conservation and scholarship. The potential of the systems described has yet to be fully explored in this field, though further efforts together with more inexpensive and user-friendly computers will almost certainly make this technology more attractive and viable in the future.[13]

The rapid pace of technological change today is often cited as a cause for the neglect and decay of our material culture: glass and steel replace brick and carved stone in our cities just as fiberglass and aluminum replace wood and canvas in our marinas and boat yards. Yet here, at least, is an ever improving opportunity to turn new technology towards the active preservation and dissemination of the old. Computers are increasingly proving to be highly useful and efficient tools in the service of heritage conservation and in the documentation and study of small craft. We simply ought not to "miss the boat."

Notes

1. For examples of how quantitative and qualitative performance data has been developed, see Colin Palmer's "Sail and Hull Performance," *WoodenBoat* 92 (1990): 76; Ole Crumlin-Pederson and Max Vinner's *Sailing into the Past: The International Ship Replier Seminar, Roskilde 1984* (Roskilde, Denmark: Viking Ship Museum, 1986); Sean McGrail and J. Eric McKee, *The Building and Trails of the Replica of an Ancient Brat: The Gokstad Faering,* Parts I–II, *Maritime Monographs and Reports* No. 11, (Greenwich, England: National Maritime Museum, 1974).

2. The market is competitive enough that most inquiries will be answered quickly and enthusiastically. Some companies even provide sample disks. Good, reliable follow-up and consultation service should also be considered an essential issue in the decision to purchase any software package.

3. The author is indebted to John Bell, supervising photogrammetric engineer with Heritage Recording and Technical Data Services Section of the Ministry of the Environment, for providing a written description of the stereo-photogrammetric system used by his section. Much of the information on the stereo photogrammetric system provided in this chapter was adapted from Mr. Bell's submission, though any errors of interpretation are entirely the fault of the author. For further information on the stereo-photogrammetry measurement system used by the Canadian Parks Service and copies of its reports and lines plans, contact: Robin Letellier, head, Heritage Recording Services Section, Restoration Services Division, Engineering & Architecture Branch, Canadian Parks Service, Ottawa, Ontario, Canada K1A 0H3.

4. United States Yacht Racing Union, *Hull Measurement Instructions for Use with the USYRU Hull Measurement Instrument,* US Version 1.1 (Newport: USYRU, 1988), 1.

5. Ibid.

6. A successful, experimental, and low-cost hull measuring device, based on the same principles as the USYRU's HMI, but requiring slightly more pencil, paper, and keypunch data entry-storage, was developed by John S. Letcher, Jr., and presented to the New England Sailing Yacht Symposium of the Society of Naval Architects and Marine Engineers on March 22, 1980. Serious inquiries about the "Maine Yankee Hull Measuring Device" can be made to: John Letcher, Sr., AEROHYDRO Inc., Box 684, Southwest Harbor, Maine 04679.

7. A USYRU-trained measurer in Toronto, Ontario, contacted in the spring of 1989 quoted a hourly rate of $50 (Canadian) for HMI measuring services.

8. Project members included Maurice Smith, executive director of the Marine Museum; David Walker, naval architect; Steve Killing, yacht designer and computer consultant; and the author, a historian and archaeologist.

9. Garth S. Wilson, "The Great Lakes Historic Ships Research Project: Computers and the Analysis and Documentation of Small Craft," *Fresh Water* 2: 2 (1987): 9–13. Portions of this chapter are adapted from this article.

10. GLHSRP is indebted to Mr. Steever for his kind permission to use his plans for this purpose.

11. An example of this can be found in, Wilson, "The Great Lakes Historic Ships Research Project."

12. Garth S. Wilson, "The Great Lakes Historic Ships Research Project: An Innovative Approach to the Documentation and Analysis of Historic Hull Design," *International Journal of Maritime History* 1: 2 (1989): 199–238.

13. The Society of Naval Architects and Marine Engineers has a panel that attempts to keep abreast of improvements in Yacht-CAD systems. For more information, write SNAME, 601 Pavonia Avenue, Jersey City, New Jersey 07306.

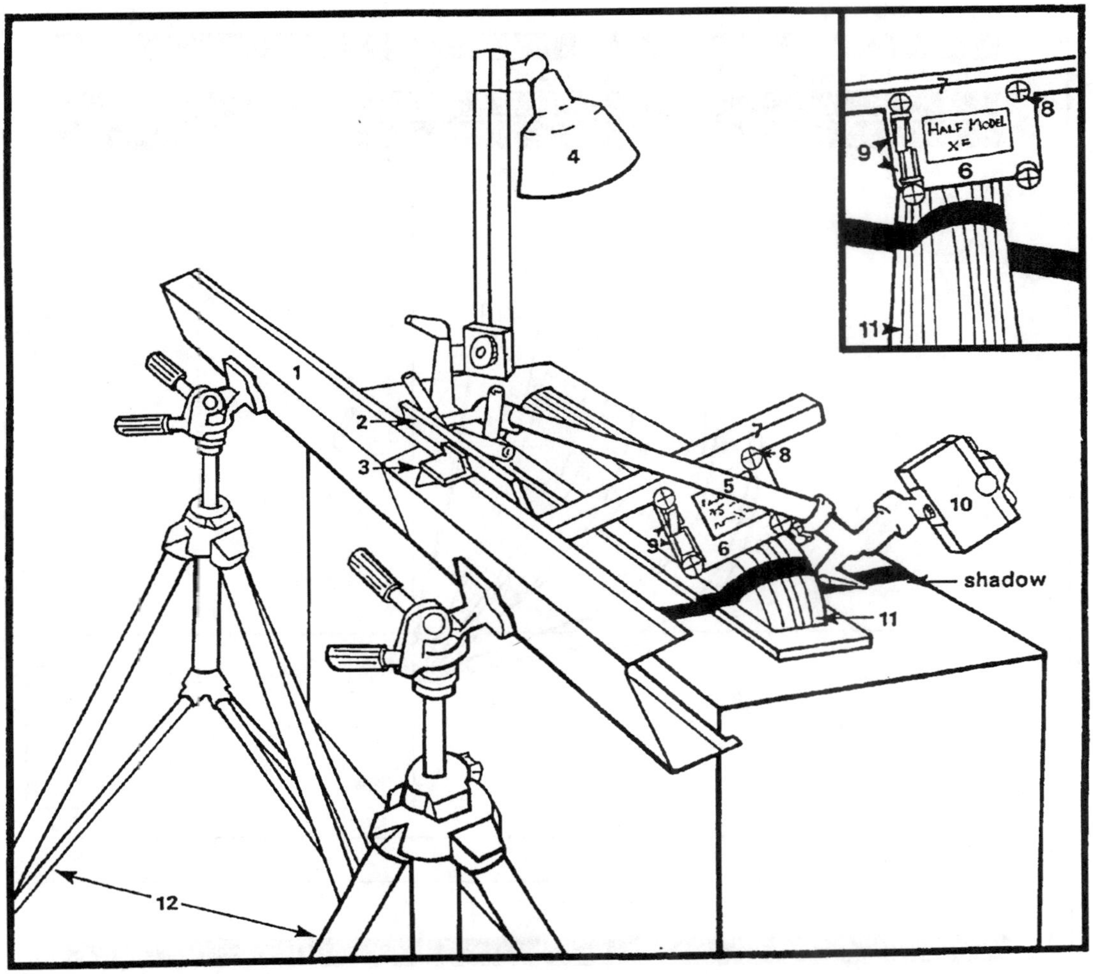

Figure 16.1. The GLHSRP photo-measuring device. (1) Track. (2) Carriage. (3) Pointer. (4) Light source. (5) Camera arm. (6) Flag. (7) Flag mast. (8) Target-datum point. (9) Focusing tabs. (10) Camera. (11) Half-model. (12) Tripods.

Flag (6): In measuring a hull, the distance between the targets on the flag must never be smaller than half the larger of either the vessel depth or its half-breadth. Thus, for example, if measuring a vessel with a half-breadth of 4 feet and a depth of 3 feet, the square defined by the targets on the flag should be no less than 2 feet by 2 feet. To reduce the weight of the flag, it could be made as a metal grid with a light-weight center. A similar separate flag of the same size is required for recording the profile of the vessel's stem and stern, although this flag requires neither raised targets nor focusing tabs. However, a different type of mast would be needed, ideally one that would hold the flag on the centerline while at the same time allowing for vertical alignment.

Light source (4): The crispness of the shadow being cast can be maintained by ensuring that the distance between the light source and the flag mast is equal to the distance between the flag mast and the hull being measured. The quality of the shadow can also be maximized by ensuring that the filament is crisp (focused). The strength (wattage) of the light source should also be increased in keeping with the increase in the flag (perhaps up to 1,000 watts), although the need for higher wattage can be somewhat offset by operating in low-light conditions.

Camera arm (5): The targets on the flag and the shadow cast on the hull make up the two essential components of every station photograph taken. Therefore, in order to ensure a complete image, a larger flag would also require a longer arm to hold the camera.

Measuring longer vessels: Two tripods are required. The maximum workable length of track is probably 16 feet (the current GLHSRP device has 8 feet of track). Lengthening the track beyond 16 feet would require another tripod, which could prove awkward when attempting to level the device.

As an alternative, vessels longer than 16 feet could be measured in two halves: from bow to midsection and from midsection to stern. While the centerline reference remains constant, the vertical reference is lost when the device is moved. To compensate for this, the midsection is photographed twice. Then, in the digitizing process, the midsection station from the two halves are aligned until they match perfectly. Once this is accomplished, the two segments are merged to form a complete body plan.

However, in cases where a vessel being measured in two segments has no clear centerline reference (that is, no prominent keel), then a straight, even length of tubing must be positioned above the vessel in such as way as to ensure that the shadow is also cast upon it. Accordingly, this bar must be long enough to be included in each station photograph. It will then serve as a point of reference common to each photograph. While it is not absolutely necessary that the tubing be parallel with the centerline or waterline, the better aligned the tubing is, the easier the digitizing process will be.

To digitize a vessel measured in this manner, the shadow cast on the bar is also digitized, perhaps as an "L" or crescent shape. In accordance with standard procedure, the operator can digitize the station curve as it appears in the image. However, in this instance, when the curve has been traced, the "line" is concluded, and the shadow cast on the bar is digitized as a separate line on the same plane. So, for each station in the body plan being produced, there will be a curve representing the hull shape and, above the curve, an independent "L" or crescent-shaped reference line.

When the body plan half-segments are completed, the operator can yaw, pitch, and heel them separately, until the reference lines for each half-segment are coincident. Once this is done, the two segments are merged together to form one complete body plan. Finally, the reference lines are removed from the finished body plan simply by using the "delete curve" option.

Accuracy in digitizing: Human error in placing the pen along the image line can be minimized to approximately .02 inches. In other words, repeatability is within .02 of an inch. Thus, when digitizing a half-model with a 1/8-inch beam and projecting full-size an image on the digitizing tablet, the possible error will be .02 inches. However, when digitizing an 8-inch projected image that represents a full-size dimension of 24 inches, the possible error at full size will be .06 inches (24 inches divided by 8 inches times .02 inches). In most instances, this resolution is completely adequate.

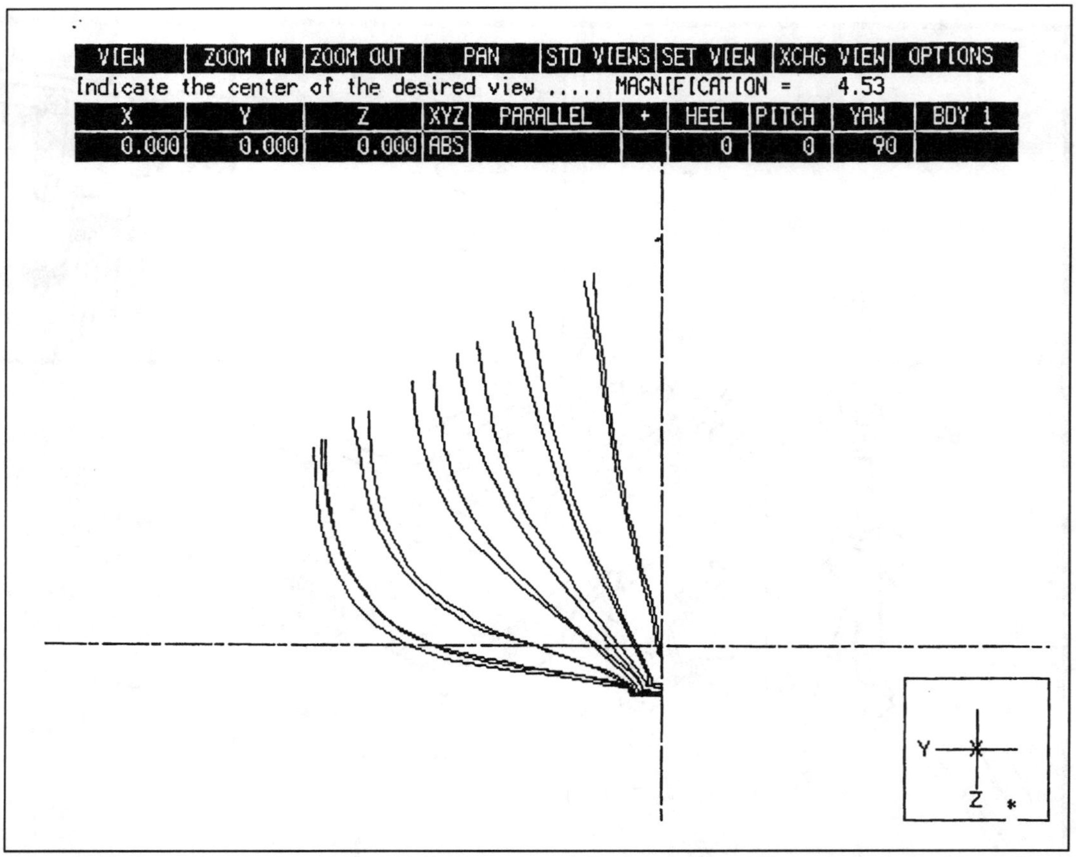

Figure 16.2. Annie, body plan view, digitized stations, "Fast Yacht" monitor image.

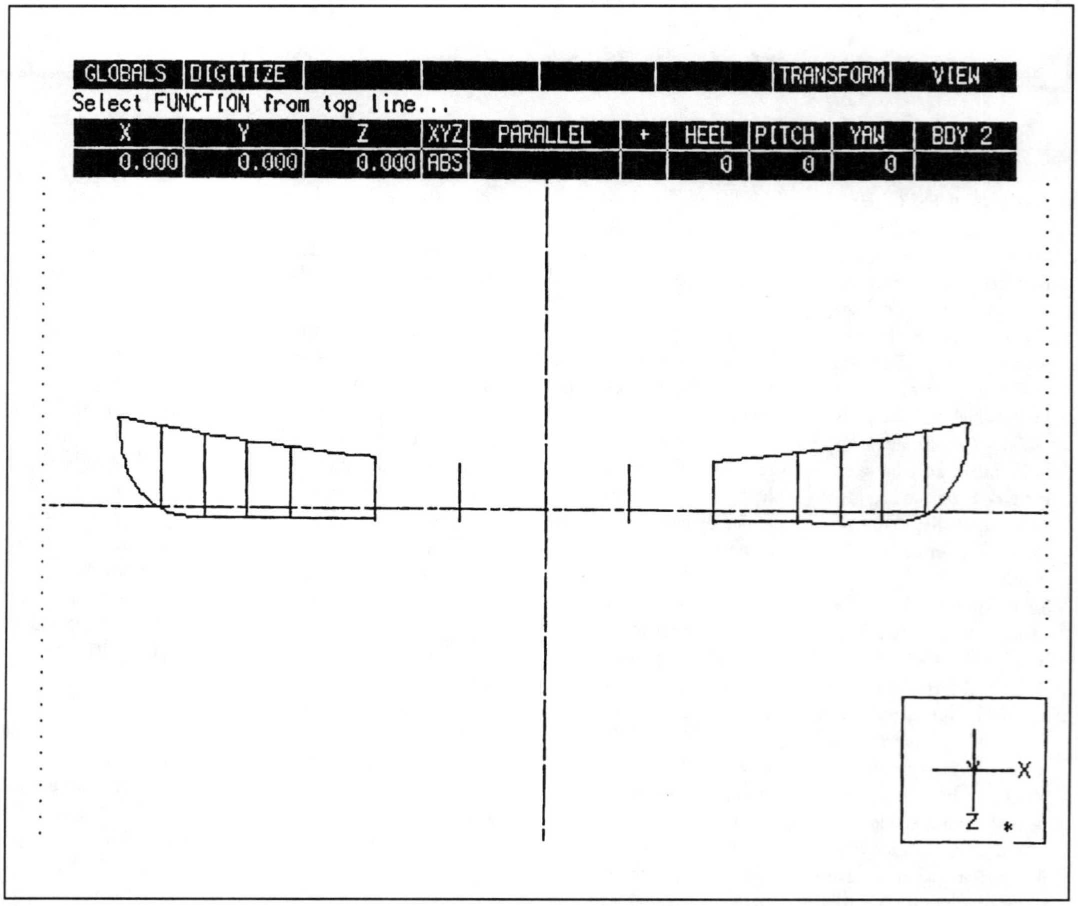

Figure 16.3. Annie, profile view, digitized stations and profile, "Fast Yacht" monitor image.

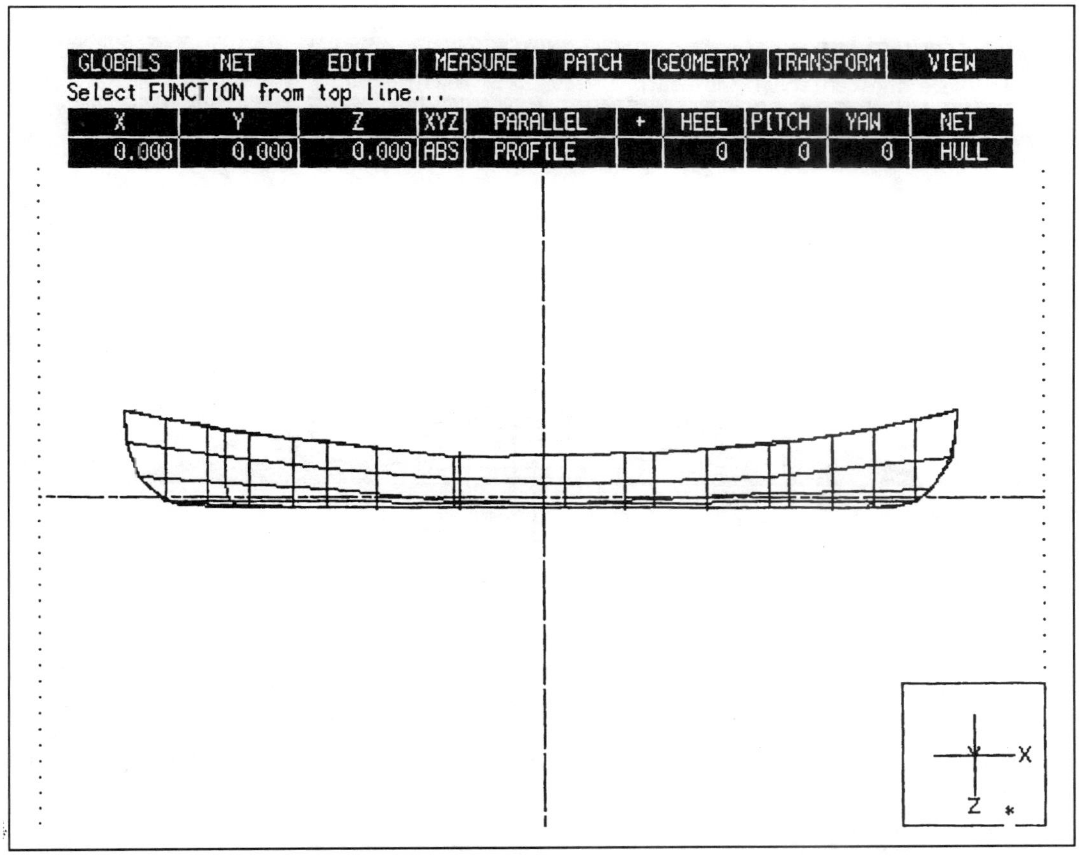

Figure 16.4. Annie, profile view, net, "Fast Yacht" monitor image.

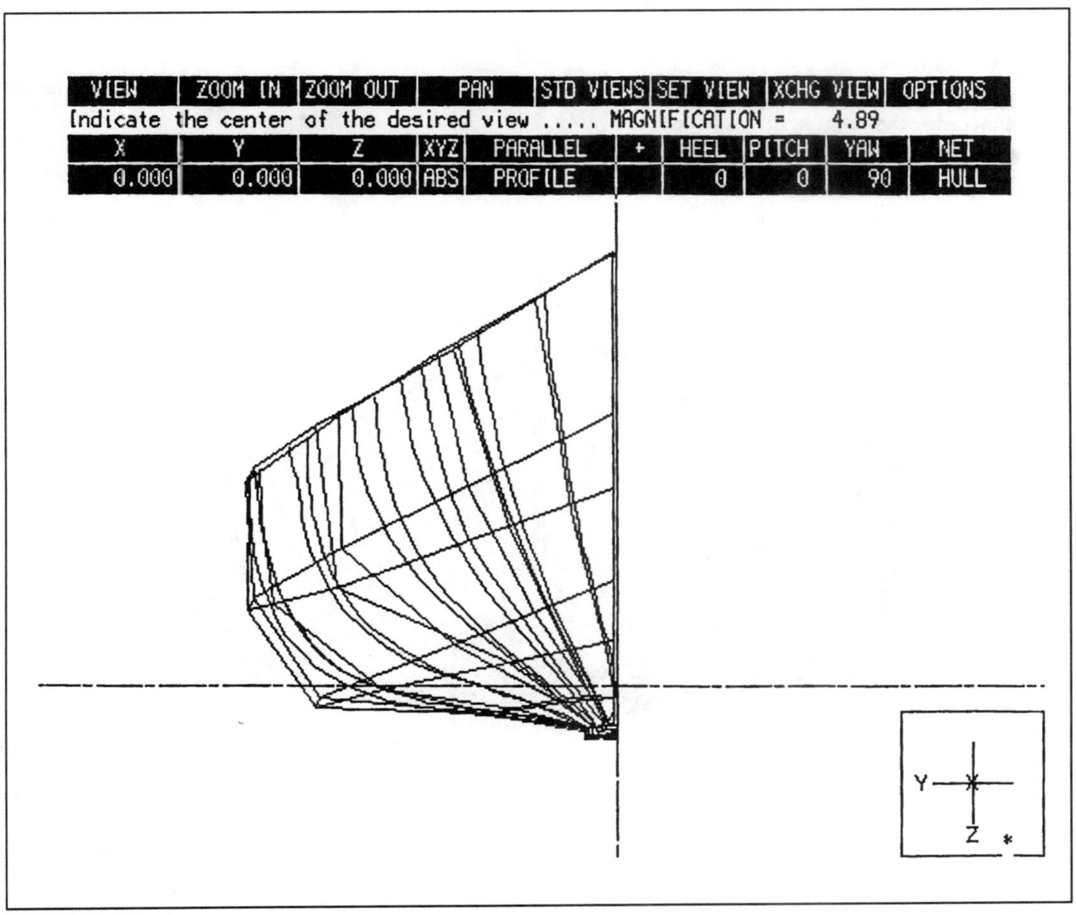

Figure 16.5. Annie, body plan view, net, "Fast Yacht" monitor image.

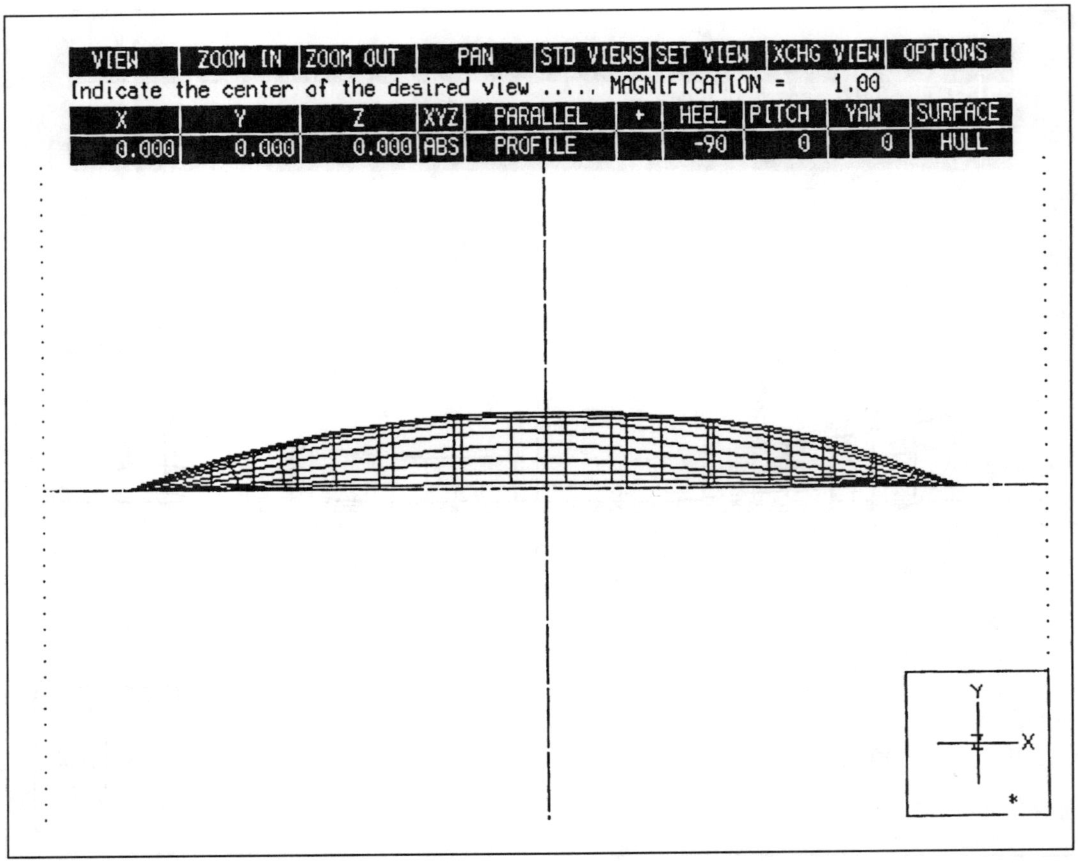

Figure 16.6. *Annie*, plan view, surface, "Fast Yacht" monitor image.

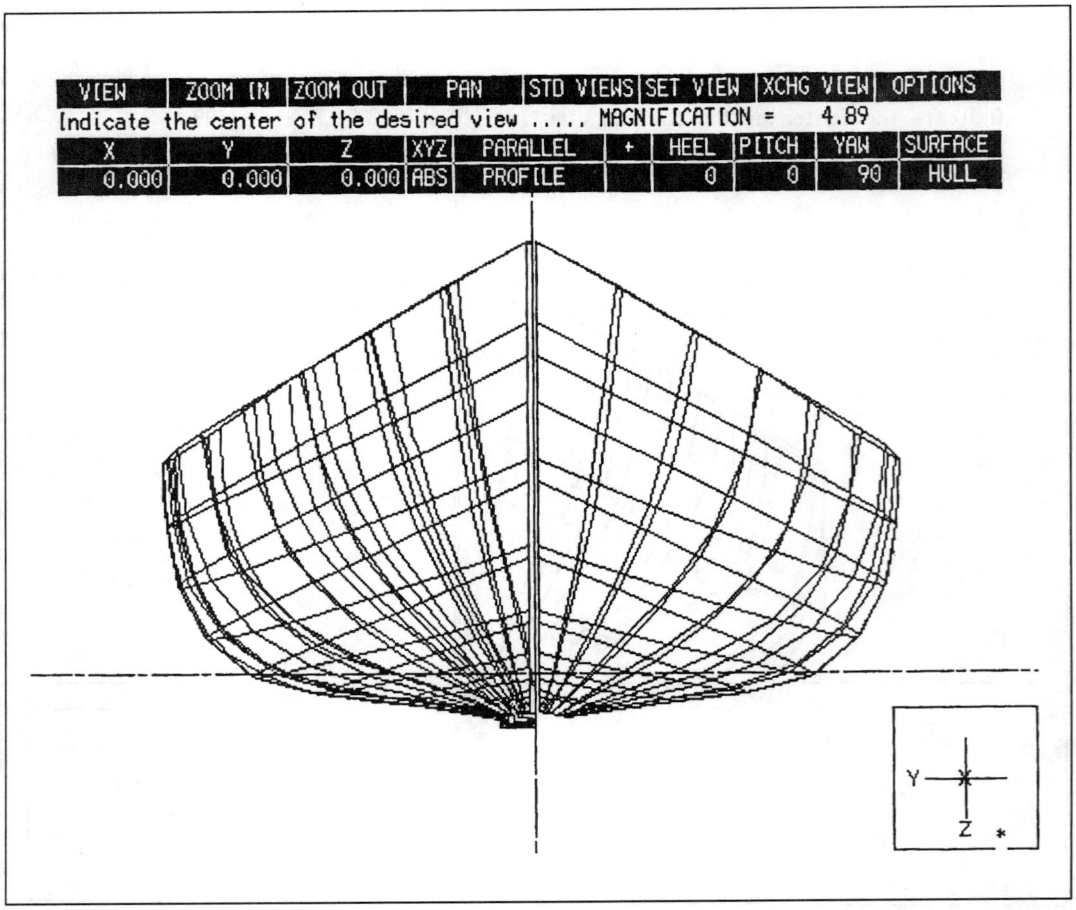

Figure 16.7. *Annie*, body plan, surface with mirror, "Fast Yacht" monitor image.

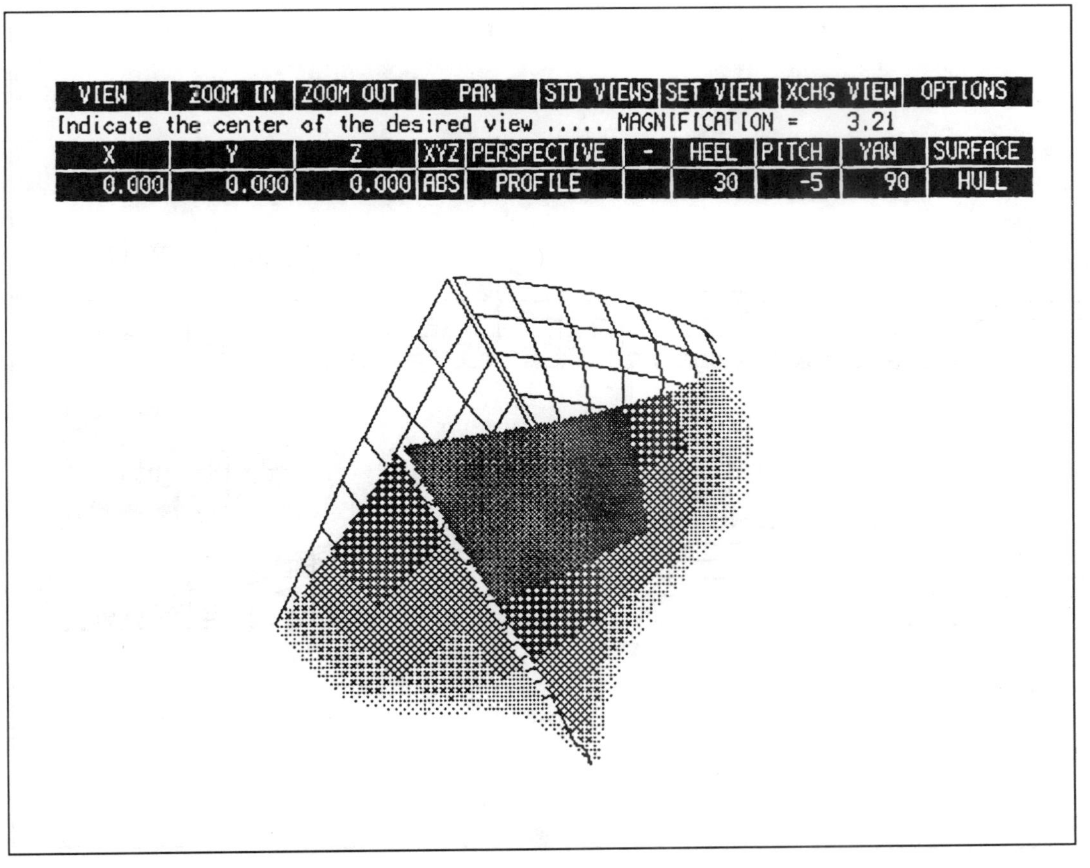

Figure. 16.8. *Annie*, rotated perspective, solid, "Fast Yacht" monitor image.

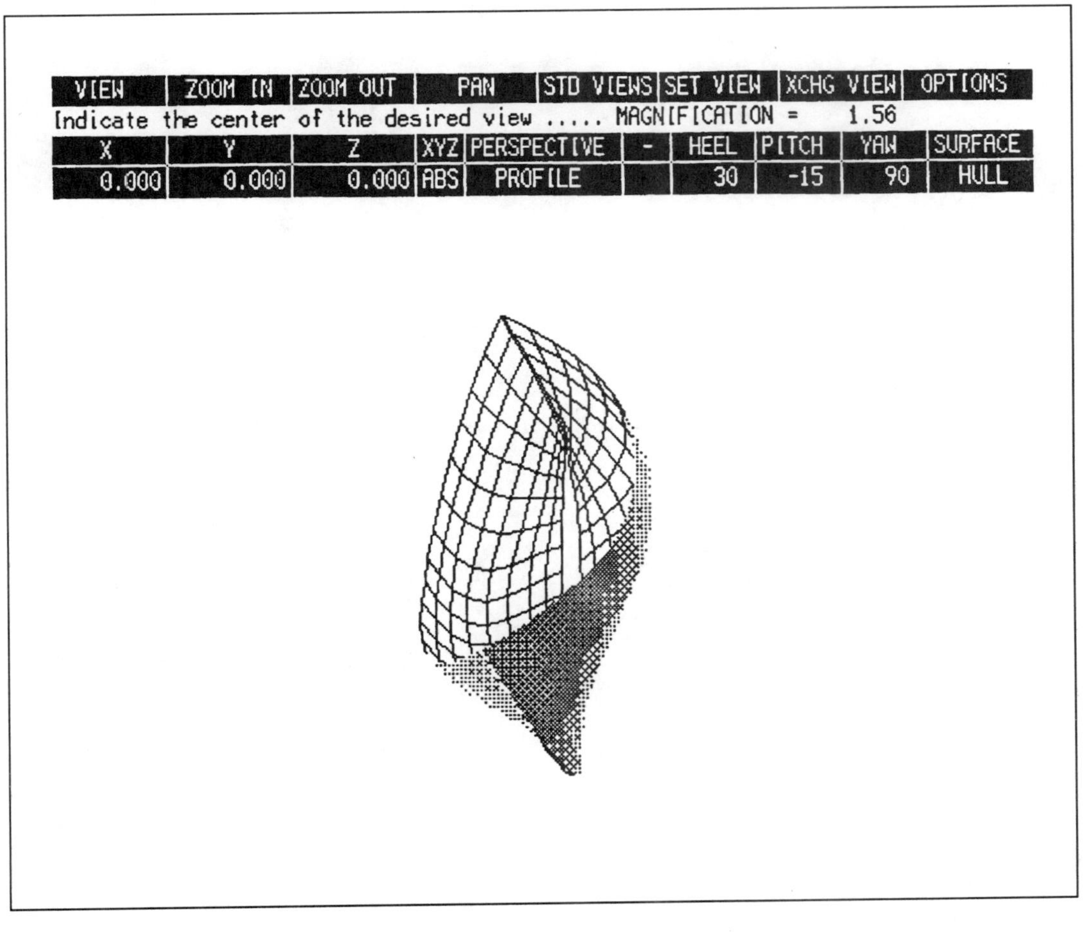

Figure 16.9. *Annie*, rotated perspective, solid, "Fast Yacht" monitor image.

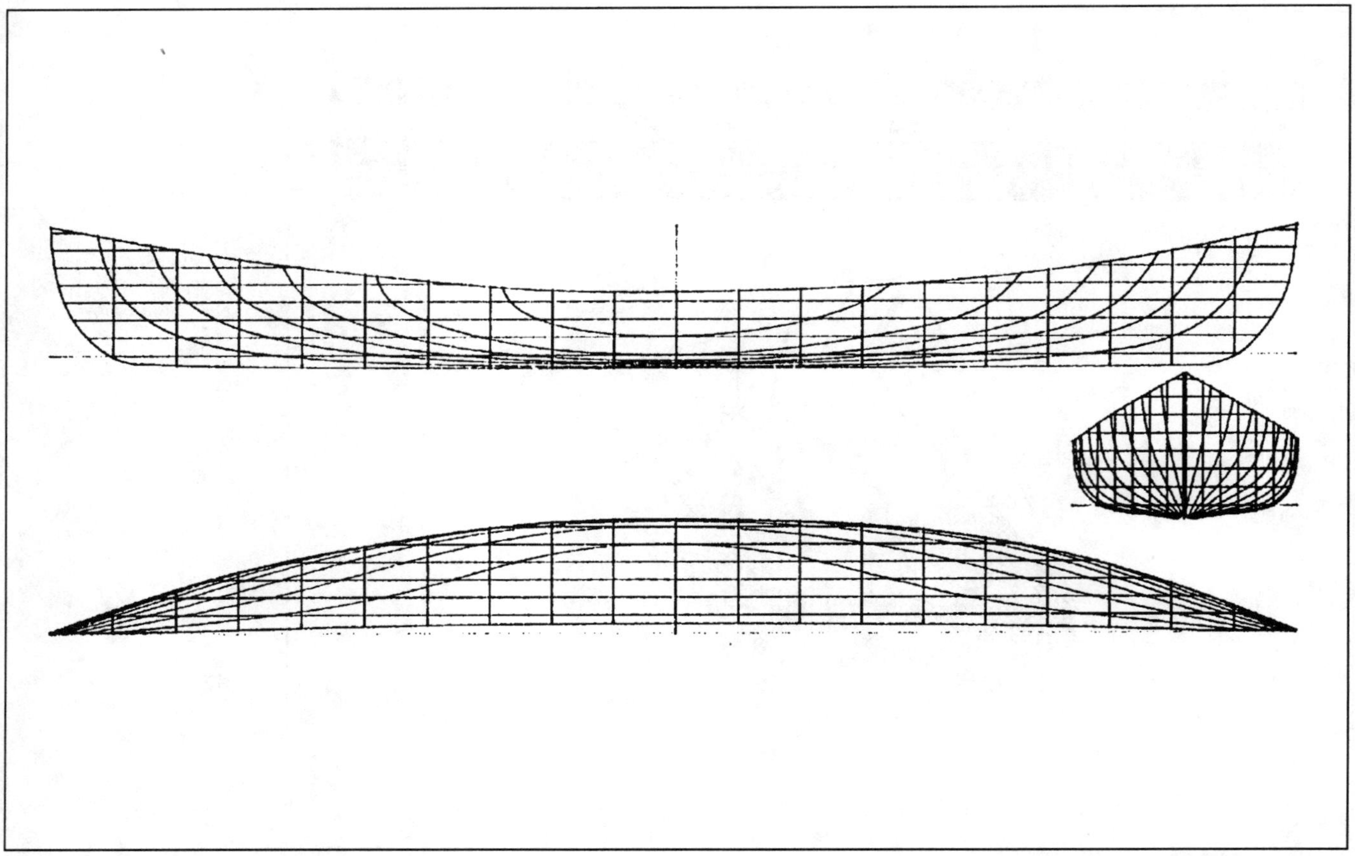

Figure 16.10. Annie, computer generated conventional lines plans, reduced.

Contributors

Willits D. Ansel was a ship's carpenter with the duPont Shipyard at Mystic Seaport Museum until his retirement in 1989. His writings include the books *Restoration of the Smack EMMA C. BERRY at Mystic Seaport* (1973) and *The Whaleboat: A Study of Design Construction and Use from 1850-1970* (1983).

Maynard Bray is a contributing editor for WoodenBoat Magazine and a trustee and former supervisor of the duPont Shipyard at Mystic Seaport Museum. His many published works include *Mystic Seaport Watercraft* (1979), *Herreshoff of Bristol* (1989), and numerous articles as technical editor of *WoodenBoat* from 1978-1988.

Kevin J. Crisman is a New World Nautical Archaeologist at the Institute for Nautical Archaeology. He contributed to *Ships and Shipwrecks of the Americas* (Bass, 1988) and wrote *Of Sailing Ships and Sidewheelers: the History and Nautical Archaeology of Lake Champlain* (1986).

David W. Dillion was an engineering draftsman before establishing a career as a freelance boat documentation specialist. He has measured and drawn over more than seventy boats up to a hundred feet in length and has taught lines-taking at the WoodenBoat School and half a dozen maritime museums across the United States. His plans have been published in *WoodenBoat* and other periodicals.

Benjamin A. G. Fuller is the former chief curator of Mystic Seaport Museum. He is the past Secretary General of the International Congress of Maritime Museums, and is currently a teacher, writer, and consultant on historic maritime preservation and museum computerization. His writings have appeared in *The Log of Mystic Seaport*, *WoodenBoat*, and other journals.

Paula J. Johnson is Maritime History Specialist for the Division of Transportation of the National Museum of American History at the Smithsonian Institution. She is editor of *Working the Water: the Commercial Fisheries of Maryland's Patuxent River* (1988) and *Historical Tours through Southern Maryland: Broome's Island* (1983).

Paul R. Lipke is a wood technologist and maritime preservation consultant. Before he raised the funds, managed and co-edited this manual, his major works were *Plank On Frame* (1981) and *The Royal Ship of Cheops* (1984). He has contributed articles to *Archaeology*, *WoodenBoat*, *Small Boat Journal*, *The Christian Science Monitor*, *The Boston Globe*, and many other publications.

William N. Peterson is curator of collections at Mystic Seaport Museum. His publications include *"The Care of Historic Photographs" The Log of Mystic Seaport*,(1979) and *"Mystic Built": Ships and Shipyards of Mystic Connecticut, 1784-1919* (1989).

Peter A. Schmid is an architect practicing in Worcester, Massachusetts. As a marine illustrator, he has contributed drawings to *The Royal Ship of Cheops* (1984), *National Geographic Magazine* (April 1988), and *WoodenBoat* (May 1986).

David A. Taylor is a folklife specialist with the American Folklife Center at the Library of Congress. He is the author of *Boat Building in Winterton, Trinity Bay, Newfoundland* and the *Documenting Maritime Folklife.*

Peter T. Vermilya is Mystic Seaport Museum's associate curator for the Yachting and Small Craft Collections and founding president of the Museum Small Craft Association. He is working on a book on the construction of Concordia Beetle Cat for *WoodenBoat* Publications.

Garth S. Wilson is curator of marine transportation at the National Museum of Science and Technology, (Ottawa).

Appendix A

List of Helpful Organizations

Adirondack Museum
The Adirondack Historical Association
Blue Mountain Lake, NY 12812

Allie Ryan Maritime
Collection of Maine Museum
Box C-1 Maine Maritime Academy
Castine, ME 04420

American Merchant Marine Museum
U. S. Merchant Marine Academy
Kings Point, NY 11024

Antique and Classic Boat Society
P. O. Box 831
Lake George, NY 12845

Antique Boat Museum
750 Mary Street
Clayton, NY 13624

Baltimore Seaport and
Baltimore Maritme Museum
Pier Four, Pratt Street
Baltimore, MD 21202

Battleship *Texas* Historical Park
3527 Battleground Road
La Porte, TX 77571

C. A. N. A. L.
36 Lakeview Avenue
Lincoln, RI 02865

C & O Canal of Cumberland,
MD, Inc.
300 Bel Air Drive
Cumberland, MD 21502

Cairo Museum
P. O. Box 349
Vicksburg, MS 39180-0349

Calvert Maritime Museum
P. O. Box 97
Solomons, MD 20688

Canal Museum
Hugh Moore Park
P. O. Box 877
Easton, PA 18042

The Canal Museum
318 Erie Boulevard
Syracuse, NY 13202

Canastota Canal Town Museum
122 Canal Street
Canastota, NY 13032

Cape Ann Historical Association
27 Pleasant Street
Gloucester, MA 01903

Cape Hatteras National Seashore
Route 1, Box 675
Manteo, NC 27945

Chesapeake Bay Maritime Museum
Navy Point
St. Michaels, MD 21663

City Island Historical Nautical Museum
190 Fordham Street, City Island
Bronx, NY 10464

Coastal Heritage Society
One Fort Jackson Road
Savannah, GA 31404

Coast Artillery Museum
at Fort Worden
433 Ranger Drive, SE
Olympia, WA 98503

Coast Guard Museum
of the North West
1519 Alaskan Way S
Seattle, WA 98134

Columbia River Maritime Museum
1792 Marine Drive
Astoria, OR 97103

Confederate Naval Museum
P. O. box 1022
Columbus, GA 31902

Connecticut River Foundation
at Steamboat Dock, Inc.
P. O. Box 261
Essex, CT 06426

Coquille River Museum
P. O. Box 737
390 SW First Street
Bandon, OR 97411

Cruiser *Olympia*
Submarine *Becuna* Association, Inc.
P. O. Box 928
Philadelphia, PA 19905

Door County Maritime Museum
6427 Green Bay Road
Sturgeon Bay, WI 54235

Dossin Great Lakes Museum
100 Strand/Belle Isle
Detroit, MI 48207

Dukes County Historical Society
Box 827
Edgartown, MA 02539

East Hampton Town
Maritime Museum
Bluff Road, P. O. Box 858
Amagansett, NY 11930

Flagship *Niagara*
P. O. Box 1026
Harrisburg, PA 17108

Galveston Historical Foundation
2016 Strand
Galveston, TX 77550

Gloucester Fisherman,s Museum
Rogers and Porter Street
Box 159
Gloucester, MA 01930

Grand Banks Schooner
Museum Trust
110 Roads End
Boothbay Harbor, ME 04538

Great Lakes Historical Society
Museum
480 Main Street, P. O. Box 435
Vermillion, OH 44089

Great Lakes Maritime Institute
Belle Isle
Detroit, MI 48207

Great Lakes Naval and
Maritime Museum
P. O. Box A-3785
Chicago, IL 60690

Hampton Roads Naval Museum
Pennsylvania H. G-29
Norfolk Naval Base
Norfolk, VA 23511

Head of the Lakes
Maritime Society
P. O. Box 775
Superior, WI 54880

Heritage Boat Club
2554 W. Pensacola Avenue
Chicago, IL 60610

Herreshoff Maritime Museum
P. O. Box 450
18 Burnside Street
Bristol, RI 02809

Howard Steamboat Museum
Clark County Historical Society
1101 E. Market Street, P. O. Box 606
Jeffersonville, IN 47130

Hudson River Maritime Center
One Roundout Landing
Kingston, NY 12401

Hull Lifesaving Museum
6 Circuit Avenue
Hull, MA 02045

Inland Rivers Library Collection
Cincinnati and Hamilton County
Public Library
8th and Vine Street
Cincinnati, OH 45202

M. I. T. Museum and Historical Collection
Hart Nautical Collection
265 Massachusetts Avenue
Cambridge, MA 02139

Maine Maritime Museum
963 Washington Street
Bath, ME 04530

Manitowoc Maritime Museum
809 South 8th Street
Manitowoc, WI 54220

Marine Corps History
and Museums Division
Building 58, Navy Yard
Washington, DC 20374

The Mariners' Museum
Newport News, VA 23606

Maritime Museum of the Atlantic
1675 Lower Water Street
Halifax, Nova Scotia
Canada B3J 1S3

Marquette Maritime Museum
1515 Lynn Street
Marquette, MI 49855

Maryland Historical Society
201 West Monument Street
Baltimore, MD 21201

Missouri River History Museum
P. O. Box 124
Brownville, NE 68321

Moosehead Marine Museum
P. O. Box 1151
Pritham Avenue
Greenville, ME 04441

Museo de Historical Military y
Naval de Puerto Rico
Fort Saint Jerome POB 4184
San Juan, PR 00905

Museum of Coastal History
610 Beachview Drive
St. Simons Island, GA 31522

Mystic Seaport Museum, Inc.
Greenmanville Avenue
Mystic, CT 06355

Nantucket Historical Association
P. O. Box 1016
Nantucket, MA 02554

National Liberty Ship Memorial, Inc.
Fort Mason—Bldg. 57
Washington Navy Yard
Washington, D. C. 20374

National Maritime Museum, San Francisco
Golden Gate National Recreation Area
San Francisco, CA 94123

National Association of Marine Surveyors
(NAMS)
305 Springhouse Lane
Morristown, NJ 08057

National River Hall of Fame
Dubuque County Historical Society
Box 305
Dubuque, IA 52001

Naval Undersea Warfare
Museum Engineering Station
NUWES Code 05m
Keyport, WA 98345

Naval War College Museum
Coasters Harbor Island
Newport, RI 02841

Navy Memorial Museum
Washington Navy Yard
Building 76
Washington, DC 20374

Navy, Marine Corps & Coast Guard
Museum of the Pacific
Bldg. 1, Treasure Island
San Francisco, CA 94130

Navy Supply Corps Museum
Navy Supply Corps School
Athens, GA 30606

New Bedford Whaling Museum
18 Johnny Cake Hill
New Bedford, MA 02740

New York Historical Association
Lake Road
Box 800
Cooperstown, NY 13326

North Carolina Maritime Museum
315 Front Street
Beaufort, NC 28516

Northwest Michigan Maritime Museum
324 Main Street
P. O. Box 389
Frankfort, MI 49635

North West Seaport
P. O. Box 2865
Seattle, WA 98111

North Wind Undersea Institute
610 City Island Avenue
Bronx, NY 10464

Old Darmouth Historical Society
Whaling Museum
18 Johnny Cake Hill
New Bedford, MA 02740

Old Lighthouse Museum
P. O. Box 512
Michigan City, IN 46360

Old Lighthouse Museum,
Stonington Historical Society
P. O. Box 103
Stonington, CT 06378

Old Presque Isle
Lighthouse Museum
6282 Blackbass Bay
Alpena, MI 49707

Oxford Maritime Museum
P. O. Box 178
Oxford, MD 21654

Pacific Fleet Submarine
Memorial Association, Inc.
11 Arizona Memorial Drive
Honolulu, HI 96818

Patriots Point Naval
and Maritime Museum
P. O. Box 986
Mt. Pleasant, SC 29464

Peabody Museum of Salem
East India Marine Hall
Salem, MA 01970

Penobscot Marine Museum
Lincoln Colcord Memorial Library
Church Street
Searsport, ME 04974

Philadelphia Maritime Museum
321 Chestnut Street
Philadelphia, PA 19106

Plimoth Plantation, Inc.
P. O. Box 1620
Plymouth, MA 02360

Port Isabel Lighthouse
State Historic Structure
P. O. Box 863
Port Isabel, TX 78578

Portsmouth Lightship Museum
London Slip at Water Street
P. O. Box 850
Portsmouth, VA 23704

Radcliffe Maritime Museum
Maryland Historical Society
201 West Monument Street
Baltimore, MD 21201

Rockport Apprenticeshop/The Artisans School
Box 539
Rockport, ME 04856

Sag Harbor Whaling and
Historical Museum
P. O. Box 1327
Sag Harbor, NY 11963

San Diego Maritime Museum
1306 North Harbor Drive
San Diego, CA 92101

Sandy Hook Museum
Gateway National Recreation Area
P. O. Box 437
Highlands, NJ 07732

Ships of the Sea Maritime Museum
503 East River Street
Savannah, GA 31401

Sleeping Bear Dunes Seashore,
National Park Service
400 Main Street
Frankfort, MI 49635

Smithsonian Institution
Curator of Naval History
National Museum of American History
Washington, DC 20560

Smithsonian Institution
Curator of Maritime History
National Museum of American History
Washington, DC 20560

South Street Seaport Museum
207 Front Street
New York, NY 10038

Southampton Historical Museum
P. O. Box 303
Southampton, NY 11968

Southold Historical Society
and Maritime Museum
Main Road and Maple Lane
Southold, NY 11971

St. Clements Island,
Potomac Museum
General Delivery
Colton Point, MD 20626

St. Marys City Commission
P. O. Box 39
St. Mary's City, MD 20686

Staten Island Ferry
Maritime Museum
New York Dept. of Maritime
St. George Ferry Terminal
Staten Island, NY 10306

Strawbery Banke, Inc.
P. O. Box 300
Portsmouth, NH 03801

Submarine Memorial Museum,
USS *Ling*
P. O. Box 395
Hackensack, NJ 07602

Suffolk County Historical Society
300 West Main Street
Riverhead, NY 11901

U. S. Coast Guard Museum
U. S. Coast Guard Academy
New London, CT 06320

U. S. Department of the Navy
Naval Historical Center, Bldg. 57
Washington Naval Yard
Washington, DC 20374

U. S. Frigate *Constellation*
Constellation Dock
Baltimore, MD 21202

U. S. Naval Academy Museum
Annapolis, MD 21402

U. S. Navy Memorial Museum
Bldg. 76
Washington Navy Yard
Washington, DC 20390

USS *Alabama* Battleship
Memorial Park
P. O. Box 65
Mobile, AL 36601

USS *Constitution* Museum
Building 22
Charlestown Navy Yard
Boston, MA 02129

USS *Massachusetts* Memorial Committee
Battleship Cove
Fall River, MA 02721

Vancouver Maritime Museum
1905 Ogden Street
Vancouver, British Columbia
Canada V6J 3J9

Virginia Beach Historical Museum
P. O. Box 24
Virginia Beach, VA 23458

Whale Museum
P. O. Box 25
Cold Spring Harbor, NY 11724

Whatcom Museum of History and Art
121 Prospect Street
Bellingham, WA 98225

The H. Lee White Marine Museum
Foot of East First Street
P. O. Box 387
Oswego, NY 13126

Henry Francis du Pont Winterthur
Museum
Winterthur, DE 19735

Appendix B

Union List of North American Watercraft: Descriptive Summary and Reporting Forms

A project of the Museum Small Craft Association

The goal of this Museum Small Craft Association project is to create a comprehensive computer database listing and detailing all watercraft (with no size limitations) in North American museum collections.

The list will be invaluable as a finding aid, research tool, and collections management tool. It will also provide essential information for future watercraft preservation efforts, by providing an overall view of what is out there. The database format will make the list easy to update, while also enabling individual museums to manipulate information on their own collections for management purposes (such as printing accession records, surveying the collection, and research). The list will also be available in hard copy form.

The core of the current Union List is the watercraft information from the 1984 Maritime Historic Resources survey, conducted by the National Trust for Historic Preservation and White Elephant Management. That data was passed also to the MSCA in 1987, and was loaded into a specially designed database. To gather information missing from the original survey and to ensure the consistent use of terms, new, more comprehensive data sheets were prepared. These were sent back to the original twelve responding museums plus three others.

The costs of this initial effort have included designing the dBase III computer program, working the "bugs" out of the database, loading and processing existing data, and sending that data out for review by participating institutions. Mystic Seaport Museum has underwritten the costs of the first stage, which comes in at a remarkable three thousand dollars. The work has been master-minded by WoodenBoat Publications, whose staff has produced effective and user-friendly programs for data entry and retrieval.

The original group of museum watercraft information (including the largest collections in the country) has been converted to MSCA format.

The second stage will be to solicit and enter new watercraft data from another dozen museums that did not respond to the original White Elephant survey. We are now developing a target list of those museums. Making the list available to researchers and museum users is also a priority.

We are convinced that the Union List is an essential and valuable survey effort, one whose time has come. As more and more institutions and individuals become involved in watercraft studies, the need for a comprehensive finding aid has become ever more apparent. Such a list can influence preservation planning by identifying "endangered species" of watercraft; it can aid research projects of all kinds, coordinate museum collecting; and, perhaps most importantly, provide the field with a "thesaurus" of boats, a consistent method for describing watercraft and a reference of cataloguing and description.

Eventually, the Union List will encompass all significant watercraft collections in the United States and Canada. The value of this project to the maritime and preservation worlds is apparent, and we are optimistic that many more museums will become involved as the MSCA moves into the second stage.

Your participation and help will be appreciated.

MSCA Vessel Record - Complete

Vessel Data:

Vessel Name [ ] Accession No. [ ]

Vessel Type [ ] ☐ class boat

Vessel SubType [ ] ☐ reproduction

Designer [ ] ☐ attributed

Original Use [ ]

LOA ft		Hull Material		Rig Type	
LOA in		Planking		Rig Shape	
Beam ft		Bottom Shape		Engine Type	
Beam in		End Shape		Engine Fuel	
Draft ft		Keel Type		Propulsion	
Draft in		Deck Layout			

Propulsion Notes:

Condition ○ unknown ○ fair ○ excellent ○ poor ○ good

Tender? ○ yes ○ no

Rowing Stations [ ]

Tender to [ ]

Other Uses

☐ education ☐ guiding ☐ livery ☐ recreation ☐ utility ☐ workboat
☐ fishing, commercial ☐ hunting ☐ logging ☐ rescue ☐ warfare
☐ fishing, recreational ☐ indian ☐ racing ☐ transport ☐ waterfowl

Builder's Data:

Builder [ ] ☐ attributed Hull No [ ]

City [ ] ☐ possibly

State [ ] Country [ ]

Year Built [ ] ☐ circa Build Date Range [ ]

Archival Info

Owned by: [ ] Code [ ]

Addl Info ☐ drawings ☐ manuscripts ☐ photographs ☐ sketches ☐ interviews ☐ models ☐ publications

General Comments:

Coding for Union List Data

Propulsion Data Keys:

RIG:
- BA = bark
- CK = cat ketch
- CR = cat-rig
- CS = cat shooner
- CY = cat yawl
- CU = cutter
- KE = ketch
- SC = schooner
- SH = ship
- SL = sloop
- YA = yawl
- Other: specify in full on data sheet

RIG SHAPE:
- GA = gaff
- GU = gunter
- LA = lateen
- LU = lug
- MA = marconi
- SP = sprit
- SQ = square
- Other: specify in full on data sheet

OAR = Number of rowing stations

PADDLE:
- D = double blade paddle
- S = single blade paddle

POLE = Yes (if applicable)

ENGINE:
- IN = inboard
- OU = outboard

FUEL:
- DI = diesel
- DE = diesel-electric
- GA = gas
- NA = naptha
- ST = steam
- Other: specify in full on data sheet

Shape and Construction Data Keys:

MATERIAL:
- BA = bark
- CO = composite, wood planking and backbone with metal frames
- FG = fiberglass
- RE = reed
- ST = steel
- SB = steel/bronze
- WC = wooden hulls sheathed with canvas
- WO = wood
- WS = wood frame covered with skin or fabric
- Other: specify in full on data sheet

PLANKING:
- BS = batten-seam carvel
- CA = carvel
- CM = cold-molded
- DE = deadrise construction
- DY = dory construction
- DD = double diagonal
- DB = double planked

PLANKING *(cont.)*:
- DU = dugout
- FL = flush lap
- LA = lapstrake
- LP = lapstrake plywood
- LO = log
- MP = molded plywood
- RA = raft
- SH = sheet (except MP and SP)
- SP = sheet plywood
- SK = skiff construction
- SF = skin or fabric over framework
- ST = strip
- Other: specify in full on data sheet

LATERAL PLANE:
- CB = centerboard
- DA = daggerboard
- FK = fin keel with separated rudder
- KC = keel with centerboard
- KL = keel
- LB = leeboard (s)
- Other: specify in full on data sheet

BOTTOM SHAPE:
- AR = arc bottom
- FL = flat bottom
- RD = round bottom
- VB = V-bottom
- Other: specify in full on data sheet

END SHAPE:
- DE = double ended
- RE = round ended
- RS = round sterned
- SQ = square ended
- TS = transom sterned
- Other: specify in full on data sheet

DECK LAYOUT:
- CA = decked with cabin
- CO = decked with cockpit only
- FD = flush deck with cabin
- OP = open or open with end deck
- PA = pilothouse aft
- PF = pilothouse forward
- Other: specify in full on data sheet

Appendix C

Informant Consent Form and Artifact Loan Form

Informant Consent Form

Thank you for participating in the ______________ Project. By signing the form below, you give your permission to include any tapes and/or photographs made during the ______________ Project in a public archive, where they will be available to researchers and the public for scholarly and educational purposes. By giving your permission, you do not give up any copyright or performance rights that you may hold.

- -

I agree to the uses of these materials described above, except for any restrictions listed below.

____

(sign)

____

(date)

____

(researcher sign)

Restrictions:

Artifact Loan Form

I hereby give permission to the ________________ Project to borrow the following artifact(s) belonging to me:

I understand that the Project is borrowing my artifact(s) for the purpose of:

____

(owner sign)

____

(date)

I, ________________, acknowledge receipt of the artifact(s) listed above on behalf of the ______________ Project. The Project agrees to return the loaned artifact(s) in undamaged condition on or before ______________ (date).

____

(project representative)

____

(date)

Bibliography

A Guide to Published Boat Documentation Projects

Benjamin A. G. Fuller

Looking at the publications produced in projects will help you select among the various tools and approaches outlined in this manual. It also tells you what has been done and what needs doing. Surveys that present drawings and brief descriptions of boats have been done in both Europe and the United States. American boat building has its roots in both European and aboriginal North American traditions, thus familiarity with examples from both areas is useful.

For this country, Howard Chapelle's *American Small Sailing Craft* (1951) is an essential starting point. While some descriptions and drawings are better than others, the book shows the great variety of pre-engine American working craft. Chapelle and Adney's survey of native American water craft (1964) is the best place to start for aboriginal craft. Drawings used in both of these books are available as blue line prints from the Division of Transportation in the Smithsonian Institution's National Museum of American History. The Historic American Merchant Marine Survey, recently published in a series of volumes (Jackson, M. H., 1984) also contains many examples of small craft drawings. The source material upon which these works are based is also, in some cases, available at the Smithsonian.

Several museums have published surveys of watercraft in their collections. Maynard Bray's 1979 *Mystic Seaport Museum Watercraft* describes 220 watercraft then in the collection. The second edition (1986) lists an additional 124 craft, and a second volume is in preparation. The Chesapeake Bay Maritime Museum has published a comparable guide to its collection (Dodds and Lesher 1992), and the Adirondack Museum and the North Carolina Maritime Museum have catalogs being written.

John Gardner's books *Building Small Craft* volumes 1 and 2 (1977, 1984) are basic reading for anyone considering documentation for reproductions. His drawings were created expressly for builders to use. Each boat is accompanied by an essay on its history and construction. Some designs have been adapted from older ones or have been modified to suit newer building methods and materials; these are so indicated.

Eric McKee's *Working Craft of Britain* (1983) presents an environmental approach for British craft, showing how they are designed and built to fit their local geography and purposes. His book has relatively few examples of completely drawn boats but is filled with explanatory sketches and drawings showing boats in use. His analysis and comparison of the rich variety of British types based on dimensions and construction is also highly valuable for those struggling to understand the relationships among the various vernacular or folk names for a boat.

In contrast is Edgar March's *Inshore Craft of Great Britain* (1970), a two-volume trip around the British Isles looking at indigenous water craft. It is much closer to Chapelle's American study but provides more historical information about boat usage and includes sketches of equipment and first-person accounts of use. For Scandinavian boats, two somewhat similar volumes on Norway (Christensen 1979) and Denmark (Nielsen 1980) include research data that was collected prior to 1950 and show what can be done with good data or manuscripts.

The National Maritime Museum of Greenwich, England, has also published an excellent monograph series. A number of these are fine examples of documenting traditional small craft. One of special interest to those thinking of documenting, replicating, and then testing a boat, is the account of building and sailing a reproduction of the Gokstad faering (McGrail and McKee 1974). This project was one of the earliest examples applying experimental archaeology to watercraft research. Such work has continued in Europe as evidenced by the many cases cited in *Sailing into the Past* (Crumlin-Pedersen, et al. 1986), which presents a methodology for this type of documentation and recounts the problems.

The only works linking documentation with analysis in America are David Zimmerly's various articles and monographs (1979, 1983, 1986) on skin boats. Not only does he present detailed drawings, photographs, and building methods, he also shows how the boats were built and used, and their place in Eskimo culture. He applies the techniques of the naval architect to attempt to measure the performance of the boats under various conditions of load. Study of Inuit craft have continued with recent findings published by the Canadian Museum of Civilization (Arima et al., 1991).

The best examples of documentation that combine technical analysis and cultural analysis in any language are Scandinavian. Gunner Eldjarn and Jan Godal's four-volume study of Norse boats, *Nordlansbåten og Åfjordsbåten* (1988, 1990), records building and use and distribution of types and provides a technical analysis using the tools of a modern naval architect. A similar Danish study by Bent and Erik Andersen, *Råsejlet–Dragens Vinge*, published in 1989, presents some of the results of technical investigations of several of the Skuldelev vessels. Both show that there is ample potential for this type of work in North America.

Perhaps the finest example of what an artist-turned-ethnographer can do when documenting small craft and their users is the work of the Dutch artist Peter Dorleijn (1982, 1983). His volumes combine interviews, photographs, and magnificent drawings to present a picture of the life of fishermen and their boats on the Zuiderzee, before it was closed off and turned into the fresh-water Ijsselmeer. A selection of his drawings is published in *Zuiderzeevisserij in Beeld* (1987). The language is Dutch, but the many drawings make language no barrier. Sam Manning's drawings in *The Dory Book* (Gardner 1987) and in *The Evolution of the Wooden Ship* (Greenhill and Manning 1989) should also be studied, as should the unpublished drawings of the *Charles W. Morgan* by Kathy Bray in the Mystic Seaport Museum collection.

Good examples of in-depth studies of particular boats abound. Most of the major Northeastern types have at least one monograph or article. Dories are found in *The Dory Book* (Gardner 1987), but no monograph exists for the peapod, another characteristic New England type, except for one early work (Brooks 1942) and chapters in Chapelle's and Gardner's books. Also emphasizing boats from Maine are the canoe-building studies of Jerry Stelmok (1980) and later, with Rollin Thurlow (1987). Both of these discuss the history and the process of building as much as they present lines of particular craft. Richard Lunt's 1975 dissertation is a scholarly study of the development of the typical Maine lobster boat, while builder Royal Lowell discusses in detail the building process (1977). David Taylor's 1982 study of Winterton, Nova Scotia, more completely illustrates boat building and its cultural context. Many of the drawings by David Dillion, this volume's principal contributor, are published in Richard Post's book on Tancook whalers (1985) and have been published by Philadelphia Maritime Museum, Mystic Seaport Museum, Maine Maritime Museum, and the Rockport Apprenticeshop. Whaleboats and their use, history, and equipment are detailed by Willets Ansel (1983). Barry Thomas's studies of the Herreshoff dinghy (1977) and the Crosby catboat (1989) show what can be achieved by actually building a reproduction as part of the documentation process. Thomas's work illustrates documentation beyond a written and drawn description to a study of how the original builders built the boats, a theme emphasized by Ansel as well.

Relatively scarce are "source" or period documentation studies by contemporary designers and builders. Most notable are W. P. Stephens's *Canoe and Boatbuilding for Amateurs* of 1885 and 1898, whose drawings have been reprinted by Mystic Seaport Museum (1987), and C. P. Kunhardt's 1891 *Small Yachts,* also reprinted in part (1985). *Standard Designs of the United States Navy* (Highborn 1900) provides a fine example of how detailed small craft drawing and specifications can be for boats that need to be able to be built by any builder.

Regions outside of New England have fared less well in small craft documentation. For the Adirondacks, Kenneth and Helen Durant's *The Adirondack Guide-Boat* (1980) is an admirable study of a regionally important type. H. Atwood Manley's *Rushton and His Times in American Canoeing* (1968) is an excellent biography of an important American builder showing how his work fit into the first serious American interest in recreational small craft. It is also a good example of flawed lines and inadequate construction drawings.

South of New York City, there is Peter Guthorn's study of the Sea Bright Skiff of the Jersey Shore (1971) and a brief article on the Delaware Ducker (Fuller 1982, Lee 1982). The Chesapeake has several excellent books; yet to be surpassed is Marion Brewington's volume on log canoes and bugeyes (1963). Chapelle's articles on crabbing skiffs (1943) and skipjacks (1944) have been reprinted by the Chesapeake Bay Maritime Museum (1979, 1981), but a complete study of the skipjack and deadrise boats needs to be done.

South of the Chesapeake is fertile ground for documentation projects. Chapelle's monograph *Migrations of an American Boat Type* (1961) provides additional insight into the New Haven sharpie not presented in *American Small Sailing Craft* and shows how this boat moved to North Carolina. Michael Alford of the North Carolina Maritime Museum has in press *North Carolina Boats and Boatbuilding.* For the South there is only Rusty Fleetwood's book on Georgia (1982). There is ample opportunity for further research.

The Pacific Coast, the Mississippi Basin, and the Great Lakes are areas with great research potential. Scholars like Orbach (1977) and Gilmore (1986) have taken anthropological approaches to their subjects. Seattle's Center for Wooden Boats has published monographs on the Davis Skiff (Loken 1983) and Poulsbo boats (Beard and Loken 1981). Brief reviews of catalogs such as Chapelle's work on the Smithsonian's watercraft collection (1976) shows that there are many types and associated cultures worth exploring.

Studying drawings of small craft can also provide insight into how documentation problems can be solved. Few original drawings exist of boats built before the middle of the nineteenth century as naval architects primarily were concerned with large vessels. Most of these drawings are unpublished; major institutions holding study collections are listed below and can provide copies.

Drawings for boats built for the United States government are not easily obtainable, despite the fact that many examples of the historical fleet have survived. Government boats were built of wood up until the 1960s, and they represent some of the best examples of the boat builder's craft. Besides *Boats of the U. S. Navy* (Highborn 1900), there are a few articles available on Life Saving Service and Coast Guard boats (McLellan 1901, Adams and Hunnewell 1929, Noble 1976). Unfortunately, original plans for these boats are difficult to find in government archives and turn up more frequently in museum plans collections. The Mariners' Museum has made a special effort to collect life-saving boat plans. In addition, The Mariners' Museum has plans of many of its boats, as well as a fine collection of material on the motorboats of the pre–World War II era, notably the work of Chris-Craft and Hacker.

In New England, Mystic Seaport Museum has plans for many of the boats in its collection, as well as the work and archives of many designers and recorders, such as W. P. Stephens, L. F. Herreshoff, Philip Rhodes, William Garden, and the firm of Sparkman and Stephens. The Hart Nautical Museum at the Massachusetts Institute of Technology preserves the plans of the Herreshoff Manufacturing Co., George Owens, and Starling Burgess, among others. The Peabody Museum of Salem, Massachusetts, has several plans collections, including the work of B. B. Crowninshield. The Maine Maritime Museum has documented many watercraft in its own collection and has acquired plans of others. Also in Maine, the Artisans School, formerly the Rockport Apprenticeshop, has a plans collection, as does *WoodenBoat* magazine.

Farther south, regional museums with plans collections include the Philadelphia Maritime Museum with plans for a number of Delaware Bay small craft as well as plans from some of the Bay's ship and boat yards. Both the Chesapeake Bay Maritime Museum and the Calvert Marine Museum have plans collections of craft of the Chesapeake. The former has also many manuscripts and pencil plans of Howard Chapelle. The North Carolina Maritime Museum has a list of its plans available.

On the West Coast, plans collections can be found at the San Francisco Maritime National Historic Park, at the San Diego Maritime Museum, and through the Center for Wooden Boats in Seattle.

Finally, in the Smithsonian's National Museum of American History are most of the plans drawn by Chapelle, together with the plans and notes compiled as part of the Historic American Merchant Marine Survey. A list is available from the Division of Transportation's maritime history section.

Bibliography

Acheson, James M. "Anthropology of Fishing." *Annual Review of Anthropology* 10 (1981): 275-316.

Acheson, James M. *The Lobster Gangs of Maine.* Hanover, N.H.: University Press of New England, 1988.

Adams, R.B., and F.A. Hunnewell. "Typical Vessels and Boats of the U.S. Coast Guard." *United States Naval Institute Proceedings* 55:5 (May 1929): 401-405.

Adkins, Jan. *Wooden Ship: The Building of a Wooden Sailing Vessel in 1870.* Boston: Houghton Mifflin Company, 1978.

Adney, Edwin T., and Howard I. Chapelle. *The Bark Canoes and Skin Boats of North America.* 1964. Reprint. Washington, D.C.: Smithsonian Institution Press, 1983.

Albion, Robert Greenhalgh. *Naval and Maritime History: An Annotated Bibliography.* Mystic, Conn.: The Marine Historical Association, Inc., 1972.

Alford, Michael. *North Carolina Boats and Boatbuilding.* Chapel Hill, N.C.: University of North Carolina Press, in press.

Allen, Barbara, and Lynwood Montell. *From Memory to History: Using Oral Sources in Local Historical Research.* Nashville, Tenn.: American Association for State and Local History, 1981.

Andersen, Bent, and Erik Andersen. *Råsejlet - Dragens Vinge.* Roskilde, Denmark: Vikingskibshallen, 1989.

Anderson, Richard K. *Guidelines for Recording Historic Ships.* Washington, D.C.: National Park Service, U.S. Department of the Interior, Historic American Engineering Record, 1988.

Ansel, Willits D. *The Whale Boat: A Study of Design, Construction and Use from 1850 to 1970.* Mystic, Conn.: Mystic Seaport Museum, Inc., 1983.

Arima, E.Y., ed. *Contributions to Kayak Studies.* Canadian Ethnological Service, Mercury Series, Paper no. 122. Hull, Quebec: Canadian Museum of Civilization, 1991.

Ashley, Clifford W. *The Ashley Book of Knots.* Garden City, N.Y.: Doubleday & Company, 1944.

Association for Preservation Technology, "Maritime Preservation." *APT Bulletin* 11:1 (1987).

Baker, William A., Sam Svensson, and George P.B. Naish. *The Lore of Sail.* New York: Facts On File Publications, 1984.

Baker, William A. *The Mayflower and Other Colonial Vessels.* London: Conway Maritime Press Ltd.; Annapolis, Md.: Naval Institute Press, 1983.

Bankston, J. *An Introductory Bibliography of Maritime References for the New World and Spanish Empire Therein, 1492-1821.* Bisbee, Ariz.: Terrenate Associates, 1988.

Bartis, Peter. *Folklife and Fieldwork: A Layman's Introduction to Field Techniques.* Publications of the American Folklife Center, No. 3. Rev .ed. Washington, D.C.: Library of Congress, 1990.

Bass, George F. *Archaeology Under Water.* New York and Washington, D.C.: Praeger, 1966.

Bass, George F., ed. *A History of Seafaring: Based on Underwater Archaeology.* London: Thames and Hudson Ltd., 1972.

_____. *Ships and Shipwrecks of the Americas: A History Based On Underwater Archaeology.* London: Thames and Hudson Ltd., 1988.

Baynes, Ken, and Francis Pugh. *The Art of the Engineer.* Woodstock, N.Y.: The Overlook Press, 1981.

Beard, Barrett T., Sr. "American Maritime Historic Preservation: Practice and Policy." Master's Thesis, Western Washington University, 1986.

Beard, Thomas, and Marty Loken. *The Poulsbo Boat.* Seattle: The Center for Wooden Boats, 1981.

Beebe, Robert P. "The Art of Taking Off Lines." *Rudder,* November, 1966, 42-47.

Biddlecomb, George. *The Art of Rigging, Containing an Explanation of Terms and Phrases and the Progressive Method of Rigging Expressly Adapted for Sailing Vessels.* Salem, Mass.: Marine Research Society, 1925.

Borchers, Perry E. *Photogrammetric Recording of Cultural Resources.* Washington, D.C.: Technical Preservation Services, National Park Service, 1977.

Bowker, R. M., and S.A. Budd. *Make Your Own Sails.* New York: St. Martin's Press, 1962.

Bray, Anne and Cynthia Curtis, comps. *The Directory of Wooden Boat Builders.* Brooklin, Maine: WoodenBoat Publications, 1986.

_____. *Boatbuilding Woods: A Directory of Suppliers.* Brooklin, Maine: WoodenBoat Publications, 1987.

Bray, Anne, comp. *The WoodenBoat Index: Issues 1-100.* Brooklin, Maine: WoodenBoat Publications, 1991.

Bray, Maynard, ed. *Taking Care of Wooden Ships.* Walpole, Maine: University of Maine, Sea Grant, 1978.

Bray, Maynard. *Mystic Seaport Museum Watercraft.* 2d ed. Mystic, Conn.: Mystic Seaport Museum, Inc., 1986.

Bray, Maynard, and Carlton Pinheiro. *Herreshoff of Bristol: A Photographic History of America's Greatest Yacht and Boat Builders.* Brooklin, Maine: WoodenBoat Publications, 1989.

Bray, Maynard, Sam Manning, and Richard Lunt. "Guidelines for Documentation." Washington, D.C.: National Trust for Historic Preservation, 1983.

Brewington, M.V. *Chesapeake Bay Log Canoes and Bugeyes.* Cambridge, Md.: Cornell Maritime Press by arrangement with The Mariners Museum, Newport News, Va., 1963.

Brooks, Alfred A. "The Boats of Ash Point, Maine." *The American Neptune* 12:4 (October 1942): 307-23.

Brouwer, Norman J. *International Register of Historic Ships.* Annapolis, Md.: Naval Institute Press, 1985.

Brown, Bruce C. *Watershots: How To Take Better Photos On and Around the Water.* Camden, Maine: International Marine Publishing Co., 1988.

Bunting, William H. *Portrait of a Port: Boston, 1852-1914.* Cambridge, Mass.: Harvard University Press, Belknap Press, 1971.

_____. *Steamers, Schooners, Cutters & Sloops: Marine Photographs of N.L. Stebbins taken 1884 to 1887.* Boston: Houghton Mifflin Co., 1974.

Byington, Robert H., ed. *Working Americans: Contemporary Approaches to Occupational Folklife.* Smithsonian Folklife Series, No. 3. Los Angeles: California Folklore Society, 1978.

Carey, George. *Maryland Folklore and Folklife.* Cambridge, Md.: Tidewater Publishers, 1970.

_____. *A Faraway Time and Place: Lore of the Eastern Shore.* Washington D.C., and New York: Robert B. Luce, 1971.

Carter, Thomas, and Carl Fleischhauer. *The Grouse Creek Cultural Survey: Integrating Folklife and Historic Preservation Field Research.* Publications of the American Folklife Center, No.13. Washington, D.C.: American Folklife Center, Library of Congress, 1988.

Casson, Lionel. *Ships and Seamanship in the Ancient World.* Princeton, N.J.: Princeton University Press, 1971.

Cederlund, Carl Olaf, ed. *Postmedieval Boat and Ship Archaeology: Papers Based on Those Presented to an International Symposium on Boat and Ship Archaeology in Stockholm in 1982.* Report No. 20. Stockholm: Swedish National Maritime Museum; BAR International Series 256. Oxford: British Archaeological Reports 1985.

Chapelle, Howard I. *Boatbuilding: A Complete Handbook of Wooden Boat Construction.* New York: W. W. Norton & Co., 1941.

_____. *American Small Sailing Craft: Their Design, Development, and Construction.* New York: W. W. Norton & Co., 1951.

_____. *Migrations of an American Boat Type.* U.S. National Museum Bulletin 228, Washington, D.C.: Government Printing Office, 1961.

_____. *Yacht Designing and Planning for Yachtsmen, Students, & Amateurs.* New York: W. W. Norton & Co., 1971.

_____. *The American Fishing Schooners 1825-1935.* New York: W. W. Norton & Co., 1973.

_____. *The National Watercraft Collection.* 2d ed. Washington, D.C.: Smithsonian Institution Press; Camden, Maine: International Marine Publishing Co., 1976.

_____. "Chesapeake Bay Crabbing Skiffs." *Yachting,* June, October 1943. Reprint. St. Michaels, Md.: Chesapeake Bay Maritime Museum, 1979.

_____. "Notes On Chesapeake Bay Skipjacks." *The American Neptune* 14:4 (October 1944). Reprint. St. Michaels, Md.: Chesapeake Bay Maritime Museum, 1981.

Chapman, Fredrik Henrik af. *Architectura Navalis Mercatoria: A Facsimile of the Classic Eighteenth-Century Treatise on Shipbuilding.* New York: Sheridan, 1985.

Chiaramonte, Louis J. *Craftsman-Client Contracts: Interpersonal Relations in a Newfoundland Fishing Community.* St. John's, Newfoundland: Institute of Social and Economic Research, 1970.

Christensen, Arne Emil, ed. *Inshore Craft of Norway: From a Manuscript by Bernhard and Øystein Færøyvik.* Oslo, Norway: Grøndahl & Søn Forlag, A.s., 1979.

Clifford, Harold B. *Charlie York, Maine Coast Fisherman.* Camden, Maine: International Marine Publishing Co., 1974.

Colleta, Paolo E. *A Bibliography of American Naval History.* Annapolis, Md.: Naval Institute Press, 1981.

Collier, John, Jr., and Malcolm Collier. *Visual Anthropology: Photography as a Research Method.* Rev. ed. Albuquerque: University of New Mexico Press, 1986.

Comeaux, Malcolm L. *Atchafalaya Swamp Life: Settlement and Folk Occupations.* Baton Rouge: School of Geoscience, Louisiana State University, 1972.

Constantine, Albert J., Jr. *Know Your Woods: A Complete Guide to Trees, Woods and Veneers.* New York: Charles Scribner's Sons, 1987.

Crowley, William. *Rushton's Rowboats and Canoes: The 1903 Catalog in Perspective.* Blue Mountain Lake, N.Y.: The Adirondack Museum; Camden, Maine: International Marine Publishing Co., 1983.

Crumlin-Pedersen, Ole, and Max Vinner, eds. *Sailing into the Past: The International Ship Replica Seminar, Roskilde 1984.* Roskilde, Denmark: The Viking Ship Museum, 1986.

Darington, Glenn. "Computer Graphics in Archaeology." *INA (Institute for Nautical Archaeology) Newsletter* 16:1 (March 1989): 12-13.

Darrah, William C. *The World of Stereographs.* Gettysburg, Pa.: W.C. Darrah, 1977.

Day, Cyrus L. *The Art of Knotting and Splicing.* 3d ed. Annapolis, Md.: Naval Institute Press, 1970.

Dean, Jeff. *Architectural Photography: Techniques for Architects, Preservationists, Historians, Photographers, and Urban Planners.* Nashville, Tenn.: American Association for State and Local History, 1982.

Dean, Martin. *Guidelines on Acceptable Standards in Underwater Archaeology.* Fife, Scotland: The Scottish Maritime Development Association for the Scottish Institute of Maritime Studies, 1988.

Delgado, James P., et al. *Nominating Historic Vessels and Shipwrecks to the National Register of Historic Places.* National Register Bulletin No. 20. Washington, D.C.: National Park Service, 1986.

Delgado, James P., and Candace Clifford, eds. *1990 Inventory of Large Preserved Historic Vessels.* Washington D.C.: National Park Service, National Maritime Initiative, 1990.

Desmond, Charles. *Wooden Ship-building.* New York: Vestal Press, 1984.

Dodds, Richard J.S., and Pete Lesher, eds. *A Heritage in Wood: The Cheaspeake Bay Maritime Museum's Small Craft Collection.* St. Michaels, Md.: Chesapeake Bay Maritime Museum, 1992.

Dorleijn, Peter. *Van gaand en staand want, De zeilvisserij voor an na de afsluiting van de Zuiderzee.* 3 vols. Bussum, Netherlands: Van Kampen, 1982, 1983.

Dorleijn, Peter. *Zuiderzeevisserij in beeld.* Zutphen, Netherlands: De Walburg Pers, 1987.

Durant, Kenneth, and Helen Durant. *The Adirondack Guide-Boat.* Blue Mountain Lake, N.Y.: Adirondack Museum, 1980.

Easterly, Diana Eames. *Early One Design Sailboats.* New York: Charles Scribner's Sons, 1979.

Edlin, Herbert L. *What Wood Is That? A Manual of Wood Identification.* 5th ed. London: Thames and Hudson Ltd.; New York: Penguin USA, 1981.

Eldjarn, Gunnar, and Jon Godal. *Nordlandsbåten og Afjordsbåten.* 4 vols. Lesja, Norway: A. Kjellands Forlag A.s., 1988: Vols. 2 and 4, Rissa, Norway: Båtstikka D/A, 1990.

Estep, H. Cole. *How Wooden Ships Are Built: A Practical Treatise on Modern American Wooden Ship Construction with a Supplement on Laying Off Wooden Vessels.* 1918. Reprint. New York: W. W. Norton & Co., 1983.

Fleetwood, William C., Jr. *Tidecraft: An Introductory Look at the Boats of Lower South Carolina, Georgia and Northeastern Florida: 1650-1950.* Savannah, Ga.: Coastal Heritage Society, 1982.

Fleischhauer, Carl, ed. *The Ninety-Six: A Cattle Ranch in Northern Nevada.* Visual Media from the American Folklife Center, No. 1. Laser videodisc and accompanying booklet. Washington, D.C.: Library of Congress, 1985.

Fleischhauer, Carl. "Sound Recording and Still Photography in the Field." In *Handbook of American Folklore,* edited by Richard M. Dorson, 384-390. Bloomington, Ind.: Indiana University Press, 1986.

Forrest, John. *Lord, I'm Coming Home: Everyday Aesthetics in Tidewater North Carolina.* Ithaca, N.Y.: Cornell University Press, 1988.

Fuller, Benjamin A.G. "Delaware Duckers: A Brief History, The Ducker in Use." *WoodenBoat* 48 (Sept./Oct. 1982): 78-86.

Gardner, John. Building Classic Small Craft. 2 vols. Camden, Maine: International Marine Publishing Co., 1977-1984.

_____. "Taking off Lines Allows Duplication of Existing Boats." *National Fisherman,* April 1986, 44-45, and May 1986, 58.

_____. "Triangulation Method is Well Suited to Lifting Lines." *National Fisherman,* August 1986, 65-67.

_____. *The Dory Book.* Mystic, Conn.: Mystic Seaport Museum, Inc. 1987.

Giesecke, Anne G. "The Best in State Historic Shipwreck Programs." In *Proceedings of the Sixteenth Conference on Underwater Archaeology,* edited by Paul. F. Johnston, 138-141. Special Publication No. 4. Boston: The Society for Historical Archaeology, 1985.

Gillmer, Thomas C., and Bruce Johnson. *Introduction To Naval Architecture.* Annapolis, Md.: Naval Institute Press, 1982.

Gilmore, Janet C. *The World of the Oregon Fishboat: A Study in Maritime Folklife.* Ann Arbor, Mich.: UMI Research Press, 1986.

Gittings, Kirk. *Introduction to Photographing Historic Properties.* Information Series No. 42. Washington,D.C.: National Trust for Historic Preservation, 1988.

_____. *Monuments of Adobe: The Religious Architecture of Northern New Mexico.* Albuquerque: Univ. of New Mexico Press, in press.

Glassie, Henry. *Pattern in the Material Folk Culture of the Eastern United States.* Philadelphia: University of Pennsylvania Press, 1969.

Goode, George Brown, ed. *The Fisheries and Fishery Industries of the United States.* 7 vols. Washington, D.C.: U.S. Government Printing Office, 1884-87.

Gorden, Raymond L. *Interviewing: Strategy, Techniques and Tactics.* Homewood, Ill.: The Dorsey Press, 1969.

Green, Ben. *Finest Kind: A Celebration of a Florida Fishing Village.* Macon, Ga.: Mercer University Press, 1985.

Greenhill, Basil. *Archaeology of the Boat: A New Introductory Study.* Middletown, Conn.: Wesleyan University Press, 1976.

_____. *The Evolution of the Wooden Ship.* New York: Facts On File, 1988.

Guthorn, Peter. J. *The Sea Bright Skiff and Other Jersey Shore Boats.* New Brunswick, N.J.: Rutgers University Press, 1971.

Haddon, A.C., and James Hornell. *Canoes of Oceania.* Special Publications No. 27, 28, and 29. Reprint. Honolulu: Bishop Museum Press 1975.

Hamilton, Donald L. "INA Enters the SHARPS Era." *INA (Institute for Nautical Archaeology) Newsletter* 15:2 (1988): 6-7.

Hasslöf, Olof, Henning Henningsen, and Arne Emil Christensen, Jr., eds. *Ships and Shipyards, Sailors and Fishermen: Introduction to Maritime Ethnology.* Copenhagen, Denmark: Rosenkilde and Bagger, 1972.

Hayward, Mary Ellen. "Final Survey Report: Maryland's Surviving Traditional Craft." Annapolis, Md.: Radcliffe Maritime Museum of the Maryland Historical Society, 1985.

Highborn, Philip. *Standard Designs for Boats of the United States Navy: Specification, Schedule of Materials, Weights and Cost.* Washington, D.C.: Government Printing Office, 1900.

Hoadley, R. Bruce. *Understanding Wood: A Craftsman's Guide To Wood Technology.* Newtown, Conn.: Taunton Press, 1981.

Horenstein, Henry. *Black and White Photography.* Boston: Little, Brown, 1983.

Hornell, James. *Water Transport.* Cambridge, England: Cambridge University Press, 1946.

Horridge, Adrian. *Outrigger Canoes of Bali and Madura, Indonesia.* Honolulu: Bishop Museum Press, 1987.

Hufford, Mary. *One Space, Many Places: Folklife and Land Use in New Jersey's Pinelands National Reserve.* Publications of the American Folklife Center, No. 15. Washington, D.C.: Library of Congress, American Folklife Center, 1986.

INA Newsletter, College Station, Texas: Institute for Nautical Archaeology, 1974-present.

International Correspondence Schools Reference Library: Marine Engines, Marine Pumps. Vol. 170. Scranton, Pa.: International Textbook Co., 1907.

International Library of Technology. *Marine Boilers, Refrigeration.* Vol. 9b. Scranton, Pa.: International Textbook Co., 1907.

_____. *Drawing, Ship Drafting, Projection, Developments.* No. 154. Scranton, Pa.: International Textbook Co., 1918.

Ives, Edward D. *The Tape-Recorded Interview: A Manual for Fieldworkers in Folklore and Oral History.* Knoxville: University of Tennessee Press, 1980.

Jackson, Bruce. *Fieldwork.* Urbana and Chicago: University of Illinois Press, 1987.

Jackson, Melvin H. ed. *Historic American Merchant Marine Survey.* 7 vols. Salem, N.H.: Ayer Co., 1984.

Johnson, Paula J., ed. *Working the Water: The Commercial Fisheries of Maryland's Patuxent River.* Charlottesville: University Press of Virginia, 1988.

Jolly, Brad. *Videotaping Local History.* Nashville, Tenn.: American Association for State and Local History, 1982.

Karara, H.M., ed. *Handbook of Non- topographic Photogrammetry.* Falls Church, Va.: American Society of Photogrammetry, 1979.

Katzev, Michael, "Voyage of Kyrenia II," *INA (Institute of Nautical Archaeology) Newsletter* 16:1 (March 1989): 4-10.

Keefe, Laurence E., and Inch, Dennis. *The Life of a Photograph: Archival Processing, Matting, Framing, Storage.* Boston: Focal Press, 1984.

de Kerchove, René. *International Maritime Dictionary: An Encyclopedic Dictionary of Useful Terms and Phrases, Together with Equivalents in French and German.* 2d ed. New York: Van Nostrand Reinhold Co., 1961.

Kipping, Robert. *Sails and Sailmaking.* London: The Technical Press, 1936.

Kunhardt, C.P. *Small Yachts: Their Design and Construction.* 1891. Reprint. Brooklin, Maine: WoodenBoat Publications, 1985.

Landberg, Leif, comp. *A Bibliography for the Anthropological Study of Fishing Industries and Maritime Communities and Supplement.* Kingston: University of Rhode Island, 1973 and 1979.

Leather, John. *Clinker Boatbuilding.* New York: Sheridan, 1987.

_____. *Dixon Kemp's Manual of Yacht and Boat Sailing.* 1878. Reprint. New York: Sheridan, 1988.

_____. "Taking Off The Lines." *Classic Boat,* June 1989, 29-33.

_____. *Gaff Rig.* 2d ed. Camden, Maine: International Marine Publishing Co., 1989.

_____. *Spritsails and Lugsails.* 2d ed. Camden, Maine: International Marine Publishing Co., 1989.

Lee, Lance R. "Delaware Duckers: Why the Ducker Today?" *WoodenBoat* 78 (Sept./Oct. 1982): 86-89.

Lever, Darcy. *The Young Sea Officer's Sheet Anchor, or a Key to the Leading of Rigging and to Practical Seamanship.* 1918. Reprint. New York: Edward W. Sweetman Co., 1963.

Lipke, Paul. *The Royal Ships of Cheops.* Archaeological Series No. 9. Greenwich, England: National Maritime Museum; BAR International Series No. 225. Oxford: British Archaeological Reports, 1984.

Loken, Marty. *The Davis Skiff.* Seattle: The Center for Wooden Boats, 1983.

Lowell, Royal. *Lobsterboat Building Downeast.* Camden, Maine: International Marine Publishing Co., 1977.

Lund, Jens. "Fishing as a Folk Occupation in the Lower Ohio Valley." Ph.D. diss., Indiana University, 1982.

Lunt, C. Richard K. "Lobsterboat Building on the Eastern Coast of Maine: A Comparative Study." Ph.D. diss., Indiana University, 1976.

McLellan, C. H. "The U.S. Lifesaving Service Self Bailing Water Ballast Surfboat." *Marine Engineering* 6:3 (March 1901): 94-99.

MacKean, H. Ray. *Catch a Line: A Practical Method of Taking Off the Lines of Small Boats.* Fredericton: New Brunswick Heritage Publication, 1980.

Manley, Atwood, with the assistance of Paul F. Jamieson. *Rushton and His Times in American Canoeing.* Syracuse, N.Y.: The Adirondack Museum/Syracuse University Press, 1968.

March, Edgar. *Inshore Craft of Great Britain in the Days of Sail and Oar.* 2 vols. Newton Abbott, England: David & Charles; Camden, Maine: International Marine Publishing Co., 1970.

Maril, Robert Lee. *Texas Shrimpers: Community, Capitalism, and the Sea.* College Station: Texas A & M University Press, 1983.

Martin, Kenneth R., and Nathan R. Lipfert. *Lobstering and the Maine Coast.* Bath: Maine Maritime Museum, 1985.

McGrail, Sean, and J. Eric McKee. *The Building and Trials of the Replica of an Ancient Boat: The Gokstad Faering. Parts I-II.* Maritime Monographs and Reports No. 11. Greenwich, England: National Maritime Museum, 1974.

McGrail, Sean, and Eric Kentley, eds. *Sewn Plank Boats: Archaeological and Ethnographic Papers Based on Those Presented to a Conference at Greenwich in November, 1984.* Archaeological Series No. 10. Greenwich, England: National Maritime Museum; BAR International Series No. 276. Oxford: British Archaeological Reports, 1985.

McGrail, Sean. *Ancient Boats in N.W. Europe: The Archaeology of Water Transport to A.D.1500.* London and New York: Longman, 1987.

McIntosh, David C. *How To Build a Wooden Boat.* Brooklin, Maine: WoodenBoat Publications, 1987.

McKee, Eric. *Working Boats of Britain: Their Shape and Purpose.* London: Conway Maritime Press, 1983.

McQuaid, James, ed. *An Index to American Photographic Collections.* Boston: G.K. Hall & Co., 1982.

Moonsammy, Rita Zorn, David S. Cohen, and Lorraine E. Williams, eds. *Pinelands Folklife.* New Brunswick, N.J.: Rutgers University Press, 1987.

Muckelroy, Keith. *Maritime Archaeology.* Cambridge: Cambridge University Press, 1978.

_____. *Discovering a Historic Wreck.* Handbooks in Nautical Archaeology. No. 1. Greenwich, England: National Maritime Museum, 1981.

Mullen, Patrick B. *I Heard the Old Fishermen Say: Folklore of the Texas Gulf Coast.* Austin and London: University of Texas Press, 1978.

National Maritime Initiative. "Computerized Inventory of Historic Maritime Resources." Washington, D.C.: National Park Service, National Maritime Initiative.

National Maritime Initiative. *Standards for Historic Vessel Preservation Projects with Guidelines for Applying the Standards.* San Francisco: National Park Service, 1987.

National Museum of American History. *Ship Plan List, Maritime Collection.* Washington, D.C.: Division of Transportation, Smithsonian Institution, 1984.

National Trust for Historic Preservation. *Wooden Shipbuilding and Small Craft Preservation.* Washington, D.C.: National Trust for Historic Preservation, 1976.

_____. *Directory of Maritime Preservation Resources.* Washington, D.C.: National Trust for Historic Preservation, 1984.

_____. *Task Force Report on the Preservation of America's Maritime Heritage, 1985-1990.* Washington, D.C.: National Trust for Historic Preservation, 1984.

_____. *List of Small Craft in 68 Maritime Collections.* Washington, D.C.: National Trust for Historic Preservation, 1987.

_____. *The Preservation of Historic Maritime Resources: A Planning Guide for Communities.* Washington, D.C.: National Trust for Historic Preservation, in press.

Newhall, Beaumont. *The History of Photography.* 5th ed. New York, N.Y.: Museum of Modern Art, 1982

Nielsen, Christian. *Wooden Boat Designs: Classic Danish Boats Measured and Described.* New York: Charles Scribner's Sons, 1980.

Noble, Dennis L. *A Legacy: The United States Life-Saving Service: A Bicentennial Publication from the Coast Guard Historian.* Washington D.C.: U.S. Coast Guard, 1976.

Norgrove, Ross. *Cruising Rigs and Rigging.* Camden, Maine: International Marine Publishing Co., 1982.

Northeast Marine Advisory Council. *A Sport Divers Handbook for Historic Shipwrecks.* Durham, N.H.: Northeast Marine Advisory Council, 1982.

Orbach, Michael. *Hunters, Seamen, and Entrepreneurs: The Tuna Seinermen of San Diego.* Berkeley: University of California Press, 1977.

Oughtred, Iain. "Variations on the Lugsail: Small Boat Rigs Worth a Second Consideration, Part I." *WoodenBoat* 73 (Nov./Dec. 1986): 80-85.

_____. "The Gunter Rig: Small Boat Rigs Worth a Second Consideration, Part II." *WoodenBoat* 74 (Jan./Feb. 1987): 100-103.

_____. "The Gaff Rig: Small Boat Rigs Worth a Second Consideration, Part III." *WoodenBoat* 75 (March/April 1987): 80-86.

_____. "The Sprit Rig: Small Boat Rigs Worth a Second Consideration, Part IV." *WoodenBoat* 76 (May/June 1987): 80-86.

_____. "The Bermudian Rig: Small Boat Rigs Worth a Second Consideration, Part V." *WoodenBoat* 77 (July/Aug. 1987): 44-51.

_____. "Exotic Rigs: Small Boat Rigs Worth a Second Consideration, Part VI." *WoodenBoat* 78 (Sept./Oct. 1987): 66-75.

Pare, Richard. *Photography and Architecture, 1839-1939.* Montreal: Canadian Centre for Architecture, 1982.

Pearson, John C., ed. *The Fish and Fishery Resources of Colonial North America: A Documentary History of the Fishery Resources of the United States and Canada.* 11 vols. Washington, D.C.: U.S. Fish and Wildlife Service, Department of the Interior, 1972.

Pilcher, William W. *The Portland Longshoremen: A Dispersed Community.* New York: Holt, Rinehart, and Winston, 1972.

Post, Robert C. *The Tancook Whalers: Origins, Rediscovery, and Revival.* Bath: Maine Maritime Museum, 1985.

Rabl, Sam S. *Ship and Aircraft Fairing and Development: For Draftsmen, Loftsmen and Sheet Metal Workers.* New York: Cornell Maritime Press, 1941.

_____. *Boatbuilding in Your Own Backyard.* 2d ed. Centerville, Md.: Cornell Maritime Press, 1958.

Rath, Frederick L., Jr., and Merrilyn Rogers O'Connell, eds. *A Bibliography on Historical Organization Practices: Historic Preservation.* Nashville, Tenn.: American Association for State and Local History, 1975.

Reproduction of Steamboats and Modern Steam Launches: A Magazine Published During the Years 1961, 1962, 1963. Burbank, Calif.: Howell-North Books, 1981.

Ritzenthaler, Mary Lynn, Gerald J. Munoff, and Margery S. Long. *Archives & Manuscripts: Administration of Photographic Collections.* Chicago: Society of American Archivists, 1984.

Roberts, Owain T.P. "An Index for Flat-Bottomed Boats." *International Journal of Nautical Archaeology* 12:4 (1983): 325-31.

Robinson, Wendy S. *First Aid for Marine Finds.* Handbooks in Nautical Archaeology No. 2. Greenwich, England: National Maritime Museum, 1981.

Ronnberg, E.A.R., Jr. "Imagery and Types of Vessels." In *Paintings by Fitz Hugh Lane* edited by John Wilmerding, 61-106. Washington, D.C.: National Gallery of Art, 1988.

Skene, Norman L., and Francis S. Kinney, eds. *Skene's Elements of Yacht Design.* 8th ed. New York: Dodd, Mead and Co., 1973.

Spectre, Peter H. [Editorial]. *WoodenBoat* 84 (Sept./Oct. 1988): 2.

Speltz, Robert. *The Real Runabouts.* 6 vols. Albert Lea, Minn.: R. G. Speltz/Graphic Publishing, 1977-1987.

Spitzer, Nicholas R., ed. *Louisiana Folklife: A Guide to the State.* Baton Rouge: Louisiana Folklife Program, Office of Cultural Development, Department of Culture, Recreation and Tourism, 1985.

Steffy, J. Richard. "The Kyrenia Ship: An Interim Report on its Hull Construction." *American Journal of Archaeology* 89 (1985): 711-01.

_____. *Reconstructing Wooden Ships and Boats.* In press.

Stelmok, Jerry. *Building The Maine Guide Canoe.* Camden, Maine: International Marine Publishing Co, 1980.

Stelmok, Jerry, and Rollin Thurlow. *The Wood and Canvas Canoe: A Complete Guide to Its History, Construction, Restoration, and Maintenance.* Gardiner, Maine: Harpswell Press, 1987.

Stephens, William P. *Fifty Plates from William Picard Stephens' "Canoe and Boatbuilding: A Complete Manual for Amateurs."* 1885. Reprint. Mystic, Conn.: Mystic Seaport Museum Press, 1987.

Sterling, Frank Ward, ed. *Marine Engineers Handbook.* New York: McGraw-Hill Book Co., 1920.

Steward, Robert M. *Boatbuilding Manual.* 3d ed. Camden, Maine: International Marine Publishing Co., 1987.

Sussman, Aaron. *The Amateur Photographer's Handbook.* 8th ed. New York: Crowell Co., 1973

Szarkowski, John. *Mirrors and Windows.* New York: Museum of Modern Art, 1978.

Taylor, David A. *Boat Building in Winterton, Trinity Bay, Newfoundland.* Canadian Centre for Folk Culture Studies, Paper No. 41. Ottawa: National Museums of Canada, 1982.

_____. "A Survey of Traditional Systems of Boat Design Used in the Vicinity of Trinity Bay, Newfoundland, and Hardanger Fjord, Norway." Ph.D. diss., Memorial University of Newfoundland, 1990

_____. *Documenting Maritime Folklife: An Introductory Guide.* Publications of the American Folklife Center, No. 18. Washington, D.C.: Library of Congress, 1992.

Thomas, Barry. *Building the Herreshoff Dinghy.* Mystic, Conn.: Mystic Seaport, Inc., 1977.

_____. *Building the Crosby Catboat.* Mystic, Conn.: Mystic Seaport Museum, Inc., 1989.

Throckmorton, Peter, ed. *The Sea Remembers: Shipwrecks and Archaeology from Homer's Greece to the Rediscovery of the* Titanic. New York: Weidenfield & Nicolson, 1987.

Toelken, Barre. *The Dynamics of Folklore.* Boston: Houghton Mifflin Co., 1979.

Toss, Brion. *The Rigger's Apprentice.* Camden, Maine: International Marine Publishing Co., 1984.

Trask, Edgar P., and John P. Comstock. *Ship Drafting.* Scranton, Pa.: International Textbook Co., 1938.

Tryckare, Tre, et al. *The Lore of Ships.* New York: Crescent Books, 1973.

Tzalas, Harry E., ed. *Proceedings: 1st International Symposium on Ship Construction in Antiquity, Piraeus 1985.* Athens, Greece: Hellenic Institute for the Preservation of Nautical Tradition, 1989.

U.S. Department of Forestry. *The Encyclopedia of Wood.* New York: Sterling Publishing Co.,1989.

U.S. Department of the Interior. "Archeology and Historic Preservation: Secretary of the Interior's Standards and Guidelines." *Federal Register* (Sept. 29, 1983): 44,716-44,742.

_____. *Historic American Engineering Photographs, Reduced Copies of Measured Drawings, Written Historical and Descriptive Data, Field Procedures for the Schooner* Wawona. HAER No. WA-14. Washington, D.C.: National Park Service, 1986.

Underhill, Harold A. *Masting and Rigging the Clipper Ship and Ocean Carrier: With Authentic Plans, Working Drawings and Details of the Nineteenth and Twentieth Century Sailing Ship.* Glasgow: Brown, Son and Ferguson, Ltd., 1946.

United States Yacht Racing Union. *Hull Measurement Instructions for Use with the USYRU Hull Measurement Instrument.* U.S. Version 1.1. Newport, R.I.: United States Yacht Racing Union, 1988.

Vaitses, Alan. *Lofting.* Camden, Maine: International Marine Publishing Co., 1980.

Walker, David A. *A Guide to the Maintenance of* Theresa E. Conner *and* Cape Sable. Lunenberg, Nova Scotia: Fisheries Museum of the Atlantic, 1984.

Warren, James Peter. "The Historic American Merchant Marine Survey." M.A. thesis, Cornell University, 1986.

Warren, Nigel. *Metal Corrosion in Boats.* Camden, Maine: International Marine Publishing Co., 1987.

Weinstein, Robert A., and Larry Booth. *Collection, Use, and Care of Historical Photographs.* Nashville, Tenn.: American Association for State and Local History, 1977.

Welling, William. *Collectors Guide to Nineteenth-Century Photographs.* New York: Collier Books, 1976.

Weston, Edward. "Seeing Photographically." In *Edward Weston on Photography* edited by Peter C. Bunnell. Salt Lake City: Smith, 1983.

Wilkes, Bill St. John. *The Handbook of Underwater Exploration.* New York: Stein and Day, 1971.

Wilson, Garth S. "The Great Lakes Historic Ships Research Project: Computers and the Analysis and Documentation of Small Craft." *Fresh Water* 2:2 (Autumn 1987): 9-13.

Wood, Virginia Steele. *Live Oaking: Southern Timber for Tall Ships.* Boston: Northeastern University Press, 1981.

WoodenBoat Magazine Brooklin, Maine: WoodenBoat Publications, 1974- present.

LOFTING:

Emmett, Jim. "Lofting Aids." *WoodenBoat* 20 (Jan./Feb. 1978): 68-69.

Lugner, David. "Saving Time in the Mold Loft, Pick'Em up with Precision." *WoodenBoat* 31 (Nov./Dec. 1979): 66-68.

MacNaughton, Thomas. "Simplicity for the Amateur Builder, Part One." *WoodenBoat* 23 (July/August 1978): 53-58.

Manning, Sam. "Some Thoughts on Lofting." *WoodenBoat* 11 (July/August 1976): 43-47.

_____. "Lofting the Lubec Boat." *WoodenBoat* 12 (Sept./Oct. 1976): 44-52.

McIntosh, Bud. "On Laying Down and Taking Off." *WoodenBoat* 11 (July/August 1976): 48-51.

_____. "To Fit a Bulkhead." *WoodenBoat* 43 (Nov./Dec. 1981): 61-63.

Meriam, James L. "The Careful Creation of Mele Kai, Part 1." *WoodenBoat* 63 (March/April 1985): 106-110.

Morris, Ralph. "Lofting by the Numbers." *WoodenBoat* 52 (May/June 1983): 69-72.

Porter, Dave. "Saving Time in the Mold Loft, Pick 'em Up Fast." *WoodenBoat* 31 (Nov./Dec. 1979): 64-65.

Rössel, C. Greg. "Lofting Demystified, Part I." *WoodenBoat* 110 (Jan./Feb. 1993): 66-77.

_____. "Lofting Demystified, Part II." *WoodenBoat* 111 (March/April 1993): 35-46.

LINES TAKING:

Dillion, D.W. "Taking Lines Off Small Craft." *WoodenBoat* 107 (July/August 1992): 71-79.

Farmer, Weston. "Understanding a Boat Plan." *WoodenBoat* 21 (March/April 1978): 30-34.

Littleton-Taylor, David. "Taking the Lines." *WoodenBoat* 19 (Nov./Dec. 1977): 42-45.

Thomas, Barry with Clark Posten and Bret Laurent. "Taking Lines off a Boat." *Woodenboat* 115 (Nov./Dec. 1993) 71-73.

PERFORMANCE:

Palmer, Colin. "Sail and Hull Performance." *WoodenBoat* 92 (Jan./Feb. 1990): 76-89.

Zimmerly, David W. *Hooper Bay Kayak Construction.* Canadian Ethnology Service, Mercury Series, Paper No. 53. Ottawa: National Museum of Man, 1979

_____. "Building the One-Hole Aleut Bidarka, Part I." *Small Boat Journal* 29 (Feb./March 1983): 26-31.

_____. "Building the One-Hole Aleut Bidarka, Part II." *Small Boat Journal* 30 (April/May 1983): 78-83.

_____. *QAJAQ: Kayaks of Siberia and Alaska.* Juneau: Division of State Museums, 1986.

Index

Bold faced entries lead to illustrations or definitions of terms.

accommodations *See* arrangement, interior
accuracy: and size/scale of drawings, 224; checking and correcting, 31-33, 156, 313; defined by the boatbuilder, 28; of computer documentation, 360-361; of 3-D digitizing, **365**; range of in measuring, 28-29, 31-**32**
adapters, plug and socket, 36
Adirondack guideboat, **61**, **97**, **107**, 389
adze, 312
aerial photography, 359
Ames Lettering Guide, 44, **49**, 232
analytical plotters *See* plotters, analytical
appraisals, 16
apron, **71**
archaeo-botanical analysis, 306
archaeological approach to documentation, 305-315, **304**, **316-320**; analysis of hull, 307-308, 313-315; drafting site/wreck plans, 313, **317**; mapping scattered timbers, 308-311; measuring/recording procedures, 307-313; planning, 306-307, 309-310, 313; publication of report, 307, 312, 315; significance of hull, 304-305; tools and materials, 308-309, 311
archaeological plans, 242
archaeological record: boat's construction history as, 52
archaeology, experimental, 6, 388
architectural significance, 306
architectural skills: compared to documentation skills, 3-4
arrangement, interior: of cockpit, 158, **168**; plotting, 158, **168**; profile/plan, 214, (example) **298**; taking off/measuring, **110**, (selective approach) 215
artifact: small craft as, 2, 338
artifacts: borrowing of, 344, (loan form) 385; recording, 308, 313

backbone: variations in small boats, **67**; variations in larger boats, **68**
backstays, **193-194**, **206**; running, **204**
ballast, **68**
Barnegat Bay Sneakbox, **99**
base plane, 116, **134-135**, 212
baseline, 116; above the boat, 122, 124, **138**; and locations of measuring equipment, 124-125, **139**, 156; establishing for deteriorated hull, 310; plotting, 159, **169**, 212, 215; setting up, 118-123, **137**; squaring station locations off, 40-41
bateaux, 305 307-**310**, **305**, **317**
batten, lacing, **188**
batten-seam construction, **63**
battens, fairing *See* fairing splines/battens
battens, sail *See* sail battens/pockets
battens: for "ghosting in" missing components (archaeological reconstructions), 315
beading: and detecting repairs, 59
beam: measuring, 213, **219**
beam compass, 44, **48**
bearding line, **70**
bedlog (horn timber), **63**, **76**
bee holes/blocks, **186**, **188**, **192**
belaying point, 181
bent on (sails), 181
Bermuda pilot gig, 99
beveled gain, **86**
bevel gauge, sliding, 41, **47**
bevels: in lapstrake planking, **86**
bibliography: of published documentation projects, 388-390; of previous ethnographic research, 341; of recommended reading, 391-402
billethead, **68**
bitter end (of rope), **207**
blocking, 12-13, **17**; centerline, 12, **17**; in deck framing, **90-91**, **112**
blocks, **206**; in running rigging, **205**; in standing rigging, **188**
boat builders, traditional, 338-339, **353**; business practices, 344, 349; information sources, 341-342; interviews with, 343, 345, 347-349, 351; transmission of knowledge, **348**-349 *See also* workmanship
boatbuilding, traditional: information sources, 341-342; legal constraints, 342 *See also* shipbuilding, workmanship
boat covers/covering, 13
boat nail, **97**
boat owners: interviews with, 349
boat to be documented: analyzing, 57-60, 307-308, 313-315, (computers) 358-364, **365-370**; assessing condition of, 10, 52; cultural context of, 2, 5-6, 10, 20-21, 338-356; deteriorated, 6, 21, (archaeological approach) 305-315, **304**, **316-320**; economic vs. historic value, 12; environmental (geographic) context of, 10, 306-308, 313, 342, 388; identifying rig, 180; labeling and inventory, 10; moving and trailering, 10-12, 37; restoration and, 6; size in relation to take-off procedure, 55-56, (computer documentation) **365**; storage, 12-16; support, 13-15 *See also* taking off/measuring
boat type/types: cultural contexts of, 338; defined, 52; descriptions of, from local informants, 348-349; filing photographs by, 336; hydrostatic analysis of (computer), 363; regional migrations/evolution, 2, 5-6, 20-21, 363; surveying for, 210 *See also specific types and regions*
bobstay, **191-192**; evidence of, **23**
body plan: drawing (round hull), 233, **266**; computer-generated (digitized), 359, 361-362, **365-366**; examples, **286**, **293**; plotting and drawing, 228-229, **254**; finished drawings, 231
body section planes, 116-117, **134-135**, 232; "eggcrate" model, 232-233, **264-265**
body sections: and overall hull shape, 20, 117; and construction, 21; planes, 116-117, **134-135**, 232 *See also* midship section, sectional view
body sections, drawing lines from field notes/patterns (round hull): (Method 1) level and plumb

bob, 235-236, **276**; (Method 2) leveling board and triangulation, 236, **276**; (Method 3) vertical staff and level measurements, 236, **277**; (Method 4) vertical staff and triangulation, 236; (Method 5) vertical staff with horizontal arm, 236, **278**; (Method 6) vertical staff with horizontal arm and diagonal, 236, **278**; (Method 7) sloped staff, (with triangulation) 236, **279**, (with square measurements) 236, **279**; (Method 8) outside of plank with joggle stick, **288**; (Method 9) inside of plank with joggle stick, **289**
body sections, taking off/measuring, 123-129, (example) **247**; (Method 1) level and plumb bob, 125-126, 130, **140-141**; (Method 2) leveling board and triangulation, 126, 130, **142**; (Method 3) vertical staff and level measurements, 126, 130-131, **143**; (Method 4) vertical staff and triangulation, 127, 130, **143**; (Method 5) vertical staff with horizontal arm, 127, 131, **144**; (Method 6) vertical staff with horizontal arm and diagonal, 127-128, 131, **145**; (Method 7) sloped staff, 128, 131, **145**; (Method 8) outside of plank with joggle stick and scrive board, 128-129, 131, **146-148**; (Method 9) inside of plank with joggle stick, 129, 131, **148**; (Method 10) inside and outside simultaneously, 129, 131, **148**; archaeological approach, 311; selective (quick) documentation approach, 210-211, 214-215
bolts: carriage, **97**; drift, **97**; hand-forged, 305; hanger, **97**; heads of, **97**; machine, **97**; stove, **97**
boomkin, **193**
booms: goosenecks, **190**, **192**; lacing batten, **188**; taking off/measuring, **185**
bow profile, 32; complex bow construction, **176-177**; example, **275**; plotting/sketching, 161, **175-177**, (selective approach) 212, **216**; taking off, 131, **149-151**
bow: shape/entrance, 20; spoon, **22** *See also* bow profile
bowline (on sail), **182**
bowsprit band **192**
bowsprit bitt, **66**
bowsprit plate, **192**
bowsprits: hogged, **191**; measuring and recording, **191-192**
braces (on sails), **182**, **204**
bracing: in cradles, 14; in propping or shoring, 37
brails (on sails), **204**
breasthooks, 57-58, **65**, **89**
bridge deck, 23, **66**
Britain: indigenous watercraft of, 388
broadaxe, 311
bulkheads: evidence of alteration, **23**; measuring and recording, **110**, (selective approach) **218**
bullseyes, **206**
bulwark stanchion (timberhead), **80**
bungs: as guide for locating deckbeams, **109**; distinguishing from treenails, **97**
butt blocks/joints: in planking, 21-23, **87**
buttock lines, 232; drawing and fairing (round hull), 233-235, 237, **266**, **269-271**; taking off, 124-125
buttocks, run of, **271**

cabin: evidence of alteration, **23**; measuring, **109**
cabin sole *See* sole, cabin/cockpit
cabintop: measuring crown of, 212, **218**
CAD (computer assisted design) systems, 358-363, **366-369**
calipers, 39, **207**; using extension ruler as, 38, 46
camber, deck, 23; measuring, **111**
canoes, 15, 59, **61**, 389; lapstrake, **69**; log, 129, **148**; Peterborough, **96**; planking variations, **96**; Rushton, 58-59, **92**; sailing, **61**, **184**; wood-and-canvas, **96**
carlins, **66**, **82**, **89**
carvel planking, 61, 305; determining plank widths, 125, **139**
catboats, 91, **193**, 389
caulking: of seams, 57, **75-76**, 117
ceiling, **63**
cement patches, 22
centerboard case/trunk, **63**, **73**; dimensioning, 242, **291**
centerboards, 22; measuring, 131; variations in construction, **73-74**
centerboard slot, **24**, **63**, **81**; two ways of cutting, **73**
centerline plane, 116, **134-135**
centerline reference, 116, **365**; setting up, 120-121
chafing strip, **87**
chainplates, 22, 23, **206**
chalk, 39, **47**
chalkline, 121
Chesapeake: ethnographic study of maritime community, 338, 347, **351-355**; work boats of, 389-390
chine: plotting and drawing, 228, **253-254**; plotting heights and half-breadths, 227, **247**, **250**
chine construction, **82-83**, **88**
chine logs, **82-83**
Chinese lugsail, **182**
chisel point, **97**
chocked joint, **79**
chocks, anchor, 213, **219**
clamp/beam/shelf: construction details, **63**, **66**, **89**, **107**, **112**; sagging, **114**
clamps, 37
cleaning: of boat, tools for, 15; of deteriorated hull/timbers (archaeological approach), **309-310**; of historical photographs, 335;
cleats: bottom, **88**; deck, **219** *See also* thumb cleats
clew, **164**
clout, **97**
club (sail), **182**, **201**; running rigging for, **203-204**
coaming, **63**, **82**; measuring, **110**
cockpit: arrangement, 158, **168**; construction details, **94**; framing, **91**; measuring and recording, (selective approach) **218**
codline, 207
collections: as identification resources, 180; cataloging by computer, 358, 380; disposition of,

336; National Register nomination requirements, 5; of historical photographs, 332, 336; of plans, 390; surveys of, 388
combination square, 41, **47**
companionway: construction/framing details, **95**; slide, **95**
compass, 43-44, **48**; beam, 44, **48**; sharpening leads, 244, **301**
components (parts of boat), 307, 312; measuring and recording, 311, 325; missing, clues of, 307, 311-314
compressive shrinkage, 13
computer applications: cataloging/managing museum collections, 358, 380; cataloging small craft designs, 358, 380; documentation and hydrostatic analysis, 358-364, **365-370**, (cost estimates) 359-361, 364, (time estimates) 359; hull design, 358, 360; software and equipment, 358-360, 364
concrete blocks, 37, **45**
construction: taking off/measuring, 52-60, **61-114**, (archaeological approach) 307-313, **317-319** *See also* boatbuilding, shipbuilding
construction plans/drawings, 30, 224, 241; computer-generated, 359; developing before lofting, 224; drawing technique, 225, 229-230; examples, **292**, **296**; finished, 231, (example) **262**; judging accuracy, 33; rough, 229-231, **255-256**; size/scale, 224-225, 244, **247** *See also* plans/drawings of traditional watercraft
construction profile, 241; rough, 225, 229, **255**; finished, 231, 241, **258**
construction sections/detail views: drawing, 229-230, 241, 243; examples **65**, **106**; in archaeological reconstructions, **304**, 315 *See also* perspective drawings
context (level of data), 307-308
conversions: work boat to yacht, 24
copper sheathing, evidence of, 23
coracles, 15
cordage: definition, **207**; sizes, **207**
counter-sterned boats: note taking/plotting, 160, **172-174**; plumbing, 118; taking off/measuring, 132, **152**
covering boards, 58, **76**, **91**, **93**
coves: measuring, 212, **217**
cradles, 12-15, **17**, 37-38; in museums, 13-15; measuring boat in, **273**
cranes, mast, **188**
cringles, **200**
crow bars, 37-38
crown: of cabintop, 212, **218**; of deck, 23, **111**, 212, **217-218**; of transom, 132-133, **154**, 238, **281**
cultural context, 2, 5-6, 10, 20-21; and archaeological documentation, 306-307, 313; documentation of (ethnographic research), 333-334, 338-356; photography and, 333-334, 350 *See also* ethnographic approach to documentation
cultural significance, 306
curve: measuring, **110**
cutter rig, **193**
cutwater (false stem), **69**

daggerboards: measuring, 131-132, **152**; variations in construction, **73-74**
damage/wear marks, 311-312
dashed lines (showing hidden lines/edges), 243, **300**, 322
datum level/lines, horizontal, 311, **318**
datum points, 310, **316-317**; in computer documentation, 360, **365**
deadeyes, **195**, **206**
deadwood, 22, **67**
deck beams, 22, **66**, **89**; cut ends, **23**; locating, **109**
deck details and hardware, recording, 22, **219**
deck framing, **66**; camber (crown), **111**; construction variations, **89-91**, **112**; field notes/sketches, 158, **167**
deck house, **66**; construction details, **94**; measuring, **111**
decking (deck planking): fore-and-aft (straight-laid), **93**; signs of sag, **114**; sprung, **93**; strip, **92**; tongue-and-groove, **93**
deck layout/structures: detecting alterations, 22
deck plan, 213, **219-221**
deck profile: measuring, 127, **144**, (selective approach) 212, **217-218**
declivity board: for measuring out-of-plumb/level boat, 117, 129-130, **149**; for developing lines from field notes, 240
deep-keeled boats, 131, 151
Delaware River gunning skiffs, 91
design: and ethnographic documentation, 349; legal constraints on, 342; new, based on documented boat, 210; of small craft, 2, 4; relation to function/environment, 20-21, 342, 349, 358, 388
designs *See* plans/drawings of traditional watercraft
detail drawings/sketches, 229-230, 241, **301**; scale of, 244
deteriorated (derelict) hulls, 6, 21; as source of data, 305-306 *See also* archaeological approach to documentation
diagonals, 232; determining line endings, 235, **272**; drawing and fairing (round hull), 235, 237, **272**
digitizing, 3-D, 360-363, **365-366**; time estimates for, 362
dimensional (3-D) shapes: translating flat surfaces into, 3; translating into drawings, 224, 322 *See also* digitizing, 3-D
dimensioned drawings, 242
dimension lines, **259**
dimensions, defined, 53
dismantling, of wooden hull: archaeological approach, 31
dividers, 33, 43-44, **48**, 244, 310
documentation: archaeological approach, 305-315, **304**, **316-320**; attention to detail, 4; computer applications, 358-364, **365-370**; disposition of records/data, 307, 358; ethnographic approach, 338-356; importance of the boat itself, 10; levels

and purposes, 4-6, (computer documentation) 358, (deteriorated hull) 306-307, (ethnographic approach) 339, 354-355, 389; manuals, 2-3; professional advice, 6-7, 306, 311, 340; publication of report, 307, 313, 315; published projects, 388-390; research and information sources, 6-7, 341-342, 346-348; role in ensuring permanence, 4; resources, 6-7; selective (quick) approach, 210-215, **216-222**; skills/training for, 4 *See also* boat to be documented
documenting *See* taking off/measuring
dory/dories, **61**, 389; bottom, **88**; dory construction, **69**; Grand Banks, **82**; transom, **75**
dory lap, **86**
double-ended boats, **61**, **71**; locating stations on, 123; measuring, 132; plumbing, 117-118, **136**
drafting: estimates of time, 210; making corrections, 234; perspective drawings, 322-324, **327-329**; techniques and tips, 229, 231-232, 234, 243-244, **300**, (selective approach) 215, **222**; tools for, 36, 42-49 *See also* sketches/sketching, lofting
drafting media, 42, 215
drafting table, 42
drag, 21-**22**
drawing/lettering aids, 43, **48-49**, 232
drawings: dimensioned, 242; measured, 242 See also construction plans/drawings, construction sections/drawings, detail drawings/sketches, illustrations, lines plans/drawings, perspective drawings
drills/drilling, 39
drive line, 55
drop slide, companionway, **95**
drying out of hull, 13, 15, 60
ducks (lead weights), 44, **49**; using, 227-228, 243, **253-254**
dugouts, 129, **148**
dumb sheave, **192**
dutchmen, 22

electrical systems, 55; taking off, **104**
end fittings: of rigging, **194-195**, **205-206**
endpost (measuring equipment), **169**
engine and bearers: evidence of, 23-24, **25**; taking off, **102-104**
engine beds, 24
engine installation, 55
engine stringers, **83**, **104**
entrance *See* bow
environmental (geographic) context, 10, 305-307, 312, 342, 388
ethnographic approach to documentation, 338-356; acknowledgments, 344-345, 352-354; background research, 341-342, 346-348; cataloging, 340; documentation levels/purposes, 339, 354-355; documenting scenes, 349-350, **351**; ethics, 344-345; field data and disposition, 340, 345, 349, 352; fieldworkers, insider/outsider pros and cons, 346-348; funding, 340-341; informants and interviews, 340-344, 346-349, 352, (consent form) 384; institutional involvement, 340-341; interviewing techniques, 345-349, 352; making local contacts, 342-344; participant observation, 350; photography, 349-352; planning, 339-340; post-fieldwork, 352-355; publication of data (reports/books/tapes/films), 340, 344, 354, 389; sources of published information/documents, 341-342; standard procedures, 339; understanding boatbuilding and design, 349
"Exercise Boat" (drawing plans), 232-235, **266-269**
exploded diagrams, 314, 322, 324, **327**
extension cords, 36
extension lines, **259**
eye splice: marrying, **207**; measuring, **195**

fairing lines (on drawings), 241, 242, **299**; flat-bottomed (straight-sided) hull, 227-228, **253-254**; round-bottomed hull, 232-235, **269-271**
fairing splines/battens: use of, 44, **49**, 227-228, 243-244, **253-254**
fairleads, **101**
fairness: of hull, 57, 59; of lines (drawings), 228, 234, **254**
fantail stern, 77
"Fast Yacht" CAD system (GLHSRP), 359-363, **366-369**
fasteners/fastenings, 20-21, 53, 58-59, 243; and estimating boat's age, 306; brass, 58; bronze, 13, 58; copper-based, 58-59; galvanized iron, 13; galvanized steel, 58; iron, 58; locating (probes), **112**; longevity, 13; manganese bronze, 58; silicon bronze, 58; taking samples of (archaeological approach), 309, 311; uniformity, 29 *See also individual types*
fastening patterns/locations, 53, **92**, 311-315; and distinguishing bungs from treenails, **97**
feeler gauge, 39, **112**
field notes: archaeological approach, 307-312; avoiding measuring/plotting errors, 31, 156-157, **162**, **185**, 211, **216**; drawing and lofting from, 224-244, **245-301**; drawing/sketching materials, 38, (water/weatherproof) 309; ethnographic approach, 345; index sheet for, 157, **166**; organizing, 156-158, **162-168**, (archaeological approach) 311-312, (ethnographic approach) 352-354; plotting, 158-161, **169-177**; scantlings list, 157-158, **166**; selective (quick) approach, 211; sketches, 323
fin-keeled boats, **68**, **114**, 131, **151**
fish plate, **68**
fisherman's snotter, **204**
fisherman's staysail, **183**; running rigging for, **203**
fishing, commercial, U.S.: **346**-347, **351**, **354**; information sources, 341-342; regulations, 342
fittings: end (for standing rigging), **195**; "patent," **206**; various types, **206** *See also specific types*
flags, reference (in computer documentation) *See* reference flags
flare of hull sides: and fairing lines, 228
flat-bottomed boats, 57, 61, **88**; construction variations, **82**; drawing and fairing lines, 227-228,

253-254; taking off/measuring, 125, **140**
floor, **68** *See also* sole, cabin/cockpit
floors: variations in construction, **80-81**
floor timbers, **24**
flying jib/jib topsail, **183**
foot (of sail), **164**
footwell/cockpit, **66**; construction details, **94**
fore staysails: running rigging for, **203**
forefoot, cutaway, 20
fore sail, **183**
forestay, **193**
fractions: and grid system (used in plotting), 159-160
frame, measuring, 158, **169-170**; using, 120
frames, 21, 24; bent, **78**, **81**; boxed, **78**; broken, 21, 58-59; double sawn, **80**; evidence of repairs, 58, **79**; jogged, **78**; knuckled, 239; measuring and recording, 58-59, **78-81**, (archaeological approach) 309-310, 311; replacement, 28; single sawn, **79**; sister, **24**; variations in construction, **78-81**
framing: deck, **66**, **89**; **90-91**; housetop, **66**, **95**
framing square, carpenter's, 41
freeboard, 20
Friendship sloops, **81**, **91**
full siding, **70**
funding: for ethnographic research, 340
futtock, **79**, **80**

gaff-headed sails, **183-184**; running rigging for, **202**
gaff jaws, **189**
gaff rig, 180, **183**
gaff sails, seam patterns, **199**
gaff topsail, running rigging for, **203**
gaffs, measuring and recording, **183**
gains, in planking, **86**
garboard plank/seam, 22, 58, **63**, **72**, **271**; extreme shape (example), 234, **271**; protection, **87**
gauge pins, 54, **62**
gauges: feeler, 39; sliding bevel, 41, **47**
girth, 57, 241
GLHSPR (Great Lakes Historic Ships Research) photo documentation system, 358-362, 364, **365**
Gokstad faering, 388
Gold Cup Racer, **104**
gooseneck fittings, **190**, **192**
Great South Bay catboats, **91**
grid system: checking squareness, 225-226; division of squares into fractions, 159-160; net (used in computer documentation), 362, **367**; subdividing cubes (perspective drawings), 324, **329**; used in drawing rough lines plan, 225-226, **248**; used in mapping scattered timbers (archaeological approach), 310-311, **317**; used in perspective drawing, 323-324, **328-329**; used in plotting field notes, 157-160, **165**, **168**, **170**, **274**
guards (guardrail, rubrail), **63**, **82**; construction variations, **107**; measuring, 212, **217**
gunter sail/rig, sliding, **182**, **199**; running rigging for, **204**

half models, 4; measurement of by computer, 359-for, 204
half models, 4; measured by computer, 359-360
half-breadths: plotting and drawing (body sections), 228-229, **254**, (sheer and chine), 227, 231, **247**, **250**, **258**; showing in table of offsets, 231, **258**
half-breadths plan, 227, **231**, 237, **250**, **258**; examples, **286**, **294**
halyard, main, **202**
hambroline, **207**
handicapping of yachts: by computer, 359-360
hanging knees: drawing (rough construction plan), 229, **255-256**; recording/sketching (field notes), **106**
hanks, sail, **201**
hardware, 28-29; deck, 22; for hanging rudder, **99-100**
harpin, 77
hatch framing: measuring/sketching, 158, **167**
hatches, measuring and recording, **219-221**
head (of sail), **164**
headboard (on sail), **201**
headledges, **73**
headstay, **183**, **194**
heel (butt) of mast, **184**
heights: plotting, 227, **247**, **250**; showing in table of offsets, 231
helm details, **214**
hinges, strap, 41, **47**
historic ships/boats: of Europe, 304; of Mediterranean, 305, **319-320**; of North America, 304-305; measurement of by computer, 360-361
historic significance, 5, 305-307
historical photographs *See* photographs, historical
hog/hogging, 13, 20, 57, **113-114**; removing in lines drawings, 240
hogged (bowsprit), **191**
hollows (in hull lines), defined, 234
hood ends, **69**, **114**
horizontal arm: used in measuring, 127-128, 131, **144-145**
horn timber, **76** *See also* bedlog
horse (rigging), **192**, **205**
horses, three-legged, 12, **17**
housetop, **66**; framing, **66**
housline, **207**
hull measuring devices, electronic: USYRU Hull Measuring Instrument (HMI), 359-360; Maine Yankee Hull Measuring Device, 364
hull profile (sheer plan): 20; drawing and fairing (round hull), 233, 237, **266**; examples, **286**, **294**; in computer documentation, 362, **366-369**; plotting/sketching, 158, **169**; taking off/measuring, 131-132, **149-151**, (selective approach) 211-212, **216-219**; underwater, 21, 212, **216**
hull shape, 5-6, 10, 20-21, 31, 52; and "knowing the boat," 28, 30; changes in (due to improper storage, damage, structural failure, etc.) 13, 57, 59, **113-114**, 117; fairness, 57, 224; noticing/detecting irregularities, 117, 224;

take-off/measuring process, 116-133, **134-154**
humidity: and dimensional changes in wood, 28, **33**; in storage environment, 13
hydrodynamics *See* hydrostatic analysis
hydrostatic analysis (projection of stability and performance), 124; by computer, 358, 360-363; traditional, 358, 364

illustrations, 315, 389; techniques and tips, 322-324
inboard, 54
inboard profile, **66**
integrity: historic, 5; structural, 10, 12, 52, 311
interior of boat: measuring, 130
intersections of lines (crossings): drawing tips, 244, **301**; scant crossings, 234, **269**; strong crossings, 234, **269**
interviews, recorded: 340-341; equipment, 345-346, 352; technique and strategy, 343-346, 348-349
Inuit skin boats, 388-389
irregular shapes: measuring, **108**
isometric lines drawing ("quickplot"), computer-generated, 363

jackline, **201**
jacks, 37-38
jackyard topsail: running rigging for, **203**
jaws: boom, **190**; gaff, **189-190**; gunter yard, **190**
jib-headed sail, **183**; measuring, **197**; seam patterns, **199**
jibs, **183**; measuring, **197-198**, 213; running rigging for, **203**
jibstay, **193-194**
joggle stick: measuring body sections, 128-129, **131**, **146-148**; developing patterns taken off with, 240, **288-289**
joinerwork, deck and interior: taking off/measuring, 52, 54-55
joints, 28, 57; in frames, **78-79**; lapped, **79**; taking off/measuring, 56, **112**
joints, types of: box (finger), **98**; dovetail, **98**; fishtail plate, **98**; half-lap, **98**; lap, **79**; miter, **98**; mortise-and-tenon, **98** *See also* scarfs
jumper strut, **193**

kayaks, 15
keel: drag, 21-**22**; hogging of, 13, 20, **113-114**; rabbet, **64**, 212; rocker, 57; sagging of, 13, **113-114**; sections, **72**; taking off/measuring, 131, **151**, (archaeological approach) 309; used for plumbing/leveling boat, 122, **136**, (archaeological approach) 310
keel, false, 22, **24**
keel batten, **67**, **71**
keels, structural: blocking and storage, 12-13; boats without, 15
keelson, **79**-80
keel types: deep, 131, **151**; fin, **68**, **114**, 131, **151**; rockered, **67**; sprung (bent), **67**, **102-104**, **114**
ketch, **183**
key plans (examples): open rowing boat, **65**; 30' cruiser, **66**
king plank, 58, **93**
knees, 57-58; in deck framing, **90**; plotting/sketching layout, **168**
knightheads, **68**
knuckle (of transom), **77**; projecting height of, 229, **254**; drawing line (rough construction plan), 230
Kyrenia wreck, 305, 315, **319**

Lake George bateaux, **304**, 305-308
laps, plank, **86**; lap and lay (lapstrake planking), 125, **141**, **288**
lapstrake (clinker-built) boats, 57, **78**, **85-86**, 304; construction plans, 240, **287**; determining plank widths/thicknesses, 125, **139**, **141**; developing hull sections, 239, **285**; drawing rough lines plan, 239-240, **284-286**; taking off construction, **86**
lateen rig, **182**
launchings: photographing, 350-351, **352**
laws and regulations affecting boat building/design/use, 342
lazy jacks, **202**
lead (rigging), 181
lead ballast, **68**
lead patches, 22
lead weights *See* ducks
leaking, 22
leeboards: variations in construction, **73-74**
lettering (on lines/construction drawings), 232; style, 157, 322
lettering guide, 44, **49**, 232
level, carpenter's, 40-41, **47**, 56
leveling board: use of, 41, **47**, 159
leveling the boat, 117-119, 122, **136-137**; archaeological approach, 310
levels, 40-41, **47**
levers, 37-38
lifting beckets, **75**
lighting: in computer photo-measuring device, **365**; in measuring area, 36
line endings, determining, 235, 241, 272
line level, mason's, 40-41, **47**
line weight/quality, 224, 231-232, **261-262**, 322
lines plans/drawings, 224-226; archaeological reconstruction, 314-315; computer-generated, 359, 363, **370**; developing before lofting, 224; insufficient for exact reproduction, 5-6; isometric, 363; parametric, 363; resulting from selective documentation, 210, 214-215, **222**; size/scale, 224-225 *See also* drafting
lines plans/drawings, finished, 225, 231-232, **262**; examples, **260**, **263**; layout, 231, **258** *See also* plans/drawings of traditional watercraft
lines plans/drawings, rough, 225, 229-231, **250**, **255-256**; drafting technique and tips, 243-244, **299-301**; example, **266**; for flat-bottomed hull, 225, **250**; for round-bottomed hull, 233-237, **263**, **266**, **268-279**
list, angle of, 311
lobster boats, 389
lodging knees, **66**, **90**
lofting, 32, 36, 42-49; as distinguished from draw-

ing, 224; body sections, (from field notes) 237, (from joggle stick patterns) 288-289; by computer, 363; for reconstruction (archaeological approach), 314
lofting surface, 42
log rail, **82**
long lines, defined, 241
long tapes, 38
loose-footed sail, running rigging for, **205**
luff, **164**; offset (gunter), **199**
lugsails, 180; standing, **182**

machinery, 28-29
Mackinaw boat, **69**
main sail, **183**
main topsail, **183**
mapping techniques: used in archaeological documentation, 308-311
maps and charts: used in archaeological documentation, 308; used in ethnographic documentation, 342
maritime preservation community *See* museums and maritime organizations
marline, **207**
marry, **207**
mast band, **186**, **194**
masthead, **193**; details, **186**
mast partners, **184**; construction variations, **89-90**, **112**
mast saddle, **189**
mast step, **66**, **70**
mast tenon, **184-186**
masts, 2, 180-181; construction variations, **187**; details, **186-188**; grooved for halyard, **187**; measuring and recording, **184**; rake of, estimating, 213; unstayed, **193**
measured drawings, 242
measurements: additive, 31, **33**; approximate (selective approach), 211; averaging, 31, 33; defined, 53; digital, 127; missing, compensating for, 243, **300**, (archaeological approach) 307, 311-315; of irregular shapes, **108**; overall, 33; recording method, 31; resolving discrepancies, 32-33; running, 31, **33**; square 128, **145**; straight-line, 131, **151**; triangulated, **110**, 126-128, 130, **142-143**, **146**; units of (archaeological approach), 308 *See also* accuracy, verification/checking
measurer's job, 29-31
measuring *See* taking off/measuring
measuring equipment, 36-41, 211, 308-309, 311, 359-361; location/setup, 124-125, **139**, 156, **169**
measuring staffs, 40, 56, **109**, 119-122, 131, **137**, **150-151**, **169**; sloped, 128-131, 145-146, **150**; vertical, 126-128, **143-146**
measuring tape (seamstress's), 31
measuring team/jobs, 29-31; in archaeological documentation, 311
mechanical and electrical systems, 55, **102-104**
mechanical plans, 241-242
Mediterranean: documentation of wrecks, 305
midsection *See* construction sections
midship section: sketch of, 158, **167**
mildew, 15
mirror option (in computer design/documentation), 363, **369**
missing data, compensating for: in measuring (archaeological approach), 307, 311-314; in drawing/reconstructing lines, 243, **300**, 314-315; in scale model, 315
mitered sail, **198**
mizzen, **183**
model: defined, 52
modeling, scale: tools and materials for, 314
models: "eggcrate," 232-233, **264-265**; half, 4, 359; scale, 314-315, **320**; used in ethnographic research, 341; various types, 315
molding: defined, 53
mooring cleat, **192**
moving a boat: cautions/pros and cons, 10-11, 37-38; deteriorated hull, 306; methods of, 11-12
Muscongus Bay sloop, 22-24
Museum Small Craft Association (MSCA): Union List of North American Watercraft (database), 6-7, 380-382
museum storage: indoor, 15-16; in-water, 16
museums and maritime organizations: advice on documentation, 6-7, 306; cataloging and identification, 180; donating boats to, 16; funds for restoration, 16; interpretation, 16; involving in ethnographic research, 340; list of, 374-378; policies for accepting boats/collections, 16; sources for maritime images (photographs), 334-335
Mylar (drafting film), 36, 42, 215, 225, 231, 244, 309, 314, 363
Mystic Seaport Museum: Vessel Authority file, 6

nails: boat, **97**; clench, **97**; ring, **97**
National Register nomination: qualifications for, 5-6
natural crooks, 21, 56-58; in floors, **81**; in stem, **69**
naval architecture, 4
negatives: glass plate, 332; storage, 326
net (grid), used in computer documentation, 362, **367**
New England: indigenous watercraft of, documentation projects, 389-390
New Haven sharpie, **61**, **99**, 390
nibbed ends: of deck planking, **93**
notetaking: boat's condition when found, 10; notetaker's job, 29-31 *See also* field notes

oarlock blocks/pads, 56, **63**, **65**; recording/sketching (field notes), **106**
offsets: how to read/write, 231; plotting, 233; table of, 31, 33, 224-225, 231, **258**, **267**, **295**, (computer-generated) 359-360; taken from reconstruction model, 315
orthagonal projection, 323
out-of-plumb/level boat See twist
outboard, 54
outboard profile, **287**

outhaul, **202**
overhanging ends: supporting, 15; unsupported (sagging), **114** *See also* counter-sterned boats

Pacific Northwest: small craft documentation projects, 390
pad eye, **206**
paint lines, 22, **23**, 24, **25**
paper: for drafting, 36, 42; for field notes and sketching, 38, 322; gridded, for plotting, 157; perspective, 322
parallel lines: drawing technique, 226, **251-252**
parrel beads, **189**
patch function (in computer design/documentation), 362
patterns, converting to drawings/loftings, 237; curved, raked transom, 238-239, **282**; sections taken with joggle stick, 240, **288-289**
patterns, used in measuring, 56; body sections, 124, 128-129, **146-148**; transom, 132, **153**
peapod, 389
pencils, leads, and erasers: for drafting and lofting, 42-43, 224; for field notes/sketches, 38, 309; for illustrations, 322; technique/drawing hints, 231, 243
pens and ink, 43, 322; pen sizes and drafting technique, 224, 231-232, 243, **261**
performance, in relation to hull shape, 5-6, 20-21; hydrostatic analysis of (by computer), 358, 363
perpendicular lines: drawing technique, 226, **251**
perpendiculars: erecting, 225, **249**; locating, 119; measuring between, 122-123; plotting, **169**
perspective drawings, 156, 158; computer images, 360, 362, **369**; cutaway sections, **63**; drawing techniques, 322-324, **327-329**
photographs, historical, 332-336; and ethnographic research, 341; care and storage of, 335-336; cataloging of collections, 332, 336; dating and identification of, 333-**334**; disposition of collection, 336; labeling, 336; evolution of processes, 332; estimating dimensions in, 334; photographers, 332; sources/repositories, 332-335; user fees, 334
photography, 32-33, 324-326; and computer measurement systems, 359-361, **365**; archaeological approach, 306-308, 311; archival quality, 325; contact sheets, 37, 325; equipment, 36-37, 181, 211, 324-325, 340, 350; ethnographic research, 349-352; film, 37, 211, 324-326; historic processes, 332; lighting, 31, 325-326; limitations of in measuring, 56; recommended views, 326, (selective approach), 213; of boat "in situ," 10, 306-308; of rig, 181, **208**; prints, 213, 324-325; selective (quick) approach, 210, 213; standards of, 324-325; storage of photographs/negatives, 335-336; technique and composition, 31, 325-326
pin, **206**
pintles and gudgeons, **99-100**
pipes: routing of, 55
pirogues: log, 53-54, **62**; plank, 54, **62**
pitch: recording, **216**
planes of a hull, 116-117, **134-135**, 232-233, **264-265**; reference planes (selective approach), 211-212, **216**
planking, deck, 22, 58; various methods, **92-93**
planking, hull, 57; bottom, **88**; detecting repairs, 23, **87**, 312; determining thickness, 39, 54, **62**; determining widths, 125, **139**; drawing plans (round hull), 234-237, **276-279**; faired into hull, 125, **141**; fastening patterns, 53, **92**; layout, 58; signs of sag, **114**; strake diagrams, 315
planking lines/seams: 21, 56-57, **76**, 160, **171**; drawing and fairing (lapstrake hull), 239, **285**; drawing and fairing (round hull), 234-237, **276-279**
planking methods: 21, **84-85**; Ashcroft, **85**; batten-seam, **85**; carvel, **84**; double, **85**; feather lap, **84-85**, **87**; lapstrake, 57, **78**, **85-86**; ship-lapped, **83**; strip, **84**
plank keel, **72**
planksheer *See* covering board
plank thickness, variation in, 28
plans, set of: summary list, 241-242 *See also* body plan, construction plans/drawings, lines plans/drawings, sail plans
plans/drawings of traditional watercraft: collections, 390; published, 388-390; sources for, 4, 388-390
plan view, **65**, 214; archaeological approach, **304**, 310
platform: for preventing damage to deteriorated hull, 309-**310**
plotters, analytical, 361
plotting/sketching, 32-33; by computer, 361, 363; field vs. final, 156; field notes, (on gridded paper) 157-160, **165**, **168**, **170**, (on plain paper) 160-161, **169**, **172-174**; offsets and, 233; plotter's job, 29-31; procedure, (field notes) 159-161, **168-174**, (rough lines drawings) 226-229, **250-252**; tools and materials, 156 *See also specific profiles/views*
plumb bobs, 39, **47**; using, 56-57
plumb height, defined, 226
plumbing the boat, 117-118, 122
point of view: in perspective drawings, 323, **328**
poppets, 12-14, **17**, 38; Brownell type, 12
portholes: measuring, 212, **218-219**
powerboats: plotting/sketching, 161; V-bottomed/chine construction, **83**, **88**
preservation of hull: lake mud and, 308
probes: using to measure angles/joints, **69**, **72**, **112**
production boats, 4, 58-59, **78**
profile view, **65**, 214, 323; archaeological approach, **304** *See also* bow profile, hull profile, stem profile, stern profile
project/projection: defined, **227**; projecting from table of offsets, 233-235, **269**
propeller shafts/tubes: fittings, **104**; taking off, **102-104**
protractors, 43, **48**
pry bars, 37-38
publication/dissemination of data, 307, 313, 315;

335, 340, 344, 354
publications: guide to published documentation projects, 388-390; sources of ethnographic information, 341-342

quarter knees, **63**, **65**, **75**; drawing (rough construction plan), 230, **256-257**, **262**; recording/sketching, **106**
Quebec City bateaux, 304, **308-310**, **316**
"quickplot" (computer-generated lines drawing), 363

rabbet: at plank keel, **72**; keel, **64**, 214; stem, **64**, **69-70**
rabbet half-breadth, 214
rabbet line, 57, **63**, **70**, 212; plotting, 226, **251**
rake, measuring: mast, 213; stem, 131, **150-151**; transom, 132-133, **154**, 212, 238, **281**
Rangeley Lakes Boat, **69**
reconstructions, 307; compensating for missing data, 307, 311-314; computers and, 358; graphic form, 314; reasons for, 242; scale model, 314-315, **316**; various forms, 305
recording *See* taking off/measuring
recreational boating, 2, 4, 389
reef bands, **164**, **196**
reefing, **202**
reef points, **200**
reef tackle, **202**
refastening: evidence of, 21, 23
reference flags (in computer documentation), 360, **365**
reference lines, 40, 116, 156; and locations of measuring equipment, 124-125, **139**, 156; and note-taking, 156-157; auxiliary reference line, 56, **109**; horizontal reference, 156-157, **159**, **170**, 309; in computer measurement systems, **365**; plotting, 158, **169**, **172-173**; setting up, 121-122, **137-138**; top reference line, 56, **109**; vertical reference, 156-157, **159**, **170**, **365**
reference tags (archaeological approach), 309
reflector lights, 36
regional surveys (of boat types within a region), 5-6, 363
repairs/alterations: as historical record of boat, 16, 21-24; evidence of, 20, 22-25, 28, 58-59, **79**, 311 *See also specific areas of hull*
replicas/replications, 5, 358, 388; bateaux, 305
reproductions (of documented watercraft), 5-6, 388-389
restoration, 16; funds for, 16; impact on documentation, 6
reverse curve, 28, **33**
rig (sails, spars, rigging): detecting changes, 22; measuring and recording, 180-181, **182-208**; stepped (for measuring), 180 *See also specific rigs*
rigging, 180 *See also* running rigging, standing rigging
riser, **63**, **65**
rivets, **97**
roach, **164**; measuring, **196-198**
rope: definitions, **207**; sizes, **207**
round-hulled boats, 57; drawing from table of offsets, 232-235, **266**, **268-271**; drawing from field notes, 235-237, **273-279**; fairing lines, 228, 233; plotting, 159-160, **171**, **175**
rudders: attachment to boat, **99-100**; photographing details, photographing details, **215**; taking off/measuring, **108**, 212, **216**; various types, **99**
rudder trunk, **68**; built-up, **76**
rulers, folding extension, 38, **45-46**
running backstay, **204**
running rigging, 180-181; measuring and recording, **202-205**
rust, 4; fastenings, 21

sag/sagging, 13, **113-114**
sail battens/pockets, **200**
sail plans: drawing, 225; examaples, **293**, **297**; of small craft, **182-185**; single-sail, **182**
sails, 22, 181; bending on, **201**; design and fabrication, **196**; details, **200**; panels and seam patterns, **198-199**; sailcloth, **196**; storage, 16; taking off/measuring, 157, **164**, **196-199**, (selective approach) 213
sail track and slide, **187**, **201**
sanitary facilities (head), 55
saw horses, 37; three-legged (for blocking and storage), 12, **17**
scale (beveled scale, "scale rule"), 39, 43, **46-47**, 225
scale, black-and-white (footbar): used in photographs, 211, 325
scale, graphic: used in drawings, 231, 242, **259**, **291**
scale drawings, 156
scale models *See* models
scale off, defined, **225**
scale plan: of archaeological site, 310, 313, **316-317**
scales, drawing: commonly used, 224-225, **246**; effect on drawing size, 225, **247**
scaling: in photographs, 56, 181, 211
Scandinavia: indigenous watercraft of, 388
scant crossing (angle of line intersection), 234, **269**
scantlings: defined, 53; ranges of accuracy for measuring, 28; taking off, 53
scantlings list: examples, **105-106**, **262**; field notes, 157-158, **166**; selective approach, 213-214
scarfs, 58, **87**, 59; in feather-lap planking, **87**; nibbed, **87**
schematics, 55, **104**
schooner rig, **183**, **193**
Scotch cut sails, **199**
scow, **61**
screws: lag, **97**; machine, **97**; ratio of head to shank diameter, **97**; wood, **97**
scrive board, used in measuring/patternmaking, 128-129, **131**, **146-148**
sculling notch, **75**
seats, measuring and recording, 213, **220-221**
sectional views, 314; plotting/sketching, 158-159, **170-171**

sections *See* body sections, construction sections
seizing, **207**
Serce Limani wreck, 305, **320**
shackle, **206**
shaft logs, **102-104**
sharpie *See* New Haven sharpie
sheaves (in mast, boom, or gaff), **101**, **187**; dumb sheave, **192**
sheer, 20, 57, 59-60, **113**, 117; sagged, **114**
sheer heights and half-breadths: measuring, 133, 212, **216**; plotting, 226-227, **247**, **250**; showing in table of offsets, 231, **258**
sheerline: fairing, 227-228, **243**, **253**; plotting and drawing, 227; representing in perspective drawings, 324
sheer strake, 58, **63**; measuring, 39; molded, compensating for, 125, **141**
sheets (running rigging), **182**, **202**; variations, **205**
shells, rowing, 15
shim ends, deck planking, **93**
ship-lap (rabbeted gain), **83**, **86**
ship's curves, 43, **48**; using, 243, **299**
shipbuilding, traditional, 3-4
shrimp net: pattern for, **355**
shrouds, **193-194**, **206**
shutter plank, **84**
sidewinder, 211
siding: defined, 53; full, **70**
significance, historic, 5
sills, **94**
site/wreck plans: mapping and drafting, 308-311, 313, **317**
size of boat, 20-21; estimating in historical photographs, 334; in relation to take-off method, 55-56, **365**
Skuldelev wreck, 305, 315
skegs, **67**, 118, **216**
sketches/sketching, 29, 56, 322-324; in field notes, 158, 211, 323; size/level of detail, 158
skiff: taking off construction, 55-56, **61**, **105-106**; drawing and lofting lines, 225, 227-228, **250**, **253-254**;
slide rail, **95**
sliding gunter *See* gunter sail
sloop rig, 193 *See also* Friendship sloop, Muscongus Bay sloop
small craft, historical: as artifacts, 2, 338; backbone variations, **67**; computer measurement systems for, 359-363; construction, 4; defined, 2; design, 2, 4; documentation, 2-7; in historical photographs, 332-336; sources of plans/designs, 4, 388-390
"small stuff" (cordage), **207**
snotters, **186**, **204**
software *See* computer applications: software and equipment
sole, cabin/cockpit (floor), **66**; measuring, **109**
solid option (in computer design/documentation), 363, **369**
spars, 2, 180; details, **186-189**; measuring and recording, 180, **185**, (selective approach) 213; storage, 16; unstayed, **193**
spikes, hand-forged, 306
spiling board, use of, 215
splices, in rigging: marry, **207**; measuring tips, **194**
splines *See* fairing splines/battens
spreaders, **193**
springstay, **193**
sprit: measuring and recording, **186**
sprit rig, **182**
sprit sails, 180; seam pattern, **199**
square sail, **182**
squares, framing/combination, 41, **47**
St. Lawrence River skiff, 362-363, **366-370**
stability, hydrostatic analysis of (by computer), 358, 363
stabilization (of hull), 10-17; defined, 10
staffs *See* measuring staffs
standing rigging, 180-181, **193**; ends/end fittings, **194-195**; measuring, **194-195**
station lines, squaring off baseline, 40-41
station reference, in measuring body sections, 124, **138**, 212
stations: digitized (traced by computer), 362, **366**; locating and recording, 123-124, 212, **217**; plotting, 158, **169** *See also* body plan
staving, **94**
stays, **191-194**
staysails, measuring, **198**
stealers, 23, **87**
steel/iron hulls, 4
steering methods: hiking stick, **101**; oar, 101; tiller, **101**; wheel, **101**; yoke, **101**
stem, false (cutwater), **69**, **72**
stem face, plotting, 226, **251**
stem profile, **72**
stem shoe, **70**
stems, **22-23**, 59-60, **63**, **65**; rabbet, **64**, **69-70**; recording/sketching (field notes), **106**; taking off/measuring, 131, **149-151**; variations in construction, **69-72**
stepped hulls (powerboats), **83**, 124
stereographs, 332-**333**
stereo-photogrammetry, 359, 361, 364
stern gland and packing, **76**
sternpost, **66-67**
stern profile: plotting/sketching, 160, **172-174** (selective approach) 212, **216**; taking off/measuring, 132-133, **152**, **154**
stern seats: drawing (rough construction plan), 229-230, **255-256**; recording/sketching (field notes), **106**
sterns: and plumbing the boat, 118, **136**; overhanging (counter), 160, **172-174**; photographing details, **215**; round and elliptical, 77; transom, **76**; variations in construction, **68**, **75-77**
stopwaters, 22
storage: improper, changes resulting from, 59, 117; in museum, 15-16; minimizing dirt, 15; of photographs, 335; long-term, 13-16; short-term, 10, 12-13; site selection, 10-11; structural integrity and, 12; temperature/humidity considerations,

13, 15-16
straight edges, 42
strake diagrams, 315
strap eye, **206**
strap hinges, 41, **47**
stresses on hull: avoiding point loads, 14
string, nylon, 40
strong crossings (angle of line intersection), 234, **269**
structural failure, signs of: in frames, 58-59
structure (level of data), 307
strut bearing, **104**
stuffing box, **102**
support of hull, 12-15, **17**, 37-38
surface design: using "Fast Yacht" software, 362-363, **368**
surveyors, marine, 10
sweet (lines), 59
symmetry, in a hull, 58
systems *See* mechanical and electrical systems

table of offsets, 31, 33; computer-generated, 359-360; drawing, 224-225, 231, **258**; examples, **267**, **295**; projecting lines from, 233-235
tabling (sails), **196**, **200**
tack (sail), **182**, **202**
tacks, copper, 2, **97**
taking off/measuring: boat in the water, 214; body sections, 123-129, **140-148**, **247**; by computer, 358-364, **365-370**; common errors to avoid, 31, 185, 211, **216**; construction, 52-114; cost, 6, 30; crown/camber, 132-133, **154**; digital measuring, 127; estimates of time, 6, 28, 30, 210, (selective approach) 212-213, (archaeological approach) 313; interior of boat, 130; irregular shapes, **108**; joinerwork, 52, 54-55; overall procedure, 29-33, (archaeological approach) 307-313; rake, 131-132, **150-151**, 154; ranges of accuracy, 28-29, 31-**32**; recording methods, 31; rigs, 180-181, **182-208**; scantlings, 53; selective (quick) approach, 210-215, **216-222**; setting up, 29-30, **273**; steps and procedure, 116-133; tools and materials, 36-41, 211, 307-308, 310; underwater (archaeological approach), 309, 311 *See also specific areas of hull and rig*
tangs, **188**, **194-195**
tape, masking, 40
tape measures: reading, 31; setup, 55-56, **105**; types, 38, 307-308
tax deduction, for donated boat, 16
temperature, in storage environment, 13
templates, 44, **49**; using, 243-244
terminal points, defined, 214
thole pins, **65**
throat of sail, **164**, **195**
through-hulls, evidence of, 23
thumb cleats, **188**, **194-195**
thwarts: and plumbing the boat, 118; defined, 54, **63**, **65**; drawing (rough construction plan), 229-230, **255-256**; knees, **107**, **114**; plotting/sketching layout, **168**; signs of sag, **114**; taking off/measuring, 56-57, **105-107**
tick strip, 3
tie rods, **83**, **90**
tiller assemblies, **101**
timberhead *See* bulwark stanchion
timbers, detached, 305, 307-308; tracing and recording, 312, **319**
toe rail, 23
tool marks and identification, 311-312
tools and materials: for drafting/lofting, 42-49; for plotting, 156; for taking off/measuring, 36-41, (archaeological approach) 308-310, (selective approach) 211 *See also* computer applications: software and equipment
top timbers (in frames), **79-80**
topping lift, **202**
topsails, running rigging for, **203**
tracing: in drawing finished lines plans, 243; of timbers (archaeological documentation), 312, **319**
track and slide *See* sail track and slide
trailers/trailering, 12
trammel heads, 44, **48**
transom, curved and raked: converting field notes/pattern to drawing/lofting, 237-239, **282**; determining arc-of-face radius, (algebraic formula) 238, **281**, (geometric solution) 238, 281; developing in body plan, 237-239, **282-283**; developing in profile and half-breadths plan, 239, **283**
transom, top of: plotting, 229; drawing (rough construction plan), 230, **257**
transom beam, **68**
transom face (expanded transom): plotting and developing, 226-227, 238-239, **247**, **252**, **280-282**
transom frame, **76**
transoms: detecting changes, 58; evidence of repairs, **75**; location, 24-**25**; maintenance, 15; plank, **75**; planking seams, **75**; plotting and drawing (round hull), 233, 236-237, **275**, **280**; plotting face, 226-227, **247**, **252**; plotting top, 229; projecting to body plan, **280**; taking off/measuring, 132-133, **153-154**; thickness of planking, 39; variations in construction, **75**
transoms, types of: dory, **75**; raked, **33**, 132-133, **154**; flat, 132-133, **153**
transom-sterned boats, **67**; locating stations on, 123; plumbing, 118, **136**; variations in construction, **76**
travelers, **192**, **201**, **205**
treenails (trunnels), 306; distinguishing from bungs, **97**
triangles, 43, **48**; "sliding" technique (used in plotting), 226-227, **251-252**
triangular sails, **182-183**
triangulation (measuring technique), **110**, 126-127, 130, **142-143**, 236, **279**, 310, **316-317**, 359
trim details, exterior: measuring and recording, 212, **217**
tripods: in computer photo-measuring device, **365**; used in photography, 325-326
truck bed, securing boat to, 12

tumblehome, 22, **76**
turn of the bilge, 21
turnbuckles, **195**, **206**; measuring, **205-206**
twist, in hull: compensating for, 118; considerations in drawing lines plans from field notes, 240-241, **290**; measuring for, 126, **141**; removing, 18;
type *See* boat type

Union List of North American Watercraft (museum database of historic watercraft), 6-7, 380; recording forms for, 381-382
unstayed masts and spars, **193**
USYRU (United States Yacht Racing Union) hull measurement system, 358-360

V-bottomed (chine) boats, 57, **82-83**, **88**; plotting/sketching, 158-159, **170**, **174**; taking off/measuring, **141**
vanishing points: in perspective drawings, 323-324, **328**
vellum, 36, 38, 42
ventilation: in storage area, 13, 15-16
ventilation and heating systems (in a boat), 55
verification and checking, 31-33, 156, 313-314; in measuring rigs, **185**
vernacular ("traditional" or "folk art") boats/names, 2, 54, 388
videotaping, 352

wands, electronic (used in computer measurement systems), 359-360, 364
waterline, load: recording breadth of, 214-215, **221**
waterline, painted, 118; recording breadth of, 214, 221; used as reference line/plane, 211-212, 215; used in leveling boat, **136**
waterline plane, 14
waterlines, 233; drawing and fairing (round hull), 233-235, 237; taking off, 124-125
web slings, 15
wedges, 37, **45**
whaleboats, 305, 389
whipping, **207**
Whitehall, **78**
windows, measuring, 212-213, **218-219**
wire rope, **195**
wiring, 55; color codings, 104; runs, **104**
wood: as hull material, 13, 20, 58; cracks in, **112**; different species used (evidence of repairs), 23; dimensional changes due to climate, 28, **33**; dimensional changes due to workmanship, 28; durability/longevity, 4; species identification, 53, 57, 311-312; waterlogged, 312
wood grain, 21, 28, **33**, 56, 58, **70**, **97**, **112**; shown in drawings, 232, 322
Woods Hole Spritsail Boat, 52, **60**, **63**, **70**, **167**
work boats: British, 388; North American, 338-339, 341-342, 349, **355**, 388-390; rigs, 180 *See also specific types and regions*
workmanship, 20-21, 28, 311; different, as evidence of repairs, 22-24
work site selecting, 10-11

wrecks: documenting, 304-320

X-bracing, in cradles, 14

yards, **182**; running rigging for, **203**
yawl, 183

zoom function (in computer design/documentation), 362